SOL-GEL SCIENCE

The Physics and Chemistry of
Sol-Gel Processing

SOL-GEL SCIENCE

The Physics and Chemistry of Sol-Gel Processing

C. Jeffrey Brinker

Sandia National Laboratories
Albuquerque, New Mexico

George W. Scherer

E. I. du Pont de Nemours & Company
Wilmington, Delaware

ACADEMIC PRESS, INC.

An Imprint of Elsevier

Boston San Diego New York
London Sydney Tokyo Toronto

ACADEMIC PRESS, INC.
An Imprint of Elsevier
1250 Sixth Avenue, San Diego, CA 92101

United Kingdom Edition published by
ACADEMIC PRESS LIMITED.
An Imprint of Elsevier
24–28 Oval Road, London NW1 7DX

Library of Congress Cataloging-in-Publication Data

Brinker, C. Jeffrey.
  Sol-gel science : the physics and chemistry of sol-gel processing
/ C. Jeffrey Brinker, George W. Scherer.
    p.   cm.
  Includes bibliographical references.
  ISBN-13: 978-0-12-134970-7   ISBN-10: 0-12-134970-5
  1. Ceramic materials.   2. Colloids.   I. Scherer, George W.
  II. Title.
  TP810.5.B75   1990
  666—dc20   89-15631
                                                CIP
ISBN-13: 978-0-12-134970-7
ISBN-10: 0-12-134970-5

To Stephanie and Lina
and
to Marty

Contents

Preface

Our goal is to present the physical and chemical principles of the sol-gel process at a level suitable for graduate students and practitioners in the field. We define *sol-gel* rather broadly as the preparation of ceramic materials by preparation of a sol, gelation of the sol, and removal of the solvent. The sol may be produced from inorganic or organic precursors (e.g., nitrates or alkoxides) and may consist of dense oxide particles or polymeric clusters. We expand the definition of ceramics to include organically modified materials, often called ORMOSILs or CERAMERs. The emphasis of our treatment is on the science, rather than the technology, of sol-gel processing. Although a chapter on applications is included, more detailed discussion is available in proceedings of conferences and in the recent collection of articles, *Sol-Gel Technology for Thin Films, Fibers, Preforms, Electronics, and Specialty Shapes* (Noyes, Park Ridge, N.J., 1988), edited by Professor Lisa Klein.

Sol-Gel Sciences was motivated by our perception that the explosion of literature in this field has made it difficult for newcomers to catch up to the state of the art, and threatens to overwhelm researchers already involved in the subject. The problem is manifested by publications that unwittingly duplicate previous work, and interpretations based on discarded theories. There is clearly a need—which the present volume is intended to fill—for a text presenting a coherent account of the principles of sol-gel processing. The absence of such a book was justified until fairly recently by the immaturity of the discipline. We feel that there has now been enough progress to enable us to present a view that is well-founded and unlikely to be overturned (except with respect to details) by future discoveries. Of course, there are many topics about which little is known (e.g., reaction kinetics in multicomponent systems or fracture mechanics of wet gels), and rapid progress in such areas can be expected to make this book require revisions within a few years.

The first chapter introduces basic terminology, provides a brief historical sketch, and identifies some excellent texts for background reading.

Thereafter, the structure of the book parallels the steps of the process. Chapters 2 and 3 discuss the mechanisms of hydrolysis and condensation for nonsilicate and silicate systems, respectively. In each chapter, aqueous systems are considered first and then contrasted with the nonaqueous systems that are the main focus of this book. The growth of polymeric clusters or small (few nm) primary particles is the endpoint of those chapters and the beginning for Chapter 4, which deals with stabilization and gelation of sols. The growth of identically sized "monospheres" is discussed at length. Chapter 5 reviews theories of gelation and examines the predicted and observed changes in the properties of a sol in the vicinity of the gel point. Chapter 6 describes the changes in structure and properties that occur during aging of a gel in its pore liquor (or some other liquid). The stiffening and coarsening obtained in this way have important consequences for the drying and sintering steps that follow. The discussion of drying is divided into two parts, with the theory concentrated in Chapter 7 and the phenomenology in Chapter 8. Readers who are not theoretically inclined can skip most of the first and they will still be able to understand the second. The structure of dried gels is explored in Chapter 9. The gel is quite different from the corresponding oxide, because of the abundance of hydroxyl and organic groups *on* and *in* the solid phase. In addition, of course, gels have a huge interfacial area (typically $300-1000 \, m^2/g$). As shown in Chapter 10, these factors present the possibility of using the gel as a substrate for chemical reactions or of modifying the *bulk* composition of the resulting ceramic by performing a *surface* reaction (such as nitridation) on the gel. Chapter 11 reviews the theory and practice of sintering, describing the mechanisms that govern densification of amorphous and crystalline materials, and showing the advantages of avoiding crystallization before sintering is complete. The properties of gel-derived and conventional ceramics are discussed in Chapter 12, where it is shown that the uniqueness of the former diminishes as the temperature of heat treatment is increased. The preparation of films is such an important aspect of sol-gel technology that the fundamentals of film formation are treated at length in Chapter 13. Films and other applications are briefly reviewed in Chapter 14.

Each chapter begins with an outline of its contents and ends with a summary. When new terms are introduced they are set in italics. All variables and abbreviations are defined when they first appear in each chapter; a table of frequently used abbreviations is provided in Chapter 1. Extensive references (typically more than 100) are provided in each chapter.

Acknowledgments

I.

Writing this book has been a far more educational experience than I wanted. I thank all those colleagues who patiently explained their work, and even provided original figures for our use. I am particularly indebted to those who took the time to write reviews of drafts of the chapters, including Jean Phalippou, Egon Matijević, Jim Martin, Paul Meakin, Tom Wood, John Ramsay, Terry Ring, Steve Wallace, Tom Gallo, Wander Vasconcelos, and Bulent Yoldas. Of course, they can't be held responsible for the final product, since I didn't necessarily understand what they told me. I also thank the management at DuPont, particularly Uma Chowdhry, Ralph Staley, and Lloyd Guggenberger, for encouraging me in this project. The photography of the numerous figures was done with great care and skill by Milt Nuttall.

My final and greatest debt is, of course, to my wife Marty, who has now suffered through my second and last book. Every evening and weekend for a year and a half was invested in this project, during which time she patiently took responsibility for every aspect of the care and feeding of the author. If you hate this book, feel free to tell me, but please tell Marty it's terrific.

George W. Scherer
Wilmington, DE

II.

As reflected in the following chapters, the topic "Sol-Gel Science" is quite diverse and interdisciplinary. I relied on the expertise of many colleagues in assimilating the information required for this project. The most important contributions to the success of this book were made by those who reviewed original drafts of the chapters, including Jacques Livage, Clement Sanchez, Alan Hurd, Tom Wood, Thomas Bein, Karin Möller, Walter Klemperer, Chip Zukoski, Helmut Schmidt, Doug Smith, and Jean Phalippou. I must also acknowledge my co-author, George Scherer, whose dedication, encouragement, and—yes—occasional harassment provided the requisite motivation to complete this book.

The "conception and birth" of this book coincided with that of my daughter, Lina. The often incompatible roles of father and author led to many lonely nights and weekends for my wife, Stephanie, whom I now thank in writing for her patience and encouragement.

Finally, I acknowledge the provocative and productive environment of Sandia National Laboratories—discussions with Alan Hurd, Keith Keefer, Jim Martin, and Dale Schaefer—encouragement by my management, Dan Doughty, Ron Loehman, and Bob Eagan, secretarial support by Julie Walker, and wonderful graphic art by Mona Aragon.

C. Jeffrey Brinker
Albuquerque, NM

Fig. 1.

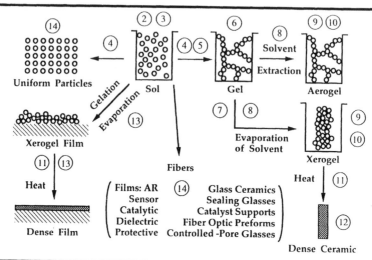

Overview of the sol-gel process with steps labelled by numbers of corresponding chapters of this book.

CHAPTER 1

Introduction

Our goal is to present the fundamental principles of sol-gel processing. This requires that we explore areas of physics (e.g., fractal geometry and percolation theory), chemistry (mechanisms of hydrolysis and condensation), and ceramics (sintering and structural relaxation) that may be unfamiliar to many readers, including those involved in sol-gel research. Therefore we introduce each topic on an elementary level and develop it only so far as necessary to make our point; numerous references are provided for those who want more detail. In this chapter we begin by stepping through the stages of the process, which parallel the structure of the book, and by introducing the terminology. Then we provide a brief historical sketch of the evolution of this field of study plus a list of "essential reading."

Figure 1 presents a schematic of the routes that one could follow within the scope of sol-gel processing, with each stage labelled by the number of the relevant chapter in this book. This illustration recurs as a "locator" at the beginning of each chapter to provide a perspective on the place of that step in the overall process.

1. _____

SOL-GEL PROCESSING

A *colloid* is a suspension in which the dispersed phase is so small (~1–1000 nm) that gravitational forces are negligible and interactions are dominated by short-range forces, such as van der Waals attraction and surface charges. The inertia of the dispersed phase is small enough that it exhibits *Brownian motion* (or *Brownian diffusion*), a random walk driven by momentum imparted by collisions with molecules of the suspending medium. A *sol* is a colloidal suspension of solid particles in a liquid. An *aerosol* is a colloidal suspension of particles in a gas (the suspension may be called a *fog* if the particles are liquid and a *smoke* if they are solid) and an *emulsion* is a suspension of liquid droplets in another liquid. All of these types of colloids can be used to generate polymers or particles from which ceramic materials can be made. A *ceramic* is usually defined by saying what it is *not*: it is nonmetallic and inorganic; some would also say it is not a chalcogenide. We thus include all metal oxides, nitrides, and carbides, both crystalline and noncrystalline. In the sol-gel process, the *precursors* (starting compounds) for preparation of a colloid consist of a metal or metalloid element surrounded by various *ligands* (appendages *not* including another metal or metalloid atom). For example, common precursors for aluminum oxide include *inorganic* (containing no carbon) salts such as $Al(NO_3)_3$ and *organic* compounds such as $Al(OC_4H_9)_3$. The latter is an example of an *alkoxide*, the class of precursors most widely used in sol-gel research. An *alkane* is a molecule containing only carbon and hydrogen linked exclusively by single bonds, as in *methane* (CH_4) and *ethane* (C_2H_6); the general formula is C_nH_{2n+2}. An *alkyl* is a ligand formed by removing one hydrogen (proton) from an alkane molecule producing, for example, *methyl* ($\bullet CH_3$) or *ethyl* ($\bullet C_2H_5$) (where the dot $\bullet$ indicates an electron that is available to form a bond). An *alcohol* is a molecule formed by adding a *hydroxyl* (OH) group to an alkyl (or other) molecule, as in *methanol* (CH_3OH) or *ethanol* (C_2H_5OH). An *alkoxy* is a ligand formed by removing a proton from the hydroxyl on an alcohol, as in *methoxy* ($\bullet OCH_3$) or *ethoxy* ($\bullet OC_2H_5$). A list of the most commonly used alkoxy ligands is presented in Table 1.

Metal alkoxides are members of the family of *metalorganic* compounds, which have an organic ligand attached to a metal or metalloid atom. The most thoroughly studied example is silicon tetraethoxide (or tetraethoxysilane, or tetraethyl orthosilicate, TEOS), $Si(OC_2H_5)_4$. *Organometallic* compounds are defined as having direct metal–carbon bonds, not metal–oxygen–carbon linkages as in metal alkoxides; thus, alkoxides are not organometallic compounds, although that usage turns up frequently in the

Table 1.

Commonly Used Ligands.

Alkyl		Alkoxy	
methyl	•CH$_3$	methoxy	•OCH$_3$
ethyl	•CH$_2$CH$_3$	ethoxy	•OCH$_2$CH$_3$
n-propyl	•CH$_2$CH$_2$CH$_3$	n-propoxy	•O(CH$_2$)$_2$CH$_3$
iso-propyl	H$_3$C(•C)HCH$_3$	iso-propoxy	H$_3$C(•O)CHCH$_3$
n-butyl	•CH$_2$(CH$_2$)$_2$CH$_3$	n-butoxy	•O(CH$_2$)$_3$CH$_3$
sec-butyl	H$_3$C(•C)HCH$_2$CH$_3$	sec-butoxy	H$_3$C(•O)CHCH$_2$CH$_3$
iso-butyl	•CH$_2$CH(CH$_3$)$_2$	iso-butoxy	•OCH$_2$CH(CH$_3$)$_2$
tert-butyl	•C(CH$_3$)$_3$	tert-butoxy	•OC(CH$_3$)$_3$

Other

acetylacetonate   H$_3$COC(•O)CH$_2$(O•)COCH$_3$

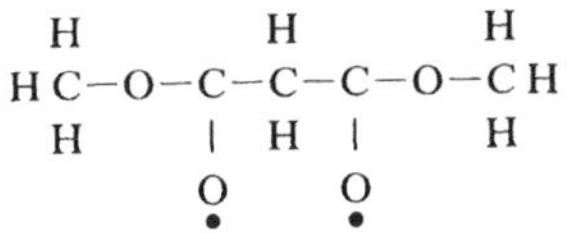

```
      H        H        H
   H C—O—C—C—C—O—C H
      H    | H |    H
           O    O
           •    •
```

acetate   •OOCCH$_3$

```
      H
   H C—C—O •
      H  ‖
         O
```

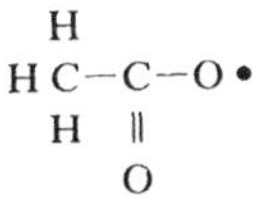

Dot (•) indicates bonding site. Parentheses indicate atom with available bond.
n = normal (meaning a linear chain), *sec* = secondary, *tert* = tertiary.

literature. Metal alkoxides are popular precursors because they react readily with water. The reaction is called *hydrolysis*, because a hydroxyl ion becomes attached to the metal atom, as in the following reaction:

$$Si(OR)_4 + H_2O \rightarrow HO\text{--}Si(OR)_3 + ROH. \tag{1}$$

The R represents a proton or other ligand (if R is an alkyl, then •OR is an alkoxy group), and ROH is an alcohol; the bar (–) is sometimes used to indicate a chemical bond. Depending on the amount of water and catalyst present, hydrolysis may go to completion (so that all of the OR groups are replaced by OH),

$$Si(OR)_4 + 4H_2O \rightarrow Si(OH)_4 + 4ROH \tag{2}$$

or stop while the metal is only partially *hydrolyzed*, $Si(OR)_{4-n}(OH)_n$. Inorganic precursors can also be hydrolyzed, as discussed in Chapters 2 and 3.

Two partially hydrolyzed molecules can link together in a *condensation reaction*, such as

$$(OR)_3Si\text{–}OH + HO\text{–}Si(OR)_3 \rightarrow (OR)_3Si\text{–}O\text{–}Si(OR)_3 + H_2O \qquad (3)$$

or

$$(OR)_3Si\text{–}OR + HO\text{–}Si(OR)_3 \rightarrow (OR)_3Si\text{–}O\text{–}Si(OR)_3 + ROH. \qquad (4)$$

By definition, condensation liberates a small molecule, such as water or alcohol. This type of reaction can continue to build larger and larger silicon-containing molecules by the process of *polymerization*. A *polymer* ("many

Fig. 2.

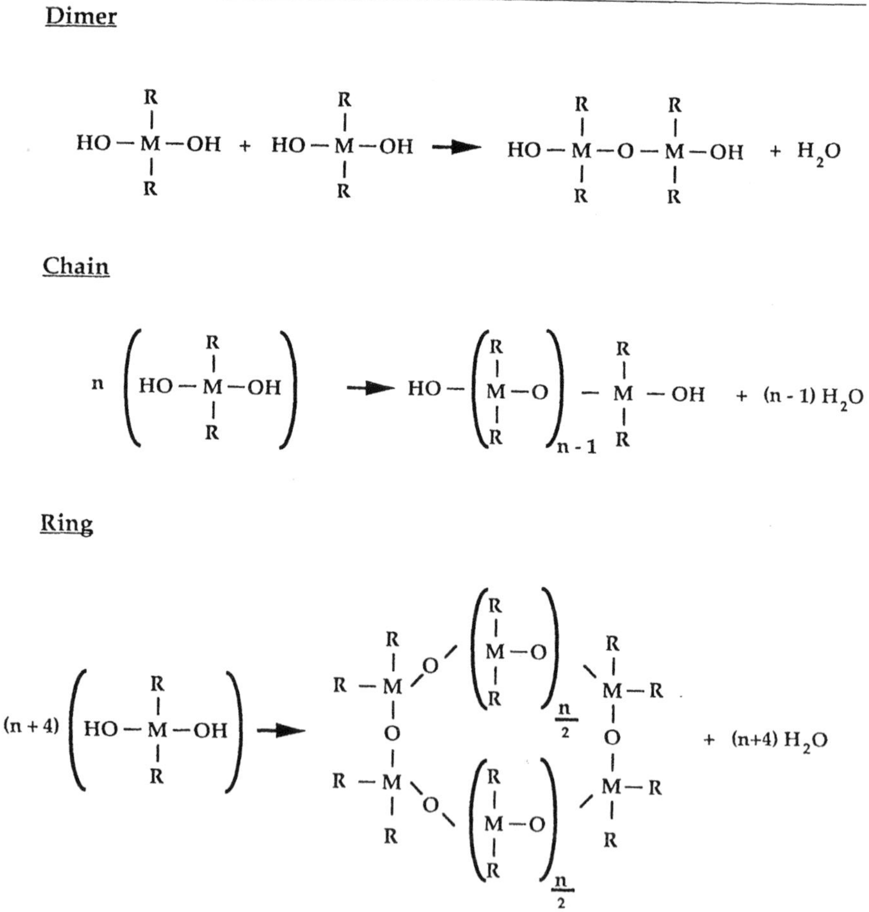

Dimer

Chain

Ring

Formation of rings and chains by bifunctional ($f = 2$) monomer: (a) dimer, (b) chain, and (c) ring.

member") is a huge molecule (also called a *macromolecule*) formed from hundreds or thousands of units called *monomers* that are capable of forming at least two bonds. An *oligomer* is a molecule of intermediate size—much larger than "mono," but much less than "macro." The number of bonds that a monomer can form is called its *functionality*, f; typical oxide monomers are *bifunctional* ($f = 2$), *trifunctional* ($f = 3$), or *tetrafunctional* ($f = 4$), any of which may be called *polyfunctional* (f arbitrary). Consider a metal atom, M, with four ligands, $MR_2(OH)_2$, of which two are unreactive *R* groups and two are reactive (*labile*) hydroxyls. Such a compound can polymerize only into linear chains or rings, as shown in Fig. 2. If a polyfunctional unit with $f > 2$ is present, the chains can be joined by *crosslinks* to form a three-dimensional structure. Polymerization of silicon alkoxide, for instance, can lead to complex *branching* of the polymer (as in Fig. 3), because a fully hydrolyzed monomer $[Si(OH)_4]$ is tetrafunctional. On the other hand, under certain conditions (e.g., low water concentration) fewer than four ligands will be capable of condensation, so relatively little branching will

Fig. 3.

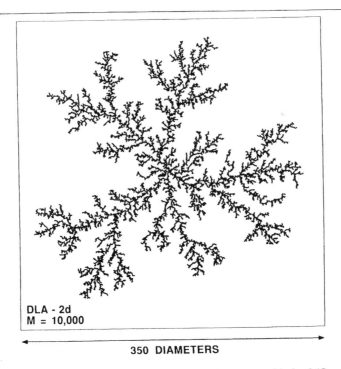

DLA - 2d
M = 10,000

350 DIAMETERS

Fractal polymer made by branching of polyfunctional monomer with $f > 2$ (Computer simulation of two-dimensional aggregation by P. Meakin [1]).

occur. The mechanisms of hydrolysis and condensation, and the factors that bias the polymer structure toward linear or branched structures are discussed in detail in Chapters 2 and 3.

Since a polymer can form a three-dimensional structure, one could extend the definition of polymer to include most solids; for example, diamond could be regarded as a polymer of a tetrafunctional C monomer. This clearly trivializes the concept (if everything is a polymer, there is no point in using the word!), so we follow Flory [2] in restricting the term polymer (in the case of 3-d networks) to structures with *random* branching. This excludes crystalline solids from the category of polymers; to exclude oxide glasses, one could further require that the average size of closed loops within the network be much greater than atomic size. For example, in silica glass one could start from a given atom and trace a continuous path of bonds back to the starting point, and the average path would contain about six silicon atoms. The diameter of such a loop would be ≤ 1 nm, so we exclude it from the class of polymers. Admittedly, this is not a rigorous definition, and it is not the last bit of ambiguity that we must face.

If we prepare a solution of monomers and allow it to condense into cross-linked polymers with an average size of several nanometers, do we have a solution or a sol? Most of us would agree that the answer is sol, but what if the polymer is a linear chain that is coiled into a ball of that size? Or if it is extended? The term *colloid* was originally coined to describe macromolecules that could not pass through porous membranes [2], so all of the preceding systems could legitimately be described as sols. In the past, we and others have used the term *colloidal* to describe sols consisting of dense oxide particles, and used *polymeric* for suspensions of branched macromolecules. We now recognize that as a poor choice, since both systems are colloids (sols), so we follow Flory [3] and Rabinovich [4] in adopting the term *particulate* to describe sols in which the dispersed phase consists of nonpolymeric solid particles. Our original motivation for introducing that terminology was to distinguish the particulate silica sols that are typically formed in *aqueous* solutions (in which the only liquid present is water) from the *polymeric* silica sols generally obtained from hydrolysis of alkoxides in *nonaqueous* solutions (containing solvents other than or in addition to water). The value of that distinction is diminished when one looks beyond the silica system. Whereas silica tends to form polymeric sols except under extreme conditions (viz., high pH and excess water), most other oxides prefer to form particles. Moreover, some aqueous systems yield ''particles'' with diameters on the order of 1 nm—so small that they might better be called oligomers. Since polymeric silicates do not have particles even on that scale, we define a polymeric sol as one in which the solid phase contains no dense oxide particles larger than 1 nm, which is at the lower limit of the colloidal range. The preparation of particulate sols and gels is the subject of Chapter 4. Most of the systems of interest contain dense particles larger than 5 nm, so there is no ambiguity about their nature. The topics discussed in Chapter 4 include the growth of colloidal particles into *monospheres* (spherical particles of uniform size) and use of particles made by vapor methods.

When a polyfunctional ($f > 2$) monomer forms bonds at random, or when a particulate sol aggregates, it is common to form *fractal* structures. (An example is the polymeric cluster in Fig. 3.) A *mass fractal* is distinguished from a conventional *Euclidean* object by the fact that the mass (m) of the fractal increases with its radius (r) according to

$$m \propto r^{d_f} \tag{5}$$

where d_f is called the *mass fractal dimension* of the object. For a Euclidean object, $m \propto r^3$, but for a fractal $d_f < 3$, so its density ($\rho \propto m/r^3$) decreases as the object gets bigger. A tree is an example of a mass fractal, because its branches become wispier as one moves away from the trunk, so the mass of the tree increases more slowly than the cube of its height. A *surface fractal* has a surface area, S, that increases faster than r^2:

$$S \propto r^{d_s} \tag{6}$$

where d_s is called the *surface fractal dimension*. One could think of a piece of paper crumpled into a ball as a surface fractal, because it is "all surface," with an area that increases as the radius of the ball cubed ($d_s = 3$); this object is not a mass fractal, because its mass also increases as r^3.

As discussed in Chapters 2–5, Euclidean objects (viz., dense spheroidal particles) are most likely to form in systems in which the particle is partially soluble in the solvent; then monomers can dissolve and reprecipitate until the equilibrium structure (having minimal interfacial area) is obtained. In nonaqueous systems, such as alkoxide–alcohol–water solutions, the solubility of the solid phase is so limited that condensation reactions are virtually irreversible. Consequently, bonds form at random and cannot convert to the equilibrium configuration, and this generally leads to the formation of fractal polymeric clusters. The fractal dimension can be measured by small-angle scattering of X-rays or neutrons from the sol: $-d_f$ is the slope of a plot of ln(scattered intensity) versus $\ln(\sin(\theta)/\lambda)$, where θ is the scattering angle and λ is the wavelength of the scattered radiation. If the slope is more negative than -3, the sol does not contain mass fractal objects, but may contain surface fractals; dense smooth particles give a slope of -4. A slope between -3 and -4 could be used as a defining characteristic of a particulate sol; however, such a slope could be obtained from a truly particulate sol if the range of particle size were wide enough.

If a monomer can make more than two bonds, then there is no limit on the size of the molecule that can form. If one molecule reaches macroscopic dimensions so that it extends throughout the solution, the substance is said to be a *gel*. The *gel point* is the time (or degree of reaction) at which the last bond is formed that completes this giant molecule. Thus a gel is a substance that contains a continuous solid skeleton enclosing a continuous liquid phase. The continuity of the solid structure gives elasticity to the gel (as in the familiar gelatin desert). Gels can also be formed from particulate sols, when attractive dispersion forces cause them to stick together in such a way as to form a network. The characteristic feature of the gel is obviously not the type of bonding: polymeric gels are covalently linked, gelatine gels form by entanglement of chains, and particulate gels are established by van der Waals forces. The bonds may be reversible, as in the particulate systems (which can often be redispersed by shaking), or permanent, as in polymeric systems. What exactly is or is not a gel? As noted by Henisch [5], a gel "has been defined as a 'two-component system of a semisolid nature, rich in liquid,' and no one is likely to entertain illusions about the rigor of such a definition." What is semisolid? That restriction would help to eliminate porous sandstone from the category of gels, but it would also seem to eliminate silica gel, which can be quite rigid. We interpret a gel to consist of continuous solid and fluid phases of colloidal dimensions. Continuity means that one could travel through the solid phase from one side of the sample to the other without having to enter the liquid; conversely, one could make the same trip entirely within the liquid phase. Since both phases are of colloidal dimension, a line segment originating in a pore (which may be filled with liquid or vapor) and running perpendicularly into the nearest solid surface must re-emerge in another pore less than 1 μm away. (See Fig. 4.) Similarly, a segment originating within the solid phase and passing perpendicularly through the pore wall must re-enter the solid phase within a distance of 1 μm.

The formation of gels and the effect of gelation on the properties of the parent sol are discussed in detail in Chapter 5. It is generally found that the process begins with the formation of fractal aggregates that grow until they begin to impinge on one another, then those clusters link together as described by the theory of *percolation*. That is, near the gel point bonds form at random between the nearly stationary clusters (polymers or aggregates of particles), linking them together in a network. The gel point corresponds to the *percolation threshold*, when a single cluster (called the *spanning cluster*) appears that extends throughout the sol; the spanning cluster coexists with a sol phase containing many smaller clusters, which gradually become attached to the network. Gelation can occur after a sol is cast into a mold, in which case it is possible to make objects of a desired shape. If the smallest dimension of the gel is greater than a few millimeters, the object is generally

Fig. 4. _____

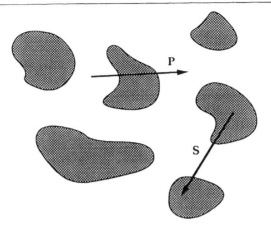

Schematic illustration of gel structure. Arrow P originates in a pore, passes perpendicularly into the solid phase (shaded) and re-emerges in another pore; arrow S similarly passes from solid to solid. Since the gel is a three-dimensional structure with a microstructure on a colloidal scale, any such arrow is less than 1 μm long.

called a *monolith*. Alternatively, gelation can be produced by rapid evaporation of the solvent, as occurs during preparation of films (Chapter 13) or fibers (Chapter 14).

Bond formation does not stop at the gel point. In the first place, the network is initially compliant, so segments of the gel network can still move close enough together to allow further condensation (or other bond-forming processes). Moreover, there is still a sol within the gel network and those smaller polymers or particles continue to attach themselves to the network. The term *aging* is applied to the process of change in structure and properties after gelation. As explained in Chapter 6, aging may involve further condensation, dissolution and reprecipitation of monomers or oligomers, or phase transformations within the solid or liquid phases. Some gels exhibit spontaneous shrinkage; called *syneresis*, as bond formation or attraction between particles induces contraction of the network and expulsion of liquid from the pores.

Shrinkage of a gel, either during syneresis or as liquid evaporates during drying, involves deformation of the network and transport of liquid through the pores. The driving forces and transport mechanisms are discussed at length in Chapter 7. A theory is introduced that permits calculation of the stresses and strains in shrinking gels. This is of particular importance for a detailed understanding of drying, but it constitutes more detail than most readers will desire, so it is separated from the qualitative discussion of drying in Chapter 8. Drying by evaporation under normal conditions gives rise to capillary pressure that causes shrinkage of the gel network. The resulting dried gel, called a *xerogel* (*xero* means dry), is often reduced in volume by a factor of 5 to 10 compared to the original wet gel. If the wet gel is placed in an autoclave and dried under supercritical conditions, there is no interface between liquid and vapor, so there is no capillary pressure and relatively little shrinkage. This process is called *supercritical* (or *hypercritical*) *drying*, and the product is called an *aerogel*. These may indeed be mostly air, having volume fractions of solid as low as ~1%.

Xerogels and aerogels are useful in the preparation of dense ceramics, but they are also interesting in themselves, because their high porosity and surface area make them useful as catalytic substrates, filters, and so on. The structure of the network of dried gels and the evolution of that structure during heating are the subjects of Chapter 9. It is found that the dried gel contains many labile sites that offer opportunities for new chemical reactions. Chapter 10 explains in detail how the reactivity of gels can be exploited, particularly for modification of the composition of the gel through chemisorption of ammonia.

Most gels are *amorphous* (noncrystalline), even after drying, but many crystallize when heated. If the objective of the processing is to produce a pore-free ceramic, it is necessary to heat the gel to a high enough temperature to cause *sintering*. As explained in Chapter 11, sintering is a process of collapse of pores driven by surface energy. In amorphous materials, transport of atoms occurs by viscous flow and the process is called *viscous sintering*; in crystalline materials, sintering involves diffusion. Since the latter is a slower process, it is important to understand the relative rates of sintering and crystallization; densification is easiest if sintering can be completed before crystals appear.

Once a gel has been densified, is it equivalent to a ceramic made by conventional means? This question is examined in Chapter 12, where it is shown that unique structures and properties can be achieved when all of the processing steps are performed at low temperatures. After a gel has been melted, however, it "forgets" its manner of preparation and acquires the equilibrium structure dictated by thermodynamics; in most cases, in fact, it is necessary only to heat slightly above the glass-transition temperature.

The unique structures (e.g., fibers), microstructures, and compositions that can be made by sol-gel methods open many possibilities for practical applications, which are discussed in Chapter 14. It is the nature of an expanding field such as this that the list presented there is outdated before the ink dries. Readers are invited to contribute to the obsolescence of this book by inventing applications of their own. Preparation of thin films is by far the most important use of sols and gels, and as such is deserving of a chapter of its own, Chapter 13 provides a detailed look at the microstructural control made possible by the polymeric nature of alkoxide-derived sols, which allows films to be made with almost any composition and degree of porosity.

2.

HISTORICAL SKETCH

The first metal alkoxide was prepared from $SiCl_4$ and alcohol by Ebelmen [6], who found that the compound gelled on exposure to the atmosphere. However, these materials remained of interest only to chemists (see the detailed account by Bradley *et al.* [7]) for almost a century. It was finally recognized by Geffcken [8] in the 1930s that alkoxides could be used in the preparation of oxide films. This process was developed by the Schott glass company in Germany and was quite well understood, as explained in the excellent review by Schroeder [9].

Inorganic gels from aqueous salts have been studied for a long time. Graham [10] showed that the water in silica gel could be exchanged for organic solvents, which argued in favor of the theory that the gel consisted of a solid network with continuous porosity. Competing theories of gel structure regarded the gel as a coagulated sol with each of the particles surrounded by a layer of bound water, or as an emulsion. The network structure of silica gels was widely accepted in the 1930s, largely through the work of Hurd [11], who showed that they must consist of a polymeric skeleton of silicic acid enclosing a continuous liquid phase. The process of supercritical drying to produce aerogels was invented by Kistler [12] in 1932, who was interested in demonstrating the existence of the solid skeleton of the gel, and in studying its structure.

Around the same time, mineralogists became interested in the use of sols and gels for the preparation of homogeneous powders for use in studies of phase equilibria [13,14]. This method was later popularized in the ceramics community by Roy [15,16] for the preparation of homogeneous powders. That work, however, was not directed toward an understanding of the mechanisms of reaction or gelation, nor the preparation of shapes (monoliths). Much more sophisticated work, both scientifically and technologically, was going on in the nuclear-fuel industry, but it was not published until later [17,18]. The goal of this work was to prepare small spheres (tens of μm in diameter) of radioactive oxides that would be packed into fuel cells for nuclear reactors. The advantage of sol-gel processing was that it avoided generation of dangerous dust, as would be produced in conventional ceramics processing, and facilitated the formation of spheres. The latter was accomplished by dispersing the aqueous sol in a hydrophobic organic liquid, so that the sol would form into small droplets, each of which would subsequently gel.

The ceramics industry began to show interest in gels in the late sixties and early seventies. Controlled hydrolysis and condensation of alkoxides for preparation of multicomponent glasses was independently developed by Levene and Thomas [19] and Dislich [20]. Ceramic fibers were made from metalorganic precursors on a commercial basis by several companies [21–23]. However, the explosion of activity that continues today can be dated from the demonstration by Yoldas [24,25] and Yamane *et al.* [26] that monoliths could be produced by careful drying of gels. This is a bit ironic in retrospect, as it is evident that monoliths are the least technologically important of the potential applications of gels. However, the allure of a room-temperature process for the preparation of bricks and windows was irresistible for research directors around the world, and enormous effort has been devoted to demonstrating that that objective is nonsense.

As is usually the case, the technology preceded the science of sol-gel processing, but great strides have been made in the past few years in under-standing the fundamental aspects of preparing homogeneous multicompo-nent ceramics (crystalline and amorphous) from alkoxide-derived gels. The increase in the volume and the quality of published work has been extraordinary—and is the main motivation for writing this book: we recog-nize that it is an overwhelming task for a newcomer to this field to catch up with the state of the art. Therefore, we have tried to pull together the best scientific work on the physics and chemistry of sol-gel processing to provide a foundation for understanding future developments. We take the process step-by-step in the following chapters and present our best understanding of the most important topics, but we are well aware of two defects in our product: first, the field is so broad and deep that we can give only cursory attention to many important subjects; second, there is controversy (duly noted in the text) with respect to many of the ideas that we present, as is to be expected in a developing field. A complete education can therefore not be obtained within these covers, so we provide a reading list in the following section. It includes the classics of the sol-gel literature, along with books and major reviews of topics on the fringes of material science, such as percolation theory and fractal geometry. In addition, abundant references are provided within each chapter.

3.

BACKGROUND READING

3.1. Journals and Conferences

For coverage of current sol-gel research, the most widely used journals are the *Journal of Non-Crystalline Solids, Journal of Materials Research*, and the journals of the American Ceramic Society and the Ceramic Society of Japan (*Yogyo Kyokai-shi*). The *Journal of Non-Crystalline Solids* carries the proceedings of the International Workshop on Gels, which have been held every other year since 1982 (volumes **48, 63, 82,** and **100** to date). Two other series of topical meetings are of major importance. The Materials Research Society has sponsored meetings under the title *Better Ceramics Through Chemistry* and the proceedings have been published in book form:

Better Ceramics Through Chemistry, eds. C. J. Brinker, D. E. Clark, and D. R. Ulrich (North-Holland, N.Y., 1984).
Better Ceramics Through Chemistry II, eds. C. J. Brinker, D. E. Clark, and D. R. Ulrich (Mat. Res. Soc., Pittsburgh, Pa., 1986).
Better Ceramics Through Chemistry III, eds. C. J. Brinker, D. E. Clark, and D. R. Ulrich (Mat. Res. Soc., Pittsburgh, Pa., 1988).

The proceedings of a biennial meeting with the title *Ultrastructure Processing* have been published as:

Ultrastructure Processing of Ceramics, Glasses, and Composites, eds. L L. Hench and D. R. Ulrich (Wiley, N.Y., 1984).
Science of Ceramic Chemical Processing, eds. L. L. Hench and D. R. Ulrich (Wiley, N.Y., 1986).
Ultrastructure Processing of Advanced Ceramics, eds. J. D. Mackenzie and D. R. Ulrich (Wiley, N.Y., 1988).

There is now a biennial series of meetings on aerogels. The proceedings to date have been published as

Aerogels, ed. J. Fricke (Springer-Verlag, N.Y., 1986).
2nd International Symposium on Aerogels, *Revue de Physique Appliquée*, **24** [c4] (1989).

Other relevant conferences include those devoted to preparation of powders, much of which involves sol-gel methods; for example,

Advances in Ceramics, Vol. 21: Ceramic Powder Science, eds. G. L. Messing, K. S. Mazdiyasni, J. W. McCauley, and R. A. Haber (Am. Ceram. Soc., Westerville, Ohio, 1987).

3.2. Chemistry

The chemistry of hydrolysis of inorganic salts is discussed in

C. F. Baes and R. E. Messmer, *The Hydrolysis of Cations* (Wiley, N.Y., 1976).

The colloid chemistry of clays is the subject of the excellent monograph

H. van Olphen, *An Introduction to Clay Colloid Chemistry*, 2d ed. (Wiley, N.Y., 1977).

Every aspect of the chemistry of silica in aqueous systems, including polymerization, gelation, gel structure, and applications is discussed in a classic text that is frequently cited in our book:

R. K. Iler, *The Chemistry of Silica* (Wiley, N.Y., 1979).

The chemistry of alkoxides (with, unfortunately, little reference to hydrolysis) is the subject of

D. C. Bradley, R. C. Mehrotra, and D. P. Gaur, *Metal Alkoxides* (Academic Press, N.Y., 1978).

Silicate chemistry is discussed in a number of excellent books:

K. A. Andrianov, *Organic Silicon Compounds* (State Scientific Publ. House for Chem. Lit., Moscow, 1955). Translation 59-11239, U.S. Dept. of Commerce, Washington, D.C.
C. Eaborn, *Organosilicon Compounds* (Butterworths, London, 1960).
Soluble Silicates, ed. J. S. Falcone, Jr. (Am. Chem. Soc., Washington, D.C., 1982).
M. G. Voronkov, V. P. Mileshkevich, and Y. A. Yuzhelevski, *The Siloxane Bond* (Consultants Bureau, N.Y., 1978).

The chemistry and physics of organic polymers (much of which is relevant to inorganic polymers as well), including the "classical" theory of gelation, is the subject of the classic text

P. J. Flory, *Principles of Polymer Chemistry* (Cornell Univ. Press, Ithaca, N.Y., 1953).

3.3. Stability of Sols

Excellent texts dealing with stabilization by electrostatic and steric repulsion, respectively, are

R. J. Hunter, *Zeta Potential in Colloid Science* (Academic Press, N.Y., 1981).
D. H. Napper, *Polymeric Stabilization of Colloidal Dispersions* (Academic Press, N.Y., 1983).

3.4. Aggregation, Fractals, and Percolation

The formation of clusters of particles or monomers by random aggregation processes is a popular topic in modern physics that is discussed in lengthy reviews, such as

P. Meakin, *Ann. Rev. Phys. Chem.*, **39** (1988) 237–267.

and in books derived from topical meetings on the subject, such as

On Growth and Form, eds. H. E. Stanley and N. Ostrowsky (Martinus Nijhoff, Dordrecht, Netherlands, 1986).

Fractal geometry is discussed in the classic book (beautifully illustrated) by the inventor of the concept:

B. B. Mandelbrot, *The Fractal Geometry of Nature* (W. H. Freeman, N.Y., 1983).

The connection of fractals with problems in materials science is revealed in

J. Feder, *Fractals* (Plenum Press, N.Y., 1988).

A delightfully lucid introduction to percolation theory, with reference to gelation, is provided in

R. Zallen, *The Physics of Amorphous Solids* (Wiley, N.Y., 1983).

3.5. Characterization of Structures

We make reference to many methods of characterization of structures in this text, some of which may be unfamiliar to ceramists. Useful discussions of these techniques are provided in the references suggested below.

Methods for characterization of the porosity and surface area of porous materials:

T. Allen, *Particle Size Measurement*, 3d ed. (Chapman and Hall, N.Y., 1981).
S. Lowell and J. E. Shields, *Powder Surface Area and Porosity*, 2d ed. (Chapman and Hall, N.Y., 1984).
S. J. Gregg and K. S. W. Sing, *Adsorption, Surface Area, and Porosity*, 2d ed. (Academic Press, N.Y., 1982).
Characterization of Porous Solids, eds. K. K. Unger, J. Rouquerol, K. S. W. Sing, and H. Kral (Elsevier, N.Y., 1988).

Nuclear magnetic resonance spectroscopy (NMR):

Basic Principles and Progress, Vol. 17: ^{17}O and ^{29}Si, eds. P. Diehl, E. Fluck, and R. Kosfeld (Springer-Verlag, Berlin, 1981).
NMR of Newly Accessible Nuclei, Vols. 1, 2, ed. P. Laszlo (Academic Press, London, 1983).

Infrared and Raman spectroscopy:

Vibrational Spectroscopy of Molecules on Surfaces, eds. J. P. Yates, Jr., and T. E. Madley (Plenum, N.Y., 1987).

J. Wong and C. A. Angell, *Glass Structure by Spectroscopy* (Marcel Dekker, N.Y., 1977).

Small-angle scattering of X-rays and neutrons:

B. J. Berne and R. Pecora, *Dynamic Light Scattering* (Wiley, N.Y., 1976).

C. F. Bohren and D. R. Huffman, *Absorption and Scattering of Light by Small Particles* (Wiley, N.Y., 1983).

A. Guinier and G. Fournet, *Small Angle Scattering of X-rays* (Wiley, N.Y., 1955).

S. W. Lovesay, *Theory of Neutron Scattering from Condensed Matter, Vol 1: Nuclear Scattering* (Clarendon Press, Oxford, England, 1984).

A particularly readable review of the theory of small-angle scattering, with reference to the growth and gelation of fractal clusters, is

J. E. Martin and A. J. Hurd, *J. Appl. Cryst.*, **20** (1987) 61–78.

Thermal analysis:

M. E. Brown, *Introduction to Thermal Analysis* (Chapman and Hall, N.Y., 1988).

3.6. Applications

A review of the principal applications of sol-gel technology is

Sol-Gel Technology for Thin Films, Fibers, Preforms, Electronics, and Speciality Shapes, ed. L. C. Klein (Noyes, Park Ridge, N.J., 1988).

4.
GLOSSARY OF ABBREVIATIONS

Many terms appear frequently enough to justify the use of abbreviations. We have tried to define each at its first use in a given chapter, but a list is provided in Table 2 for convenient reference.

Table 2.

Frequently Used Abbreviations.

Term	Meaning
AcAc	Acetyl acetone, acetyl acetones [e.g., $M(AcAc)_x(OR)_y$]
Bu	Butyl (e.g., BuOH = butanol)
Bu^n, Bu^s, Bu^i, Bu^t	*n*-Butyl, *sec*-Butyl, *iso*-Butyl, *tert*-Butyl
CERAMER	Ceramic–polymer (generalization of ORMOSIL)
CRH	Constant rate of heating

Table 2—*continued*

Term	Meaning
CRP	Constant rate period (during drying)
DCCA	Drying control chemical additive
DLA	Diffusion-limited aggregation or aggregate
DLCCA, DLCA	Diffusion-limited cluster–cluster aggregation or aggregate
DLMCA	Diffusion-limited monomer-cluster aggregation
DLVO	Derjaguin-Landau-Vervey-Overbeek (originators of theory of colloidal stability)
DSC	Differential scanning calorimetry
DTA	Differential thermal analysis
Et	Ethyl (e.g., EtOH = ethanol)
FRP	Falling rate period (during drying)
IEP	Isoelectric point
IR	Infrared
Me	Methyl (e.g., MeOH = methanol)
MOR	Modulus of rupture
NMR	Nuclear magnetic resonance
ORMOSIL	Organically modified silicate
Pr	Propyl (e.g., PrOH = propanol)
Pr^n, Pr^i	*n*-Propyl, *iso*-Propyl
PZC	Point of zero change
QELS	Quasi-elastic light scattering
RDF	Radial distribution function
RLA	Reaction-limited aggregation or aggregate
RLCCA, RLCA	Reaction-limited cluster–cluster aggregation or aggregate
RLMCA	Reaction-limited monomer-cluster aggregation or aggregate
SANS	Small-angle neutron scattering
SAXS	Small-angle X-ray scattering
SEM	Scanning electron microscopy
TEM	Transmission electron microscopy
TEOS	Tetraethoxysilane, $Si(OC_2H_5)_4$
TGA	Thermogravimetric analysis
TMOS	Tetramethoxysilane, $Si(OCH_3)_4$

REFERENCES

1. P. Meakin, *Ann. Rev. Phys. Chem.*, **39** (1988) 237–267.
2. P.J. Flory, *Principles of Polymer Chemistry* (Cornell Univ. Press, Ithaca, N.Y., 1953).
3. P.J. Flory, *Faraday Disc. Chem. Soc.*, **57** (1974) 7–18.
4. E.M. Rabinovich, in *Sol-Gel Technology for Thin Films, Fibers, Preforms, Electronics, and Speciality Shapes*, ed. L. C. Klein (Noyes, Park Ridge, N.J., 1988), pp. 260–294.
5. H.K. Henisch, *Crystal Growth in Gels* (Penn. State Univ. Press, University Park, Pa., 1970), p. 41.
6. J.J. Ebelmen, *Ann.*, **57** (1846) 331.

7. D.C. Bradley, R.C. Mehrotra, and D.P. Gaur, *Metal Alkoxides* (Academic Press, N.Y., 1978).

8. W. Geffcken and E. Berger, German Patent 736 411 (May 1939).

9. H. Schroeder, *Phys. Thin Films*, **5** (1969) 87–141.

10. T. Graham, *J. Chem. Soc.*, **17** (1864) 318–327.

11. C.B. Hurd, *Chem. Rev.*, **22** (1938) 403–422.

12. S.S. Kistler, *J. Phys. Chem.*, **36** (1932) 52–64.

13. R.H. Ewell and H. Insley, *J. Res. NBS*, **15** (1935) 173–186.

14. R.M. Barrer and L. Hinds, *Nature*, **166** (1950) 562.

15. R. Roy, *J. Am. Ceram. Soc.*, **39** [4] (1956) 145–146.

16. R. Roy, *J. Am. Ceram. Soc.*, **52** [6] (1969) 344.

17. R.M. Dell, in *Reactivity of Solids*, eds. J. S. Anderson, M.W. Roberts, and F.S. Stone (Chapman and Hall, N.Y., 1972), pp. 553–566.

18. J.L. Woodhead, *Silicates Ind.*, **37** (1972) 191–194.

19. L. Levene and I.M. Thomas, U.S. Patent 3,640,093 (February 8, 1972).

20. H. Dislich, *Angewandt Chemie*, **10** [6] (1971) 363–370.

21. E. Wainer, German Patent 1,249,832 (April 11, 1968).

22. H.G. Sowman, U.S. Patent 3,795,524 (March 5, 1974).

23. S. Horikuri, K. Tsuji, Y. Abe, A. Fukui, and E. Ichiki, Japanese Patent 49-108325 (October 15, 1974).

24. B.E. Yoldas, *J. Mater. Sci.*, **10** (1975) 1856–1860.

25. B.E. Yoldas, *J. Mater. Sci.*, **12** (1977) 1203–1208.

26. M. Yamane, A. Shinji, and T. Sakaino, *J. Mater. Sci.*, **13** (1978) 865–870.

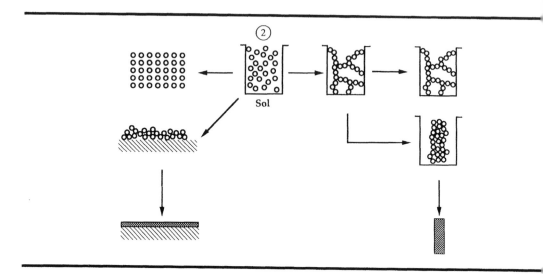

Sol

CHAPTER 2

Hydrolysis and Condensation I

Nonsilicates

Although the main focus of this book is on gels made by hydrolysis and condensation of metal alkoxide precursors, principally silicates, we begin our discussion of solution chemistry by considering the hydrolysis and condensation of non-silicates: both inorganic precursors in an aqueous solution and metal organic precursors in a variety of mixed solvents. This chapter concentrates mainly on early transition metals (e.g., Ti, V, Zr) and Group IIIB metals (B and Al). Generally speaking, these systems are distinguished from silicates by greater chemical reactivity resulting from the lower electronegativity of the metal and its ability to exhibit several coordination states, so that coordination expansion occurs spontaneously upon reaction with water or other nucleophilic reagents. An outline of the contents of this chapter follows.

1. *Transition Metals*

   1.1. *Solution chemistry of inorganic precursors* discusses hydrolysis and condensation of inorganic salts in aqueous solution. It introduces terminology associated with inorganic polymerization used throughout this chapter and in Chapter 3.

   1.2. *Solution chemistry of metal alkoxide precursors* discusses hydrolysis and condensation of metal alkoxides and chemically modified alkoxides in nonaqueous solvents.

2. *Aluminates*

   2.1. *Solution chemistry of inorganic precursors* describes the hydrolysis and condensation of aluminum salts such as nitrates in aqueous solution.

   2.2. *Metal alkoxides* discusses the hydrolysis and condensation of aluminum alkoxides resulting in alumina sols or gels.

3. *Borate Systems*

    3.1. *Aqueous borates* describes oligomeric borate species formed in aqueous solutions of boric acid at intermediate pH.

    3.2. *Alkoxide-derived borate gels* establishes criteria for gel formation from metal alkoxide precursors in nonaqueous solvents.

1.

TRANSITION METALS

Transition metal oxide gels include one of the most successful sol-gel products, IROX™ TiO_2 coatings on architectural glass [1], produced by Schott Glaswerke, Mainz, West Germany. Transition metal oxide gels are also the basis of several important thin-film ferroelectric materials such as barium titanate [2] and PLZT [3] as well as semiconducting V_2O_5 films [4], electrochromic WO_3 films [5], and magnetic ferrite films or particles [6]. Recently transition metal oxide gels have been used extensively in chemical routes to high-temperature superconducting ceramics in the $YBa_2Cu_3O_{7-x}$ and $La_{2-x}Sr_xCuO_{4-y}$ systems. (See, e.g., [7].)

    This section is largely based on an excellent review of transition metal oxide gels by Livage *et al.* [8]. We discuss *hydrolysis* and *condensation* of several important inorganic and metal organic precursors to transition metal oxides. Concepts related to inorganic polymerization are introduced that are used throughout this chapter and in the following chapter on silicates.

1.1. Solution Chemistry of Inorganic Precursors

1.1.1. HYDROLYSIS

When dissolved in pure water, metal cations, M^{z+}, often introduced as salts, are solvated by water molecules according to:

$$M^{z+} + :O\begin{matrix} \diagup H \\ \diagdown H \end{matrix} \rightarrow \left[M \leftarrow O \begin{matrix} \diagup H \\ \diagdown H \end{matrix} \right]^{z+}. \tag{1}$$

For transition metal cations, charge transfer occurs from the filled $3a_1{}^\dagger$ bonding orbital of the water molecule to the empty d orbitals of the transition metal [8]. This in turn causes the partial charge on the hydrogen to increase, making the water molecule more acidic. Depending on the water acidity and, hence, the magnitude of the charge transfer, the following equilibria are

---

†Symmetry species of bonding orbital of water.

established, which we define as *hydrolysis* [8]:

$$[M(OH_2)]^{z+} \rightleftarrows [M–OH]^{(z-1)+} + H^+ \rightleftarrows [M{=}O]^{(z-2)+} + 2H^+. \quad (2)$$

Equation 2 defines the three types of ligands present in noncomplexing aqueous media:

$$\begin{array}{ccc} M–(OH_2) & M–OH & M{=}O \\ \text{Aquo} & \text{Hydroxo} & \text{Oxo} \end{array} .$$

The rough formula of any inorganic precursor can then be written as $[MO_N H_{2N-h}]^{(z-h)+}$, where N is the *coordination number of water molecules* around M and h is defined as the *molar ratio of hydrolysis*. When $h = 0$, the precursor is an *aquo-ion*, $[MO_N H_{2N}]^{z+}$, whereas for $h = 2N$, the precursor is an *oxy-ion*, $[MO_N]^{(2N-z)-}$. If $0 < h < 2N$, the precursor can be either an *oxo-hydroxo* complex, $[MO_x(OH)_{N-x}]^{(N+x-z)-}$ $(h > N)$; a *hydroxo-aquo* complex, $[M(OH)_x(OH_2)_{N-x}]^{(z-x)+}$ $(h < N)$; or a *hydroxo* complex, $[M(OH)_N]^{(N-z)-}$ $(h = N)$.

In general, hydrolysis is facilitated by increases in the charge density on the metal, the number of metal ions bridged by a hydroxo or oxo ligand, and the number of hydrogens contained in the ligand [9]. Hydrolysis is inhibited as the number of hydroxo ligands coordinating M increases [9].

The precise nature of the complex depends on the charge, z, coordination number, N, and electronegativity, χ_M^0, of the metal and the pH of the aqueous solution [8]. It is also necessary to consider the possible effects of ligand field stabilization which are most important for d^3 and d^8 ions [10]. The typical effects of charge and pH are shown schematically in Fig. 1, where three domains corresponding to aquo, hydroxo, and oxo ions are defined [11,12]. This diagram explains in a qualitative manner why the hydrolysis of low-valent cations ($z < 4$) yields aquo, hydroxo, or aquo-hydroxo complexes

Fig. 1.

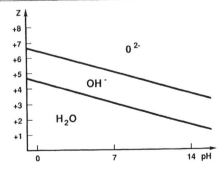

Charge versus pH diagram indicating the "aquo," "hydroxo," and "oxo" domains. From Kepert [11].

over the complete pH scale, whereas over the same range of pH high-valence cations ($z > 5$) form oxo or oxo-hydroxo complexes. Tetravalent metals are on the borderline; depending on pH they can form any of the possible complexes.

Livage and coworkers [8,13] have developed a *partial-charge model* in order to quantify the pH-charge relationship presented in Fig. 1. The basis of the model is that when two atoms combine, charge transfer occurs causing each atom to acquire a *partial positive* or *negative charge*, δ_i. According to the electronegativity equalization principle [14] (which is equivalent to the principle of chemical potential equalization in the equilibrium state), electron transfer stops when the electronegativities of all the atoms equal the mean electronegativity, $\bar{\chi}$. The partial-charge model does not account for the precise electronic structures and changes in coordination of the reactant species. However, this model provides insight into inorganic polymerization pathways, since in nucleophilic substitution reactions (S_N) and nucleophilic addition reactions (A_N) common to hydrolysis and condensation chemistry, the substituent with the largest partial negative charge, δ^-, is the *nucleophile*, and in S_N reactions the substituent with the largest partial positive charge, δ^+, is the *leaving group* or, according to the terminology of Livage *et al.* [8], nucleofugal (literally "fleeing the nucleus"). Nucleophilic reactions cease when the strongest nucleophile aquires a partial charge $\delta \geq 0$.

The partial-charge model can be used to calculate the magnitude of charge transfer between oxo, hydroxo, or aquo ligands and M^{z+} allowing calculation of the charge-pH diagram (Fig. 1). Under acidic conditions, spontaneous hydrolysis:

$$[MO_N H_{2N}]^{z+} + pH_2O \rightarrow [MO_N H_{2N-p}]^{(z-p)+} + pH_3O^+ \qquad (3)$$

is limited by the cleavage reaction of the O–H bond due to the polarizing power of M [8]. The reaction proceeds as long as $\delta(OH) > 0$ in the $[MO_N H_{2N-p}]^{(z-p)+}$ precursor. According to the partial charge model,

$$p = 1.45z - 0.45N - 1.07(2.71 - \chi_M^0)/\sqrt{\chi_M^0}. \qquad (4)$$

indicating that the number of protons removed depends on the charge, coordination number, and electronegativity of M.

Three limiting cases of Eq. 3 are encountered when applying Eq. 4:

$p = 0$, $(2N - p = 2N)$: $[M(OH_2)_N]^{z+}$ is not deprotonated, and a base must be added to cause hydrolysis. Examples include Ag^+ and Mn^{2+}.

$p > 2N$, $(2N - p < 0)$: the oxy-ion, $[MO_N]^{(2N-z)-}$, cannot be protonated by H_3O^+. A typical example is RuO_4^0.

$0 < p < 2N$, $(0 < 2N - p < 2N)$: under acid conditions two species corresponding to $h = E(p)$ and $h = E(p + 1)$ are in equilibrium, where $E(p)$

represents the integral part of p. Examples include Mn(VII), Cr(VI), V(V), Ti(IV), and Fe(III).

Under basic conditions, hydrolysis is limited by the cleavage of the M–OH bond due to the low polarizing power of the $[MO_N H_{2N-q}]^{(z-q)+}$ precursor [8]. Hydrolysis proceeds until $\delta(OH) = -1$ in $[MO_N H_{2N-q}]^{(z-q)+}$ where $2N - q$ corresponds to the number of protons that cannot be removed even at very high pH. From the partial charge model [8]:

$$q = 1 + 1.25z - 0.92(2.49 - \chi_M^0)/\sqrt{\chi_M^0}. \tag{5}$$

Two limiting cases are encountered when applying Eq. 5:

$q > 2N$, $(2N - q < 0)$: the most basic form of M is an oxy-ion $[MO_N]^{(2N-z)-}$. Examples include Ru(VIII) and Mn(VII).

$0 < q < 2N$, $(0 < 2N - q < 2N)$: two species corresponding to $h = E(q)$ and $h = E(q + 1)$ are in equilibrium at very high pH:

either oxo-hydroxo complexes such as, V(V), Ti(IV), Zr(IV), and Fe(III), or hydroxo-aquo complexes such as Mn(II) or Ag(I).

1.1.2. CONDENSATION

Condensation can proceed by either of two nucleophilic mechanisms depending on the coordination of the metal. When the preferred coordination is satisfied, condensation occurs by *nucleophilic substitution* (S_N):

$$M_1-OX + M_2-OY \rightarrow M_1-\overset{\displaystyle X}{\overset{\displaystyle |}{O}}-M_2 + OY. \tag{6}$$

When the preferred coordination is not satisfied, condensation can occur by *nucleophilic addition* (A_N):

$$M_1-OX + M_2-OY \rightarrow M_1-\overset{\displaystyle X}{\overset{\displaystyle |}{O}}-M_2-OY \tag{7}$$

with an attendant increase in the coordination number of M_2[†].

According to the partial charge model, oxo-ligands contained in oxy-ions $[MO_N]^{(2N-z)-}$ (predominant species in the high-pH/high-z domain, Fig. 1)

[†] Reactions 6 and 7 require that the coordination of oxygen increase from 2 to 3. Many examples of tri-coordinated oxygen exist, rutile (TiO_2) being the most common to the ceramist. The creation of the additional bond involves a lone pair electron on oxygen, and the bond formed may or may not be equivalent to the other two bonds.

are good nucleophiles, $\delta(0) \ll 0$, but poor leaving groups.[†] Therefore, condensation occurs only by addition reactions when at least one of the reactant species is *coordinatively unsaturated* (maximum coordination number, N, is less than the oxidation state, z). Aquo-ligands in aquo-ions $[M(OH_2)_N]^{z+}$ (predominant species in the low-pH/low-z domain, Fig. 1) are good leaving groups, $\delta(H_2O) > 0$, but poor nucleophiles. Condensation does not occur, since no attacking group is present. Hydroxo-ions present at intermediate pH and intermediate z contain both good nucleophiles (O or OH) and good leaving groups (H_2O or OH). Condensation occurs as soon as one OH is present in the coordination sphere of M. Referring to the charge-pH diagram (Fig. 1), it is generally necessary to be in the hydroxo domain to generate condensed species (except for the case of coordinatively unsaturated precursors). This is done by adding a base or oxidizing agent to aquo-ions or by adding an acid or reducing agent to oxy-ions.

1.1.2.1.   Olation

Olation is a condensation process in which a *hydroxy bridge* ("ol" bridge) is formed between two metal centers [15]. For coordinatively saturated hydroxo-aquo precursors, olation occurs by an S_N mechanism where the hydroxy group is the nucleophile and H_2O is the leaving group. According to Ardon *et al.* [16,17], olation occurs via a reaction intermediate involving H_3O_2 bridging ligands:

$$(8)$$

[†] Leaving groups require partial charges greater than 0.

Fig. 2. _____

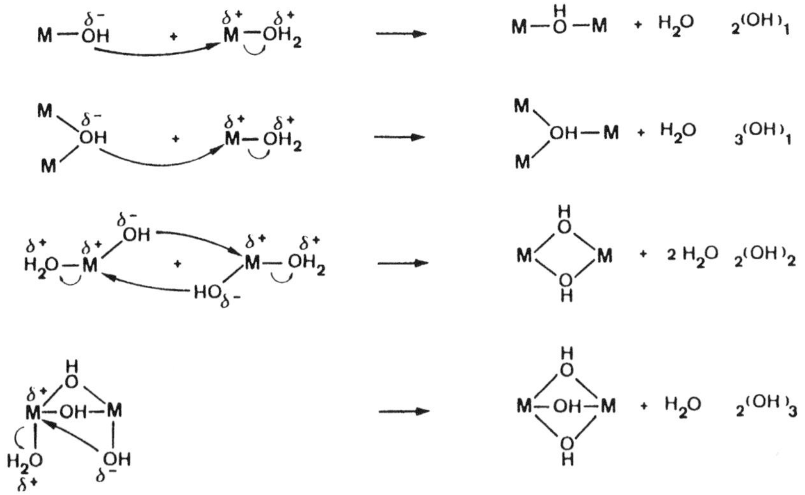

Olation mechanisms. From Livage et al. [8]. $_x(OH)_y$ terminology defined in text.

Several types of OH bridges can be formed as shown in Fig. 2 where the terminology $_x(OH)_y$, defines the number of M atoms bridged by a single OH (x) and the number of bridges formed between these x metal centers (y) [18]. Since H_2O is the leaving group, the kinetics of olation are related to the *lability* of the aquo ligand (ability to dissociate), which depends on size, electronegativity [19,20], and the electronic configuration of M [10,21]. In general, the smaller the charge and the larger the size, the greater the rate of olation. (See Fig. 3.)

Fig. 3. _____

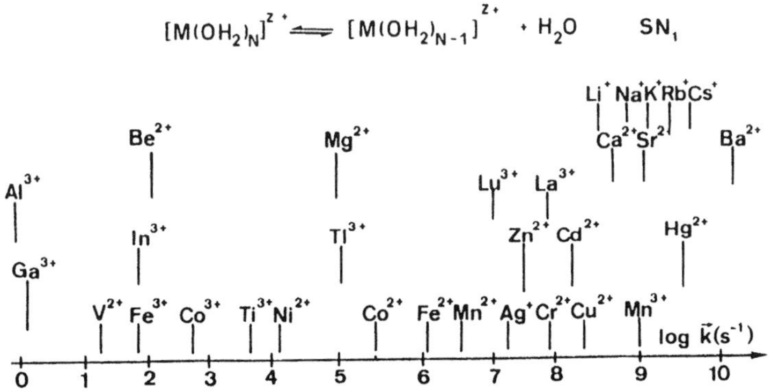

$$[M(OH_2)_N]^{z+} \rightleftharpoons [M(OH_2)_{N-1}]^{z+} + H_2O \qquad SN_1$$

Lability of some aquo-ions ranked according to the dissociative rate constant. From Livage et al. [8], based on the data of Eigen [19] and Kruger [20].

Fig. 4.

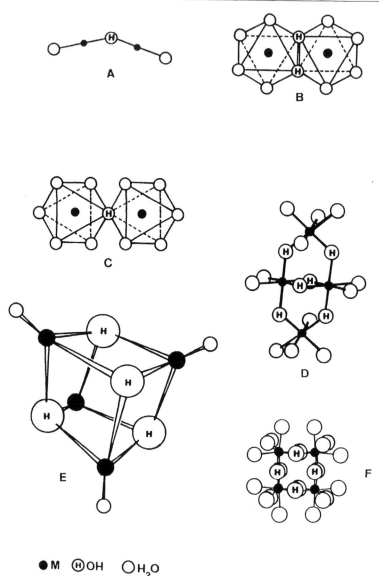

● M   (H)OH   ◯ H₂O

Transition metal polycations [8]:
(a) $[M_2(OH)(OH_2)_x]^{3+}$, M = Mn^{2+}, Co^{2+}, Ni^{2+} [10]; (b) $[M_2(OH)_2(OH_2)_x]^{(2z-2)+}$,
M = VO^{2+}, Cr^{3+}, Fe^{3+}, Ti^{3+}, Cu^{2+} [10]; (c) $[Cr_2(OH)(OH_2)_{10}]^{5+}$ [23];
(d) $[M_4(OH)_6(OH_2)_{12}]^{6+}$, M = Cr^{3+}, Co^{3+} [24]; (e)$[M_4(OH)_4(OH_2)_4]^{4+}$, M = Co^{2+},
Ni^{2+} [22]; (f) $[M_4(OH)_8(OH_2)_{16}]^{8+}$, M = Zr^{4+}, Hf^{4+}, [25,26].

Due to the S_N mechanism, olation reactions between charged cationic precursors $(z - h \geq 1)$ stop when $\delta(OH) \geq 0$ [8]. As electron-donating aquo ligands are removed during olation, $\delta(OH)$ becomes progressively less negative, approaching zero. Depending on the nucleophilic power of the starting precursor, condensation in certain pH regimes is limited to dimers $(Mn^{2+}, VO^{2+}, Fe^{3+}, Ti^{3+}, Cu^{2+})$ or tetramers $(Zr^{4+}, Hf^{4+}, Co^{3+})$. Structures of several transition metal *polycations* formed via olation are shown in Fig. 4 [10,22-26]. These *oligomers* (polymer containing a few repeat units) are the final products of hydrolysis and condensation in a narrow range of pH.

1.1.2.2. Oxolation

Oxolation is a condensation reaction in which an *oxo bridge* $(-O-)$ is formed between two metal centers. When the metal is coordinately unsaturated, oxolation occurs by nucleophilic addition (A_N) [27-29] with rapid kinetics $(>10^5 M^{-1}s^{-1}$ [30]) leading to edge- or face-shared polyhedra:

$$-M\overset{O}{\underset{O}{<}} + \overset{O}{\underset{O}{>}}M- \;\rightarrow\; -M\overset{O}{\underset{O}{<>}}M- \qquad _2(O)_2 \qquad (9)$$

$$-M\overset{O}{\underset{O}{<}} + O-M \;\rightarrow\; -M\overset{O}{\underset{O}{<}}O-M- \qquad _2(O)_3. \qquad (10)$$

For coordinatively saturated metals, oxolation proceeds by a two-step S_N reaction between oxyhydroxy precursors involving nucleophilic addition (Eq. 11) followed by water elimination to form a $M-O-M$ bond (Eq. 12):

$$M-OH + M-OH \;\rightarrow\; M-\overset{\overset{H}{|}}{O}-M-OH \qquad (11)$$

$$M-\overset{\overset{H}{|}}{O}-\overset{\overset{OH}{|}}{M} \;\rightarrow\; M-O-M + H_2O. \qquad (12)$$

(Olation rather than oxolation would preferentially occur for aquohydroxy precursors containing good H_2O leaving groups.) The first step (Eq. 11) is catalyzed by bases that deprotonate the hydroxo ligands creating stronger nucleophiles:

$$M-OH + OH^- \;\rightarrow\; M-O^- + H_2O \qquad (13)$$

$$M-O^- + M-OH \;\rightarrow\; M-O-M + OH^-. \qquad (14)$$

The second step (Eq. 12) is catalyzed by acids that protonate hydroxo ligands creating better leaving groups:

$$
\overset{\displaystyle H}{\underset{\displaystyle |}{M}}-O-M-OH + H_3O^+ \rightarrow [M-\overset{\displaystyle H}{\underset{\displaystyle |}{O}}-M-OH_2]^+ + H_2O \qquad (15)
$$

and

$$
M-O-M + H_3O^+ \xleftarrow{\ H_2O\ } [M-\overset{\displaystyle \downarrow H}{\underset{\displaystyle |}{O}}-M]^+ + H_2O. \qquad (16)
$$

Thus compared to olation, oxolation occurs over a wider range of pH, but due to the two-step process, kinetics are slower and never diffusion-controlled. Generally oxolation kinetics are minimized at the isoelectric point (IEP) where $[MO_{z-N}(OH)_{2N-z}]^0$ species are predominant and hence neither step is catalyzed according to Eqs. 13–16.

1.1.3. POLYMERIC SPECIES AND GELATION

Near the IEP, neutral precursors ($h = z$) are able to condense indefinitely via olation and/or oxolation reactions to form metal hydroxide or oxyhydroxide products, depending on $\delta(H_2O)$: when $\delta(H_2O) < 0$, hydroxide products are isolated [31]; when $\delta(H_2O) > 0$, oxyhydroxides are formed as metastable intermediates to fully condensed oxides, $MO_{z/2}$.

Whether precipitation or gelation occurs depends not only on processing factors (such as pH gradients, temperature, and speed of mixing) [32], but also on the condensation kinetics. For example, homogeneous gels can be formed by adding base to Cr^{3+} aquo-ions ($3d^3$ ions) that have a rather low dimerization rate via olation: $k = 10^{-5} M^{-1}s^{-1}$[†] at 25°C [33]. By comparison Fe^{3+} ions ($3d^5$ ions) that exhibit no ligand field stabilization[††] exhibit a high dimerization rate, $k = 450\, M^{-1}s^{-1}$ at 25°C [34] and form only precipitates [8].

Polyanions are formed at rather high pH from oxo-hydroxo precursors via oxolation reactions, for example:

$$
2[CrO_2(OH)_2]^0 \rightarrow [(HO)O_2Cr-O-CrO_2(OH)]^0 + H_2O \qquad (17)
$$

$$
\delta(OH) = -0.01 \qquad\qquad \delta(OH) = +0.04
$$

followed by deprotonation of the dimer:

$$
[Cr_2O_5(OH)_2]^0 \rightarrow [Cr_2O_6(OH)]^- + H^+ \qquad (18)
$$

$$
[Cr_2O_6(OH)]^- \rightarrow [Cr_2O_7]^{2-} + H^+. \qquad (19)
$$

[†] The units $M^{-1}s^{-1}$ are equivalent to $l\,mol^{-1}\,sec^{-1}$ appropriate for a second order rate constant.

[††] Stability conferred by ligand field splitting [31].

Fig. 5. _____

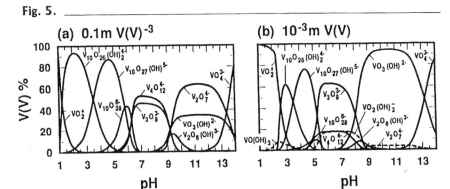

The distribution of vanadate species at 25°C in 0.1-m V(V) and 0.001-m V(V) solutions; total ion concentration $I = 1$ m. From Baes and Mesmer [10].

The acidities of these oligomers have led to a second name, polyacids. Unlike aquo-hydroxo precursors, the extent of condensation via oxolation in the first step is limited by the loss of nucleophilic power of OH with the extent of condensation. For example, polycondensation of $[VO(OH)_3]^0$ ceases after an extent of oligomerization of 10:

$$10[VO(OH)_3]^0 \rightarrow [H_6V_{10}O_{28}]^0 + 12H_2O$$

$$\delta(OH) = -0.09 \qquad \delta(OH) = +0.003. \tag{20}$$

Subsequent deprotonation at high pH produces the polyanion:

$$[H_6V_{10}O_{28}]^0 \rightarrow [V_{10}O_{28}]^{6-} + 6H^+. \tag{21}$$

Decavanadate species, written generally as $V_{10}O_{28-z}(OH)_z^{(6-z)-}$, are the predominant V(V) species at moderate acidities [10] and may also form by polymerization of $[VO_2(OH_2)_4]^+$ precursors. (See, e.g., Fig. 5.) Structures of several common polyanions are shown in Figs 6a and 6b. More compact polyanions (Fig. 6a) are formed when the kinetics of condensation via A_N or oxolation mechanisms are rapid, whereas slower kinetics result in more open structures (Fig. 6b) and generally allow the formation of clear gels when an acid is added.

1.1.4. STRUCTURAL STUDIES

The pentavalent V(V) system is a rather well studied example of gelation in aqueous media. Gels are prepared by the addition of acid to vanadate salts, hydration of V_2O_5 [64], or by proton exchange of sodium or ammonium

Fig. 6a.

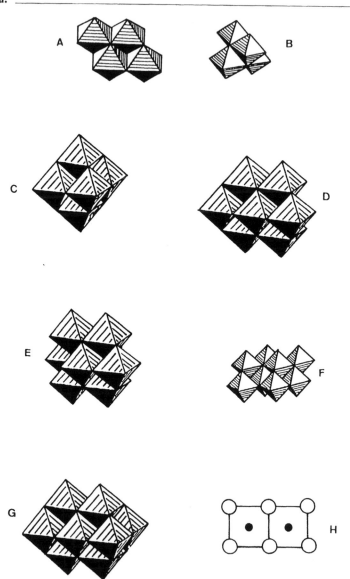

Structures of some compact polyanions [8]:
(a) $[W_4O_{12}(OH)_4]^{4-}$ [35,36]; (b) $[W_4O_{16}]^{8-}$ [35,36]; (c) $[M_6O_{19}]^{8-}$ M = Nb, Ta [37,38], $[M_6O_{19}]^{2-}$ M = W [39]; Mo [40,41]; (d) $[M_7O_{24}]^{6-}$ M = W [35,36]; Mo [42,43]; (e) β-$[Mo_8O_{26}]^{4-}$ [44–47]; (f) $[Mo_8O_{26}(OH)_2]^{6-}$ [48,49]; (g) $[M_{10}O_{28}]^{6-}$ M = V [50–52]; Nb [53]; (h) $[Au_2O_6]^{6-}$ [31].

Fig. 6b.

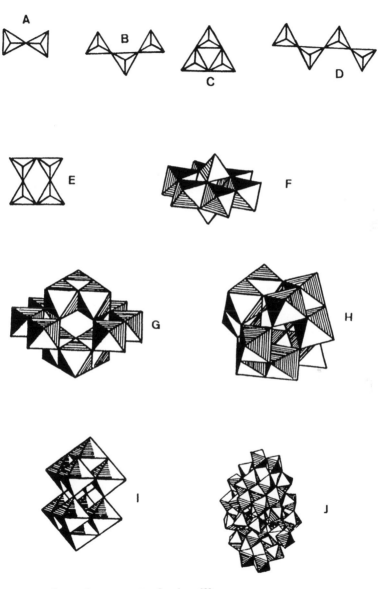

Structures of some less compact polyanions [8]:
(a) $[M_2O_7]^{2-}$ M = Cr [54]; Mo [35]; $[M_2O_7]^{4-}$ M = V [55]; (b) $[Cr_3O_{10}]^{2-}$ [56];
(c) $[V_3O_9]^{3-}$ [55]; (d) $[Cr_4O_{13}]^{4-}$ [57]; (e) $[V_4O_{12}]^{4-}$ [35,55]; (f) α-$[Mo_8O_{26}]^{4-}$ [46,57,58];
(g) $[H_2W_{12}O_{42}]^{10-}$ [59]; (h) $[H_2W_{12}O_{40}]^{6-}$ [60]; (i) $[W_{10}O_{32}]^{4-}$ [61,62];
(j) $[Mo_{36}O_{112}(OH_2)_{16}]^{8-}$ [63].

Fig. 7.

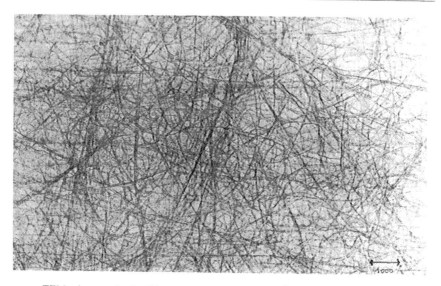

TEM micrograph of a fibrous V_2O_5 gel. Bar = 1000 Å. From Livage *et al.* [8].

metavanadate solutions [65]. To explain the fibrous structure of vanadium pentoxide gels (Fig. 7), Livage *et al.* propose the following growth model [8]. Acidification of vanadate solutions causes protonation leading to the tetrahedrally coordinated $h = 5$ $[VO(OH)_3]^0$ precursor in which V is quite electrophilic ($\delta(V) = +0.62$). Any nucleophilic ligands then cause the coordination of V to increase. In aqueous solution, nucleophilic addition of water forms a sixfold coordinated species containing a long $V-OH_2$ bond along the z-axis opposite the short $V=O$ double bond:

$$\tag{22}$$

Since water is a good leaving group ($\delta(H_2O) = +0.10$) and the hydroxyl ligands are quite nucleophilic ($\delta(OH) = -0.14$), olation reactions involving OH ligands in *trans* positions (from Latin, meaning *across*) can readily occur leading to $[VO(OH)_3(OH_2)]_n$ chains composed of octahedral units linked by

$_2(OH)_1$ bridges:

$$(23)$$

Subsequent oxolation reactions between chains convert unstable $_2(OH)_1$ bridges into stable $_3(O)_1$ bridges forming double chains:

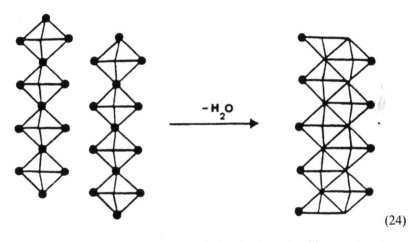

$$(24)$$

Further oxolation reactions between chains lead to the fibrous structures evident by TEM (Fig. 7). Electron and X-ray diffraction studies have shown that the fibers are actually flat ribbons ~10 nm wide and 1 nm thick [66]. These ribbons in turn are composed of fibrils ~2.7 nm wide linked side by side. Since compact decavanadate species are formed by the proton exchange method, these polyions must restructure to form the fibrous structures portrayed in Fig. 7. The mechanistic details of this restructuring process are currently not understood.

Gelation of vanadate systems depends on the solution concentration [67]. In proton exchanged systems, decavanadic acid (mol/wt. = 1000 g) predominates below 10^{-3} M. At higher concentrations, polymeric species are observed (mol. wt. = 2×10^6 g). Above 2×10^{-2} M, aggregation occurs leading to gelation if the vanadium concentration exceeds 0.1 M. Livage *et al.* report that decavanadic acid and fibrous polymeric species coexist at the gel point [8].

Bunker and Keefer [68] have investigated hydrolysis and condensation of tetravalent Zr (IV) systems. Above ~pH 0, aquo-ions spontaneously hydrolyze and undergo olation to form the $[Zr_4(OH)_8(OH_2)_{16}]^{8+}$ polycation exhibiting eightfold coordination (Fig. 4f) according to the following probable scheme:

$$4[Zr(OH_2)_6]^{4+} \rightarrow 4[Zr(OH)_2(OH_2)_4]^{2+} + 8H_3O^+ \qquad (25)$$

$$4[Zr(OH)_2(OH_2)_4]^{2+} \rightarrow [Zr_4(OH)_8(OH_2)_{16}]^{8+}. \qquad (26)$$

Corresponding small-angle X-ray scattering (SAXS) data [68] indicate the presence of small clusters, radius of gyration, $R_g \sim 0.4$ nm, consistent with these polynuclear species. At higher pH, further hydrolysis of the terminal aquo-ligands results in rapid condensation yielding gelatinous precipitates or particles depending on the solution concentration. SAXS investigations of this second stage of growth are consistent with *diffusion-limited aggregation*[†] [69] of the primary tetramer units, not classical nucleation and growth.

In a related study, Bleier and Cannon [70] observed that aging zirconyl salt solutions at 90 to 98°C produced quite uniform monoclinic ZrO_2. They proposed that tetramers (Eq. 26, Fig. 4f) aggregate to produce nuclei (comprising 12–24 tetramers), and further aging results in 3-nm crystallites. A secondary controlled aggregation process involving these crystallites produces 80 nm particles that exhibited an unusually high degree of crystallographic alignment of the primary crystallites presumably due to anisotropy in the surrounding double layers. The key features of the growth model proposed by Bleier and Cannon—viz, (a) generation of discrete solute species, in this case $[Zr_4(OH)_8(OH_2)_{16}]^{8+}$; (b) slow growth–aggregation of these species to form stable nuclei and subsequently primary crystallites; and (c) secondary aggregation of primary crystallites—may pertain to a variety of polyvalent metal oxide sols including FeOOH [71,72], TiO_2 [73,74], and GeO_2 [75] as well as the V_2O_5 system described previously. For more details concerning the structures of particulate sols, refer to the discussion of coacervates and tactoids in Chapter 4.

Trivalent Cr(III) and Fe(III) gels or precipitates can be prepared by the addition of base to sulphate, nitrate, or chloride precursors. Despite similar electronegativities and coordination of the metals, the hydrolysis kinetics are somewhat different:

$$[Cr(OH_2)_6]^{3+} + H_2O \rightarrow [Cr(OH)(OH_2)_5]^{2+} + H_3O^+$$

$$\vec{k}_1 = 1.4 \times 10^5 \, s^{-1} \, [33] \qquad (27)$$

$$[Fe(OH_2)_6]^{3+} + H_2O \rightarrow [Fe(OH)(OH_2)_5]^{2+} + H_3O^+$$

$$\vec{k}_1 = 3.0 \times 10^7 \, s^{-1} \, [34]. \qquad (28)$$

[†]Discussions of aggregation processes appear in Chapter 3 (Section 2.6.5.2.) and in Chapters 4 and 5.

These kinetic differences reflect in part the ligand field stabilization of Cr(III), a $3d^3$ ion, compared to Fe(III), a $3d^5$ ion. It is observed that slower kinetics facilitate gel formation in Cr(III) systems where the average gel stoichiometry corresponds to $[Cr(OH)_3(OH_2)_3] \cdot nH_2O$ [76]. Gelatinous precipitates form in Fe(III) systems having a composition between α-FeOOH (goethite) and α-Fe$_2$O$_3$ (hematite) [77,78].

According to Flynn [79], the hydrolysis of inorganic Fe(III) solutions consists of (1) formation of low–molecular-weight species; (2) further condensation to form a red cationic polymer; (3) aging of the polymer causing conversion to oxide phases; or (4) precipitation of oxide phases directly from low–molecular-weight precursors. These steps are illustrated schematically in Fig. 8. Possible polymerization mechanisms are as follows [71,72,77,78–88].

The $h = 1$ $[Fe(OH)(OH_2)_5]^{2+}$ precursor undergoes dimerization via olation–oxolation reactions to nucleate γ-FeOOH (lepidocrocite), a phase isomorphous with γ-AlOOH (boehmite), comprising double chains of

Fig. 8.

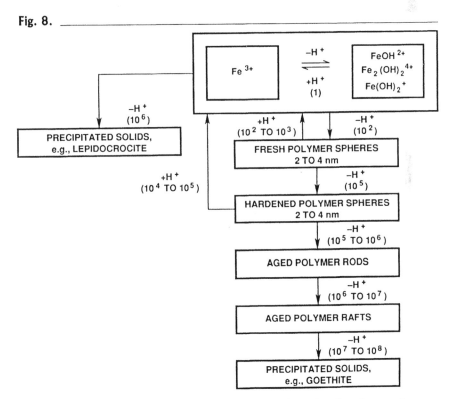

Stages of hydrolysis and condensation in Fe(III) solutions. Numbers in parentheses indicate reaction times in seconds at 25°C. From Flynn [79].

Fig. 9a.

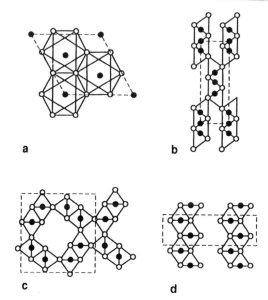

Schematic crystal structures of Fe(III) oxides and oxyhydroxides: (a) Hematite (α-Fe$_2$O$_3$); (b) Goethite (α-FeO(OH)); (c) Akaganeite (β-FeO(OH)); (d) Lepidocrocite (γ-FeO(OH)). From Flynn [79].

Fig. 9b.

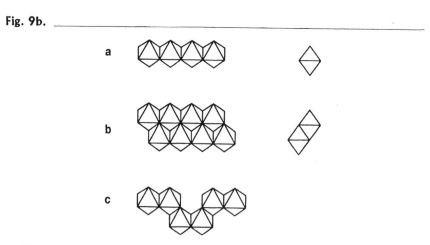

Chains of Fe(O, OH, OH$_2$)$_6$ octahedra: (a) single straight chain; (b) double straight chain; (c) single kinked chain. From Flynn [79].

Fe(O,OH)$_6$ octahedra. (See Fig. 9.) The $h = 2$ [Fe(OH)$_2$(OH$_2$)$_4$]$^{1+}$ precursor condenses to form polycationic spheres 2–4 nm in diameter, containing $\sim 10^2$ Fe(III) ions, whose composition is [Fe$_4$O$_3$(OH)$_4$]$_n^{2n+}$. Further aging or addition of base results in aggregation of these primary particles, forming needles of α-FeOOH having the same diameter as the original polycation. A second ordered aggregation process produces rods that organize into tactoids responsible for the gelatinous appearance of the precipitate.

Flynn [79] suggests that the polymer forms by oxolation reactions between chains of edge-shared Fe(O, OH, OH$_2$)$_6$ octahedra resulting in double chain structures of α or β-FeOOH. (See Fig. 9.) The formation of γ-FeOOH cannot proceed as simply from the double chains consistent with its formation only from low–molecular-weight precursors.

1.1.5. ROLE OF THE ANION

In addition to the aquo, hydroxo, and oxo ligands discussed in the previous sections, most aqueous systems also contain counterions introduced by dissolution of inorganic salts in water. These anions compete with aquo ligands for coordination to the metal centers and, in many cases, strongly affect the evolving particle morphology and stability as demonstrated many times by Matijević [89–92]. (See, e.g., Fig. 10.) Whether or not an anion,

Fig. 10. _____

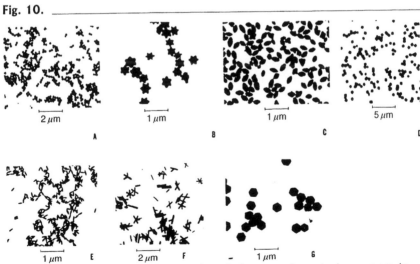

Morphologies of particles as a function of the type of counter-ions present in solution. From Matijević [89–92]. (a) Cl$^-$ (α-Fe$_2$O$_3$) [89]; (b) ClO$_4^-$ (α-Fe$_2$O$_3$) [91]; (c) NO$_3^-$ (α-Fe$_2$O$_3$) [91]; (d) Cl$^-$/EtOH (α-Fe$_2$O$_3$) [92]; (e) H$_2$PO$_4^-$ (α-Fe$_2$O$_3$) [89]; (f) Cl$^-$ (β-FeO(OH)) [89]; (g) HSO$_4^-$ (Fe$_3$(OH)$_5$(SO$_4$)$_2 \cdot$2H$_2$O) [90].

X, can coordinate M (and hence influence hydrolysis and condensation as well as the double-layer composition and solution ionic strength) has been discussed by Livage *et al.* [8] on the basis of the partial-charge model.

Complexation to form associated species $[M(OH)_h(X)(OH_2)_{N-h-1}]^{(z-h-1)+}$ can occur when positively charged precursors and negatively charged anions, X^-, are both present in solution [93–98]. Complexation occurs by nucleophilic substitution when M is coordinatively saturated. The complexing ability of X^- depends mainly on the extent of charge transfer from X to M in the M–X bond causing a charge variation, Δx, of the anion. However, since H_2O is both a solvent with a high dielectric constant ($\varepsilon = 80$) and a moderately strong nucleophile, the complex must be stable against both ionic dissociation and hydrolysis in aqueous solution in order to influence the particle morphology and stability. In general, stability is conferred when X^- is less electronegative than water resulting in a more covalent M–X bond that is stable against dissociation and when the partial charge of the conjugate acid $\delta(HX) < 0$, so that the protonated species remains attracted by the positively charged metal.

Table 1 lists the mean electronegativities of a series of common anions and the partial charges $\delta(X)$ and $\delta(HX)$ in $[Fe(X)(OH_2)_4]^{2+}$ and $Fe(OH)(HX)(OH_2)_4]^{2+}$, respectively, corresponding to the following equilibria:

Ionic dissociation:
$$[Fe(X)(OH_2)_4]^{2+}_{aq} \xrightleftharpoons{H_2O} [Fe(OH_2)_6]^{3+}_{aq} + X^-_{aq} \tag{29}$$

Hydrolysis:
$$[Fe(OH)(HX)(OH_2)_4]^{2+}_{aq} \xrightleftharpoons{H_2O} [Fe(OH)(OH_2)_5]^{2+}_{aq} + HX_{aq}. \tag{30}$$

The susceptibility toward ionic dissociation decreases as the electronegativity, $\bar{\chi}$, of the anion decreases, whereas the susceptibility toward hydrolysis increases with $\bar{\chi}$. For the Fe(III) precursor referred to in Eqs. 29 and 30, the associated complex is stable when the charge variation of the anion (Δx)

Table 1.

Partial Charges $\delta(X)$ and $\delta(HX)$ in $[Fe(X)(OH_2)_4]^{2+}$ and $[Fe(OH)(HX)(OH_2)_4]^{2+}$ Species, Respectively, as a Function of the Mean Electronegativity $\bar{\chi}$ of the Anion X^-_{aq}.

X^-	ClO_4^-	NO_3^-	HSO_4^-	HCO_3^-	Cl^-aq	CH_3COO^-
$\bar{\chi}$	2.86	2.76	2.64	2.49	2.40	2.20
$\delta(X)$	-0.92	-0.84	-0.50	-0.34	-0.06	$+0.40$
Δx	$+0.08$	$+0.18$	$+0.50$	$+0.66$	$+0.94$	$+1.40$
$\delta(HX)$	-0.52	-0.42	-0.15	$+0.02$	$+0.25$	$+0.70$

Source: Livage *et al.* [8].

Fig. 11.

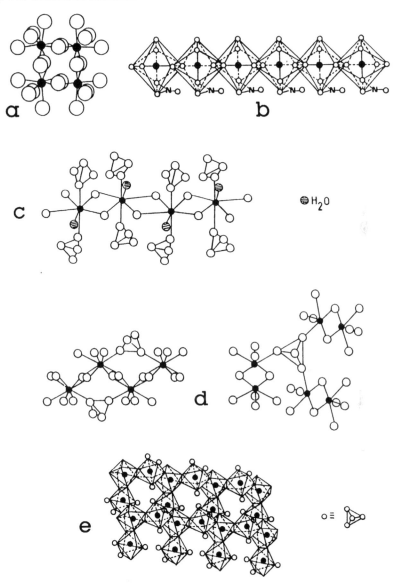

X-ray structures of some basic salts of zirconium. From Livage *et al.* [8]:
(a) $[Zr_4(OH)_8(OH_2)_{16}]^{8+}$, $8ClO_4^-$ [25,26]; (b) $Zr(OH)_2(NO_3)(OH_2)_2]_n^{n+}$, nNO_3^- [101];
(c) $[Zr(OH)_2]_n^{2n+}$, nSO_4^{2-} [101]; (d) $[Zr_3(OH)_6(CrO_4)]_n^{4n+}$, $2nCrO_4^{2-}$ [102];
$[Zr_3(OH)_6(HPO_4)]_n^{4n+}$, $4nH_2PO_4^-$ [103]; (e) $[Zr_4(OH)_6(CrO_4)]_n^{8n+}$, $4nCrO_4^{2-}$
[104,105].

resulting from charge transfer is greater than 0 and when $\delta(HX) < 0$. From Table 1, these conditions are met for ClO_4^-, NO_3^-, and HSO_4^-. More electronegative anions than ClO_4^- result in ionic dissociation, and less electronegative anions than HSO_4^- result in hydrolysis. Stability of the associated complex is further influenced by the extent of hydrolysis of the precursor and therefore the solution pH. For example, highly electronegative anions such as perchlorate are able to coordinate Fe^{3+} only under very acidic conditions where $h = 0$ (see Table 1), whereas less electronegative anions such as HCO_3^- are able to coordinate Fe^{3+} only under basic conditions where h is large.

Some anions such as sulfate have mean electronegativities close to that of H_2O ($\bar{\chi} = 2.49$) and are therefore able to form stable complexes over a wide range of pH. Such anions strongly influence the morphology of colloids and precipitates. They may be bidentate and can remain coordinated to the metal in the final condensation product as terminal or bridging ligands, forming basic salts. The structures of several basic salts of zirconium are shown in Fig. 11. It should be noted in Fig. 11 that nitrate ligands serve only as terminal ligands, whereas sulfate anions (SO_4^{2-}) can serve as network-forming ligands bridging up to three $[Zr(OH_2)]_n^{2n+}$ chains.

In addition to these electrostatic factors it is necessary to consider the stabilizing influence of entropic and resonance effects observed with chelating anions such as EDTA and α-hydroxy acids that are often used to control precipitation processes [106–110].

1.2. Solution Chemistry of Metal Alkoxide Precursors

Transition *metal alkoxides*, $M(OR)_z$, especially those of the d^0 transition metals (Ti, Zr), are widely used as molecular precursors to glasses and ceramics. Metal alkoxides are in general very reactive due to the presence of highly electronegative OR groups (hard-π donors) that stabilize M in its highest oxidation state and render M very susceptible to nucleophilic attack [8,111]. However, several factors distinguish transition metal alkoxides from Group IV silicon alkoxides ($Si(OR)_4$), the most commonly used precursors in sol-gel processing [8]: (1) The lower electronegativity of transition metals causes them to be more electrophilic and thus less stable toward hydrolysis, condensation, and other nucleophilic reactions. (2) Transition metals often exhibit several stable coordinations, and when coordinatively unsaturated, they are able to expand their coordination via olation, oxolation, alkoxy bridging, or other nucleophilic association mechanisms. For example, transition metal alkoxides dissolved in nonpolar solvents often form oligomers via *alkoxy bridging*, an A_N mechanism similar to olation:

$$2\,M-OR \rightarrow M{\overset{\displaystyle OR}{\underset{\displaystyle OR}{\diagup\!\!\!\diagdown}}}M. \tag{31}$$

In polar solvents such as alcohol, either alkoxy bridging or alcohol association can occur. By comparison for $Si(OR)_4$, $N = z$, and neither oligomerization nor alcohol association is observed. (3) The greater reactivity of transition metal alkoxides requires that they be processed with stricter control of moisture and conditions of hydrolysis in order to prepare homogeneous gels rather than precipitates. (4) The generally rapid kinetics of nucleophilic reactions cause fundamental studies of hydrolysis and condensation of transition metal alkoxides to be much more difficult than for $Si(OR)_4$. In this regard it is noteworthy that Berglund and coworkers have developed a rapid mixing apparatus to study chemical and structural changes occurring at very short times [112].

1.2.1. MECHANISMS OF HYDROLYSIS AND CONDENSATION

For coordinatively saturated metals in the absence of catalyst, hydrolysis and condensation both occur by nucleophilic substitution (S_N) mechanisms involving nucleophilic addition (A_N) followed by proton transfer from the attacking molecule to an alkoxide or hydroxo-ligand within the transition state and removal of the protonated species as either alcohol (*alcoxolation*) or water (oxolation) [8,13,113]:

$$H-\underset{\underset{H}{|}}{O} + M-OR \longrightarrow \underset{H}{\overset{H}{\diagdown}}O: \rightarrow M-OR \longrightarrow HO-M \leftarrow O\overset{\diagup R}{\diagdown H}$$

$$\qquad\qquad (a)\qquad\qquad\quad (b)\qquad\qquad (c)$$

$$\longrightarrow M-OH + ROH \tag{32}$$

$$(d)$$

hydrolysis

$$M-\underset{\underset{H}{|}}{O} + M-OR \longrightarrow M-O: \rightarrow M-OR \longrightarrow M-O-M \leftarrow O\overset{\diagup R}{\diagdown H}$$

$$\qquad (a)\qquad\qquad\qquad\quad H\ (b)\qquad\qquad (c)$$

$$\longrightarrow M-O-M + ROH \tag{33}$$

$$(d)$$

alcoxolation

$$M-\underset{\underset{\text{(a)}}{H}}{\overset{|}{O}} + M-OH \longrightarrow M-\underset{\underset{H}{\diagdown}}{O:} \overset{\nearrow}{\to} M-OH \longrightarrow M-O-M \leftarrow :O\underset{\diagdown H}{\overset{\diagup H}{}}$$

$$\text{(b)} \qquad\qquad\qquad\qquad \text{(c)}$$

$$\longrightarrow M-O-M + H_2O. \qquad\qquad (34)$$

$$\text{(d)}$$

oxolation

When $N - z > 0$, condensation can occur by olation:

$$M-OH + M \leftarrow O\underset{\diagdown R}{\overset{\diagup H}{}} \longrightarrow M-\underset{\overset{|}{M}}{\overset{H}{O}}-M + ROH$$

$$M-OH + M \leftarrow O\underset{\diagdown H}{\overset{\diagup H}{}} \longrightarrow M-\underset{\overset{|}{M}}{\overset{H}{O}}-M + H_2O.$$

$$(35)$$

The thermodynamics of hydrolysis, alcoxolation, and oxolation are governed by the strength of the entering nucleophile, the electrophilicity of the metal, and the partial charge and stability of the leaving group. These reactions are favored when $\delta(O) \ll 0$, $\delta(M) \gg 0$, and $\delta(H_2O)$ or $\delta(ROH) > 0$ [8]. Calculation of the charge distribution within the transition state of a titanate dimer pertaining to either oxolation or alcoxolation $[Ti_2(OEt)_6(OH)_2]$ indicates that protonation of OEt produces a more positively charged leaving group ($\delta(EtOH) = +0.02$) than protonation of OH ($\delta(HOH) = -0.25$) [8]. Thus alcoxolation should be the favored condensation reaction between partially hydrolyzed, coordinatively saturated titanate precursors. It should be noted that although the same trend is predicted for silicates [8], oxolation rather than alcoxolation is the favored reaction between partially hydrolyzed monomers based on the results of ^{29}Si NMR investigations [114,115]. Perhaps solvent interactions that are not accounted for by the partial-charge model influence the quality of the leaving group.

Since the transition state involves an associative mechanism accompanied by a proton transfer, kinetics are governed by the extent of coordination unsaturation of the metal, $N - z$, and the transfer ability of the proton. Larger values of $(N - z)$ and greater acidities of the protons reduce the associated activation barriers and enhance the kinetics. The thermodynamics of olation depend on the strength of the entering nucleophile and the electrophilicity of the metal. The kinetics of olation are systematically

Table 2.

Positive Partial Charge $\delta(M)$ for Metals in Various Alkoxides.

Alkoxide	$Zr(OEt)_4$	$Ti(OEt)_4$	$Nb(OEt)_5$	$Ta(OEt)_5$	$VO(OEt)_3$	$W(OEt)_6$	$Si(OEt)_4$
$\delta(M)$	+0.65	+0.63	+0.53	+0.49	+0.46	+0.43	+0.32

Source: Livage *et al.* [8].

fast because $(N - z) > 0$ and because no proton transfer occurs in the transition state.

Transition metals are very electropositive. Values of $\delta(M)$ calculated from the partial charge model are listed in Table 2 where they are compared to $\delta(Si)$. Livage *et al.* use Table 2 to help explain why the hydrolysis and condensation kinetics of transition metal alkoxides are much faster than for $Si(OR)_4$:[†] literature values of the hydrolysis rate for $Si(OEt)_4$ range from $k_h = 10^{-4}$ to $10^{-6} \, M^{-1} \, s^{-1}$ at pH = 3 [114–121], which can be extrapolated to about $5 \times 10^{-9} \, M^{-1} \, s^{-1}$ at pH 7; a rough estimate of the minimum hydrolysis rate constant for $Ti(OR)_4$ [122–126] at pH 7 is $k_h = 10^{-3} \, M^{-1} \, s^{-1}$ [122], more than five orders of magnitude greater! A comparison of the condensation rates reveals $k_c(Si(OEt))_4 = 10^{-4} \, M^{-1} \, s^{-1}$ [127–128], whereas a global second-order rate constant of $k_c = 30 \, M^{-1} \, s^{-1}$ was observed for TiO_2 precipitated form $Ti(OEt)_4$ [126].

Another factor that influences reaction kinetics is the extent of oligomerization (*molecular complexity*) of the metal alkoxides. The molecular complexity depends on the nature of the metal atom. Within a particular group, it increases with the atomic size of the metal (see Table 3), thus explaining the tendency of divalent transition metal alkoxides (Cu, Fe, Ni, Co, Mn) to polymerize rendering them insoluble [129]. Molecular complexity

Table 3.

Molecular Complexity (Number of Metal Atoms per Osmotic Molecule) of Some Transition Metal Ethoxides as a Function of Metal Size.

Compound	$Ti(OEt)_4$	$Zr(OEt)_4$	$Hf(OEt)_4$	$Th(OEt)_4$
Covalent radii (Å)	1.32	1.45	1.44	1.55
Molecular complexity	2.9	3.6	3.6	6.0

Source: Livage *et al.* [8].

[†] As discussed earlier, Ti may be coordinatively unsaturated, allowing coordination expansion via olation, oxolation, and alkoxy bridging which also contribute to enhanced hydrolysis and condensation kinetics.

Fig. 12.

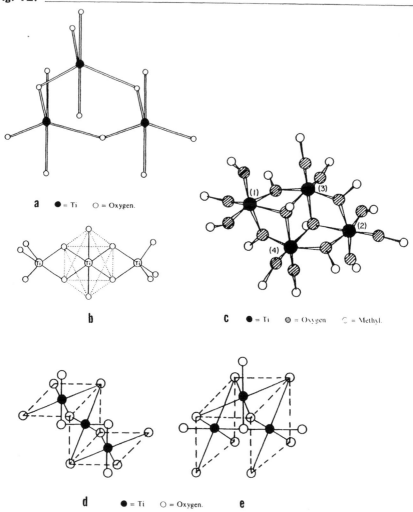

Possible structures of Ti(OR)$_4$ oligomers. From Bradley *et al.* [130]:
(a) [Ti(OEt)$_4$]$_3$; (b) Ti$_3$(OEt)$_{12}$; (c) [Ti(OMe)$_4$]$_4$; (d) [Ti(OEt)$_4$]$_3$; (e) [Ti(OEt)$_4$]$_3$.

also depends on the alkoxide ligand; for example, Ti(OEt)$_4$ exhibits an oligomeric structure, whereas Ti(OPr^i)$_4$ remains monomeric. Structures of several titanium alkoxide oligomers are shown in Fig. 12 [130].

It is observed that alkoxy bridges are more stable toward hydrolysis than associated solvent molecules and in some cases terminal OR ligands. Therefore, starting from a particular alkoxide, the kinetics and resulting structure can be controlled by appropriate choice of solvent [8]. For example, partial hydrolysis of $Zr(OPr^n)_4$ dissolved in the polar, protic solvent,[†] n-PrOH, results in a precipitate, whereas homogeneous gels are obtained by hydrolysis of $Zr(OPr^n)_4$ dissolved in nonpolar, aprotic solvents, such as cyclohexane [131]. These differences reflect the influence of molecular complexity on the hydrolysis kinetics: alkoxy-bridging occurs in cyclohexane allowing controlled hydrolysis, whereas alcohol association (rather than alkoxy bridging) occurs preferentially in n-PrOH resulting in rapid hydrolysis and formation of a highly condensed product. A similar difference is observed for titanium alkoxides: $Ti(OEt)_4$ dissolved in EtOH exhibits an oligomeric structure, and hydrolysis results in precipitation of monosized particles [123,124]. $Ti(OPr^i)_4$ dissolved in Pr^iOH is monomeric and hydrolysis results in rapid precipitation of a polydisperse product [123–125].

The size and electron-providing or -withdrawing characteristics of the organic ligand also affect the hydrolysis and condensation kinetics. For a series of titanium n-alkoxides ($Ti(OR^n)_4$), the hydrolysis rate decreases with the alkyl chain length [122,132] consistent with the steric effect expected for an associative S_N reaction mechanism. In addition, Livage *et al.* show a trend of decreasing $\delta(Ti)$ and $\delta(H)$ with alkyl chain length that should also contribute to slower kinetics [8]. The influence of the alkyl chain length on condensation is illustrated qualitatively by observations that precipitation cannot be avoided, even under mild hydrolysis conditions, when $R = Et$, Pr^n, or Pr^i [123–125], whereas stable sols are obtained when $R = Bu^n$ (mean molecular weight M = 5600 g) [122,133,134] or Am^t (M = 3800 g) [135]. The molecular weights correspond to Ti species containing at most several tens of Ti atoms. This suggests that, as observed in inorganic systems, the initial condensation products are oligomeric species that subsequently aggregate to form gels or precipitates. The oligomer size depends on R: the larger the size of R, the smaller the oligomer. The R group also exerts an influence on the morphology (particle size and surface area) and crysallization behavior of the resulting gel [136,137], perhaps by altering the size and structure of the primary oligomeric building blocks.

1.2.2. ROLE OF THE CATALYST

Acid or base catalysts can influence both the hydrolysis and condensation rates and the structure of the condensed product. Acids serve to protonate negatively charged alkoxide groups, enhancing the reaction kinetics by

---

[†] A solvent containing a labile proton such as alcohol or water. A discussion of solvent effects appears in Chapter 3 (Section 2.4.3.).

producing good leaving groups,

$$M-OR + H_3O^+ \longrightarrow M^+ \leftarrow :O\overset{\displaystyle H}{\underset{\displaystyle R}{\diagup}} + H_2O, \qquad (36)$$

and eliminating the requirement for proton transfer within the transition state. Hydrolysis goes to completion when sufficient water is added. The relative ease of protonation of different alkoxide ligands can influence the condensation pathway as demonstrated by consideration of a typical partially hydrolyzed polymer [8]:

$$
\begin{array}{cccc}
\text{OR} & \text{O} & \text{OR} & \text{OR} \\
\text{HO-Ti-O-} \cdots \text{-O-Ti-O-} \cdots \text{-O-Ti-O-} \cdots \text{-O-Ti-OR.} \\
\text{OR} & \text{OR} & \text{OR} & \text{OR} \\
\text{(A)} & \text{(B)} & \text{(C)} & \text{(D)}
\end{array}
$$

$$(37)$$

$\delta(OR)$ for the sites A–D calculated from the partial charge model are listed in Table 4. The ease of protonation decreases as $D \gg A > C \gg B$, which reflects the electron-providing power of the ligands, which decreases as alkoxy, hydroxo, oxo. Therefore acid-catalyzed condensation is directed preferentially toward the ends rather than the middles of chains, resulting in more extended, less highly branched polymers. This tendency is consistent with the observation that acid catalysts combined with low r values (water : metal ratio in the alkoxide) often result in monolithic gels [138,139] or spinnable sols [140,141]. High acid concentrations ($H^+/Ti \to 1$) severely retard the condensation kinetics [8]. Protonation of hydroxo ligands

Table 4.

Charge Distribution According to the
Partial-Charge Model within a Titanium
Oxo-Polymer.

Site	$\delta(OR)$	$\delta(Ti)$
A	-0.01	$+0.70$
B	$+0.22$	$+0.76$
C	$+0.04$	$+0.71$
D	-0.08	$+0.68$

Source: Livage *et al.* [8].

becomes possible resulting in the same aquo-hydroxo species observed in inorganic systems.

Alkaline conditions produce strong nucleophiles via deprotonation of hydroxo ligands:

$$L-OH + :B \rightarrow L-O^- + BH^+ \tag{38}$$

where L = M or H and B = OH^- or NH_3. Although base additions promote the hydrolysis of $Si(OR)_4$, Bradley [142] observed that the hydrolysis rate of $Ti(OBu^s)_4$ was less in basic conditions (NaOH) than in acidic or neutral conditions, perhaps because nucleophilic addition of OH^- reduces $\delta(Ti)$ [8]. Condensation kinetics are systematically enhanced under basic conditions. Based on values of $\delta(Ti)$ calculated from the partial-charge model for sites A–D on a typical partially hydrolyzed polymer (Table 4), the order of reactivity toward nucleophilic attack should decrease as $B \gg C \approx A > D$. Thus base-catalyzed condensation (as well as hydrolysis) should be directed toward the middles rather than the ends of chains, leading to more compact, highly branched species.

1.2.3. STRUCTURE OF CONDENSED PRODUCTS

The structure of condensed products depends on the relative rates of the four reactions: hydrolysis, oxolation, alcoxolation, and olation. The contributions of each of these reactions depend in turn on internal parameters such as the nature of the metal atom and alkyl groups and the molecular complexity as well as external parameters such as r, the choice of catalyst, concentration, solvent, and temperature. Unfortunately, due to the fast kinetics of the hydrolysis and condensation reactions, relatively little information is available concerning progressive structural evolution in transition metal oxide systems [112].

Hydrolysis and condensation of transition metal alkoxides is quite facile and hydroxide or hydrated oxide precipitates readily occur when $r \geqslant 2$ [130]. However for low r values and carefully controlled hydrolysis conditions, alcoxolation and alcolation result in oxo-alkoxide products that can be isolated as single crystals and identified by X-ray crystallography. Oxo-alkoxides are the organic counterparts of polyanions and polycations obtained in aqueous solution and many structural similarities exist. For example, $Ti_7O_4(OEt)_{20}$ is isostructural with $Mo_7O_{24}^{6-}$, and $Nb_8O_{10}(OEt)_{20}$ is composed of the same structural units as paratungstate $[H_2W_{12}O_{42}]^{10-}$. The structures of several oxo-alkoxides are shown in Fig. 13, where it is observed that the metals have acquired their maximum coordination numbers.

Fig. 13.

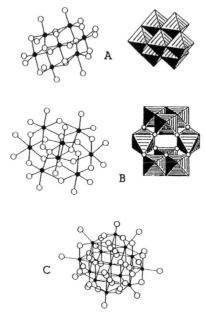

X-ray structures of some transition metal oxo-alkoxides. From Livage *et al.* [8]:
(a) $Ti_7O_4(OEt)_{20}$ [143]; (b) $Nb_8O_{10}(OEt)_{20}$ [144]; (c) $Zr_{13}O_8(OMe)_{36}$ [145].

Bradley *et al.* have determined the molecular complexity of oxo-alkoxide oligomers as a function of *r* [146] and have proposed several structural models to account for their observations [147]. Models I–IV summarized in Table 5 (where *h* is equivalent to *r*) assume that the coordination of the metal alkoxides increases to six via alcolation and/or alcohol association and that hydrolysis and condensation occur without modification of the oligomeric or monomeric building blocks. Model I, which pertains to $Ti(OEt)_4$ (shown to be trimeric) predicts that the initial isolable condensation product ($x = 1$ and $h = 0.67$ in Table 5) is $Ti_6O_4(OEt)_{16}$. *r* values between 0 and 0.67 should result in mixtures of $Ti_3(OEt)_{12}$ and $Ti_6O_4(OEt)_{16}$. Above $r = 0.67$, the percentage of $Ti_6O_4(OEt)_{16}$ decreases until $r = 0.89$, where pure $Ti_9O_8(OEt)_{20}$ ($x = 3$) is expected. When $r = 1.33$, the formation of an insoluble, infinitely long triple-chain polymer $Ti_3O_4(OEt)_4$ is expected. Above $r = 1.33$, more highly condensed products are formed leading ultimately to TiO_2 at $r = 2$.

Bradley *et al.* [130] claim that the maximum yield of $Ti_6O_4(OEt)_{16}$ was obtained approximately at $r = 0.67$, that the product at $r = 0.89$ was $Ti_9O_8(OEt)_{20}$, and that the product at $r = 1$ was $TiO(OEt)_2$ ($x = 3$) consistent with Model I. However XANES-EXAFS experiments [148,149]

Table 5.

Structural Models for Metal Alkoxides.

System	General Formula	Molecular Complexity (n)	Variation of n with h	Variation of x with h	Molecular Formula of the Infinite Polymer
Model I	$M_{3(x+1)}O_{4x}(OR)_{4(x+3)}$	$3(x+1)$	$n = 12/(4-3h)$	$x = 3h/(4-3h)$	$[M_3O_4(OR)_4]_x$
Model II	$M_{2(x+1)}O_{3x}(OR)_{2(x+4)}(ROH)_{2(x+1)}$	$2(x+1)$	$n = 6/(3-2h)$	$x = 2h/(3-2h)$	$[M_2O_3(OR)_2(ROH)_2]_x$
Model III	$M_{(x+1)}O_{3x}(OR)_{(4-2x)}(ROH)_{2(x+1)}$	$(x+1)$	$n = 3/(3-h)$	$x = h/(3-h)$	$[M_3O_6(ROH)_6]$
Model IV	$M_{(x+1)}O_{2x}(OR)_{(5+x)}(ROH)_{(x+1)}$	$(x+1)$	$n = 2/(2-h)$		$[MO_2(OR)\cdot(ROH)]_x$

Source: Bradley et al. [130].

n = number of Ti atoms per osmotic molecule, i.e. molecular complexity.
h = number of H_2O molecules added per atom of Ti.

Model I: $Ti_3(OR)_{12}$ R = Et, Pr^n, Bu^n
Model II: M = Zr, Ce, Ti
Model III: M = Ti, Zr, Ce
Model IV: M = Ta, Nb

suggest that Ti is 5-coordinated in the unhydrolyzed oligomer, and X-ray data indicate that the first hydrolysis product is $Ti_7O_4(OEt)_{20}$ [143]. (See Fig. 13.) Despite these quantitative differences, the basic premise of Bradley's model, viz. that condensation occurs between well-defined oligomeric units, appears to be qualitatively correct for many transition metal alkoxide systems.

Based on the partial-charge model, Livage *et al.* [8] predict that the first two steps of hydrolysis of $Ti(OPr^i)_4$ (known to be monomeric) occur easily leading to condensed products via alcoxolation ($\delta(Pr^iOH) > 0$). When $r = 1$, condensation could theoretically result in chain polymers $[MO(OR)_2]_n$ as proposed by Boyd and Winter for $Ti(OBu^n)_4$ systems [122,133] and Kamiya *et al.* [140–141] for spinnable titanate sols.[†]

When $r > 2$, $\delta(OR)$ becomes positive, potentially causing proton transfer to become rate-limiting (in neutral conditions). Consequently, hydrolysis may not go to completion even when $h = 4$ consistent with experiment [122,124,133–135]. Under these conditions, condensation via oxolation becomes competitive with alcoxolation, and olation is the preferable condensation pathway for coordinatively unsaturated precursors.

Large values of r favor highly condensed products. Using an excess of water, monodispersed powders, Ti, Zr, and Ta hydroxides or oxo-hydroxides, have been obtained from $Ti(OEt)_4$ [123], $Zr(OPr^n)_4$ [151], and $Ta(OEt)_5$ [152]. It is likely that the spherical particles obtained in these syntheses form by a controlled aggregation of much smaller oligomeric building blocks in a manner analogous to the inorganic zirconate systems discussed in Section 1.1.4. For more details concerning this controlled aggregation process, refer to Chapter 4.

1.2.4. CHEMICAL MODIFICATION

Chemical modification of transition metal alkoxides with alcohols, chlorides, acids or bases, chelating ligands, etc. is commonly employed to retard the hydrolysis and condensation reaction rates in order to control the condensation pathway of the evolving polymer [8,113]. In most cases, the modification occurs by an S_N reaction between a nucleophilic reagent (XOH) and the metal alkoxide to produce a new molecular precursor [113]:

$$XOH + M(OR)_z \rightarrow M(OR)_{z-x}(OX)_x + xROH \qquad (39)$$

[†] It should be noted that based on ^{29}Si NMR results, there is no evidence for strictly linear or ladder polymers in spinnable silicate sols prepared with $r = 1.5$ or 1.67 [150]. By analogy, claims that linear polymers form in spinnable titanate systems may be unjustified.

or when M is coordinatively unsaturated $(N - z > 1)$ by nucleophilic addition (A_N):

$$XOH + M(OR)_z \rightarrow M(OR)_z(XOH)_{N-z}. \qquad (40)$$

The reactivity of the alkoxide toward this modification increases when $\delta(M) \geqslant 1$ and $(N - z) > 1$, and when XOH is a strong nucleophile [8]. For a given group, the quantity $(N - z)$ increases when going down a column of the periodic table.

The hydrolysis and condensation behavior of the modified precursor depends, of course, on the stability of the modifying ligands, which is not always possible to predict on the basis of the stability of the parent compound, $M(OX)_z$. In general, less electronegative ligands are removed preferentially during hydrolysis, whereas more electronegative ligands (normally the modifying ligands) are removed more slowly (if at all) during condensation [8]. In substituted precursors (Eqs. 39 and 40), stable modifying ligands cause the effective functionality toward condensation to be reduced, resulting in less highly condensed products and promoting gelation.

Alcohol exchange reactions occur readily with metal alkoxide precursors:

$$M(OR)_z + xR'OH \rightarrow M(OR)_{z-x}(OR')_x + xROH \qquad (41)$$

causing alcohol exchange to be a common method of alkoxide synthesis [130]. Exchange is facilitated when $\delta(M) \geqslant 1$ and when R' is less sterically bulky than R. For example, alcohol exchange rates decrease as MeOH $>$ EtOH $>$ Pr^{i}OH $>$ Bu^{t}OH [8]. However, since hydrolysis rates decrease with steric bulk of the alkoxy ligands, chemical modification of transition metal precursors normally involves exchange of a bulky ligand for a less bulky one, for example:

$$VO(OPr^i)_3 + xHOAm^t \rightarrow VO(OPr^i)_{3-x}(OAm^t)_x + xPr^iOH. \qquad (42)$$

Because V is very electrophilic in $VO(OR)_3$ $(\delta(V) \approx +0.46)$ [8], this reaction occurs within seconds at room temperature without catalyst. By comparison, exchange of OPr^i for OEt in $Si(OEt)_4$ $(\delta(Si) = +0.32)$ takes about twenty hours with acid catalyst [116].[†]

Alcohol exchange can significantly alter the hydrolysis behavior of transition metal alkoxides. Hydroxo or oxyhydroxide precipitates always occur when $Ti(OEt)_4$ or $Ti(OPr^i)_4$ are hydrolyzed with excess water $(r > 2)$, whereas stable sols are obtained with $Ti(OAm^t)_4$. It should be noted that in the mixed alkoxide, $Ti(OPr^i)_{4-x}(OAm^t)_x$, (OAm^t) is less electronegative

---

[†] Exchange of OPr^i for OEt in TEOS apparently does not occur in the absence of water, suggesting that steric factors are important and/or that the exchange mechanism involves a partially hydrolyzed transition state. See [84] in Chapter 3.

than (OPr^i), causing preferential hydrolysis of (OAm^i) ligands and rapid gelation [113]. This reinforces the statement that the stability of mixed alkoxides cannot be judged on the basis of the parent compounds.

Chloride-modified transition metal alkoxides can be prepared by reaction of metal alkoxides with halides or hydrogen halides or by reaction of metal chlorides with alcohols [130]:

$$TiCl_4 + 2EtOH \rightarrow TiCl_2(OEt)_2 + 2HCl. \tag{43}$$

The reactivity of the metal chloride toward alcoholysis decreases with $\delta(M)$. Therefore, although SiCl$_4$ reacts completely in excess alcohol to form Si(OR)$_4$, under similar conditions Ti and Zr react only partially to form TiCl$_2$(OEt)$_2\cdot$EtOH [153] and ZrCl$_3$(OEt)$\cdot$EtOH [154]. ThCl$_4$ forms only an addition compound, e.g., ThCl$_4\cdot$4EtOH [130]. Unlike SiCl$_4$, chlorine-modified transition metal alkoxides can be quite stable toward hydrolysis. For example, although it is difficult to prepare homogeneous gels from Nb(OR)$_5$ or NbCl$_5$, the chlorine-modified alkoxide, NbCl$_2$(OR)$_3$, can be hydrolyzed to form a clear gel in excess water [155].

Metal alkoxo-acetates can be formed by the reaction of acetic acid (HOAc) with metal alkoxides [156–158]. Whereas the addition of HOAc decreases the gel time of silicon alkoxide systems [159], the reverse effect is observed for Ti(OR)$_4$ and Zr(OR)$_4$ [113,158]. To understand this behavior in acetic acid-modified Ti(OR)$_4$ systems, Livage and coworkers [160] and Sanchez and coworkers [148] have performed careful experiments combining XANES,[†] ^{13}C NMR, IR, and EXAFS.[††]

Ti(OPr^i)$_4$ is monomeric and the Ti atom exhibits a coordination number, $N = 4$. The addition of glacial acetic acid causes an exothermic reaction, and XANES experiments indicate that N increases to 6. ^{13}C NMR shows that the chemical environments of the carboxylic and methyl carbons in HOAc are altered after reaction with Ti(OPr^i)$_4$ due to bonding with titanium and that both terminal and bridging OPr^i groups are present. The IR spectrum of the reaction product exhibits absorption bands due to the titanium alkoxide together with acetic acid and the propyl acetate ester. Also, there are two strong bands at ~1500 cm^{-1} assigned to the symmetric and antisymmetric stretching vibrations of carboxylic groups: $v_s(CO_2^-) = 1450$ cm^{-1} and $v_a(CO_2^-) = 1580$ cm^{-1}. The frequency difference ($\Delta v = 130$ cm^{-1}) is typical of an acetate ion acting as a *bidentate* (bifunctional) ligand, but it is not possible to determine on the basis of Δv whether the bidentate ligand is bridging two titaniums or chelating one. IR confirms the presence of both terminal (1084 cm^{-1}) and bridging (1034 cm^{-1}) OPr^i ligands. This evidence

---

[†] XANES is the abbreviation of X-ray absorption near edge spectroscopy.

[††] EXAFS is the abbreviation of extended X-ray absorption fine structure spectroscopy.

combined with a Ti–Ti correlation of 3.05 Å determined from EXAFS is consistent with a dimer composed of bridging and terminal OPr^i ligands and bridging CH$_3$COO$^-$ ligands [160]:

$$
\begin{array}{c}
CH_3 \\
| \\
C \\
\end{array}
$$

(44)

During hydrolysis, ^{1}H and ^{13}C NMR and IR spectroscopy indicate that OPr^i groups are preferentially hydrolyzed [160], whereas the bridging acetate ligands remain bonded to titanium throughout much of the condensation process. Since they are not hydrolyzed, bridging CH$_3$COO$^-$ ligands effectively alter the condensation pathway perhaps promoting the formation of linear polymers composed of edge-shared octahedra.

An equimolar mixture of acetylacetone (acac) and Ti(OPr^i)$_4$ results in an exothermic reaction and a clear yellow solution is obtained [148]. ^{1}H and ^{13}C NMR indicate a 3 : 1 ratio of OPr^i : acac and the presence of two types of OPr^i groups with a ratio of 2 : 1. The ^{1}H and ^{13}C chemical shifts corresponding to acac indicate that acac is not free, but bonded to titanium. A single CH resonance is observed indicating only one type of acac is present. The corresponding IR spectrum (Fig. 14) has a broad absorption band near 620 cm^{-1} assigned to v(Ti–O–Pr^i) and bands at 1590 and 1530 cm^{-1} assigned to acac groups bonded to Ti. The XANES spectrum (Fig. 15) shows a single pre-peak, excluding octahedral Ti coordination. The intensity of this pre-peak

Fig. 14.

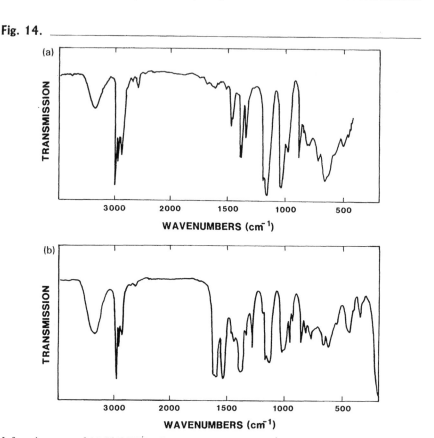

Infrared spectra of (a) $Ti(OPr^i)_4$; (b) acac-modified $Ti(OPr^i)_4$. From Sanchez *et al.* [161].

indicates that the acac reaction caused the coordination of Ti to increase from 4 to 5. EXAFS data show no Ti–Ti correlation. Two Ti–O distances with an intensity ratio 3 : 2 are observed corresponding to OPr^i and acac ligands bonded to Ti, respectively. These data are consistent with a *chelated* titanate precursor, $Ti(OPr^i)_3acac$:

$$Ti(OPr^i)_4 + acac \rightarrow Pr^i-O-Ti-O-Pr^i + Pr^iOH. \tag{45}$$

Fig. 15. _____

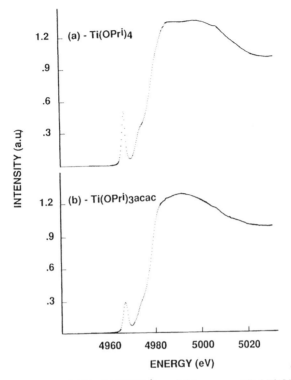

K-edge XANES spectra of Ti in (a) Ti(OPr^i)$_4$ and (b) acac-modified Ti(OPr^i)$_4$. From Sanchez *et al.* [161].

A 2 : 1 mixture of acac : Ti(OPr^i)$_4$ results in an octahedrally coordinated *dichelated* precursor:

$$\text{Ti(OPr}^i\text{)}_4 + 2\,\text{acac} \longrightarrow \text{Pr}^i\text{O}-\text{Ti}-\text{O}-\text{Pr}^i + 2\text{Pr}^i\text{OH}. \qquad (46)$$

Fig. 16. _____

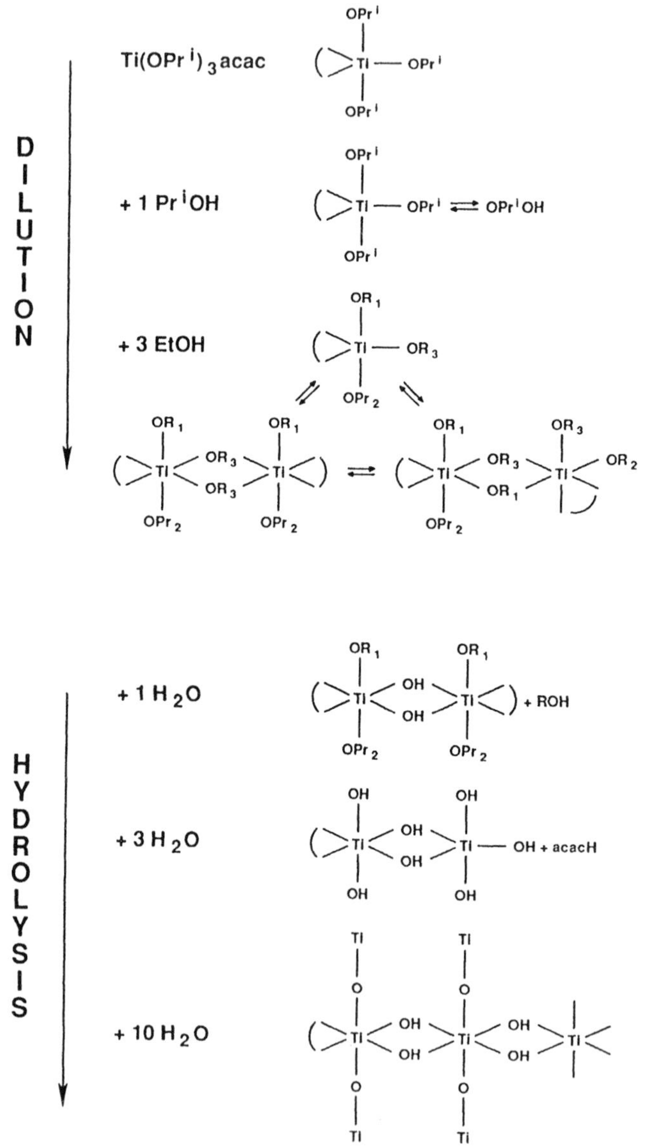

Structural evolution during dilution and hydrolysis of Ti(OPr^i)$_3$ acac. From Babonneau *et al.* [162].

The addition of water to the $Ti(OPr^i)_3acac$ precursor causes OPr^i ligands to be preferentially hydrolyzed [148]. When $r \geq 3$, the 620-cm^{-1} $v(Ti-O-Pr^i)$ band completely disappears, and, although bands at 1700 and 1720 cm^{-1} associated with free acac begin to appear, the doublet associated with acac bonded to Ti is still evident [148]. Even a large excess of water ($r = 20$) does not hydrolyze all the acac ligands bonded to titanium. The XANES spectrum of the hydrolyzed product ($r = 3$) shows a triplet pattern characteristic of octahedrally coordinated Ti, and EXAFS reveals Ti–Ti correlations at 3.2 Å consistent with condensed species. A generalized scheme for the initial stages of $Ti(OPr^i)_3acac$ hydrolysis and condensation is portrayed in Fig. 16. Excess water leads finally to a stable, three-dimensional, colloidal sol comprising ~5.0-nm particles with structures similar to anatase [162], whereas hydrolysis of $Ti(OPr^i)_2acac_2$ is said to lead to one-dimensional polymers suitable for fiber drawing [163]. By comparison, hydrolysis of $Ti(OPr^i)_4$ results in a polydisperse sol composed of 10–20-nm particles.

2.

ALUMINATES

Aluminum, the second most abundant metal in the earth's crust, exhibits only the trivalent state in compounds and in solution [10]. The hydrolysis chemistry of aluminum has been of interest for over a century because of its relationship to soil chemistry and because the most common mineral source of aluminum, bauxite, is composed largely of aluminum hydroxides and oxo-hydroxides. The hydrothermal digestion of bauxite in the Bayer process relies on the solubility of Al(III) as the anionic species, $[Al(OH)_4]^-$, at high pH. More recently, aluminum hydroxides and oxo-hydroxides have become of interest as precursors to transition aluminas used as catalyst supports and adsorbents as well as α-Al_2O_3 used commonly as a structural ceramic. Furthermore, the discovery by Yoldas [164,165] in the late 1970s that hydrolysis and condensation of aluminum alkoxides could result in mono-lithic alumina gels was largely responsible for the explosion in sol-gel research that continues to this day. This section explores the hydrolysis and condensation behavior of Al(III) in both aqueous, inorganic solutions and non-aqueous solutions derived from metal alkoxides.

2.1.   Solution Chemistry of Inorganic Precursors

Al^{3+}, with an ionic radius of 0.5 Å, has a coordination number of water of $N = 6$ and exists as the unhydrolyzed species $[Al(OH_2)_6]^{3+}$ below pH 3 [10]. With increasing pH, $[Al(OH_2)_6]^{3+}$ can be hydrolyzed extensively:

$$[Al(OH_2)_6]^{3+} + hH_2O \rightarrow [Al(OH)_h(OH_2)_{6-h}]^{(3-h)+} + hH_3O^+ \quad (47)$$

$$hH_3O^+ + hOH^- \rightarrow 2hH_2O \quad (48)$$

where h is defined as the molar ratio of hydrolysis, which is equivalent to the OH : Al ratio according to the net reaction (Eq. 47 + Eq. 48). Subsequent condensation (via olation or oxolation) results in polynuclear hydroxides or oxo-hydroxides, which can be stable indefinitely, but which are in fact metastable with respect to the precipitation of bayerite α-$Al(OH)_3$ [166]. (See Fig. 17.) Metastable precipitates can form and redissolve slowly, the species distribution being very sensitive to the precise conditions of the hydrolysis procedure, e.g., time-temperature history and method of base addition [10].

Based on potentiometric studies and colorimetric titrations [10,167], the stable mononuclear species in solution correspond to the series, $h = 0$–4 (Eq. 47), although as shown in Fig. 18, the $h = 2$ and $h = 3$ species are significant only in very dilute solutions. The potentiometric results of Baes and Mesmer [10,167] are consistent with two small polynuclear species at low equivalent additions of base, $[Al_2(OH)_2(OH_2)_4]^{4+}$

Fig. 17.

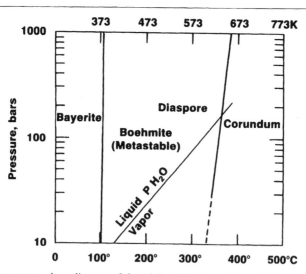

Pressure-temperature phase diagram of the Al_2O_3–H_2O binary. From Wefers and Misra [166].

Fig. 18.

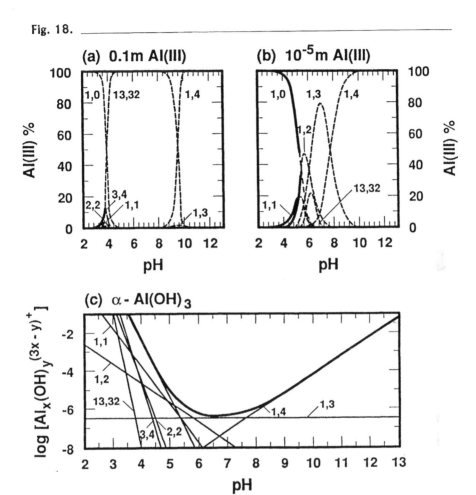

Distribution of hydrolysis products (x, y) in $[Al_x(OH)_y]^{(3x-y)+}$ at $I = 1$ m and $25°C$ in (a) 0.1 m Al(III); (b) 10^{-5} m Al(III); and (c) solutions saturated with α-Al(OH)$_3$. From Baes and Mesmer [10].

and $[Al_3(OH)_4(OH_2)_9]^{5+}$ and, at higher base additions, one large polynuclear species, $[AlO_4Al_{12}(OH)_{24}(OH_2)_{12}]^{7+}$, the Al_{13} ion. The existence of both the dimer and the Al_{13} ion has been confirmed by X-ray crystallography of the corresponding sulfate salts. The Al_{13} ion has also been identified by SAXS [168,169] and by a narrow ^{27}Al NMR resonance at ca 63.5 ppm (relative to $[Al(OH_2)_6]^{3+}$ (Fig. 19) attributed to the single tetrahedrally coordinated aluminum located at the center of three layers of edge-sharing octahedra. (See Fig. 20.)

Fig. 19.

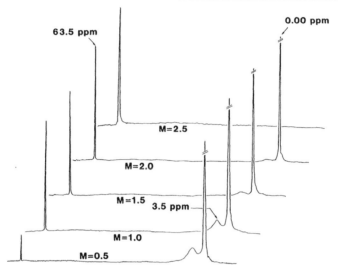

²⁷Al NMR spectra of 0.25 M Al(III) solutions hydrolyzed with aqueous bicarbonate, where the hydrolysis ratio M (equivalent to h) varies from 0.5 to 2.5. From Wood [170].

Fig. 20.

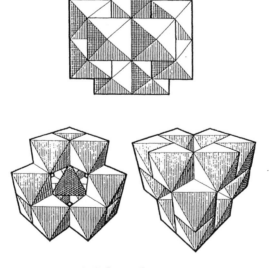

THE "Al-13" ION

The Al_{13} ion. From Wood [170].

Fig. 21.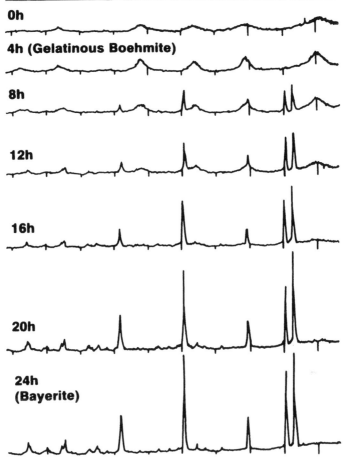

X-ray diffraction patterns of an aluminum hydroxide gel as a function of aging time at pH 9 and 300K. From Wefers and Misra [166].

Above $h = 2.46$ (theoretical value for Al_{13}), rapid precipitation of more highly condensed, amorphous, or weakly crystalline phases occurs [10]. (See Fig. 18c.) The common crystalline phase is pseudoboehmite (or gelatinous boehmite). The X-ray diffraction pattern of pseudoboehmite shows broad lines that coincide with the major reflections of well-crystallized *boehmite*, γ-AlO(OH), but are shifted to varying degrees to higher *d*-spacings (see Fig. 21) consistent with intercalation or association of water. The degree of crystalline order, particle size, and chemical composition of the gelatinous aluminas depends critically on temperature, rate of precipitation, final pH, ionic composition, concentration of starting solutions, and time of aging [10,171]. Both amorphous and psuedo crystalline phases convert eventually to α-Al(OH)$_3$ upon aging at intermediate pH. (See Fig. 21.)

2.1.1.   CONDENSATION MECHANISMS

The dimer consisting of two edge-shared octahedra forms via an olation reaction between two singly hydrolyzed monomers ($h = 1$):

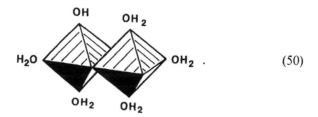

$$2\left[Al(H_2O)_6^{3+} - H^+ \rightarrow Al(H_2O)_5OH^{2+}\right] - 2H_2O \rightarrow Al_2(OH)_2(H_2O)_8^{4+}$$

$$(49)$$

The structures of more condensed products can be understood qualitatively on the basis of an aluminum hydrolysis model developed by Wood and coworkers [170] which predicts that the initial hydrolytic interactions of two aluminum cations will proceed along the reaction pathway that maximizes their hydrogen bonding. According to this premise, preferred condensation sites are those that maximize the interactions between lone pair electrons on a bound hydroxide ligand on one aluminum species with a proton on a water bound to another aluminum species. The most acidic water ligand in the dimer is *cis* to (on the side of) existing bridging hydroxides. Therefore, deprotonation generates the following species:

$$(50)$$

The addition of the $h = 1$ monomer, [Al(OH)(OH$_2$)$_5$]$^{2+}$, occurs preferentially to form the $[Al_3(tri-\mu-OH)(OH)_3(OH_2)_9]^{5+}$ trimer:

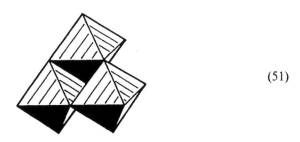

(51)

where tri-μ-OH emphasizes that one OH ligand bridges three aluminums and distinguishes this species from the linear species composed of three edge-sharing octahedra:

.   (52)

The more compact, tri-μ-OH trimer (51) forms preferentially because it maximizes the strong hydrogen-bond interactions between bridging or terminal hydroxides and bound waters in the transition (three versus a maximum of two in the case of the linear species). Similar arguments dictate that the reaction of two dimers occurs by a "side-on" reaction that allows a maximum of four strong hydrogen bond interactions between bridging or terminal hydroxides and bound water ligands. The reactions to form the proposed trimers or tetramers involve an activated complex containing $H_3O_2^-$ bridging ligands which have been shown to exist in both solution and the solid state [16,17].

Once a species is generated containing a tri-μ-OH ligand, further growth is dominated by the nucleophilic character of this ligand following deprotonation. The smallest congener (member of a family) containing the tri-μ-OH ligand is the trimer (51). Deprotonation produces the cation $[Al_3(tri-\mu-O)(OH)_3(H_2O)_9]^{4+}$, which contains a very nucleophilic electron pair on the tri-μ-OH ligand. Nucleophilic attack by this species on a monomeric aluminum species initiates the condensation process that generates the Al_{13} ion: because of the bulky nature of the tri-μ-OH trimer, this monomer is then forced to adopt tetrahedral coordination. The addition of two more tri-μ-OH trimers creates the Al_{13} ion (Fig. 20).

This scenario is consistent with ^{27}Al NMR investigations by Akitt and Farthing [171] that show only monomer, dimer, and Al$_{13}$ in solutions prepared by rapid, dropwise addition of Na$_2$CO$_3$ (1.125 M) to boiling AlCl$_3$ (1.67 M): due to the extremely high reactivity of the tri-μ-O ligand, Al$_{13}$ ions form rapidly without producing measurable (by ^{27}Al NMR) concentrations of trimer.[†] Akitt and Farthing [171] proposed an alternative condensation pathway to explain the NMR results, viz., that this hydrolysis procedure produces locally high concentrations of base favoring the formation of [Al(OH)$_4$]$^-$ anions around which six dimers rapidly condense, forming the Al$_{13}$ ion. This mechanism has been criticized for the following reasons:

1. the dimer and [Al(OH)$_4^-$] exhibit very different pK$_a$ values and would more naturally interact rapidly via an acid–base reaction involving proton transfer from the dimer to the monomer [170].
2. Al$_{13}$ forms under weakly basic conditions where [Al(OH)$_4^-$] is an improbable solution species [170].

However, it is noteworthy that more homogeneous hydrolysis procedures, e.g., *in situ* generation of base, do not result in large concentrations of Al$_{13}$ ions. Either the Akitt and Farthing model is correct or these conditions favor consumption of monomers to produce dimers, causing the tetramer to be the more probable condensation product [170]. The tetramer is the fundamental building block of the ubiquitous double ribbon structure common to boehmite and its Fe(III) and Cr(III) isomorphs:

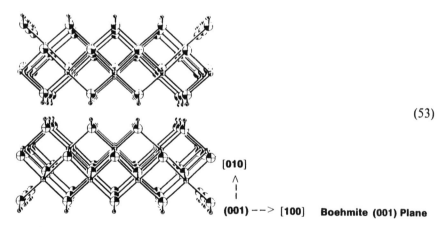

(53)

[010]

∧
⋮
⋮

(001) --> [100]   **Boehmite (001) Plane**

[†] Due to quadrupole broadening of ^{27}Al NMR resonances, there is always a finite amount of Al that is unaccounted for in the ^{27}Al NMR spectra attributable to polynuclear species larger than dimers.

This accounts perhaps for the tendency to form pseudoboehmite (or gelatinous boehmite) as the initial precipitation product over wide ranges of temperatures and pressures.

2.1.2. KINETICS

Based largely on potentiometric studies, the following five conclusions were drawn by Baes and Mesmer [10] concerning the kinetics of hydrolysis and condensation.

1. Mononuclear hydrolysis products are formed rapidly and reversibly.
2. Small polynuclear species, $[Al_2(OH)_2(OH_2)_8]^{4+}$ and $[Al_3(OH)_4(OH_2)_9]^{5+}$, are formed less rapidly.
3. The Al_{13} ion is formed more slowly still.
4. The precise conditions of hydrolysis (viz., the compositions of the solutions that are combined, the rate of combination, the amount of agitation, and the temperature) determine the nature and quantity of transient polymeric species, colloidal particles, or amorphous solid phases formed.
5. The rate at which these transient species are converted either to the Al_{13} ion or to crystalline $Al(OH)_3$ depends on the temperature, pH, the amount of $Al(OH)_3$ already present, and perhaps the available anions.

2.2. Metal Alkoxides

The Yoldas [164,165] process outlined schematically in Fig. 22 is the most common method of forming transparent, monolithic alumina gels. (See Fig. 23.) It consists of hydrolyzing an *aluminum alkoxide*, $Al(OR)_3$, normally $Al(OBu^s)_3$, in a large excess of water ($r = 100$–200) at 80–$100°C$, resulting in precipitation of fibrillar boehmite, followed by peptization with a mineral acid (HNO_3) to yield a stable particulate sol. Cold-water hydrolysis forms an amorphous precipitate that converts to bayerite upon aging through a dissolution–recrystallization process. Gelation is generally achieved by concentration of the sol via boiling or evaporation. Yoldas observed that a minimum gel volume occurred with an acid : Al mole ratio of 0.07. (See Fig. 24.) Several studies have shown that variations of the value of r, the acid content, the hydrolysis temperature, etc. are manifested as microstructural changes in the resulting xerogel products [173–175].

Fig. 22.

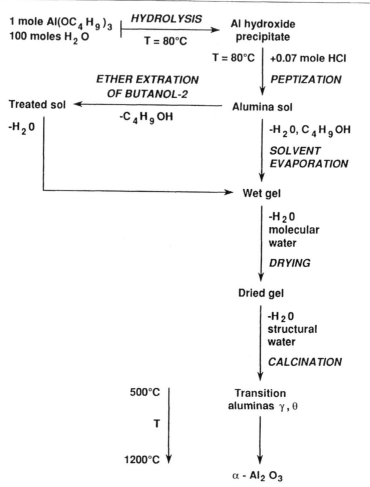

Flow chart of the Yoldas process for preparing alumina. From Assih *et al.* [172].

2.2.1. MECHANISMS

As discussed in Section 1, hydrolysis and condensation proceed differently in alkoxide systems than in inorganic systems in two fundamental ways: (1) the unhydrolyzed monomers often associate to form oligomers via alcolation and (2) OR ligands must be replaced by OH ligands for condensation to proceed further, so the effective functionality toward condensation may be reduced by employing low values of *r*. The extent of oligomerization

Fig. 23.

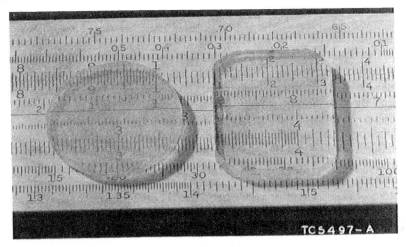

Monolithic alumina xerogels produced by the Yoldas process and heated to 500°C. From Yoldas [164].

Fig. 24.

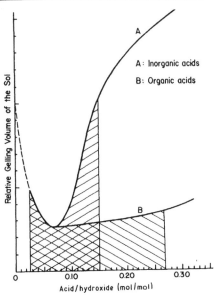

Relative gelling volumes versus the acid-to-aluminum ratio for alumina sols prepared by the Yoldas process. Shaded areas denote the range of acid-to-aluminum ratios yielding transparent, monolithic gels or films: Curve A, inorganic acids; Curve B, organic acids. From Yoldas [165].

(molecular complexity) of aluminum alkoxides depends mainly on the steric bulk of the alkoxide ligand. Steric crowding is reduced for alkoxide ligands exhibiting –O–CH$_2$– bonding to the metal (R = n-C$_4$H$_9$, i-C$_4$H$_9$, n-C$_5$H$_{11}$, i-C$_5$H$_{11}$, etc.), resulting in symmetric molecules corresponding to structure D in Fig. 25a. Sterically bulky OR groups (secondary or tertiary) impede the conversion of tetrahedral Al atoms to pentacoordinated or octahedral Al atoms, leading to the formation of less constrained linear species, either dimers or trimers (structures A and B). Based on ^{27}Al NMR investigations, structure B is proposed for Al (OBu^s)$_3$ [176].

Under neutral conditions, it is expected that both hydrolysis and condensation occur by nucleophilic addition followed by proton transfer and elimination of either water or alcohol in a manner analogous to transition metal alkoxides (Eqs. 32–34). Likewise, both of these reactions are catalyzed by addition of either acid or base (Eqs. 36 and 38): acids protonate OR or OH ligands creating good leaving groups and eliminating the requirement for proton transfer in the intermediate; bases deprotonate water or OH ligands creating strong nucleophiles. Although the hydrolysis kinetics are not well documented, since aluminum alkoxides may be coordinatively unsaturated and are able to adopt three stable coordination numbers, kinetic pathways of nucleophilic reactions should be quite facile. Consequently the rates of hydrolysis (and condensation) are greater than for silicon alkoxides which are coordinatively saturated and normally exhibit only one stable coordination number.

Fig. 25a. _____

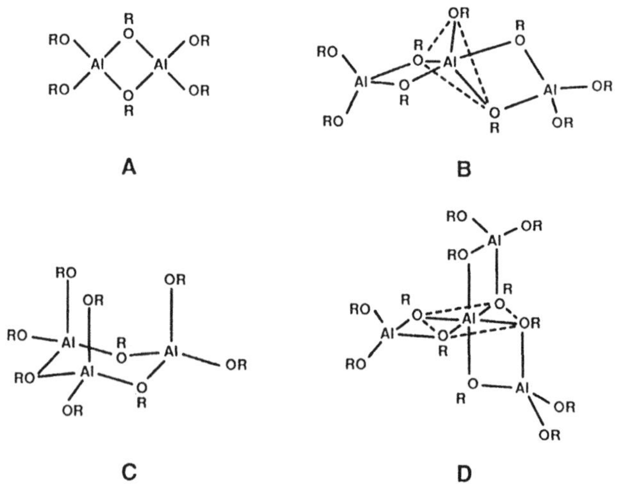

Various [Al(OR)$_3$]$_n$ oligomers. From Kriz et al. [176].

Fig. 25b. _____

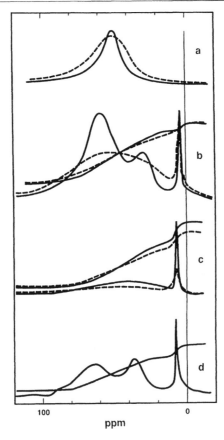

^{27}Al NMR spectra (in benzene) of (a) dimeric [Al(OBu^t)$_3$]$_2$; (b) trimeric [Al(OPr^i)$_3$]$_3$; (c) tetrameric [Al(OPr^i)$_3$]$_4$; and (d) tetrameric [Al(OEt)$_3$]$_4$. Solid lines denote spectra recorded at 70°C, dashed lines at 22°C. Spectrum of Al(OBu^s)$_3$ is similar to spectrum (b). From Kriz *et al.* [176].

2.2.2. STRUCTURAL EVOLUTION

Structural changes that occur during the Yoldas process (and variations of this general procedure) have been characterized by several groups employing ^{27}Al NMR [177,178] or ^{27}Al NMR in combination with Raman [172] and IR [174] spectroscopy, gel permeation chromatography [175], or viscosity measurements [173]. Olson and Bauer [178] closely followed the Yoldas procedure (Fig. 22) and observed ^{27}Al NMR spectra (Fig. 26) consistent with

Fig. 26.

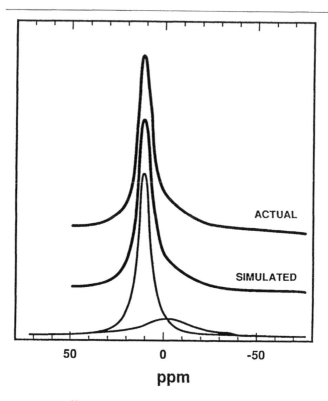

Actual and simulated ^{27}Al NMR spectra of an alumina sol prepared by the Yoldas process. From Olson and Bauer [178].

the presence of only octahedral Al species (0–8-ppm resonances relative to AlCl$_3$). No peaks near 60 ppm attributable to tetrahedrally coordinated Al species were resolved. Spectral deconvolution revealed a broad resonance centered at ca 0 ppm tentatively assigned to higher–molecular-weight species containing asymmetrical Al sites, and a narrow resonance at ca 5 ppm assigned to more symmetric Al sites in the sol polymers. These resonances were remarkably stable with respect to temperature (implying fairly rigid structures) and aging time. A second addition of acid (H$^+$/Al = 0.08) led to gel formation, accompanied by an increase in the intensity of the 0-ppm resonance with a corresponding decrease in intensity of the 8-ppm resonance (see Fig. 27) revealing a narrow band at 1.5 ppm assigned to the *aquated* monomer, [Al(OH$_2$)$_6$]$^{3+}$. Base additions (rather than acid) also led to gelation. The main spectral change observed under basic conditions was a decrease in the intensity of the 8-ppm resonance without the appearance of

Fig. 27.

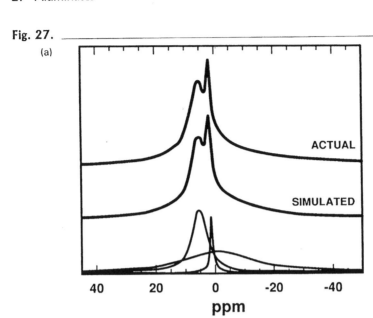

(a)

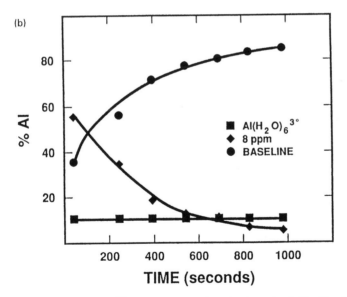

(b)

(a) Actual and simulated ^{27}Al NMR spectra of an alumina sol as in Fig. 26, after a second addition of HNO$_3$. (b) Temporal changes in relative intensities of Al resonances after second addition of acid. The Al (OH$_2$)$_6^{3+}$ resonance is the narrow band at ca 1 ppm in Fig. 27a. The baseline resonance refers to the very broad band centered at ca 0 ppm in Fig. 27a. From Olson and Bauer [178].

monomer. Based on these results, Olson and Bauer [178] concluded that both acid- and base-catalyzed gelation occurred by condensation reactions between more-symmetric species to form less-symmetric, higher–molecular-weight species.

In contrast to these results, lower hydrolysis temperatures ($\leq 80°C$) and/or higher acid concentrations (≥ 0.28 HNO$_3$/Al) result in substantial

Fig. 28.

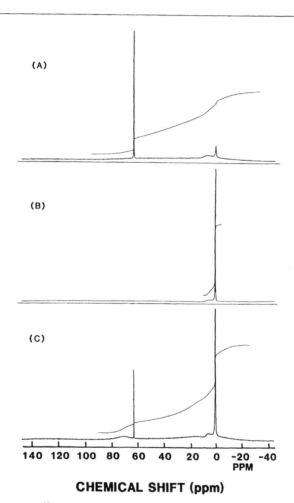

CHEMICAL SHIFT (ppm)

Room-temperature ^{27}Al NMR spectra of alumina sols: (a) 0.5 M HNO$_3$ sol, hydrolyzed and aged at 20°C for 8 h; (b) 0.5 M HNO$_3$ sol, hydrolyzed and aged at 90°C for 14 h; (c) 0.5 M HNO$_3$ sol, hydrolyzed and aged at 20°C for 14 h, followed by aging at 90°C for 14 h. From Nazar and Klein [177].

concentrations of tetrahedral aluminum atoms identified by ^{27}Al NMR resonances at 60–70 ppm (relative to aqueous solutions of $AlCl_3$ or $Al(NO_3)_3$) [172,177]. Nazar and Klein [177] observed that hydrolysis of $Al(OBu^s)_3$ ($r = 100$) at 20°C using acid ratios between ~0.3 and 1.0 mol HNO_3/Al resulted in clear sols in which the primary ^{27}Al NMR resonance was a line at 62.5 ppm (Fig. 28). On the basis of the extremely narrow bandwidth, Nazar and Klein assigned this band to the central, tetrahedrally coordinated Al in the symmetric Al_{13} ion (Fig. 20). The narrow resonance at 0 ppm and the two broad resonances at ca 4.2 and 7.6 ppm were assigned respectively to the octahedrally coordinated Al species: $[Al(OH_2)_6]^{3+}$, the aquated dimer, and a larger polymer (possibly tetramer) resulting from dimer condensation. Acid ratios lower than 0.3 resulted in milky sols exhibiting a broad ill-defined resonance centered at ca 9 ppm. According to GPC results by Olson and Bauer [175], milky sols are characterized by a very broad molecular weight distribution compared to clear sols.

High-temperature (90°C) hydrolysis resulted in alumina sols containing only octahedrally coordinated Al species for all acid ratios investigated [177] consistent with the observations of Olson and Bauer [178]. Due to molecular-weight broadening effects, only about 20% of the aluminum species are observable by ^{27}Al NMR, indicating a greater extent of polymerization compared to the low-temperature conditions. It is interesting to note that high-temperature aging (90°C) of the low-temperature sols does not cause rapid conversion to the high-temperature form; instead, a new tetrahedral resonance is observed at 71.2 ppm and a new octahedral resonance at 10–12 ppm accompanied by a gradual reduction of the Al_{13} tetrahedral band (Fig. 28). These new resonances were also observed by Akitt and Farthing [171] in aqueous aluminum systems hydrolyzed with Na_2CO_3. Nazar and Klein [177] attribute these features to higher–molecular-weight products of Al_{13} condensation.

Assih *et al.* [172] combined Raman spectroscopy and ^{27}Al NMR to follow the hydrolysis, peptization, and concentration steps of the Yoldas process (at 80°C). In contrast to the results of Olson and Bauer [178] on sols prepared at 100°C, both octahedral and tetrahedral aluminum species were observed in the 80°C peptized sol (0.07 HNO_3/Al). The ^{27}Al NMR spectra were simulated by three resonances at -7, 0, and 60–70 ppm (Fig. 29), however the 60–70 ppm resonance assigned to tetrahedral Al species was very broad compared to the low-temperature, high-acid sols investigated by Nazar and Klein (Fig. 28). The corresponding Raman spectra showed a peak at 360 cm^{-1}, which correlated with the 0-ppm ^{27}Al NMR resonance, assigned to a bending vibration of octahedrally coordinated Al species in poly-crystalline boehmite. The absence of the 1060-cm^{-1} Al-O-C stretching vibration indicated that hydrolysis was complete.

Fig. 29.

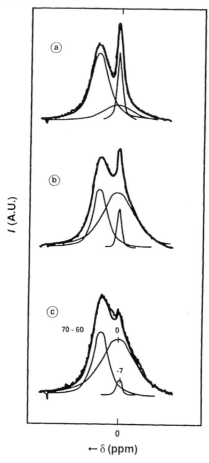

²⁷Al NMR spectra of (a) an alumina sol prepared according to the Yoldas process at 80°C with concentration C_0; (b) a concentrated alumina sol (concentration C_1); and (c) a concentrated gel (concentration C_2), with $C_0 < C_1 < C_2$. From Assih *et al.* [172].

The NMR and Raman changes accompanying concentration via solvent evaporation are shown in Figs. 29 and 30. The 360-cm⁻¹ Raman band becomes very intense while the broad 0-ppm resonance grows at the expense of the narrow −7-ppm resonance assigned to $[Al(OH_2)_3]^{3+}$. No obvious change occurs in the intensity or bandwidth of the 60–70-ppm resonance. New Raman bands at 451, 494, and 677 cm⁻¹ emerge from the background. The Raman spectrum of the concentrated sol (gel) corresponds closely to that of boehmite.

Fig. 30.

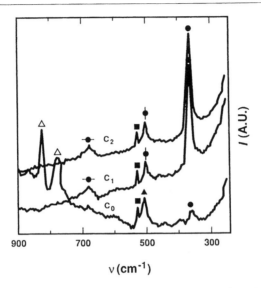

Evolution of the Raman spectra of alumina sols with concentration. $C_0 < C_1 < C_2$ as in Fig. 29. ●: pseudo-boehmite. △: sec-butanol. ■: laser. From Assih *et al.* [172].

Based on these observations, Assih *et al.* postulate that boehmite-like structures form during concentration by consumption of $[Al(OH_2)_6]^{3+}$ [172]. The 60–70-ppm ^{27}Al NMR resonance is assigned to the central tetrahedral Al site in the Al_{13} ion; however, its resonance appears to be much too broad to be explained on the basis of the isolated polycation. (Compare, e.g., Figs. 28 and 29.) The breadth of the resonance is in fact reminiscent of the 60–70-ppm resonance observed in the unhydrolyzed trimer (Fig. 25b), suggesting that under these conditions hydrolysis may occur without destroying the original oligomeric structure. Assih *et al.* argue that Al_{13} ions form at pH 4–4.3 during peptization and adsorb on the surfaces of the boehmite causing electrostatic stabilization of the sol. Adsorption that reduces the Al_{13} ion symmetry is an alternate explanation of the peak broadening. Adsorption of extensively solvated polynuclear cations on particle surfaces during acid peptization of concentrated boehmite sols ($HNO_3/Al = 0.02$) is consistent with the elastic and often thixotropic properties observed by Ramsay *et al.* [179], ascribed to short-range inter-particle repulsion. The stabilizing role of polynuclear ions at the boehmite surface is also consistent with the powerful coagulating effects of anions, such as IO_3^-, F^-, and SO_4^{2-} which form insoluble complexes with hydrolyzed aluminum ions, e.g., $Na[Al_{13}O_4(OH_2)_{12}(SO_4)_4]$ [180].

2.3. Summary

Hydrolysis and condensation studies of both aqueous and organic aluminum systems indicate that the condensation pathway is very sensitive to the precise conditions of hydrolysis, aging, peptization, etc. Due to molecular-weight broadening effects associated with the quadrupolar Al nucleus, it is not possible to unambiguously identify Al species other than Al_{13}, monomer, dimer, possibly tetramer, and higher–molecular-weight polymer. Therefore, the sequential processes comprising structural evolution in alumina sols remain uncertain.

Due to the high $H_2O : Al(OR)_3$ ratios employed in the Yoldas process, which promote OR hydrolysis, it is not surprising that many similarities exist between the aqueous and organic routes. There appears to be little evidence that under these aggressive hydrolysis conditions, the oligomeric structure of the $Al(OR)_3$ precursors is maintained. Therefore, the influence of organic versus inorganic precursors on the evolving structure appears to be minimized. Little work has been done to characterize $Al(OR)_3$ hydrolysis at low r values. Perhaps as observed by Bradley for $Ti(OR)_4$, the oligomeric structure will influence the structure of the condensed product.

3.

BORATE SYSTEMS

B_2O_3 forms a glass, and borates are commonly added to silicate glasses to reduce the melting temperature while maintaining high chemical durability, as demonstrated by the well-known PYREX[R] borosilicate glasses. A Group III metal like Al, B is trivalent; however, because B has no available d orbitals, the boron atom is trigonal coplanar with sp^2 hybridization in the majority of cases. In this configuration, boron is electron-deficient in the sense of the Lewis octet theory, causing the trigonal boron atom to be very electrophilic and thus quite susceptible to nucleophilic attack compared, for example, to silicon. This section examines the hydrolysis and condensation behavior of borates in both aqueous and nonaqueous solution.

3.1. Aqueous Borates

The small boron (III) atom forms no simple cations in solution [10]. The least hydrolyzed form of boron in aqueous solution is boric acid, $B(OH)_3$ [10]. The principal hydrolysis reaction involving boric acid is the rapid and

Fig. 31.

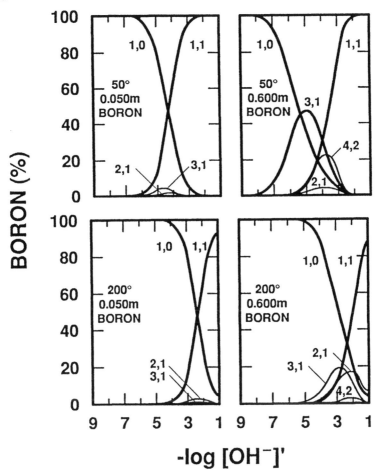

Distribution of borate species for 0.05- and 0.6-m B(III) in 1-m KCl at 50 and 200°C. The species are represented by the notation x, y corresponding to the formula $B_x(OH)_{3x+y}^{y-}$. From Baes and Mesmer [10].

reversible formation of the borate ion in which boron exhibits tetrahedral coordination with sp^3 hybridization:

$$B(OH)_3 + H_2O \rightarrow B(OH)_4^- + H^+. \tag{54}$$

This reaction becomes significant above pH 7 as shown in Fig. 31 [10], and above pH 11 $B(OH)_4^-$ is the predominant solution species. At intermediate pH and boric acid concentrations greater than about 0.05 M, polyborate anions are formed rapidly and reversibly by reactions between boric acid and borate ion [181], e.g.:

$$2B(OH)_3 + B(OH)_4^- \rightarrow B_3O_3(OH)_4^- + 3H_2O \tag{55}$$

or generally [10]:

$$xB(OH)_3 + (y - z)H_2O \rightarrow B_xO_z(OH)_{3x+y-2z}{}^{y-} + yH^+. \tag{56}$$

Polymerization is not observed above pH 11 because borate anions repel each other and because tetrahedrally coordinated boron without any available d orbitals is coordinatively saturated and unable to participate in the $sp^3\ d$ intermediate required for condensation via an S_N mechanism [181]. Below pH 4 polymerization is inhibited because the OH ligands on boric acid are not sufficiently strong nucleophiles to have an affinity for other boric acid molecules [181].

Depending on the pH, the cation present, and the temperature, a variety of hydrated polyborates can appear as saturating solid phases [10]. In fact, many hydrated polyborates occur naturally [10]. The structures of polyborate anions in solution appear to conform to a list of "rules" formulated by Edwards and Ross [182] by analogy to the structures of crystalline hydrated polyborates:

Fig. 32. _____

AQUEOUS BORATE SPECIES

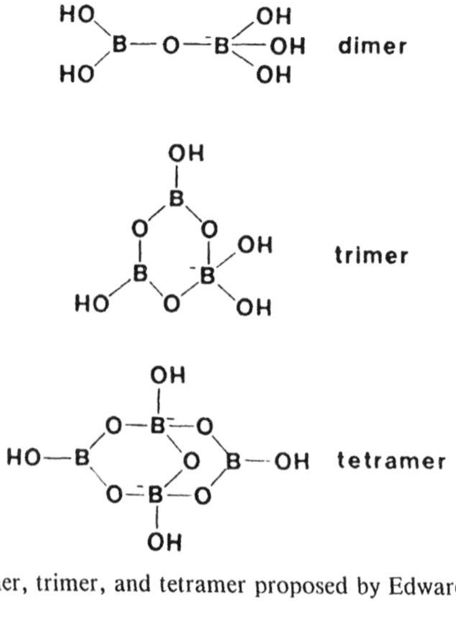

Aqueous borate dimer, trimer, and tetramer proposed by Edwards and Ross [182].

Fig. 33. _____

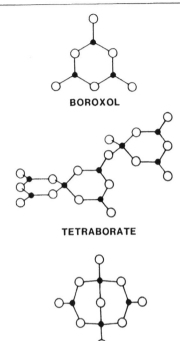

Examples of alkali-borate structural units postulated to exist in alkali borate glasses by Krogh-Moe [183].

(1) Boron atoms occur in trigonal BO_3 and tetrahedral BO_4 groups that have a formal charge of 0 and -1, respectively.

(2) The basic structure of the polyborates is a six-membered ring of alternating boron and oxygen atoms.

(3) To be stable, a ring must contain one or two tetrahedral boron atoms.

(4) Rings may be fused at tetrahedral boron atoms.

(5) Long-chain polyanions may be formed from the rings by repeated dehydration.

Based on these rules, the structures of the stable dimer, trimer, and tetramer are shown in Fig. 32. It is interesting to note that similar structures composed of three-membered rings (Fig. 33) were proposed by Krogh-Moe [183] to exist in alkali borate glasses where alkali ions (rather than protons) compensate the charges on the tetrahedral borate anions.

Despite rule 5, there appears to be no evidence for gel formation in aqueous borate systems . This argues that B–O–B bonds linking together the primary polyborate units are only metastable with respect to hydrolysis so that infinite (spanning) polymers are statistically unlikely. The stability of B–O–B bonding is discussed in detail in the next section where it is proposed as the primary criterion for gel formation in nonaqueous systems.

3.2. Alkoxide-derived Borate Gels

The lower alkoxides of boron are highly soluble compounds that can be distilled at atmospheric pressure to yield essentially monomeric derivatives [130]. The first synthesis of an alkoxide-derived borate gel was reported in 1984 by Toghe *et al.* [184]. Since then, Brinker and coworkers [185–187] have investigated the hydrolysis and condensation of boron alkoxides in lithium borate systems (nominally, xLi_2O–$(1 - x)B_2O_3$) using ^{11}B NMR, IR, and SAXS and have established the primary criterion for gelation to be the kinetic stability of borate bonds toward dissociative attack. (See Eq. 64.) This section is based largely on these later investigations.

3.2.1. BORON COORDINATION

In $B(OR)_3$ boron is trigonal coplanar with sp^2 hybridization:

$$\begin{array}{c} RO \\ \diagdown \\ B\!-\!OR. \\ \diagup \\ RO \end{array} \tag{57}$$

In this configuration boron is very electrophilic, and, in the presence of a sufficiently strong nucleophile, undergoes a nucleophilic addition reaction to form a tetrahedrally coordinated species with sp^3 hybridization:

$$LiOR' + \begin{array}{c} RO \quad\; OR \\ \diagdown \diagup \\ B \\ | \\ OR \end{array} \rightleftharpoons Li^+ \left[\begin{array}{c} OR \\ | \\ RO\!-\!B\!\bullet\!\bullet\!\bullet OR \\ | \\ OR' \end{array} \right]^{-} . \tag{58}$$

Figure 34 shows the fraction of tetrahedrally coordinated borons, N_4, versus x for mixtures of lithium methoxide and $B(OBu^n)_3$ in a mixed methanol, 2-methoxyethanol solvent with H_2O : OR ratios, $w = 0$–0.4. N_4 varies as $N_4 = (0.43)x + 0.104$ regardless of the initial value of w and remains constant during subsequent aging steps in 100% relative humidity (RH) water vapor. FTIR investigations of the pure solution components and mixtures of

Fig. 34.

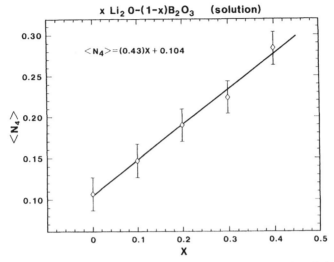

Fraction of tetrahedrally coordinated borons, N_4, as a function of x in solutions of xLiOMe $(1 - x)$ B(OBu^n)$_3$, including all investigated $H_2O:OR$ ratios (0–0.4). From Brinker *et al.* [185,186].

components identify lithium methoxide as the nucleophile that causes boron to adopt tetrahedral coordination via nucleophilic addition (A_N) (Eq. 58). FTIR and high-resolution ^{11}B NMR show that H_2O, *n*-butanol, and methanol do not produce tetrahedrally coordinated boron species, indicating that the thermodynamics of this reaction are governed largely by the basicity of the oxygen (magnitude of $\delta(O)$). For example, carboxylic acids can stabilize tetrahedral boron, whereas aliphatic alcohols do not [188].

If all the added lithium methoxide resulted in the production of tetrahedral boron, N_4 would vary with x as $x/(1 - x)$. Instead, the linear dependence implies that an equilibrium is established in which there remains unreacted LiOCH$_3$ as confirmed by FTIR. Both ^{11}B and FTIR showed that the nonzero intercept in Fig. 34 is a result of ca 10% tetrahedral boron in the neat B(OBu^n)$_3$. This could be explained by limited oligomerization of the monomers via alcolation causing the average molecular complexity to exceed one.

3.2.2. HYDROLYSIS AND POLYMERIZATION

Hydrolysis and condensation can both occur by S_N mechanisms involving nucleophilic attack of OH or OR ligands on electrophilic, trigonal boron with concerted elimination of alcohol or water. Under neutral conditions these reactions may be represented as follows:

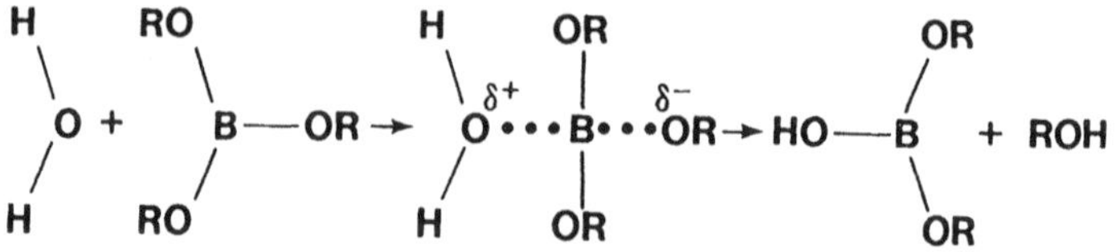

$$
\begin{array}{c}
\overset{H}{\underset{H}{\diagdown}}O \;+\; \overset{RO}{\underset{RO}{\diagup}}B{-}OR \;\rightarrow\; \overset{H}{\underset{H}{\diagdown}}O{\cdots}\overset{\delta+}{B}{\cdots}\overset{\delta-}{O}R \;\rightarrow\; HO{-}\overset{OR}{\underset{OR}{\diagup}}B \;+\; ROH \qquad (59)
\end{array}
$$

Hydrolysis

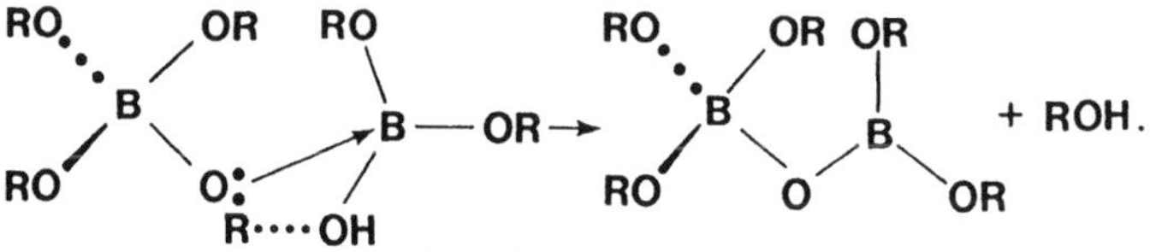

$$
(60)
$$

Condensation

It is also conceivable that for trigonal borons, condensation could occur by nucleophilic addition with a concomitant increase in coordination. Due to the lack of sp^3d hybridization required for the intermediates in Eqs. 59 and 60, we expect nucleophilic reactions involving tetrahedral borons to be unlikely. Thus, tetrahedrally coordinated borons should be kinetically stable toward hydrolysis or alcoholysis, condensation reactions between two tetrahedrally coordinated boron species are unexpected, and, in reactions between tetrahedrally and trigonally coordinated borons, the oxygen residing on the tetrahedrally coordinated boron should be incorporated in the B–O–B bond (Eq. 60). Consistent with these ideas, high-resolution ^{11}B NMR indicates that the addition of water to an unhydrolyzed solution of LiOMe and B(OBu″)$_3$ causes a rapid change in the chemical environment of trigonally coordinated borons due to hydrolysis, while the environment of tetrahedrally coordinated borons initially remains unchanged [185].

Due to the proposed stability of cyclic species containing at least one tetrahedral boron (Fig. 32), it is expected that the initial stage of growth involves the condensation of partially hydrolyzed trigonal boron monomers with unhydrolyzed tetrahedral boron monomers to form primary units composed of six-membered rings, for example:

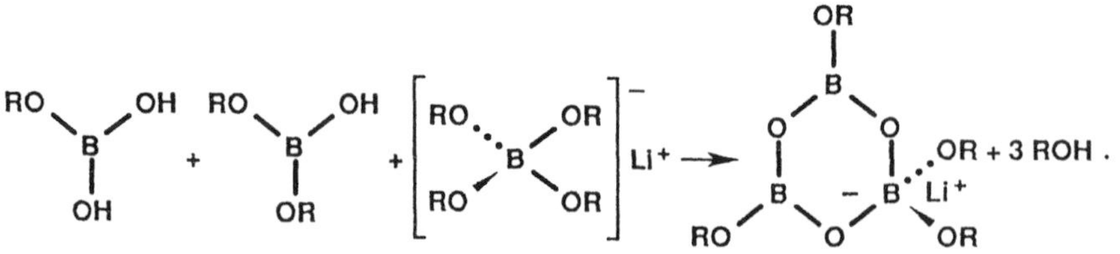

$$
(61)
$$

This reaction requires on average a minimum $H_2O : OR$ ratio of $w = 0.3$. Additional hydrolysis of OR ligands bonded to trigonal borons is required to form extended polymers in which the primary units are linked by B–O–B bonds, for example:

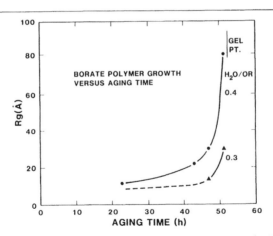

$$\text{(62)}$$

This establishes $w = 0.4$ as the minimum average value required for polymer growth.[†] This hydrolysis criterion is supported by SAXS investigations of the $x = 0.3$ sol (Fig. 35) that show, for $w < 0.4$, growth (if any) of polyborates is limited to species less than ~ 10 Å in radius, consistent with the formation of six-membered rings. Measurable growth is observed only during aging in 100% RH water vapor when w exceeds 0.4.

Fig. 35.

Borate polymer growth (Rg: radius of gyration) versus aging time in 100% RH water vapor for 0.3 LiOMe–0.7 B(OBu^n)$_3$ sols prepared with initial $H_2O : OR$ ratios of 0.4 or 0.3. Gel point for the 0.4 sample indicated. From Brinker *et al.* [185].

[†] Eqs. 59–62 are general, depicting formation and linkage of six-membered rings. The actual concentration and distribution of primary units (e.g., Fig. 32) depends on the extent of hydrolysis, the fraction of tetrahedrally coordinated borons, and the relative stabilities of the various primary units.

Fig. 36. _____

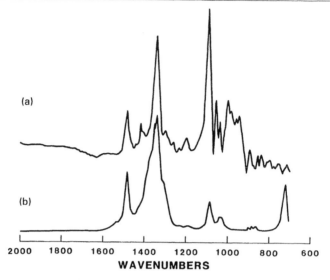

(a) FTIR solution spectrum of the 0.3 LiOMe–0.7 B(OBu″)₃ sol after 24 h aging in 100% RH water vapor (solvents substracted). (b) FTIR spectrum of trimethoxyboroxine. From Brinker *et al.* [185].

Further evidence in support of the general growth model just described is derived from IR studies of the $x = 0.3$ lithium borate sol. Figure 36 compares the FTIR difference spectrum (sol spectrum minus solvent spectrum) of the $w = 0.3$ sol after 24 hours of aging in 100% RH water vapor to the FTIR spectrum of trimethoxyboroxine (63):

$$\qquad\qquad(63)$$

The similarities in the ~1300–1500-cm^{-1} region, assigned to trigonal B–O asymetric stretching [189], suggest that after 24 hours of aging the environments of the trigonal borons in the sol and in trimethoxyboroxine are similar, presumably because in both cases the trigonal borons are contained in six-membered rings. Differences in the ~850–1100-cm^{-1} region, assigned to tetrahedral B–O asymetric stretching, reflect that $N_4 = 0.22$ (from ^{11}B NMR data) for the lithium borate sol ($x = 0.3$), whereas $N_4 = 0$ for trimethoxyboroxine. The difference spectrum, 120–72 hours of aging, for the

Fig. 37.

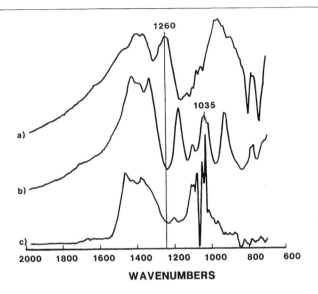

FTIR spectra of (a) anhydrous crystalline lithium tetraborate; (b) partially hydrolyzed crystalline lithium tetraborate; and (c) difference spectrum of lithium borate sols 120–72 hours of aging in 100% RH water vapor for the 0.3 LiOMe-0.7 B(OBu")$_3$ sol. From Brinker *et al.* [185].

$w = 0.3$ sol, is compared to the spectra of anhydrous and partially hydrolyzed crystalline lithium tetraborate (Li$_2$O-4 B$_2$O$_3$) in Fig. 37. The similarities in relative intensities of the envelope of vibrations centered at $\sim 1400\,\mathrm{cm^{-1}}$ and the vibration at $1035\,\mathrm{cm^{-1}}$ indicate that the products formed during the later stages of aging are similar in structure to partially hydrolyzed lithium tetraborate.

3.2.3. CRITERIA FOR GEL FORMATION

In addition to a minimum hydrolysis ratio, Brinker and coworkers [185–187] observed that there was also a minimum fraction of tetrahedrally coordinated borons, N_4, that must be exceeded in order to form gels. For example, in 2-methoxyethanol no polymer growth was observed by SAXS for $N_4 \leq 0.14$, and only very weak gels were formed for $N_4 \leq 0.19$, whereas stiff gels were obtained for $N_4 = 0.22$ or 0.28. In the aprotic solvent, THF, stiff gels were observed for $N_4 \geq 0.17$. These experimental results can be explained on the basis of the stability of B–O–B bonds toward dissociative reactions of the type:

$$ROH + \quad \begin{array}{c} \\ B-O-B \\ \end{array} \rightleftharpoons \begin{array}{c} \overset{R}{\underset{\overset{O-H}{\cdots}}{\vphantom{.}}} \\ B-\overset{\cdots}{\underset{\cdot\cdot}{O}}-B \\ \end{array} \rightleftharpoons \begin{array}{c} \overset{R}{\overset{O}{|}} \\ B \\ / \backslash \end{array} + \begin{array}{c} \overset{H}{\overset{O}{|}} \\ B \\ / \backslash \end{array}. \quad (64)$$

intermediate

Because boron cannot utilize sp^3d hybridization, this nucleophilic mechanism is not possible for 4-coordinated borons. Thus, the kinetic stability of B–O–B bonds is expected to decrease in the following order:

$$[\equiv B\text{-}O\text{-}B\equiv] > [\equiv B\text{-}O\text{-}B=] > [=B\text{-}O\text{-}B=].$$

This is consistent with hydrolysis mechanisms proposed by Bunker to explain the aqueous corrosion of alkali borosilicate glasses [181].

For all systems investigated, $N_4 \leq 0.27$. Thus, there are no possible structural models that both exclude all linkages between two trigonal borons ($=B\text{-}O\text{-}B=$) and fully incorporate boron in an "infinite" network. Since B–O–B bonds exist both within primary structural units (Figs. 32 and 33) and between primary units, the N_4 criterion may reflect the kinetic stability of either of these types of borate bonds. For vitreous B_2O_3 (composed completely of trigonal borons residing in boroxol units), Krogh-Moe [183] determined that $=B\text{-}O\text{-}B=$ bonds that link units together are more susceptible to hydrolysis than $=B\text{-}O\text{-}B=$ bonds within units. Thus, in the solution environment employed in gel processing, it is expected that gel formation will be limited by the kinetic stability of borate bonds that link together primary structural units.

In order to test this hypothesis and determine an appropriate structural model for solution-derived lithium borate networks (gels), Brinker *et al.* performed FTIR investigations of crystalline alkali borates and partially hydrated crystalline alkali-borates (with $N_4 = 0.25$ to 0.50). A prominent spectral feature (see, e.g., Fig. 37a) of the anhydrous tetra- and triborate compounds is the ~ 1260-cm^{-1} vibration due to oxygens bridging between trigonal borons contained in separate primary units [183] (Fig. 38a). Anhydrous lithium diborate is composed exclusively of diborate units linked by oxygens bridging between 3- and 4-coordinated borons and exhibits no 1260-cm^{-1} vibration [185]. Partial hydrolysis of the tetra- (and triborates) causes a dramatic reduction in the relative intensity of the 1260-cm^{-1} vibration (Fig. 37b).

The preceding observations prove that oxygens bridging trigonal borons residing in separate primary units (e.g., tetraborates or triborates) are unstable in aqueous (and presumably alcoholic) environments. Thus, any reasonable structural model proposed to explain borate gel networks formed in mixed alcohol–water solutions must exclude $=B\text{-}O\text{-}B=$ bonds between

Fig. 38.

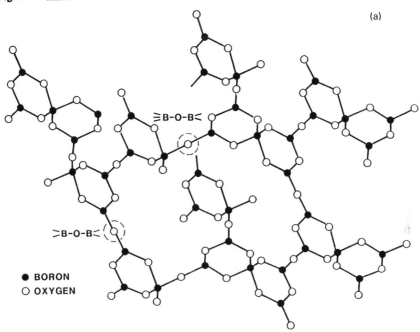

(a)

≡B-O-B<

≥B-O-B<

● BORON
○ OXYGEN

Tetraborate Network
Excluding ⟩B-O-B⟨ Between Units

(b)

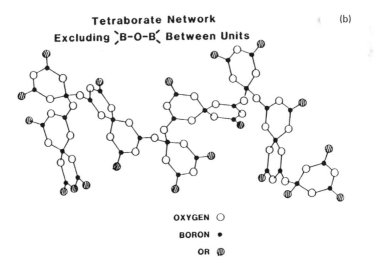

OXYGEN ○

BORON ●

OR ⦶

(a) A portion of the crystalline tetraborate network showing both =B-O-B= and =B-O-B≡ linkages between primary units. (b) Tetraborate network that excludes =B-O-B= bonding between primary units. From Brinker *et al.* [185].

units. In support of this idea, difference spectra obtained during the latter stages of aging (120–72 hrs, Fig. 37c), where considerable growth occurs, show no evidence of the 1260-cm^{-1} band. Because condensation reactions between two 4-coordinated borons are forbidden, the linkages responsible for polymer growth and gelation are oxygens bridging 3- and 4-coordinated borons. Due to the increased relative intensity of the 1035-cm^{-1} band (Fig. 37c) during the latter stages of aging, this vibration was assigned to the asymetric B–O stretch of $\equiv$B–O–B$=$ linkages between units. This vibration is also prominent in the spectrum of partially hydrolyzed (partially depolymerized) lithium tetraborate (Fig. 37b).

The similar shapes of the envelopes of vibrations associated with trigonal (1300–1500-cm^{-1}) and tetrahedral (850–1100-cm^{-1}) borons in partially hydrolyzed lithium tetraborate (Fig. 37b) and the 120–72 hour product species (Fig. 37c) indicate that the polyborate species formed during the latter stages of aging are structurally similar to partially hydrolyzed lithium tetraborate. The strong bands at about 930 and 1175 cm^{-1}, present only in the partially hydrolyzed lithium tetraborate spectrum, were assigned to B–O–H out-of-plane and in-plane bending, respectively, on the basis of D_2O and $H_2{}^{18}O$ hydrolyses [185].

To arrive at a structural model for lithium borate gels ($x = 0.3$) based on a tetraborate primary unit, Brinker *et al.* excluded all $=$B–O–B$=$ linkages between units and terminated nonbonded sites with alkoxides (Fig. 38b) [185]. This tetraborate borate network represents the lowest average mole fraction of tetrahedral borons ($N_4 = 0.25$) in which a continuous network can be constructed which excludes $=$B–O–B$=$ between units and includes all borons. Replacing 2-methoxyethanol with the cyclic ether, THF (which due to its nonlabile protons does not react according to Eq. 64), broadens the gel-forming region to include $x \geq 0.15$. This reduced requirement for N_4 reflects the reduced concentration of alcohols that can undergo dissociative reactions with B–O–B linkages (64).

4.

SUMMARY

Gel formation in organoboron systems involves the partial hydrolysis of the borate precursor, condensation of monomers to form primary units, and linkage of primary units to form extended networks. The hydrolysis and condensation processes as well as the kinetic stability of the borate network are all influenced by the electrophilic behavior of the trigonal boron atom. In solution environments containing molecules that can dissociatively react with trigonal boron (H_2O, ROH, etc.), a critical fraction of 4-coordinated borons must be exceeded in order to obtain gels. The gel network appears to be composed of primary units linked exclusively by $\equiv$B–O–B$=$ bonds.

REFERENCES

1. H. Dislich and E. Hussmann, *Thin Solid Films*, **77** (1981) 129.
2. R.G. Dosch in *Better Ceramics Through Chemistry*, Mater. Res. Soc. Symp. Proc., **32**, eds. C.J. Brinker, D.E. Clark, and D.R. Ulrich (North-Holland, New York, 1984), p. 157.
3. K.D. Budd, S.K. Dey, and D.A. Payne, *Brit. Ceram. Soc. Proc.*, **36** (1985) 107–121.
4. C. Sanchez, M. Nabavi, and F. Taulelle in *Better Ceramics Through Chemistry III*, Mater. Res. Soc. Symp. Proc., **121**, eds. C.J. Brinker, D.E. Clark, and D.R. Ulrich (Mater. Res. Soc., Pittsburgh, Pa., 1988), p. 93.
5. J. Livage and J. Lemerle, *Ann. Rev. Mat. Sci.*, **12** (1982) 103–122.
6. K. Tanaka, T. Yoko, M. Atarashiki, K. Kamiya, *J. Mat. Sci.*, **8** (1989) 83–85.
7. B.C. Bunker, J.A. Voigt, D.L. Lamppa, D.H. Doughty, E.L. Venturini, J.F. Kwak, D.S. Ginley, T.J. Headley, M.S. Harrington, M.O. Eatough, R.G. Tissot, Jr., and W.F. Hammetter in *Better Ceramics Through Chemistry III*, Mater. Res. Soc. Symp. Proc., **121**, eds. C.J. Brinker, D.E. Clark, and D.R. Ulrich (Mater. Res. Soc., Pittsburgh, Pa., 1988), p. 373.
8. J. Livage, M. Henry, and C. Sanchez, *"Sol-Gel Chemistry of Transition Metal Oxides"* in *Progress in Solid State Chemistry*, **18** (1988) 259-342.
9. B.C. Bunker, private communication.
10. C.F. Baes and R.E. Mesmer, *The Hydrolysis of Cations* (Wiley, New York, 1976).
11. D.L. Kepert, *The Early Transition Metals* (Academic Press, London, 1972).
12. C.K. Jorgensen, *Inorganic Complexes* (Academic Press, London, 1963).
13. J. Livage and M. Henry in *Ultrastructure Processing of Advanced Ceramics*, eds. J.D. Mackenzie and D.R. Ulrich (Wiley, New York, 1988), p. 183.
14. R.T. Sanderson, *Science*, **114** (1961) 670.
15. C.L. Rollinson in *The Chemistry of the Coordination Compounds*, ed. J.C. Bailar (Reinhold, New York, 1956), p. 448.
16. M. Ardon and B. Magyar, *J. Am. Chem. Soc.*, **106** (1984) 3359-3360.
17. M. Ardon, A. Bino, and K. Michelson, *J. Am. Chem. Soc.*, **109** (1987) 1986-1990.
18. V. Baran, *Coordin. Chem. Rev.*, **6** (1971) 65.
19. M Eigen, *Pure Appl. Chem.*, **6** (1963) 97.
20. H. Kruger, *Chem. Soc. Rev.*, **11** (1982) 227.
21. R.G. Pearson, *J. Chem. Educ.*, **38** (1961) 164.
22. G.B. Kolski, N.K. Kildahl, and D.W. Margerum, *Inorg. Chem.*, **8** (1969) 1211.
23. C.K. Jorgensen in *Atoms and Molecules* (Academic Press, London, 1962), p. 80.
24. E. Bang, *Acta Chem. Scand.*, **22** (1968) 2671.
25. A. Clearfield and P.A. Vaughan, *Acta Cryst.*, **9** (1956) 555.
26. T.C.W. Mak, *Can. J. Chem.*, **46** (1968) 3491.
27. M.L. Freedman, *J. Am. Chem. Soc.*, **80** (1958) 2072.
28. D.L. Kepert, *Prog. Inorg. Cehm.*, **4** (1962) 199.
29. R.H. Tytko and O. Glemser, *Adv. Inorg. Chem, Radiochem*, **19** (1976) 239.
30. G. Schwarzenback and J. Meier, *J. Inorg. Nucl. Chem.*, **8** (1958) 302.
31. A.F. Wells, *Structural Inorganic Chemistry* (Clarendon Press, Oxford, England, 1984).
32. W.Kopaczewski, *Bull. Soc. Chem. Fr.* (1950) 149.
33. D.M. Grant and R.E. Hamm, *J. Am. Chem. Soc.*, **78** (1956) 3006.

34. H. Wendt, *Inorg. Chem.*, **8** (1969) 1527.
35. M.T. Pope, *Heteropoly and Isopoly Oxometalates* (Springer-Verlag, Berlin, Heidelberg, 1983).
36. K.F. Jahr and J. Fuchs, *Angew, Chem. Int. Ed.*, **5** (1966) 689.
37. I. Lindqvist, *Arkiv Kemi*, **5** (1953) 247.
38. I. Lindqvist, *Arkiv Kemi*, **7** (1954) 49.
39. V.J. Fuchs, W. Freiwald, and H. Hartl, *Acta Cryst.*, **B34** (1978) 1764.
40. O. Nagano and Y. Sasaki, *Acta Cryst.*, **B25** (1979) 2387.
41. H.R. Allcock, E.C. Bissel, and E.T. Shawl, *Inorg. Chem.*, **12** (1973) 2963.
42. H.T. Evans, B.M. Gatehouse, and P. Leverett, *J. Chem. Soc. Dalton* (1975) 505.
43. H.T. Evans, *J. Am. Chem. Soc.*, **90** (1968) 3275.
44. I. Lindqvist, *Arkiv Kemi*, **2** (1950) 349.
45. T.J.R. Weakley, *Polyhedron*, **1** (1982) 17.
46. W.W. Day, M.F. Fredrich, W.G. Klemperer, and J.W. Shum, *J. Am. Chem. Soc.*, **99** (1977) 952.
47. A.F. Masterz, S.F. Gheller, R.T.C. Brownlee, M.J. O'Connor, and A.G. Wedd, *Inorg. Chem.*, **19** (1980) 3866.
48. E.M. McCarron, J.F. Whitney, and D.B. Chase, *Inorg. Chem.*, **23** (1984) 3275.
49. M. Isobe, F. Marumo, T. Yamase, and T. Ikawa, *Acta Cryst.*, **B34** (1978) 2728.
50. A.G. Swallow, F.R. Ahmed, and W.H. Barnes, *Acta Cryst.*, **21** (1966) 397.
51. H.T. Evans, *Inorg. Chem.*, **5** (1966) 967.
52. H.T. Evans and M.T. Pope, *Inorg. Chem.*, **23** (1984) 501.
53. E.J. Graeber and B. Morosin, *Acta Cryst.*, **B33** (1977) 2137.
54. J.K. Brandon and I.D. Brown, *Can. J. Chem.*, **46** (1968) 933.
55. R.U. Russel and J.E. Salmon, *J. Chem. Soc.* (1958) 4708.
56. D. Blum and J.C. Guitel, *Acta Cryst.*, **B36** (1968) 135.
57. P. Lofgren, *Acta Cryst.*, **B29** (1973) 2141.
58. J. Fuchs and H. Hartl, *Angew. Chem. Int. Ed.*, **15** (1976) 375.
59. J. Fuchs and E.P. Flindt, *Z. Naturforsch*, **34b** (1979) 412.
60. R. Allman, *Acta Cryst.*, **B27** (1971) 1393.
61. J. Fuchs, H. Hartl, W. Schiller, and U. Gerlach, *Acta Cryst.*, **B22** (1976) 740.
62. J. Fuchs, H. Hartl, and W. Schiller, *Angew. Chem. Int. Ed.*, **12** (1973) 420.
63. B. Krebs and I. Paulat-Boschen, *Acta Cryst.*, **B38** (1982) 1710.
64. A. Ditte, *Compt. Rend. Acad. Sci.*, **101** (1985) 698.
65. J.F. Hazel, W.M. McNabb, and R. Santini, *J. Phys. Chem.*, **57** (1953) 681.
66. J.J. Legendre and J. Livage, *J. Colloid Interface Sci.*, **94** (1983) 75.
67. N. Gharbi, C. Sanchez, J. Livage, J. Lemerle, L. Nejem, and J. Lefebvre, *Inorg. Chem.*, **21** (1982) 2758.
68. B.C. Bunker and K.D. Keefer, unpublished results.
69. T.A. Witten and L.M. Sanders, *Phys. Rev. Lett.*, **47** (1981) 1400.
70. A. Bleier and R.M. Cannon in *Better Ceramics Through Chemistry II*, Mater. Res. Soc. Symp. Proc., **73**, eds. C.J. Brinker, D.E. Clark, and D.R. Ulrich (Mater. Res. Soc., Pittsburgh, Pa., 1986), p. 71.
71. J.H.A. Van Der Woude, P.L. De Bruyn, and J. Pieters, *Colloids and Surface*, **9** (1984) 173.
72. J.H.A. Van Der Woude and P.O. De Bruyn, *Colloids and Surface*, **12** (1984) 179.
73. L.H. Edelson, K. Gaugler, and A.M. Glaeser, *Am. Ceram. Soc. Bull.*, **65** (1986) 504.
74. D.G. Pickles and E. Lilley, *J. Am. Ceram. Soc.*, **68** (1985) C-222.
75. E. Matijević, *Langmuir*, **2** (1986) 12.
76. K.K. Singh, P.R. Sarode, and P. Ganguly, *J. Chem. Soc. Dalton* (1983) 1895.
77. S. Rajendran, V.S. Rao, and H.S. Maiti, *J. Mater. Sci.*, **17** (1982) 2709.

78. S.V.S. Prasad and V.S. Rao, *J. Mater. Sci.*, **19** (1984) 3266.
79. C.M. Flynn, Jr., *Chem. Rev.*, **84** (1984) 31–41.
80. P.J. Murphy, A.M. Posner, and J.P. Quirk, *J. Colloid Interface Sci.*, **56** (1976) 270, 284, 298, 312.
81. J. Dousma and P.L. De Bruyn, *J. Colloid Interface Sci.*, **56** (1976) 527.
82. J. Dousma and P.L. De Bruyn, *J. Colloid Interface Sci.*, **64** (1978) 154.
83. J. Dousma and P.L. De Bruyn, *J. Colloid Interface Sci.*, **72** (1979) 314.
84. J.H.A. Van Der Woude and P.L. De Bruyn, *Colloids and Surface*, **8** (1983) 55.
85. J.H.A. Van Der Woude, P. Verhees, and P.L. De Bruyn, *Colloids and Surface*, **8** (1983) 79.
86. W. Schneider, *Comments Inorg. Chem.*, **3** (1984) 205.
87. C.M. Flynn, Jr., *Chem. Rev.*. **84** (1984) 31.
88. R.J. Knight and R.N. Sylva, *J. Inorg. Nucl. Chem.*, **36** (1974) 591.
89. E. Matijević, *Ann. Rev. Mater. Sci.*, **15** (1985) 483.
90. E. Matijević, R.S. Sapieszko, and J.B. Melville, *J. Colloid Interface Sci.*, **50** (1975) 567.
91. E. Matijević and P. Scheiner, *J. Colloid Interface Sci.*, **63** (1978) 509.
92. S. Hamada and E. Matijević, *J. Chem. Soc. Faraday Trans. I*, **78** (1982) 2147.
93. R.S. Sapiezsko, R.C. Patel, and E. Martijević, *J. Phys. Chem.*, **81** (1977) 1061.
94. V. Strahm, R.C. Patel, and E. Matijević, *J. Phys. Chem.*, **83** (1979) 1689.
95. P.H. Fries, N.R. Jagannathan, F.G. Hering, and G.N. Patey, *J. Phys. Chem.*, **91** (1987) 215.
96. A. Fratiello in *Inorganic Chemical Reactions Mechanisms, Part II*, ed. J.O. Edwards (Interscience, New York, 1972), p. 57.
97. M. Magini and R. Caminiti, *J. Inorg. Nucl. Chem.*, **39** (1977) 91.
98. M. Magini, *J. Inorg. Nucl. Chem.*, **40** (1978) 43.
99. A. Clearfield and P.A. Vaughan, *Acta Cryst.*, **9** (1956) 555.
100. T.C.W. Mak, *Can. J. Chem.*, **46** (1968) 3491.
101. D.B. McWhan and G. Lundgren, *Acta Cryst.*, **16** (1963) A36.
102. W. Mark, *Acta Chem. Scand.*, **26** (1972) 3744.
103. L. Baetsle and J. Pelsmaekers, *J. Inorg. Nucl. Chem.*, **21** (1962) 124.
104. G. Lundgren, *Arkiv Kemi*, **13** (1958) 59.
105. W. Mark, *Acta Chem. Scand.*, **27** (1973) 177.
106. B.J. Intorre and A.E. Martell, *J. Am. Chem. Soc.*, **82** (1960) 358.
107. B.J. Intorre and A.E. Martell, *J. Am. Chem. Soc.*, **83** (1961) 3618.
108. A.E. Martell in *Proc. 2nd Symp. Coord. Chem.*, **1** (Tihany, 1964), p. 165.
109. F. Fairbrother, D. Robinson, and J.B. Taylor in *Chemistry of the Coordination Compounds*, Symp. Pub. Div. (Pergamon, New York, 1958), p. 296.
110. F. Fairbrother and J.B. Taylor, *J. Chem. Soc.* (1956) 4946.
111. L.G. Hubert-Pfalzgraf, *New J. of Chem.*, **11** (1987) 663.
112. R.W. Hartel and K.A. Berglund in *Better Ceramics Through Chemistry II*, Mater. Res. Soc. Symp. Proc., **73**, eds. C.J. Brinker, D.E. Clark, and D.R. Ulrich (Mater. Res. Soc., Pittsburgh, Pa., 1986), p. 633.
113. C. Sanchez, J. Livage, M. Henry, and F. Babonneau, *J. Non-Cryst. Solids*, **100** (1988) 65–76.
114. R.A. Assink and B.D. Kay, *J. Non-Cryst. Solids*, **99** (1988) 359.
115. B.D. Kay and R.A. Assink, *J. Non-Cryst. Solids*, **104** (1988) 112–122.
116. J.C. Pouxviel, J.P. Boilot, J.C. Beloeil, and J.Y. Lallemand, *J. Non-Cryst. Solids*, **89** (1987) 345.
117. R. Aelion, A. Loebel, and F. Eirich, *J. Am. Chem. Soc.*, **72** (1950) 1605.
118. H. Schmidt, H. Scholze, and A. Kaiser, *J. Non-Cryst. Solids*, **63** (1984) 1.
119. E. Akerman, *Acta Chem. Scand.*, **10** (1956) 298.

120.   E. Akerman, *Acta Chem. Scand.*, **11** (1957) 298.

121.   G. Orcel and L. Hench, *J. Non-Cryst. Solids*, **79** (1986) 177.

122.   G. Winter, *Oil and Colour Chemist's Association*, **34** (1953) 30.

123.   E.A. Barringer and H.K. Bowen, *Langmuir*, **1** (1985) 414.

124.   E.A. Barringer and H.K. Bowen, *Langmuir*, **1** (1985) 420.

125.   R.W. Hartel and K.A. Berglund in *Better Ceramics Through Chemistry II*, Mater. Res. Soc. Symp. Proc., **73**, eds. C.J. Brinker, D.E. Clark, and D.R. Ulrich (Mater. Res. Soc., Pittsburgh, Pa., 1986), p. 633.

126.   B.J. Ingebrethsen and E. Matijević, *J. Colloid Interface Sci.*, **100** (1984) 1.

127.   M.F. Bechtold, R.D. Vest, and L. Plambeck, *J. Am. Chem. Soc.*, **90** (1968) 4590.

128.   M.F. Bechtold, W. Mahler, and R.A. Schunn, *J. Polym. Sci. Polym. Ed.*, **18** (1980) 2823.

129.   R.W. Adams, E. Bishop, R.L. Martin, and G. Winter, *Aust. J. Chem.*, **19** (1966) 207.

130.   D.C. Bradley, R.C. Mehrotra, and D.P. Gaur, *Metal Alkoxides* (Academic Press, London, 1978).

131.   D. Kundu and D. Ganguli, *J. Mater. Sci. Lett.*, **5** (1986) 293.

132.   E. Bistan and I. Gomory, *Chem. Zvesti*, **10** (1956) 91.

133.   T. Boyd, *J. Polym. Sci.*, **7** (1951) 591.

134.   S. Minami and T. Ishino, *Technol. Rept. Osaka Univ.*, **3** (1953) 357.

135.   N.M.S. Cullinane, J. Chard, G.F. Price, B.B. Millward, and G. Langlois, *J. Appl. Chem.*, **1** (1951) 400.

136.   L. Springer and M.F. Yan in *Ultrastructure Processing of Ceramics, Glasses and Composites*, eds. L.L. Hench and D.R. Ulrich (Wiley, New York, 1984), p. 464.

137.   M. Vallet-Regi, M.L. Veiga-Blanco, and A. Mata-Arjon, *Ann. Quim.*, **76B** (1980) 172, 177, 182, 187.

138.   M. Prassas and L.L. Hench in *Ultrastructure Processing of Ceramics, Glasses and Composites*, eds. L.L. Hench and D.R. Ulrich (Wiley, New York, 1984), p. 100.

139.   B.E. Yoldas, *J. Mater Sci.*, **21** (1986) 1087.

140.   S. Sakka and K. Kamiya, *J. Non-Cryst. Solids*, **42** (1980) 403.

141.   K. Kamiya, K. Tanimoto, and T. Yoko, *J. Mat. Sci. Lett.*, **5** (1986) 402.

142.   D.C. Bradley, *Adv. Chem. Ser.*, **23** (1959) 10.

143.   K. Watenpaugh and C.N. Caughlan, *Chem. Comm.* (1967) 76.

144.   D.C. Bradley, M.B. Hurthouse, and P.F. Rodesiler, *Chem. Comm.* (1968) 1112.

145.   B. Morosin, *Acta Cryst.*, **B33** (1977) 303.

146.   D.C. Bradley, R. Gaze, and W. Wardlaw, *J. Chem. Soc.* (1957) 469.

147.   D.C. Bradley, R. Gaze, and W. Wardlaw, *J. Chem. Soc.* (1955) 3977.

148.   C. Sanchez, F. Babonneau, S. Doeuff, and A. Leaustic in *Ultrastructure Processing of Advanced Ceramics*, eds. J.D. Mackenzie and D.R. Ulrich (Wiley, New York, 1988).

149.   F. Babonneau, S. Doeuff, A. Leaustic, C. Sanchez, C. Cartier, and M. Verdaguer, *Inorg. Chem.*, **27** (1988) 3166–3172.

150.   C.J. Brinker and R.A. Assink, *J. Non-Cryst. Solids*, **111** (1989) 48–54.

151.   B. Fegley, P. White, and H.K. Bowen, *J. Amer. Ceram. Soc.*, **68** (1985) C60.

152.   T. Oglihara, T. Ikemoto, N. Mitzutani, M. Kato, and Y. Mitarai, *J. Mater. Sci.*, **21** (1986) 2771.

153.   R.C. Paul and M.S. Makhni, *Ind. J. Chem.*, **9** (1971) 247.

154.   D.C. Bradley, F.M. Abd-El-Halim, and W. Wardlaw, *J. Chem. Soc.* (1950) 3450.

155.   C. Alquier, M.T. Vandenborre, and M. Henry, *J. Non-Cryst. Solids*, **79** (1986) 383.

156.   R.C. Mehrotra and R. Bohra, *Metal Carboxylates* (Academic Press, London, 1983).

157.   B.E. Yoldas, *Amer. Ceram. Soc. Bull.*, **54** (1975) 289.

158.   S. Doeuff, M. Henry, C. Sanchez, and J. Livage, *J. Non-Cryst. Solids*, **89** (1987) 206.

159.   E.J.A. Pope and J.D. Mackenzie, *J. Non-Cryst. Solids*, **87** (1986) 185.

160. J. Livage, C. Sanchez, M. Henry, and S. Doeuff, *Solid State Ionics*, **32/33** (1989) 633–638.

161. C. Sanchez, F. Babonneau, S. Doeuff, and A. Leaustic in *Ultrastructure Processing of Advanced Ceramics*, eds. J.D. Mackenzie and D.R. Ulrich (Wiley, New York, 1988), p. 77.

162. F. Babonneau, A. Leaustic, and J. Livage in *Better Ceramics Through Chemistry III*, Mater. Res. Soc. Symp. Proc., **121**, eds. C.J. Brinker, D.E. Clark, and D.R. Ulrich (Mater. Res. Soc., Pittsburgh, Pa., 1988), p. 317.

163. W.C. LaCourse and S. Kim in *Science of Ceramic Chemical Processing*, eds. L.L. Hench and D.R. Ulrich (Wiley, New York, 1986), p. 310.

164. B.E. Yoldas, *Amer. Ceram. Soc. Bull.*, **54** (1975) 286–290.

165. B.E. Yoldas, *J. Mat. Sci.*, **10** (1975) 1856.

166. K. Wafers and C. Misra in *Oxides and Hydroxides of Aluminum*, Alcoa Tech. Paper 19, revised 1987 (Alcoa Tech. Center, Alcoa, Pa).

167. R.E. Mesmer and C.E. Baes, Jr., *Inorg. Chem.*, **10** (1971) 2290.

168. W.V. Rausch and H.D. Bale, *J. Chem. Phys.*, **40** (1964) 3391.

169. D.W. Schaefer, R.A. Shellman, K.D. Keefer, and J.E. Martin, *Physica*, **104A** (1986) 105–113.

170. T. Wood, private communication.

171. J.W. Akitt and A. Farthing, *J. Chem. Soc. Dalton* (1981) 1617.

172. T. Assih, A. Ayral, M. Abenoza, and J. Phalippou, *J. Mat. Sci.*, **23** (1988) 3326.

173. J-Y. Chane-Ching and L.C. Klein, *J. Am. Ceram. Soc.*, **71** (1988) 83–85.

174. J-Y. Chane-Ching and L.C. Klein, *J. Am. Ceram. Soc.*, **71** (1988) 86–90.

175. W.L. Olson in *Better Ceramics Through Chemistry II*, Mater. Res. Soc. Symp. Proc., **73**, eds. C.J. Brinker, D.E. Clark, and D.R. Ulrich (Mater. Res. Soc., Pittsburgh, Pa., 1986), p. 611.

176. O. Kriz, B. Casensky, A. Lycka, J. Fusek, and S. Hermanek, *J. Mag. Reson.*, **60** (1984) 375–381.

177. L.F. Nazar and L.C. Klein, *J. Am. Ceram. Soc.*, **71** (1988) C85–C87.

178. W.L. Olson and L.J. Bauer in *Better Ceramics Through Chemistry II*, Mater. Res. Soc. Symp. Proc., **73**, eds. C.J. Brinker, D.E. Clark, and D.R. Ulrich (Mater. Res. Soc., Pittsburgh, Pa., 1986), p. 187.

179. J.D.F. Ramsay, S.R. Dash, and C.J. Wright, *J. Chem. Faraday* (1978) 65–75.

180. G. Johansson, *Acta. Chem. Scand.*, **14** (1960) 771.

181. B.C. Bunker in *Better Ceramics Through Chemistry II*, Mater. Res. Soc. Symp. Proc., vol. 73, eds. C.J. Brinker, D.E. Clark, and D.R. Ulrich (Mater. Res. Soc., Pittsburgh, Pa., 1986), p. 49.

182. J.O. Edwards and V. Ross, *J. Inorg. Nucl. Chem.*, **15** (1960) 329.

183. J. Krogh-Moe, *Phys. Chem. Glasses*, **6** (1965) 46.

184. N. Tohge, G.S. Moore, and J.D. Mackenzie, *J. Non-Cryst. Solids*, **63** (1984) 95.

185. C.J. Brinker, K.J. Ward, K.D. Keefer, E. Holupka, and P.J. Bray in *Better Ceramics Through Chemistry II*, Mater. Res. Soc. Symp. Proc., **73**, eds. C.J. Brinker, D.E. Clark, and D.R. Ulrich (Mater. Res. Soc., Pittsburgh, Pa., 1986), p. 57.

186. C.J. Brinker, K.J. Ward, K.D. Keefer, E. Holupka, P.J. Bray, and R.K. Pearson in *Aerogels*, ed. J. Fricke (Springer-Verlag, Berlin, 1986), pp. 387–411.

187. C.J. Brinker, B.C. Bunker, D.R. Tallant, and K.J. Ward, *J. de Chimie Phys.*, **83** (1986) 851–858.

188. K. Kustin and R. Pizer, *J. Am. Chem. Soc.*, **91-2** (1969) 317.

189. S.D. Ross in *The Infrared Spectra of Minerals*, ed. V.C. Farmer (Mineralogical Society, London, 1974).

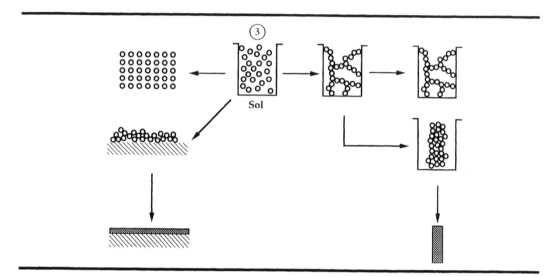

Sol

CHAPTER 3

Hydrolysis and Condensation II

Silicates

Silicon is the most abundant metal in the earth's crust, and evidence of silicate hydrolysis and condensation to form polysilicate gels and particles is seen in many natural systems. For, example, precious opal is composed of amorphous silica particles glued together by a lower-density silicate gel [1,2]. The essential ingredients required to form opals are an abundant supply of readily soluble silica and a source of water. Repeated hydrolysis and condensation steps involving the soluble silica lead to aqueous polysilicate species that, under appropriate chemical conditions, evolve into spherical particles of essentially anhydrous SiO_2. Flint, which our ancient ancestors recognized as the toughest stone available, was apparently formed from the siliceous skeletons of ancient sponges by a mysterious process involving solution transport [1]. Since the 1970s it has been recognized that soluble silica in trace amounts also plays a role in the development of mammals [1].

Manmade synthesis of polysilicate gels from alkoxide precursors closely followed the first preparation of silicon tetrachloride ($SiCl_4$) in 1824 [3]. Ebelmen [4] reacted $SiCl_4$ with ethanol to form tetraethoxysilane[†] (TEOS) as early as 1845. His subsequent publications [5,6] document the hydrolysis of TEOS to yield silicate solutions from which fibers could be drawn and the casting of amorphous gels that could be dried over a sufficiently long period of time to yield optical elements such as lenses. In the 1850s, Mendeleyev [7] conceived of the novel idea that hydrolysis of $SiCl_4$ yields a product ($Si(OH)_4$) that undergoes repeated condensation reactions to form high–molecular-weight polysiloxanes.

[†] Also commonly referred to as tetraethyl orthosilicate.

The discovery of the exceptional tendency of organosilicon compounds to form siloxane polymers containing organic side groups (silicones) caused an explosion of activity in the late 1930s that established a chemical and physical basis for understanding the processes of hydrolysis and condensation. More recently, the "rediscovery" of monolithic gel formation and the low-temperature conversion of gels to glasses without melting [e.g., 8,9] has caused renewed interest in the topic of the hydrolysis and condensation of silicates.

This chapter reviews the hydrolysis and condensation of silicates in both aqueous and organic systems used in sol-gel processing of silica glass. To complement the treatise *The Chemistry of Silica* published by Iler in 1979 [1], we emphasize more recent results derived from in situ methods of analysis—such as nuclear magnetic resonance (NMR) spectroscopy, vibrational spectroscopy, and small-angle scattering of X-rays (SAXS), neutrons (SANS), or light—and we concentrate on organic rather than inorganic systems. The topics discussed in this chapter are outlined as follows:

1. *Aqueous Silicates* provides a *concise* review of hydrolysis and condensation of aqueous silicate systems in which Iler's views [1] are augmented by more recent studies that employ ^{29}Si NMR.
2. *Hydrolysis and Condensation of Silicon Alkoxides*
    2.1 *General Trends* presents an overview of silicate polymerization in alkoxide-based systems.
    2.2 *Precursor Molecules* identifies common silicate precursors and their synthesis.
    2.3. *Hydrolysis* addresses the effects of processing parameters, such as catalysts, H_2O : Si ratio, and solvent, as well as steric and inductive factors on the mechanism of hydrolysis of silicon alkoxides and the reverse reaction, esterification.
    2.4. *Condensation* discusses the effects of catalyst and solvent and the influence of steric and inductive factors on the mechanisms of the condensation reaction and the reverse reactions, siloxane bond hydrolysis or alcoholysis.
    2.5. *Sol Gel Kinetics* presents the results of experiments from which rates of concurrent hydrolysis and condensation are derived.
    2.6. *Structural Evolution* follows the growth of silicate polymers from their genesis to very near the gel point, using various in situ methods such as NMR, vibrational spectroscopy, and small-angle scattering. The section also presents growth models to explain structural evolution.
    2.7. *Summary.*
3. *Multicomponent Silicates* briefly discusses synthetic schemes employed to make multicomponent gels and presents the results of several structural studies that examine polymer growth in binary silicate systems.

1.

AQUEOUS SILICATES

The $+4$ *oxidation state* $(z = 4)$ is the only important one in the chemistry of silicon in naturally occurring systems [10], and the *coordination number* of silicon, N, is most often four. Compared to transition metals discussed in the previous chapter, silicon is generally less electropositive, e.g., the *partial positive charge*[†] on silicon $\delta(Si)$ in $Si(OEt)_4$ is $+0.32$, whereas $\delta(Ti) = +0.63$ and $\delta(Zr) = +0.65$ in $Ti(OEt)_4$ and $Zr(OEt)_4$, respectively [11]. The reduced $\delta(M)$ makes silicon comparatively less susceptible to *nucleophilic attack*, and since $N = z$, *coordination expansion* does not spontaneously occur with nucleophilic reagents. These factors make the kinetics of hydrolysis and condensation considerably slower than observed in transition metal systems or in Group III systems.

Silicon is hydrolyzed even in dilute acid [10], as expected from its small ionic radius (0.42 Å), and $Si(OH)_4$ is the predominant mononuclear solution species below ca pH 7. (See Fig. 1.) Above pH 7, further hydrolysis produces anionic species:

$$Si(OH)_4(aq) \rightarrow SiO_x (OH)_{4-x}^{x-} + xH^+ \tag{1}$$

where $SiO(OH)_3^-$ ($x = 1$ in Eq. (1)) is the predominant mononuclear species above ca pH 7. Because $SiO(OH)_3^-$ is a very weak acid, $SiO_2(OH)_2^{2-}$ is observed in appreciable quantities only above pH 12. (See Fig. 1.)

Freundlich [12] made silicic acid by acidifying a soluble silicate:

$$Na_2SiO_3 + H_2O + 2HCl \rightarrow Si(OH)_4 + 2NaCl \tag{2}$$

or hydrolyzing the ester:

$$Si(OEt)_4 + H_2O(excess) \rightarrow Si(OH)_4. \tag{3}$$

He observed that silicic acid diffuses easily through parchment or animal membranes and has a molecular weight (by freezing-point depression) corresponding to the monomer. Soon the molecular units become larger and pass through membranes slowly and then not at all [12]. Iler [1] has considered two alternatives to explain these observations, viz., monomers or other small

---

[†] For a definition of the italicized terms and a discussion of the partial charge model, refer to Section 1 in Chapter 2.

Fig. 1.

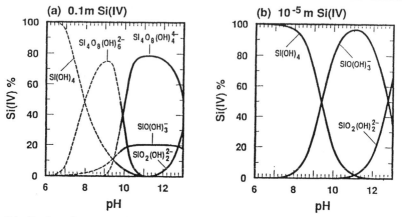

Distribution of aqueous silicate species at 25°C in (a) 0.01-m Si(IV) and (b) 10^{-5}-m Si(IV). Ionic strength, I = 3 m. From Baes and Mesmer [10].

primary particles aggregate or individual particles increase in size and decrease in number.

Since silicic acid solutions slowly thicken and finally *gel*, appearing outwardly like organic gels, most researchers prior to 1979 believed that $Si(OH)_4$ polymerized into siloxane chains that branched and cross-linked like many organic polymers [1]. However Iler states, "There is no relation or analogy between silicic acid polymerized in an aqueous system and condensation-type organic polymers." Instead, silicic acid polymerizes into discrete particles that in turn aggregate into chains and networks as first recognized by Carmen [13]. According to Iler [1], polymerization occurs in three stages:

1. Polymerization of monomer to form particles.
2. Growth of particles.
3. Linking of particles into chains, then networks that extend throughout the liquid medium, thickening it to a gel.

The formation of particles in the first stage has been investigated by potentiometric methods [14], trimethylsilylation [15], molybdic acid reagent [16], paper chromatography [17], and most recently ^{29}Si NMR. (See, e.g. Refs. 18–21.) The potentiometric results were explained on the basis of the cyclic tetramers, $Si_4O_6(OH)_6^{2-}$ and $Si_4O_8(OH)_4^{4-}$, along with the three mononuclear species previously discussed. (See Fig. 1.)

Using a high field spectrometer and samples enriched in the magnetically active ^{29}Si nucleus along with various "spin-perturbation" techniques, Knight and coworkers [18,19] have positively identified at least 12 species in moderately concentrated potassium silicate solutions [18]. In 0.5M

Fig. 2. _____

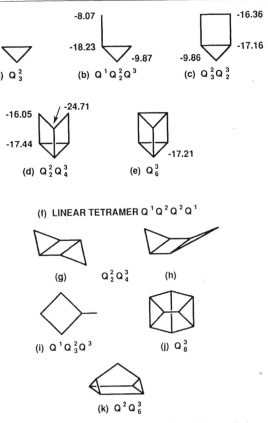

Some possible silicate species present in aqueous potassium silicate solution with atomic ratio $K:Si = 1.0$ and concentration $\sim 1.4\,M$ identified by ^{29}Si NMR. Vertices in diagrams represent positions of silicon atoms that are bridged by oxygens. Additional oxygen (or OH groups) are present as required for valency fulfillment. Chemical shifts, given in ppm with respect to the monomer resonance, are included for positively identified species. Subscripts denote the number of individual Q^n species contained in each structure. After Harris _et al._ [18].

$K_2O–SiO_2$ at $22°C$, the _monomer_ (Q^0), _dimer_ (Q_2^1), _cyclic trimer_ (Q_3^2), _linear trimer_ $(Q_2^1 Q^2)$, _substituted cyclic trimer_ $(Q^1 Q^3 Q_2^2)$, _cyclic tetramer_ (Q_4^2), and _prismatic hexamer_ (Q_6^3) account for 80% of the silicon in solution.[†] (See Fig. 2.)

[†] In Q_x^n notation, the superscript n denotes the number of bridging oxygens (–OSi) surrounding the central silicon, and the sum of the subscripts, x, equals the number of silicons comprising the silicate species, e.g., dimer $(\sum x = 2)$, trimer $(\sum x = 3)$, etc., where Q^n species without subscripts are counted as $x = 1$.

These NMR results largely support Iler's view (see Fig. 3) that condensation takes place in such a fashion as to maximize the number of Si–O–Si bonds and minimize the number of terminal hydroxyl groups through internal condensation. Thus rings are quickly formed to which monomers add, creating three-dimensional particles. These particles condense to the most compact state leaving OH groups on the outside. (See Fig. 4.). According to Iler the three-dimensional particles serve as nuclei. Further growth occurs by an *Ostwald ripening* mechanism whereby particles grow in size and decrease in number as highly soluble small particles dissolve and reprecipitate on larger, less soluble nuclei. Growth stops when the difference in solubility between the smallest and largest particles becomes only a few ppm. Due to greater solubility, growth continues to larger sizes at higher temperatures, especially above pH 7.

Fig. 3.

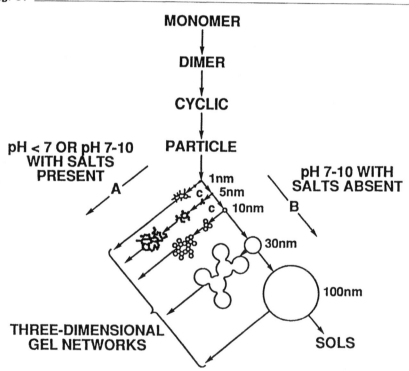

Polymerization behavior of aqueous silica. In basic solution (B) particles grow in size with decrease in number; in acid solution or in the presence of flocculating salts (A), particles aggregate into three-dimensional networks and form gels. From Iler [1].

Fig. 4.

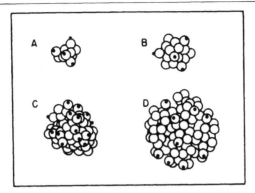

Models of (A) cyclic trisilicic, (B) cubic octasilicic acids, and (C) and (D) the corresponding theoretical colloidal particles formed by condensing monomers to form closed rings until the original species is surrounded by one layer of deposited silica bearing silanol groups. When formed above pH 7, the inner silica contains few silanol groups. Different kinds of incompletely condensed oligomers could form the cores of colloidal particles. There is no evidence that A and B are specifically involved. Spheres are oxygen atoms, black dots, hydrogen atoms. Silicon atoms are not visible. From Iler [1].

1.1. pH-Dependence

Iler divides the polymerization process into three approximate pH domains: $<$pH 2, pH 2–7, and $>$pH 7. pH 2 appears as a boundary, since the *point of zero charge* (PZC), where the surface charge is zero, and the *isoelectric point* (IEP), where the electrical mobility of the silica particles is zero, both are in the range pH 1–3. pH 7 appears as a boundary because both the silica solubility and dissolution rates are maximized at or above pH 7 and because the silica particles are appreciably ionized above pH 7 so that particle growth occurs without aggregation or gelation. (See Fig. 5.)

1.1.1. POLYMERIZATION pH 2–7

Since the gel times decrease steadily between pH 2 and ca pH 6, it is generally assumed that above the IEP the condensation rate is proportional to $[OH^-]$ as in the following reaction sequence:

$$\equiv\text{Si-OH} + \text{OH}^- \xrightarrow[\text{fast}]{} \equiv\text{Si-O}^- + \text{H}_2\text{O} \qquad (4)$$

$$\equiv\text{Si-O}^- + \text{HO-Si} \xrightarrow[\text{slow}]{} \equiv\text{Si-O-Si}\equiv + \text{OH}^-. \qquad (5)$$

Fig. 5.

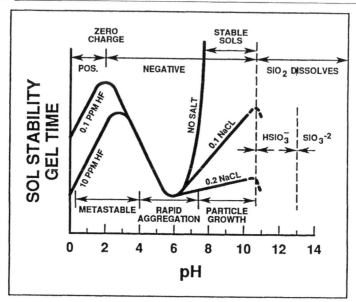

Effects of pH in the colloidal silica-water system. From Iler [1].

In any given distribution of silicate species, the most acidic silanols (and hence the most likely to be deprotonated according to Eq. 4) are those contained in the most highly condensed species [1]. Therefore, condensation according to Eq. 5 occurs preferentially between more highly condensed species and less highly condensed, neutral species. This means that the rate of dimerization is low, but once dimers form they react preferentially with monomers to form trimers which in turn react with monomers to form tetramers. At this point cyclization is rapid due to the proximity of the chain ends and the substantial depletion of the monomer population. Cyclic trimers may also form, but the reduced Si–O–Si bond angles and the associated strain make them much less stable in this pH range. (They are, however, quite stable above ca pH 12 [22] as discussed in the following section.)

Further growth occurs by continued addition of lower–molecular-weight species to more highly condensed species (either conventional polymerization or ripening) and by aggregation of the condensed species to form chains and networks. Near the IEP where there is no electrostatic particle repulsion, the growth and aggregation processes occur together and may be indistinguishable. In any case, since the solubility of silica is low in this pH range, particle growth stops when the particles reach 2–4 nm where the solubility and size-dependence of solubility is greatly reduced. (See Fig. 6.)

1.1.2. POLYMERIZATION ABOVE pH 7

Above pH 7, polymerization occurs by the same nucleophilic mechanism (Eqs. 4 and 5). However, because all the condensed species are more likely to be ionized and therefore mutually repulsive, growth occurs primarily by the addition of monomers to more highly condensed particles rather than by particle aggregation. Particles 1–2 nm in diameter are formed in a few minutes above pH 7. Above ca pH 12, where most of the silanols are deprotonated, the primary building blocks are composed primarily of cyclic trimers and tetramers (Fig. 2). Cyclic trimers are stable in this pH range because the planar, cyclic configuration (D_{3h} symmetry) permits the greatest separation of charge between the deprotonated sites [22].

Due to the greater solubility of silica and the greater size-dependence of solubility above pH 7 (Fig. 6), growth of the primary particles continues by Ostwald ripening. Particles grow rapidly to a size that depends mainly on

Fig. 6a.

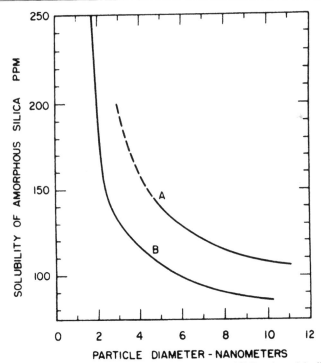

Relation between solubility of amorphous silica in water at 25°C and particle diameter. (A) particles made at 80–100°C at pH 8. (B) particles made at 25–50°C at pH 2.2. From Iler [1].

Fig. 6b.

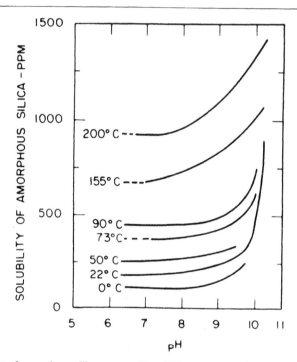

Solubility of amorphous silica versus pH at different temperatures. From Iler [1].

the temperature. (Higher temperatures produce larger particles due to greater silica solubility. See Table 1.) Since growth occurs by the dissolution of smaller particles and deposition of soluble silica on larger particles, the growth rate depends on the particle-size distribution.

In the absence of salt, no chaining or aggregation occur, because the particles are mutually repulsive. Stable sols of large particle sizes can be prepared and concentrated for industrial use [1]. The addition of salt reduces the thickness of the double layer at a given pH, dramatically reducing the gel times. (See Fig. 5.) Factors governing the stability of sols are discussed in the following chapter.

1.1.3   POLYMERIZATION BELOW pH 2

As discussed previously, pH 2 represents a metastability region where the observed gel times are quite long (Fig. 5). Below pH 2, the polymerization rate is proportional to $[H^+]$ [1]. Although Iler and others propose that the

Table 1.

Growth of Silica Particles by Heating a 4% Sol of Silicic Acid at pH 8–10.

Mole Ratio $SiO_2 : Na_2O$	Time	Temperature (°C)	Specific Surface Area $(m^2 g^{-1})$	Estimated Particle Diameter (nm)
100	1 hr	80	600	5
64	6 hr	85	510	6
100	5 hr	95	420	7
78	6 hr	98	406	7
80	30 min	100	350	8
85	3 hr	160	200	15
85	3.25 min	270	200	15
85	0.9 min	250	225	15
90	3.1 min	200	271	10
85	10 min	200	228	12
85	10 min	295	78	36
85	30 min	295	—	64
Very high[a]	3 hr	340	—	88
Very high[a]	6 hr	340	—	105
Very high[a]	3 hr	350	20	150

Source: Iler [1].

[a] Traces of sodium ions remaining in the starting particles after deionization of the sol before autoclaving resulted in a pH of about 8 in the final sol.

acid-catalyzed polymerization mechanism involves a siliconium ion intermediate ($\equiv Si^+$):

$$\equiv Si-OH + H_3O^+ \rightarrow \equiv Si^+ + 2H_2O \tag{6}$$

$$\equiv Si^+ + HO-Si \rightarrow \equiv Si-O-Si \equiv + H^+, \tag{7}$$

we shall show in the following section on alkoxide polymerization that the condensation process more likely proceeds via an associative $\equiv SiOR(OH_2)^+$
(with H above) intermediate.

In the absence of fluoride ion, the solubility of silica below pH 2 is quite low (except $\ll$ pH 0, [23]), and at moderate acidities (pH 0–2) the silicate species should not be highly ionized. For these reasons it is likely that the formation and aggregation of primary particles occur together and that ripening contributes little to growth after the particles exceed 2 nm in diameter. Gel networks are formed composed of exceedingly small primary particles (Fig. 3). Traces of F^- or the addition of HF decrease the gel times (Fig. 5) and produce gels similar to those formed above pH 2. As discussed in the following section, F^- and OH^- are of similar size and have the same influence on the polymerization behavior.

2.

HYDROLYSIS AND CONDENSATION OF SILICON ALKOXIDES

2.1.   General Trends

Silicate gels are most often synthesized by hydrolyzing monomeric, tetrafunctional *alkoxide* precursors employing a mineral acid (e.g., HCl) or base (e.g., NH_3) as a catalyst. At the functional group level, three reactions are generally used to describe the sol-gel process:

hydrolysis

$$\equiv Si - OR + H_2O \quad \rightleftharpoons \quad \equiv Si - OH + ROH \qquad (8)$$

esterification

alcohol condensation

$$\equiv Si - OR + HO - Si\equiv \quad \rightleftharpoons \quad \equiv Si - O - Si\equiv \, + ROH \qquad (9)$$

alcoholysis

water condensation

$$\equiv Si - OH + HO - Si\equiv \quad \rightleftharpoons \quad \equiv Si - O - Si\equiv \, + H_2O \qquad (10)$$

hydrolysis

where R is an alkyl group, C_xH_{2x+1}. The *hydrolysis reaction* (Eq. 8) replaces alkoxide groups (OR) with hydroxyl groups (OH). Subsequent *condensation reactions*[†] involving the silanol groups produce siloxane bonds (Si–O–Si) plus the by-products alcohol (ROH) (Eq. 9) or water (Eq. 10). Under most conditions, condensation commences (Eqs. 9 and 10) before hydrolysis (Eq. 8) is complete. Because water and alkoxysilanes are immiscible (see, for example, Fig. 7), a mutual solvent such as alcohol is normally used as a homogenizing agent. However, gels can be prepared from silicon alkoxide-water mixtures without added solvent [25], since alcohol produced as the by-product of the hydrolysis reaction is sufficient to homogenize the initially phase separated system. It should be noted that alcohol is not simply a solvent. As indicated by the reverse of Eqs. 8 and 9, it can participate in *esterification* or *alcoholysis* reactions.

[†]Nomenclature used in the organo-silicate literature, *alcohol condensation* and *water condensation*, is equivalent to the terminology, *alcoxolation* and *oxolation*, used in the discussion of inorganic polymerization in the previous chapter.

Fig. 7.

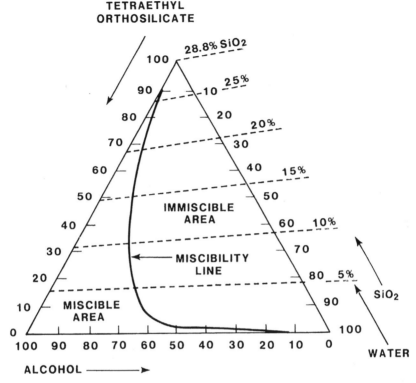

TEOS, H_2O, Synasol (95% EtOH, 5% water) ternary-phase diagram at 25°C. For pure ethanol the miscibility line is shifted slightly to the right [24].

The H_2O : Si *molar ratio* (r) in Eq. 8 has been varied from less than one to over 50, and the concentrations of acids or bases have been varied from less than 0.01 [26] to 7 M [27], depending on the desired end product. Typical gel-synthesis procedures used to produce bulk gels, films, fibers, and powders are listed in Table 2.

Because water is produced as a by-product of the condensation reaction, an r value of 2 is theoretically sufficient for complete hydrolysis and condensation to yield anhydrous silica as shown by the net reaction:

$$n\text{Si(OR)}_4 + 2n\text{H}_2\text{O} \rightarrow n\text{SiO}_2 + 4n\text{ROH}. \tag{11}$$

However, even in excess water ($r \gg 2$), the reaction does not go to completion. Instead, a spectrum of intermediate species ($[\text{SiO}_x(\text{OH})_y(\text{OR})_z]_n$; where $2x + y + z = 4$) are generated. Figure 8 shows a sequence of ^{29}Si NMR spectra representing the temporal evolution of silicate species in an

Table 2.

Sol-Gel Silicate Compositions for Bulk Gels, Fibers, Films, and Powders.

SiO₂ Gel Types	mole %					
	TEOS	EtOH	H₂O	HCl	NH₃	H₂O/Si(r)
Bulk						
1-step acid [9]	6.7	25.8	67.3	0.2	—	10
1-step base [9]	6.7	25.8	67.3	—	0.2	10
2-step acid–base [26]						
1st step-acid	19.6	59.4	21.0	0.01	—	1.1
2nd step-acid (A2)	10.9	32.8	55.7	0.6	—	5.1
2nd step-base (B2)	12.9	39.2	47.9	0.01	0.016	3.7
fibers [28]	11.31	77.26	11.31	0.11	—	1.0
films [29]	5.32	36.23	58.09	0.35	—	10.9
Monodisperse spheres [27]	0.83	33.9	44.5	—	20.75	53.61

Source: Nogami and Moriya [9], Brinker *et al.* [26], Stöber *et al.* [27], Sakka [28], Sakka *et al.* [29].

acid-catalyzed silicate solution prepared with $r = 2$ [30]. Clearly the reaction to form anhydrous silica has not gone to completion as evidenced by the complex distribution of Q^0 through Q^4 species even at very long times.

Numerous investigations have shown that variations in the synthesis conditions (for example, the value of r, the catalyst type and concentration, the solvent, temperature, and pressure) cause modifications in the structure and properties of the polysilicate products. For example, Sakka and coworkers [28,29,31] observed that hydrolysis of TEOS utilizing r values of 1 to 2 and 0.01 M HCl as a catalyst yields a viscous, *spinnable* sol (capable of being drawn into a fiber) when aged in open containers exposed to the atmosphere. Subsequent studies showed that spinnable solutions exhibit a strong concentration-dependence of the intrinsic viscosity and a power-law–dependence of the reduced viscosity on the number averaged molecular weight:

$$[\eta] = kM_n^{\alpha}. \tag{12}$$

Values of α range from 0.5 to 1.0, which indicate linear or chain molecules [32]. More recently, Kamiya *et al.* [33] postulated ladder polymers to account for spinnability.

By comparison, hydrolyses utilizing r values greater than 2 and/or base catalysts produced solutions that were not spinnable at equivalent viscosities [28,31]. Values of α in Eq. 12 equalled 0.1–0.5, indicating sphere or disk-shaped particles. The latter results are consistent with the structures that

Fig. 8.

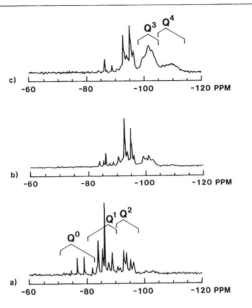

^{29}Si NMR spectra of acid-catalyzed TEOS ($r = 2$) after a) 3 hours, b) 3 days, c) 14 days [30]. Chemical shift, δ, in ppm, relative to TMS.

emerge under the conditions employed in the so-called "Stöber process" [27] for preparing SiO_2 powders: hydrolysis of TEOS with r values ranging from 20 to over 50 and concentrations of ammonia ranging from ~1 to 7 M results in monodisperse, spherical particles.

Based on the results of in situ SAXS investigations, Schaefer, Keefer, and coworkers [26,34,35] have demonstrated that variations in the hydrolysis and condensation conditions profoundly affect the structures of sol-gel silicates on intermediate-length scales ~1–20 nm. A two-step hydrolysis procedure ($1H_2O$/Si under acidic conditions followed after 90 min by an additional ~3 or $4H_2O$/Si under acidic or basic conditions) produced ramified, weakly branched structures characterized by a *mass fractal dimension*, d_f [36].[†] Single-step base-catalyzed hydrolysis produced fractally rough [36] particles, characterized by a *surface fractal dimension*, d_s,[††] and base-catalyzed hydrolysis and condensation under aqueous conditions produced smooth colloidal particles (nonfractal).

[†] For mass fractals d_f relates the polymer mass, M, to its radius, r, according to $M \propto r^{d_f}$, where in three dimensions $d_f \leq 3$.

[††] The surface fractal dimension, d_s, relates the area of an object to its size. In three dimensions, d_s varies from 2 for a pristine surface to 3 for a highly convoluted surface.

Thus a consistent trend is apparent. Acid-catalyzed hydrolysis with low $H_2O:Si$ ratios produces weakly branched "polymeric" sols, whereas base-catalyzed hydrolysis with large $H_2O:Si$ ratios produces highly condensed "particulate" sols. Intermediate conditions produce structures intermediate to these extremes.

It is evident from Eqs. 8–10 that the structure of sol-gel silicates evolves sequentially as the product of successive hydrolysis and condensation reactions (and the reverse reactions: esterification and alcoholic or hydrolytic depolymerization). Since structural variations can only result from a different sequence of these three basic reactions, the fundamental questions surrounding hydrolysis and condensation in silicates concern what chemical and physical factors determine the precise sequence of these reactions under different processing conditions.

The next section discusses precursor molecules used in silicate gel syntheses. In the following three sections, the mechanisms and kinetics of the hydrolysis and condensation reactions (and the reverse reactions) are discussed with respect to catalytic, steric, inductive and solvent effects. The remaining sections document structural changes leading to gelation based on the results of in situ methods of analysis. We relate chemistry occurring on short-length scales to macroscopic models of kinetic growth processes in order to gain physical insight into structural evolution in silicate systems.

2.2. Precursor Molecules

The most common *tetraalkoxysilanes* used in the sol-gel process are *tetraethoxysilane* ($Si(OC_2H_5)_4$) and *tetramethoxysilane* ($Si(OCH_3)_4$), which are abbreviated in the literature as TEOS and TMOS, respectively. The traditional method of preparing tetraalkoxysilanes is by reacting tetrachlorosilane with alcohol [3]. When anhydrous ethanol is used, the product is TEOS with hydrogen chloride as a by-product:

$$SiCl_4 + EtOH \rightarrow Si(OEt)_4 + 4HCl. \tag{13}$$

Table 3 lists the formulae and properties of several tetraalkoxysilanes used in sol-gel processing [37].

In order to reduce the functionality (potential number of sites able to form Si–O–Si bonds) of the alkoxide precursor, impart organic character to or derivatize the siloxane network, it is also possible to use *organotrialkoxysilane* or *diorganodialkoxysilane* precursors ($R'Si(OR)_3$ or $R'_2Si(OR)_2$, respectively) in which the R' represents a nonhydrolyzable organic substituent. Table 4 lists the formulae and properties of some organoalkoxysilanes that have been used in sol-gel processing. Of course, the silicone

Table 3

Physical Properties of Typical Tetraalkoxysilanes.

Name	MW	bp	n_D (20°)	d (20°)	η (ctsks)	Dipole Moment	Solubility
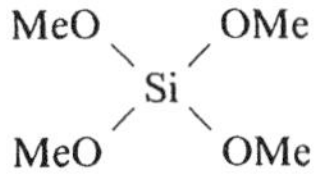 MeO—Si—OMe (MeO, OMe) $Si(OCH_3)_4$ tetramethoxysilane TMOS	152.2	121	1.3688	1.02	5.46	1.71	alcohols
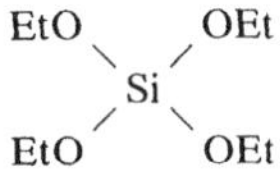 EtO—Si—OEt (EtO, OEt) $Si(OC_2H_5)_4$ tetraethoxysilane TEOS	208.3	169	1.3838	0.93	—	1.63	alcohols
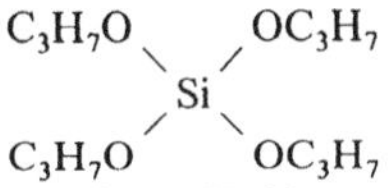 C_3H_7O—Si—OC_3H_7 (C_3H_7O, OC_3H_7) $Si(n\text{-}C_3H_7O)_4$ tetra-n-propoxysilane	264.4	224	1.401	0.916	1.66	1.48	alcohols
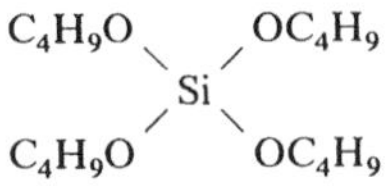 C_4H_9O—Si—OC_4H_9 (C_4H_9O, OC_4H_9) $Si(n\text{-}C_4H_9O)_4$ tetra-n-butoxysilane	320.5	115	1.4126	0.899	2.00	1.61	alcohols
$(MeOCH_2CH_2O)_4Si$ tetrakis(2-methoxyethoxy) silane	328.4	179	1.4219	1.079	4.9	—	alcohols

Source: Anderson *et al.* [37].

industry is based on polydiorganosiloxanes and many possibilities exist for the organic substituents. For a more complete listing of commercially available organoalkoxysilane precursors, consult reference [37].

Silicate gels have also been synthesized using *oligomeric* precursors.[†] Ethyl Silicate 40 is a commercial form of ethoxypolysiloxane (ethyl poly-silicate) that results when the ethanol used in the production of TEOS contains some water (industrial spirit) [24]. Water plus the hydrogen chloride by-product (Eq. 13) result in the partial hydrolysis and condensation of the

---

[†] *Oligomer* is defined as a polymer composed of a few monomers. We use this term to include species ranging from dimers to octamers.

Table 4

Physical Properties of Typical Organoalkoxysilanes.

Name	Formula	MW	pb (°C)	d (g-cm^{-3})	Dipole Moment (debyes)	n_D	η (millipoise)
Methyltriethoxysilane	MeSi(OEt)$_3$	178.3	141	0.895	1.72	1.3832	—
Methyltrimethoxysilane	MeSi(OMe)$_3$	136.22	102	0.955	1.6	1.3696	0.5
Methyl tri-n-propoxysilane	MeSi(n-OPr)$_3$	220.4	83	0.88	—	—	—
Phenyltriethoxysilane	PhSi(OEt)$_3$	240.37	112	0.996	1.85	1.4718	—
Vinyltriethoxysilane	H$_2$C=C-Si(OEt)$_3$ | H	190.31	160	0.903	1.69	1.396	0.7

Source: Anderson *et al.* [37].

tetraethoxysilane. In practice, reaction conditions are chosen that give on ignition SiO$_2$ equivalent to 40 wt%, which corresponds to a mixture of ethoxysiloxanes with an average of five silicon atoms per oligomer.

Klemperer and coworkers [38,39] have pioneered the use of specific oligomers as "molecular building blocks" for the construction of silicate networks with precisely controlled architectures. Precursors studied include hexamethoxydisiloxane (14), octamethoxytrisiloxane (15), and a methoxylated, cubic octamer (Si$_8$O$_{12}$)(OCH$_3$)$_8$ (16):

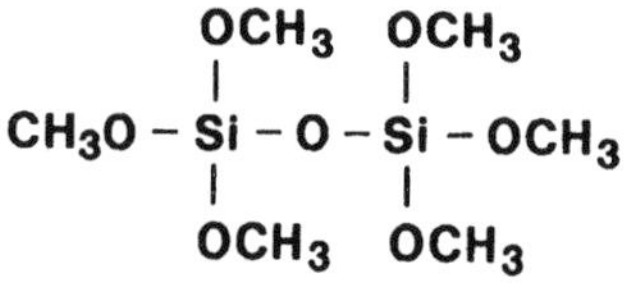

(14)

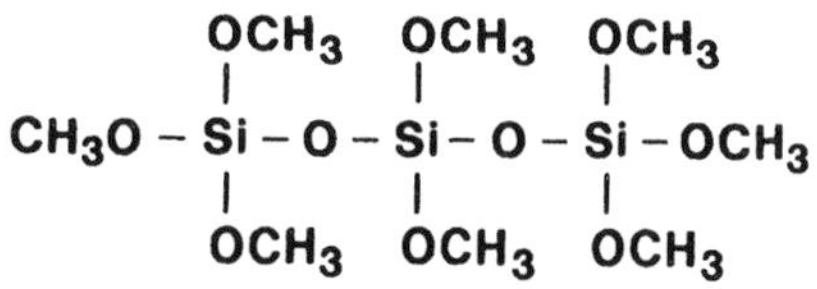

(15)

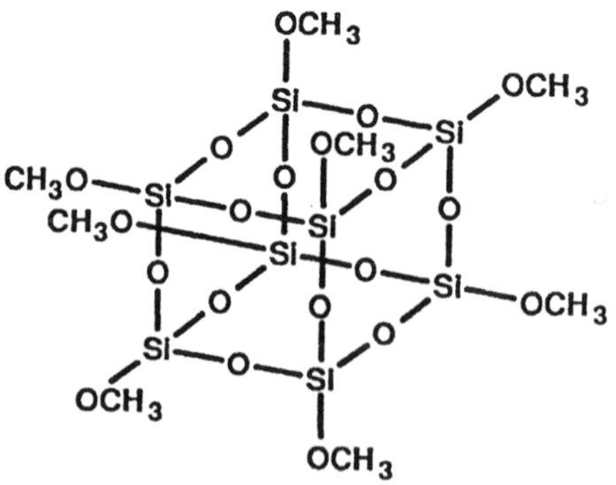

(16)

The octamer was prepared using the following two-step reaction sequence [40]:

$$[Si_8O_{12}]H_8 + 8Cl_2 \rightarrow [Si_8O_{12}]Cl_8 + 8HCl \qquad (17)$$

$$[Si_8O_{12}]Cl_8 + CH_3ONO \rightarrow [Si_8O_{12}](OCH_3)_8 + 8NOCl. \qquad (18)$$

With regard to monomeric precursors, the steric bulk and the electron-providing or -withdrawing characteristics of the alkoxide or organic substituents attached to silicon will largely determine the kinetics of the hydrolysis and condensation reactions [3]. Therefore the use of specific precursors is often mandated by kinetic considerations or compatibility with precursors of other network-forming elements in multicomponent silicate gel syntheses. (See Section 3.) Oligomeric precursors are desirable when it is necessary to increase the silicate content of a solution (e.g., in refractory binders) or when tailor-made substructure is required. The latter concept holds the promise of designing inorganic silicates with novel properties.

Organically modified silicates (ORMOSILs [41,42] or CERAMERs [43,44]) represent hybrid systems in which several precursor types are combined. Schmidt and coworkers [41,42] have developed silicate products with unique properties by combining tetraalkoxysilanes with alkyl-substituted and organofunctional alkoxysilanes: $Si(OR)_4 + R_2Si(OR)_2 + YR'Si(OC_2H_5)_3$. R is an alkyl group, R' is an alkylene, and Y is an organofunctional group such as $-(CH_2)_3NH_2$, $-(CH_2)_3NHCO-O-NH_2$, $-(CH_2)_3S(CH_2)_2CHO$, etc. If Y is a polymerizable ligand such as an epoxide, it is possible to form an organic network in addition to the inorganic one [45]:

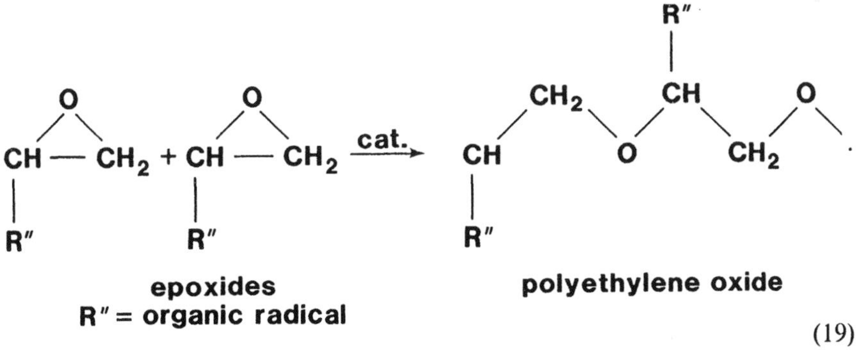

epoxides
R" = organic radical

polyethylene oxide

$$(19)$$

In principle such hybrid systems allow an unlimited number of chemical and structural modifications to be performed. Choice of specific precursors may be made on the basis of solubility or the thermal stability of the organofunctional substituents.

2.3. Hydrolysis

Hydrolysis occurs by the nucleophilic attack of the oxygen contained in water on the silicon atom as evidenced by the reaction of isotopically labelled water with TEOS that produces only unlabelled alcohol in both acid- and base-catalyzed systems [46]:

$$\text{Si-OR} + \text{H}^{18}\text{OH} \rightleftharpoons \text{Si-}^{18}\text{OH} + \text{ROH} . \tag{20}$$

The same behavior is observed in organoalkoxysilanes, $R_xSi(OR)_{4-x}$, where $x = 1$, 2, or 3 [47].

Tetraalkoxysilanes, organotrialkoxysilanes, and diorganodialkoxysilanes hydrolyze upon exposure to water vapor [47]. Hydrolysis is facilitated in the presence of homogenizing agents (alcohols, dioxane, THF, acetone, etc.) that are especially beneficial in promoting the hydrolysis of silanes containing bulky organic or alkoxy ligands, such as phenylphenoxysilane, which, when neat (undiluted), remains unhydrolyzed upon exposure to water vapor [47]. It should be emphasized, however, that the addition of solvents may promote esterification or depolymerization reactions according to the reverse of Eqs. 8 and 9. (See discussion of reverse reactions in Sections 2.3.6 and 2.4.5.)

2.3.1. EFFECTS OF CATALYSTS

Hydrolysis is most rapid and complete when *catalysts* are employed [47]. Although mineral acids or ammonia are most generally used in sol-gel processing, other known catalysts are acetic acid, KOH, amines, KF, HF, titanium alkoxides, and vanadium alkoxides and oxides [47]. Many authors report that mineral acids are more effective catalysts than equivalent concentrations of base. However, neither the increasing acidity of silanol groups with the extent of hydrolysis and condensation (acidic silanols may neutralize basic catalysts [48]) nor the generation of unhydrolyzed monomers via base-catalyzed alcoholic or hydrolytic depolymerization processes have generally been taken into acount.

McNeil and coworkers [49] and Pohl and Osterholtz [50] have studied the hydrolysis of alkyltrialkoxysilanes in buffered aqueous solution. These studies do not suffer from complications due to pH excursions, polymerization, or mixed organic-aqueous solvent systems. In aqueous solution (excess water) the pseudo first-order rate constants (measured spectrophotometrically or by extraction) were extrapolated to zero buffer concentration to yield the spontaneous first-order rate constants at a particular pH. The rate constants thus obtained by Pohl and Osterholtz for γ-glycidoxypropyltrialkoxysilane are plotted versus pH in Fig. 9. The hydrolysis appears to be both *specific acid* (hydronium ion) and *specific base* (hydroxyl ion) catalyzed, because the slopes of the plot above and below the minimum rate around pH 7 are drawn to be +1 and −1, respectively. Similar behavior was observed by McNeil *et al.* [49] for the hydrolysis of tris(2-methoxyethoxy)phenylsilane. However a pH independent hydrolysis rate constant was observed for bis(2-methoxyethoxy)diphenylsilane above pH 10 [49], which may indicate general base catalysis.

Aelion *et al.* [51,52] investigated the hydrolysis of TEOS under acidic and basic conditions using several cosolvents: ethanol, methanol, and dioxane. The extent of hydrolysis (Eq. 8) was determined by distillation of the ethanol by-product. Karl Fischer titration was used to follow the consumption of water by hydrolysis (Eq. 8) and its production by condensation (Eq. 10). Quantitative information (for example, specific rate constants) derived from these investigations is unreliable due to the inadvertent consumption of silanol groups by the Karl Fischer reagent, but it is worthwhile to present here and in the following sections some of the qualitative features of their

Fig. 9.

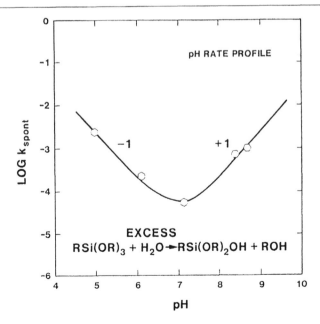

pH rate profile for the hydrolysis of γ-glycidoxypropyltrialkoxysilane in aqueous solution [50].

results. Aelion *et al.* observed that the rate and extent of the hydrolysis reaction was most influenced by the strength and concentration of the acid or base catalyst. Temperature and solvent were of secondary importance. They found that all strong acids behaved similarly, whereas weaker acids required longer reaction times to achieve the same extent of reaction. From a plot of the logarithm of the hydrolysis rate constants versus log [HCl], they obtained a slope of one and concluded that the reaction was first-order in acid concentration. Although general acid catalysis could not be completely ruled out, the hydrolysis reaction mechanisms that were postulated both involved hydrogen ions implying specific acid catalysis.

As under acidic conditions, the hydrolysis of TEOS in basic media was a function of the catalyst concentration [51]. The order of the reaction was determined by comparing the times required to complete specific degrees of hydrolysis. In very dilute solutions, the hydrolysis reaction was found to be first-order in [NaOH]. However when the concentration of TEOS was increased, the reaction no longer followed a simple order but apparently became complicated by secondary reactions. The weaker bases, ammonium hydroxide and pyridine, produced measurable speeds of reaction only if they were present in large concentrations. Compared to acidic conditions, the hydrolysis kinetics were more strongly affected by the nature of the solvent.

The effects of a variety of catalysts on the overall hydrolysis and condensation rates, as judged by the times required for gelation, have been summarized by Pope and Mackenzie [53] for TEOS hydrolyzed with four equivalents of water in ethanol ($r = 4$). Their results, listed in Table 5, not only show the effects of hydronium ion and hydroxyl ion on the gel times but also the effects of the conjugate base, most notably F^-. F^- is about the

Table 5.

Gel Times and Solution pH for TEOS Systems Employing
Different Catalysts.

Catalyst	Concentration (mol.: TEOS)	Initial pH of solution	Gelation time (h)
HF	0.05	1.90	12
HCl	0.05	0.05[a]	92
HNO$_3$	0.05	0.05[a]	100
H$_2$SO$_4$	0.05	0.05[a]	106
HOAc	0.05	3.70	72
NH$_4$OH	0.05	9.95	107
No catalyst	—	5.00	1000

Source: Pope and Mackenzie [53].
[a] Between 0.01 and 0.05.

same size as OH^- and has the ability to increase the coordination of silicon above four, for example in $R_3SiF_2^-$ [54]. Many of the properties of HF catalyzed gels are similar to those of base-catalyzed gels, which suggests that the roles of OH^- and F^- are similar. Andrianov [55] proposes that the catalytic effect of F^- involves the displacement of an OR^- via a *bimolecular nucleophilic* (S_N2-Si) mechanism in which the nucleophile, F^-, attacks Si followed by preferential hydrolysis of the Si–F bond. However Corriu and coworkers [56] have shown that the first step is fast reversible formation of a pentavalent intermediate

$$\begin{array}{c} \text{OR} \\ \text{F}^- \diagdown \; | \\ \text{RO} - \text{Si} - \text{OR} \\ | \\ \text{OR} \end{array} \longrightarrow \left[\begin{array}{c} \text{OR} \\ | \\ \text{F} - - \text{Si} - -\text{OR} \\ \diagup \diagdown \\ \text{OR} \quad \text{OR} \end{array} \right]^{(-)} \tag{21}$$

that stretches and weakens the surrounding Si–OR bonds. The subsequent rate-determining step is the nucleophilic attack of water on the hypervalent silicon leading to nucleophilic substitution by proton transfer and elimination of ROH.

$$\left[\begin{array}{c} \text{OR} \\ | \\ \text{F} --- \text{Si} --- \text{OR} \\ \diagup \diagdown \\ \text{OR} \quad \text{OR} \end{array} \right]^{(-)} \xrightarrow{(+H_2O)} \left[\begin{array}{c} \text{OR} \quad \text{OR} \\ \diagdown \diagup \\ \text{F} --- \text{Si} --- \text{OH}_2 \\ \diagup \diagdown \\ \text{OR} \quad \text{OR} \end{array} \right]^{(-)} \longrightarrow \begin{array}{c} \text{OR} \\ | \\ \text{OH} - \text{Si} - \text{OR} + \text{HOR} + \text{F}^-\cdot \\ | \\ \text{OR} \end{array}$$
$$\tag{22}$$

As shown by Coltrain *et al.* [57], the reduction in gel time resulting from acetic acid (HOAc) catalyst compared to HCl, HNO_3, or H_2SO_4 is not due to a catalytic effect of the acetate ion but to a reduction in the acidity of HOAc when in alcohol. Figures 10a and b compare gel times versus pH of the water solution used for hydrolysis and pH of the reacting sol (pH*). When compared on the basis of pH*, the curves coincide, indicating no catalytic effect of the counterions.

2.3.2. STERIC AND INDUCTIVE EFFECTS

Steric (spatial) factors exert the greatest effect on the hydrolytic stability of organoxysilanes [47]. Any complication of the alkoxy group retards the hydrolysis of alkoxysilanes, but the hydrolysis rate is lowered the most by

Fig. 10a.

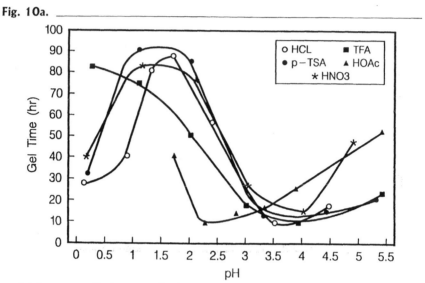

Gel time versus pH of water used to hydrolyze TEOS ($r = 4$). p-TSA: *p*-toluenesulfonic acid. HOAc: acetic acid. TFA: trifluoroacetic acid. From Coltrain *et al.* [57]

Fig. 10b.

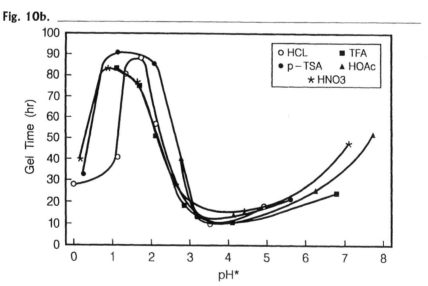

Gel time versus pH* measured in the reacting sols prepared as in Fig. 10a. From Coltrain *et al.* [57].

branched alkoxy groups [47]. The effects of alkyl chain length and degree of branching observed by Aelion *et al.* [51] are illustrated in Tables 6a and 6b for the acid hydrolysis of tetraalkoxysilanes. Figure 11 compares the hydrolysis of TEOS and TMOS under acidic and basic conditions. The retarding effect of the bulkier ethoxide group is clearly evident.

According to Voronkov *et al.* [47], in the case of mixed alkoxides, $(RO)_x(R'O)_{4-x}Si$ where $R'O$ is a higher (larger) alkoxy group than RO, if the $R'O$ has a normal (i.e., linear) structure, its retarding effect on the hydrolysis rate is manifested only when $x = 0$ or 1. If $R'O$ is branched, it has a retarding effect even when $x = 2$. Gas-liquid chromatography investigations of TEOS hydrolyzed in *n*-propanol, using the two-step acid- or base-catalyzed procedure outlined in Table 2, indicate substantial replacement of ethoxide groups with *n*-propoxide groups during the first (acid-catalyzed) hydrolysis

Table 6a.

Rate Constant k for Acid Hydrolysis of Tetraalkoxysilanes $(RO)_4Si$ at 20°C.

R	k 10^2 ($1\ mol^{-1}\ s^{-1}\ [H^+]^{-1}$)
C_2H_5	5.1
C_4H_9	1.9
C_6H_{13}	0.83
$(CH_3)_2CH(CH_2)_3CH(CH_3)CH_2$	0.30

Source: Aelion *et al.* [51].

Table 6b.

Rate Constants k (10^2 ($1\ mol.^{-1}\ s^{-1}\ [H^+]^{-1}$)) for Acid Hydrolysis of Alkoxyethoxysilanes $(RO)_{4-n}Si(OC_2H_5)_n$ at 20°C.

	R			
n	C_6H_{13}	$CH_3CH(CH_3)CH_2$ \ CH / H_3C	$CH_3(CH_2)_5$ \ CH / H_3C	$CH_3CH(CH_3)CH_2$ \ CH / $CH_3CH(CH_3)CH_2$
0	0.8	—	—	0.030
1	1.1	—		—
2	5.0	0.15	0.095	0.038
3	5.0	—		—

Source: Aelion *et al.* [51].

Fig. 11. _____

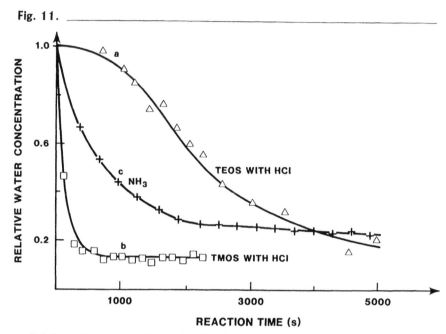

Relative water concentration versus time during acid-or base-catalyzed hydrolysis of TEOS or TMOS [58].

step [59]. The hydrolysis of the n-propoxide groups was observed to be slower than the ethoxide groups during the second hydrolysis step under both acidic and basic conditions. This result suggests that a retarding effect of higher, normal alkoxide groups is realized regardless of the extent of substitution.

Substitution of alkyl groups for alkoxy groups increases the electron density on the silicon. Conversely, hydrolysis (substitution of OH for OR) or condensation (substitution of OSi for OR or OH) decreases the electron density on silicon. (See Fig. 12.) Alkyl substitution and hydroxyl or bridging oxygen substitution therefore increase the stability of positively and negatively charged transition states, respectively (discussed further in Section 2.3.5.1). If inductive effects are important, this should enhance the kinetics of reactions involving these positively and negatively charged transition states.

Inductive effects are evident from investigations of the hydrolysis of methylethoxysilanes [61], $(CH_3)_x(C_2H_5O)_{4-x}Si$ where x varies from 0 to 3. Figure 13 shows that under acidic (HCl) conditions, the hydrolysis rate increases with the degree of substitution, x, of electron-providing alkyl groups, whereas under basic (NH₃) conditions the reverse trend is clearly observed. Figure 13 also shows the accelerating effect of methoxide

Fig. 12. _____

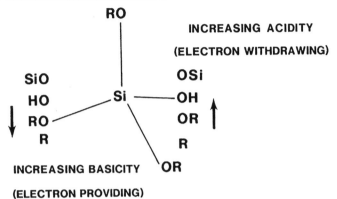

Inductive effects of substituents attached to silicon, R, OR, OH, or OSi [60].

substitution on the hydrolysis rate (TMOS versus TEOS). The consistent acceleration and retardation of hydrolysis with increasing x under acidic and basic conditions, respectively, suggests that the hydrolysis mechanism is sensitive to inductive effects and (based on the consistent trends) is apparently unaffected by the extent of alkyl substitution. Because increased stability of the transition state will increase the reaction rate, the inductive effects are evidence for positively and negatively charged transition states or intermediates under acidic and basic conditions, respectively.

This line of reasoning leads to the hypothesis that under acidic conditions, the hydrolysis rate decreases with each subsequent hydrolysis step (electron withdrawing), whereas under basic conditions, the increased electron-withdrawing capabilities of OH (and OSi) compared to OR may establish a condition in which each subsequent hydrolysis step occurs more quickly as hydrolysis and condensation proceed. From the standpoint of organically modified alkoxysilanes, $R_xSi(OR)_{4-x}$, the inductive effects indicate that acid-catalyzed conditions are preferable [61], because acid is effective in promoting hydrolysis both when $x = 0$ and when $x > 0$.

2.3.3. H_2O : Si RATIO, r

As indicated in Table 2, the hydrolysis reaction has been performed with r values ranging from <1 to over 25 depending on the desired polysilicate product, for example, fibers, bulk gels, or colloidal particles. From Eq. 8, an increased value of r is expected to promote the hydrolysis reaction. Aelion et al. [51,52] found the acid-catalyzed hydrolysis of TEOS to be first-order in [H_2O]; however, they observed an apparent zero-order dependence of

Fig. 13.

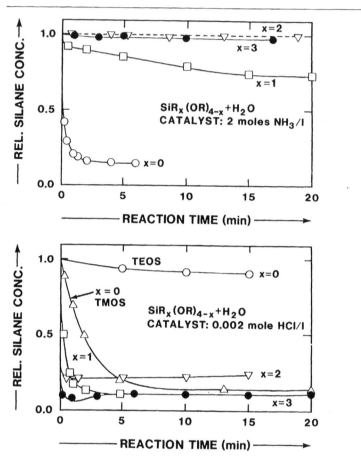

Relative silane concentration versus time during acid- and base-catalyzed hydrolysis of different silanes in ethanol (volume ratio silane to EtOH = 1 : 1). ●: $(CH_3)_3SiOC_2H_5$. ▽: $(CH_3)_2Si(OC_2H_5)_2$. □: $CH_3Si(OC_2H_5)_3$. ○: $Si(OC_2H_5)_4$. △: $Si(OCH_3)_4$ [61].

the water concentration under base-catalyzed conditions. As explained in Section 2.4.5, this is probably due to the production of monomers by siloxane bond hydrolysis and redistribution reactions. Pouxviel and coworkers [62] have used ^{29}Si NMR to investigate the acid-catalyzed hydrolysis of TEOS for three systems in which $r = 0.3$ (low water), 4 (medium water), or 10 (high water). Figure 14 compares the temporal evolution of Q^0 species for the medium- and high-water systems. The most obvious effect of the increased value of r is the acceleration of the hydrolysis reaction (Eq. 8): for the $r = 4$ system, measurable quantities of the unhydrolyzed

Fig. 14.

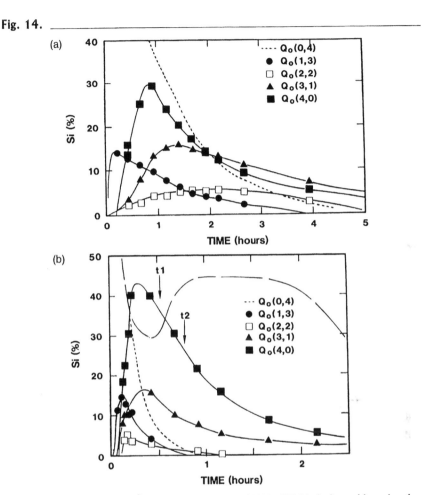

Temporal evolution of Q^0 (x, y) silanol species $(Si(OH)_x(OEt)_y)$ during acid-catalyzed hydrolysis as determined by ^{29}Si NMR. a: H_2O/Si = 3.8, b: H_2O/Si = 10 [62].

monomer $(Q_0(0, 4)^†$ in Fig. 14) persisted after four hours of reaction time, whereas when r = 10, no unhydrolyzed monomer remained after one hour. This result apparently contradicts earlier results that show a retarding effect of increased H_2O/Si on the hydrolysis rate under acidic conditions [58].

It is also evident from comparisons of the concentrations of the fully hydrolyzed monomers, $Q_0(4, 0)$, that the higher value of r caused more

---

†According to the nomenclature used by Pouxviel *et al.* [62], subscripts rather than superscripts are used to distinguish between possible Q species (0–4), and (0, 4) indicates the values of x and y, respectively, in $Si(OH)_x(OR)_y$ monomers.

complete hydrolysis of the monomers before significant condensation occurred. According to Eqs. 9 and 10, differing extents of monomer hydrolysis should affect the relative rates of the alcohol- or water-producing condensation reactions. Generally with understoichiometric additions of water ($r \ll 2$), the alcohol-producing condensation mechanism is favored [63], whereas the water-forming condensation reaction is favored when $r \geq 2$ [63]. (See Section 2.5.)

Although increased values of r generally promote hydrolysis, when r is increased while maintaining a constant solvent : silicate ratio, the silicate concentration is reduced. This in turn reduces the hydrolysis and condensation rates, causing an increase in the gel time. This effect is evident in Fig. 15 which shows gel times for acid-catalyzed TEOS systems as a function of r and the initial alcohol : TEOS molar ratio [64].

Fig. 15. _____

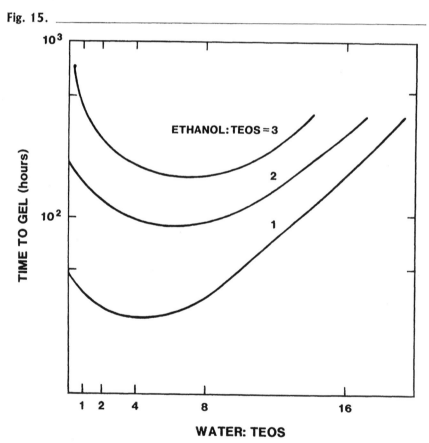

Gel times versus H_2O : TEOS ratio, r, for three ratios of EtOH to TEOS [64].

As indicated by the alcohol–water–TEOS ternary-phase diagram shown in Fig. 7, large values of r cause liquid–liquid immiscibility; however both alcohol produced as the by-product of the hydrolysis reaction and partial hydrolysis of the TEOS precursor lead to homogenization. Finally, because water is the by-product of the condensation reaction (Eq. 10), large values of r promote siloxane bond hydrolysis (the reverse of Eq. 10). The effects of reverse reactions are discussed extensively in Sections 2.3.6 and 2.4.5.

2.3.4. SOLVENT EFFECTS

Traditionally, solvents are added to prevent liquid–liquid phase separation during the initial stages of the hydrolysis reaction (Figure 7 shows the region of immiscibility in the TEOS–H_2O–ROH ternary system) and to control the concentrations of silicate and water that influence the gelation kinetics. More recently, the effects of solvents have been studied primarily in the context of *drying control chemical additives* (DCCA) used as cosolvents with alcohol in order to facilitate rapid drying of monolithic gels without cracking [65].

Solvents may be classified as *polar* or *nonpolar* and as *protic* (containing a removable or "labile" proton) or *aprotic*. Physical properties of some common solvents used in sol-gel processing are listed in Table 7. With regard to solvating power, several important characteristics of solvents are (1) polarity, (2) dipole moment, and (3) the availability of labile protons. The polarity largely determines the solvating ability for polar or nonpolar species. More polar solvents (e.g., water, alcohol, or formamide) are normally used to solvate polar, tetrafunctional silicate species used in sol-gel processing. Less polar solvents such as dioxane or tetrahydrofuran (THF) may be used in alkyl-substituted or incompletely hydrolyzed systems. Ether alcohols such as methoxyethanol or ethoxyethanol exhibit both polar and nonpolar character and are useful when a distribution of polar and nonpolar species is present in solution. As discussed in Chapter 4, the dipole moment of a solvent determines the length over which the charge on one species can be "felt" by surrounding species. The lower the dipole moment, the larger this length becomes. This is important in electrostatically stabilized systems (for example, electrostatic stabilization is more effective in systems with low dipole moments) and when considering the distance over which a charged catalytic species, for example an OH^- nucleophile of H_3O^+ electrophile, is attracted to or repelled from potential reaction sites, depending on their charge.

The availability of labile protons (protic versus aprotic solvents in Table 7) determines whether anions or cations are solvated more strongly through hydrogen bonding. Because hydrolysis is catalyzed either by hydroxyl

Table 7.

Physical Properties of Typical Solvents.

	MW	bp	ρ	n_D	ε	η	μ
Protic							
water H_2O	18.01	100.00	1.000	1.333	78.5	10.1	1.84
methanol CH_3OH	32.04	64.5	0.791	1.329	32.6	5.4	1.70
ethanol C_2H_5OH	46.07	78.3	0.785	1.361	24.3	10.8	1.69
2-ethoxyethanol $C_4H_{10}O_2$	90.12	135	0.93	1.408	—	—	2.08
formamide CH_3ON	45.04	193	1.129	1.448	110	33.0	3.7
Aprotic							
dimethylformamide C_3H_7NO	73.10	152	0.945	1.430	36.7	7.96	3.86
dioxane 1,4 $C_4H_8O_2$	88.12	102	1.034	1.422	2.21	10.87	0
tetrahydrofuran C_4H_8O	72.12	66	0.889	1.405	7.3	—	1.63

bp: °C. n_D: 20°. η: millipoise.
ρ: g/cm³ at 20°C. Dielectric constant, ε: at 25°C. Dipole moment, μ: debyes.

(pH > 7) or hydronium ions (pH < 7), solvent molecules that hydrogen bond to hydroxyl ions or hydronium ions reduce the catalytic activity under basic or acidic conditions, respectively. Therefore, aprotic solvents that do not hydrogen bond to hydroxyl ions have the effect of making hydroxyl ions more nucleophilic, whereas protic solvents make hydronium ions more electrophilic [66][†]. Hydrogen bonding may also influence the hydrolysis mechanism. For example, hydrogen bonding with the solvent can sufficiently activate weak leaving groups to realize a bimolecular, nucleophilic (S_N2–Si) reaction mechanism [47].

The availability of labile protons also influences the extent of the reverse reactions, reesterification (reverse of Eq. 8) or siloxane bond alcoholysis or hydrolysis (reverse of Eqs. 9 and 10). Aprotic solvents do not participate in reverse reactions such as reesterification or hydrolysis, because they lack

[†] Corria *et al.* [56] point out that in mixed organic/H_2O solvents (e.g., H_2O/ROH) the organic solvents modify the structure of water so that the acidity or basicity often decreases compared to that in pure water.

sufficiently electrophilic protons and are unable to be deprotonated to form sufficiently strong nucleophiles (e.g., OH^- or OR^-) necessary for these reactions:

$$\begin{array}{c} R \\ \diagdown \\ \underset{\cdot\cdot}{O} - H \\ \vdots \\ \equiv Si - O - Si \equiv \; \rightleftharpoons \; \equiv Si - OR + HO - Si \equiv. \end{array} \qquad (23)$$

Therefore compared to alcohol or water, aprotic solvents such as THF or dioxane are considerably more "inert" (i.e., they do not formally take part in sol-gel processing reactions), although, as discussed previously, they may influence reaction kinetics by increasing the strength of nucleophiles or decreasing the strength of electrophiles.

When cosolvents, such as formamide, are added as DCCAs in order to influence the structure or drying behavior of the gel network, it is also necessary to consider the surface tensions, vapor pressures, and chemical reactivities of the additives. The effects of surface tension and vapor pressure are addressed in the chapter on drying. The remainder of this section will consider chemical effects of DCCAs.

Numerous accounts of the use of formamide, $HCONH_2$, as a DCCA appear in the literature [e.g., 65,67]. Using ^{29}Si NMR and Raman spectroscopy, Jonas [68], Artaki et al. [69], and Orcel and Hench [70] have studied the effects of formamide on the hydrolysis of TMOS in methanol under neutral conditions. They find that, compared to using pure methanol as a solvent, formamide causes a reduction in the hydrolysis rate constant ($\sim$ a factor of 5, based on the disappearance of monomer), whereas the condensation-rate constant is increased. Using similar methods, Hench [67] observed that, under acidic conditions, replacement of 50% of the methanol with formamide caused the reverse effects: increased hydrolysis and reduced condensation rates.

The effects of formamide observed under neutral conditions were generally explained on the basis of hydrogen bonding and solvent viscosity [69,70]. Although both methanol and formamide can act as donors or acceptors in hydrogen bonding, formamide exhibits higher values of both ε and μ (see Table 7), and is therefore expected to hydrogen bond more strongly to protons and hydroxyls under acidic and basic conditions, respectively, reducing the effective catalyst activity [70]. In addition, the viscosity of formamide is approximately seven times greater than methanol at 25°C. (See Table 7.) Because the hydrolysis mechanism may require a reorientation of ligands surrounding the nucleus to achieve a maximum separation of charge between the attacking and leaving groups, it was postulated that the increased viscosity may reduce the ability of the nucleus to reorient [69].

Increased viscosity would also reduce the diffusion coefficients of reactive species within the solution, further reducing the hydrolysis rate.[†]

The preceding explanations, however, do not take into account possible reactions of formamide with the solution components. Rosenberger *et al.* [72] used ^{1}H NMR to investigate the effects of formamide addition on the acid-catalyzed hydrolysis of TEOS. They concluded that formamide does not react chemically with TEOS or ethanol but is hydrolyzed to produce NH_3 plus formic acid:

$$HC \overset{\displaystyle O}{\underset{\displaystyle NH_2}{\diagdown}} \xrightarrow{\ H_2O,\ H^+,\ NO_3^-\ } HC \overset{\displaystyle O}{\underset{\displaystyle O^-(NH_4)^+}{\diagdown}} \rightleftarrows HC \overset{\displaystyle O}{\underset{\displaystyle OH}{\diagdown}} + NH_3. \qquad (24)$$

These reactions lead to a progressive increase of the solution pH with time[††]. (See Fig. 16.) Because the hydrolysis reaction is generally more rapid and complete under acidic conditions, whereas the average condensation rate is maximized near pH 4 (see Fig. 10b), formamide additions to acid-catalyzed solutions should allow efficient hydrolysis (low pH) followed by efficient condensation ($\sim$pH 4). At pH 4, the solubility of silica and the dissolution rate are quite low [1]. Therefore, one effect of formamide additions may be to strengthen the gel by maximizing the extent of condensation without reorganization into a particulate structure. This is consistent with observations that formamide increases the microhardness of wet gels[†††] and correspondingly the pore sizes of dried gels while maintaining a narrow pore size distribution [67].

2.3.5. MECHANISMS

The previous sections have documented that the hydrolysis reaction of tetraalkoxy- and organoalkoxysilanes is influenced by steric and inductive effects and appears to be specific-acid-(H_3O^+) and specific-base-(OH^-) catalyzed. The reaction order with respect to water and silicate is observed to be two and one, resulting in third- and second-order overall kinetics, respectively. Based on these factors, it is generally argued that hydrolysis proceeds by bimolecular nucleophilic displacement reactions (S_N2–Si reactions) involving pentacoordinate intermediates or transition states

---

[†]Zerda and Hoang combined high-pressure Raman spectroscopy and viscosity measurements to show that the viscosity does not significantly alter the hydrolysis kinetics [71].

[††]Formamide has little effect on pH under neutral or basic conditions, C.S. Ashley, unpublished.

[†††]S.H. Wang and L.L. Hench in *Better Ceramics Through Chemistry*, Eds. C.J. Brinker, D.E. Clark, and D.R. Ulrich (Elsevier, NY, 1984), p.71.

Fig. 16.

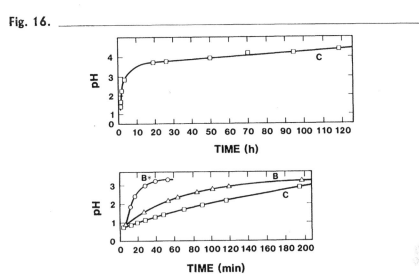

pH versus time during B: acid-catalyzed hydrolysis of TMOS at room temperature in methanol–formamide solutions, $TEOS:EtOH:H_2O:HNO_3:CH_3NO = 1:3:10:0.3:3$; $B* = 60°C$; C: acid-catalyzed hydrolysis of formamide, $EtOH:H_2O:HNO_3:CH_3NO = 3:10:0.3:3$. Upper curve: 60°C; lower curve: room temperature [72].

[e.g., 49,50,60], although by analogy to carbon chemistry *siliconium ions*, $Si(OR)_3^+$, have also been suggested as possible intermediates [61].

2.3.5.1. Acid-catalyzed Hydrolysis

Under acidic conditions, it is likely that an alkoxide group is protonated in a rapid first step. Electron density is withdrawn from silicon, making it more electrophilic and thus more susceptible to attack by water. Pohl and Osterholz [50] favor a transition state with significant S_N2-type character. The water molecule attacks from the rear and acquires a partial positive charge. The positive charge of the protonated alkoxide is correspondingly reduced, making alcohol a better *leaving group*. The transition state decays by displacement of alcohol accompanied by *inversion* of the silicon tetrahedon:

$$S_N2\text{-Si HYDROLYSIS} \tag{25}$$

Consistent with this mechanism, the hydrolysis rate is increased by substituents that reduce steric crowding around silicon (Fig. 11 and Table 6). Electron-providing substituents (e.g., alkyl groups) that help stabilize the developing positive charges should also increase the hydrolysis rate (Fig. 13) but to a lesser extent, because the silicon acquires little charge in the transition state.

Keefer [48] has discussed possible consequences of inversion of the silicon tetrahedron with regard to retarding the hydrolysis of silicate species that are contained in polymers. Using optically active monomers $R'_x Si^*(OR)_{4-x}$, Sommer and coworkers [73–75] have proven that inversion occurs during hydrolysis of several monomers including $R_3 Si^* OCH_3$. As a general rule, inversion occurs in displacement reactions with good leaving groups such as Cl^- or $OCOR^-$ whose *conjugate acids* (corresponding protonated anions) have $pK_a < 6$, regardless of the nature of the solvent provided that the attacking reagent furnishes an entering group more basic than the leaving group [75]. For poorer leaving groups such as H or OR whose conjugate acids have $pK_a > 6$, retention or inversion may occur depending on the nature of the catalyst cation and the solvent polarity. For example, hydrogen bonding of the solvent may facilitate inversion by activating poor leaving groups [47]:

$$
\text{H} - \text{O} \cdots \underset{\underset{\text{H}}{|}}{\overset{\diagup\diagdown}{\text{Si}}} \cdots \text{O} \cdots \text{H} - \text{OR.}
$$

$$\underset{\text{H}}{|} \qquad \underset{\text{R}}{|}$$

(26)

Klemperer and coworkers [38] have shown that, under neutral conditions, constrained oligomers (cubic octamers) hydrolyze without inversion. (See pathway A, Fig. 17.) This suggests that retention or inversion is further influenced by specific bonding configurations.

Several investigators have proposed hydrolysis mechanisms involving flank-side attack without inversion of the silicon tetrahedron [48,76]. A possible acid-catalyzed mechanism is the following:

$$
\underset{\text{RO}}{\overset{\overset{\displaystyle OR}{|}}{\text{RO}\,\text{I}\text{IIIII}\,\text{Si} - \text{OR}}} \quad \xrightarrow[\text{H}_2\text{O}]{\text{H}^+} \quad \underset{\underset{\text{H}}{|}}{\overset{\overset{\displaystyle OR}{|}}{\underset{\text{RO}}{\overset{\text{RO}}{\text{Si}}} \overset{\text{H}}{\underset{\oplus}{-}\text{O} - \text{R}}}} \quad \longrightarrow \quad
\begin{array}{l}(RO)_3\,SiOH\ +\\ HOR + H^+.\end{array}
$$

(27)

This mechanism is subject to both steric and inductive effects. Compared to the S_N2 mechanism described previously, electron-providing substituents

Fig. 17.

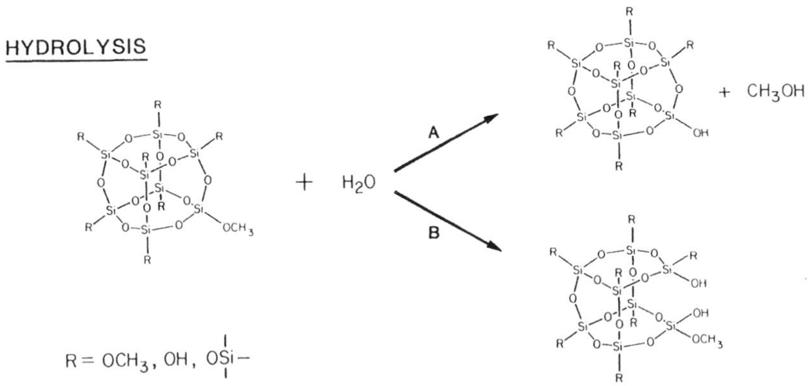

HYDROLYSIS

+ CH_3OH

+ H_2O

A

B

R = OCH_3, OH, $O\overset{|}{\underset{|}{Si}}-$

Possible pathways for hydrolysis of methoxy terminated Q^3 species contained in cubic octamers: A) with retention of configuration, B) with siloxane bond hydrolysis [38].

should have a greater effect, because the silicon acquires more charge in the transition state.

Timms [77] has proposed an acid-catalyzed hydrolysis mechanism involving a siliconium ion ($\equiv Si^+$). An alkoxide group is rapidly protonated followed by a slower step in which a siliconium ion is formed by the removal of alcohol:

$$\equiv Si-OR + H^+ \rightleftharpoons \equiv Si-\underset{H^+}{OR} \xrightarrow{\text{SLOW}} \equiv Si^+ + ROH, \qquad (28)$$

$$\equiv Si^+ + ROH \overset{H_2O}{\rightleftharpoons} \equiv Si-OH + ROH + H^+. \qquad (29)$$

Water reacts with the siliconium ion to form a silanol and the proton is regenerated (Eq. 29). Consistent with the observed behavior, this reaction should have third-order overall kinetics and should be accelerated by electron-providing substituents attached to silicon (Figs. 12 and 13). The identical reaction has been proposed by Jada [78] to explain the acid-catalyzed reaction of TEOS and water generated by the reaction of acetic acid with ethanol:

$$H_3C-COOH + C_2H_5OH \overset{H^+}{\rightleftharpoons} H_3C-\overset{O}{\overset{\|}{C}}-O-C_2H_5 + H_2O. \qquad (30)$$

Several factors argue against mechanisms involving siliconium ion intermediates, however. McNeill and coworkers [49] measured the deuterium solvent isotope effect and the activation parameters ΔH^* and ΔS^*. From rate constants measured in hydrochloric-acid solutions using the stopped flow method, a $k_{H_3O^+}/k_{D_3O^+}$ value of 1.24 was obtained. Rather than a dissociative mechanism, this deuterium isotope effect is in accord with an associative mechanism (as in Eq. 25 or 27) in which the silicate monomer is protonated (deuterated) in a rapid first step followed by attack of water and generation of alcohol in subsequent, slower, rate-limiting steps. From a plot of $\ln(k_{obsd}/T)$ versus $1/T$ (where k_{obsd} is the observed first-order rate constant at temperature T), the entropy of activation was calculated to be $-39\,cal\,deg^{-1}\,mol^{-1}$. This large negative entropy of activation suggests a highly ordered A-2–type transition state with at least one water molecule associated with it [49].

Swain et al. [79] compared the hydrolysis behavior of triphenylmethyl fluoride (TMF) and triphenylsilyl fluoride (TSF) under similar conditions. They found that, whereas the hydrolysis of TMF was consistent with a positively charged carbonium ion intermediate, the hydrolysis of TSF was not consistent with a positively charged siliconium ion intermediate but rather an intermediate in which silicon is less positively charged than in the original molecule. They concluded that pentacoordinate intermediates are easy pathways for displacements on silicon that are not available for carbon which cannot expand its valence to include more than eight electrons.

Corriu and Henner [80] have examined the siliconium-ion question and have failed to prove the existence of siliconium ions by physiochemical methods, the preparation of stable salts, or identification of reaction intermediates. They conclude that the high affinity of silicon for nucleophiles explains the failure to prove the presence of such ions, but they point out that it is not possible to completely exclude their existence.

Using ^{29}Si NMR and Raman spectroscopy, Jonas [68], Artaki et al. [81], and Zerda and Hoang [71] have investigated the hydrolysis of TMOS (pH ~ 5–7.5) at pressures up to 5 kbar. Their observations that pressure increases the rates of reaction without affecting the distributions of hydrolyzed and/or condensed species are consistent with an associative mechanism involving a pentacoordinate intermediate (transition-state volume, ΔV^*, is negative) rather than a dissociative mechanism in which the rate-determining step is the formation of a tricoordinated siliconium ion plus alcohol (ΔV^* positive).

2.3.5.2. Base-catalyzed Hydrolysis

Under basic conditions it is likely that water dissociates to produce nucleophilic hydroxyl anions in a rapid first step. The hydroxyl anion then attacks the silicon atom. Iler [1] and Keefer [48] propose an S_N2–Si mechanism in which OH^- displaces OR^- with inversion of the silicon tetrahedron:

$$RO\overset{RO}{\diagdown}\quad\quad RO\diagdown\overset{\delta^-}{}\diagup OR\quad\quad OR\overset{OR}{\diagup}$$

$$HO^- + Si\!\!-\!\!OR \rightleftharpoons HO\cdots \overset{\delta^-}{Si}\cdots OR \rightleftharpoons HO\!-\!Si + OR^-. \quad (31)$$

$$\underset{RO}{\diagup}\qquad\qquad \underset{OR}{\big|}\qquad\qquad \underset{OR}{\diagdown}$$

As discussed for acid-catalyzed hydrolysis, this mechanism is affected by both steric and inductive factors; however steric factors are more important because silicon acquires little charge in the transition state.

Pohl and Osterholtz [50] favor an S_N2^{**}–Si or S_N2^*–Si mechanism involving a stable 5-coordinated intermediate.[†] The intermediate decays through a second transition state in which any of the surrounding ligands can acquire a partial negative charge:

$$S_N2^{**}\text{-Si}\qquad S_N2^{*}\text{-Si}$$

$$OH^- + Si(OR)_4 \rightleftharpoons \left[\overset{\overset{\delta^-}{OH}}{\underset{OR}{RO\!-\!\overset{\delta^-}{Si}\!\!\!\diagdown\!\!\overset{OR}{OR}}} \right] \longleftarrow \overset{OH}{\underset{OR}{RO\!-\!\overset{-}{Si}\!\!\!\diagdown\!\!\overset{OR}{OR}}} \rightleftharpoons \quad (32)$$

$$\text{T.S. 1}$$

$$\left[\overset{OH}{\underset{OR}{RO\!-\!\overset{\delta^-}{Si}\!\!\diagup\!\!\overset{OR}{\underset{\delta^-}{OR}}}} \right] \longrightarrow Si(OR)_3OH + RO^-.$$

$$\text{T.S. 2}$$

Hydrolysis occurs only by displacement of an alkoxide anion, which may be aided by hydrogen bonding of the alkoxide anion with the solvent (as in Eq. 26).

[†] The asterisks distinguish between the formation or the decay of the transition states, T.S.1 and T.S.2 as the rate-determining steps, S_N2^{**} and S_N2^*, respectively.

Because the silicon acquires a formal negative charge in the transition state, S_N2^{**}–Si or S_N2^*–Si mechanisms are quite sensitive to inductive as well as steric effects. Electron-withdrawing substituents such as –OH or –OSi should help stabilize the negative charge on silicon, causing the hydrolysis rate to increase with the extent of OH substitution, whereas electron-providing substituents should cause the hydrolysis rate to decrease (in accord with Fig. 13). Because inversion of configuration is not implicit in Eq. 32, the hydrolysis rate may also increase with the extent of condensation.

The hydrolysis kinetics are expected to be first-order in [OH$^-$] and second-order with respect to water and silicate (third-order overall kinetics) for all three mechanisms, S_N2–Si, S_N2^{**}–Si, and S_N2^*–Si. Therefore, it is generally not possible to distinguish between these mechanisms based on the reaction order.

2.3.6. TRANSESTERIFICATION, REESTERIFICATION, AND HYDROLYSIS

The hydrolysis reaction (Eq. 8) may proceed in the reverse direction, in which an alcohol molecule displaces a hydroxyl group to produce an alkoxide ligand plus water as a by-product. This reverse process, *reesterification*, presumably occurs via mechanisms similar to those of the forward reactions, i.e., by bimolecular nucleophilic substitution reactions: S_N2–Si, S_N2^{**}–Si, or S_N2^*–Si. An S_N2^{**}–Si mechanism is shown in Eq. 33 for an acid-catalyzed reesterification reaction:

T.S. 1   Pentacoordinate Intermediate

T.S. 2   (33)

Voronkov [47], however, proposes the formation of an active six-membered transition complex that contains two alcohol molecules.

Observations of the extent of reesterification of polysiloxanes [e.g., 26,59] indicate that it proceeds much further under acidic conditions than under basic conditions. This led Keefer [48] to the conclusion that the base-catalyzed mechanism involves inversion of configuration, while the acid-catalyzed mechanism does not. It is likely that the first step of the acid-catalyzed reesterification reaction involves the protonation of a silanol group, whereas under base-catalyzed conditions the first step is the deprotonation of an alcohol to form the nucleophile, OR^-. Therefore, the tendency for reesterification to be more complete under acidic than basic conditions may also result from the greater ease of protonation of silanol groups under the acidic conditions normally employed in sol-gel processing (pH ~ 1–3)[†] than deprotonation of alcohols under the weakly basic conditions normally employed (pH ~ 8–10).

Reesterification is quite important during drying of gels, because in many solvent systems (e.g., ethanol or propanol) excess water can be completely removed via the azeotrope, which has a higher vapor pressure than the neat solvent or water. Keefer [48] has calculated the number of silanols reesterified for several common gel-forming conditions and degrees of condensation. According to these calculations, it is easy to understand the common and contradictory observations that although hydrolysis readily goes to completion with slight excesses of water under acidic conditions, the dried gel may be substantially esterified [26,59].

Transesterification, in which an alcohol displaces an alkoxide group to produce an alcohol molecule:

$$R'OH + Si(OR)_4 \rightarrow Si(OR)_3OR' + ROH \qquad (34)$$

has been extensively studied, because it is the common method of producing various alkoxides of silicon [82]. As discussed by Voronkov *et al.* [47], a wide range of catalysts have been employed, although information on the relative catalytic activity is sparse.

In sol-gel processing, transesterification often occurs when alkoxides are hydrolyzed in alcohols containing different alkyl groups. For example, Brinker *et al.* [59] observed substantial ester exchange during the acid-catalyzed hydrolysis of TEOS in *n*-propanol. Transesterification will also be important in multicomponent systems that employ several alkoxides with differing alkoxide substituents. After transesterification has occurred, subsequent hydrolysis kinetics will depend on the steric and inductive characteristics of the exchanged alkoxide. For example, displacement of ethoxide by *n*-propoxide groups in TEOS was shown to reduce the overall hydrolysis rate [59].

[†] pH is not well defined in nonaqueous solutions. Often it is a measured value (via a nonaqueous pH electrode) rather than exactly $-\log[H^+]$.

It is suggested that transesterification proceeds by a type of nucleophilic displacement reaction (either S_N2–Si, S_N2^{**}–Si, or S_N2^*–Si) similar to reesterification except that an alcohol molecule rather than a water molecule is displaced. As such, transesterification should be subject to both steric and inductive effects. For example, Table 8 indicates that the extent of transesterification is severely reduced when bulky secondary or tertiary alcohols are employed. Investigations by Uhlmann *et al.* [84] and Peace *et al.* [83] show that under acidic conditions, transesterification occurs more readily after partial or complete hydrolysis of the silicate species has occurred, suggesting possible steric constraints. Similar to the case of reesterification, Yamane *et al.* [85] observed transesterification of TMOS under acid-catalyzed conditions but not under base-catalyzed conditions.

From studies of optically active organosilicon compounds, it has been established that transesterification can proceed with retention or inversion of configuration depending on the nature of the leaving group, the solvent, and the catalyst [47]. Because alkoxide substituents are poor leaving groups (alcohols have a $pK_a > 10$), Voronkov *et al.* proposed that only in very polar solvents can there be sufficient separation of charge in the transition state to realize the S_N2–Si mechanism that proceeds with inversion of configuration [47]:

$$
\begin{array}{ccccc}
\text{RO} & & \text{OR} & & \\
\diagdown & & \diagup & & \\
\text{R}'' - \text{O} \cdots \text{Si} \cdots \text{O} \cdots \text{H} - \text{OR}''. \\
| & | & | & & \\
\text{H} & \text{OR} & \text{R} & &
\end{array}
\qquad (35)
$$

In less polar solvents, retention of configuration is generally observed. In sol-gel systems, we expect that transesterification of highly condensed species will proceed generally with retention of configuration.

Table 8.

Extent of Hydrolysis and Alcohol Exchange for Acid-Catalyzed
Hydrolysis of TEOS in Different Solvents.

Solvent	Hydrolysis (%)	Exchange (%)
Methanol	76	44.7
Cellosolve (ethylene glycol mono ethyl ether)	63	45.9
Isopropanol	63	35.2
t-butanol	67	0

Source: Peace *et al.* [83].

Isotopic labelling investigations have confirmed that silanols can be hydrolyzed. When triethylsilanol is hydrolyzed in oxygen-labelled water in the presence of acid or base catalyst or under neutral conditions, the oxygen of the silanol is completely replaced by the water oxygen [47].

$$H_2{}^{18}O + (RO)_3SiOH \rightarrow (RO)_3Si{}^{18}OH + H_2O. \qquad (36)$$

This reaction is generally not important in influencing structural development in sol-gel systems; however it does emphasize that all substituents attached to silicon are quite labile and will depend in an equilibrium sense on the changing concentrations of alcohol and water as well as the solvent, nature of the catalyst, and extent of condensation.

2.4. Condensation

Polymerization to form siloxane bonds occurs by either an alcohol-producing condensation reaction (Eq. 9) or a water-producing condensation reaction (Eq. 10). The latter reaction has been discussed in detail by Iler [1] with regard to forming silicate polymers and gels in aqueous media. Engelhardt and coworkers [21] employed ^{29}Si NMR to investigate the condensation of aqueous silicates at high pH. Their results indicate that a typical sequence of condensation products is monomer, dimer, linear trimer, cyclic trimer, cyclic tetramer, and higher-order rings. As discussed in Section 1, the rings form the basic framework for the generation of discrete colloidal particles commonly observed in aqueous systems [1].

This sequence of condensation requires both depolymerization (ring opening) and the availability of monomers, which are in solution equilibrium with the oligomeric species and/or are generated by depolymerization (reverse of Eqs. 9 and 10). However in alcohol–water solutions normally employed in sol-gel processing, the depolymerization rate is lower than in aqueous media, especially at low pH [1]. Under these conditions, Iler [1] proposed:

Where depolymerization is least likely to occur so that the condensation is irreversible and siloxane bonds cannot be hydrolyzed once they are formed, the condensation process may resemble classical polycondensation of polyfunctional organic monomer resulting in a three dimensional molecular network. Owing to the insolubility of silica under these conditions the condensation polymer of siloxane chains cannot undergo rearrangement into particles.

This is quite a prophetic statement with regard to the sequence of condensation reactions in sol-gel systems where, depending on conditions, a complete spectrum of structures ranging from molecular networks to colloidal particles may result.

2.4.1. EFFECTS OF CATALYSTS

Although the condensation of silanols can proceed thermally without involving catalysts, their use especially in organosilanes is often helpful. Numerous catalysts have been employed: generally compounds exhibiting acid or base character but also neutral salts and transition metal alkoxides, e.g. $Ti(OEt)_4$ [86]. In sol-gel systems mineral acids, ammonia, alkali metal hydroxides, and fluoride anions are most commonly used. As discussed in the section on hydrolysis, the understanding of catalytic effects is often complicated by the increasing acidity of silanol groups with the extent of hydrolysis and polymerization, and by the effects of reverse reactions that become increasingly important with greater concentrations of water and/or base.

Pohl and Osterholtz [50] used ^{13}C and ^{29}Si NMR to investigate the condensation of alkylsilanetriol to bis-alkyltetrahydroxydisiloxane in buffered aqueous solutions as a function of pD ($-\log[D_3O^+]$). The second-order rate constants for the disappearance of the triol are plotted in Fig. 18a as a pD rate profile. The slopes of the plot above and below the rate minimum at pD 4.5 are $+1$ and -1, respectively, indicating that the condensation is specific acid- and base-catalyzed. Assink [30] investigated the acid-catalyzed condensation of TMOS over a narrow range of pH. He observed that the condensation rate was proportional to $[H_3O^+]$, which is also consistent with specific acid catalysis.

It is interesting to compare the results in Fig. 18a to the pH-dependence of the gel time (Fig. 18b), which is often used as a measure of the overall condensation kinetics for sol-gel systems (gel time $\propto 1/$(average condensation rate)). According to Fig. 18b, the overall condensation rate is minimized at about pH 1.5 and maximized at intermediate pH. Stable (nongelling) systems are obtained under more basic conditions. The minimum at $\sim$pH 2 corresponds to the isoelectric point of silica: surface silanol groups are protonated and deprotonated at lower and higher pH values, respectively. Because silanols become more acidic with the extent of condensation of the siloxane network to which they are attached, the shift in the minimum rate from $\sim$pD 4.5 (condensation of monomers, $RSi(OH)_3$) to pH 2 (condensation of polydisperse, higher-order polymers leading to gelation) presumably reflects the increasing acidity of silanols with the degree of condensation. Regardless of the cause, the pH-dependence suggests that for more highly cross-linked systems, protonated and deprotonated silanols are involved in the acid- and base-catalyzed condensation mechanisms at pH < 2 and pH > 2, respectively. Under more basic conditions where the gel times are observed to increase (see, e.g., Fig. 5), condensation reactions proceed but gelation does not occur. In this pH regime, particles are formed that, after

Fig. 18.

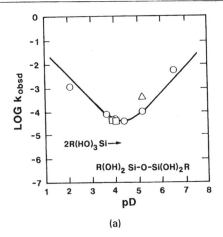

(a)

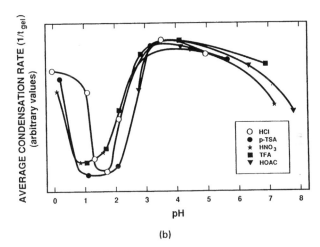

(b)

(a) pD-rate profile for the condensation of alkylsilanetriol in buffered aqueous solution [50]. (b) average condensation rates (1/gel times) for TEOS hydrolyzed with solutions of various acids: (*p*-TSA) *p*-toluenesulfonic acid; (HOAc) acetic acid; (TFA) tri-fluoroacetic acid. After Coltrain *et al.* [57].

reaching a critical size, become stable toward gelation due to mutual repulsion effects. This high-pH region represents the conditions in which so-called Stöber silica particles [27] are formed.

In both aqueous and mixed alcohol–water systems, traces of HF have a remarkable catalytic effect on the polymerization rate [87,88]. (See Table 5.) Below pH 2, Iler [1] observed that the polymerization rate is proportional to the concentration of H^+ and F^-. He proposed that the condensation mechanism may involve a bimolecular intermediate in which the fluorine anion temporarily increases the coordination of one silicon from four to five or six just as in the case of the OH^- ion. Rabinovich and Wood [87] proposed a mechanism in which F^- displaces an OH^-, causing localized attractions to other silanol species, thereby increasing the condensation rate. Because F^- is more electron-withdrawing than OH^-, an alternate argument is that F^- substitution for OH^- reduces the electron density on Si, making it more susceptible to nucleophilic attack, consistent with the view of Corriu et al. [56].

2.4.2. STERIC AND INDUCTIVE EFFECTS

During sol-gel processing, condensation can proceed by two different reactions (Eqs. 9 and 10) that can occur between substantially different solution species (monomers, oligomers, etc.), which have undergone different extents of hydrolysis. Therefore, steric and inductive effects are not well documented for tetraalkoxides. According to Voronkov et al. [47], the condensation rate of triorganosilanols decreases with increase in the length or branching of the chain of the alkyl radical, or, if aromatic groups are present, with increase in their number. Likewise in tetrafunctional alkoxides normally employed in sol-gel processing, we expect that substituents that increase steric crowding in the transition state will retard condensation. Voronkov et al. [47] also say that the condensation rate increases with an increase in the number of silanols on the silicon atom (increasing silanol acidity). This result may be explained on the basis of steric, inductive, or statistical effects.

The previous section indicates that the acid- and base-catalyzed condensation mechanisms involve protonated and deprotonated silanols, respectively. In organosilanes organic substituents influence the acidity of silanols involved in condensation. Electron-providing alkyl groups reduce the acidity of the corresponding silanol. This should shift the isoelectric point toward higher pH values as observed in Fig. 18a, significantly influencing the pH-dependence of the condensation mechanism. Conversely, electron-withdrawing groups ($-OH$ or $-OSi$) increase the silanol acidity, and the minimum condensation rate for oligomeric species occurs at about pH 2 (as judged from gel times versus pH, Fig. 18b). Thus, the extent of both hydrolysis and condensation and, in organoalkoxysilanes [$R_x Si(OR)_{4-x}$], the value of x determine the reaction mechanism and define what is meant by acid- or base-catalyzed condensation. In general we refer to base-catalyzed condensation as occurring when the pH is above about 2.

Although inductive effects are clearly important, Voronkov *et al.* state that in acid-catalyzed condensation of dialkysilanediol, steric effects predominate over inductive effects [47]. Therefore, in tetrafunctional alkoxide precursors the inductive effects resulting from longer-chain alkyl substituents are probably even less important.

2.4.3. EFFECTS OF SOLVENT

As discussed in the section on hydrolysis (Section 2.4.3), solvent effects have been primarily evaluated in the context of drying control chemical additives added to facilitate the rapid drying of monolithic gels without cracking. Solvents (or additives) may be either protic or aprotic and may vary in their polarity. (See Table 7.) Depending on the pH, either protonated or deprotonated silanols are involved in the condensation mechanism. Because protic solvents hydrogen bond to nucleophilic deprotonated silanols and aprotic solvents hydrogen bond to electrophilic protonated silanols, protic solvents retard base-catalyzed condensation and promote acid-catalyzed condensation. Aprotic solvents have the reverse effect.

Unfortunately, few data are available to determine unambiguously the effect of solvent type on the condensation rate. Artaki *et al.* [89] investigated the effects of protic and aprotic solvents on the growth rates of polysilicates prepared from TMOS ($r = 10$) with no added catalyst (pH measured immediately after mixing the reactants equalled 6–7). Using Raman spectroscopy and molybdic acid reagent methods (see, e.g., ref. 1), qualitative information related to the size of the polysilicate species was obtained as a function of the reduced time, t/t_{gel}, for five solvent systems: methanol, formamide, dimethylformamide, acetonitrile, and dioxane. Table 9 lists the stabilized

Table 9.

Relative Raman Intensity of 830-cm^{-1} Band, I^{tot},
at $t/t_{gel} = 2$; Inverse Relative Molybdic Acid Rate
Constant, K^{-1}, and Gel Times for TMOS Hydrolyzed
in Different Solvents.

Solvent	I^{tot}	K^{-1}	t gel (h)
Methanol	1	1	8
Formamide	1.3	1.3	6
Dimethyl-formamide	1.4	1.3	28
Acetonitrile	1.7	1.9	23
Dioxane	2.1	1.9	41

Source: Artaki *et al.* [89].

final intensity, I^{tot}, of the 830-cm^{-1} band (a value proportional to the number of siloxane bonds in the scattering volume[†]) evaluated at $t/t_{gel} = 2$, the reciprocal molybdenum rate constant, k^{-1} (proportional to the polymer size) evaluated at $t/t_{gel} = 0.8$, and the gel time, t_{gel}, for the five solvent systems.

These solvents may be categorized as polar protic (water, methanol, formamide), polar aprotic (acetonitrile, dimethylformamide) and non-polar aprotic (dioxane). (See Table 7.) Artaki *et al.* [89] explained that under base-catalyzed condensation conditions (pH > 2.5), the aprotic solvent, dioxane, is unable to hydrogen bond to the SiO$^-$ nucleophile. In addition, because it is nonpolar, it does not tend to stabilize the reactants with respect to the activated complex. Therefore, dioxane should result in a significant enhancement of the condensation rate and cause an efficient condensation leading to the formation of large, compact spherical particles. The polar, aprotic solvents, dimethylformamide and acetonitrile, also do not hydrogen bond to the silicate nucleophile involved in the condensation reaction. However, due to their polarity the anionic reactants are stabilized with respect to the activated complex slowing down the reaction to some extent.

Both methanol and formamide are protic solvents and as such can hydrogen bond to SiO$^-$, making it less nucleophilic. However, according to Artaki *et al.* [89], formamide is capable of forming significantly stronger hydrogen bonds to the reactant species (dipole moment, μ, equals 3.7 and 1.7 for formamide and methanol, respectively). (See Table 7.) Therefore, compared to methanol, formamide should provide more extensive steric shielding around silicon preventing efficient condensation. This line of reasoning appears inconsistent with the results in Table 9 that show more extensive condensation in the formamide system. This apparent contradiction may be explained by the partial hydrolysis of formamide to produce ammonia plus formic acid [72]. Because base-catalyzed silanol condensation (pH > 2.5) is generally observed to be first-order in [OH$^-$], ammonia should increase the condensation rate. To elucidate the effects of solvent type on condensation, further work should be performed in buffered solutions as a function of pH.

A second and important effect of the solvent, one that tends to be ignored, is its ability to promote depolymerization, for example, by the reverse of Eqs. 9 and 10. Iler [1] insightfully suggested that under conditions in which depolymerization is suppressed, condensation may lead to molecular

---

[†] Note added in proof: Due to a problem with the spectrometer, the 830-cm^{-1} band was probably caused by light scattering rather than Raman scattering. However, since light scattering also is a measure of particle growth, the conclusions of this work are qualitatively correct.

networks, whereas under conditions in which depolymerization is promoted, restructuring occurs ultimately resulting in the formation of highly condensed colloidal particles. Because the nucleophile (OH^-) is involved in the base-catalyzed hydrolysis of siloxane bonds (reverse of Eq. 10), aprotic solvents, which are unable to hydrogen bond to OH^- and therefore make OH^- a stronger nucleophile, promote restructuring leading to more highly condensed species. Enhanced restructuring resulting from the activation of the nucleophile in aprotic systems may further explain the results of Artaki *et al.* [89]. Clearly, further work needs to be performed to elucidate both the effects of solvent reactivity (e.g., the hydrolysis of formamide) and the tendency of the solvent to promote restructuring via enhanced depolymerization. The effects of depolymerization on structural evolution are discussed further in Section 2.6.

2.4.4. MECHANISMS

2.4.4.1. Base-catalyzed Condensation

The most widely accepted mechanism for the condensation reaction involves the attack of a nucleophilic deprotonated silanol on a neutral silicate species as proposed by Iler [1] to explain condensation in aqueous silicates systems[†]:

$$SiO^- + Si(OH)_4 \leftrightarrows Si-O-Si + OH^-. \qquad (37)$$

This reaction pertains above the isoelectric point of silica (>pH 2–4.5, depending on the extent of condensation of the silicate species), where surface silanols may be deprotonated depending on their acidity. The acidity of a silanol depends on the other substituents on the silicon atom. When basic OR and OH are replaced with OSi, the reduced electron density on Si increases the acidity of the protons on the remaining silanols [66]. Therefore, Iler's mechanism favors reactions between larger, more highly condensed species, which contain acidic silanols, and smaller, less weakly branched species. The condensation rate is maximized near neutral pH where significant concentrations of both protonated and deprotonated silanols exist. A minimum rate is observed near the isoelectric point. (See Fig. 18b.)

Pohl and Osterholtz [50] and Voronkov *et al.* [47] propose essentially the same mechanism to account for deuteroxide (hydroxyl) anion and

[†] Johnson *et al.* recently proposed a condensation mechanism involving a hydrogen-bonded intermediate that contains a five-coordinate $Si(OH)_5^-$ anionic species. (*J. Am. Chem. Soc.* **111** (1989) 3250).

general base-catalyzed condensation of alkylsilanetriol and alkylsilanediol, respectively.

$$R\text{-Si(OH)}_3 + OH^- \underset{k_{-1}}{\overset{k_1}{\rightleftharpoons}} R\text{-Si(OH)}_2O^- + H_2O$$

fast

$$R\text{-Si(OH)}_2O^- + RSi(OH)_3 \underset{k_{-2}}{\overset{k_2}{\rightleftharpoons}} R\text{-Si(OH)}_2\text{-O-Si(OH)}_2R + OH^-$$

slow

$$-d\left[\text{silanetriol}\right]/dt = k_1 k_2/k_{-1} \left[RSi(OH)_3\right]^2 \left[OH^-\right] \tag{38}$$

According to Pohl and Osterholtz [50], deuteroxide (hydroxyl) anion reversibly reacts with silanetriol in a rapid first step leading to an equilibrium concentration of *silanolate anion*, $Si(OH)_3O^-$. Silanolate anion reacts with neutral triol in a slower rate-determining step resulting in dialkyl-tetra-hydroxydisiloxane and regeneration of hydroxyl anion. Consistent with this mechanism, the condensation rate is observed to be first-order in deuteroxide anion and second-order in triol, $-d$[silanetriol]$/dt = (k_1k_2/k_{-1})[RSi(OD)_3]^2$ [OD$^-$]. Further condensation of the disiloxane was not observed at short reaction times, presumably due to steric effects.

It is generally believed that the base-catalyzed condensation mechanism involves penta- or hexacoordinated silicon intermediates or transition states [79,81,90–92]. For silicic acid polymerization, Okkerse [90] proposed a bimolecular intermediate involving one hexacoordinated silicon:

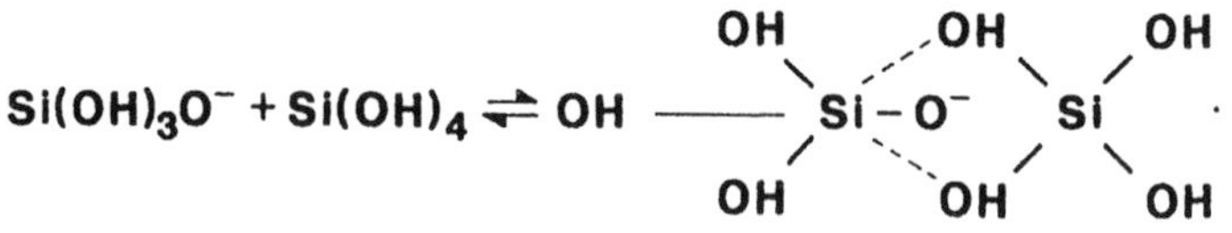

$$Si(OH)_3O^- + Si(OH)_4 \rightleftharpoons OH \quad\text{———}\quad \tag{39}$$

Grubbs [91] proposed that the condensation of trimethylsilanol in methanol occurs by an S_N2–Si mechanism in which the nucleophile approaches the backside of the silicon which subsequently undergoes displacement of its hydroxyl anion. Swain *et al.* [79] have proposed that silicon forms stable pentacoordinate intermediates: either S_N2^{**}–Si or S_N2^*–Si mechanisms:

$$S_N2^{**} \text{-SI} \quad S_N2^* \text{-SI} \quad \text{CONDENSATION}$$

$$\equiv SiOH + OH^- \longrightarrow \equiv SiO^- + HOH$$

$$\equiv SiO^- + Si(OR)_4 \longrightarrow$$

An associative condensation mechanism involving a penta- or hexa-coordinated intermediate is also consistent with the enhanced condensation kinetics observed at high pressures by Artaki *et al.* [81]. (See Fig. 19.) In order to explain these effects, Artaki *et al.* analyzed the activation volumes associated with a base-catalyzed condensation mechanism involving a hexa-coordinated intermediate (Eq. 39). They concluded that due to rearrangements of solvent molecules around the anionic nucleophile, SiO^-, and the smaller volume of the transition state compared to the volume of the reactant species, both dissociation of the silanol species and the formation of the transition state contribute to a reduction in the activation volume. Thus, both reactions should be accelerated by pressure. This same reasoning is applicable to mechanisms involving pentacoordinate intermediates.

Using a semi-empirical molecular orbital technique, Davis and Burggraf [92] have examined base-catalyzed silanol polymerization. Their calculations show that water is more easily eliminated when the coordination at silicon can change from hexavalent to pentavalent rather than from pentavalent to tetravalent. These results argue in favor of a condensation mechanism involving a hexacoordinate silicon intermediate. The same reasoning appears

Fig. 19.

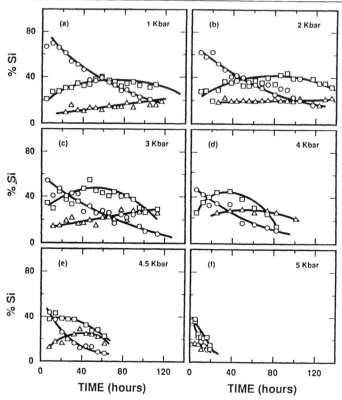

Temporal evolution of all observable condensed species during acid-catalyzed hydrolysis of TMOS. (a) 1 bar, (b) 2 kbar, (c) 3 kbar, (d) 4 kbar, (e) 4.5 kbar, (f) 5 kbar ($r = 10$); ○: Q^0; □: $(Q^1)_2$, $(Q^1)_3$; △: $Q^1Q^2Q^1$, $(Q^2)_4$ [81].

to apply to base-catalyzed hydrolysis where it may be easier to displace alcohol from a dimer through a hexacoordinated intermediate than from the monomer, implying concerted hydrolysis and dimerization.

2.4.4.2. Acid-catalyzed Condensation

Because in aqueous silicate systems, gel times are observed to decrease below the isoelectric point of silica (Fig. 10b), it is generally believed that the acid-catalyzed condensation mechanism involves a protonated silanol species. Protonation of the silanol makes the silicon more electrophilic and thus more susceptible to nucleophilic attack. The most basic silanol species (silanols contained in monomers or weakly branched oligomers) are the most likely to be protonated. Therefore, condensation reactions may occur preferentially between neutral species and protonated silanols situated on monomers, end groups of chains, etc. (See also Section 1.2.2 in Chapter 2.)

Pohl and Osterholtz [50] proposed that below about pD 4.5, the increased condensation rate of alkylsilanetriol (Fig. 18a) also involved a deuterated (or protonated) silanol:

$$
\text{R-Si(OH)}_3 + \text{H}^+ \underset{\substack{k_{-1}\\ \text{fast}}}{\overset{k_1}{\rightleftharpoons}} \text{R-Si(OH)}_2
$$

(41)

$$
\text{R-Si(OH)}_2 + \text{RSi(OH)}_3 \underset{\substack{k_{-2}\\ \text{slow}}}{\overset{k_2}{\rightleftharpoons}} \text{R-Si-O-Si-R} + \text{H}_3\text{O}^+
$$

$$
-d\left[\text{silanetriol}\right]/dt = k_1\, k_2/k_{-1}\, \left[\text{RSi(OH)}_3\right]^2\, \left[\text{H}^+\right].
$$

Consistent with this mechanism, the condensation rate was observed to be first-order in deuterium ion and second-order in silanetriol. From these results it is not possible to distinguish between S_N2–Si, S_N2^{**}–Si, or S_N2^*–Si mechanisms.

Because the proposed reaction mechanisms for both base- and acid-catalyzed condensation involve penta- or hexacoordinate transition states or intermediates, the condensation reaction kinetics will be influenced by both steric and inductive factors. Substituents attached to silicon that reduce steric crowding in the transition state or intermediate will enhance the condensation kinetics. Bulky groups attached to silicon or perhaps partial condensation of the silicon involved in the condensation reaction are expected to retard the condensation process. Replacement of more electron-providing OR groups with progressively more electron-withdrawing OH and OSi groups stabilizes the negative charge on the anionic nucleophile involved in the base-catalyzed condensation reaction and therefore should enhance the kinetics. Similar reasoning leads to the hypothesis that extensive hydrolysis and condensation should destabilize the positively charged intermediate or transition state involved in the acid-catalyzed condensation reaction and thus retard the condensation kinetics.

As discussed for the hydrolysis reaction, the specific mechanism may be influenced by the local environment of the silicon undergoing condensation, the solvent, and nature of the catalyst. For example, Klemperer *et al.* [38] proved that two Q^3 silicon species contained in separate cubic octamers undergo condensation under neutral conditions with essentially complete retention of configuration (see pathway A in Fig. 20), whereas condensation reactions between monomers may result in the inversion of one of the silicate tetrahedra.

2.4.5.   EFFECTS OF REVERSE REACTIONS

Alcoholysis and hydrolysis of siloxane bonds (reverse of Eqs. 9 and 10) provide a means for bond breakage and reformation allowing continual restructuring of the growing polymers. The rate of hydrolysis of siloxane bonds (dissolution of silica) exhibits a strong pH-dependence as shown in Fig. 21. Between about pH 3 and 8, the dissolution rate increases by over three orders of magnitude in aqueous solution. Partial replacement of water (pH 9.5) with methanol decreases the solubility by over a factor of 20 as shown in Table 10.

Fig. 20. _____

CONDENSATION

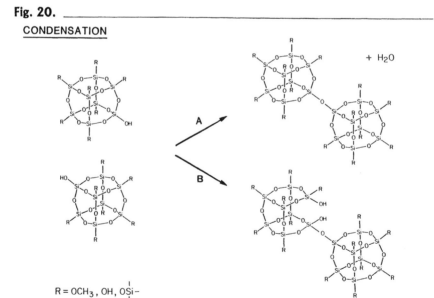

$R = OCH_3, OH, O\overset{|}{\underset{|}{Si}}-$

Possible pathways for condensation of Q^3 species contained in cubic octamers. A) With retention of configuration. B) With siloxane bond hydrolysis [38].

Fig. 21.

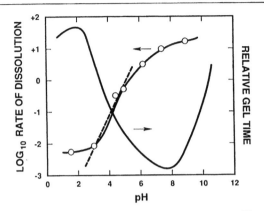

pH-dependence of the dissolution rate and gel times for aqueous silicates [1].

Klemperer and coworkers [93] have shown that alcoholysis occurs under basic conditions, leading to a redistribution of siloxane bonds:

$$ROH + (RO)_3\ Si -- O -- Si(OR)_3 \overset{OH^-}{\rightleftharpoons} Si(OR)_4 + Si(OR)_3\ OH$$

$$Si(OR)_3\ OH + (RO)_3\ Si -- O -- Si(OR)_3 \rightleftharpoons trimer + ROH$$

$$2\ dimers \overset{OH^-}{\underset{MeOH}{\rightleftharpoons}} \mathbf{REDISTRIBUTION} \atop trimer + unhydrolyzed\ monomers.$$

(42)

Table 10.

Silica Solubility at 25°C in Water–Methanol Solutions.

Wt.% Methanol	Solubility at 25°C (mg l^{-1})
0	140
25	75
50	40
75	15
90	5

Source: Iler [1].

They propose that this reaction accounts for the common observation that under base-catalyzed conditions, unhydrolyzed monomers persist past the gel point even with overstoichiometric additions of water. (See, e.g., ref. 59.) Based on capillary gas chromatography and ^{29}Si NMR results, Klemperer and coworkers [93,94] have shown that the redistribution reactions result in an "inverted" molecular-weight distribution in which high- and low-molecular-weight species are maximized with respect to intermediate-molecular-weight species. As discussed in Section 2.6, this may be explained on the basis of classical nucleation and growth or ripening theories.

According to Iler [1], the dissolution of amorphous silica above pH 2 is catalyzed by OH$^-$ ions that are able to increase the coordination of silicon above four weakening the surrounding siloxane bonds to the network. This general nucleophilic mechanism could presumably occur via S$_N$2-Si, S$_N$2**Si, or S$_N$2*-Si transition states or intermediates and could equally well explain alkoxide ion- or fluorine ion-catalyzed depolymerization mechanisms.

2.5. Sol-Gel Kinetics

Thus far we have discussed the hydrolysis and condensation reactions separately at a rudimentary level that largely ignored how the various functional groups, (OR), (OH), and (OSi), are distributed on the silicon atoms. At this level only three reactions and three rate constants are necessary to describe the functional group kinetics (k_h, $k_{cw}/2$, and $k_{ca}/2$ in equations 43–45), where h is hydrolysis, cw is condensation of water, and ca is condensation of alcohol:

$$\text{SiOR} + \text{H}_2\text{O} \xrightarrow{k_h} \text{SiOH} + \text{ROH} \tag{43}$$

$$2\text{SiOH} \xrightarrow{k_{cw}/2} 2(\text{SiO})\text{Si} + \text{H}_2\text{O} \tag{44}$$

$$\text{SiOH} + \text{SiOR} \xrightarrow{k_{ca}/2} 2(\text{SiO})\text{Si} + \text{ROH.} \tag{45}$$

In practice, hydrolysis and condensation occur concurrently, and at the nearest functional group level there are 15 distinguishable local chemical environments. Kay and Assink [63,95–97] have represented the 15 silicate species in matrix form as shown in Fig. 22, where the ordered triplet (X, Y, Z) represents the number of –OR, –OH, and –OSi functional groups attached to the central silicon: Si(OR)$_X$(OH)$_Y$(OSi)$_Z$, and $X + Y + Z = 4 =$ coordination of silicon. At this more sophisticated level, there are 165 distinguishable rate coefficients: 10k_h, 55k_{cw}, and 100k_{ca}, considering only the forward reactions. At the next-to-nearest functional group level, there are 1365 distinct local silicon environments requiring 199,290 rate coefficients [95]!

Fig. 22.

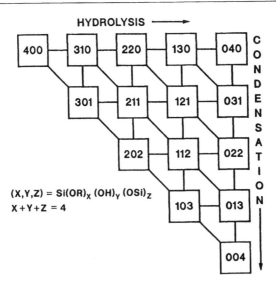

Chemical speciation at next-to-nearest–neighbor level represented in matrix form [63,95,96].

Assink and Kay [63,95–97] have used ^{1}H and ^{29}Si NMR to determine the values of k_h, k_{cw}, and k_{ca} during the initial stages of the acid-catalyzed hydrolysis of TMOS and have considered several kinetic models to explain the temporal evolution of functional groups surounding silicon. The following discussion is largely based on their line of reasoning.

^{1}H NMR is used to measure the relative amounts of methoxy functional groups and methanol molecules as a function of reaction time. Two limiting cases exist. If the hydrolysis rate is much larger than either condensation rate, then the methoxy functional group concentration will quickly be reduced to the value corresponding to complete hydrolysis without condensation. (Reduction in methoxy functional group concentration equals the concentration of added water.) Further reduction in the methoxy functional group concentration occurs at a lower rate commensurate with the overall rate of the condensation reactions. If the hydrolysis rate is much smaller than either condensation rate, hydrolysis will be followed by immediate condensation. In this case, the reduction in methoxy functional group concentration will occur at a rate proportional to the hydrolysis rate. Figure 23 shows the relative methoxy functional-group concentration as a function of time for several values of the initial H_2O/TMOS mole ratio, r. For r values up to 2, the reaction proceeds quickly (in less than three minutes) to the stoichiometry corresponding to complete hydrolysis without condensation.

Fig. 23.

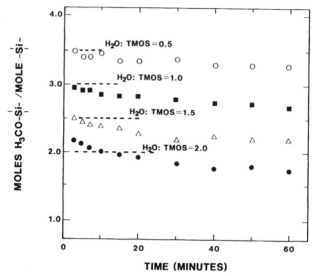

Moles $SiOCH_3$/mole Si versus time for H_2O/Si values ranging from 0.5 to 2.0. Dashed lines are the theoretical values corresponding to complete hydrolysis without condensation [63].

This corresponds to the limiting case in which the initial rate of the hydrolysis reaction is much larger than the sum of the rates of the condensation reactions so that [SiOH] equals the consumption of H_2O by hydrolysis. Based on second-order hydrolysis kinetics (constant $[H_3O^+]$), a lower limit for the hydrolysis-rate coefficient was established as 0.2 l/mole-min) [63].

^{29}Si may be used to determine the rate of formation of Si–O–Si bonds. Again, two limiting cases may be considered based on the equation:

$$d[(SiO)Si]/dt = k_{cw}[SiOH]^2 + k_{ca}[SiOH][SiOR]. \qquad (46)$$

If k_{cw} is much greater than k_{ca}, the condensation rate will be proportional to $[SiOH]^2$. If k_{cw} is much smaller than k_{ca}, the condensation rate will be proportional to [SiOH][SiOR]. By measuring the initial overall condensation rate as a function of the initial water concentration, it is possible to determine if either of these limiting cases is applicable.

In the case where the initial overall condensation reaction is negligible with respect to the initial hydrolysis rate and the initial hydrolysis reaction is complete (see Fig. 23), Eq. 46 may be re-expressed as:

$$\frac{d[(SiO)Si]/dt}{\langle[SiOH]\rangle} = (k_{cw} - k_{ca})\langle[SiOH]\rangle + k_{ca}[SiOMe]_o. \qquad (47)$$

This expression is valid at early times when the concentration of Si–O–Si is small compared to the initial methoxy functional-group concentration. According to this equation, k_{cw} and k_{ca} are determined from a plot of the initial condensation rate divided by $\langle[SiOH]\rangle$ versus $\langle[SiOH]\rangle$ where $\langle[SiOH]\rangle$ is the average value of the silanol group concentration over the measurement window. (See Fig. 24). k_{cw} and k_{ca} determined from the slope and intercept of this plot are 0.006 and 0.001 l/mole-min, respectively. It is important to note that both the hydrolysis- and condensation-rate coefficients determined for TMOS are significantly greater than corresponding values reported by Pouxviel *et al.* [62] for TEOS. Because increased electron provision by the alkoxide substituents (OEt versus OMe) would be expected to increase the acid-catalyzed hydrolysis- and condensation-rate constants if inductive effects were important, this opposite effect suggests that for tetrafunctional alkoxysilanes, steric rather than inductive factors are more important in determining sol-gel kinetics.

Fig. 24. _____

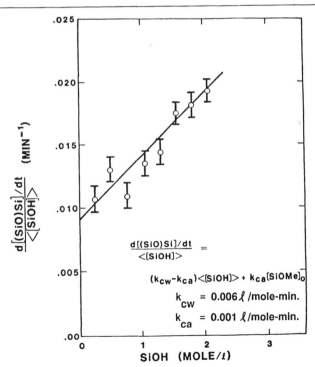

Initial condensation rates divided by the initial silanol concentrations versus the initial silanol concentration. This plot should be linear with a slope of $k_{cw} - k_{ca}$ and intercept of $k_{ca}[SiOR]_o$ [63].

Fig. 25.

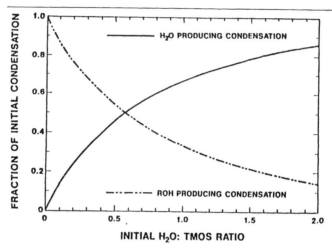

Fraction of condensation occurring by the water-forming reaction or the alcohol-forming reaction versus the initial H_2O : TMOS ratio [63].

Figure 25 shows the fraction of the initial condensation occurring by water- or alcohol-producing reactions as a function of r. The curves are calculated using Eq. 46 and the values of k_{cw} and k_{ca} determined from Fig. 24. It is evident that for small values of r, the alcohol-producing condensation reaction dominates, whereas the water-producing condensation reaction becomes more important when r exceeds about 0.5. Clearly both reactions must be considered to describe condensation kinetics accurately.

To explain the temporal evolution of functional groups surrounding silicon, Kay and Assink [95,63] have considered several kinetic models. When each hydrolysis- and condensation-rate coefficient is much greater than the rate coefficient for the step preceding it, only two of the matrix species will be present, (004) and (400), the relative concentrations of which depend on the H_2O : Si ratio, r. Based on the steric and inductive effects expected for nucleophilic attack of OH^- (hydrolysis) and $\equiv Si-O^-$ (condensation) on silicon, this situation might occur in base-catalyzed systems [60]. Conversely, when each subsequent rate coefficient is much smaller than the rate coefficient preceding it, the reaction proceeds down the diagonal elements of the matrix [95,63]. Based on the importance of inductive effects, it has been suggested that this situation may occur under acidic conditions, although steric considerations, especially for higher alkoxy groups, may outweigh any inductive effects [60].

A third model assumes that steric and inductive effects are unimportant and that the speciation is governed merely by statistics [95,63]. This model utilizes two simplifying statistical assumptions: the hydrolysis- and condensation-rate constants depend only on the functional group reactivity, not the local silicon chemical environment; and the rate coefficient for a particular species undergoing one of the three reactions is simply the product of a statistical factor and the appropriate functional group rate coefficient, k_h, k_{cw}, or k_{ca}. For example, the hydrolysis-rate coefficient for species (400) is four times the rate coefficient for species (130), and the water-forming condensation-rate coefficient for species (040) with itself is $4 \times 4 = 16$ times greater than that of species (310) with itself. This reduces the number of rate coefficients required to describe the kinetics at the functional group level from 165 to 3 (k_h, k_{ca}, and k_{cw}).

Kay and Assink used the experimentally determined values of k_h, k_{ca}, and k_{cw} in the statistical model to calculate both the equilibrium distribution of species as a function of r at long reaction times and the temporal evolution of species, during acid-catalyzed hydrolysis of TMOS. The results of these calculations are presented in Figs. 26 and 27 where they are compared to values determined experimentally using ^{29}Si NMR. Agreement between theory and experiment is very good. Because the statistical model implicitly ignores steric and inductive effects as well as reverse reactions such as reesterification, good agreement with experiment suggests that these effects are unimportant for the early stages of hydrolysis and condensation of TMOS at low r values.

Fig. 26. _____

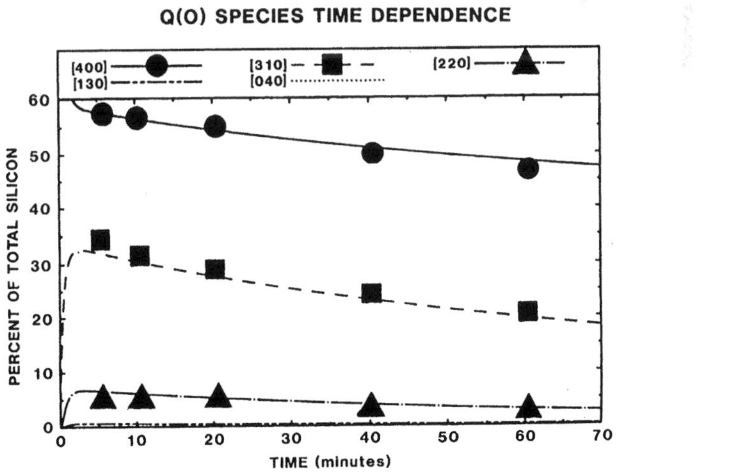

Temporal distribution of Q^0 species during acid-catalyzed hydrolysis of TMOS ($r = 0.5$). Lines are drawn according to the statistical model. Points are experimentally determined by ^{29}Si NMR [96].

Fig. 27.

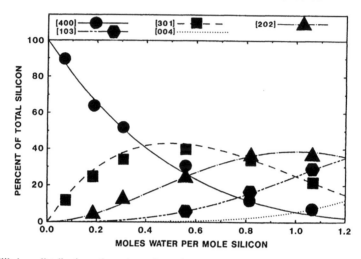

COMPARISON OF EXPERIMENTAL AND
STATISTICAL EQUILIBRIUM DISTRIBUTIONS

Equilibrium distribution of species at long times: lines drawn according to statistical model; points determined experimentally by ^{29}Si NMR [95].

^{29}Si NMR investigations of the hydrolysis and condensation of TEOS reveal inconsistencies with the statistical model, however. Pouxviel *et al.* [62] report that the relative hydrolysis rate coefficients are 1:5:12:5 compared to 4:3:2:1 (statistical model) for the species (400), (310), (220), and (130), respectively. Lin and Basil [86] do not observe the dimer, (301)–(301), at early times during acid-catalyzed hydrolysis of TEOS, whereas the statistical model predicts its concentration to be ~23% of the total dimer concentration and ^{29}Si NMR shows that it accounts for ~19% of the dimer concentration at early times during the hydrolysis of TMOS [98]. Based on numerical simulations of ^{29}Si NMR data, Pouxviel and Boilot [99] show that the condensation rate decreases with the extent of condensation and that reesterification is significant especially for large values of r.

The rate-constant results [62,86] suggest that as the size of the alkoxide substituents increases or, as the reacting species become more highly condensed, steric effects become increasingly important as observed, for example, in organoalkoxysilanes [47].

Assink (in [100]) has introduced a multiplicative factor ($R < 1$) in the statistical model to account for the reduction in condensation kinetics with the extent of condensation. This modified statistical approach is qualitatively consistent [101] with the rate-constant data of Pouxviel and Boilot.

Although the absence of the (301)–(301) dimer can be rationalized by steric or inductive arguments (the less acidic species, (400) and (310), prefer to condense with the more acidic species, (200), (130), etc., rather than with themselves), the appearance of this species during the latter stages of the reaction, when the water concentration is greatly reduced [86], may indicate that for unknown reasons this species hydrolyzes extremely rapidly compared to other species in the TEOS and TMOS systems.

Using the kinetic model of Kay and Assink [95], Doughty and coworkers [102] evaluated the temporal evolution of hydrolysis and condensation products of the dimer, hexamethoxydisiloxane, to derive the hydrolysis- and condensation-rate constants defined in Eqs. 43–45. Using ^{29}Si NMR, two types of measurements were performed. The amount of dimer and the extent to which it was hydrolyzed were determined at early times by integration of the Q^1 resonances. The extent of condensation was determined by comparing the integrals of the Q^1 and Q^2 resonances.

The initial rate of formation of siloxane bonds and the average value of the silanol functional-group concentration over the measurement time window determined from the ^{29}Si NMR investigations were used in Eq. 47 to derive the values of the alcohol- and water-producing condensation-rate constants, k_{ca} and k_{cw}, respectively. Calculated values of k_{ca} and k_{cw} are listed in Table 11 where they are compared to the respective values obtained for the monomer, TMOS. The observed rate constants for the alcohol-producing condensation reaction are approximately the same for monomer and dimer (0.001 and 0.0007 l/mol-min, respectively), whereas the rate constants for the water-producing condensation reaction are significantly different (0.006 and 0.0011 l/mol-min, respectively). The differences in the values of k_{cw} obtained for monomer and dimer were explained on the basis of steric and inductive arguments [102]. Bulky OSi groups attached to the silicon undergoing condensation retard the kinetics. Replacement of electron-providing OR groups with more electron-withdrawing OSi groups decreases the stability of the positively charged intermediate also reducing the condensation rate.

Table 11.

Condensation Rate Constants for
Methoxy-substituted Monomers and Dimers.

	Q^0 Monomers	Q^1 Dimers
k_{ca} (l/mol-min)	0.001	0.0007
k_{cw} (l/mol-min)	0.006	0.0011

Source: Doughty *et al.* [102].

If the first step in the acid-catalyzed alcohol-producing condensation reaction is the protonation of an alkoxide group, which subsequently becomes the leaving group, MeOH, then the increase in the rate-constant ratio, k_{ca}/k_{cw}, in changing from monomer to dimer condensation (Table 11) suggests that MeOH has become a better leaving group than water as predicted by the partial-charge model [11], or that the electron provision by the R-group predominates over the other (electron-withdrawing) substituents attached to silicon [102].

The statistical model of Kay and Assink appears to be the only available model that predicts the temporal evolution of silicate species during the early stages of the hydrolysis and condensation reactions. For it to be more generally applicable, it must be improved upon to account for steric and inductive effects as well as reverse reactions such as reesterification and siloxane bond hydrolysis.

2.6. Structural Evolution

It is necessary to discuss the structure of silicate solution species on scales of several lengths. On the shortest length scale, the nearest neighbor of silicon may be an alkoxide group (OR), a hydroxyl group (OH), or a bridging oxygen (OSi). On intermediate length scales, oligomeric species (dimers, trimers, tetramers, etc.) may be linear, branched, or cyclic. On length scales large with respect to the monomer and small with respect to the polymer, structures may be dense with well-defined solid–liquid interfaces, uniformly porous, or tenuous networks characterized by a mass or surface fractal dimension, d_f or d_s, respectively [36].

The methods of choice for determining structure on these different-length scales are nonintrusive in situ methods such as nuclear magnetic resonance (NMR) spectroscopy, Raman and infrared spectroscopy, and X-ray, neutron, and light scattering. In the following subsections, structural information obtained by these and other in situ methods are presented in the order of increasing length scales.

2.6.1. NMR INVESTIGATIONS

[1]H and [29]Si nuclear magnetic resonance spectroscopy (NMR) have been employed extensively to elucidate the extent and kinetics of the hydrolysis and condensation reactions accompanying gelation, and the speciation of silicate solutions during the early stages of polymerization of TMOS, TEOS,

and several oligomeric species, $Si_2O(OR)_6$, $Si_3(OR)_8$, and $Si_8O_{12}(OR)_{12}$, where $R = CH_3$ [38,39,102,103] or $R = C_2H_5$ [86]. In addition, ^{17}O and ^{13}C NMR have been used to monitor the water and solvent content in several systems [104].

Typical ^1H spectra are shown in Fig. 28 for TEOS after different times during acid-catalyzed hydrolysis ($r = 1$). The quartet of peaks centered at 3.90 ppm is the resonance of methylene ($-CH_2-$) protons associated with a silicon, and the quartet centered at 3.70 ppm is the resonance of methylene protons associated with ethanol. In general, the extent of hydrolysis of TMOS or TEOS is readily quantified by integration of the resonances of methyl ($-CH_3$) and methylene ($-CH_2-$) protons associated with silicon or the

Fig. 28a.

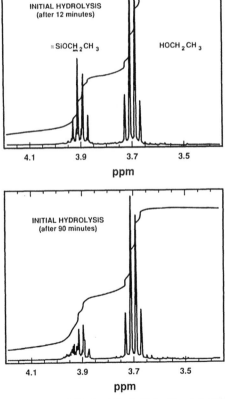

^1H NMR spectra after 12 min and 90 min of the initial acid-catalyzed hydrolysis of TEOS ($r = 1$) [59].

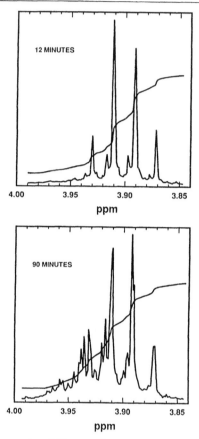

^{1}H NMR multiplet centered at 3.90 ppm after 12 min and 90 min of the initial acid-catalyzed hydrolysis of TEOS ($r = 1$). The complex spectrum obtained after 90 min is a superposition of many quartets resulting from the hydrolysis and condensation products [59].

alcohols (methanol and ethanol, respectively) [97]. Figure 29 shows the fraction of methylene protons associated with silicon relative to the total number of methylene protons for two-step acid- or base-catalyzed hydrolyses of TEOS. It is evident that after 1000 min of the first step (1 mole H_2O/Si with acid catalyst) the reaction:

$$nSi(OEt)_4 + nH_2O \rightarrow [SiO(OEt)_2]_n + 2EtOH \qquad (48)$$

has not gone to completion (point C in Fig. 29), which suggests that an equilibrium distribution of species containing uncondensed silanols is slowly

Fig. 29. _____

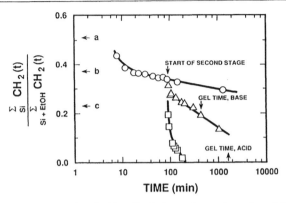

Fraction of CH_2 groups associated with a Si relative to the total number of CH_2 groups for first (O) and second (□ acid, △ base) stages of reaction as a function of time. For the first stage, a = starting composition, b = complete hydrolysis, no condensation, and c = complete hydrolysis and condensation [97].

approached. When, after 90 min of the first step, additional water plus acid or base are added so that $r = 4-5$, it is observed [97] that the hydrolysis reaction goes to completion in acid long before the gel point, whereas under basic conditions gelation occurs prior to complete hydrolysis (Fig. 29). According to the NMR/GC results of Klemperer and Ramamurthi [93], incomplete hydrolysis with base catalyst is a consequence of the redistribution reaction (Eq. 42), that produces unhydrolyzed monomers.

The speciation of polysilicate solutions was originally investigated in basic, aqueous systems by Engelhardt and coworkers [21]. According to their ^{29}Si NMR results, a common condensation sequence is monomer, dimer, trimer, cyclic trimer, cyclic tetramer, higher-order rings. They also observed an equilibrium distribution of species at long reaction times that was independent of the starting materials (mono-, di- or trisilicic acids), indicating that complete restructuring occurs under these conditions. A similar reaction pathway was proposed by Iler [1] for the formation of aqueous silicate colloids: highly condensed cage structures form the precritical nuclei for subsequent colloid growth. (See Section 1 of this chapter.)

Although the ^{29}Si NMR spectra of aqueous silicate systems are quite complex, the speciation of polysilicates formed from tetraalkoxysilanes is more complicated, because hydrolysis and condensation occur concurrently. At the nearest functional-group level there are 15 distinguishable local chemical environments, which Kay and Assink have represented in matrix form in Fig. 22. The ordered triplet (X, Y, Z) represents the number of –OR, –OH, and –OSi functional groups attached to the central silicon [63,97].

According to the Q notation adopted by Engelhardt *et al.* [21], from top to bottom, the five rows correspond to Q^0 through Q^4 species (increasing extents of condensation). From left to right, columns represent different extents of hydrolysis. At the nearest functional-group level, the aqueous silicate species investigated by Engelhardt *et al.* [21] and Harris and Knight [105] are represented by the right-hand column.

Figure 30 compares the spectra of TMOS and TEOS in methanol and ethanol, respectively, after hydrolysis with a silicon : water : acid ratio of 1 : 2 : 0.02. The major features of ^{29}Si NMR spectra of TMOS systems (Fig. 30a) can be understood in terms of an approximately $+2$-ppm shift for each $-OH$ that is substituted for $-OCH_3$ and an approximately -8-ppm shift for each $-OSi$ that is substituted for $-OCH_3$, relative to the reference TMOS peak (~ -79.0 ppm). Thus, the simple hydrolysis products of TMOS, $Si(OCH_3)_x(OH)_{4-x}$, will show resonances at approximately [107]:

$$x = \quad 4 \qquad 3 \qquad 2 \qquad 1 \qquad 0$$
$$\quad -79.0 \quad -77.0 \quad -75.5 \quad -74.2 \quad -73.1 \text{ ppm.}$$

The simplest condensation products are Q^1 species, $Si(OSi)(OCH_3)_y(OH)_{3-y}$, in which the silicon of interest is bonded to a single other OSi. These species show resonances approximately 8 ppm lower than the corresponding

Fig. 30.

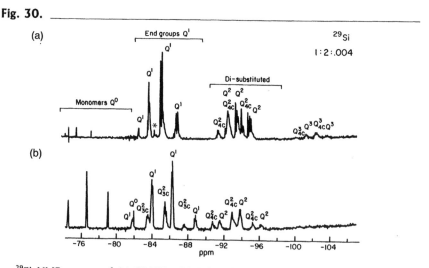

^{29}Si NMR spectra of (a) TMOS + MeOH and (b) TEOS + EtOH at $t_o + 2$ hours with a silicon : water : acid ratio of 1 : 2 : 0.02. Note the presence of ring compounds (Q^2_{3c}) in TEOS compared to TMOS within the Q^1 region. From Kelts and Armstrong [106].

uncondensed species [107]:

$$y = \begin{array}{cccc} 3 & 2 & 1 & 0 \\ -86.2 & -84.8 & -83.5 & -82.1 \text{ ppm}. \end{array}$$

Similar behavior with respect to the effects of hydrolysis and condensation on silicon resonances is observed in TEOS systems [86]. Tables 12a and b list assignments for many of the Q resonances observed in TMOS and TEOS sol-gel systems. From Table 12a we observe that it is possible to reliably resolve differences in the chemical shift of one Q^1_2 silicon nucleus (a Q^1 species contained in a dimer) that result from the extent of hydrolysis of the other Q^1_2 nucleus to which it is attached (fourth-nearest-neighbor

Table 12a.

^{29}Si Chemical Shifts, δ, in ppm Relative to TMS, for Q^0 and Q^1 species.

| | | Q^0 | $Si(OR)_x(OH)_{4-x}$ | | |
| | | | x | | |
	4	3	2	1	0
TMOS[a]	−78.5[c]	−77.0	−75.5	−74.2	−73.1
TEOS[b]	−81.95	−79.07	−76.58	−74.31	—

			$(OR)_y(OH)_{3-y}$		
		Q^1_2	$\underline{Si}(OSi)(OR)_x(OH)_{3-x}$		
			x		
		3	2	1	0
TMOS y	3	−85.81[d]	−83.99[d]	−82.62[d]	−81.6[a]
	2	−85.76[d]	−84.02[d]	−82.69[d]	—
	1	−85.72[d]	−84.04[d]	—	—
	0	—	—	—	79.93[e]
		3	2	1	0
TEOS y	3	−88.85[b]	−86.27[b]	−83.92[b]	−81.72[b]
	2	−88.75[b]	−86.17[b]	−83.80[b]	−81.62[b]
	1	−88.64[b]	−86.06[b]	−83.72[b]	−81.55[b]
	0	—	−85.96[b]	−83.64[b]	−79.93[e]

[a] Ref. 108.
[b] Ref. 86.
[c] Ref. 63.
[d] Ref. 102.
[e] Ref. 105.

Table 12b.

^{29}Si Chemical Shifts, δ, in ppm Relative to TMS, for Monomers and Oligomers.

	Q^0	Q^1	Q^2	Q^3
TMOS				
–OCH$_3$ substituted				
Monomer Q^0	-78.47^c	—		
Dimer $(Q^1)_2$	—	-85.81^d		
Trimer $Q^1Q^2Q^1$		-85.99^d	-93.69^d	
Cyclic trimer $(Q^2)_3$			-83.3^a(possibly hydrolyzed)	
Cyclic tetramer $(Q^2)_4$			-92.9^a	
Linear tetramer $Q^1Q^2Q^2Q^1$		-85.98^d	-93.9^d	
Branched tetramer $Q^1Q^3Q^1Q^1$		-86.19^d		-102.10^d
TEOS				
–OC$_2$H$_5$ substituted				
Monomer Q^0	-81.95			
Dimer $(Q^1)_2$		-88.85^b		
Trimer $Q^1Q^2Q^1$		-88.99^b	-96.22^b	
Aqueous silicates				
–OH substituted				
Monomer Q^0	-73.1^e			
Dimer $(Q^1)_2$		-79.92^e		
Linear trimer $Q^1Q^2Q^1$		-79.485^e	-88.19^e	
Cyclic trimer $(Q^2)_3$			-82.02^e	
Prismatic hexamer $(Q^3)_6$				-90.4^e
Cubic octamer $(Q^3)_8$				-99.3^e

Source: References the same as in Table 12a.

effect). It is also possible to resolve differences in Q^1 silicon resonances depending on whether it is contained in a dimer, trimer, or tetramer, Q^1_2, Q^1_3, or Q^1_4 species, respectively.

Resonances attributable to Q^2 species, which include chains and rings, are generally observed between -91 and -95 ppm. Small rings, such as cyclic trimers or tetramers, however, cause the silicon resonance to shift toward more positive values. For example, the Q^2_3 resonance for the cyclic trimer is observed in the Q^1 region of the spectrum ($\sim$10-ppm shift) [108], and the Q^2_4 resonance of the cyclic tetramer is observed at -92.9 ppm compared to -93.90 ppm observed in the linear tetramer [108]. The magnitude of this positive shift is correlated with the reduction in size of the Si–O–Si bridging angle, ϕ, resulting from the formation of the ring [109].

Q^3 species are observed between -99 and -103 ppm and Q^4 species are observed as a broad band around -110 ppm. The breadth of the Q^2–Q^4

resonances is due to the overlap of a large number of single resonance lines produced as a result of the highly varied environments of the silicon nuclei of condensed oligomeric species. For example, depending on the extent of hydrolysis, a total of 48 chemically different linear trimers and 81 chemically different cyclotetramers are possible. For this reason the use of ^{29}Si NMR spectroscopy as a structural tool is generally limited to the study of the initial stages of the condensation process [110].

Knowing the distribution of Q species within the solution is obviously not sufficient to uniquely identify the multitude of possible polysilicate trimers, tetramers, pentamers, hexamers, etc. that may form during the course of gelation. Klemperer and coworkers [93,94] have combined capillary gas chromatography, mass spectroscopy (MS), and ^{29}Si {^{1}H} NMR techniques (1-pulse, 1-D-INADEQUATE, and 2-D-INADEQUATE [94]) to identify specific species and distinguish between isomers containing up to six silicons. Capillary gas chromatography separates the various oligomers according to their diffusivities. The capillary gas chromatogram of a partially hydrolyzed TMOS solution is shown in Fig. 31 [94]. Using MS a formula can be assigned to the species responsible for each peak. ^{29}Si {^{1}H} NMR techniques are used to create a "connectivity map" that identifies the type of Q species to which each silicon comprising the oligomer is attached. In this fashion it is possible

Fig. 31. _____

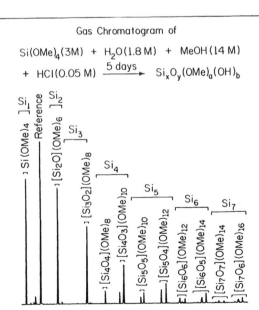

Gas Chromatogram of

$$Si(OMe)_4(3M) + H_2O(1.8\,M) + MeOH(14\,M)$$
$$+ HCl(0.05\,M) \xrightarrow{5\ days} Si_xO_y(OMe)_a(OH)_b$$

Gas chromatogram of the polysilicate ester solution prepared as indicated at the top of the figure. From Klemperer *et al.* [94].

Fig. 32.

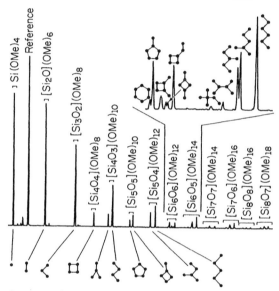

Capillary Gas Chromatographic Analysis
of Hydrolyzed TMOS

$$Si(OMe)_4(3M) + H_2O(1.8M) + HCl(0.05M) + MeOH(14M) \xrightarrow[\text{R.T.}]{\text{5 days}} [Si_xO_y](OMe)_a(OH)_b$$

Complete assignments of the gas chromatogram shown in Fig. 31, where the principal peaks are assigned using the following structural nomenclature: lines represent siloxane oxygens, dots represent silicon centers, and tetravalence at silicon is satisfied by adding the necessary number of OMe groups. From Klemperer *et al.* [94].

to distinguish between isomers as shown by the assignments of the gas chromatogram (Fig. 32). This work represents the current state of the art with regard to structural identification of polysilicate solution species, although from a practical standpoint these techniques are limited to rather low-molecular-weight structures.

In addition to the identification of specific species comprising the solution at any stage of the gelation process, ^{29}Si NMR is useful in the determination of structural trends that result under varying synthesis conditions. Figure 33 compares the ^{29}Si NMR spectra of TMOS systems near the gel point ($t/t_{gel} \sim 0.9$) for three different pH conditions. It is apparent that acidic conditions promote hydrolysis (thus allowing polymerization to occur) but inhibit condensation: at pH 1 the system is 100% polymerized but only 26% tetrasubstituted (Q^4). More basic conditions have the reverse effect:

Fig. 33.

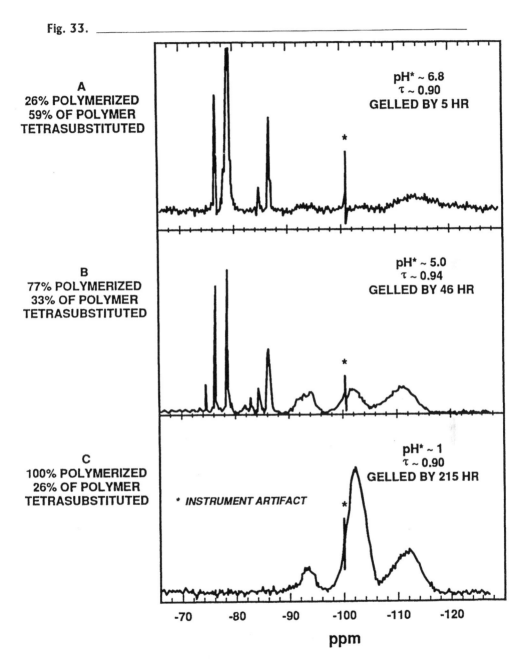

^{29}Si NMR spectra during hydrolysis of TMOS at pH ~6.8, 5.0, and 1; $t/t_{gel}(\tau) \sim 0.9$, $r = 2.6$ [111]. *: instrument artifact.

Fig. 34.

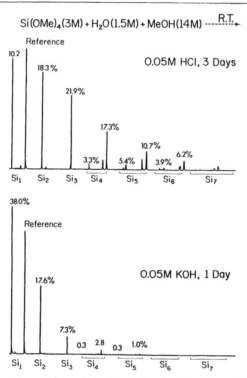

Capillary gas chromatographic analysis of hydrolyzed and condensed TMOS ($r = 0.5$) under acidic and basic conditions [93].

at pH 6.8 the system is only 26% polymerized, but 59% of the polymeric species are tetrasubstituted. Unhydrolyzed monomers primarily account for the Q^0 species [59]. Klemperer and Ramamurthi [93] have investigated these trends using capillary gas chromatography and ^{29}Si NMR. Figure 34 shows the distribution of oligomeric species resulting from acid- or base-catalyzed hydrolysis of TMOS ($r = 0.5$) at long reaction times. The acid-catalyzed system is composed almost completely ($\Sigma = 97\%$) of a "normal" (monotonic) distribution of oligomers containing up to seven silicons.[†] The oligomeric fraction of the base-catalyzed system contains primarily monomers and dimers ($\sim56\%$), but higher–molecular-weight species, containing more than seven silicons, account for 36% of the total silicate. Thus an "inverted" (bimodal) molecular-weight distribution is inferred.[††]

[†] Mole percent of silicon present as a function of degree of polymerization shows a maximum near the number average degree of polymerization.

[††] Most of the silicon is accounted for by low and high molecular-weight species.

Fig. 35.

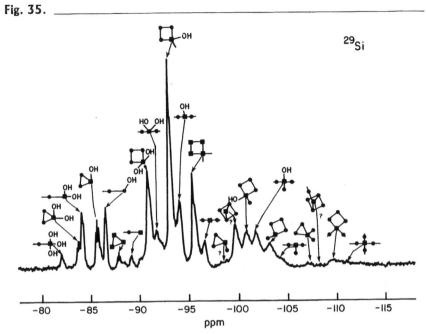

^{29}Si NMR spectrum of TEOS + EtOH with a silicon : water : acid ratio of 1 : 2 : 0.02 at t_0 + 15 h with assignments. Chemical shift distinctions are made between cyclic trimers and tetramers and di-, tri-, and tetrasubstituted silicons in linear species. The broadening observed in these peaks is due to the higher–molecular-weight oligomers. Alkoxy groups are implied where not shown to make the Si 4-coordinate. Also, oxygen atoms are assumed between all Si atoms. The Si atom referred to in the assigned resonance is represented by a square; all other Si atoms in the molecule are circles [106].

Kelts and Armstrong [106] compared the distributions of cyclic species formed during acid-catalyzed hydrolysis of TMOS and TEOS (r = 2). Figure 35 shows cyclic assignments in the Q^1 and Q^2 portions of the ^{29}Si NMR spectrum of the TEOS system after 15 hours of reaction time. Quite remarkably cyclic tri- and tetrasiloxane species are much more prominent in TEOS and higher alkoxide systems than in TMOS systems. (See Fig. 36.) This suggests that cyclic species are favored based on steric considerations. It should be noted that species containing cyclic tetramers account for the majority of polysilicate solution species in TEOS systems, and, after about 70 hours of reaction time, cyclic and linear species are present in approximately equal concentrations in the TMOS system. Since cyclic species are excluded in the Flory–Stockmayer [112,113] theory of condensation of polyfunctional monomers, the utility of this theory in adequately describing silicate speciation during sol-gel processing is questionable.

Fig. 36.

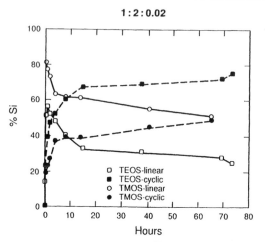

Graph of linear and cyclic species calculated from ^{29}Si NMR spectra of TEOS + EtOH and TMOS + MeOH with a silicon : water : acid ratio of 1 : 2 : 0.02 [106].

The hydrolysis and condensation of oligomeric species, for example, dimers, trimers, and cubic octamers (Structures 14–16) have been investigated by several groups. Klemperer *et al.* [38,39] observed no degradation of the trimer (15) or octamer (16) during hydrolysis and condensation under neutral conditions employing 20 and 8 equivalents of water, respectively ($r = 2.5$ and 1). However, production of monomer during neutral hydrolysis of the dimer (14) is clear evidence of siloxane bond hydrolysis (reverse of Eq. 10). The increased stability of the trimer and octamer compared to the dimer may be attributable to steric factors.

As an extension of this work, Klemperer and Ramamurthi investigated acid- and base-catalyzed redistribution reactions of the dimer, (MeO)$_3$SiOSi-(OMe)$_3$ [93]:

$$Si_2O(OMe)_6 \ (1.5 \ M) + MeOH \ (14 \ M) \xrightarrow[0.05 \ M \ KOH]{0.05 \ M \ HCl} . \tag{49}$$

Figures 37a and b show that after sufficiently long reaction times the species distributions resulting from redistribution reactions are nearly identical to the distributions resulting from the hydrolysis of TMOS with one-half equivalent of water ($r = 0.5$) employing 0.05 M HCl or KOH as a catalyst. (See Fig. 34.) This indicates that for dimers, siloxane bond formation is reversible. Furthermore, since redistribution under acidic and basic conditions reproduces the same types of polysilicate distributions generated by hydrolysis–condensation, Klemperer and Ramamurthi concluded that

Fig. 37.

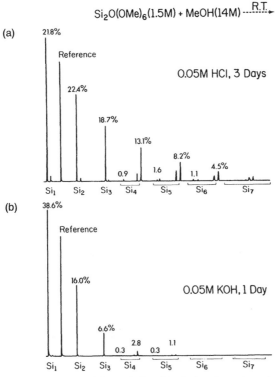

$Si_2O(OMe)_6$(1.5M) + MeOH(14M) $\xrightarrow{R.T.}$

(a) acidic conditions: 0.05M HCl, 3 Days

(b) basic conditions: 0.05M KOH, 1 Day

Capillary gas chromatographic analysis of hydrolyzed–condensed $Si_2(OMe)_6$ under (a) acidic and (b) basic conditions [93].

thermodynamic rather than kinetic factors are primarily responsible for silicate speciation, because different kinetic schemes are involved in each case. However, the hydrolytic stability of the trimer and cubic octamer shows that siloxane bond formation is not necessarily reversible, so that kinetic control is probable after larger oligomers appear. Thermodynamic versus kinetic control is further discussed in Section 2.6.5.

Balfe and Martinez [108] investigated the acid- and base-catalyzed hydrolysis of TMOS, and the dimer (15) and linear trimer (16) (H_2O/OCH_3 = 0.5–1.0). They observed cyclic tetramer in the condensation products of the monomer and dimer but not in the products of trimer polymerization. From this they concluded that hydrolytic decomposition of the trimer is kinetically unimportant under their reaction conditions. Similarly they observed a resonance attributable to cyclic trimer in the condensation products of the monomer or trimer but not in the products of dimer polymerization. Thus in contrast to the results of Klemperer *et al.* [38,39], Balfe and Martinez [108] observed no detectable evidence of siloxane bond hydrolysis during the acid- or base-catalyzed hydrolysis of the dimer or trimer. As we shall see, this has important implications with respect to kinetic versus thermodynamic control in silicate polymerization pathways.

Lin and Basil [86] investigated the acid-catalyzed hydrolysis of TEOS and the corresponding dimer and linear trimer for $r = 1$. They observed that the initial rates of hydrolysis of Q^0, Q^1_2, and Q^1_3 groups were similar, but the terminal ethoxide groups (Q^1_3) of the trimer hydrolyze more rapidly than the middle (Q^2_3) ethoxide groups (evidence for steric or inductive effects). Detectable amounts of all possible monomers and dimers except $Si(OH)_4$ and $Si_2O(OH)_6$ were observed, but no branched (Q^3) species were present during the first four hours and the unhydrolyzed dimer was not detectable during the initial stage of the reaction.

2.6.2.   IR AND RAMAN SPECTROSCOPIC INVESTIGATIONS

Infrared and Raman spectroscopy have been used both in combination with NMR to identify specific oligomeric species in solution and to follow the evolution of inorganic frameworks by comparisons with model compounds of known structures. Lippert *et al.* [107] and Mulder and Damen [114] have combined Raman spectroscopy and ^{29}Si NMR to investigate the hydrolysis and condensation of TMOS and TEOS, respectively. Lippert *et al.* [107] were able to assign eight of the Raman bands observed during the gelation of TMOS by positive correlations with the temporal behavior of ^{29}Si NMR resonances. (See Figs. 38 and 39 and Table 13.) The intermediates a–d represent Q^0 species and e–h represent Q^1 through Q^4 species, respectively.

Mulder and Damen [114] hydrolyzed TEOS with water under acidic conditions ($r = 1$). Raman and ^{29}Si NMR spectral comparisons of the non-OH substituted dimers, trimers, and tetramers isolated by fractional distillation allowed assignments of some of the Raman bands observed during the initial stages of the hydrolysis and condensation reactions. The principal symmetric SiO_4 stretch vibrations of TEOS and the non-OH substituted dimer, linear trimer, and linear tetramer are 654, 600, 576, and 545 cm^{-1}, respectively. No evidence of the cyclic trimer was observed, in contrast to the results of Balfe and Martinez [108] and Kelts and Armstrong [106].

Balfe *et al.* [115] combined Raman and Fourier transform infrared spectroscopy to determine the hydrolysis behavior of model linear and cyclosilicate compounds under neutral conditions. They observed that all linear as well as cyclotetra- and cyclopentaorganosiloxanes (which have little or no bond strain resulting from reduced Si–O–Si bond angles, ϕ, or

Fig. 38.

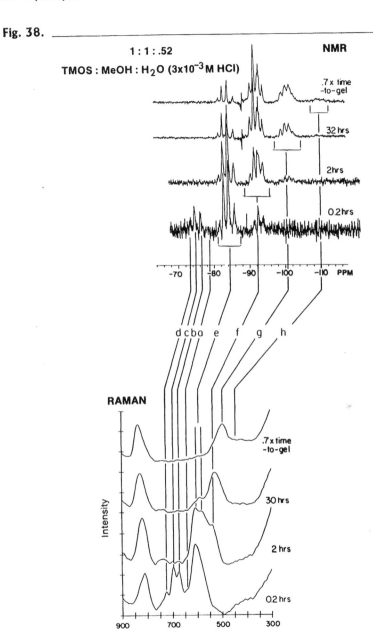

^{29}Si NMR and Raman spectra obtained at various times during the sol-gel reaction in a solution containing 1:1:0.24 (vol) TMOS, MeOH, and 3×10^{-3} M aqueous HCl (1:3.7:2.0 (mol)). The bands indicated by a–h are described in Table 13 [107].

Fig. 39.

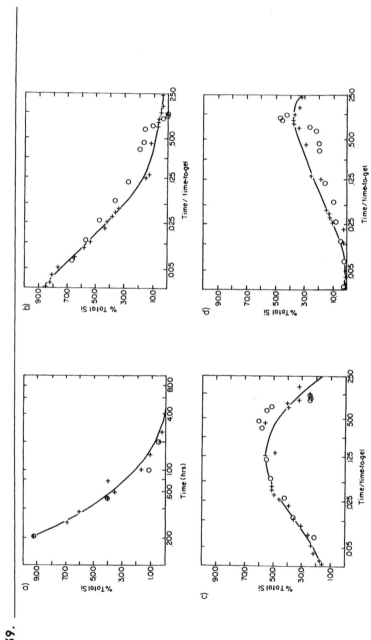

Time-dependence of several NMR and Raman bands in the sol-gel reaction of $1:3.7:2.0$ (mol) TMOS, MeOH, and 3×10^{-3} M aqueous HCl. NMR intensities (normalized to % total Si) are shown by (O), while Raman peak heights (adjusted to best fit the NMR data) are shown by (+). a) Comparison of intensities of bands labelled "c" in Fig. 38. b) Bands labelled "e", all NMR bands between -81 and -87 ppm were added. c) Bands labelled "f", all NMR bands between -90 and -96 ppm were added. d) Bands labelled "g", all NMR bands between -98 and -104 ppm were added [107].

Table 13.

Structural Assignments of Raman Bands Observed during
TMOS Hydrolysis.

a.	Si–(OCH$_3$)$_4$	(646 cm^{-1})	monomer
b.	Si–(OCH$_3$)$_3$ $\|$ OH	(673 cm^{-1})	$\Big\}$
c.	Si–(OCH$_3$)$_2$ $\|$ (OH)$_2$	(697 cm^{-1})	hydrolyzed
d.	Si–OCH$_3$ $\|$ (OH)$_3$	(725 cm^{-1})	
e.	$\equiv$Si–O–Si–(OH)$_n$–(OCH$_3$)$_{3-n}$	(609 and 589 cm^{-1}) end groups
f.	$\equiv$Si–O–Si–(OH)$_m$–(OCH$_3$)$_{2-m}$ $\|$ O–Si$\equiv$	(525 cm^{-1}) chains, rings
g.	$\equiv$Si–O–Si–OH $\|$ (O–Si$\equiv$)$_2$	(487 cm^{-1}) trisubstituted
h.	$\equiv$Si–O–Si–(O–Si$\equiv$)$_3$	(435 cm^{-1}) tetrasubstituted

Source: Lippert *et al.* [107].

elongated Si–O bonds) are relatively inert to siloxane bond hydrolysis under neutral conditions. Cyclotrisiloxane rings, which exhibit a characteristic symmetric Si–O breathing vibration at 585 cm^{-1} (Raman) and are predicted to be mildly strained ($\phi = 136°$ compared to 150° for fused silica [116]), hydrolyze at a rate of 3.8×10^{-3} min^{-1}, which is approximately 75 times greater than unstrained vitreous silica. Limited infrared data for the organo-cyclodisiloxanes, which are highly strained ($\phi = 91°$ [116]), suggest that they are hydrolyzed at least four times faster than the cyclotrisiloxanes [115]. These results suggest that cyclodisiloxanes and cyclotrisiloxanes are not important solution species in normal sol-gel processing conditions ($r > 2$, pH < 12). However, cyclotetrasiloxanes are quite stable, which is consistent with their prominence in numerous sol-gel silicate systems. (See, for example, Fig. 35.)

Jonas, Artaki, Zerda and co-workers [68,110,117] have combined Raman spectroscopy, ^{29}Si NMR, and the molybdic acid reagent technique to monitor the polymerization process of silicate systems as a function of solution pH (1–9) and solvent composition (methanol or methanol plus formamide). Raman and ^{29}Si NMR investigations were also performed at pressures ranging from ambient to 5 kbar. As in the previously cited examples, it was possible to correlate some of the Raman bands with specific ^{29}Si resonances. Table 14 lists the assignments of Raman bands observed in the 600–1050 cm^{-1} region of the spectrum [110].

Table 14.

Assignments of Raman Bands Observed between 600 and 1050 cm^{-1} During TMOS Hydrolysis.

ν, cm^{-1}	Assignment	
610	formamide	
645	Si(OCH$_3$)$_4$	Si–O–C
675	Si(OCH$_3$)$_3$OH	Si–O–C
695	Si(OCH$_3$)$_3$(OH)$_2$	Si–O–C
720	Si(OCH$_3$)(OH)$_3$	Si–O–C
795	silica dimers	Si–O–Si
830	silica network	Si–O–Si
1029	methanol	C–O

Source: Zerda *et al.* [117].

By assuming that the Raman intensity reflects the number of bonds in the scattering volume and that, for $r = 10$, the silicate species are uniform and grow spherically, Zerda *et al.* [71] related the Raman intensity of the 830-cm^{-1} band, assigned to the silica network,[†] to the particle radius, r:

$$I_{830\,\text{cm}^{-1}} \propto (8/3)\pi r^3 - 2\pi r^2. \qquad (50)$$

Using this relationship the growth of the silica network was monitored for times ranging from $t/t_{gel} = 0.8$ to >30. Quantitative determinations of particle sizes is impossible by this method [68]. Qualitative information was derived by using methanol as an internal standard after hydrolysis was determined to be complete. Figures 40 and 41 show the normalized intensities of the 830-cm^{-1} band versus t/t_{gel} for TMOS systems prepared under different pH conditions. It is apparent that for all values of pH, the normalized Raman intensity continues to increase long after the gel point. This is consistent with the idea that gelation occurs by a percolation process, in which case as little as 20% of the silicate is incorporated in the infinite spanning cluster at the gel point [118]. Continued condensation of oligomeric species with the spanning cluster accounts for the increased Raman intensities for $t/t_{gel} > 1$.

By assuming that the shape and structure (described, for example, by the fractal dimension) of the silicate species or network are similar under varying pH conditions, Zerda *et al.* concluded that neutral or basic conditions promote the formation of larger, more highly condensed particles prior to

[†]Note added in proof: Due to a problem with the spectrometer, the 830-cm^{-1} band was probably caused by light scattering rather than Raman scattering. However, since light scattering is also a measure of particle growth, the conclusions of this study are qualitatively correct.

Fig. 40.

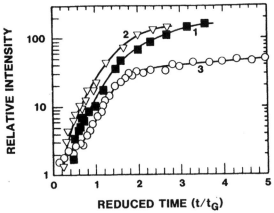

Normalized intensities of the 830-cm^{-1} Raman band of TMOS-derived SiO$_2$ gels of different pH values versus reduced time, t/t_{gel}. The gelation times are 820 h at pH = 1 (■); 1050 h at pH = 2 (▽); 150 h at pH = 3 (○). The relatively short gelation times for pH = 3 reflect the fact that pH = 3 is already above the isoelectric point [117].

Fig. 41.

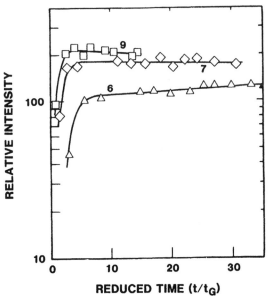

Normalized intensities of the 830-cm^{-1} Raman band of SiO$_2$ gels of different pH values versus reduced time. The gelation times are 8.5 h at pH = 6 (△); 11 h at pH = 7 (◇); 17 h at pH = 9 (□) [117].

gelation compared to acidic conditions [117]. Without detailed information concerning the fractal dimension of the solution species prior to gelation, this conclusion is merely speculative. However it appears justified to conclude that under neutral or basic conditions, a greater number of bridging oxygens are present per unit volume compared to the acid-catalyzed systems. Substitution of formamide for some of the methanol in a TMOS system ($r = 10$, no added catalyst) had an effect similar to the addition of base, i.e. it promoted the formation of more highly condensed silicates [69]. This might be explained by the hydrolysis of formamide to produce NH_3 plus formic acid, which Rosenberger *et al.* [72] observed to cause an increase in solution pH. (See Fig. 16.)

The results of the high-pressure Raman study (Fig. 42) show that pressure appreciably accelerates the condensation process. The gelation of a TMOS system ($r = 10$, pH $= 3.5$) was decreased from 168 hours at ambient pressure to 6 hours at 3.5 kbar. Compared to ambient pressure conditions, a significantly enhanced condensation rate was observed at 3.5 kbar for normalized times up to $t/t_{gel} \sim 2$.

Lin and Basil [86] combined Fourier transform infrared spectroscopy with size exclusion chromatography (FTIR/SEC) to follow the initial stages of

Fig. 42.

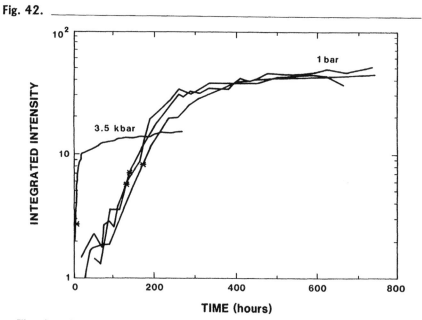

Time dependence of normalized intensities of the 830-cm^{-1} band of TMOS-derived SiO_2 gels at different pressures ($r = 10$; pH ~ 3.5). Asterices denote gel times [68].

the hydrolysis and condensation of TEOS ($r = 1$, with HNO_3 catalyst). SEC separates the various monomeric and oligomeric species according to their molecular weights. Commercial ethylsilicate 40 was used as a five-point calibration standard for the molecular-weight range 208–744. Figures 43 and 44 show an SEC chromatogram after 60 min of reaction and the corresponding SEC–FTIR spectra in which the solution components are separated according to molecular weights. The appearance of the 3675-cm^{-1} band, corresponding to silanol O–H stretching, clearly identifies the successive elution of the disiloxane (Q^1) and monomeric (Q^0) silanols at elution times of 6.3 and 7.3 minutes, respectively. The chromatogram shows that the addition of tetraethyltitanate (TET) results in complete elimination of all silanol species. This indicates that TET is a very effective siloxane-condensation catalyst. (Further discussion of this topic appears in Section 3.)

2.6.3 PHOTOPROBE INVESTIGATIONS

Photophysical or photochemical probe molecules, e.g., pyrene or 1,3-di(1-pyrenyl)propane, incorporated in a medium provide structural information related to the surroundings of the probe molecule on a length scale of one to several nanometers. Kaufman, Avnir, and coworkers [119,120] used photoprobes to investigate structural development during the sol-to-gel-to-xerogel transitions. A photophysical probe molecule such as pyrene (Py) exhibits changes in its emission spectrum in response to changes in its environment. The addition of photoprobes to a reacting sol-gel system therefore provides a means to monitor structural changes in situ, with the proviso

Fig. 43. _____

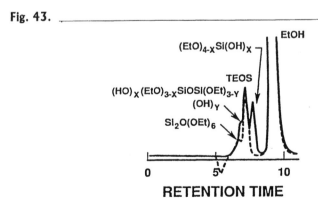

RETENTION TIME

SEC chromatogram of TEOS hydrolysis solution at 60 min. (——) and after adding TET tetraethyltitanate (---). TEOS : H_2O : HNO_3 : EtOH mole ratios $= 1:1:0.00056:4.5$ [86].

Fig. 44.

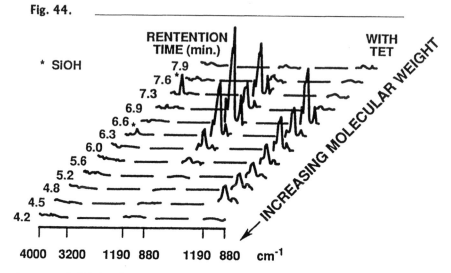

FTIR data for the 60-min TEOS hydrolysis solution of Fig. 43 [86].

that the added probe does not interfere with or otherwise alter the course of the hydrolysis and condensation reactions in its surroundings. Pyrene (Py):

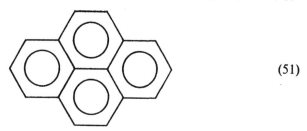

(51)

is a commonly used photophysical probe in surface science by virtue of three characteristics:

Its ability to form an excimer, i.e., a complex between an excited state Py* and a ground state Py. (The excimer, (PyPy*), has a characteristic emission different from the emission of Py*. See Fig. 45.)

Its monomer emission, which has a typical vibronic structure. (The ratio of peak 1 to peak 5 (I_1/I_5, Fig. 45A) increases with an increase in environmental polarity.)

Its singlet lifetime is long, in many cases exceeding 100 ns [120].

The following discussion relies on the excimer formation process.

Fig. 45.

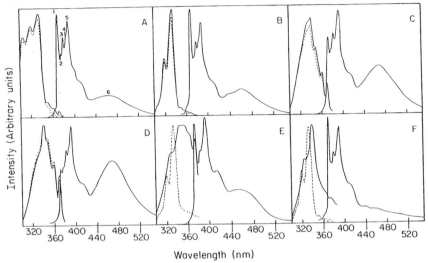

Emission (right) and excitation (left, λ_{ex} = 470 nm and λ_{ex} = 392 nm) spectra of pyrene (1×10^{-3} M) at six stages of the sol-gel process (neutral hydrolysis of TMOS). The figures show the following sequence in time: A) time "zero"; B) 48 hours; C) 80 hours; D) 100 hours, maximum excimer emission intensity; E) 172 hours; F) 288 hours, pyrene molecules are trapped and isolated [120].

TMOS or TEOS were hydrolyzed in methanol or ethanol, respectively, with no added catalyst. The concentration of pyrene was 1×10^{-3} M. The solutions were slowly dried to form gels and then xerogels. Figure 45 shows the Py emission and excitation spectra during the sol-to-gel-to-xerogel transition. The PyPy* excimer emission intensity (peak 6 in Fig. 45) is observed to change dramatically relative to the monomeric emission (peaks 1–5). I_1/I_5 is also observed to change. Figure 46 plots the excimer emission intensity and I_1/I_5 as a function of time.

The changes in excimer emission are believed to reflect changes in geometry, surface irregularity, and/or porosity. Both an increase in Py concentration and Py trapping in small pores increase the probability of Py + Py* → (PyPy*). Therefore, the increase in excimer emission is attributable to a gradual increase in the surface irregularity or number of small pores (which causes an effective increase in local Py concentration), on length scales up to several nm, during the sol-to-gel-to-xerogel transition. The excimer emission increases mainly after the gel point (which is not a special point on the length scales probed) and then decreases to practically zero. The decrease is interpreted as monomer trapping within small pores which precludes excimerization. The differences observed between TMOS and

Fig. 46.

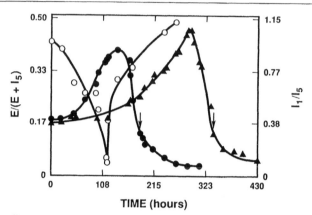

The relative fluorescence intensity of pyrene excimer (E/T + M) as a function of gelation time for silica gel glasses from ● TMOS and ▲ TEOS (no added catalyst). Arrows indicate isostructural points of the two gels. ○: the change of the $I_1 : I_5$ ratio as function of gelation time [120].

TEOS reflect the differences in kinetics through the gel point. However, the systems behave similarly during drying. Kaufman and Avnir [120] claim that for times greater or equal to those indicated by the arrows in Fig. 46, the TMOS and TEOS systems are isostructural due to complete hydrolysis.

The value of r has a more pronounced effect on the excimer emission than does pH. For low r values, inferred geometric irregularities increased gradually throughout the complete process, whereas for large r values the initial stage of growth resulted in smooth particles, both at low and high pH (low excimer emission), that subsequently aggregated increasing the excimer emission.

2.6.4.   SMALL-ANGLE–SCATTERING INVESTIGATIONS

Small-angle–scattering investigations utilizing neutrons (SANS), X-rays (SAXS), or visible light (static light scattering or quasi-elastic light scattering (QELS)) have been employed to investigate the growth and topology of macromolecular networks that precede gelation, the aggregation of colloids, and the structures of porous gels and aerogels. Martin and Hurd [121], Schaefer [122], and Schaefer and Keefer [34,123] have published excellent reviews of this topic. See also the discussion in Chapter 4.

In small-angle scattering, an incident light beam (light, neutrons, or X-rays) impinges on a sample and the angular dependence of the scattered intensity is measured. Figure 47 shows a schematic small-angle–scattering

Fig. 47.

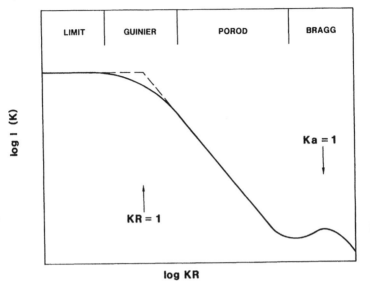

Schematic small-angle X-ray curves (log intensity versus log KR) from a dilute macromolecular solution [123].

curve from a dilute macromolecular solution in which the scattered intensity is plotted as a function of the Fourier spatial frequency, K:

$$K = (4\pi/\lambda) \sin(\theta/2) \tag{52}$$

where λ is the wavelength of the incident radiation. K relates a characteristic length scale, $l = 2\pi/K$, to the scattering angle, θ, through Bragg's law. By scanning θ, one effectively studies an object at different length scales. Using a combination of light, X-ray, and neutron scattering, it is possible to probe length scales from 0.1 nm to 1 micron [122].

Schaefer and Keefer [123] have divided the scattering curve into several regions depending on the set of lengths (K^{-1}, R, a) where R is the *radius of gyration*, a length related to the size of the polymer or colloid, and a is the bond length. (See Fig. 47.) Information is obtained from scattering experiments on length scales proportional to $1/K$.

At large scattering angles ($Ka \sim 1$) information is obtained concerning interatomic spacings. This is the *Bragg region*. In amorphous systems diffuse peaks may be observed from which radial distribution curves can be derived.

At low scattering angles ($KR \sim 1$) the scattered intensity is exponentially related to R [124]:

$$I(K) \sim e^{-K^2 R^2/3}. \tag{53}$$

This is the *Guinier region*. It provides information concerning polymer mass or radius, which is determined by the intercept and slope of a plot of log $I(K)$ versus K^2 [124].

At intermediate angles $(R \gg K^{-1} \gg a)$ the scattered intensity decays as a power law:

$$I(K) \sim K^{-X} \tag{54}$$

which does not depend on R or a. This is called the *Porod region* [125]. Porod showed that for systems with sharp boundaries, for example, dispersions of dense, colloidal particles, $X = 4$. Recently a more general expression for the exponent, $-X$, has been derived [122]:

$$-X = -2d_f + d_s = P \tag{55}$$

where P is called the *Porod slope*, d_f is the mass fractal dimension $(0 \le d_f \le 3)$, and d_s is the surface fractal dimension $(2 \le d_s \le 3)$. For uniform (nonfractal) objects $d_f = 3$, $d_s = 2$, and P reduces to -4. For mass fractal objects, $d_f = d_s$, and $P = -d_f$. In this case the mass fractal dimension is obtained directly from the slope. Mass fractal dimensions of various objects are shown in Table 15. For surface fractal objects, $d_f = 3$ and $P = d_s - 6$. Schmidt has shown, however, that for a polydisperse system of randomly oriented, independently scattering pores with a number distribution of pore diameters having the form of a power law, the scattered intensity also obeys a power law [126].[†] Therefore, without knowledge of the extent of polydispersity, it may not be possible to derive meaningful information concerning fractal dimensions from analyses of the Porod slope.

Table 15.

Porod Slopes for Various Structures.

Linear ideal polymer (random walk)	-2	
Linear swollen polymer (self-avoiding walk)	$-5/3$	
Randomly branched ideal polymer	$-16/7$	Mass fractals
Swollen branched polymer	-2	Slope $= -d$
Diffusion-limited aggregate	-2.5	
Multiparticle diffusion-limited aggregate	-1.8	
Percolation (single cluster)	-2.5	
Fractally rough surface	-3 to -4	Surface fractals
Smooth surface (non-fractal)	-4	Slope $= d_s - 6$

Source: Schaefer and Keefer [34].

[†] The same reasoning applies to a power law size distribution of particles.

The following subsections present structural information obtained for silicate systems from the Guinier and Porod regions of small-angle scattering. In the following section this information is rationalized on the basis of several kinetic growth processes pertinent to silicate polymerization.

2.6.4.1. Guinier Region

Figures 48a and b show growth information obtained from the Guinier region for TEOS-derived silicate systems ($r \sim 4$–5) prepared using a two-step acid- or two-step base-catalyzed synthesis, respectively [26,59]. The acid-catalyzed system prepared at pH ~ 1.0 (sample A2 in refs. 26 and 59) shows a strong concentration-dependence of the measured Guinier radius: dilutions of at least 10:1 with ethanol prior to the SAXS experiment result in an apparent increase in the value of the radius [26,123] (approximately a factor of 4). This indicates that in the more concentrated system the solution species are highly overlapped soon after condensation begins [26,123]. The measured radius therefore is an interchain correlation length, ζ. Dilution disentangles the polymers, causing ζ to approach the true radius of gyration, R.[†] The base-catalyzed system prepared at pH ~ 8 (sample B2 in refs. 26 and 59) shows no concentration-dependence of the Guinier radius.

Fig. 48a.

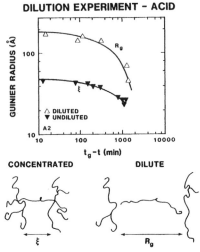

Time-and concentration-dependences of the Guinier radius for two-step acid-catalyzed hydrolysis of TEOS ($r = 5$). Dilution (10:1) disentangled the polymers causing an apparent increase in size [123].

[†] An alternative explanation is that the polymers are swelling upon dilution [121].

Fig. 48b.

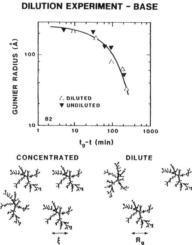

Time- and concentration-dependence of the Guinier radius for two-step acid- and base-catalyzed hydrolysis of TEOS. Dilution (10 : 1) had no effect on the apparent radius [123].

This indicates that the solution species are compact, more highly branched clusters that do not overlap [26,123].

Mandelbrot [127] considered the conditions required for structures to interpenetrate. He has shown that if two rigid structures of radius R are each placed independently of the other in the same region of space, the number of intersections, $M_{1,2}$, is expressed as:

$$M_{1,2} \propto R^{d_{f,1}+d_{f,2}-d} \tag{56}$$

where $d_{f,1}$ and $d_{f,2}$ are the respective fractal (or Euclidian) dimensions and d is the dimension of space. If $d = 3$ and each structure has a fractal dimension less than 1.5, $M_{1,2}$ decreases indefinitely with R, which would allow the structures to interpenetrate freely as their concentration was increased or, conversely, disentangle as their concentration was decreased. The structures are *mutually transparent*. However if $d_{f,1}$ and $d_{f,2}$ are greater than 1.5, the probability of intersection increases algebraically with R: the structures are *mutually opaque*. Although these concepts of mutual transparency or opacity are based on rigid structures that stick irreversibly on contact (not the case for silicate polymers), this idea may be used to rationalize the dilution results. The A2 system must be less highly branched than the B2 system (lower d_f) and/or exhibit a lower condensation rate. Decreased branching reduces the probability of intersection. A reduced condensation rate reduces the probability of sticking if an intersection were to occur.

2.6.4.2. Porod Region

Figure 49 shows power law information derived from the Porod region of scattering for a variety of silicate systems investigated by Schaefer *et al.* [35]. The alkoxide-derived silicates exhibit fractal behavior. Samples A and B, prepared by a two-step acid- or base-catalyzed process (identified as A2 and B2, respectively, in the preceding discussion of the Guinier region) are mass fractals with fractal dimensions, $d_f = 1.9$ and 2.0, respectively. Sample C prepared by one-step base-catalyzed hydrolysis of TEOS ($r = 1$) is a mass fractal ($d_f = 2.8$), whereas sample D prepared by one-step base-catalyzed hydrolysis of TEOS ($r = 2$) is a surface fractal, $d_s = P + 6 = 2.7$. LUDOX®, an aqueous silicate colloid, exhibits $P = -4$. These results illustrate that sol-gel–derived silicates are generally not dense, colloidal particles, and by variation of the processing conditions it is possible to generate a spectrum of silicate species that are structurally quite different on the ~1–20-nm–length scales.

Fig. 49.

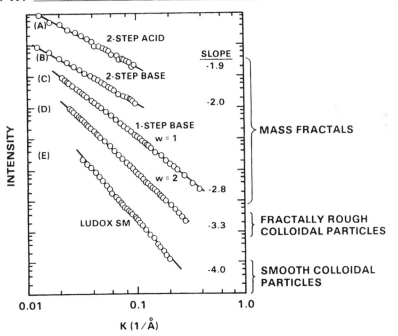

Porod plots of a variety of silicate solution species investigated by Schaefer *et al.* [35]. (A) two-step acid-catalyzed TEOS system. (B) two-step acid- and base-catalyzed TEOS system. (C) one-step base-catalyzed TEOS system ($r = 1$). (D) one-step base-catalyzed TEOS system ($r = 2$). (E) aqueous silicate system, LUDOX®.

Fig. 50.

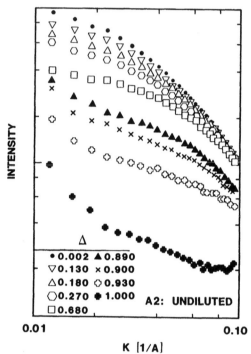

Δ

• 0.002	▲ 0.890
▽ 0.130	× 0.900
△ 0.180	◌ 0.930
◯ 0.270	✦ 1.000
▢ 0.680	

A2:   UNDILUTED

0.01   **0.10**

K [1/A]

Development of the scattering profile for two-step acid-catalyzed TEOS system, A2 ($r = 5$). Δ is the normalized time increment to gelation: $\Delta = (t_{gel} - t)/(t_{gel} - t_0)$ where t_0 is the initiation time and t is the observation time [123].

Figure 50 shows the Porod region of scattering for the A2 systems as a function of the normalized time from gelation [123]. The fractal dimension, which is obtained from the limiting value of the slopes at high K, appears to be unaffected as $t \rightarrow t_{gel}$. The gel point does not appear as a special point in the Porod region, because gelation depends on the critical connectivity on length scales that are very large compared to the ~1–10-nm–length scale probed by the SAXS experiment. Comparisons of Fig. 50 with schematic representations of the development of scattering curves during polymerization (Fig. 51), indicate that, during the latter stages, polymerization proceeds mainly by growth of a fixed number of solution species (scatterers).

Martin and coworkers [128,129] have studied the structure of silicate gels prepared from TMOS on length scales: 34 nm < $1/K$ < 440 nm arbitrarily close to the gel point by using dynamic and static light scattering on very dilute systems. These experiments examine the regime of growth associated with the connectivity transition on length scales that are large compared to

Fig. 51. _____

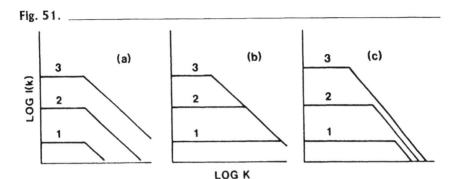

Development of the scattering curves during polymerization: (a) increasing number of scatterers without increase in size; (b) increasing size of a constant number of scatterers (big clusters growing at the expense of small); (c) simultaneously increasing size and number of scatterers [123].

the correlation length of the undiluted gel (~10 nm). Figure 52 shows Porod scattering for base-catalyzed TMOS gels ($r = 4$) as a function of dilution prior to analysis [128]. Two salient features of the scattering functions are:

At higher concentrations the intensity per unit concentration is smaller, which indicates significant overlap of the clusters.

Coupled with this lower intensity is a less negative slope.

Figure 53 shows an extrapolation of the dependence of the scattering exponents to zero concentration for acid- and base-catalyzed systems. The linear extrapolation yields a slope of 1.58 (average of three preparations) which is in close agreement with the percolation prediction of 1.59. Similar results were obtained for two-step acid–base-catalyzed systems.

Comparison of these results with the results of Schaefer *et al.* [34,35] (slope = 1.6 vs. 2.0) suggests that the factors that control structure on the short-length scale probed by SAXS differ from those that control structure on the larger-length scale examined by light scattering. According to Martin and Keefer [128], the short-range structure is determined by chemical effects far from the gel point, and there is no reason to expect chemically controlled growth to be described by percolation. The structure of silicate systems near the gel point is discussed further in Chapter 5.

2.6.5. GROWTH MODELS

The previous section has documented that under most conditions of sol-gel synthesis, the evolving structures on the 1–200-nm scale are not uniform

Fig. 52.

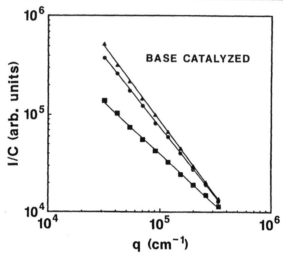

Elastic light-scattering measurements from the 1.0 M base-catalyzed sol show a strong dependence on concentration. The lowest curve is a 10:1 dilution, the intermediate curve is a 50:1 dilution, and the highest curve is a 250:1 dilution. q is the scattering vector [128].

Fig. 53.

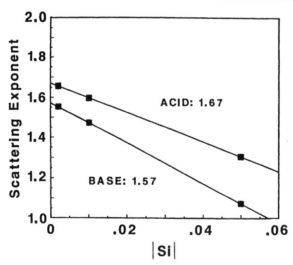

The elastic scattering exponents derived from the slopes in Fig. 52 are plotted versus concentration. Although the linear extrapolation is without theoretical justification, it fits the data quite well for all samples [128].

objects described by Euclidian geometry but tenuous structures appropriately described by a mass or surface fractal dimension. This section examines several equilibrium and kinetic models proposed to describe polymer growth in silicate systems as reviewed by Martin [130] and Schaefer [122]. We shall see that fractal structures generally emerge unless growth can occur predominately by reaction-limited addition of monomers or particles to a growing cluster (analogous to classical nucleation and growth).

2.6.5.1. Equilibrium Growth Models

The first theory that attempted to describe the divergences in cluster mass and average radius accompanying gelation is that of Flory [112] and Stockmayer [113]. In their model, bonds are formed at random between adjacent nodes on an infinite Cayley tree or Bethe lattice. (See Fig. 54.) The Flory–Stockmayer (FS) model is qualitatively successful because it correctly describes the emergence of an infinite cluster at some critical extent of reaction and provides good predictions of the gel point. In addition, for the Porod region of scattering the FS model predicts $I(K) \sim K^{-2}$ [130], which is in reasonable agreement with Porod slopes observed in acid-catalyzed systems (See, for example, Fig. 49.) Klemperer and Ramamurthi [93] have used the FS model to explain the "normal" molecular-weight distributions observed in acid-catalyzed TMOS systems (see Fig. 34) and the tendency for polysilicate esters, $Si_xO_y(OMe)_z$, to avoid dense, highly branched structures and prefer extended, unbranched structures. According to Klemperer and Ramamurthi, the preference for unbranched structures parallels the greater conformational entropy of unbranched isomers, where internal rotations generate far more distinct conformations than for branched isomers [93]. However, if conformational entropy were the primary factor in dictating structure, it is unlikely that cyclic species (see Figs. 35 and 36) would be so prevalent during acid-catalyzed polymer growth. The FS model has also

Fig. 54.

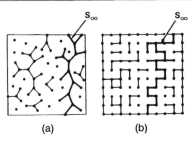

Gelation theories (schematic). (a) Flory's model ($f = 3$). (b) Percolation on a square lattice. In each case, a spanning s_∞ gel molecule is indicated embedded in the sol [131].

been criticized becaused of the unphysical nature of the Cayley tree. Because cyclic configurations are avoided, the purely branched clusters formed on the Cayley tree are predicted to have a fractal dimension of 4'($M \sim R^4$).

Because of the inherent problems with the FS approach, Stauffer [132] and de Gennes [133] advanced *bond percolation* as a description of polycondensation. In the percolation model, bonds are formed at random between adjacent nodes on a regular or random d-dimensional lattice. (See Fig. 54.) In this approach, cyclic clusters are allowed and excluded volume effects are directly accounted for [130]. As described in the previous section, Martin has shown that the percolation model accounts for the static structure of the polymer ensemble on the 25–400-nm–length scale near the gel point; however divergences of the average cluster mass and radius are stronger than predicted by the percolation model (and much stronger than predicted by the FS model). This topic is discussed further in Chapter 5, "Gelation."

2.6.5.2. Kinetic Growth Models

Most kinetic growth processes produce objects with *self-similar* fractal properties, i.e., they look self-similar under transformation of scale such as changing the magnification of a microscope [122]. According to a review by Meakin [134], the origin of this *dilational symmetry* may be traced to three key elements describing the growth process: 1) the reactants (either monomers or clusters), 2) their trajectories (Brownian or ballistic), and 3) the relative rates of reaction and transport (diffusion or reaction-limited conditions). The effects of these elements on structure are illustrated by the computer-simulated structures shown in the 3 × 2 matrix in Fig. 55.

2.6.5.2.1. Monomer–Cluster Growth

Depending on conditions, growth in silicate systems may occur predominately by the condensation of monomers with growing clusters (defined here as oligomers or polymers) or by condensation reactions of clusters with either monomers or other clusters. These two types of growth processes are identified as *monomer–cluster* and *cluster–cluster*, respectively in Fig. 55.

Monomer–cluster growth requires a continual source of monomers which, for either chemical or physical reasons, condense preferentially with clusters rather than each other. Examination of the ^{29}Si NMR spectra of TEOS-derived silicates (Fig. 33) shows that substantial concentrations of monomer are available even at $t/t_{gel} = 0.9$ for systems prepared at pH 5 and 7, whereas the system prepared at pH 1 is composed almost completely of Q^2-Q^4 species. (Monomers are consumed by $t/t_{gel} = 0.01$ [111].) ^{29}Si NMR investigations of more alkaline systems have shown a trend of maximization of Q^0 and Q^4 species at the expense of Q^3 and Q^2 species [30]. Thus under neutral and

Fig. 55.

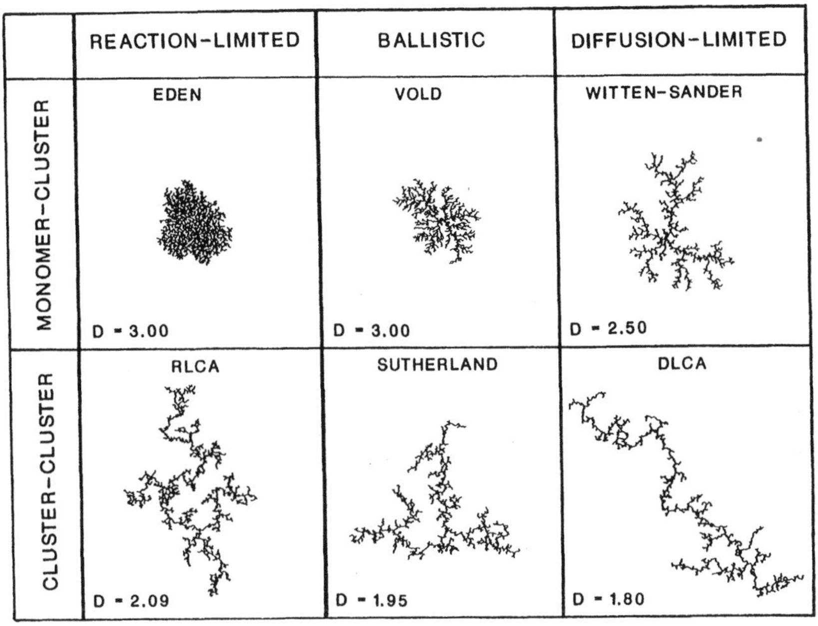

	REACTION-LIMITED	BALLISTIC	DIFFUSION-LIMITED
MONOMER-CLUSTER	EDEN D = 3.00	VOLD D = 3.00	WITTEN-SANDER D = 2.50
CLUSTER-CLUSTER	RLCA D = 2.09	SUTHERLAND D = 1.95	DLCA D = 1.80

Simulated structures resulting from various kinetic growth models [122]. Fractal dimensions are listed for 3-d clusters even though their 2-d analogs are displayed. Each cluster contains 1,000 primary particles. Simulations by Meakin [134].

basic conditions a source of monomers is available, which is a necessary condition for monomer–cluster growth.

There are several explanations for the large concentration of monomers present under neutral and basic conditions. From Fig. 21 we see that the rate of hydrolysis of siloxane bonds increases by over three orders of magnitude between pH 4 and 7. Because hydrolysis occurs preferentially at less highly condensed Q^1 sites [1], monomers are the primary by-product of siloxane bond hydrolysis. In a related study, Klemperer and Ramamurthi [93] have shown that siloxane bonds are broken by redistribution reactions under basic conditions (Eq. 42) that produce unhydrolyzed monomers as a by-product. In addition, from the pH-dependence of the hydrolysis reaction (Fig. 9), we see that the hydrolysis rate is minimized at neutral pH. Because the rate constant of the alcohol-producing condensation reaction is less than that of the water-producing reaction [63,95], unhydrolyzed or partially hydrolyzed monomers may persist in solution past the gel point. Presumably these combined factors contribute to the large concentrations of monomers observed under neutral and basic conditions.

As discussed in Section 2.4.4, above pH ∼ 2.0 the condensation reaction occurs preferentially between a nucleophilic deprotonated silanol species and a neutral species. The most highly condensed species are the most acidic and therefore the most likely to be deprotonated. Conversely the least highly condensed species, monomers, are the least likely to be deprotonated. Thus the base-catalyzed condensation mechanism (pH > 2.0) biases the growth process toward monomer–cluster growth.

Monomer–cluster aggregation can occur under diffusion-limited or reaction-limited conditions. *Diffusion-limited monomer–cluster aggregation* (DLMCA) is simulated by the Witten and Sander model [135]. (See Fig. 55.) In this model monomers are released one by one from sites arbitrarily far from a central cluster. The monomers travel by random walks and stick irreversibly at first contact with the growing cluster. Because of their Brownian trajectories, which simulate diffusion, monomers cannot penetrate deeply into a cluster without intercepting a cluster arm. The arms effectively screen the interior from the flux of incoming monomers; therefore growth occurs preferentially at exterior sites resulting in mass fractal objects whose density decreases radially from the center of mass (in three dimensions, $d_f = 2.45$).

Reaction-limited monomer–cluster growth (RLMCA) is distinguished from DLMCA in that there is a barrier to bond formation. The effect of this barrier is to reduce the "sticking" probability (condensation rate), so that many collisions between monomer and cluster are required to successfully form a bond [136]. In this process all potential growth sites are sampled by the monomers. The probability of attachment to a particular site per encounter is dictated by the local structure rather than the large-scale structure, which governs the probability that a monomer will encounter a given site [136]. Keefer simulated reaction-limited monomer–cluster growth by the Eden Model [136]. In this model originally developed to simulate the growth of cell colonies, unoccupied perimeter sites are selected randomly and occupied with equal probability. Because all sites are accessible and filled with equal probability, this model leads to compact, smooth clusters ($d_f = 3 =$ dimension of space).

In silicates, the condensation-rate constant is sufficiently small that growth is assumed to occur under reaction-limited conditions [136]. Thus neutral and basic conditions may be expected to result in compact smooth structures as a consequence of reaction-limited monomer–cluster growth. Keefer [136,137] used SAXS to study silicate growth in base-catalyzed systems prepared from TEOS ($r = 1$–4 and $NH_4OH = 0.01$ M). Porod scattering curves shown in Fig. 56 indicate that the structures change from mass fractals to progressively smoother surface fractals as r increases. Keefer developed a "poisoned" Eden growth model to simulate the observed behavior [136].

Fig. 56.

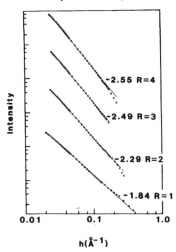

Small-angle scattering curves (slit smeared, P = slope $-$ 1) from partially hydrolyzed TEOS in alkaline solution. R is the molar ratio of water to TEOS. The slope for $R = 1$ is consistent with mass fractals with $d_f = 2.84$, and the slopes for $R = 2, 3$, and 4 are consistent with surface fractals with $d_s = 2.81, 2.51$, and 2.45, respectively [136,137].

In this model a certain fraction of sites (representing alkoxy groups) are prohibited from being occupied, i.e., in terms of silicate polymerization these sites do not undergo condensation. Depending on the number of poisoned sites and their distribution, the poisoned Eden growth model generates structures that vary from being uniformly porous (nonfractal) to surface fractals to mass fractals (see Fig. 57) consistent with the SAXS experiments.

The poisoned Eden model is relevant to silicate condensation under neutral conditions where the hydrolysis rate is minimized. Ether-forming (ROR) condensation is forbidden, and the rate of the alcohol-forming condensation reaction is lower than that of the water-forming condensation reaction [6,95]. Therefore, unhydrolyzed alkoxide sites, which are most abundant at neutral pH, act as "poisons" by inhibiting condensation (at least temporarily). Keefer argues that once unhydrolyzed sites are incorporated in the cluster, their hydrolysis may be impeded due to steric factors [136]. If so, the poisoned Eden model would be physically and chemically relevant.

Increasing the solution pH above 7 and/or increasing the value of r causes the hydrolysis reaction to be more complete for all condensed species (Q^1-Q^3). Thus all cluster sites are reactive, and (because a source of

Fig. 57.

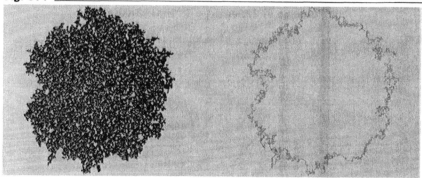

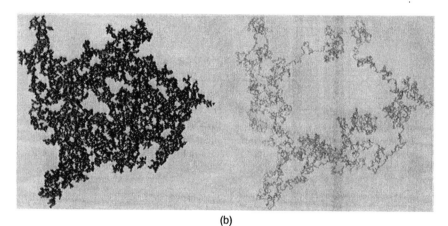

(b)

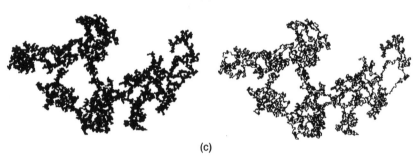

(c)

(a) A cluster of mass 25,600 generated by Eden growth from a $1:1:1$ mixture of 4, 3, and 2 functional monomers and its perimeter. The cluster is porous but is neither a mass nor a surface fractal. (b) A cluster of mass 25,600 generated from a $3:3:3:1$ mixture of 4, 3, 2, and 1 functional monomers. This is a surface fractal with $d_s = 1.3$ and $d_f = 2$. (c) A cluster of mass 10,000 generated for a $1:3:3:1$ mixture of 4, 3, 2, and 1 functional monomers. This is a mass fractal with $d_f = 1.82$. Note that almost every site is on the perimeter [136].

monomers is available) the Eden model should pertain. Porod slopes of -4, indicative of smooth uniform particles, are observed for LUDOX®, a commercial particulate silicate synthesized at high pH. (See E in Fig. 49.) According to Iler [l], LUDOX® is formed by classical nucleation and growth. At a critical degree of supersaturation, nuclei form and subsequently grow by the addition of monomer. The monomer (monosilicic acid, $Si(OH)_4$, in aqueous systems) is initially supplied by the supersaturated solution. As the degree of supersaturation diminishes, oligomers that have been in solubility equilibrium with the higher monomer concentration depolymerize to form monomers. Because monomeric species are less acidic than more highly condensed species, further condensation (according to Eq. 37) occurs preferentially between monomers and clusters rather than between monomers.

Smooth monosized SiO_2 spheres may be formed from TEOS by the Stöber process ($r = 7.5$ to >50 and $[NH_3] = 1$–7 M, see Table 1) [27]. However TEM investigations [138] reveal that growth does not occur simply by the addition of monomers to a finite number of clusters (classic nucleation and growth) as required by the Eden model. Bogush and Zukoski have used TEM to examine the temporal evolution of silicate species produced by the Stöber process [138]. The sequence of micrographs shown in Fig. 58 illustrates that small silica particles emerge at early times and that these particles are present throughout the growth stage.[†] Bogush and Zukoski concluded that there is obviously not a single nucleation event. According to classical nucleation theory, nucleation is predicted to stop when the soluble silica concentration drops below a critical supersaturation value. By analysis of the number density of particles, their size, and the molar volume of silica, Bogush and Zukoski [139] determined that the critical supersaturation value was exceeded during the complete course of the process and that nucleation proceeds for a substantial fraction of the reaction period.

To account for the observed behavior, Bogush and Zukoski [139] proposed a nucleation and aggregation model. Although negatively charged, the initial primary particles (≤ 10 nm) are unstable due to their small size, causing aggregation. The colloidally stable aggregates then sweep through the suspension picking up freshly formed primary particles and smaller aggregates. Monodispersity of the final precipitate is achieved through size-dependent aggregation rates.

Since this model does not take into account the solubility of positively curved silicate surfaces at high pH (see Fig. 1 in Chapter 4), an alternate explanation of the observed behavior consistent with particle growth in aqueous conditions (see Section 1 in this chapter) is that the initial aggregates

[†] It is presently unclear whether any of the structure observed in Fig. 58 is an artifact of drying the sol prior to the TEM investigations. J.K. Bailey and M.L. Mecartney are employing cryo-TEM to address this question.

Fig. 58. _____

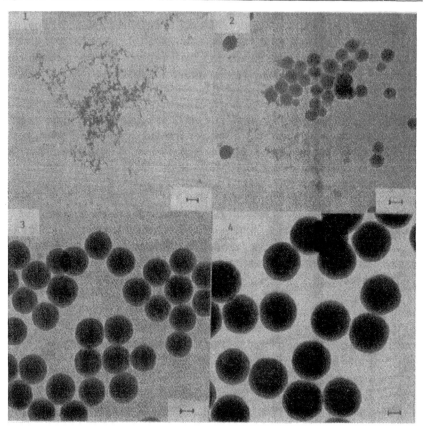

Silica particles precipitated from solution of 0.17 M tetraethyl orthosilicate, 1.3 M ammonia, and 2.0 M H$_2$O in ethanol at 25°C. Grids taken 2 min, 8 min, 30 min, and 20 hrs after initiation of reaction for Figs. 1–4, respectively. Aggregation of the small particles probably occurs as they are concentrated on the grid by drying. Bar = 100 nm [138].

restructure through dissolution-reprecipitation to form larger, more stable particles, thereby consuming the small primary particles (*Ostwald ripening*). However, at room temperature, the difference in solubility due to curvature effects practically vanishes when the particles exceed about 5 nm in diameter, a size smaller than even the primary particles observed by Bogush and Zukoski. Thus it is unlikely that Ostwald ripening could account for the relatively rapid growth (<1 day) of 45–250 nm particles. In support of the aggregation model there are many other examples of growth occurring by nucleation and controlled aggregation of primary particles to form monodisperse particles as discussed in Chapters 2 and 4. The aggregate structure apparently causes a collective reduction in surface area of the particles comprising it. The concomitant reduction in surface energy is the driving force for the ordered aggregation.

Growth of monosized silica spheres by the Stöber process therefore represents reaction-limited monomer–cluster growth on two length scales. The smaller primary particles form by nucleation and growth and ripening as in aqueous silicate systems. Repeated dissolution-reprecipitation insures that reaction-limited conditions exist and that a source of monomers is available. Because the value of r can exceed 25, it is likely that the growing primary particles are fully hydrolyzed. Thus the Eden model should pertain. The larger monosized aggregates grow in effect by reaction-limited monomer–cluster aggregation where the "monomers" are the primary particles. Reaction-limited conditions must pertain since particles add in an orderly fashion maintaining a minimum liquid-aggregate surface area. This certainly requires that the primary particles sample many potential growth sites before reacting at the most energetically favored ones.

In situ SAXS investigations of particles formed by the Stöber process yield Porod slopes ranging from -3 to -4 [140] indicative of more compact structures: either surface fractals or uniform (nonfractal) species. Solvent evaporation results in Porod slopes of -4 [141], which indicates that surface roughness, if any, is collapsed by the capillary pressure during drying.

In summary, reaction-limited monomer–cluster growth models, either the Eden or poisoned Eden model, appear to adequately described silicate growth under neutral or basic conditions. Of course it is unlikely that growth occurs exclusively by the addition of monomers to growing clusters, so that structural deviations from the predictions of the models may be anticipated. Under neutral and basic conditions, important sources of monomers are the by-products of depolymerization reactions: either siloxane bond hydrolysis and alcoholysis reactions or redistribution reactions. Siloxane bond hydrolysis insures the realization of reaction-limited monomer–cluster growth conditions for several reasons. First, it provides for the continued supply of monomers. Second, repeated depolymerization–repolymerization in essence insures that reaction-limited conditions are achieved: depoly-merization occurs preferentially at less stable sites (e.g., Q^1 sites), so repeated depolymerization–repolymerization forms stable configurations at the expense of unstable ones. Individual monomers have the ability to sample many potential growth sites, finally "sticking" at the most favored ones rather than the first encountered sites [136]. Depolymerization is also esssential to Ostwald ripening in which many smaller particles are reorganized into fewer larger particles with a corresponding reduction in the surface energy [1].

2.6.5.2.2.   Cluster–Cluster Growth Models

A second class of growth models is cluster–cluster aggregation [142]. (See Fig. 55.) These models describe growth that results when a "sea" of monomers undergoes random walks, forming a collection of clusters that continue to grow by condensing with each other and with remaining monomers. Under diffusion-limited conditions (DLCA) clusters stick irreversibly on first contact, whereas under reaction-limited conditions (RLCA) the sticking probability is less than unity. Compared to monomer–cluster aggregation, the strong mutual screening of colliding clusters creates very open fractal structures even under reaction-limited conditions [121]. The fractal dimension for DLCA, 1.80 [142], is in fact only slightly less than for RLCA [134], 2.09. As indicated by the simulated structures in Fig. 55, in contrast to monomer–cluster growth, cluster–cluster growth produces objects with no obvious centers.

Cluster–cluster growth is expected to pertain when there is not a continuous source of monomers and when there is no mechanism favoring condensation predominantly between low- and high–molecular-weight species. Re-examination of the ^{29}Si NMR spectra shown in Fig. 33 shows that these conditions may exist under acidic conditions. At pH 1 the solution is depleted of monomers at $t/t_{gel} = 0.9$, yet the concentration of Q^4 species is much less than under neutral conditions. In fact the monomers are essentially depleted for $t/t_{gel} \geq 0.01$ [111]. Under strongly acidic conditions the hydrolysis reaction is complete long before the gel point (see Fig. 29), which indicates that the rate of condensation is low with respect to the rate of hydrolysis. Below about pH 3, the depolymerization reaction is minimized. Below about pH 2, the isoelectric point of silica, condensation occurs by a mechanism involving a protonated silanol or alkoxide species.

These combined factors suggest that for large values of r, hydrolysis is complete at an early stage of the reaction. Since all species are hydrolyzed, they can all condense to form dimers and other low–molecular-weight oligomeric species, rapidly depleting the solution of monomers (e.g., at $t/t_{gel} \leq 0.01$ [111]). Condensation involves the most basic silicate species, i.e., monomers (Q^0) or end groups on chains (Q^1). When monomers are depleted, condensation occurs preferentially between Q^1 species and more acidic Q^2 and Q^3 species (chain ends react with chain middles) leading to open, randomly branched structures. Because the rate of siloxane bond hydrolysis is low and redistribution reactions are suppressed, there are no monomers available to fill in voids after the original source of monomers is depleted. Thus, under acidic conditions, reaction-limited cluster–cluster (RLCA) growth predominates at an early stage of the condensation process resulting in very open fractal structures ($d = 2.09$).

RLCA is also assured by two-step processes (see Table 2) in which the first step involves acid-catalyzed hydrolysis with understoichiometric additions of water ($r < 2$). Under these conditions, Assink and Kay [97] have shown that condensation commences before hydrolysis is complete. All the water is rapidly consumed to produce a distribution of partially hydrolyzed monomers which subsequently condense to form low–molecular-weight oligomeric species. (See Fig. 29.) Growth must cease when all available water and silanols are reacted. Addition of water plus catalyst (acid or base) in a second step ($r \geq 4$) causes all remaining alkoxide sites to hydrolyze, rendering all sites approximately equally reactive [34,122]. Subsequent condensation is forced to occur primarily between oligomeric species (cluster–cluster growth). Porod slopes obtained for two-step acid- and two-step base-catalyzed silicates, -1.9 and -2.1, respectively [34,35], are consistent with the RLCA prediction of -2.09. (See Figs. 49 and 55.) This indicates that for the two-step base conditions employed ($r = 3.7$ and pH ~ 8) restructuring via dissolution-reprecipitation or redistribution is not sufficiently extensive prior to gelation to achieve primarily monomer–cluster growth conditions. However the more negative Porod slope, -2.1, obtained for two-step base conditions is consistent with more compact structures that do not swell upon dilution compared to two-step acid conditions ($P = -1.9$).

With regard to structural evolution, perhaps the most important feature of the acid-catalyzed growth mechanism is that for most oligomeric species condensation is virtually irreversible. Tenuous, fractal structures are therefore kinetically stabilized, because there is no mechanism for restructuring and there is little monomer available to fill in voids. Unlike basic conditions in which dissolution-reprecipitation and the condensation process (monomer–cluster growth) naturally result in an "inverted" Q distribution (primarily Q^0 and Q^4 species), acidic conditions produce a distribution of Q^1–Q^4 species as expected for classic polycondensation of tetrafunctional monomers. (See Fig. 33.) The importance of irreversible condensation to the evolving silicate structures was predicted by Iler [1].

2.6.6. RHEOLOGICAL INVESTIGATIONS

Although rheological measurements characterize the bulk properties of a solution (for example, the viscosity or storage modulus), the dependence of rheological properties on concentration, molecular weight, or shear rate can be used to infer structural information on rather short length scales. Qualitative rheological investigations have been performed on numerous sol-gel silicate systems. For example, the gel point is often identified by examination to be the time at which the solution loses fluidity, i.e., no flow

Fig. 59.

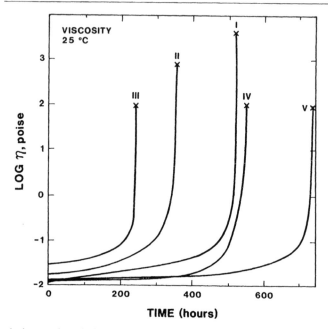

Temporal changes in solution viscosity for acid- or base-catalyzed TEOS systems. Crosses indicate gel points [31]. Samples I–V are identified in Table 17.

of the gel is observed when its container is tipped [143]. Numerous quantitative viscosity measurements have also been performed during the course of the sol-to-gel transition [e.g., 31,144]. In these investigations, the gel point is often defined as the time at which the viscosity is observed to increase abruptly [31]. (See Fig. 59.) Viscosity versus time measurements are used to identify the aging conditions required to draw fibers or thin sheets directly from solution. For example, Sakka and coworkers [28,31] reported that fiber formation was possible when the viscosity was ~1–100 Pa-s [29], and sheet formation was possible when the viscosity was ~1–10 Pa-s. However according to Sacks and Sheu [145], most of the measured viscosities were not absolute, because narrow-gap viscometers were not used and, hence, shear rates were poorly defined.

Sacks and Sheu [145] prepared acid- and base-catalyzed silicate systems in which the molar ratios of the components were varied as indicated in Table 16. Acid-catalyzed, low-water systems (compositions 1 and 2) are suitable for drawing fibers; composition 3 ($r = 20$) is suitable for bulk gels, and the dilute, base-catalyzed system, composition 4 ($r = 20$), results in the formation of particles. Shear stress and viscosity versus shear rate measurements

Table 16.

Compositions Investigated by Sacks and Sheu [145].

Solution	Water/TEOS Molar Ratio, r	Ethanol/TEOS Molar Ratio	Catalyst/TEOS Molar Ratio	Aging Temperature
1	2	5.8	0.1 HNO_3	25°C
2	1.5	1.91	0.0027 HNO_3	25°C
3	20	0.3	0.01 HNO_3	25°C
4	20	3.82	0.01 NH_4OH	50°C

for composition 1 show Newtonian flow behavior (shear-rate–independent viscosity) for aging times up to 1176 h. This is followed by a period of shear-thinning behavior (viscosity decreases with increasing shear rate). Further aging results in yield behavior and hysteresis in the shear stress versus shear rate curve indicating thixotropic flow behavior.

These changes are similar to those observed in marginally stable particle-liquid systems as the volume fraction of particles is increased [145]. At low particle concentrations the viscosity is rather unaffected by particle–particle interactions and Newtonian behavior is observed. Aging leads to aggregated structures, causing the viscosity to increase due to liquid immobilized within the aggregate, which in effect increases the apparent solids loading. As the shear rate is increased, these tenuous aggregates break down, releasing immobilized liquid and thus reducing the viscosity. This corresponds to shear-thinning behavior. Further aging leads to extensive network formation imparting elastic character to the system as indicated by the yield stress in the shear stress versus shear rate curve. After the yield stress is exceeded shear-thinning behavior and hysteresis are observed (thixotropic flow behavior).

Spinnability (the ability to draw fibers from the solution) was observed for compositions 1 and 2 but not for compositions 3 and 4 in Table 16 [145]. The best spinnability was observed when the viscosity behavior was highly shear thinning but not thixotropic. Shear-thinning behavior however is not sufficient for spinnability. All systems exhibited transformations from Newtonian to shear-thinning to thixotropic. Compositions 1 and 2 are distinguished from compositions 3 and 4 in that their viscosities were much higher when the transformation to shear-thinning behavior occurred. Apparently a very high viscosity is necessary to prevent the drawn fiber from breaking up into droplets. High viscosities required for stable fiber formation are generally achieved by concentration of the sol through solvent evaporation [31].

Sakka and coworkers [28,29,31,146] and Kamiya et al. [147] have investigated the rheology of silicate systems prepared from TEOS (see Table 17) in conjunction with fiber formation (spinnability). Acid- and

Table 17.

Compositions and Behavior of Sol-Gel Systems Investigated by Sakka *et al.* [28,29,31].

Solution	Si(OC$_2$H$_5$)$_4$ (g)	H$_2$O (g)	C$_2$H$_5$OH (g)	Mole Ratio (r) of H$_2$O to Si(OC$_2$H$_5$)$_4$	Catalyst[a]	Time for Gelling (h)	Spinnability
I	169.5	14.7	239.7	1	HCl	525	yes
II	382.0	33.0	83.4	1	HCl	360	yes
III	169.5	292.8	37.5	20	HCl	248	no
IV	50	3.8	47.6	1	NH$_4$OH	565	no
V	50	7.6	47.6	2	NH$_4$OH	742	no

[a] Mole ratio of HCl or NH$_4$OH to the Si(OC$_2$H$_5$)$_4$ is 0.01.

base-catalyzed systems prepared with r values ranging from 1 to 20 were evaluated by determining the concentration-dependence of the reduced viscosity, η_{sp}/C, and the molecular-weight–dependence of the intrinsic viscosity, $[\eta]$. Figure 60 compares the concentration dependence of η_{sp}/C for composition I (acid-catalyzed, $r = 1$, see Table 17) to that of LUDOX® (spherical silicate colloids) and sodium metasilicate (chain-like silicates) after various periods of aging in open containers. The reduced viscosity of a solution of noninteracting spherical particles (e.g., LUDOX®) is independent of concentration, C [148]:

$$\eta_{sp}/C = k/\rho \qquad (57)$$

where k is a constant and ρ is the density of the particles. Therefore the silicate species present at $t/t_{gel} = 0.34$ (see Fig. 60) are inferred to be compact and noninteracting. This is consistent with the Newtonian behavior observed by Sacks and Sheu [145] at early stages of aging.

Further aging causes a progressively larger dependence of η_{sp} on C. According to the Huggins equation, chainlike or linear polymers (e.g., metasilicate) show a concentration-dependence of the reduced viscosity [149]:

$$\eta_{sp}/C = [\eta] + k[\eta]^2 C \qquad (58)$$

where $[\eta]$ is the intrinsic viscosity and k is a proportionality constant. Therefore the progressively larger dependence of η_{sp}/C on C with aging time (Fig. 60) may be explained as arising from a gradual progression of the silicate structure from small noninteracting species to extended, weakly branched polymers. This corresponds to the shear thinning region observed by Sacks and Sheu [145].

Figure 61 plots log$[\eta]$ versus log number–averaged molecular weight, M_n, for acid-catalyzed silicate systems in which r was varied from 1 to 20.

Fig. 60.

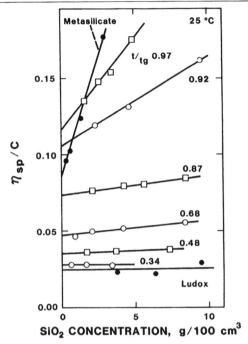

Concentration-dependence of the reduced viscosity, η_{sp}/C, of solution I (acid-catalyzed TEOS, $r = 1$, see Table 16) as a function of t/t_{gel}. Data obtained for LUDOX® and sodium metasilicate are shown for comparison [31].

For organic polymer solutions it is known that $[\eta]$ is related to M_n according to [32]:

$$[\eta] = kM_n^{\alpha} \tag{59}$$

where k is a constant that depends on the kind of polymer, solvent, and temperature, while α depends on the polymer structure: $\alpha = 0$ for rigid spherical particles; $\alpha = 0.5$–1.0 for flexible, chainlike, or linear polymers; and $\alpha = 1.0$–2.0 for rigid, rodlike polymers [32]. For example, for high molecular-weight polymethylsiloxanes, $\alpha = 0.5$ for linear polymers, $\alpha = 0.21$–0.28 for branched polymers, and $\alpha = 0.3$ for spherical particles [149]. From this relationship the results presented in Fig. 61 indicate that the spinnable systems ($r = 1$ or 2) are composed of flexible chainlike or linear polymers ($\alpha = 0.64$–0.75), whereas the nonspinnable systems are composed of more highly branched structures. Based on fractal geometry Eq. 59 may be re-expressed as follows [122]:

$$\eta \sim R_g^3/M_n \sim M_n^{(3/d_f)-1} \tag{60}$$

Fig. 61.

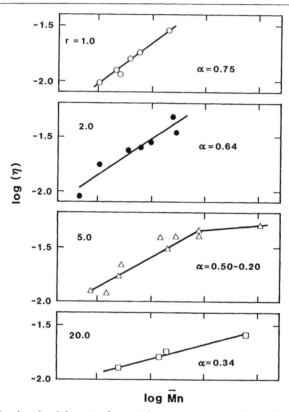

Log intrinsic viscosity $[\eta]$ versus log number average molecular weight, M_n, for trimethylsilylated silicates prepared by acid-catalyzed hydrolysis of TEOS and $r = 1$–20 [28]. α is defined in Eq. 59.

where R_g is the Guinier radius and d_f is the mass fractal dimension. According to Eq. 60, α in Eq. 59 is equivalent to $(3/d_f - 1)$. Therefore α values of 0.64–0.75 observed for spinnable systems correspond to mass fractal dimensions, $d_f = 1.83$ to 1.71, consistent with structures ranging from linear swollen polymers to swollen branched polymers (Table 15). It is not possible to distinguish between these various structures on the basis of d_f alone.

Comparing the results of Sakka and coworkers [28,29,31,146] to those of Sacks and Sheu [145], we see that a consistent set of requirements for spinnability are high viscosity without premature gelation, acid catalyst, and low values of r (≤ 2). High viscosity kinetically stabilizes the fibers from spheridization. According to Eqs. 58–60, for any particular concentration and extent of condensation, weakly branched "extended" structures, e.g., chains or rods ($d_f \ll 3$), are more efficient than compact structures, e.g., uniform particles ($d_f = 3$), in increasing the viscosity. Overlapped extended structures interact at lower concentrations which explains the concentration- and molecular weight–dependence of the reduced and intrinsic viscosities, respectively. As we have discussed in Section 2.6.5, acid-catalyzed conditions lead to the formation of weakly branched, extended structures that exhibit a strong concentration dependence of R_G, suggesting significant overlap prior to gelation. The combination of acid-catalyzed condensation and low values of r allows the concentration (and therefore viscosity) to be greatly increased by solvent evaporation without premature gelation, both because the condensation rate is low under acidic conditions and because more ethoxide groups are retained on the siloxane backbone, further reducing the condensation rate. Further solvent evaporation that accompanies fiber drawing presumably causes a sufficient increase in viscosity to stabilize the drawn fiber due to the strong concentration-dependence of viscosity.

In order to determine if linear or rigid rod polymers are necesssary for spinnability, Sakka *et al.* investigated the hydrolysis and condensation of difunctional, dimethyldiethoxysilane and trifunctional, methyltriethoxy-silane [150]. These molecules are potential precursors for linear and ladder polymers, respectively. It was observed that the hydrolysis of dimethyl-diethoxysilane resulted preferentially in the formation of the cyclic tetramer leading to immiscibility. Based on this observation it was suggested that rigid-rod (e.g., ladder) polymers are necessary for fiber formation, because flexible, linear polymers would tend to cyclize [150] leading to compact structures. Ladder polymers have also been associated with the inflection in the viscosity-versus-time behavior observed in spinnable systems hydrolyzed with the theoretical quantity of water required to form ladder polymers, $r \cong 1.7$. (See Fig. 62.)

Several factors argue against the existence of rigid-rod ladder polymers in spinnable systems, however. First, according to Eqs. 59 and 60, rigid-rod polymers should have a mass fractal dimension, $d_f = 1$, resulting in an α value of 2 rather than 0.75 or 0.64 as observed (Fig. 61). Second, solutions containing rigid-rod polymers should exhibit Porod slopes of -1. Preliminary SAXS experiments show that the spinnable systems are highly polydisperse [151]: no power law region is observed, which implies that if rigid-rod polymers are present, they do not constitute the predominant solution species. Finally, it is unlikely that random-growth of tetrafunctional precursors results in such highly ordered species as ladder polymers.

The general synthetic approach to the formation of ladder polymers is acid-catalyzed hydrolysis of trifunctional silanes (e.g., trichlorophenylsilane)

Fig. 62.

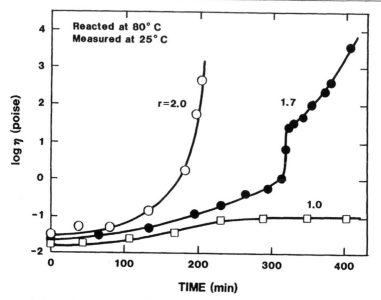

Temporal change in viscosity of acid-catalyzed TEOS systems exposed to the atmosphere at 80°C [28].

to yield incompletely condensed oligomers [152]. Higher–molecular-weight ladder polymers:

$$(61)$$

are then obtained by equilibrating the partially condensed oligomers with alkali rearrangement catalysts, e.g., KOH, at temperatures between 110 and 150°C. It is reported that alkali-catalyzed condensation and equilibration of the hydrolysis products of trifunctional silanes usually occur entirely randomly except in the case where the organic substituents are phenyl groups.

If ladder polymers were the primary product of the tetrafunctional silicate systems that exhibit spinnability, ^{29}Si NMR investigations would reveal

Fig. 63.

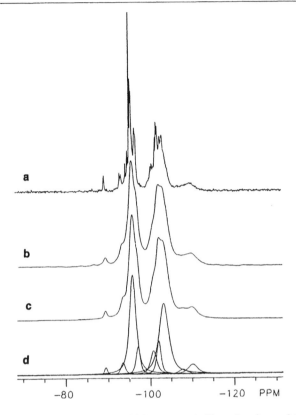

^{29}Si NMR spectrum of spinnable TEOS system (acid-catalyzed $r = 1.5$) [153]. (a) Experimental, 1.0 Hz exponential line broadening; (b) experimental, 30 Hz exponential line broadening; (c) computer simulation; (d) resonance components of computer simulation.

mainly Q^3 species (double chain ladder polymers) or Q^3 and Q^4 species in molar ratio $2:1$ (triple chain ladder polymers). A ^{29}Si NMR spectrum of a spinnable silicate solution prepared from TEOS ($r = 1.5$) is shown in Fig. 63 [153]. Q^1–Q^4 species are evident, with approximately equal concentrations of Q^2 and Q^3 species. Clearly if ladder polymers are present, they do not comprise the predominant solution species. Analysis of the data according to a statistical model (Table 18) [63,95] shows that the species distribution can be accounted for by a random (statistical) growth model modified to take into account steric and inductive effects [100]. Thus it is more likely that the species distribution shown in Fig. 63 is the product of a random rather than an ordered growth process.

Table 18.

Observed and Calculated Q Distribution of an Acid-Catalyzed Silicate Sol Prepared with $r = 1.5$.

		Theoretical Distribution		
Speciation	Experimental Distribution	$H_2O = 1.5$	$H_2O = 1.31$	$H_2O = 1.31$ $R = 0.35$
Q^0	0.0	0.4	1.4	0.0
Q^1	1.0	4.7	10.8	0.8
Q^2	41.9	21.1	30.6	41.5
Q^3	50.3	42.2	38.8	52.5
Q^4	6.8	31.6	18.4	5.2
Standard deviation		7.5	4.9	0.6

Note: Calculated values are derived from the statistical reaction model [63,95] modified by a multiplicative factor, R, to account for steric and inductive effects [153].

2.6.7. STRUCTURAL SUMMARY

To summarize trends in silicate growth, it is informative to consider the pH-dependences of the hydrolysis, condensation, and depolymerization (dissolution) rates shown schematically in Fig. 64 for an arbitrary value of r.[†]

2.6.7.1. Low-pH Conditions

Below ca pH 2, both hydrolysis and condensation occur by bimolecular nucleophilic displacement reactions (either S_N2–Si, S_N2^*–Si, or S_N2^{**}–Si) involving protonated alkoxide groups. Under these conditions, the rate of hydrolysis is large compared to the rate of condensation. For r values greater than about 4, we expect that hydrolysis will be essentially complete at an early stage of the reaction. After monomers are depleted (e.g., $t/t_{gel} \geq 0.01$), condensation between completely hydrolyzed species occurs by reaction-limited cluster–cluster aggregation leading to weakly-branched structures characterized by a mass fractal dimension, $d_f \sim 2$. Because the rate of dissolution (siloxane bond hydrolysis) is low for trimers and higher polysilicate species, the condensation reaction is essentially irreversible. Thus these tenuous, nonequilibrium structures are kinetically stabilized: because bond breakage does not occur, they are unable to restructure and there is no source of monomers available to fill in voids.

[†]The concept of pH is ill defined in nonaqueous solutions. pH often refers to the experimentally measured value, pH*, rather than strictly $-\log[H^+]$.

Fig. 64.

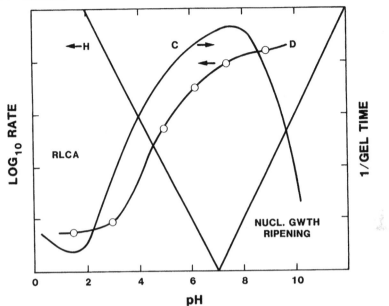

Schematic representation of the pH-dependences of hydrolysis (H), condensation (C), and dissolution (D) for an arbitrary value of r (1.5) [60].

When the hydrolysis reaction is performed with understoichiometric amounts of water ($r < 4$), condensation commences before hydrolysis is complete. Condensation between incompletely hydrolyzed species is also expected to occur by a cluster–cluster process; however, because the alcohol-producing condensation rate is less than the water-producing condensation rate, the pattern of condensation reflects the pattern of hydrolysis. Unhydrolyzed OR groups effectively reduce the functionality, promoting the formation of more weakly-branched structures. Weakly-branched "extended" structures increase the concentration-dependence of the reduced viscosity, the M_n-dependence of the intrinsic viscosity, and the concentration-dependence of the Guinier radius. Because the ether-forming (ROR) condensation reaction is unlikely, unhydrolyzed extended structures can be highly concentrated without gelling. These concentrated systems exhibit high viscosities required for spinnability.

If additional water is added in a second hydrolysis step, complete hydrolysis quickly occurs, rendering all sites approximately equal in reactivity. Condensation of completely hydrolyzed oligomers occurs by reaction-limited cluster–cluster aggregation.

2.6.7.2. High-pH Conditions

Above pH 7, hydrolysis and condensation occur by bimolecular nucleophilic displacement reactions (either S_N2-Si, S_N2^*-Si, or S_N2^{**}-Si) involving OH^- and SiO^- anions, respectively. For r values greater than 4, the hydrolysis of all polymeric species is expected to be complete. Dissolution reactions that occur preferentially at weakly-branched Q^1 sites provide a continual source of monomers. Redistribution reactions can produce unhydrolyzed monomers. Because condensation occurs preferentially between weakly acidic species that tend to be protonated and strongly acidic species that are deprotonated, growth occurs primarily by monomer–cluster aggregation. Dissolution and redistribution reactions provide a source of monomers required for monomer–cluster growth and insure that reaction-limited conditions are achieved. RLMCA is equivalent to nucleation and growth and leads to compact near-equilibrium structures that are nonfractal.

Understoichiometric additions of water ($r \ll 4$) cause unhydrolyzed sites to be incorporated in the growing clusters. The probability of condensation at these sites is less than at hydrolyzed sites. Under these conditions growth is described by a "poisoned" Eden model which, depending on the number and distribution of the poisoned sites, may result in mass or surface fractals or uniformly porous objects. Additional water added in a second hydrolysis step is expected to completely hydrolyze the clusters and further growth should be described by the Eden model.

Under the conditions used by Stöber et al. [27], primary particles formed by nucleation, growth, and ripening undergo an ordered aggregation process to form monosized spherical particles. This process represents reaction-limited monomer–cluster aggregation occurring concurrently on two length scales.

2.6.7.3. Intermediate-pH Conditions

Intermediate values of pH (3–8) represent conditions in which a spectrum of transitional structures might be expected. The dissolution and condensation rates smoothly increase with pH, whereas the hydrolysis rate goes through a minimum at approximately neutral pH. Increased $H_2O : Si$ ratios, r, increase the dissolution and hydrolysis rates at any particular pH value.

Consider, as an example, neutral pH. For $r \ll 4$, the condensation reaction initially proceeds between incompletely hydrolyzed species. Hydrolysis is rate-limiting, so the pattern of condensation reflects the pattern of hydrolysis. According to Eq. 37, the base-catalyzed condensation reaction occurs preferentially between acidic, deprotonated species and less acidic protonated species. Thus the hydrolyzed species are more likely to condense with a larger cluster than with themselves. Poisoned Eden growth should pertain, resulting in structures ranging from mass fractals to surface fractals to porous uniform clusters, depending on the number and distribution of unhydrolyzed sites. Depolymerization and redistribution reactions provide an additional source of monomers (see Fig. 33) that likewise condense preferentially with clusters. Increasing the value of r causes an increase in the hydrolysis and dissolution rates, both of which promote the formation of more compact structures. At sufficiently large r the Eden model should obtain.

Below pH 7 the hydrolysis rate increases and the rates of dissolution and condensation decrease. Hydrolysis occurs by an acid-catalyzed mechanism involving a basic, protonated alkoxy substituent. Condensation occurs by a base-catalyzed mechanism involving an acidic depronated silanol. Therefore hydrolysis occurs preferentially on monomers and weakly branched oligomers that subsequently condense preferentially with clusters. However, the availability of monomers at later stages in the reaction decreases with pH; therefore, the predominant growth mechanism changes from monomer–cluster to cluster–cluster with decreasing pH and increasing time of reaction. Below pH 4 the condensation process becomes essentially irreversible and more weakly branched structures predominate.

From pH 7 to pH 8, the dissolution and hydrolysis rates increase. Eden growth should obtain especially for large r values. Restructuring reactions provide a continual supply of monomers and insure that reaction-limited conditions are achieved. Obviously at sufficiently high r, the process is akin to aqueous silicate polymerization, and smooth colloidal particles should form. It should be noted that on larger-length scales silicate particles formed at intermediate pH will rapidly aggregate by RLCA. Thus when discussing structure it is necessary to specify length scales.

Gelation, which represents a divergence in the critical connectivity at large length scales, most likely occurs by a percolation process involving substructures formed by the kinetic growth processes described previously.

2.7. Summary

It is generally agreed that both hydrolysis and condensation occur by acid- or base-catalyzed bimolecular nucleophilic substitution reactions involving, e.g., S_N2–Si, S_N2^{**}–Si, or S_N2^*–Si transition states or intermediates. The acid-catalyzed mechanisms are preceded by rapid protonation of the OR or OH substituents bonded to Si, whereas under basic conditions hydroxyl or silanolate anions attack Si directly. Statistical and steric effects are probably most important in influencing the kinetics; however, inductive effects are certainly evident in the hydrolysis of organoalkoxysilanes.

With respect to structural evolution, many of the observed trends (e.g., extent of branching, particle versus polymer gels, occurrence of fractality, and spinnability) may be understood by considering the pH and $[H_2O]$-dependence of the hydrolysis, condensation, *and* dissolution rates. Particulate sols (nonfractal on short length scales) occur only when there is a continuous source of fully functional monomers and condensation is reaction-limited. Cluster–cluster growth or reduced functionality results in weakly branched mass or surface fractals. The divergence in critical connectivity at large length scales associated with gelation is best described by percolation as will be described in Chapter 5.

3.

MULTICOMPONENT SILICATES

It is certainly beyond the scope of this book to discuss in any detail the solution chemistry pertinent to the large number of multicomponent silicate systems described in the literature.[†] However, it is necessary to address at the minimum the synthetic strategies employed in multicomponent gel synthesis. In aqueous systems coprecipitation is so commonly used that it warrants no further discussion here. In alkoxide systems, there are two general approaches: 1) hydrolysis of mixed-alkoxide or metal organic precursors and 2) sequential addition of alkoxides to partially hydrolyzed precursors.

The first method was invented by Dislich [154,155], whose procedure for preparing an eight-component glass composition is illustrated schematically in Fig. 65. The idea is to form a complex via alcolation that contains all the metals in the proper stoichiometry. Hydrolysis converts the complex to an oxy-hydroxide or oxide that is homogeneous at the atomic level. Although it is improbable that an eight-membered oligomeric species could form and hydrolyze as a single unit as implied by Fig. 65, the formation of double alkoxides such as $MgAl_2(OR)_8$ is well documented [82,156–160] and Mehrotra and coworkers [161–165] have synthesized and isolated tri- and tetrametallic alkoxides. However, since the synthesis of the mixed-metal alkoxides was originally motivated by the need for volatile precursors for chemical vapor deposition [82], little attention was given to the mechanisms of mixed-metal alkoxide hydrolysis. Riman's work on the hydrolysis of $SrTi(OR)_4$ [157] and the work of Jones *et al.*

[†] For a partial listing of the investigated compositions, refer to Table 7 in Chapter 9.

Fig. 65. _____

$$m \, Si(OMe)_4 + n \, Al(Osec.Bu)_3 + o \, P_2O_5 + p \, LiOEt +$$
$$q \, Mg(OMe)_2 + r \, NaOMe + s \, Ti(OBu)_4 + t \, Zr(OPr)_4$$

$\underset{\xrightarrow{\hspace{2cm}}}{\textit{complexation}}$

$$[(Si_m \, Al_n \, P_o \, Li_p \, Mg_q \, Na_r \, Ti_s \, Zr_t)(OR)_{4m+3n+p+2q+r+4s+4t} \, (OH)_{3o}]$$

$\underset{\xrightarrow{\hspace{2cm}}}{\textit{hydrolysis}}$

$$[(Si_m \, Al_n \, P_o \, Li_p \, Mg_q \, Na_r \, Ti_s \, Zr_t)(OH)_{4m+3n+p+2q+r+4s+4t+3o}]$$

$\underset{\xrightarrow{\hspace{2cm}}}{\textit{condensation}}$

$$[(Si_m \, Al_n \, P_o \, Li_p \, Mg_q \, Na_r \, Ti_s \, Zr_t \, OR_{4m+3n+3o+p+2q+r+4s+4t}]$$

Synthesis of a glass ceramic

Synthesis of an eight-component oxide via complexation followed by hydrolysis and condensation. From Dislich [155].

on $Mg[Al(OR)_4]_2$ [158] appear to be the only thorough investigations of this topic. Although Jones et al. [158] present evidence that the magnesium aluminum double alkoxide remains intact during hydrolysis, further work is required to support or refute the general reaction scheme proposed in Fig. 65.

The second approach, sequential additions of different alkoxides to partially hydrolyzed precursors, was invented by Thomas [166] and popularized by Yoldas [167,168]. In this process alkoxides are added in the reverse order of their respective reactivities (least reactive precursors first) and a partial hydrolysis step is performed after each addition. The idea is that the newly added, unhydrolyzed alkoxides will condense with partially hydrolyzed sites on the polymeric species formed by the preceding hydrolysis steps (*heterocondensation*) rather than with themselves (*homocondensation*). Unlike the previous method where molecular-level homogeneity is predicted, the homogeneity of the product will depend on the size of the polymeric species to which the last component is added. Two sequential hydrolysis schemes to prepare a multicomponent borosilicate glass composition are shown in Fig. 66 where alkali and alkaline earth metals are introduced as salts or alkoxides.

The purpose of this section is to examine structural evolution during hydrolysis and condensation of multicomponent silicate systems. Instead of presenting an exhaustive review of all the work in this area, we present several illustrative examples representing structural evolution in three common glass-forming binaries: Al_2O_3–SiO_2, B_2O_3–SiO_2, and TiO_2–SiO_2.

Fig. 66.

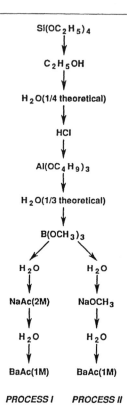

PROCESS I   PROCESS II

Synthesis of a five-component oxide by sequential additions of alkoxides (or salts) to partially hydrolyzed condensates. From Brinker and Mukherjee [169].

3.1. The Al$_2$O$_3$–SiO$_2$ System

Although Al$_2$O$_3$–SiO$_2$ gels have been widely investigated from the standpoint of preparing mullite (3Al$_2$O$_3$–2SiO$_2$) at low temperatures [e.g., 170–178], there have been relatively few fundamental investigations of the hydrolysis and condensation process [175–178]. Pouxviel and coworkers [175] used ^{27}Al and ^{29}Si NMR and SAXS to examine the structures of aluminosilicates prepared by hydrolysis of the aluminosilicate ester, (Bu^{s}O)$_2$–Al–O–Si–(OEt)$_3$. ^{29}Si NMR of the neat ester showed Si–O–Si resonances, attributable to partial condensation, and evidence of transesterification between the (OBu^s) and (OEt) ligands. The ^{27}Al NMR spectrum of the precursor showed a narrow, intense resonance at 51 ppm assigned to a symmetric, tetrahedral

Al environment consistent with the dimer:

$$(62)$$

where $R = {}^sBu$ and $R' = Si(OEt)_3$. Figure 67 shows the temporal evolution of the ^{27}Al NMR spectra during hydrolysis of $(Bu^sO)_3-Al-O-Si-(OEt)_3$ at room temperature ($H_2O/Si = 10$, pH = 2.5). After five minutes, three ^{27}Al resonances are observed: the sharp line at 51 ppm and two broad bands centered at 56 and 7 ppm (intensity ratio 5:1) assigned to tetrahedral and octahedral aluminum sites formed during hydrolysis and condensation. With time, the sharp line decreases in intensity, and at $t/t_{gel} \sim 0.96$ only the two broad bands are observed. From the peak intensities at $t/t_{gel} \sim 0.96$, 62% of the Al atoms are tetrahedrally coordinated and 38% octahedrally coordinated. These results indicate that the hydrolysis and condensation process proceeds with a conversion of tetrahedrally coordinated aluminums to octahedrally coordinated aluminums and that the Al environments become more varied, as is evident from the broadening of the 56- and 7-ppm resonances.

Fig. 67.

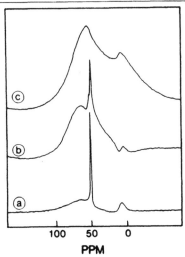

^{27}Al NMR spectra (ppm from $Al(OH_2)_6^{3+}$ of $(Bu^sO)_2Al-O-Si(OEt)_3$ hydrolysis (pH 2.5, $r = 10$) in ${}^iPrOH$ (gel time = 24 hours). (a) After 5 minutes. (b) After 7 hours. (c) After 23 hours. From Pouxviel *et al.* [175].

In contrast to the ^{27}Al results, the ^{29}Si NMR spectra remain relatively unchanged after 23 hours of reaction apart from partial hydrolysis of the alkoxide ligands. This is a surprising result when considering the substantial changes in the environment of Al to which the Q^1 silicon species are bonded in the original precursor. Pouxviel *et al.* [175] concluded from these results that the inorganic polymerization is due primarily to the formation of Al–O–Al linkages, first involving tetrahedrally coordinated aluminum species and later octahedrally coordinated aluminum species, leading to three-dimensional particle growth.

Irwin *et al.* [176] investigated the hydrolysis of the same ester, $(Bu^sO)_2Al-O-Si(OEt)_3$, with no acid catalyst. Using solid-state MAS ^{27}Al NMR on xerogels dried at 40°C, they observed a resonance at 29 ppm, assigned to pentacoordinate aluminum, in addition to the tetrahedrally and octahedrally coordinated aluminum resonances observed by Pouxviel *et al.* [175]. The corresponding ^{1}H–^{29}Si MAS NMR spectrum showed only one broad peak at ca -94 ppm with no resolution of Q^2, Q^3, or Q^4 sites. The ^{29}Si chemical shift is in the range expected for silicon bonded (through oxygen) to two aluminums, Si(2Al), rather than Si(4Al) expected from the stoichiometry of the aluminosilicate ester. ^{1}H–^{29}Si cross-polarization indicates that the remaining two ligands surrounding silicon are silanols. These results appear to support the conclusion of Pouxviel *et al.* that polymerization occurs primarily through Al–O–Al bonding, indicating that the gel is inhomogeneous on the molecular scale.

Irwin *et al.* [176] also investigated aluminosilicate gels (Al:Si ratio = 1:14) prepared by the addition of $Al(OBu^s)_3$ to partially hydrolyzed TEOS (acid-catalyzed) followed by a second hydrolysis step. The solid-state ^{27}Al MAS NMR spectrum of the dried gel (Fig. 68a) is similar to those observed

Fig. 68a.

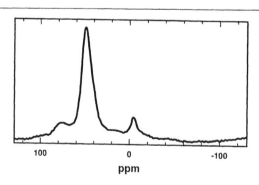

Solid-state ^{27}Al MAS NMR spectrum of an aluminosilicate gel made by the sequential hydrolysis method after drying at 40°C. From Irwin *et al.* [176].

Fig. 68b.

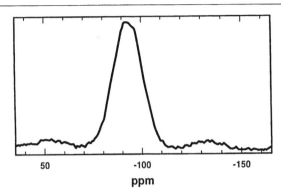

Solid-state ^{29}Si MAS NMR spectrum of the aluminosilicate gel in Fig. 68a. From Irwin *et al.* [176].

by Aukett *et al.* [179] for ZSM-5 zeolites, where Al exists primarily in tetrahedral coordination within the silicate framework. The broader peak at 52 ppm is due to aluminum in tetrahedral environments and the less intense, narrower resonance at one ppm is due to octahedrally coordinated aluminum. The corresponding ^{1}H–^{29}Si cross-polarization MAS spectrum (Fig. 68b) shows a very broad resonance centered at ca 100 ppm. This spectrum is similar to those observed in gels prepared as zeolite precursors. The breadth of the resonance has been attributed to the wide range of chemical environments in the second coordination sphere of silicon caused by Si–O–Al bonding [180].

In addition to conventional alkoxide-based syntheses, Williams and Interrante [177] and Pouxviel *et al.* [175] investigated aluminosilicate gels prepared from organometallic precursors. Williams and Interrante reacted diketonate (chel) aluminum alkoxides with trimethylacetoxysilane forming heterocondensed products according to the following scheme:

$$[Al(chel)(OR)_2]_x + nSi(CH_3)_3OAc$$

$$\rightarrow [Al(chel)(OR)_{2-n}]_x[OSi(CH_3)_3]_n + nROAc. \qquad (63)$$

This class of precursors (which readily dissolve in organic solvents to give viscous solutions suitable for film formation) convert cleanly to aluminosilicates at 450°C in air while maintaining the initial Al : Si ratio.

Pouxviel *et al.* [175] prepared a similar organometallic precursor according to the reaction:

$$Al(OPr^i)_3 + nSi(CH_3)OAc$$

$$\rightarrow Al(OPr^i)_{3-n}(OSiCH_3)_n + nAcOPr^i. \qquad (64)$$

Fig. 69.

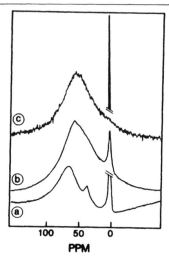

^{27}Al NMR spectra (ppm from Al(OH$_2$)$_6$$^{3+}$) of Al(O^{i}Pr)(OSiMe$_3$)$_2$. (a) In CH$_2Cl_2$ (saturated solution). (b) In HO^{i}Pr (0.1 M) after 4 hours of hydrolysis in an atmosphere of 70% RH (gel time = 24 hours). (c) After 19 hours of hydrolysis. From Pouxviel *et al.* [175].

The ^{27}Al NMR spectrum of the pure aluminosiloxane (n = 2) in CH$_2$Cl$_2$ (Fig. 69) is composed of an intense narrow band at 3.5 ppm assigned to octahedrally coordinated aluminum, a broad band at 64 ppm assigned to tetrahedrally coordinated aluminum, and a rather weak, broad band at 38.9 ppm assigned to pentacoordinate aluminum. Based on the structures of aluminum alkoxides [181], Pouxviel *et al.* [175] proposed that the alumino-siloxane solution is a mixture of the dimer, trimer, and tetramer:

(65)   (66)   (67)

$$R = {}^i Pr, \qquad R' = SiMe_3$$

in the approximate molar ratio $2:1:2$. During aging in 70% relative humidity, the 3.5-ppm band disappears along with the 38.9-ppm band (Fig. 69). Just prior to the gel point the ^{27}Al NMR spectrum consists of a single, intense, broad resonance centered at 58 ppm assigned to tetrahedrally coordinated aluminum. Corresponding SAXS investigations indicated that the aluminosilicate polymers are mass fractals with $d_f = 1.8$. From these results Pouxviel et al. concluded that polymer growth occurs by a chain polymerization mechanism involving tetrahedrally coordinated aluminums.

3.2. Borosilicate Systems

In conventional glass processing, B_2O_3 is added to silicate glasses to reduce the melting temperature without severely degrading the chemical durability. PYREX® is a well-known example. B_2O_3 additions to silicate gels likewise reduce the temperature required for viscous sintering, motivating many researchers to synthesize binary and multicomponent borosilicate gels [e.g., 154,169,182–185]. However few of these studies examine structural evolution during hydrolysis and condensation.

In the only thorough investigation of this topic, Irwin et al. [184] prepared borosilicate compositions by the addition of trimethylborate, $B(OMe)_3$, to partially hydrolyzed solutions of TMOS under acidic, neutral, or basic conditions. (See Table 19 for details.) ^{11}B and ^{29}Si NMR and IR spectroscopy

Table 19.

Sample Preparation Procedures for $20B_2O_3$–$80SiO_2$ Compositions.

Type I		Type II		Type III	
TMOS	1 mol	TMOS	1 mol	TMOS	1 mol
H₂O	2 mol	H₂O	2.2 mol	H₂O (0.15 M HCl)	2 mol
methanol	3 × vol of TMOS	methanol	0.46 × vol of TMOS	methanol	0.1 × vol of TMOS
	↓		↓		↓
	50–60 h room temp.		reflux 1 h		1 h room temp
	↓		↓		↓
trimethyl borate 0.43 mol		trimethyl borate 0.43 mol		trimethyl borate 0.43 mol	
					↓
				stir 1 h 50°C	
				↓	
				H₂O 2 mol	

Source: Irwin et al. [184].

Fig. 70.

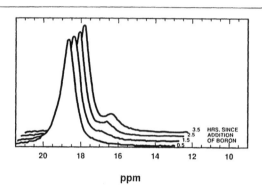

ppm

^{11}B NMR (expanded scale) of Type I borosilicate sample (Table 19) for first 3.5 hours
after addition of trimethyl borate. All spectra are plotted on the same intensity scale.
From Irwin *et al.* [184].

were used to examine structural changes that occurred during gelation. The
temporal evolution of the ^{11}B NMR spectra for a Type I sample without
catalyst (Table 19) is shown in Fig. 70. The large, broad peak centered at
18.3 ppm (relative to $BF_3-O(C_2H_5)_2$) is a superposition of the resonances
of trimethylborate and boric acid $(B(OH)_3)$. The second smaller peak
appearing with time is assigned to trigonal boron in borosiloxane bonds
$(=^{11}B-O-Si\equiv)$. The behavior of this latter band under different processing
conditions is shown in Fig. 71. In all cases a reduction in borosiloxane bonding is observed at long times and zero borosiloxane bonds are observed

Fig. 71.

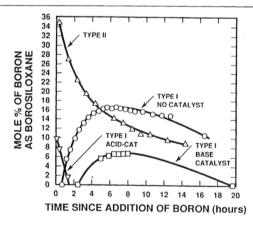

Time evolution of borosiloxane bonds after addition of trimethyl borate to hydrolyzed
silicate solutions. See Table 19 for details. From Irwin *et al.* [184].

at the gel point ($\geqslant 20$ hours). Corresponding ^{29}Si NMR spectra showed that the Type II system was much more hydrolyzed than the Type I system prior to addition of B(OMe)$_3$ and that B(OMe)$_3$ additions accelerate TMOS hydrolysis.

Since trigonal boron is very electrophilic compared to silicon, it preferentially hydrolyzes in the presence of water, rapidly forming boric acid [186], and is very susceptible to nucleophilic attack. (See Section 3.1 in Chapter 2.) Therefore Irwin *et al.* explained their results on the basis of the following reactions involving boric acid:

$$\equiv \text{Si-O-R} \quad \rightarrow \ \equiv \text{Si-O-B} = \ + \ \text{ROH} \tag{68}$$
$$=\text{B-O-H}$$

$$\equiv \text{Si-O-H} \quad \rightarrow \ \equiv \text{Si-O-B} = \ + \ \text{H}_2\text{O} \tag{69}$$
$$=\text{B-O-H}$$

$$\equiv \text{Si-O-R} \quad \rightarrow \ \equiv \text{Si-O-H} \ + \ =\text{B-O-R}. \tag{70}$$
$$\text{H-O-B} =$$

The induction period observed in the formation of borosiloxane bonds in the Type I systems (Table 19) without catalyst or with base catalyst parallels the hydrolysis of TMOS, indicating that borosiloxane bonds form preferentially by the reaction of boric acid with silanol groups (Eq. 69). The apparent absence of an induction period in the Type I system with acid catalyst and the Type II system (Fig. 71) is explained by the extensive hydrolysis of these systems prior to the addition of B(OMe)$_3$: both acid additions (Type I, acid catalyst) and refluxing (Type II) promote the hydrolysis of TMOS. The decay in the borosiloxane concentrations at longer times parallels the production of water as the by-product of silanol condensation reactions. Water promotes borosiloxane bond hydrolysis (reverse of Eq. 69) as is evident from the fact that there appear to be no known examples in the chemical literature where borosiloxane bonds are formed in other than anhydrous conditions [187,188]. In addition to refluxing, the low concentration of methanol used in the Type II process also contributes to the high initial borosiloxane concentration, since high methanol concentrations (Type I system) promote borosiloxane alcoholysis reactions (reverse of Eq. 68).

Based on the discussion in Section 3.1 of the preceding chapter, we expect that kinetically stable borosiloxane bonds can only be formed between tetrahedrally coordinated boron and silicon: tetrahedrally coordinated boron is coordinatively saturated and lacks the d orbitals necessary for the sp^3d intermediate required for the borosiloxane hydrolysis or alcoholysis reactions. Irwin *et al.* [183] observed no evidence of tetrahedrally coordinated boron under any of the conditions investigated: neither water nor alcohol is sufficiently nucleophilic to cause an increase in the boron coordination. This is consistent with investigations of alkali borate gels [189] where the concentration of 4-coordinated borons was unaffected by water concentration.

3.3. TiO$_2$–SiO$_2$ Systems

Sol-gel processing of TiO$_2$–SiO$_2$ systems is of significant technological interest for the low-temperature formation of refractory, ultralow thermal expansion (ULE®) glasses (~ 8 mole% TiO$_2$) as well as the preparation of graded refractive index (GRIN) optics. (For details, refer to Chapter 14.) The homogeneity of sol-gel processed TiO$_2$–SiO$_2$ compositions has been considered by Yoldas [190] and Basil and Lin [86,191,192]. Yoldas argued that the addition of Ti(OR)$_4$ to partially hydrolyzed silicon alkoxides should result in homogeneous glasses, because homocondensation of the silicates is slow with respect to the heterocondensation reaction:

$$(R'O)_3Si–OH + RO–Ti(OR)_3 \rightarrow (R'O)_3Si–O–Ti(OR)_3 \qquad (71)$$

(where R′ in this case could be H, Si, or an alkyl); therefore dissimilar constituents tend to become neighbors, i.e., silicon-oxygen-titanium rather than silicon-oxygen-silicon.

However since Ti(OR)$_4$ is known to catalyze silanol condensation [193], Basil and Lin tested Yoldas's hypothesis by observing the effects of Ti(OEt)$_4$ additions (1–50 mole%, based on the theoretical silicon concentration) on silanol concentrations in partially hydrolyzed solutions of TEOS, or the ethoxylated dimer or trimer. Figure 72 presents a summary of the species observed by ^{29}Si NMR after various additions of Ti(OEt)$_4$ (TET) to solutions containing TEOS, H$_2$O, HNO$_3$, and ethanol in the molar ratios, 1:1:0.006:4.5, reacted two hours prior to the TET addition. The amount of unreacted TEOS present is unaffected by TET additions, but concentrations of the silanol containing monomer and dimer decrease disproportionately with the amount of added TET. Sixty-two percent of the silanols were removed with only 0.025 equivalents of TET, and with 10% TET no silanol groups were detectable. Accompanying the loss of silanol containing species is the production of higher–molecular-weight, fully condensed silicon ethoxides, particularly Si$_2$O(OEt)$_6$ and the linear trimer and tetramer. There was no evidence for reactions between singly hydrolyzed silicate monomers and TET to form products containing (EtO)$_3$Si–O–Ti≡ groups. Based on these observations Basil and Lin concluded that TET primarily serves to catalyze silanol condensation reactions. When added to a partially hydrolyzed solution, TET is incorporated only in oligomers

Fig. 72.

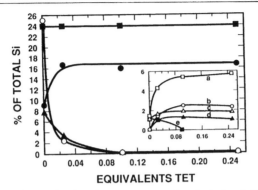

Relative concentrations of selected components of a hydrolyzed TEOS solution (Si : H$_2$O : HNO$_3$: EtOH = 1 : 1 : 0.006 : 4.5, reacted 2 hours at 66°C) after addition of various amounts of Ti(OEt)$_4$ (TET). ■: unhydrolyzed TEOS. O: Si(OEt)$_3$OH. ●: (EtO)$_3$Si–O–Si(OEt)$_3$. ▲: (EtO)$_2$(OH)$\underline{Si}$–O–Si(OEt)$_3$. Inset: (a) Q$^1_3$ + Q$^1_4$; (b) cyclo-Q$^2_4$; (c) Q$^2_3$; (d) cyclo-Q$^2_6$; (e) Q$^1_3$(OH) + Q$^1_4$(OH). From Basil and Lin [191].

with a Si : Ti ratio approaching 10 : 1, rather than condensing rapidly with available silanols to produce 1 : 1 condensation products such as (OEt)$_3$Si–O–Ti(OEt)$_3$ plus unreacted silanols (the expected result, if the heterocondensation rate is much greater than the rate of silicate homo-condensation). This is a significant finding with regard to the preparation of homogeneous TiO$_2$–SiO$_2$ gels by the sequential addition scheme. To insure homogeneity it may be necessary to synthesize mixed-metal precursors or complexes composed of Ti–O–Si bonds.

3.4. Summary

The SiO$_2$–Al$_2$O$_3$, B$_2$O$_3$–SiO$_2$, and TiO$_2$–SiO$_2$ systems are illustrative of several of the possible consequences of multicomponent silicate gel processing. In the Al$_2$O$_3$–SiO$_2$ system, the coordination of aluminum and the relative amounts of hetero- versus homocondensation (and therefore molec-ular-scale homogeneity) are influenced by both the choice of precursors and the processing conditions. The B$_2$O$_3$–SiO$_2$ system is an example where homogeneity is dictated by the hydrolytic or alcoholic stability of the mixed-metal, borosiloxane bond. In alcohol–water solution there is no evidence of Si–O–B bonding at the gel point. Homogeneity is achieved only during subse-quent thermal processing. (See Chapter 9.) In the TiO$_2$–SiO$_2$ system, Ti(OR)$_4$ catalyzes silanol condensation, promoting homocondensation of the silicate species rather than uniform incorporation of Ti at the molecular level.

REFERENCES

1. R.K. Iler, *The Chemistry of Silica* (Wiley, New York, 1979).
2. J.V. Sanders, *Journal de Physique*, **C3** (1985) 1-8.
3. K.A. Andrianov, *Organic Silicon Compounds* (State Scientific Publishing House for Chemical Literature, Moscow, 1955), translation 59-11239 U.S. Dept of Commerce, Washington, D.C.
4. M. Ebelmen, *Ann. Chim. Phys.* **15** (1845) 319.
5. M. Ebelmen, *Ann. Chim. Phys.*, **16** (1846) 129.
6. M. Ebelmen, *Comptes Rend. de l'Acd des Sciences*, **25** (1847) 854.
7. D.I. Mendeleyev, *Khim. Zhur. Sok. i. Eng.*, **4** (1860) 65.
8. B.E. Yoldas, *J. Mater. Sci.*, **12** (1977) 1203-1208.
9. M. Nogami and Y. Moriya, *J. Non-Crystalline Solids*, **37** (1980) 191-201.
10. C.F. Baes and R.E. Mesmer, *The Hydrolysis of Cations* (Wiley, New York, 1976).
11. J. Livage, M. Henry, and C. Sanchez, *Sol-Gel Chemistry of Transition Metal Oxides*, Progress in Solid State Chemistry (Plenum, to be published in 1989).
12. H. Freundlich, *Colloid and Capillary Chemistry* (Methuen, London, 1926) (Engl. transl.).
13. P.C. Carmen, *Trans. Faraday Soc.*, **36** (1940) 964.
14. G. Lagerström, *Acta Chem. Scand.*, **13** (1959) 722.
15. H.P. Calhoun and C.R. Masson, *Rev. Silicon, Germanium, Lead Tin Compd.*, **5** (1981) 153.
16. D. Hoebbel and W. Wieker, *Z. Anorg. Allg. Chem.*, **400** (1973) 146.
17. W. Wieker and D. Hoebbel, *Z. Anorg. Allg. Chem.*, **366** (1969) 139.
18. R.K. Harris, C.T.G. Knight, and W.E. Hull in *Soluble Silicates*, ed. J.S. Falcone, Jr., ACS Symp. Series **194** (American Chemical Society, Washington, D.C., 1982), pp. 79-93.
19. C.T.G. Knight, R.J. Kirkpatrick, and E. Oldfield, *J. Mag. Reson.*, **78** (1988) 31-40.
20. H.C. Marsmann, *Chem. Zeitung*, **97** (1973) 128.
21. V.G. Engelhardt, W. Altenburg, D. Hoebbel, and W.Z. Wieker, *Anorg. Allg. Chem.*, **418** (1977) 43.
22. L.S. Dent-Glasser and E.E. Lachowski, *J. Chem. Soc. Dalton Trans.* (1980) 393, 399.
23. R. Yu Sheinfain, O.P. Stas, and T.F. Makovskaya, *Koll. Zh.*, **34** (1972) 869 (Eng. trans.).
24. H.D. Cogan and C.A. Setterstrom, *Chem. and Eng. News*, **24** (1946) 2499.
25. D. Avnir and V.R. Kaufman, *J. Non-Crystalline Solids*, **192** (1987) 180-182.
26. C.J. Brinker, K.D. Keefer, D.W. Schaefer, and C.S. Ashley, *J. Non-Crystalline Solids*, **48** (1982) 47-64.
27. W. Stöber, A. Fink, and E. Bohn, *J. Colloid and Interface Sci.*, **26** (1968) 62-69.
28. S. Sakka in *Better Ceramics Through Chemistry*, eds. C.J. Brinker, D.E. Clark, and D.R. Ulrich (North-Holland, New York, 1984), p. 91.
29. S. Sakka, K. Kamiya, K. Makita, and Y. Yamamoto, *J. Non-Crystalline Solids*, **63** (1984) 223-235.
30. R.A. Assink, unpublished.
31. S. Sakka and K. Kamiya, *J. Non-Crystalline Solids*, **48** (1982) 31-46.
32. H. Tsuchida, *Science of Polymers* (Baihukan, Tokyo, 1975), p. 85 (in Japanese).
33. K. Kamiya, Y. Iwamoto, T. Yoko, and S. Sakka, *J. Non-Crystalline Solids*, **100** (1988) 195-200.
34. D.W. Schaefer, J.E. Martin, and K.D. Keefer in *Physics of Finely Divided Matter*, eds. N. Bocarra and M. Daoud (Springer-Verlag, Berlin, 1985), p. 31.
35. D.W. Schaefer and K.D. Keefer in *Fractals in Physics*, eds. L. Pietronero and E. Tosatti (North-Holland, Amsterdam (1986), pp. 39-45.

36. B.B. Mandelbrot, *Fractals, Form, and Chance* (Freeman, San Francisco, 1977).
37. R. Anderson, B. Arkles, and C.L. Larson, *Petrarch Systems Silanes and Silicones* (Petrarch Systems, 1987).
38. W.G. Klemperer, V.V. Mainz, and D.M. Millar, in *Better Ceramics Through Chemistry II*, eds. C.J. Brinker, D.E. Clark, and D.R. Ulrich (Mat. Res. Soc., Pittsburgh, Pa., 1986), p. 3.
39. W.G. Klemperer, V.V. Mainz, and D.M. Millar, in *Better Ceramics Through Chemistry II*, eds. C.J. Brinker, D.E. Clark, and D.R. Ulrich (Mat. Res. Soc., Pittsburgh, Pa., 1986), p. 15.
40. V.W. Day, W.G. Klemperer, V.V. Mainz, and D.M. Millar, *J. Am. Chem. Soc.*, **107** (1985) 8262-8264.
41. H. Schmidt in *Better Ceramics Through Chemistry*, eds. C.J. Brinker, D.E. Clark, and D.R. Ulrich (North-Holland, New York, 1984), pp. 327-335.
42. G. Philipp and H. Schmidt, *J. Non-Crystalline Solids*, **63** (1984) 283-92.
43. H. Huang, B. Orler, and G.L. Wilkes, *Macromolecules*, **20** (1987) 1322-1330.
44. G.L. Wilkes, B. Orler, and H. Huang, *Polym. Prepr.*, **26** (1985) 300-302.
45. G. Phillip and H. Schmidt, *J. Non-Crystalline Solids*, **82** (1986) 31-36.
46. I.G. Khaskin, *Dokl. Akad. Nauk SSSR*, **85** (1952) 129.
47. M.G. Voronkov, V.P. Mileshkevich, and Y.A. Yuzhelevski, *The Siloxane Bond* (Consultants Bureau, New York, 1978).
48. K.D. Keefer in *Better Ceramics Through Chemistry*, eds. C.J. Brinker, D.E. Clark, and D.R. Ulrich (North-Holland, New York, 1984), pp. 15-24.
49. K.J. McNeill, J.A. DiCaprio, D.A. Walsh, and R.F. Pratt, *J. Am. Chem. Soc.*, **102** (1980) 1859.
50. E.R. Pohl and F.D. Osterholtz in *Molecular Characterization of Composite Interfaces*, eds. H. Ishida and G. Kumar (Plenum, New York, 1985), p. 157.
51. R. Aelion, A. Loebel, and F. Eirich, *J. Am. Chem. Soc.*, **72** (1950) 5705-5712.
52. R. Aelion, A. Loebel, and F. Eirich, *Recueil Travaux Chimiques*, **69** (1950) 61-75.
53. E.J.A. Pope and J.D. Mackenzie, *J. Non-Crystalline Solids*, **87** (1986) 185.
54. R.J.P. Corriu and J.C. Young, chapter 20 in *The Chemistry of Organic Silicon Compounds*, eds. S. Patai and Z. Rappoport (Wiley, New York, 1989), pp. 1241-1288.
55. K.A. Andrianov, *Metal Organic Polymers* (Wiley, New York, 1965).
56. R.J.P. Corriu, D. LeClercq, A. Vioux, M. Pauthe, and J. Phalippou in *Ultrastructure Processing of Advanced Ceramics*, eds. J.D. Mackenzie and D.R. Ulrich (Wiley, New York, 1988), pp. 113-126.
57. B.K. Coltrain, S.M. Melpolder, and J.M. Salva, *Proceedings of the IVth Int'l. Conference on Ultrastructure Processing of Ceramics, Glasses, and Composites, Feb. 19-24, 1989, Tucson, AZ*, eds. D.R. Uhlmann and D.R. Ulrich (Wiley, New York, to be published).
58. H. Schmidt, A. Kaiser, M. Rudolph, and A. Lentz, in *Science of Ceramic Chemical Processing*, eds. L.L. Hench and D.R. Ulrich (Wiley, New York, 1986), pp. 87-93.
59. C.J. Brinker, K.D. Keefer, D.W. Schaefer, R.A. Assink, B.D. Kay, and C.S. Ashley, *J. Non-Crystalline Solids*, **63** (1984) 45-59.
60. C.J. Brinker, *J. Non-Crystalline Solids*, **100** (1988) 30-51.
61. H. Schmidt, H. Scholze, and A. Kaiser, *J. Non-Crystalline Solids*, **63** (1984) 1-11.
62. J.C. Pouxviel, J.P. Boilet, J.C. Beloeil, and J.Y. Lallemand, *J. Non-Crystalline Solids*, **89** (1987) 345.
63. R.A. Assink and B.D. Kay, *J. Non-Crystalline Solids*, **99** (1988) 359.
64. L.C. Klein, *Ann. Rev. Mater. Sci.*, **15** (1985) 227-248.
65. L.L. Hench, G. Orcel, and J.L. Nogues in *Better Ceramics Through Chemistry II*, eds. C.J. Brinker, D.E. Clark, and D.R. Ulrich (Mat. Res. Soc., Pittsburgh, Pa., 1986), p.35.

66. R.T. Morrison and R.N. Boyd, *Organic Chemistry* (Allyn & Bacon, Boston, 1966).

67. L.L. Hench in *Science of Ceramic Chemical Processing*, eds. L.L. Hench and D.R. Ulrich (Wiley, New York, 1986), pp. 52–64.

68. J. Jonas in *Science of Ceramic Chemical Processing*. eds. L.L. Hench and D.R. Ulrich (Wiley, New York, 1986), p. 65.

69. I. Artaki, M. Bradley, T.W. Zerda, J. Jonas, G. Orcel, and L.L. Hench in *Science of Ceramic Chemical Processing*, eds. L.L. Hench and D.R. Ulrich (Wiley, New York, 1986), pp. 73–80.

70. G. Orcel and L.L. Hench, *J. Non-Crystalline Solids*, **79** (1986) 177–194.

71. T.W. Zerda and G. Hoang, *J. Non-Crystalline Solids*, **109** (1989) 9–17.

72. H. Rosenberger, H. Burger, H. Schutz, G. Scheler, and G. Maenz, *Zeitschrift fur Physikalische Chemie Neue Folge*, **153** (1987) 27–36.

73. L.H. Sommer and C.F. Frye, *J. Am. Chem. Soc.*, **82** (1960) 3796.

74. L.H. Sommer, C.F. Frye, M.C. Muslof, G.A. Parker, P.G. Rodewald, K.W. Michael, Y. Okaya, and P. Pepinski, *J. Am. Chem. Soc.*, **83** (1961) 2210.

75. L.H. Sommer, G.A. Parker, N.C. Lloyd, C.L. Frye, and K.W. Michael, *J. Am. Chem. Soc.*, **89** (1967) 857–860.

76. D.R. Uhlmann, B.J. Zelinski, and G.E. Wnek, in *Better Ceramics Through Chemistry*, eds. C.J. Brinker, D.E. Clark, and D.R. Ulrich (North-Holland, New York, 1984), pp. 59–70.

77. R.E. Timms, *J. Chem. Soc., A* (1971) 1969–1974.

78. S.S. Jada, *J. Am. Ceram. Soc.*, **70** [11] (1987), C298–C300.

79. C.G. Swain, R.M. Esteve, and R.H. Jones, *J. Am. Chem. Soc.*, **11** (1949) 965.

80. R.J.P. Corriu and M. Henner, *J. of Organometal. Chem.*, **74** (1974) 1–28.

81. I. Artaki, S. Sinha, A.D. Irwin, and J. Jonas, *J. of Non-Crystalline Solids*, **72** (1985) 391–402.

82. D.C. Bradley, *Metal Alkoxides* (Academic Press, London, 1978).

83. B.W. Peace, K.G. Mayhan, and J.F. Montle, *Polymer*, **14** (1973) 420–422.

84. D.R. Uhlmann, B.J. Zelinski, L. Silverman, S.B. Warner, B.D. Fabes, and W.F. Doyle in *Science of Ceramic Chemical Processing*, eds. L.L. Hench and D.R. Ulrich (Wiley, New York, 1986), pp. 173–183.

85. M. Yamane, S. Inoue, and A. Yasumori, *J. of Non-Crystalline Solids*, **63** (1984) 13–21.

86. C.C. Lin and J.D. Basil in *Better Ceramics Through Chemistry II*, eds. C.J. Brinker, D.E. Clark, and D.R. Ulrich (Mat. Res. Soc., Pittsburgh, Pa., 1986), p. 585.

87. E.M. Rabinovich and D.L. Wood in *Better Ceramics Through Chemistry II*, eds. C.J. Brinker, D.E. Clark, and D.R. Ulrich (Mat. Res. Soc., Pittsburgh, Pa., 1986), p. 251.

88. R. Winter, J.-B. Chan, R. Frattini, and J. Jonas, *J. Non-Crystalline Solids*, **105** (1988) 214–222.

89. I. Artaki, T.W. Zerda, and J. Jonas, *J. Non-Crystalline Solids*, **81** (1986) 381.

90. C. Okkerse in *Physical and Chemical Aspects of Adsorbents and Catalysts*, ed. B.G. Linsen (Academic Press, New York, 1970).

91. W.T. Grubbs, *J. Am. Chem. Soc.*, **76** (1954) 3408.

92. L.P. Davis and L.W. Burggraf in *Ultrastructure Processing of Advanced Ceramics*, eds. J.D. Mackenzie and D.R. Ulrich (Wiley, New York, 1988), pp. 367–378.

93. W.G. Klemperer and S.D. Ramamurthi in *Better Ceramics Through Chemistry III*, eds. C.J. Brinker, D.E. Clark, and D.R. Ulrich (Mat. Res. Soc., Pittsburgh, Pa., 1988), pp. 1–14.

94. W.G. Klemperer, V.V. Mainz, S.D. Ramamurthi, and F.S. Rosenberg in *Better Ceramics Through Chemistry III*, eds. C.J. Brinker, D.E. Clark, and D.R. Ulrich (Mat. Res. Soc., Pittsburgh, Pa., 1988), pp. 15–24.

95. B.D. Kay and R.A. Assink, *J. Non-Crystalline Solids*, **104** (1988) 112.
96. R.A. Assink and B.D. Kay, *J. Non-Crystalline Solids*, **107** (1988) 35–40.
97. R.A. Assink and B.D. Kay in *Better Ceramics Through Chemistry*. eds. C.J. Brinker, D.E. Clark, and D.R. Ulrich (North-Holland, New York, 1984), pp. 301–306.
98. R.A. Assink and B.D. Kay, unpublished.
99. J.E. Pouxviel and J.P. Boilot, *J. Non-Crystalline Solids*, **94** (1987) 374–386.
100. C.J. Brinker and R.A. Assink, *J. Non-Crystalline Solids*, **111** (1989) 48–54.
101. R.A. Assink, unpublished results.
102. D.H. Doughty, R.A. Assink, and B.D. Kay, *J. Non-Crystalline Solids*, submitted.
103. H.D. Marsmann, E. Meyer, M. Vongehr, and E.F. Weber, *Macromol. Chem.*, **184** (1983) 1817.
104. C.W. Turner and K.J. Franklin in *Science of Ceramic Chemical Processing*, eds. L.L. Hench and D.R. Ulrich (Wiley, New York, 1986), pp. 81–86.
105. R.K. Harris and C.T.G. Knight, *J. Chem. Soc.*, Faraday Trans. 2, **79** (1983) 1525.
106. L.W. Kelts and N.J. Armstrong in *Better Ceramics Through Chemistry III*, eds. C.J. Brinker, D.E. Clark, and D.R. Ulrich (Mat. Res. Soc., Pittsburgh, Pa., 1988), p. 519.
107. J.L. Lippert, S.B. Melpolder, and L.W. Kelts, *J. Non-Crystalline Solids*, **104** (1988) 139–147.
108. C.A. Balfe and S.L. Martinez in *Better Ceramics Through Chemistry II*, eds. C.J. Brinker, D.E. Clark, and D.R. Ulrich (Mat. Res. Soc., Pittsburgh, Pa., 1986), pp. 27–33.
109. R. Oestrieke, W.H. Yang, R.J. Kirkpatrick, R.L. Herrig, A. Navrotsky, and B. Montez, *Geochim. Cosmochim. Acta*, **51** (1987) 2199.
110. I. Artaki, M. Bradley, T.W. Zerda, and J. Jonas, *J. Phys. Chem.*, **89** (1985) 4399–4404.
111. L.W. Kelts, N.J. Effinger, and S.M. Melpolder, *J. Non-Crystalline Solids*, **83** (1986) 353–374.
112. P.J. Flory, *J. Am. Chem. Soc.*, **63** (1941) 3083; *J. Phys. Chem.*, **46** (1942) 132.
113. W.H. Stockmeyer, *J. Chem. Phys.*, **11** (1943) 45.
114. C.A.M. Mulder and A.A.J.M. Damen, *J. Non-Crystalline Solids*, **93** (1987) 169–178.
115. C.A. Balfe, K.J. Ward, D.R. Tallant, and S.L. Martinez in *Better Ceramics Through Chemistry II*, eds. C.J. Brinker, D.E. Clark, and D.R. Ulrich (Mat. Res. Soc., Pittsburgh, Pa., 1986), p. 619.
116. M. O'Keeffe and G.V. Gibbs, *J. Chem. Phys.*, **81** (1984) 876.
117. T.W. Zerda, I. Artaki, and J. Jonas, *J. Non-Crystalline Solids*, **81** (1986) 365–379.
118. R. Zallen, *The Physics of Amorphous Solids* (Wiley, New York, 1983).
119. V.R. Kaufman, D. Levy, and D. Avnir, *J. Non-Crystalline Solids*, **82** (1986) 103–109.
120. V.R. Kaufman and D. Avnir, *Langmuir*, **2** (1986) 717–722.
121. J.E. Martin and A.J. Hurd, *J. Appl. Cryt.*, **20** (1987) 61–78.
122. D.W. Schaefer, *MRS Bulletin*, **8** (1988) 22–27.
123. D.W. Schaefer and K.D. Keefer in *Better Ceramics Through Chemistry*, eds. C.J. Brinker, D.E. Clark, and D.R. Ulrich (Elsevier North-Holland, New York, 1984), 1–14.
124. G. Guinier, C. Fournet, C.B. Walker, and K.L. Yudovitch, *Small Angle Scattering of X-Rays* (Freeman, New York, 1955).
125. G. Porod, *Kolloid Z.*, **124** (1951) 83.
126. P.W. Schmidt, *J. Appl. Cryst.*, **15** (1982) 567–569.
127. B.B. Mandelbrot, *The Fractal Geometry of Nature* (Freeman, San Francisco, 1982).
128. J.E. Martin and K.D. Keefer, *Phys. Rev. A*, **34** [67] (1986) 4988–4992.
129. J.E. Martin, J. Wilcoxon, and D. Adolf, *Phys. Rev. A*, **36** [4] (1987) 1803–1810.
130. J.E. Martin in *Proceedings of Atomic and Molecular Processing of Electronic and Ceramic Materials: Preparation, Characterization, Properties*, eds. I.A. Aksay, G.L. McVay, T.G. Stoebe, and J.F. Wager (Mat. Res. Soc., Pittsburgh, Pa., 1987), pp. 79–89.

131. J. Zarzycki in *Science of Ceramic Chemical Processing*, eds. L.L. Hench and D.R. Ulrich (Wiley, New York, 1986), pp. 21–36.

132. D. Stauffer, *J. Chem. Soc. Faraday Trans. II*, **72** (1976) 1354.

133. P.G. de Gennes, *Scaling Concepts in Polymer Physics* (Cornell Univ. Press, Ithaca, New York, 1979).

134. P. Meakin in *On Growth and Form*, eds. H.E. Stanley and N. Ostrowsky (Martinus-Nijhoff, Boston, 1986), pp. 111–135.

135. T.A. Witten and L.M. Sander, *Phys. Rev. Lett.*, **47** (1981) 1400.

136. K.D. Keefer in *Better Ceramics Through Chemistry II*, eds. C.J. Brinker, D.E. Clark, and D.R. Ulrich (Mat. Res. Soc., Pittsburgh, Pa., 1986), pp. 295–304.

137. K.D. Keefer in *Science of Ceramic Chemical Processing*, eds. L.L. Hench and D.R. Ulrich (Wiley, New York, 1986), pp. 131–139.

138. G.H. Bogush and C.F. Zukoski, *Proc. of the 44th Annual Meeting of the Electron Microscopy Society of America*, ed. G.W. Bailey (San Francisco Press, San Francisco, 1986), pp. 846–847.

139. G.H. Bogush and C.F. Zukoski in *Ultrastructure Processing of Advanced Ceramics*, eds. J.D. Mackenzie and D.R. Ulrich (Wiley, New York, 1988), pp. 477–486.

140. C.J. Brinker and A.J. Hurd, unpublished.

141. C.J. Brinker, A.J. Hurd, and K.J. Ward, in *Ultrastructure Processing of Advanced Ceramics*, eds. J.D. Mackenzie and D.R. Ulrich (Wiley, New York, 1988), pp. 223–240.

142. P. Meakin, *Phys. Rev. Lett.*, **51** (1983) 1119.

143. P. Yu, H. Liu, and Y. Wang, *J. Non-Crystalline Solids*, **52** (1981) 511–520.

144. L.C. Klein and G.T. Garvey, "Soluble Silicates," ACS Symp. Series no. 194, ed. J.S. Falcone, Jr. (Am. Chem. Soc., Washington, D.C., 1982), p. 293.

145. M. Sacks and R. Sheu, *J. Non-Crystalline Solids*, **92** (1987) 383–396.

146. S. Sakka, K. Kamiya, and T. Kato, *Yogyo-Kyokai-Shi*, **90** (1982) 79–80.

147. K. Kamiya, T. Yoko, and H. Suzuki, *J. Non-Crystalline Solids*, **93** (1987) 407–414.

148. A. Einstein, *Ann. Phys.*, **19** (1906) 289; **34** (1911) 591.

149. Y. Abe and T. Misono, *J. Polym. Sci., Polymer Chem.*, **21** (1983) 41.

150. S. Sakka, Y. Tanaka, and T. Kokubo, *J. Non-Crystalline Solids*, **82** (1986) 24–30.

151. C.J. Brinker and A.J. Hurd, unpublished results.

152. N. Yamazaki, S. Nakahama, J. Goto, T. Nagawa, and A. Hirao, in *Proc. U.S. Japan Polym. Symp., Nov. 21–26, 1980, Palm Springs, CA.* (Plenum, New York, 1984), pp. 105–113.

153. C.J. Brinker and R.A. Assink, *J. Non-Crystalline Solids*, in press.

154. H. Dislich, *Angew Chem. Int. Ed. Engl.*, **10** (1971) 363.

155. H. Dislich in *Transformation of Organometallics into Common and Exotic Materials*, ed. R. Laine, NATO ASI Series E, No. 141 (Nijhof, Dordrecht, 1988), pp. 236–249.

156. R.C. Mehrotra, *J. Non-Crystalline Solids*, **100** (1988) 1–15.

157. R.E. Riman, Ph.D. thesis, MIT, Cambridge, Mass. (1987).

158. K. Jones, T.J. Davies, H.G. Emblem, and P. Parkes in *Better Ceramics Through Chemistry III*, eds. C.J. Brinker, D.E. Clark, and D.R. Ulrich (Mat. Res. Soc., Pittsburgh, Pa., 1988), pp. 111–116.

159. S.R. Gurkovich and J.B. Blum in *Ultrastructure Processing of Glasses, Ceramics and Composites*, eds. L.L. Hench and D.R. Ulrich (Wiley, New York, 1984), p. 153.

160. B.J.J. Zelinsky, B.D. Fabes, and D.R. Uhlmann, *J. Non-Crystalline Solids*, **82** (1986) 307.

161. M. Aggrawal and R.C. Mehrotra, *Polyhedron*, **4** (1985) 845.

162. M. Aggrawal and R.C. Mehrotra, *Synth. React. Inorg. Met. Org. Chem.*, **11** (1981) 139.

163. R.K. Dubey, Ph.D. thesis, University of Rajasthan, Jaipur (1987).

164. R.K. Dubey, A. Singh, and R.C. Mehrotra, *Inorg. Chim. Acta* (in press).

165. R.C. Mehrotra, A. Singh, R.K. Dubey, and A. Shah, *J. Chem. Soc. Chem. Comm.* (in press).

166. I.M. Thomas, U.S. Patent 3,791,808 (February 12, 1974).

167. B.E. Yoldas, *J. Mat. Sci.*, **12** (1977) 1203–1208.

168. B.E. Yoldas, *J. Mat. Sci.*, **14** (1979) 1843–1849.

169. C.J. Brinker and S.P. Mukherjee, *J. Mat. Sci.*, **16** (1981) 1980–1988.

170. K.S. Mazdiyasni and L.M. Brown, *J. Am. Cer. Soc.*, **55** (1972) 548–552.

171. Y.M.M. Al-Jarsha, K.D. Biddle, A.K. Das, T.J. Davies, H.G. Emklem, K. Jones, J.M. McCollough, M.A. Mohd, Abd Rahaman, A.N.A. El-M. Sharf El Deen, and R. Wakefield, *J. Mat. Sci.*, **20** (1985) 1773–1781.

172. G. Orcel, J. Phalippou, and L.L. Hench, *Rev. Int. Hautes Temp. Refract., Fr.*, **22** (1985) 185–190.

173. B. Sonuparlak, *Adv. Ceram. Mat.*, **3** (1988) 263–267.

174. D.W. Hoffman, R. Roy, and S. Komarneni, *J. Am. Cer. Soc.*, **67** (1984) 468–471.

175. J.C. Pouxviel, J.P. Boilet, S. Sanger, and L. Huber in *Better Ceramics Through Chemistry II*, eds. C.J. Brinker, D.E. Clark, and D.R. Ulrich (Mat. Res. Soc., Pittsburgh, Pa., 1986), pp. 269–274.

176. A.D. Irwin, J.S. Holmgren, and J. Jonas, *J. Mat. Sci.*, **23** (1988) 2908–2912.

177. A.G. Williams and L.V. Interrante in *Better Ceramics Through Chemistry*, eds C.J. Brinker, D.E. Clark, and D.R. Ulrich (North-Holland, New York, 1984), pp. 151–156.

178. C.V. Edney, R.A. Condrate, Sr., W.B. Crandall, and M.E. Washburn, *Mat. Lett.*, **5** (1987) 463–467.

179. P.N. Aukett, S. Cartlidge, and I.J.F. Poilett, *Zeolites*, **6** (1986) 169.

180. C.A. Fyfe, G.C. Gobbi, and J.S. Hartman, *J. Am. Chem. Soc.*, **108** (1986) 3218.

181. O. Kriz, B. Casensky, A. Lycka, J. Fusek, and S. Hermanek, *J. Magn. Res.*, **60** (1984) 375.

182. N. Tohge, A. Matsuda, and T. Minami, *J. Am. Cer. Soc.*, **70** (1987) C-13–C-15.

183. A.D. Irwin, J.S. Holmgren, and J. Jonas, *J. Non-Crystalline Solids*, **101** (1988) 249–254.

184. A.D. Irwin, J.S. Holmgren, T.W. Zerda, and J. Jonas, *J. Non-Crystalline Solids*, **89** (1987) 191–205.

185. D.M. Haaland and C.J. Brinker in *Better Ceramics Through Chemistry*, eds. C.J. Brinker, D.E. Clark, and D.R. Ulrich (North-Holland, New York, 1984), pp. 267–273.

186. H. Steinberg, *Organoboron Chemistry*, Vol. 1 (Wiley, New York, 1964), chapter 4, p. 21.

187. E.A. Hengelein, R. Lang, and K. Sheinhost, *Makromol. Chem.*, **15** (1955) 177.

188. K. Andrianov and L.M. Volkova, *Izvest Alcad Nauk SSSR, Otdel Khim. Nauk* (1955) 303; *Chem. Abstr.*, **51** (1957) 14544.

189. C.J. Brinker, K.J. Ward, K.D. Keefer, E. Holupka, and P.J. Bray in *Better Ceramics Through Chemistry II*, eds. C.J. Brinker, D.E. Clark, and D.R. Ulrich (Mat. Res. Soc., Pittsburgh, Pa., 1986), pp. 57–68.

190. B.E. Yoldas, *J. Non-Crystalline Solids*, **38, 39** (1980) 81–86.

191. J.D. Basil and C.-C. Lin in *Ultrastructure Processing of Advanced Ceramics*, eds. J.D. Mackenzie and D.R. Ulrich (Wiley, New York, 1988), pp. 783–794.

192. J.D. Basil and C.-C. Lin in *Better Ceramics Through Chemistry III*, eds. C.J. Brinker, D.E. Clark, and D.R. Ulrich (Mat. Res. Soc., Pittsburgh, Pa., 1988), pp. 49–55.

193. W. Noll, *Chemie und Technologies der Silicone*, 2nd ed. (Verlag Chemie, Weinheim, F.R. Germany, 1968).

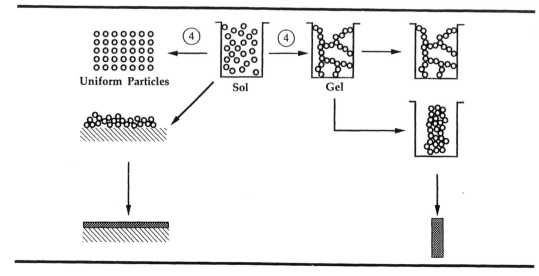

Uniform Particles   Sol   Gel

CHAPTER 4

Particulate Sols
and Gels

Although the focus of this book is on polymeric gels made from metalorganic compounds, there are many other types of gels and precursors that deserve mention. This chapter provides a brief review of the most important of these. Broadly speaking, the factor that distinguishes the systems discussed here is that they consist of dense particles, rather than polymeric clusters; the particles may be made in solution, or grown in the vapor phase and then dispersed. The particles are sometimes so small (<5 nm) that the difference between particles and polymers begins to blur. We define *polymeric* systems as having no dense (nonfractal) particles larger then 1 nm, whereas *particulate* systems have identifiable primary particles >1 nm. This size (1 nm) is generally recognized as the approximate lower bound of the colloidal range, so the systems we identify as "polymeric" do not contain dense colloidal particles. The particulate systems include some of the most successful commercial applications of sol-gel technology, such as nuclear fuel pellets and Ludox® colloidal silica. In some cases, inorganic polymers are combined with colloidal particles to great advantage. An outline of the contents of this chapter follows.

1. *Aqueous Metal Salts*
    1.1. *Sols* explains the mechanisms of electrostatic and steric stabilization that prevent agglomeration of sols, and discusses the structure of the sol particles.
    1.2. *Gels* discusses the means by which stability is eliminated, leading to the formation of aggregates and gels. The considerable information now available on the structure of particulate gels is briefly reviewed.

2. *Monodisperse Particles from Solution* examines a class of materials of widespread interest for the fabrication of ceramic bodies. Many methods of preparation have been devised. The mechanisms of growth and the structure of the particles are discussed.

3. *Other Methods of Making Particles* describes procedures including hydrolysis of aerosols, evaporative decomposition of solutions, vapor-phase reactions (by flame, plasma, or laser), and thermal decomposition of resins, among others. Both oxides and nonoxides can be prepared.

4. *Dispersion of Pyrogenic Particles* discusses methods used to prepare particulate gels of commercial significance, including high-quality silica glass for optical waveguides.

1.

AQUEOUS METAL SALTS

One of the earliest applications of sol-gel technology, already in use in the 1930s [1], was the preparation of homogeneous powders for studies of phase relations. This approach was later brought to the attention of the ceramics community by Roy and colleagues [2,3], who began with the mixing of ingredients such as salts, alkoxides, and colloidal silica. The solution was dried, or spray-dried, or gelled and dried, to obtain powder that could be sintered or melted [3]. The intimate mixing of components facilitated preparation of equilibrium phases. Solutions have now been used to make powders by a wide variety of methods. One direct approach is to spray a salt solution into a base to precipitate hydroxide powder [4]; however, as explained by Dell [5], if the base is an alkali hydroxide, it may be difficult to wash the alkali out of the precipitate, and if ammonium hydroxide is used, soluble complexes may form that result in loss of certain elements. Homogeneity is improved by allowing the solution to gel [6], by encouraging copolymerization of the component oxides [7], or by trapping the hydrous oxides in an organic gel [5,8]. These methods can produce pure, homogeneous oxides with very small particles that may be crystalline or amorphous. Some materials, because of their small particle size, appear to be amorphous when examined by X-rays, but are found to be crystalline by electron diffraction [9]. Precipitated powders tend to be highly agglomerated, which can make them difficult to sinter.

Unagglomerated uniform particles can be made in several ways. Matijević [10] has shown that conditions exist under which oxides (and some nonoxides) will precipitate in the form of (submicron) uniform spheres from salt solutions, but these methods require low concentrations and precise control of conditions. Considerable effort has been devoted to this problem, and these methods are discussed in detail in Section 2. An elegant and practical approach [11] for making free-flowing spheres from precipitated powders derives from methods used in the preparation of nuclear fuels. The idea is to disperse a salt solution as an emulsion in an organic liquid and gel the droplets to produce spheres (several microns in diameter) of hydrous oxides. Johnson [12] has reviewed many such methods of producing powders; numerous examples are discussed in Sections 2 and 3.

In this section, we take a detailed look at the stabilization and gelation of sols made from inorganic precursors. The structure and properties of these materials will be compared with polymeric gels in later chapters.

1.1. Sols

1.1.1. PRIMARY PARTICLES

The preparation of dense primary particles (1–5 nm in diameter) was discussed at length in the preceding chapters. As explained by Iler [13] the fate of these particles depends on their size, as well as on the temperature and pH of the solution. The solubility, S, of a particle is related to its radius, r, by the Ostwald–Freundlich equation:

$$S = S_0 \exp\left(\frac{2\gamma_{SL} V_m}{R_g T r}\right) \tag{1}$$

where S_0 is the solubility of a flat plate, γ_{SL} is the solid–liquid interfacial energy, V_m is the molar volume of the solid phase, R_g is the ideal gas constant, and T is the temperature. This is illustrated for silica in Fig. 1. The effect of size on solubility is most important for particles with diameters <5 nm; smaller particles in this size range will tend to dissolve and reprecipitate on larger particles. This process of particle growth, known as *Ostwald ripening*, will raise the average diameter of the silica particles to 5 to 10 nm at pH >7, whereas at low pH growth will be negligible for particles larger than 2 to 4 nm. The final particle size increases with temperature and pressure, as both factors increase the solubility of silica. Since the condensation reaction is exothermic, each Si atom tries to surround itself with four *siloxane* (i.e., $\equiv$Si-O-Si$\equiv$) bonds. For particles smaller than 5 nm, more than 50% of the Si atoms are on the surface, so they must have one or more *silanol* (i.e., $\equiv$Si-OH) bonds; nevertheless, the interiors of the colloidal particles can be regarded as dense SiO_2.

Fig. 1.

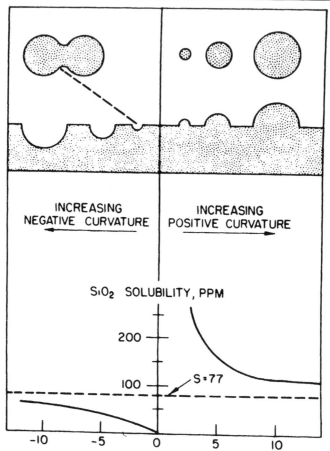

Variation in solubility of silica with radius of curvature of surface. The positive radii of curvature are shown in cross section as particles and projections from a planar surface; negative radii are shown as depressions or holes in the surface, and in the crevice between two particles. From R. K. Iler, *The Chemistry of Silica* (Wiley, New York, 1979).

A sol of hydrous oxide can evolve along a variety of paths, as illustrated in Fig. 3 of Chapter 3: polymers can remain in solution or condense into a polymeric gel; polymers can condense into particles that remain stably suspended, or aggregate into a particulate gel, or grow so large that they settle out of suspension. In the next subsection, we examine the factors that contribute to the stability of colloidal suspensions. We shall see that silica is anomalous in this, as in so many of its properties, so that the theory that accounts for most colloidal systems fails for this important material.

1.1.2. STABILITY OF SOLS

The stability and coagulation of sols is the most intensively studied aspect of colloidal chemistry; consequently, many excellent review articles [14–16] and texts [17–21] are available on the subject. Here we provide a brief overview of elementary aspects of the subject.

Since the electrons surrounding a nucleus do not constitute a spatially and temporally uniform screen, every atom is a fluctuating dipole. This effect creates an attraction between atoms, known as the *van der Waals* or *dispersion*[†] energy, which is proportional to the polarizabilities of the atoms and inversely proportional to the sixth power of their separation. The van der Waals forces result from three types of interactions: permanent dipole–permanent dipole (Keesom forces), permanent dipole–induced dipole (Debye forces), and transitory dipole–transitory dipole (London forces). It is the London forces that produce the long-range attraction between colloidal particles. When atoms are far apart, the movement of their electrons is uncorrelated; but, as shown in Fig. 2, upon closer approach the electrons will distribute so as to minimize the energy of the system. The transitory dipoles formed in this way produce a net attraction, although the net permanent dipole moment is zero. These dipoles fluctuate with a period of $\sim 10^{-16}$ s; thus, if the atoms are separated by a distance greater than ~ 30 nm, by the time the electromagnetic wave travels from one atom to another, the electron distribution of the first atom has changed. This is called *retardation* of the dispersion forces, and results in reduced attraction; only the unretarded forces are of importance in colloids.

Fig. 2.

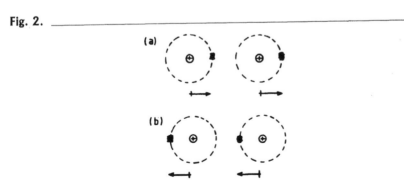

Schematic representation of the origin of the London dispersion forces for two atoms. From Napper [19].

[†] Napper [19] points out that the term *dispersion* comes from the phenomenon of optical dispersion, which results from the same electronic oscillations that produce London forces.

The attractive dispersion force is approximately additive, so that it becomes important for particles of colloidal dimension, and represents the attractive force tending to cause flocculation of colloids. The theory of dispersion force is thoroughly discussed in the monograph by Mahanty and Ninham [22]. For two infinite slabs separated by distance h, the attractive potential is

$$V_A = -A/12\pi h^2 \qquad (2)$$

where A (which has a magnitude on the order of 10^{-19} to 10^{-20} J) is a material property called the *Hamaker constant*. Note that the dependence of force on distance has changed from $V_A \propto -1/h^6$ for atoms to $V_A \propto -1/h^2$ for plates; for spheres the force involves a logarithmic function that decays more slowly than for plates at small separations (viz., comparable to the radius of the sphere) and much more rapidly at large separations. As a result of the modest dependence of V_A on h, the attractive force extends over distances of nanometers, so to prevent aggregation it is necessary to erect barriers of comparable dimensions. This can be done by creating electrostatic repulsion between particles or by adsorbing a thick organic layer (called a *steric barrier*) on the particles to prevent close approach.

The stabilization of colloids by electrostatic repulsion is successfully described by the DLVO theory (named after its principal creators, Derjaguin, Landau, Verwey, and Overbeek). The net force between particles in suspension is assumed to be the sum of the attractive van der Waals forces and the electrostatic repulsion created by charges adsorbed on the particles. The repulsive barrier depends on two types of ions that make up the *double layer*: *charge-determining ions* that control the charge on the surface of the particle and *counterions* that are in solution in the vicinity of the particle and act to screen the charges of the potential determining ions. For hydrous oxides, the charge-determining ions are H^+ and OH^-, which establish the charge on the particles by protonating or deprotonating the MOH bonds on the surfaces of the particles:

$$M\text{–}OH + H^+ \rightarrow M\text{–}OH_2{}^+ \qquad (3)$$

or

$$M\text{–}OH + OH^- \rightarrow M\text{–}O^- + H_2O. \qquad (4)$$

The ease with which protons are added or removed from the oxide (that is to say, the *acidity* of the MOH group) depends on the metal atom. The pH at which the particle is neutrally charged is called the *point of zero charge* (PZC). At pH > PZC, Eq. 4 predominates, and the particle is negatively charged, whereas at pH < PZC, Eq. 3 gives the particle a positive charge. Values of the PZC for several oxides are given in Table 1; the data are taken from an extensive tabulation by Parks [23]. The magnitude of the surface

Table 1.

Point of Zero Charge (PZC) of Selected Oxides.

Oxide type	Typical Ranges[a] PZC
M_2O	$11.5 < pH$
MO	$8.5 < pH < 12.5$
M_2O_3	$6.5 < pH < 10.5$
MO_2	$0 < pH < 7.5$
M_2O_5, MO_3	$pH < 0.5$

Oxide	Examples[b] PZC
MgO	12
FeOOH	6.7
Fe_2O_3	8.6
Al_2O_3	9.0
Cr_2O_3	8.4
SiO_2	2.5
SnO_2	4.5
TiO_2	6.0

[a] From Parks [23].
[b] Selected from Hunter [20].

potential, ϕ_o, depends on the departure of the pH from the PZC, and that potential attracts oppositely charged ions (counterions) that may be present in the solution.

Hydrous metal oxides are *hydrophilic*[†] (literally, "water-loving"), which means that layers of water molecules are strongly adsorbed at the surface of a particle, bound by hydrogen bonds and van der Waals force. Counterions are also strongly attracted by the van der Waals force, as well as by the electrostatic potential of the potential-determining ions. The force of attraction increases with the polarizability of the ion, so it increases with valence and with size, for instance from Li^+ to Cs^+ and from F^- to I^-. However, the high charge density of small cations causes them to be surrounded by a layer of hydration, so the *hydrated* form of a small cation (e.g., Li^+ + associated H_2O) may be *larger* than that of a larger ion (e.g., Cs^+). Consequently, small hydrated ions actually pack less densely near a

[†] If an oxide is heated to a high enough temperature to eliminate hydroxyl groups from the surface, it becomes hydrophobic. Consequently, oxide sols are often classified as lyophobic [17] ("solvent-fearing"); however, when hydroxyls are present (as is usually the case) they are lyophilic ("solvent-loving") in polar solvents, such as water.

Fig. 3.

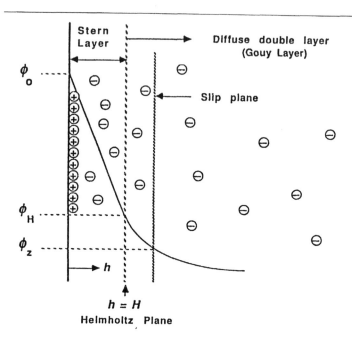

Schematic of Stern and Gouy layers. Surface charge on particle is assumed to be positive.

particle so that, for ions of a given valence, larger ions more effectively screen the surface charge.

The double layer is illustrated schematically in Fig. 3. According to the standard theory, the potential drops linearly through the tightly bound layer of water and counterions, called the *Stern layer*. Beyond the *Helmholtz plane* (at $h = H$), *in the Gouy layer*, the counterions diffuse freely; in that region the repulsive electrostatic potential of the double layer varies with distance from the particle, h, approximately according to

$$V_R \propto e^{-\varkappa(h-H)} \qquad (h \geq H) \qquad (5)$$

where $1/\varkappa$ is called the *Debye–Hückel screening length* and $\varkappa$ is given by

$$\varkappa = \sqrt{\frac{F^2 \sum_i c_i z_i^2}{\varepsilon \varepsilon_o R_g T}} \qquad (6)$$

where F is Faraday's constant, ε_o is the permittivity of vacuum, ε is the dielectric constant of the solvent, and c_i and z_i are the concentration and valence of the counterions of type i. When the screening length is large (i.e., $\varkappa$ is small) the repulsive potential extends far from the particle; this is the case when the counterion concentration is small. When counterions are present, the potential drops more rapidly with distance. Since the repulsive force is proportional to the slope of the potential,

$$F_R = \frac{dV_R}{dh} \propto \kappa e^{-\kappa(h-H)}, \tag{7}$$

the repulsive force increases with small additions of electrolyte (i.e., F_R increases with κ). Of course, large amounts of counterions will collapse the double layer (i.e., the exponential term in Eq. 7 takes over), eliminating the repulsion.

When an electric field is applied to a colloid, the charged particles move toward the electrode with the opposite charge; this transport process is called *electrophoresis*. When the particle moves, it carries along the adsorbed layer and part of the cloud of counterions, while the more distant portion of the double layer is drawn toward the opposite electrode. The *slip plane* or *plane of shear* separates the region of fluid that moves with the particle from the region that flows freely. The rate of movement of the particle in the field depends on the potential at the slip plane, known as the *zeta* (ζ) *potential*, ϕ_ζ [20]. In general, the ζ-potential is smaller than the surface potential, ϕ_o, because of the screening effect of the counterions within the slip plane. As shown in Fig. 3, the slip plane is believed to occur within the Gouy layer (beyond the Helmholtz plane). The pH at which ϕ_ζ is zero is called the *isoelectric point* (IEP), which in general is not equal to the PZC. The stability of the colloid correlates closely with the magnitude of the ζ-potential; roughly speaking, stability requires a repulsive potential $\phi_\zeta \geq 30$–50 mV.

Figure 4 shows the net effect of adding the van der Waals force and the double-layer repulsion, as required by the DLVO theory. Near the particle is a deep minimum in the potential energy produced by the van der Waals attraction; farther away is a maximum (repulsive barrier) produced by the electrostatic double layer. If the barrier is greater than $\sim 10\,kT$, where k is Boltzmann's constant, the collisions produced by Brownian motion will generally not overcome the barrier and cause aggregation. As the concentration of counterions increases, the double layer is compressed, because the same number of charges are required to balance the surface charge and they are now available in a smaller volume surrounding the particle. Since the attractive force is unchanged, while the repulsive barrier is reduced, a secondary minimum appears in the potential diagram, but it is not deep enough to bind the particles together. As the counterion concentration is increased further, the double-layer repulsion is reduced to the point that the net interparticle potential is attractive and the colloid will coagulate immediately.

Fig. 4A.

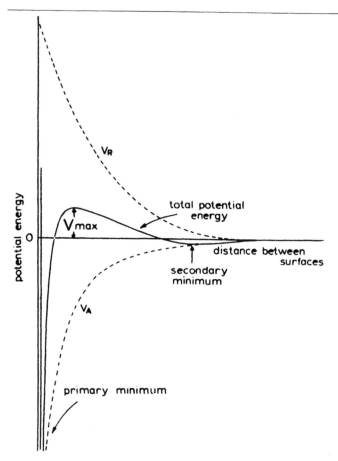

Schematic of DLVO potential: V_A = attractive van der Waals potential, V_R = repulsive electrostatic potential. From Parfitt [24].

Silica does not conform to the DLVO theory because it is apparently stabilized by a layer of adsorbed water that prevents coagulation even at the IEP. This form of stabilization is possible because of the unusually small Hamaker constant of silica. To destabilize an aqueous silica sol, it is necessary to reduce the degree of hydration. Allen and Matijević [25] showed that adding salt to the sol would produce ion exchange,

$$\equiv Si{-}OH + M^{z+} \rightarrow \equiv Si{-}OM^{(z-1)+} + H^+, \qquad (8)$$

where M^{z+} is an unhydrolyzed cation of charge z. Since the silanol groups are the adsorption sites for water, the removal of SiOH by ion exchange

Fig. 4B. _____

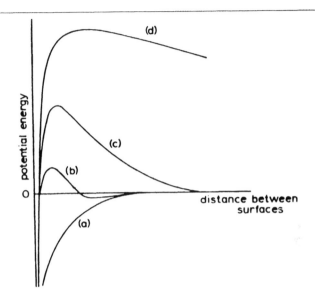

Influence of electrolyte concentration on total potential energy of interaction of two spherical particles of radius 100 nm in aqueous media: (a) $1/\kappa = 10^{-7}$ cm, (b) $1/\kappa = 10^{-6}$ cm, (c) $1/\kappa = 10^{-5}$ cm, (d) $1/\kappa = 10^{-4}$ cm. From Parfitt [24].

reduces the amount of hydration and lessens the stability of the colloid. At low pH, the amount of silanol that had to be removed to destabilize a colloid was found to be the same, regardless of the ion used for exchange ($M = Li^+, Na^+, K^+, Cs^+, Ca^{2+}, La^{3+}, Al^{3+}$). If the pH is high enough for the ions to be hydrolyzed, other surface reactions may occur, including exchange of the hydrolyzed species,[†]

$$\equiv Si-OH + La(OH)(H_2O)_5{}^{2+} \rightarrow \equiv Si-OLa(OH)(H_2O)_4{}^+ + H_3O^+ \quad (9)$$

or condensation between hydroxyls on the surface and the cation,

$$\equiv Si-OH + La(OH)(H_2O)_5{}^{2+} \rightarrow \equiv Si-OLa(H_2O)_5{}^{2+} + H_2O. \quad (10)$$

Stability may return at high salt concentrations, as the adsorbed ions reverse the charge of the particles and provide electrostatic stabilization. (See the discussion that follows.)

Figure 5 illustrates a very important point about the effect of particle size on stability: for the same surface potential, the repulsive barrier is greater for larger particles. This effect, which results from the different dependences

[†] More recent work [26] indicates that rare-earth ions have coordination numbers of 8 or 9, not 6 as indicated in Eqs. 9 and 10.

Fig. 5. _____

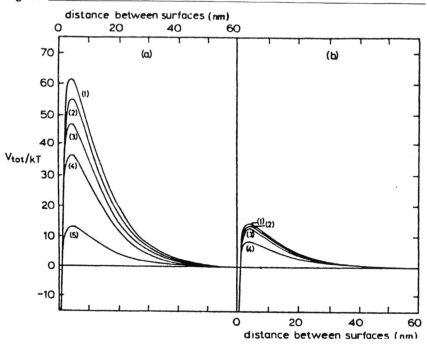

Theoretical curves of total potential energy against distance of separation of two spherical particles of radii a_1 and a_2 and equal surface potential (35.86 mV); $A = 5 \times 10^{-20}$ J, $1/\kappa = 10^{-6}$ cm^{-1}, $\varepsilon = 78.5$.

(a): $a_1 = 125$ nm, $a_2 = $ 125 nm (1), 100 nm (2), 75 nm (3), 50 nm (4), 12.5 nm (5).

(b): $a_1 = 12.5$ nm, $a_2 = 125$ nm (1), 100 nm (2), 75 nm (3), 12.5 nm (4). From Parfitt [24].

of V_A and V_R on the radius of the particle, explains why particles can grow into stable sols. The initial nuclei may be unstable, but once they have grown to sufficient size, the repulsive barrier becomes large enough to prevent coagulation (or further growth). The particle-size–dependence of the inter-particle potential has another surprising consequence [10]: if the sol contains particles varying in size, small particles may aggregate with large particles, even though the small and large particles would each be stable in a sol that was *monodisperse* (i.e., containing only one particle size).

According to the DLVO theory, coagulation results from reduction in the double-layer repulsion, which results from decreasing the surface potential (by changing pH) or from increasing the concentration of electrolyte (counterions). The effectiveness of a counterion in reducing the repulsion depends very strongly on the valence of the ion, as shown in Fig. 6. The empirical *Schulze–Hardy rule* indicates that the concentration of electrolyte

Fig. 6.

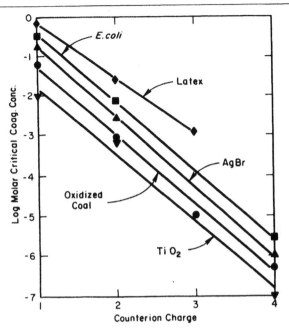

Illustration of the Schulze–Hardy rule for several sols. From Rubin [27].

required to cause flocculation, c_f, is inversely proportional to the sixth power of the charge of the ion. Although this "rule" is quite approximate, it is considered to be one of the triumphs of the DLVO theory [14] that the theory predicts [21]

$$c_f = \frac{C\varepsilon^3 k^5 T^5}{A^2 e^6 z^6} \qquad (11)$$

where C is a constant (weakly dependent on the ratio of cation to anion charge), e is the electron charge, and the other variables have been previously defined. The very strong dependence on z reflects the ability of highly charged ions to screen the surface potential. However, as noted earlier, the attraction of an ion to a particle is not simply electrostatic. The tendency for an ion to adsorb depends on its size and polarizability, in addition to its charge, and these factors contribute to the approximate nature of the Schulze–Hardy rule. In general, that rule is not obeyed by ions that chemisorb or cause chemical reactions [28].

Since the attraction is not simply electrostatic, it is possible for ions to adsorb to such an extent that they actually *reverse* the sign of the ζ-potential, so that ϕ_o and ϕ_ζ have opposite signs [21]. It has been shown [29] that charge reversal is produced by adsorption of hydrolyzed metal ions; neither unhydrolyzed cations nor insoluble hydroxides produce this effect. Since polyvalent cations can be hydrolyzed at relatively low pH [30], they can reverse the charge of the colloid and produce electrostatic stabilization over a wide range of pH. On the other hand, divalent ions, such as Zn^{2+}, hydrolyze at such high pH that insoluble hydroxides may form and prevent adsorption that would produce charge reversal. Monovalent metal ions cannot cause reversal, but large monovalent organic ions may exhibit such strong specific adsorption that they produce charge reversal at very low concentrations [29]. Adsorption effects make the electrical character of the Stern layer more complex than suggested in Fig. 3. Johnson [31] has proposed a model for calculating the ζ-potential that avoids the vague description of the electrostatic fields in the Stern layer. The model requires consideration of a set of chemical equilibria, including adsorption of electrolyte ions; since the equilibrium constants are not known a priori, they must be determined by curve fitting.

The kinetics of coagulation were first analyzed by Smoluchowski [32], who assumed that no repulsive barrier was present, and that aggregation occurred by the attachment of single particles to clusters (ignoring cluster–cluster aggregation). The theory was further developed by Fuchs [33], who showed that a repulsive potential $V(r)$ would reduce the coagulation rate by the factor W, called the *stability ratio*:

$$W = 2a \int_{2a}^{\infty} \exp(V/kT) r^{-2} \, dr \qquad (12)$$

where r is the distance from the center of a spherical particle with radius a. A complete theory of coagulation must take account of several other factors [15]. For example, during a Brownian collision it is difficult to displace the tightly bound liquid near the surfaces of the particles; a correction for this hydrodynamic effect has been derived and verified. In addition, the interaction between the approaching particles can cause distortion of the double layer or even desorption of ions. Generalizations of Smoluchowski's model have been shown to describe the kinetics and cluster-size distribution in aggregation of colloids, as found in actual experiments and computer simulations. These developments are discussed in detail in Chapter 5.

Peptization is the process of redispersing a colloid that has been coagulated. This can sometimes be done by washing, to remove counterions that caused coagulation; however, polyvalent ions that adsorb strongly to the surface may be impossible to dislodge. Alternatively, a sol may be peptized by adsorbing charge determining ions that re-establish the double layer. Clearly it is difficult or impossible to peptize a colloid that has fallen into the primary potential minimum indicated in Fig. 4, as this may require an energy of hundreds of kT. As Overbeek [15] points out, it is surprising how frequently it *is* possible to redisperse a precipitate. In many cases, particles are evidently prevented from entering the primary minimum by tightly bound layers of hydration. In the case of silica, as previously noted, the effect of hydration is so strong that colloids are stable at the isoelectric point [13].

Although aqueous systems are most commonly stabilized by an electrostatic double layer, it is also possible to prevent coagulation by use of a thick adsorbed layer of organic molecules, which constitute a *steric barrier*. This technique, which is illustrated schematically in Fig. 7, is discussed in detail in refs. [18] and [19]. The adsorbed layer discourages close approach of the particles in two ways: entropically and enthalpically. When the layers of neighboring particles overlap, the freedom of motion of the chains is diminished, which reduces the entropy of the system. At the same time, solvent molecules that surround the polymer chains are squeezed out, and this costs energy; essentially, an osmotic pressure is created that tends to suck liquid back into the space between the particles and push them apart. To provide an effective barrier, the adsorbed layer must meet the following requirements [34]: (1) the surface of the particle should be completely covered (to prevent polymer chains from attaching to both particles, causing what is called *bridging flocculation*); (2) the polymer should be firmly anchored to the surface so that it cannot be displaced during a Brownian collison; (3) the layer must be thick enough (typically >3 nm) to keep the point of closest approach outside of the range of the attractive van der Waals forces; (4) the nonanchored portion of the polymer must be well solvated by the liquid. Block copolymers are particularly effective steric barriers, because one end can be made to be strongly adsorbed at the surface, while the rest can be made to have a high affinity for the solvent. The higher the affinity

Fig. 7A.

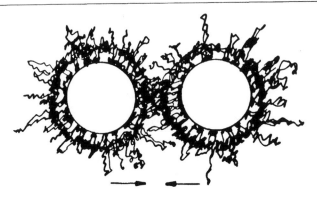

Schematic illustration of steric barrier. From Napper [19].

Fig. 7B. _____

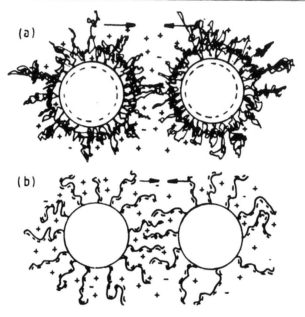

Schematic representation of electrosteric stabilization: (a) charged particles with nonionic polymers; (b) polyelectrolytes attached to uncharged particles (not to scale). From Napper [19].

of the solvent for the polymer, the greater the enthalpic contribution to stabilization. In *electrosteric* stabilization (Fig. 7B), electrostatic repulsion is combined with steric effects. Coagulation can be induced by reducing the solvency of the polymer (e.g., by adding a nonsolvent or changing the temperature). Once such a dispersion is coagulated, it can be rather easily redispersed, because the thick adsorbed layer prevents the particles from entering the primary minimum in the attractive potential. (See Fig. 4.) In some cases (to be discussed in Section 2.1) the steric barrier can be present on the particle during growth, preventing aggregation, but not becoming incorporated into the particle [35].

1.1.3. STRUCTURE OF SOLS

In a sol prepared by hydrolysis of a salt, the primary particles have diameters of ~3 to 10 nm and are nonporous. Iler [13] attributes the dense structure to the solubility of the oxide in the aqueous medium, and speculates that this might not be the case in a nonaqueous (e.g., alcoholic) solution. As we saw in the preceding chapters, he was exactly correct: the polymers that grow in alcohol–water solutions do not form dense particles, but are fractal objects. The absence of open pores in particles grown in aqueous solution is demonstrated by the work of Ramsay and Booth [36] (see also the excellent review by Ramsay [37]), who examined sols of ceria (made from a nitrate solution) and silica (such as Ludox®, made by Du Pont Co. by ion exchange of alkali silicate). The particles all were nearly spherical with a narrow size distribution; the ceria particles were 6–7 nm in diameter, and several grades of colloidal silica were used, with particle sizes ranging from 8 to 80 nm. The particle size was determined by transmission electron microscopy (TEM) and by small-angle neutron scattering (SANS) from dilute solutions; these techniques were in good agreement. In addition, the surface areas of the dried powders were determined by the BET method, and the calculated particle size agreed reasonably well with the other data. If the particles were porous, the BET area would be large and the calculated particle size would be smaller than the size seen in the TEM, so these particles were certainly dense hydrous oxides. Of course, such particles can aggregate into large porous structures, as we shall see in the next section.

Although the cores of the primary particles in aqueous sols are nonporous, it must be recalled that the particles may contain hydroxyl groups that reduce their density. An example is provided by the work of Crucean and Rand [38], who prepared zirconia by dripping zirconyl chloride into water with the pH adjusted to 4, 6, 8, or 10, and gelled the precipitate by centrifuging and drying. On heating to 390°C, the mass and pore volume of the gel decreased, as condensation of Zr–OH groups produced water that was expelled. At the same time, the skeletal density of the amorphous gel, measured by helium pycnometry, increased from ~ 4.8 to $\sim 5.1 \, g/cm^3$. This indicates that the particles of hydrous oxide in the gel shrank during dehydration, so there must have been hydroxyl groups within the particles, as well as on the surfaces of the pores. This is also characteristic of the alkoxide-derived gels, which do not contain particles. Further increases in skeletal density (to $\sim 5.5 \, g/cm^3$) occurred at higher temperatures as the amorphous zirconia gel crystallized.

The growth of dense particles is indicated in computer simulations of particle growth, as illustrated in Fig. 8 from the work of Meakin and Jullien [39]. The first figure shows a simulated aggregate of particles produced by collisions of clusters that stick firmly upon colliding and cannot rearrange. The result is a loosely packed agglomerate with a fractal dimension of $d_f \sim 1.8$. The simulation was then modified to allow the clusters to rotate around the point of contact until another contact was made. Figures 8a, b, and c represent clusters that were allowed to make one, two, or three such rotations, respectively. The increase in the density of packing on a local scale is quite profound, although the fractal dimension increases only

Fig. 8a.

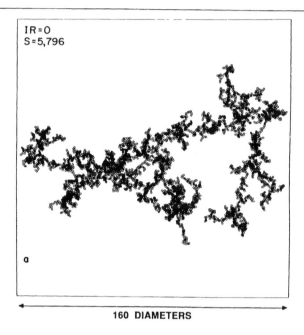

IR = 0
S = 5,796

a

160 DIAMETERS

Fig. 8b.

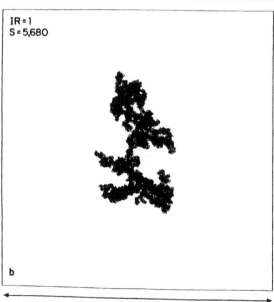

IR = 1
S = 5,680

b

160 DIAMETERS

Fig. 8c. _____

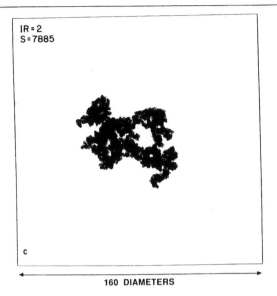

IR = 2
S = 7885

c

160 DIAMETERS

Fig. 8d. _____

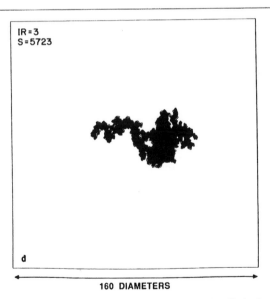

IR = 3
S = 5723

d

160 DIAMETERS

Clusters generated by 3-d polydisperse model for diffusion-limited cluster–cluster aggregation. The cluster in (a) was obtained without restructuring, while those in (b), (c), and (d) were generated in simulations allowing one, two, or three stages of rotational restructuring, respectively. Fractal dimensions of the clusters are 1.80 (a), 2.09 (b), 2.17 (c), and 2.18 (d). From Meakin and Jullien [39].

slightly (to 2.09, 2.17, and 2.18, respectively). This calculation is actually intended to be descriptive of aggregation of particles, rather than attachment of molecules to a growing particle; nevertheless, it could describe the approach of an ion surrounded by a layer of hydration that does not immediately form a chemical bond. A model that more directly simulates dissolution and reprecipitation (in two dimensions) was developed by Shih *et al.* [40]. They allowed molecules to detach from growing clusters at a rate dependent on the number of nearest neighbors; that is, the activation energy for dissolution was three times as great for a molecule bonded to three others as for a molecule attached to the cluster by a single bond. This encourages an atom to break away from the cluster and reattach repeatedly until it finds a site where it binds to several neighbors. Eventually this will lead to the formation of dense particles. The clusters grown in the simulations by Shih *et al.* were always fractal, but the fractal dimension (density of the cluster) increased as the bond energy decreased. These processes of rearrangement are analogous to the structural adjustments that could occur in an aqueous salt solution in which the oxide is significantly soluble. In contrast, the alkoxide-derived gels described in Chapters 2 and 3, because of their low solubililty, develop structures more closely resembling that in Fig. 8a, with porosity on all size scales.

It seems contradictory that dense particles can grow from solution, resulting in a stable sol. Why aren't the polynuclear ions and polymers repelled by the electrostatic double layer that stabilizes the colloidal particles? The answer is revealed in Fig. 5: the repulsive barrier increases with the size of the particles, so the nuclei may be unstable against aggregation until they reach a certain size. Kramer *et al.* [41] directly demonstrated that the rate of growth of titania particles was related to the electrostatic barrier. Particles grew to a diameter of 4 nm in ~24 h at pH 9.7, but reached 50 nm in 4 h if salts were added to compress the double layer, and they reached 6 nm in 1 h if the pH was reduced to 7 (near the IEP of titania). Of course, if the repulsive barrier is too low, the particles do not form a stable sol. According to Iler [13], silica sols made by hydrolysis of alkali silicate will precipitate if the salt concentration exceeds 0.3 N.

The arrangement of particles in a sol can be investigated using scattering methods such as SANS and SAXS (small-angle X-ray scattering). The scattered intensity is used to construct a radial distribution function (RDF), g(r), that indicates the probability that the centers of two particles will be separated by a distance r. If the particles are in an ordered arrangement, as in a crystal, the RDF consists of peaks, as in the familiar Bragg reflections in an X-ray powder pattern; at the other extreme, in a dilute solution the RDF is flat, indicating uniform density. Figure 9 shows the change in the RDF of a stable silica sol as the concentration is increased.

Fig. 9.

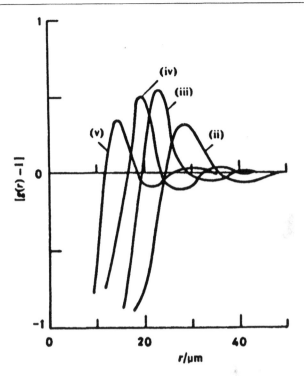

Radial distribution functions g(r) for silica gel and Ludox® S3 sols of different concentrations (g/cm³): (ii) 0.16, (iii) 0.41, (iv) 0.65, (v) ca. 1.2. From Ramsay and Booth [36].

There are few particles very near others, so $g(r) \to 0$ at short distances, but there is a large peak at longer range that reflects the onset of ordering. Computer simulations [42] suggest that the peak corresponds to particles residing in the secondary minimum in the interparticle potential. (See Fig. 4B.) As the concentration (c) increases, the average interparticle distance (i.e., the location of the main peak in the RDF) decreases according to $g(r)_{max} \propto c^{-1/3}$, as shown in Fig. 10. This indicates that the *coordination number* (i.e., the average number of nearest neighbors around any particle) is constant as the concentration increases. In this case, the coordination number remains equal to ~8, as in random dense packing of spheres, and only the average interparticle spacing changes. Because of the polydispersity in particle size (±20%) these particles never settle into the long-range order (analogous to crystallinity) that is seen in the monodisperse particles discussed in the next section. The RDF can be calculated, given an assumption for the interparticle potential. Ramsay and Scanlon [43] found

Fig. 10.

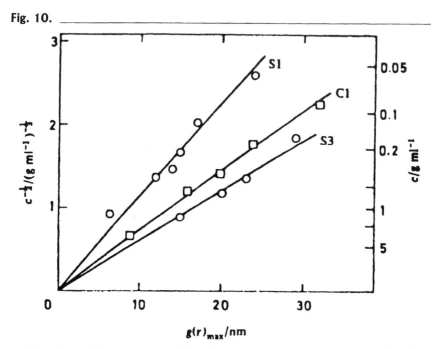

Dependence of the maximum in g(r) on the concentration, c, of sols and gels. S1 and S3 are Ludox® sols with average particle sizes of 8 and 16 nm, respectively, and C1 is a ceria sol prepared from an aqueous nitrate solution. From Ramsay and Booth [36].

excellent agreement between their SANS data and distributions calculated by assuming that the repulsive potential on silica particles consists of a hard sphere (infinite repulsive potential) surrounded by a DLVO-type electrostatic repulsion; the particle radius obtained by fitting the data to the theory agreed with that measured by TEM. If the electrostatic repulsion is ignored and the particle is assumed to behave simply as a hard sphere, the calculations do not fit the data as well, and the size of the particle required for the fit is larger than the true size [44].

These data indicate that when the concentration of a stable sol is increased, the particles pack more closely together, but remain randomly distributed. Similar behavior is observed in sols that are stabilized by a steric barrier. Van Helden and Vrij [45] stabilized monodisperse silica particles (radius ~17 nm) in cyclohexane or chloroform by chemisorbing a layer of stearyl alcohol on the surface. For volume fractions up to 0.4, the compressibility, light scattering [45], and SANS [46] from the sols were in close accord with that expected from a suspension of hard spheres, and the particle size obtained by fitting the theory to the data was in agreement with that seen by TEM. The particles act like hard spheres, because the steric barriers interact only on contact. A particularly nice illustration of the short-range nature of the steric barrier was provided by Cairns *et al.* [47], who measured the pressure required to compress a polymer latex stabilized with chains of poly(12-hydroxy stearic acid). The layer thickness would be ~9 nm, which means that the barrier layers would just touch at a volume fraction of particles of 0.53. As shown in Fig. 11, the pressure begins to rise at that point, and diverges near a volume fraction of 0.566, corresponding to a particle separation of 14.5 nm. Clearly, there is no significant repulsion until the steric barriers have interpenetrated.

Ramsay *et al.* [44] mixed stable sols of spherical titania and rod-shaped FeOOH (goethite) particles and examined the resulting sol with SANS. This technique allows independent observation of particles with different compositions, because the effectiveness of neutron-scattering varies strongly with atom type. Both types of particles were positively charged and the mixture remained stable, with the titania particles grouped around the goethite needles. Sometimes, however, sols that are independently stable

Fig. 11.

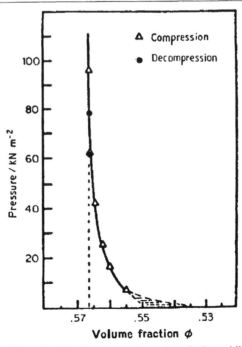

Pressure versus volume fraction for compression of sterically stabilized poly(methyl methacrylate) latex; core diameter of particles = 155 nm. From Cairns *et al.* [47]

will aggregate when mixed. Matijevic [10] points out that the attraction is greater between small and large particles, so that small particles may become attached to larger particles, even if they have the same charge. For example, when a silica sol is mixed with a suspension of polyvinyl chloride latex, the silica coats the latex, even though both are negatively charged [48].

When the double-layer repulsion between particles is reduced by addition of salt or a change of pH, the particles begin to aggregate, and the process can be followed by small-angle scattering. As illustrated in Fig. 12, a plot of scattered intensity versus scattering vector ($|Q| = 4\pi \sin(\theta)/\lambda$, where θ = scattering angle and λ = wavelength) reveals two regimes of structure. On the smallest scale ($< a_2$) the slope corresponds to the structure of the primary particles. Schaefer $et\ al.$ [49,50] examined destabilized Ludox® sols by SAXS and light scattering, with the results shown in Fig. 13. Where the data probe the size range of the primary particles ($a_2 \approx 2.7\,\text{nm}$), the slope of the scattered intensity plot is characteristic of dense particles, but on longer-length scales the data indicate the formation of fractal aggregates. Similarly, Schaefer $et\ al.$ [51] examined a sol made by raising the pH of an aqueous aluminum nitrate solution and found a fractal dimension of ~2

Fig. 12. _____

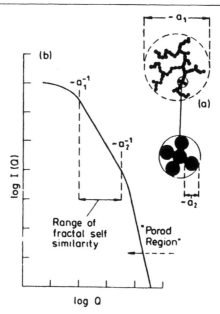

Schematic representation of a particle aggregate (a) having a range of self-similarity (fractality) between approximately a_1 and a_2. The form of the scattering expected is depicted in (b). From Ramsay and Scanlon [43].

Fig. 13.

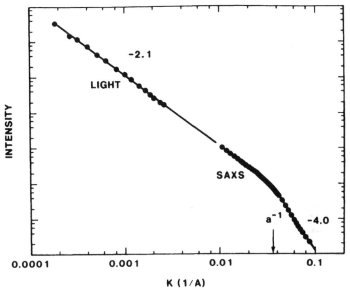

Combined SAXS and light-scattering results showing that silica aggregates have fractal structure (fractal dimension 2.1) on length scale greater than the radius (2.7 nm) of the primary particles. On shorter length scales, the Porod slope of -4 is found, indicating dense particles. From Schaefer *et al.* [50].

for aggregates with radii of gyration of ~ 140 nm. The slope increased at larger $|Q|$, indicating that the primary particles were dense particles with radii of ~ 1.2 nm. These results were obtained on systems containing 2.5 OH groups per Al ion, whereas sols containing 2 OH/Al showed evidence only for objects with radii of 0.32 nm, corresponding to a polycation containing 13 Al atoms. Axelos *et al.* [52] found the same primary particles (radius ~ 1 nm) in an experimental study of aggregation of colloidal $Al(OH)_{2.5}$ (made by partial hydrolysis of $AlCl_3$ by NaOH), but mistakenly identified them as Al_{13} polycations. The fractal dimension, d_f, of 2.1 is typical of clusters grown under conditions of slow aggregation, whereas fast aggregation often leads to $d_f \approx 1.8$. This is consistent with predictions of models for reaction-limited and diffusion-limited aggregation, respectively; such models are discussed in detail in Chapter 5. Aubert and Cannell [53] found $d_f \approx 1.8$ and 2.1 for fast and slow aggregation, respectively, but also found $d_f \approx 2.1$ in circumstances where restructuring was rapid, exactly as expected from Fig. 8b, in which restructuring occurs by rotation around points of contact.

In some cases the force field around the particles induces nonuniform ordering. As illustrated in Fig. 14, once three particles have formed a cluster,

Fig. 14.

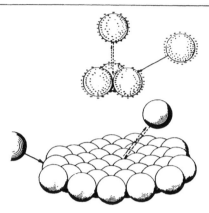

Mechanism of aggregation of particles into sheets. Particles can approach the edges of a sheet where the repulsive electrostatic force is less than on the face. From Iler, *The Chemistry of Silica* (Wiley, New York, 1979).

it is easier for the next particle to approach toward the edge than toward the face. This phenomenon has been examined by Jullien [54,55] in computer simulations of aggregation, in which he found that such clusters develop the low fractal dimension of 1.42 ± 0.05. This is believed to account for the measured fractal dimension of 1.45 found for alumina clusters by Axelos *et al.* [52]. Their data, which were obtained within minutes of the start of aggregation, indicate a much lower d_f than was obtained by Schaefer *et al.* [51] on aged sols of the same composition. The latter authors found that the sol precipitated immediately after pH adjustment and then became resuspended on aging, so it is evident that restructuring is occurring. This presumably accounts for the difference in the measured fractal dimensions obtained by these two studies. Plate-like particles of goethite (α-FeOOH) tend to aggregate in a face-to-face arrangement, and this also leads to a low fractal dimension [56]. A reduction in d_f can also result from magnetic moments on particles. Simulations by Mors *et al.* [57] predict fractal dimensions to increase continuously from 1.35 to 1.78 as the magnetic moment of the particle increases. These results nicely bracket the values measured by Kim *et al.* [58] for iron (1.34) and cobalt (1.72) particles, which have large and small moments, respectively.

There is a strong analogy between the behavior of particles in a sol and atoms in a fluid [59]. At low concentration the particles behave independently, like atoms in a gas. As the concentration increases, short-range ordering appears and the RDF of the sol is strikingly similar to that of atoms in a liquid. Further increases in concentration can lead to a cluster or gel in which the arrangement of primary particles is amorphous or crystalline [14]. Aksay and Kikuchi [60] have calculated phase diagrams for sols, introducing a parameter proportional to the square of the ζ-potential, which is analogous to temperature. When the "temperature," $\theta \propto \phi_\zeta^2$, is high, the sol is gaslike, and lowering θ leads to the formation of "glasses" or "crystals"; there is also a miscibility gap in which ordered clusters form and precipitate, leaving singlets in suspension. A sudden decrease in θ leaves no time for ordering, so an unstructured aggregate is formed, in analogy to the formation of a glass by rapid quenching of a melt. Slow reduction in θ allows ordering into crystalline structures, if the particles are monodisperse; simulations [61] indicate that the face-centered cubic structure is most likely. If there is a range of sizes, such ordering is inhibited, as observed by Ramsay and Booth [36]. The latter situation is consistent with the "principle of maximum confusion," which states that it is easier to vitrify a liquid containing many elements, because it is difficult for the constituents to organize into a crystal. This principle is exploited in the preparation of metallic alloy glasses, which crystallize unless the atoms of the alloy are sufficiently different in size. Segregation by sizes is required before crystalline order can develop. This process has been observed in polydisperse latex colloids by Hachisu *et al.* [62].

We have seen that colloidal particles can mimic the behavior of atoms in a gas or liquid and can deposit in sediments with an amorphous (disordered) or crystalline structure. In addition, colloids can exhibit phase separation in which the particles segregate into concentrated regions surrounded by dilute sol. The possibilities are summarized in Fig. 15, from an excellent review by Heller [63]. The first column shows *coacervates*, concentrated regions of particles that are attracted but not bound to one another; that is, the regions shown are still viscous suspensions. There is a sort of surface tension associated with coacervates that causes them to adopt a spheroidal shape. The second column of the figure shows *tactoids*, which contain an ordered arrangement of particles. Again, the particles are not rigidly bonded to one another, but their viscosity is non-Newtonian. The ordering is induced by anisotropy in the double layer surrounding the particles, and that anisotropy opposes the spheroidization that is typical of coacervates. If tactoids are slowly dried or the repulsive barrier is gradually reduced, the particles can become irreversibly bound into an ordered structure called a *crystalloid*, as indicated in the third column of Fig. 15. Some structures of this type are discussed in Section 2. On the other hand, rapid aggregation produces disordered clusters of the sort shown in the fourth column.

In the following subsection, we examine the ways in which the decrease in stability (or thermodynamic temperature, θ) is controlled, and the structures that result.

Fig. 15.

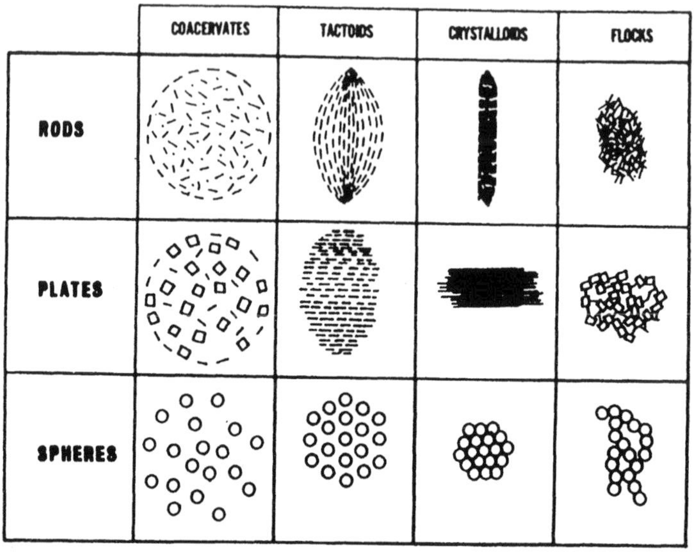

	COACERVATES	TACTOIDS	CRYSTALLOIDS	FLOCKS
RODS				
PLATES				
SPHERES				

The principal types of aggregates of rigid colloidal particles. From Heller [63].

1.2. Gels

As we have seen, aggregation results from collapse of the repulsive double layer. In dilute solution, this leads to the growth of fractal aggregates, and in more concentrated systems to the formation of gels. The theory of gelation is discussed in detail in Chapter 5. Here we discuss the methods by which aqueous sols are destabilized, and the structures of the resulting gels, both before and after drying.

Figure 5 of Chapter 3 illustrates the influence of pH and electrolyte concentration on the time (t_{gel}) to gel silica sols. At high pH the particles are stabilized by a negative charge, so t_{gel} is long; however, the effectiveness of the electrostatic barrier is sensitive to the presence of salts that compress the double layer. For most oxides, the gelation rate increases continuously as the isoelectric point is approached, but silica is anomalous: near the IEP (pH ~ 2) the stability is moderate, apparently because of protection provided by layers of bound (adsorbed) water [13]. At lower pH (<IEP), where the particles become positively charged, instability is attributed to fluoride impurities.

The phenomenon of gelation has been most ingeniously exploited in the formation of nuclear fuels, as explained in several reviews [5,64,65]. The use of aqueous sols is advantageous for commercial production of radioactive materials, because the use of sols reduces the production of dangerous dust that results from conventional ceramic-processing techniques. Moreover, the resulting oxides are pure and homogeneous. It is particularly convenient for practical reasons that sols can be prepared containing 5–50 nm particles at concentrations of 1–5 M that are stable for months. Methods have been developed for the conversion of the sols into monodisperse gel particles with diameters of 10 to 1000 microns. If three groups of particles, each differing in diameter by a factor of ~7, are mixed together, they can pack to a density of ~90% [66]. Vibratory compaction of such particles is the preferred way of preparing nuclear fuels. Emulsion methods for making monodisperse gel particles are described in the following paragraphs, while techniques for growing monodisperse particles directly in solution are discussed in the next section.

Thorium nitrate is unique in that a dispersible powder can be formed directly by heating the nitrate, which first dissolves in its own water of hydration, evolving water and nitric acid. On further heating, water evaporates, followed by oxides of nitrogen [67]. Alternatively, the nitrate can be decomposed in steam [68]. The precipitate is then peptized and the aqueous sol is dripped into an organic liquid, such as 2-ethyl hexanol. Droplets of the sol form with a size controlled by the quantity of surfactant added, and the droplets gel as the organic liquid extracts water from the sol. The sphericity of the droplets can be improved if they fall through an atmosphere of NH_3, which produces a firm "skin" on the spheres before they fall into the organic liquid [65]. Urania sols can be prepared by hydrolysis of uranyl nitrate. The aqueous sol is mixed with hexamethylene tetramine (HMTA) and the solution is dripped into hot silicone oil, where the rising temperature decomposes the HMTA, releasing ammonia which causes gelation [69]. The same procedure can be used to produce carbides, if carbon powder is included in the sol. Another method for sphere production is to drip the aqueous sol into a solution of xylene and a long-chain amine, which extracts acid from the aqueous phase; spheres of hydrous iron oxide have been prepared in this way [64]. Zirconia spheres, either pure or doped with calcia, yttria, or magnesia, have been made by dripping an aqueous sol into trichlorethane containing a long-chain amine; the amine causes gelling by deionizing the sol [67,70]. Similar methods have been used to make spheres of tin-doped indium oxide [71], Ni–Zn ferrite, and $Y_3Fe_5O_{12}$ garnet [11]. The spheres made by these techniques are gels with porosity of ~50%.

A novel method for producing gels on a large scale was developed by Shoup [72]. A sol of Ludox® particles is mixed with a solution of potassium silicate at high pH. Gelation is produced by addition of formamide ($HCONH_2$), which is hydrolyzed under basic conditions to produce

Fig. 16a.

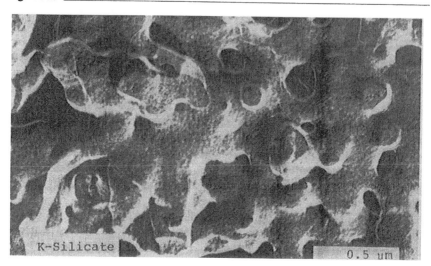

Fig. 16b.

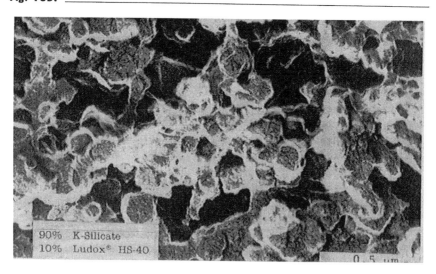

Fig. 16c.

Fig. 16d.

Controlled-pore microstructures of gel made by Shoup's method. From Shoup [72].

ammonium hydroxide, thereby lowering the pH from ~11.8 to ~10.8. The colloidal silica particles act as nuclei for the precipitation of silicate polymers from the alkali silicate solution, resulting in the unique microstructures shown in Fig. 16. The variation in pore size can be limited to ±30% of the mean size, and the mean can be varied from 10 to 360 nm by varying the ratio of potassium silicate to Ludox®. To avoid cracking during drying, the mean pore size must be >60 nm. This technique is essentially limited to the preparation of silica gels, although small amounts of other oxides can be incorporated by addition of oxides or salts to the starting solutions [73]. The greatest advantage of Shoup's technique is that large pieces (e.g., 30 × 30 × 9-cm blocks) can be dried without cracking [74].

The structure of a gel made from colloidal particles depends on the size distribution of the particles and the strength of the attractive forces between them. If the particles are spherical and monodisperse, and the repulsive barrier is reduced gradually, the sol may develop an ordered ("crystalline") structure. For example, natural opal (shown in Fig. 17) contains a face-centered cubic arrangement of silica spheres; the phenomenon of *opalescence* results from Bragg diffraction of visible light from this crystalloid. This sort of structure develops when the repulsive barrier is sufficient to allow the particles to slip into a dense packing, whereas disordered structures with high porosity result when the attractive potential is too great. This principle is clearly illustrated in Fig. 18. As the ζ-potential of the sol increases, the density of the centrifuged sediment increases. For that reason, Nelson *et al.* [75] find that gels made from aggregated sols are ~70% porous when dried, and do not sinter to full density when fired; in contrast, gels from unaggregated sols are <40% porous and densify readily at relatively low temperatures.

Ramsay and Booth [36] performed an excellent study of the structure of gels prepared from Ludox® and ceria sols. As noted earlier, the particles in concentrated sols were found to exhibit short-range order analogous to that found for atoms in a liquid. As shown in Fig. 10, the average distance between nearest neighbors, $g(r)_{max}$, varied smoothly with concentration from the sol into the gel state, with the interparticle spacing in the gel state being very close to twice the radius of the particles. This indicates that gelation does not involve a qualitative change in arrangement of the particles. In particular, there was no indication of ordering into domains of the type shown in Fig. 18, apparently because of the polydispersity of particle size (±20%) in the sols. The specific surface area of the gels, measured by nitrogen adsorption, was in good agreement with the geometric area, $S_o = 6/D\rho$, calculated using the particle diameter (D) measured by TEM and the density (ρ) of the oxide. This reveals that the primary particles were dense and nonporous. The pore size, r_p, measured by nitrogen adsorption was compared with that calculated by assuming that the primary particles were arranged in ordered

Fig. 17.

Electron micrograph of opal structure lightly etched with HF. Fossil from Cooper Pedy, Australia; photo by J.V. Sanders. From Iler, *The Chemistry of Silica* (Wiley, New York, 1979).

arrangements with coordination numbers (i.e., number of nearest neighbors) of 6, 8, and 12. As illustrated in Fig. 19, the relation between r_p and D is quite linear for $D > 16$ nm and indicates that the coordination number is ~ 8, as in random dense packing. For smaller particles, the measured pore size is constant, because they are smaller than the limit of the applicability of the technique of nitrogen adsorption; a more reliable estimate of pore size for $D < 16$ nm is probably obtained by extrapolation of the solid line in Fig. 19. The structural features of the gels examined by Ramsay and Booth are probably characteristic of the spheres made by the emulsion techniques previously discussed, since each of those spheres is a small gel.

Fig. 18.

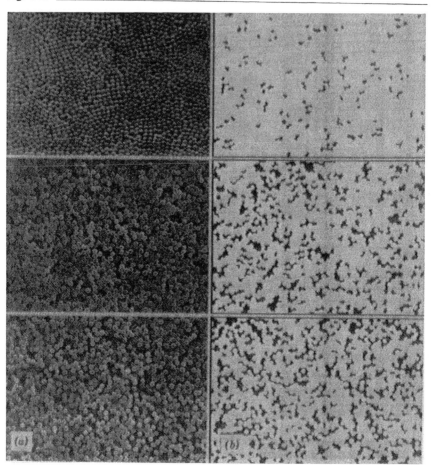

(a) Arrangement of particle domains formed by centrifugal sedimentation of SiO_2 microspheres (average diameter $0.7\,\mu$m) in H_2O at $\zeta = 110\,$mV (top), $68\,$mV (middle), and $0\,$mV (bottom). (b) Images where only the second-generation voids are highlighted as dark regions in order to illustrate the continuous variations in the domain size with ζ-potential. From Aksay and Kikuchi, "Structures of Colloidal Solids," pp. 513–521 in *Science of Ceramic Chemical Processing* (Wiley, New York, 1986).

Fig. 19.

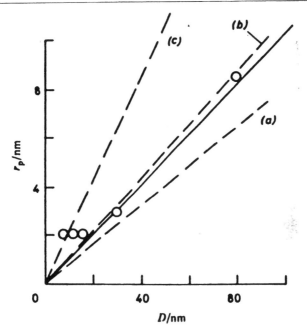

Dependence of mean pore radius, r_p, found by nitrogen adsorption, on mean particle size of silica, D, determined by TEM. Broken lines show calculated relationship of r_s to D for regular packings of spheres with coordination numbers of (a) 12 (hexagonal close packing), (b) 8 (body-centered cubic or random dense packing), and (c) 6 (simple cubic). r_s is radius of largest sphere that will pass between particles. From Ramsay and Booth [36].

2.

MONODISPERSE PARTICLES FROM SOLUTION

The need for ceramics with improved mechanical properties has led to tremendous interest in methods for preparation of superior powders. In particular, many studies have examined methods for making monodisperse particles from solution. This is based on the argument that the ideal powder should be pure, stoichiometric, dense, equiaxed (i.e., spheroidal), and nearly monodisperse. Uniform particle size is said to facilitate preparation of stable dispersions, as well as dense, uniform powder compacts [76]. However, if the particles are well dispersed, a denser body can be made with a range of particle sizes, as the smaller ones fit into the spaces between the larger ones. It is sometimes argued that the best sintering behavior will be obtained when the particles are arranged in ordered arrays, as in the top photo in Fig. 18a. However, both Fig. 17 and Fig. 18 show that the ordered arrays of particles contain stacking faults analogous to dislocations in crystals, so the pores occur on two scales: interparticle voids within ordered regions (called *domains*) and interdomain voids. (See Fig. 18b). When the gel is sintered, the densely packed domains sinter readily, but the defects (interdomain voids) disappear slowly and may remain as strength-limiting flaws in the sintered ceramic [77]. For sintering to high density it is more important to have a uniform *pore* size than a uniform *particle* size. In highly disordered gels, the pore size may actually be more uniform, because the absence of ordered domains means that no small intradomain pores exist. To achieve a densely packed, but disordered, structure that will sinter without retaining defects, Ring [77] recommends using sols with a variation in particle size in the range of 10 to 30% and shows that such particles can be prepared continuously in a plug flow or stirred tank reactor. When the polydispersity is on this order, computer simulations [78] show that ordered domains are eliminated and the particles pack with liquid-like order.

In this section we examine many of the methods that have been developed to make monodisperse materials. Whether or not they are ideal for ceramics fabrication, monodisperse particles are certainly important as model systems for testing theories of colloidal stability, light scattering, drying, and sintering.

2.1. Preparation of Spheres

Matijević and his colleagues have demonstrated that an enormous variety of materials can be grown from solution as monodisperse spheres (or polyhedra). Several widely applicable techniques are described in a series of excellent reviews [10,79–81]. The method of *forced hydrolysis* promotes deprotonation of hydrous metal ions,

$$n[M(H_2O)_p]^{z+} \rightarrow [M_n(H_2O)_{np-m}(OH)_m]^{(nz-m)+} + mH^+ \qquad (13)$$

followed by polymerization. Of course, hydrolysis and polymerization can be accomplished simply by neutralizing a sol with a base, but that does not give control over particle morphology. The best results are obtained under mild conditions and low concentration. For instance, a compound such as formamide, which thermally decomposes to yield ammonia, can be used to raise the pH gradually. Alternatively, the solution can simply be aged at elevated temperature. Matijević [79] points out that this technique

Fig. 20.

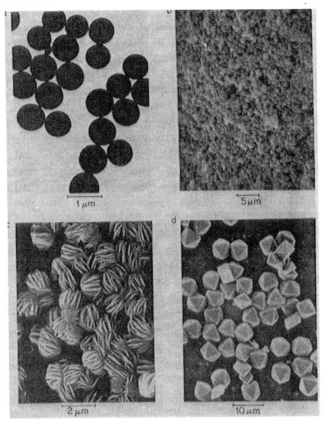

(a) TEM of aluminum hydroxide particles obtained by keeping a 0.002 M solution of $Al_2(SO_4)_3$ at 97°C for 48 h. (b) SEM of hematite, α-Fe_2O_3, particles from solution of 0.032 M $FeCl_3$ + 0.005 M HCl aged at 100°C for 2 weeks. (c) SEM of boehmite, α-AlOOH, particles from 0.0030 M solution of $Al(ClO_4)_3$ aged at 125°C for 12 h. (d) SEM of alunite, $Fe_3(SO_4)_2(OH)_5 \cdot 2H_2O$, particles from solution of 0.18 M $Fe(NO_3)_3$ + 0.27 M$(NH_4)_2SO_4$ aged at 80°C for 1.5 h. From Matijević [79].

reproducibly yields uniform particles, but is very sensitive to such factors as salt concentration, pH, nature of the anion, and temperature. Figure 20 shows examples of particles prepared in this way that are very uniform in size, though not necessarily spherical. Note that these particles are grown in very dilute solutions. The method is not limited to particles with one type of metal ion. For example, ferrites ($\leq 2\%$ Sr) can be made by aging a solution of $FeCl_2$ and $SrCl_2$; the particles are cubic with a narrow size distribution and mean size between 0.08 and 0.5 μm [82].

A given cation may yield different particle morphologies, depending on the conditions of growth. The anion used in the solution is remarkably important, as revealed by comparison of parts a and c in Fig. 20. Another striking example of the influence of the anion is the observation that chromium hydroxide grows as spherical amorphous particles from solutions containing sulfate or phosphate ions, but no particles grow from solutions containing chloride, nitrate, or acetate ions. Matijević [79] speculates that ions such as sulfate and phosphate may promote polymerization by coordinating with the hydrous oxide complexes. Polymerizing complexes tend to form amorphous particles, whereas discrete complexes (as in ferric sulfate) crystallize. Forced hydrolysis sometimes produces spherical crystalline particles (e.g., TiO_2 in the form of rutile, and Fe_2O_3 as hematite, as in Fig. 20b). The lack of faceting in such cases is attributed to the fact that the spheres are made of much smaller crystallites (e.g., 4-nm crystallites in the hematite spheres).

Another widely applicable method of preparation of monospheres is *thermal decomposition of complexes* formed by metal ions with chelating agents, such as triethanolamine, nitrilotriacetic acid, and (ethylenedinitrilo) tetraacetic acid [79]. The chelated ions are dissolved in a strongly basic solution, so they are hydrolyzed at a rate determined by the rate of decomposition of the complex. This approach allows a broader range of experimental conditions than forced hydrolysis, including addition of oxidizing or reducing agents, and control of reaction rate through the choice of chelating agent. This method also requires dilute solutions; for example, to obtain a narrow particle size distribution of barium titanate, the concentration of the titanium precursor (viz., titanium isopropoxide) must be <0.02 M [83].

In addition to a variety of single-component oxides (e.g., Fe_2O_3, CuO, V_2O_5, ZnO), this method has been used to make composite particles, as well as several types of nonoxides. Garg and Matijević [84], who coated spindle-shaped magnetite particles with chromia, cite many other examples of particles with cores of one composition surrounded by shells of another. Metallic particles (e.g., Ni) are prepared by adding a reducing agent, such as hydrazine, to the solution. Cadmium sulfide particles are grown [79] by slow decomposition of thioacetamide (TAA) in a solution of cadmium nitrate (0.0010 M $Cd(NO_3)_2$ + 0.0050 M CH_3CSNH_2 at pH 0.75, 26°C for 36 h). The particles are spherical with an average diameter of ~1 μm. Mixed sulfides and selenides can be made by using TAA and selenourea [10]. Ferric phosphate particles can be prepared by aging a very dilute solution of ferric perchlorate (0.00080 M) and phosphoric acid (0.030). Many other examples are cited by Sugimoto [81].

A widely used method for preparing monodisperse silica spheres, developed by Stöber, Fink, and Bohn (SFB) [85], is to hydrolyze tetraethylorthosilicate (TEOS) in a basic solution of water and alcohol. The overall hydrolysis and condensation reactions, respectively, can be written as

$$Si(OC_2H_5)_4 + 4H_2O \rightarrow Si(OH)_4 + 4C_2H_5OH \qquad (14)$$

and

$$Si(OH)_4 \rightarrow SiO_2 + 2H_2O. \qquad (15)$$

Actually, as discussed in depth in the preceding chapter, hydrolysis and condensation are concurrent. The purpose of writing out these schematic reactions is to point out that the stoichiometric amount of water for hydrolysis is 4 moles for every mole of silicon alkoxide, or 2 moles if condensation goes to completion. However, in the preparation of particles, the ratio of water to TEOS is typically more than 20/1 and the pH is very high, and both of these factors promote condensation. This encourages the formation of compact structures, rather than extended polymeric networks of the kind generally found in alkoxide-derived gels.

As practiced by SFB, alcohol, ammonia, and water are mixed, then TEOS is added, resulting in visible opalescence within ~10 minutes. Examples of the resulting particles are shown in Fig. 18; typically <5% of the particles differ by more than 8% from the mean size. The particle size depends on the reactant concentrations, as shown in Fig. 21. The influence of reaction

Fig. 21.

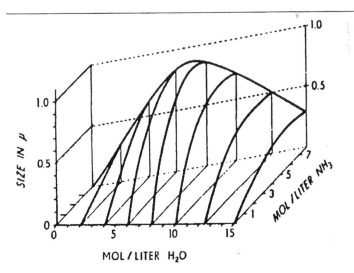

Final particle sizes as obtained by reacting 0.28 M TEOS with various concentrations of water and ammonia in ethanol. From Stöber et al. [85].

conditions on particle size has been further studied by others [86–88] with similar results. With TEOS as the source of silicon, the largest mean size that can be produced with a narrow size distribution is ~0.7 μm. The breadth of the distribution increases and the sphericity of the particles deteriorates when the concentration of TEOS exceeds ~0.2 M [86]. In addition to TEOS and ethanol, SFB examined alkoxides with larger and smaller alkoxy groups, and used mixtures of alcohols as solvents. In general, smaller particles and narrower size distributions are obtained with smaller alcohols as solvent; faster reactions and smaller particles are obtained with smaller alkoxides. Particles up to ~2 μm can be made using tetrapentylorthosilicate [85], by adding more alkoxide after the particles have formed [87], or by performing the reaction at low temperature [88]. If it is desired to produce small particles (radius ~20–35 nm) with uniform size, Ludox® can be introduced into the reaction mixture as a nucleant [89].

This method has now been extended to compositions other than pure silica. Jubb and Bowen [90] doped SiO_2 spheres with ~2% B_2O_3 by adding boron alkoxide during hydrolysis of the TEOS. To reduce the hydrolysis rate of the boron compound, so that it would be comparable to that for TEOS, it was necessary to use a larger alkoxy group (tri-*n*-butyl borate). Fegley and Barringer [91] have reviewed work in their group that has led to preparation of monospheres of pure and doped SiO_2, TiO_2, ZrO_2, and ZrO_2-Al_2O_3 from alkoxides. In the latter case, the alumina was added as a (conventional) powder; otherwise, all of the particles were spheroidal, submicron, uniform, and amorphous. Heistand *et al.* [92,93] used ethyl *tert*-butoxy zinc to make monospheres of ZnO with diameters of ~0.2.

The quality (i.e., sphericity and monodispersity) of the particles made by these methods is sensitive to the choice of reactants and reaction conditions. For example, Fegley *et al.* [94] found that unagglomerated monodisperse spheres could be made from Zr *n*-propoxide, but not from Zr isopropoxide; yttria-stabilized zirconia monospheres were successfully prepared by using ethanol as a solvent, but agglomerated particles resulted if isopropanol was used. Even better zirconia monospheres were made by Ogihara *et al.* [95] using Zr butoxide with ethanol as solvent. The best results were obtained in a narrow range of composition (near 0.1 M alkoxide and 0.1–0.2 M water) with a remarkably small (substoichiometric!) water : alkoxide ratio. Similarly, Ogihara *et al.* [96] made high-quality monospheres of Ta_2O_5 from Ta pentaethoxide in ethanol, but obtained agglomerated dumbbell-shaped particles with butanol as solvent. In this case, the best water : alkoxide molar ratio was ~5 : 1.

Although the solutions used in the SFB-type processes are more concentrated than those studied by Matijević *et al.*, they are still too dilute for production of commercial amounts of material. If appears, however, that the rate of production can be substantially increased, and that particles with very narrow size distributions can be made by conventional chemical-engineering techniques [77]. A particularly interesting development is the discovery that titania particles can be grown in the presence of hydroxypropyl cellulose, which acts as a steric stabilizer during growth, but does not become incorporated into the particles [35,97,98]. This allows growth without aggregation in more concentrated solutions.

2.2. Structure of Spheres

Monospheres of various compositions made by hydrolysis of alkoxides ("SFB spheres") have several microstructural features in common [91]: (1) the skeletal density measured by helium pycnometry is $<80\%$ of that of the corresponding oxide; (2) the BET surface area is much greater than the geometric area calculated from the particle diameter measured by TEM; (3) the weight loss on heating is ~ 8–20%, most of which is water, with $<1\%$ of the weight of the dried particles being residual organics. The low skeletal density is confirmed by the observation [87] that silica spheres shrink $\sim 5\%$ in diameter (corresponding to $\sim 15\%$ increase in density) when subjected to intense radiation in the TEM. Fegley and Barringer [91] attribute the high surface area to precipitates that form on the surfaces of the particles during drying, and note that much lower areas result if the washing is done with alcohol instead of water. However, the low skeletal density suggests that pores are present within the particles, and such pores were seen in nitrogen adsorption–desorption isotherms by Lecloux et al. [99] in a thorough study of silica spheres. They prepared particles with diameters ranging from 8 to 200 nm, some of which are shown in Fig. 22. The fraction of the pore volume contributed by *micropores* (i.e., those with diameters <2 nm) increased from $\sim 2\%$ of the total porosity in 200-nm particles to $\sim 50\%$ in the smallest particles. When the spheres were pressed into slabs, they found that the difference between the total pore surface area and the micropore area equalled the geometric area of the interparticle pores.

The observations on the surface area of spheres by Fegley and Barringer and by LeCloux et al. are not necessarily contradictory. Van Helden and Vrij [100] studied SFB spheres (diameter 110 nm) by light scattering and found that the index of refraction within the particle varied drastically with radius, from ~ 1.40 at the center to ~ 1.46 at the periphery. This suggests that the density of the particle increases as it grows, beginning from the highly porous state of the 8-nm particles in Fig. 22c and evolving toward larger particles with nonporous surfaces, as in Fig. 22a.

Fig. 22a.

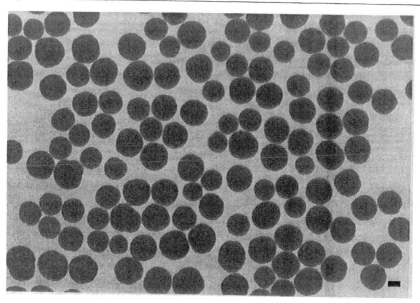

Fig. 22b.

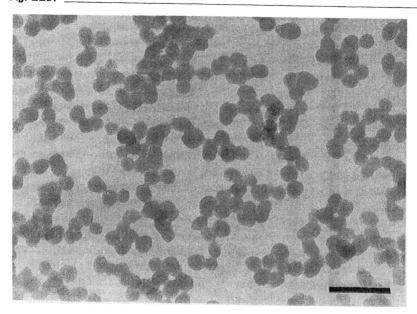

Fig. 22c. _____

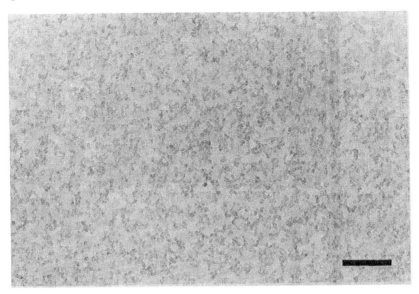

TEM of silica particles with diameters of (a) 206 nm, (b) 40 nm, (c) 8 nm. Note difference in scale: bar = 0.1 μm in each photo. From Lecloux _et al._ [99].

The ZnO powder made by Heistand _et al._ [93] actually looks fuzzy, as shown in Fig. 23. The dark-field TEM reveals that the 200-nm spheres are made of crystallites with diameters of ~20 nm. The surface area of the spheres decreases by a factor of ~3 if the solution is aged at room temperature for 19 days [92], indicating that a process of dissolution and reprecipitation is occurring. This process may account for the variation in density with radius seen in silica spheres: as growth proceeds, the supersaturation of the solution is reduced, facilitating rearrangement processes of the sort discussed in connection with Fig. 8. Consequently, the porosity and surface area are reduced and the density increases.

2.3. Mechanism of Growth

The growth of monodisperse particles is generally explained in terms of the theory of LaMer and Dinegar [101]. The concept is illustrated schematically in Fig. 24: (1) the supersaturation of the hydrous oxide is increased continuously (e.g., by a change in temperature or pH) until the critical concentration, c_N, is reached, where nucleation is extremely rapid;

Fig. 23. _____

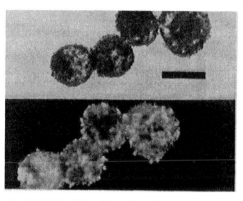

TEM of ZnO powder. (a) Bright field. (b) Dark field. From Heistand *et al.*, "Synthesis and Processing of Submicrometer Powder," pp. 482–496 in *Science of Ceramic Chemical Processing* (Wiley, New York, 1986).

(2) precipitation of particles reduces the supersaturation below the point c_o, where further nucleation is unlikely; (3) growth on the existing nuclei continues until the concentration is reduced to the equilibrium solubility, c_s. By producing a single burst of nucleation that consumes the excess solute, a single particle size can be achieved; if new nuclei form during the growth period, then a range of sizes results. Once nucleated, the rate of growth of the particles may be controlled by the diffusive flux of molecules to the particle, or by the rate of the condensation reaction between the particle and the solute. The kinetics of growth resulting from these processes have been analyzed by several authors and are discussed in the monograph by Nielsen [102].

Fig. 24. _____

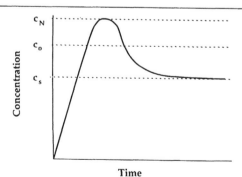

Schematic illustration of concentration of precipitating species before and after nucleation. After LaMer and Dinegar [101].

This mechanism may not seem to account for the porous structure and low density of the spheres described in the previous section. That is, if growth were occurring by diffusion of molecules to the surface of the sphere, one might expect the particle to be nonporous. In fact, the natural product of this mechanism of growth would seem to be polyhedral crystalline particles, of the sort described by Matijević [79]. However, the fact that amorphous oxides are frequently obtained reflects the fact that molecules may bond irreversibly on contact with the particle and be unable to reorient into the position required for a crystalline structure. Only if the attachment is weak enough to allow the molecule to redissolve and reattach is it possible to grow a crystal. This is much more likely in a dilute aqueous solution than in concentrated or alcoholic solutions of the kind used in the SFB process. Nevertheless, there are good reasons to doubt that the model of LaMer and Dinegar is applicable to the growth of such particles, as explained shortly.

Jean and Ring [97] examined the kinetics of growth of titania monospheres (made by hydrolysis of tetraethoxytitanate) and found that the data fit the theoretical curve for growth under diffusion control. However, the diffusion coefficient determined from the fit was $\sim 10^{-9}$ cm^2/s, which is of the order of the Brownian diffusion coefficient for a particle with a diameter of $\sim 4\,\mu$m, more than 10 times larger than the particles they grew! This enormous discrepancy was tentatively attributed to the retarding effect of the outward flow of water produced by the condensation reaction. Later Dirksen and Ring [103] reinterpreted these results in terms of aggregation of primary particles, as noted shortly.

Bogush and Zukoski [87] made a very thorough study of the growth kinetics of silica spheres according to the SFB process. They calculated the nucleation rate for the growth conditions used in their experiments and found that the concentration of silica in the solution remained above c_o (see Fig. 24) during most of the reaction period. Therefore, nuclei should have been created continuously, and a broad range of particle sizes would be expected, according to the theory of LaMer and Dinegar [101]; nevertheless, the spheres were monodisperse. The kinetics of growth were found to fit the theoretical curve for diffusion-controlled growth, but the diffusion coefficient was $\sim 10^{-12}$ cm^2/s, corresponding to Brownian diffusion of a particle with a diameter of one *millimeter*! Moreover, the kinetics were equally well fit by the theory for particle growth controlled by a surface reaction. Further studies [104] directly demonstrated that nucleation was continuing throughout the period of growth of the monospheres. The most impressive evidence against the classical theory of growth is electron micrographs showing the coexistence of large spheres with rough surfaces and large numbers of tiny particles which are apparently aggregating. (See Fig. 25.)

Fig. 25a.

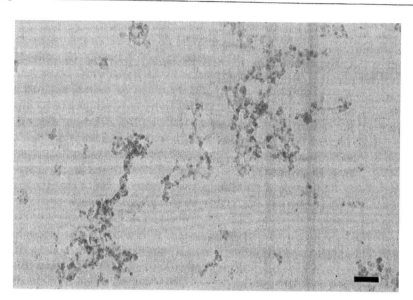

Fig. 25b.

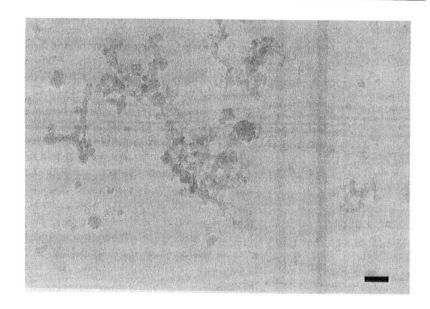

Fig. 25c.

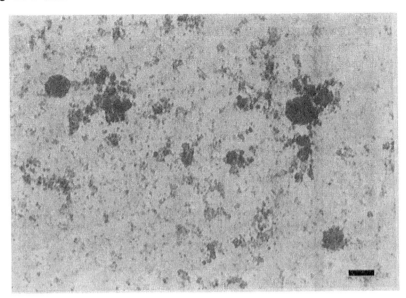

Fig. 25d.

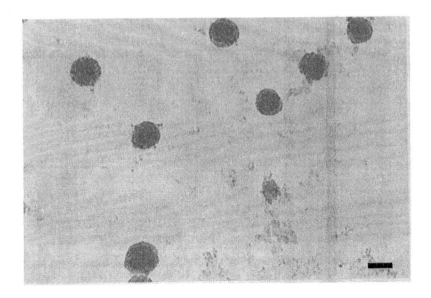

Fig. 25e.

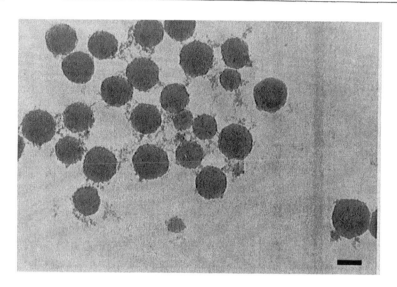

Fig. 25f.

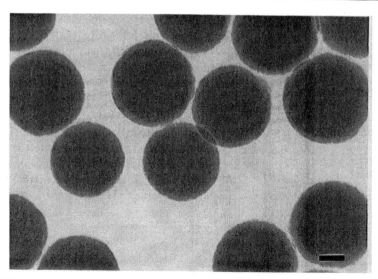

TEM showing growth of spheres by SFB method occurring by aggregation of spheres. Samples taken from solution of 0.17 M TEOS, 1.3 M NH$_3$, and 3.3 M H$_2$O at various times after mixing (a) 2 min, (b) 4 min, (c) 6 min, (d) 8 min, (e) 10 min, (f) >3 h. Bar = 100 nm. From Bogush and Zukoski [87].

This mechanism of growth is by no means limited to silica. Photos similar to Fig. 25 show that monospheres of iron oxides also grow by aggregation of smaller particles [105,106], with the largest spheres occurring near the isoelectric point of the oxide. Titania pigments, made by hydrolysis of titanium sulfate, consist of spheres 0.2–0.5 μm in diameter, but the spheres are made up of 6–8-nm crystallites [107]. A particularly remarkable type of growth is exhibited by ceria [108], in which aggregation produces uniform hexagonal particles made up of tiny crystallites, as shown in Fig. 26. These are crystalloids of the type shown schematically in Fig. 15. Similar behavior is observed for ferric oxides. Murphy *et al.* [109] showed that the product of hydrolysis of ferric salt is a spherical polycation 1.5–3 nm in diameter; those primary particles link into rods of two to four spheres. During aging the rods may grow to 20 nm long and 3 nm thick, and the individual spheres become indistinguishable as a result of dissolution and reprecipitation. For aggregates to form perfect spheres, there must be sufficient short-range repulsion to allow the particles to settle into the configuration of lowest

Fig. 26. ──────────────────────────────────────

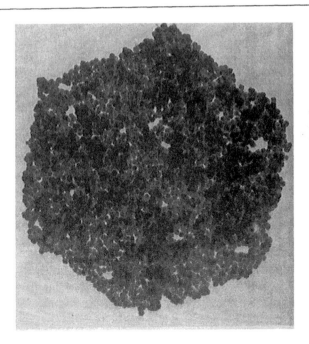

TEM of aggregate obtained by aging at 90°C for 12 h a solution of 0.015 M $(NH_4)_2Ce(NO_3)_6$, 0.064 M H_2SO_4, and 0.16 M Na_2SO_4, pH 1.1. From Hsu *et al.* [108].

energy, as in a coacervate. To form chains or polyhedra, there must also be anisotropy in the repulsive double layer.

Thus, the most plausible mechanism for the growth of monospheres is that small particles nucleate and aggregate. The growth of the primary particles may obey the classical model (Fig. 24), but the aggregation into large coacervates or polyhedra is expected to be described by Smoluchowski's equation. (See Chapter 5.) This would account not only for the kinetics of growth, but also for the rougher texture seen on smaller spheres (as in Fig. 22a and Fig. 23), as well as for the high surface areas and low skeletal densities that have been measured. The increase in density with radius of the particles [100] and the reduced microporosity in larger spheres [99] suggest that rearrangement on a molecular scale is concurrent with aggregation and accelerates as the supersaturation is reduced.

3.

OTHER METHODS OF MAKING PARTICLES

3.1. Aerosols

An *aerosol* is a colloidal dispersion of liquid droplets in a vapor. Aerosols can be used to make oxide powders in a variety of ways. For example, an aqueous sol can be sprayed into alcohol, where the aerosol droplets gel to produce spheres [110]. Subsequent firing produces perfect dense oxide spheres whose size depends on the concentration of the initial sol. Schwartz *et al.* [4] prepared powders of lead lanthanum zirconate titanate by spraying a salt solution into ammonium hydroxide and then spray-dried the resulting powder to produce uniform spheres 0.5–2 μm in diameter. In such complex compositions, aerosols are advantageous, because each droplet is a small "reaction vessel," and heterogeneity cannot occur on a scale larger than the size of the droplet.

A widely practiced aerosol technique, generally known as *evaporative decomposition of solutions* (EDS) [111], involves spraying of salt solutions into a furnace, where the droplets dry and the salts decompose into oxides. The equilibrium decomposition temperature of salts is generally below 550°C, but the reaction might be completed at much higher temperatures during flash evaporation; furnace temperatures for EDS are typically ~900–1000°C. This method is now practiced on an industrial scale [112]. Dell [5] found that solutions that can gel (e.g., those based on Al_2O_3 or Fe_2O_3) tend to form hollow shells, while those that cannot (e.g., NiO, CoO) yield dense spheres. This generalization may not be valid, however, as shells can be produced by crusts of precipitated salts as well as by gel layers. Gardner *et al.* [113] found that EDS produced shells and shell fragments of MgO, NiO, and ZnO when aqueous nitrate solutions were used, but dense oxide particles resulted from decomposition of acetate solutions. In the latter case, they speculated that the exothermic oxidation of the organic material shattered the shells, so that the final particles (0.1–0.3 μm in diameter) represented small fragments of shells (originally 1–3 μm). In either case, the particles were polycrystalline, with grain sizes of 15 to 34 nm. It should be noted that hollow shells are sometimes the desired product [114], for use as targets for laser fusion.

An elegant variation on this theme was performed by Visca and Matijević [115]. They passed a gas stream through a vapor of AgCl to condense nuclei and then passed that aerosol over a film of titanium ethoxide (or isopropoxide or chloride). The vapor of the titanium compound condensed on the AgCl nuclei; the droplets were then hydrolyzed at <100°C in a chamber containing water vapor and then were heated to ~150°C to complete the reaction. The resulting amorphous titania particles were nearly monodisperse with mean sizes ranging from 0.06 to 0.6 μm. A similar procedure has been used to make particles containing mixed oxides such as TiO_2–SiO_2 [116]. Mayville *et al.* [117] extended this procedure to produce titania particles with coatings of polyurea up to 0.25 μm thick, to provide a steric barrier when the particles were subsequently dispersed. They transported the aerosol of titania into a chamber containing vapor of hexamethylenediisocyanate; this vapor condensed on the particles and was then polymerized by exposure to vapor of ethylenediamine. The electrophoretic mobility of the coated particles was found to be the same as for the pure polymer. It may be possible to produce monodisperse aerosols at a rapid rate without seeding by use of a properly designed nozzle [118], but this has not yet been demonstrated experimentally.

Instead of using a furnace as a heat source, the salt solution can be sprayed into a flame or plasma. Dell [5] found that dense spheres resulted when the aerosol was directed into a propane flame. In that case, the temperature was high enough to sinter, but not to vaporize, the oxide. However, radio frequency plasmas generate temperatures of ~8000 K, so the solution decomposes into atoms and the final powder is independent of the droplet size in the original aerosol. For example, Kagawa *et al.* [119] made MgO particles with diameters of 16 to 44 nm by spraying a nitrate solution into an argon plasma. This approach has been used to produce a variety of single and multicomponent oxides [120], but care must be taken to avoid quenching the plasma.

3.2. Vapor-Phase Methods

Powders can be prepared by oxidation, reduction, decomposition, and other chemical reactions, using high temperatures produced by a furnace, laser, electron beam, plasma, or flame. The advantages of vapor methods include high purity of product, because of ease of purification of reactants and the absence of contamination by contact with a container; small particle size (typically ≪1 μm); homogeneity, because of atomic-scale mixing in the vapor phase; and ability to prepare nonoxide compositions. These processes, some of which are of great commercial importance, are briefly reviewed here. The methods are organized according to the heat source used. The use of vapor-generated powders to make gels is discussed in Section 4.

3.2.1. FURNACE

Mazdiyasni and his colleagues have long advocated the use of metalorganic precursors, including alkoxides, acetylacetonates (acac), and trifluoro-acac for the preparation of pure oxides. For example, zirconia can be prepared by decomposing zirconium *tert*-butoxide in a furnace at 325 to 500°C in an atmosphere of nitrogen [121]:

$$Zr(OC_4H_9)_4 \rightarrow ZrO_2 + \text{alcohol} + \text{olefin}. \tag{16}$$

This sort of reaction, in which a molecule is fragmented at high temperature, is called *pyrolysis*. The product of Eq. 16 was crystalline zirconia with an average particle size of 5 nm and no particle larger than 30 nm.

Oxidation of halide compounds, such as

$$SiCl_4 + O_2 \rightarrow SiO_2 + 2Cl_2, \tag{17}$$

yields similar results. At 1200°C, Eq. 17 yields silica particles in the form of amorphous spheres with diameters of 15 to 100 nm; high BET area indicates that the particles are porous or rough [122]. Oxidation of AlBr$_3$ at 950 to 1200°C produces nonporous 35–300 nm particles of alumina [123]. The most important commercial application of this type of reaction is the production of titania pigment by oxidation of TiCl$_4$, which is performed at the rate of hundreds of thousands of tons per year. The resulting powder is equiaxed and nearly monodisperse, with a primary particle size of ~0.25 micron.

3.2.2. LASER

Haggerty and his colleagues have developed the science and technology of using a laser to produce ceramic powders [124]. A carbon-dioxide laser generates an intense beam of light with a wavelength of $\sim 10.6\,\mu$m, which is strongly absorbed by a wide range of molecules. Therefore, the beam excites the molecules thermally, in much the same way as do the gases in a hot furnace. In contrast to a furnace or flame, however, the hot zone of the laser is very small and the heating rates are very high ($\sim 10^{6\circ}$C), so nucleation occurs suddenly and this makes the breadth of the particle-size distribution small. Haggerty has made silicon metal particles by pyrolysis of silane,

$$SiH_4 \rightarrow Si + 2H_2, \tag{18}$$

silicon nitride by reaction of silane and ammonia,

$$3SiH_4 + 4NH_3 \rightarrow Si_3N_4 + 12H_2, \tag{19}$$

and silicon carbide by reaction of silane with ethylene,

$$2SiH_4 + C_2H_4 \rightarrow 2SiC + 6H_2. \tag{20}$$

Typically, the largest particle is ~ 3 times as big as the smallest, and the mean size is 0.02–$0.2\,\mu$m. Although the particles seem to be aggregated when examined in the TEM, light scattering from sols reveals that the particles are not bonded together. Measurements of surface area and density indicate that the particles are nonporous.

3.2.3. ELECTRON BEAM

An electron beam can be used as a heat source to evaporate a solid oxide, producing an oxide vapor that condenses into a fine powder. The rate of production depends on the vapor pressure of the oxide, which follows from the amount of energy that can be concentrated on the source material. Ramsay and Avery [125] used this method to make unagglomerated powders of MgO, CaO, Al_2O_3, SiO_2, ZrO_2, CeO_2, and U_3O_6. The particles were small (<10 nm), nonporous, and exhibited good sintering behavior. They claim that this process is comparable in efficiency to the plasma methods discussed in the next subsection.

3.2.4. PLASMA

The use of plasma for synthesis of ceramic powders has been reviewed by Phillips and Vogt [120]. All of the reactions just discussed can be performed in a plasma. The difference is that the temperatures are so high (7–10,000 K) that the reactants are vaporized, so a wider variety of precursors (including sols, powders, and solutions, in addition to vapors) can be used. In a dc arc plasma, heat is transferred to the gas stream by physical contact with the electrodes, providing efficient use of powder, but introducing a risk of contamination of the product with the electrode material. The temperatures in such plasmas are ~15,000 K and the axial velocity of the exiting gas is ~100 m/s. Radio frequency (RF) plasmas heat the gas stream by electromagnetic interaction, so the risk of contamination is removed, but the efficiency is half that of the dc arc plasma. RF plasmas generate temperatures of ~8,000 K and a turbulent gas stream moving at ~10 m/s. A hybrid reactor uses a dc arc to produce a rapid gas stream to feed into an RF plasma. This decreases the turbulence of the RF discharge and increases the temperatures and flow velocities. The inhomogeneity of the conditions in the reaction zone lead to powders with a broad range of particle sizes and, often, a high degree of agglomeration.

Plasmas have been used to produce a wide range of materials, many of which are described in reviews by Hamblyn and Reuben [126] and by Johnson [12]. As indicated by the examples listed in Table 2, products include metals from pyrolysis of halides, oxides from oxidation of halides, and a variety of nonoxides, including carbides, nitrides, and borides. The powders have submicron particles, are generally agglomerated, and are either amorphous or crystallized into nonequilibrium phases.

3.2.5. FLAME OXIDATION

A commercially important method of making oxide powders is oxidation of halides in a flame. Equation 17 is performed in a H_2/O_2 torch to produce silica powders, which are sold as Cab-o-Sil® (Cabot Corp.) and Aerosil® (Degussa). This material is sometimes called *fumed silica*. Titania and alumina are prepared in a similar way from $AlCl_3$ and $TiCl_4$ by Degussa. The same sort of reaction, produced in a H_2/O_2 or CH_4/O_2 flame, is used in the preparation of optical waveguides for telecommunications [127]. Compositions of waveguides typically include silica, germania, and/or phosphorus pentoxide, and the total impurity content (including hydroxyl ions) must be in the low ppb range. When performed in a flame, Eq. 17 is often called *flame hydrolysis*, but this is a misnomer. Although the halides react vigorously with water at room temperature, the oxidation reaction is faster than hydrolysis at the high temperatures of the flame [128].

The growth of particles in a flame has been analyzed by Ulrich [129]. The reaction occurs in the vapor phase, and liquid droplets of the oxide nucleate rapidly, then coalesce. The morphology of the aggregates is strongly dependent on the residence time in the flame [130], as illustrated by the titania particles shown in Fig. 27. Both powders were made by flame oxidation of $TiCl_4$, but the large spheres had a longer residence time in the flame

Table 2.

Powders Made by Vapor Reactions in Plasmas.

Reaction	Precursors	Products	Reference
Oxidation	AlC_3	Al_2O_3	HR
	$TiCl_4$	TiO_2	HR
	$SiCl_4$	SiO_2	HR
	$AlCl_3$, $TiCl_4$	$Al_2O_3-TiO_2$	J
	$AlCl_3$, $CrCl_3$	$Al_2O_3-Cr_2O_3$	HR
	$AlBr_3$, $SiCl_4$	$Al_2O_3-SiO_2$	GM
	$CrCl_3$, $TiCl_4$	$Cr_2O_3-TiO_2$	BBL
	Zn	ZnO	HR
	Sb	Sb_2O_3	HR
	WO_3	WO_3	HR
	MoO_3	MoO_3	HR
Reduction	Al_2O_3, H_2, CH_4, CO	Al	HR
	$SiCl_4$	Si	HR
	SiH_4	Si	HR
	BCl_3	B	HR
Carbide	$SiCl_3CH_3$	SiC	HR, J
	BCl_3, H_2, CH_4	B_4C	HR
	TaC	TaC	J
Boride	Ti, B	TiB_2	HR
	TiO_2, B	TiB_2	HR
	TiO_2, B_2O_3, C	TiB_2	HR
	ZrB_2, B_2O_3	ZrB_2	HR
Nitrides	$SiCl_4$, NH_3	Si_3N_4	J
	B_2O_3, NH_3	BN	J

HR: original reference cited by Hamblyn and Reuben [126].
J: original reference cited by Johnson [12].
GM: M.S.J. Gani and R. McPherson, *J. Mater. Sci.*, **12** (1977) 999–1009.
BBL: T.I. Barry, R.K. Bayliss, and L.A. Lay, *J. Mater. Sci.*, **3** (1968) 239–243.

(though still on the order of milliseconds). The unaggregated spheres in Fig. 27b are believed to result from coalescence of aggregates of the sort in Fig. 27a. Regardless of the residence time, the product of flame oxidation is generally a metastable phase: amorphous silica; anatase, rather than the rutile form of titania; delta or theta, rather than alpha, alumina. It is typical of rapidly quenched liquids that they crystallize into the high-temperature modification of the crystal, presumably because the structure of that phase is nearest to the structure of the liquid from which it forms.

Fig. 27.

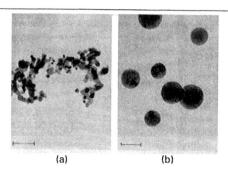

(a)   (b)

Electron micrographs of titania particles made by flame oxidation of $TiCl_4$. (a) Degussa titania. (b) Larger particles result from longer residence time in flame, allowing aggregates (a) to coalesce into single spheres (b). From Scherer [130]. Bar $= 0.1\mu m$.

The structure of aggregates of silica made by flame oxidation has been studied extensively. The results, reviewed by Schaefer [131], lead to the following physical picture. The initial aggregation process in the flame is ballistic, which means that the mean free path of the aggregating species is large compared to the cluster size. The straight-line paths allow incoming monomers to penetrate deeply into target clusters, resulting in relatively dense structures. This stage of growth continues until the particles reach diameters of ~9 nm. For powders with the highest surface areas (corresponding to very short residence times in the flame), the surfaces of these primary particles are rough; longer residence times allow the surfaces to smooth by viscous flow. Once growth has made the particles large compared to the mean free path, the trajectories change from ballistic to Brownian. The meandering path of a Brownian particle encourages attachment of the incoming cluster to the periphery of the target cluster, creating ramified fractal structures of the sort seen in Fig. 27a.

This interpretation is based on the scattering behavior of suspensions of Cab-o-Sil® (examined with visible light and SAXS [132,133]) and Aerosil® (visible light and SANS [43]). As illustrated in Fig. 12, on the smallest length scale ($< a_2$) the slope of the plot of scattered intensity reflects the nature of the primary particles. For most samples, the slope is near -4, indicating that the particles are smooth and nonfractal, but for powders with the highest surface area (~ 400 m^2/g) the slope is smaller, indicating that the particles are rough [133]. On a longer length scale, the slope of the plot approaches ~ -1.8, which is the fractal dimension characteristic of growth by diffusion-limited cluster aggregation. Remarkably similar results were obtained in all these studies: the crossover point (a_2) was very close to the primary particle size measured by TEM, and the fractal dimensions of the clusters were in the range of -1.7 to -2.0 for all of the materials studied. Ramsay and Scanlon [43] noted that the fractal dimension of the sol decreased under conditions of pH that enhanced the surface charge on the particles. Evidently, when the repulsive barrier is low, the clusters can interpenetrate to some extent, increasing their apparent density and raising the fractal dimension [134].

3.3. Other Solution Methods

The variety of approaches that have been explored for preparation of powders is enormous. Reviews by Johnson [12] and Turova and Yanovskaya [135] describe many, though only a few are cited here. Whole conferences are now devoted to this topic (e.g., ref. [136]).

If a multicomponent oxide is to be made from a salt solution, it is necessary to prevent segregation of the various salts (which inevitably differ in solubility) during drying. Several methods are based on the idea of trapping precipitated salts in an organic matrix. Pechini [8] patented a method for making multicomponent powders containing oxides of Ti, Nb, or Zr, by dissolving hydrous oxides or alkoxides together with a polyhydroxy alcohol and a chelating agent, such as citric acid. Heating to remove solvent yields an amorphous resin, which can be thermally decomposed to obtain a homogeneous oxide powder; the resin can also be applied on a substrate as a film $0.3-0.5\,\mu m$ thick. For example, one can make strontium titanate [137] by heating at 150°C to form a resin, charring at 250°C, then calcining at 700°C to form an agglomerated, but homogeneous powder. The same approach has been used to make stoichiometric powders of barium titanate, yttria-doped zirconia, and La_2NiO_4 [138]. The latter powders were amorphous after firing at 400°C in air, but crystalline if the atmosphere was oxygen (presumably because the exothermic oxidation raised the actual temperature of the powder above that of the furnace). The agglomerated powders had crystallite sizes of ~ 10 nm. A very similar method has been used by Marcilly $et\ al.$ [139] to make a variety of multicomponent oxides. Dell [5] made spheres by adding $1-5\%$ polysaccharide to an aqueous salt solution and dripping it into a solution of NaOH or NH_4OH. This produced beads with the hydrous metal oxides incorporated in an organic gel. The alkali could be washed out without removing the incorporated oxides, and the organics could be burned out to obtain oxide powder.

A few methods have been developed for making nonoxide powders in solution. Johnson $et\ al.$ [140] made sulfides by mixing a solution of a metal alkyl in toluene with a solution of toluene saturated with H_2S. They produced fine aggregated powders of ZnS from ZnR_2, where R = methyl, ethyl, or *tert*-butyl. Similar results were obtained with triethyl aluminum and diethyl magnesium. Ritter and Frase [141] describe a method for making carbides and borides by mixing metal chlorides and sodium metal in a nonpolar solvent. For example, titanium diboride is made in heptane at 25 to 160°C according to the reaction

$$10Na + TiCl_4 + 2BCl_3 \rightarrow TiB_2 + 10NaCl. \qquad (21)$$

When heated in vacuum to 700°C, the sodium chloride evaporates, leaving the crystalline boride powder behind. They have also made SiC and B_4C by this method.

4.

DISPERSION OF PYROGENIC PARTICLES

In this section we discuss methods of making gels from suspensions of particles made by flame oxidation. This class of gels is singled out because it will provide an enlightening contrast with the properties of alkoxide-derived gels, to be discussed in subsequent chapters. An advantage of this type of gel is that the particle size (typically ~ 0.05–$0.2\,\mu m$) facilitates drying without cracking, yet allows sintering at modest temperatures [130]. Some of these methods are discussed in a review by Rabinovich [142] on methods of making glass by sintering.

Ehrburger *et al.* [143] mixed Aerosil® in water, methanol, and undecane ($C_{11}H_{24}$) to find the solids content at which spontaneous gelation occurred. This material is pure silica with an aggregate structure similar to that shown in Fig. 27a; several grades are commercially available, with surface areas up to $\sim 200\ m^2/g$. They found that gels resulted when the volume of solid phase in the suspension was equal to the tap volume (the volume occupied by loose powder shaken in a dry vessel). Grades of Aerosil® with higher surface areas, having more ramified aggregates, created space-filling structures (and gelled) at lower volume fractions. Ramsay and Avery [144] found that dispersions of such materials dried to yield gels with porosities >70%, implying that the coordination number of the average particle was between 3 and 4; random dense packing gives a coordination number of ~ 8. The structure of these gels is not very different from that of optical waveguide preforms made by depositing such particles directly from the flame [145]. The high porosity of the gels has been found useful for making chromatographic columns [146].

Rabinovich *et al.* [147–150] made gels from Cab-o-Sil® flame-generated silica with a surface area of ~230 m²/g. The silica was mixed in water (sometimes with boric acid, to make B_2O_3–SiO_2 compositions) at concentrations up to ~40 wt% (24 vol%). The pH of the water dropped to 2.7, presumably because of chlorine chemisorbed on the particles (which are made according to Eq. 17), as well as dissociation of acidic silanol groups. After 1 to 2 h, the slurry would gel spontaneously, and would crack into pieces during drying. This material was then heated to 300 to 900°C, then redispersed in a high-speed blender, and allowed to gel and dry a second time. The resulting structure is shown schematically in Fig. 28. The aggregates formed in the blender have sizes on the order of microns, as do the inter-aggregate pores; the pores within the aggregates are similar in size to the particles (13–20 nm). Drying is facilitated by the presence of the large pores, which reduce the capillary pressure that causes cracking (see Chapter 8); the porosity of the dried gels was ~75%. However, when the dried gel was sintered, the smaller pores sintered first, trapping gas in the larger pores and causing the densified body to be hazy. The purpose of their work was to make materials for optical waveguides, so it was necessary to remove residual hydroxyl (which absorbs light at the wavelengths used for telecommunications) by treating the gel with chlorine gas at elevated temperatures. However, it was found that the chlorine would reboil when the glass was heated to temperatures sufficient for drawing fiber (~2000°C). This problem was eliminated by adding fluorine to the solution (as HF or NH_4F) and/or to the atmosphere during sintering [151]. The fluorine replaces

Fig. 28.

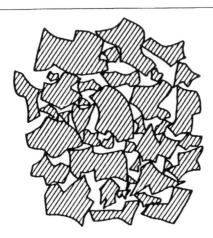

Schematic illustration of microstructure of twice-dispersed colloidal silica. From Rabinovich *et al.* [148].

Si–OH groups with Si–F, and the Si–F bond is so strong that it does not break to cause reboil during fiber drawing. This work has been reviewed by Rabinovich [152].

Scherer and Luong [153] made gels by dispersing Aerosil OX-50® in chloroform. This material has an average particle diameter of ~55 nm (BET surface area of 50 m²/g), and consists of individual spheres (as in Fig. 27b), rather than fractal aggregates (as in Fig. 27a). The particles were dispersed by an adsorbed layer of alcohol (e.g., decanol) that would hydrogen bond to the silanol on the surface of the particle, providing a steric barrier. Ordinarily, much thicker barriers are required, but the refractive index of chloroform and silica are nearly equal (which indicates that their Hamaker constants are similar), so the van der Waals attraction is relatively weak in this system. They were able to prepare suspensions containing up to ~30 vol% silica with viscosity ≤0.1 Pa·s. The sol could be molded and then gelled by exposure to a base, such as ammonia vapor. The gelling mechanism, illustrated in Fig. 29, was believed to result from charges created by deprotonation of the silanol groups: the positive ammonium ions would draw the negatively charged particles together, overcoming the weak repulsive barrier. This sort of attraction is expected in nonpolar solvents, according to calculations by Féat and Levine [154]. These gels could be dried without cracking, because they differed from those made by Rabinovich *et al.* in two essential respects: the pores were relatively large (similar in size to the particles, ~60 nm); and the surface

Fig. 29. _____

GELATION

BASE DEPROTONATES SILANOL, CHARGES CAUSE FLOCCULATION

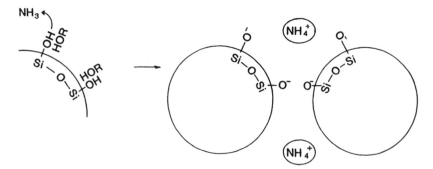

Gelation mechanism of silica dispersion in chloroform. Base deprotonates silanol. Charges cause flocculation. From Scherer and Luong [153].

tension of the solvent was ~1/4 as great as for water. Both of these factors reduced the capillary forces developed during drying. The porosity of the dried gels was 50–75%, decreasing as the solids content of the suspension increased [130]. The narrow distribution of pore sizes allowed the glass to sinter without trapping bubbles. The resulting glass could be dried by chlorine treatment without suffering reboil, presumably because the initial silanol concentration was so low that the retained concentration of chlorine did not exceed the solubility limit. The sintered glass was of such high quality that fibers drawn from it exhibited optical attenuation of <4 decibels per km at a wavelength of 0.83 μm [155]. The sintering behavior of this material is discussed in detail in Chapter 11.

Clasen [156,157] has combined the best elements of the processes just described. He stabilized Aerosil OX-50® in aqueous suspensions by use of ammonium fluoride, obtaining up to 50 wt% solids in a suspension with a viscosity of 0.05 Pa•s. This method avoids the small particle size that caused cracking of the gels made by Rabinovich et al., as well as the pollution and toxicity problems associated with the nonaqueous system used by Scherer and Luong. The fluoride increases the solubility of silica in the water, so the surface area of the powder was found to decrease from its initial value of 46 to 39 m²/g in the dried body. This solubility leads naturally to the growth of necks between particles, as silica deposits in those regions of negative curvature. (See Fig. 1.). The necks impart strength to the structure, allowing it to survive drying without cracking, so that a sample 3 cm thick could be dried in a matter of hours. The gel was then treated with chlorine gas to remove hydroxyls during sintering, resulting in a final OH content of ~20 ppb. The quality of the sintered glass was excellent. Samples have been formed by molding, extrusion, and electrophoretic deposition [157]. Clasen [158] has reported preparation of α-alumina particles with surface area of ~30 m²/g (diameter ~50 nm) by flame oxidation of $AlCl_3$, so this processing approach may be extended to preparation of crystalline materials.

A related process has been developed by Toki et al. [159] for manufacture of large plates, rods, and tubes of high-quality silica. They combine Aerosil OX-50® with TEOS in an alcohol solution and gel the system by hydrolysis and condensation of' the alkoxide; about half of the silica content is provided by the powder, and the rest by the alkoxide. The silica gel produced by the alkoxide acts as a binder, strengthening the gel so that it does not crack during drying. Pantano and colleagues [160,161] mixed Aerosil® with a titanium alkoxide and produced titania silicate glasses with zero thermal expansion. These methods are analogous to that of Shoup [72], in which colloidal particles are bound together by a gel grown from alkali silicate, and the resulting microstructure is similar.

5. _____

SUMMARY

The primary particles grown in aqueous solutions are nonporous, apparently because the solubility of the hydrous oxides allows dissolution and reprecipitation to form dense structures. This process is less facile in non-queous systems, so the gels grown in alcoholic systems are generally composed of polymeric species, rather than particles. Once particles have appeared, they will tend to aggregate because of attractive van der Waals forces, unless a barrier (either electrostatic or steric) to aggregation is erected. The arrangement of particles in a stable sol is analogous to that of atoms in a liquid, corresponding to random dense packing with a coordination number of ~8. Once destabilized, aggregates form that have a fractal structure whose dimension depends on the nature of the interparticle forces (e.g., anisotropic electrostatic or magnetic fields produce more ramified clusters).

Monodisperse particles can be formed by a variety of methods. The classical picture, that a single burst of nucleation is followed by diffusion-controlled growth, seems to apply only to the growth of primary particles (<10 nm). It appears that particles larger than ~10 nm are formed by aggregation of smaller primary particles, but it remains to be explained in detail why such aggregates become perfectly spherical. These particles are generally porous and have internal variation in density, being most dense at the periphery. This may reflect the greater ease of dissolution and reprecipitation at the end of the growth process, after the supersaturation of the solution has decreased, which allows the development of denser structures.

Many other techniques for producing particles have been developed, the most generally applicable of which are vapor-phase methods. The particles grow by nucleation and coalescence, resulting in highly ramified clusters with fractal dimensions of ~1.8. However, if the residence time at high temperature is sufficient, the clusters will coalesce into single spheres with diameters typically on the order of 0.05 to 0.1 μm. Particles of this type have been dispersed, gelled, dried, and sintered to obtain silica glass of high enough quality to make (in the laboratory) optical telecommunications fibers. These gels, being made of dense glass particles, are quite different from alkoxide-derived gels in their drying and sintering behavior, as we shall see.

REFERENCES

1. R.H. Ewell and H. Insley, *J. Res. NBS*, **15** (1935) 173–186.
2. R. Roy, *J. Am. Ceram. Soc.*, **39** [4] (1956) 145–146.
3. R. Roy, *J. Am. Ceram. Soc.*, **52** [6] (1969) 344.
4. R.W. Schwartz, D.J. Eichorst, and D.A. Payne in *Better Ceramics Through Chemistry II*, Mater. Res. Soc. Symp. Proc., **73**, eds. C.J. Brinker, D.E. Clark, and D.R. Ulrich (Mater. Res. Soc., Pittsburgh, Pa., 1986), pp. 123–128.
5. R.M. Dell in *Reactivity of Solids*, eds. J.S. Anderson, M.W. Roberts, and F.S. Stone (Chapman and Hall, New York, 1972), pp. 553–566.
6. R. Roy, *J. Am. Ceram. Soc.*, **52** [6] (1969) 344.
7. L. Levene and I.M. Thomas, U.S. Patent 3,640,093 (February 8, 1972).
8. M.P. Pechini, U.S. Patent 3,330,697 (July 11, 1967).
9. S. Komarneni, E. Breval, and R. Roy, *J. Non-Cryst. Solids*, **79** (1986) 195–203.
10. E. Matijević in *Science of Ceramic Chemical Processing*, eds. L.L. Hench and D.R. Ulrich (Wiley, New York, 1986), pp. 463–481.
11. J.L. Woodhead and D.L. Segal, *Proc. Br. Ceram. Soc.*, **36** (1985) 123–128.
12. D.W. Johnson, Jr., *Cer. Bull.*, **60** [2] (1981) 221–224, 243.
13. R.K. Iler, *The Chemistry of Silica* (Wiley, New York, 1979).
14. R.H. Ottewill, *J. Colloid Interface Sci.*, **58** [2] (1977) 357–373.
15. J.Th.G. Overbeek, *J. Colloid Interface Sci.*, **58** [2] (1977) 408–422.
16. D.H. Napper, *J. Colloid Interface Sci.*, **58** [2] (1977) 390–407.
17. *Colloid Science*, Vol. I, ed. H.R. Kruyt (Elsevier, Amsterdam, 1952).
18. T. Sato and R. Ruch, *Stabilization of Colloidal Dispersions by Polymer Adsorption* (Marcel Dekker, New York, 1980).
19. D.H. Napper, *Polymeric Stabilization of Colloidal Dispersions* (Academic Press, New York, 1983).
20. R.J. Hunter, *Zeta Potential in Colloid Science* (Academic Press, New York, 1981).
21. S. Voyutsky, *Colloid Chemistry* (Mir Publishers, Moscow, 1978).
22. J. Mahanty and B.W. Ninham, *Dispersion Forces* (Academic Press, New York, 1976).
23. G.A. Parks, *Chem. Rev.*, **65** (1965) 177–198.
24. G.D. Parfitt in *Dispersion of Powders in Liquids*, 3d ed. (Applied Science, London, 1981), pp. 1–50.
25. L.H. Allen and E. Matijević, *J. Colloid and Interface Sci.*, **31** [3] (1969) 287–296; **33** [3] (1970) 420–429; **35** [1] (1971) 66–76.
26. A. Habenschuss and F.H. Spedding, *J. Chem. Phys.*, **70** [6] (1979) 2797–2806; **70** [8] (1979) 3758–3763.
27. A.J. Rubin in *Emergent Process Methods for High-Technology Ceramics*, eds. R.F. Davis, H. Palmour, III, and R.L. Porter (Plenum, New York, 1984), pp. 45–57.
28. J.T.G. Overbeek in *Emergent Process Methods for High-Technology Ceramics*, eds. R.F. Davis, H. Palmour, III, and R.L. Porter (Plenum, New York, 1984), pp. 25–44.
29. E. Matijević in *Principles and Applications of Water Chemistry*, eds. S.D. Faust and J.V. Hunter (Wiley, New York, 1967), pp. 328–369.
30. D.L. Kepert, *The Early Transition Metals* (Academic Press, New York, 1972).
31. R.E. Johnson, Jr., *J. Colloid Interface Sci.*, **100** [2] (1984) 540–554.
32. M. Smoluchowski, *Phys. Z.*, **17** (1916) 557, 585; *Z. Phys. Chem.* **92** (1917) 129.
33. N, Fuchs, *Z. Phys.*, **89** (1934) 736.
34. Th.F. Tadros in *The Effect of Polymers on Dispersion Properties*, ed. Th.F. Tadros (Academic Press, New York, 1982), pp. 1–38.

35. T.E. Mates and T.A. Ring, *Colloids and Surfaces*, **24** (1987) 299–313.
36. J.D.F. Ramsay and B.O. Booth, *J. Chem. Soc., Faraday Trans. I*, **79** (1983) 173–184.
37. J.D.F. Ramsay, *Chem. Soc. Rev.*, **15** (1986) 335–371.
38. E. Crucean and B. Rand, *Trans. J. Brit. Ceram. Soc.*, **78** [3] (1979) 58–64.
39. P. Meakin and R. Jullien, *J. Chem. Phys.*, **89** [1] (1988) 246–250.
40. W.Y. Shih, I.A. Aksay, and R. Kikuchi, *Phys. Rev. A.*, **36** [10] (1987) 5015–5019.
41. J. Kramer, R.K. Prud'homme, and P. Wiltzius, *J. Colloid Interface Sci.*, **118** [1] (1987) 294–296.
42. G.C. Ansell and E. Dickinson, *Faraday Disc. Chem. Soc.*, **83** (1987) 167–177.
43. J.D.F. Ramsay and M. Scanlon, *Colloids and Surfaces*, **18** (1986) 207–221.
44. J.D.F. Ramsay, R.G. Avery, and L. Benest, *Faraday Disc. Chem. Soc.*, **76** (1983) 53–63.
45. A.K. van Helden and A. Vrij, *J. Colloid Interface Sci.*, **78** [2] (1980) 312–329.
46. C.G. de Kruif, W.J. Briels, R.P. May, and A.Vrij, *Langmuir*, **4** (1988) 668–676.
47. R.J.R. Cairns, R.H. Ottewill, D.W.J. Osmond, and I. Wagstaff, *J. Colloid Interface Sci.*, **54** (1976) 45–51.
48. A. Bleier and E. Matijević, *J. Chem. Soc. Faraday Trans. I*, **74** (1978) 1346–1359.
49. D.W. Schaefer, J.E. Martin, and K.D. Keefer, *J. de Physique*, **46** [3] (1985) C3-127–C3-135.
50. D.W. Schaefer, J.E. Martin, P. Wiltzius, and D.S. Cannell, *Phys. Rev. Lett.*, **52** [26] (1984) 2371–2374.
51. D.W. Schaefer, R.A. Shelleman, K.D. Keefer, and J.E. Martin, *Physica*, **140A** (1986) 105–113.
52. M.A.V. Axelos, D. Tchoubar, and R. Jullien, *J. Physique*, **47** [10] (1986) 1843–1847.
53. C. Aubert and D.S. Cannell, *Phys. Rev. Lett.*, **56** [7] (1986) 738–741.
54. R. Jullien, *Phys. Rev. Lett.*, **55** [16] (1985) 1697.
55. R. Jullien, *J. Phys.*, **A19** (1986) 2129–2136.
56. V.A. Hackley and M.A. Anderson, *Langmuir*, **5** (1989) 191–198.
57. P.M. Mors, R. Botet, and R. Jullien, *J. Phys.*, **A20** (1987) L975–L980.
58. S.G. Kim and J.R. Brock, *J. Colloid Interface Sci.*, **116** [2] (1987) 431–443.
59. R.H. Ottewill, *Langmuir*, **5** (1989) 4–11.
60. I.A. Aksay and R. Kikuchi in *Science of Ceramic Chemical Processing*, eds. L.L. Hench and D.R. Ulrich (Wiley, New York, 1986), pp. 513–521.
61. W.Y. Shih, I.A. Aksay, and R. Kikuchi, *J. Chem. Phys.*, **86** [9] (1987) 5127–5132.
62. S. Hachisu, A. Kose, Y. Kobayashi, and K. Takano, *J. Colloid Interface Sci.*, **55** [3] (1976) 499–509.
63. W. Heller in *Polymer Colloids II*, ed. R.M. Fitch (Plenum, New York, 1980), pp. 153–207.
64. D.L. Segal, *J. Non-Cryst. Solids*, **63** (1984) 183–191.
65. C.W. Turner, B.C. Catworthy, and A. Celli, *Proc. Inf. Conf. CANDU Fuel, Chalk River, Canada* (1986), pp. 155–167.
66. R.K. McGeary, *J. Am. Ceram. Soc.*, **44** [10] (1961) 513–522.
67. J.L. Woodhead, *Silicates Ind.*, **37** (1972) 191–194.
68. R.B. Matthews and M.L. Swanson, *Ceramic Bulletin*, **58** [2] (1979) 223–227.
69. V.N. Vaidya, S.K. Mukerjee, J.K. Joshi, R.V. Kamat, and D.D. Sood, *J. Nuclear Mater.*, **148** (1987) 324–331.
70. J.L. Woodhead in *Science of Ceramics*, **4**, ed. G.H. Stewart (Brit. Ceram. Soc., 1968), pp. 105–111.
71. D.L. Segal and J.L. Woodhead, *Br. Ceram. Proc.*, **38** (1986) 245–250.
72. R.D. Shoup in *Colloid and Interface Science*, **III** (Academic Press, New York, 1976), pp. 63–69.

73. R.D. Shoup, U.S. Patent 3,678,144 (July 18, 1972).
74. R.D. Shoup in *Ultrastructure Processing of Advanced Ceramics*, eds. J.D. Mackenzie and D.R. Ulrich (Wiley, New York, 1988), pp. 347-354.
75. R.L. Nelson, J.D.F. Ramsay, J.L. Woodhead, J.A. Cairns, and J.A.A. Crossley, *Thin Solid Films*, **81** (1981) 329-337.
76. E. Barringer, N. Jubb, B. Fegley, R.L. Pober, and H.K. Bowen in *Ultrastructure Processing of Ceramics, Glasses, and Composites*, eds. L.L. Hench and D.R. Ulrich (Wiley, New York, 1984), pp. 315-333.
77. T.A. Ring, *Mater. Res. Soc. Bull.*, **12** [7] (1987) 34-38.
78. E. Dickinson, S.J. Milne, and M. Patel, *Ind. Eng. Chem. Res.*, **27** (1988) 1941-1946.
79. E. Matijević, *Acc. Chem. Res.*, **14** (1981) 22-29.
80. E. Matijević in *Ultrastructure Processing of Ceramics, Glasses, and Composites*, eds. L.L. Hench and D.R. Ulrich (Wiley, New York, 1984), pp. 334-352.
81. T. Sugimoto, *Adv. Colloid Interface Sci.*, **28** (1987) 65-108.
82. X.-J. Fan and E. Matijević, *J. Am. Ceram. Soc.*, **71** [1] (1988) C60-C62.
83. P. Gherardi and E. Matijević, *Colloids and Surfaces*, **32** (1988) 257-274.
84. A. Garg and E. Matijević, *Langmuir*, **4** (1988) 38-44.
85. W. Stöber, A. Fink, and E. Bohn, *J. Colloid Interface Sci.*, **26** (1968) 62-69.
86. A.K. van Helden, J.W. Jansen, and A. Vrij, *J. Colloid Interface Sci.*, **81** [2] (1981) 354-368.
87. G.H. Bogush and C.F. Zukoski in *Ultrastructure Processing of Advanced Ceramics*, eds. J.D. Mackenzie and D.R. Ulrich (Wiley, New York, 1988), pp. 477-486.
88. C.G. Tan, B.D. Bowen, and N. Epstein, *J. Colloid Interface Sci.*, **118** [1] (1987) 290-293.
89. S. Coenen and C.G. de Kruif, *J. Colloid and Interface Sci.*, **124** [1] (1988) 104-110.
90. N.J. Jubb and H.K. Bowen, *J. Mater. Sci.*, **22** (1987) 1963-1970.
91. B. Fegley, Jr., and E.A. Barringer in *Better Ceramics Through Chemistry*, Mater. Res. Soc. Symp. Proc., **32**, eds. C.J. Brinker, D.E. Clark, and D.R. Ulrich (North-Holland, New York, 1984), pp. 187-197.
92. R.H. Heistand II and Y.-H. Chia in *Better Ceramics Through Chemistry II*, Mater. Res. Soc. Symp. Proc., **73**, eds. C.J. Brinker, D.E. Clark, and D.R. Ulrich (Mater. Res. Soc., Pittsburgh, Pa., 1986), pp. 93-98.
93. R.H. Heistand II, Y. Oguri, H. Okamura, W.C. Moffatt, B. Novich, E.A. Barringer, and H.K. Bowen in *Science of Ceramic Chemical Processing*, eds. L.L. Hench and D.R. Ulrich (Wiley, New York, 1986), pp. 482-496.
94. B. Fegley, Jr., P. White, and H.K. Bowen, *Cer. Bull.*, **64** [6] (1985) 1115-1120.
95. T. Ogihara, N. Mizutani, and M. Kato, *Ceramics International*, **13** (1987) 35-40.
96. T. Ogihara, T. Ikemoto, N. Mizutani, M. Kato, and Y. Mitarai, *J. Mater. Sci.*, **21** (1986) 2771-2774.
97. J.H. Jean and T.A. Ring, *Langmuir*, **2** (1986) 251-255.
98. J.H. Jean and T.A. Ring in *Better Ceramics Through Chemistry II*, Mater. Res. Soc. Symp. Proc., **73**, eds. C.J. Brinker, D.E. Clark, and D.R. Ulrich (Mater. Res. Soc., Pittsburgh, Pa., 1986), pp. 85-92.
99. A.J. LeCloux, J. Bronckart, F. Noville, C. Dodet, P. Marchot, and J.P. Pirard, *Colloids and Surfaces*, **19** (1986) 359-374.
100. A.K. van Helden and A. Vrij, *J. Colloid Interface Sci.*, **76** [2] (1980) 418-433.
101. V.K. LaMer and R.H. Dinegar, *J. Am. Chem. Soc.*, **72** [11] (1950) 4847-4854.
102. A.E. Nielsen, *Kinetics of Precipitation* (Macmillan, New York, 1964).
103. J.A. Dirksen and T.A. Ring, in *High Tech Ceramics Viewpoints and Perspectives*, ed. G. Kosforz (Academic Press, NY, 1989), pp. 29-39.

104. G.H. Bogush, G.L. Dickstein, P. Lee, and C.F. Zukoski IV in *Better Ceramics Through Chemistry III*, eds. C.J. Brinker, D.E. Clark, and D.R. Ulrich (North-Holland, New York, 1988).

105. K.M. Towe and W.F. Bradley, *J. Colloid Interface Sci.*, **24** (1967) 384-392.

106. T. Sugimoto and E. Matijević, *J. Colloid Interface Sci.*, **74** [1] (1980) 227-243.

107. E. Santacesaria, M. Tonello, G. Storti, R.C. Pace, and S. Carra, *J. Colloid Interface Sci.*, **111** [1] (1986) 44-53.

108. W. Hsu, L. Ronnquist, and E. Matijević, *Langmuir*, **4** (1988) 31-37.

109. P.J. Murphy, A.M. Posner, and J.P. Quirk, *J. Coll. Interface Sci.*, **56** [2] (1976) 284-297, 298-311, 312-319.

110. J.M. Fletcher and C.J. Hardy, *Chem. and Ind.*, **18** (1968) 48-51.

111. D.M. Roy, R.R. Neurogaonkar, T.P. O'Holleran, and R. Roy, *J. Am. Ceram. Soc.*, **56** [11] (1977) 1023-1024.

112. M.J. Ruthner in *Ceramic Powders*, ed. P. Vincenzini (Elsevier, Amsterdam, 1983), pp. 515-531.

113. T.J. Gardner, D.W. Sproson, and G.L. Messing in *Better Ceramics Through Chemistry*, Mater. Res. Soc. Symp. Proc., **32**, eds. C.J. Brinker, D.E. Clark, and D.R. Ulrich (North-Holland, New York, 1984), pp. 227-232.

114. R.L. Downs, M.A. Ebner, and W.J. Miller in *Sol-Gel Technology for Thin Films, Fibers, Preforms, Electronics, and Speciality Shapes*, ed. L.C. Klein (Noyes, Park Ridge, N.J., 1988), pp. 330-381.

115. M. Visca and E. Matijević, *J. Colloid Interface Sci.*, **68** [2] (1979) 308-319.

116. A. Balboa, R.E. Partch, and E. Matijević, *Colloids and Surfaces*, **27** (1987) 123-131.

117. F.C. Mayville, R.E. Partch, and E. Matijević, *J. Colloid Interface Sci.*, **120** [1] (1987) 135-139.

118. J.R. Turner, T.T. Kodas, and S.K. Friedlander, *J. Chem. Phys.*, **88** [1] (1988) 457-465.

119. M. Kagawa, M. Kikuchi, and R. Ohno, *J. Am. Ceram. Soc.*, **64** [1] (1981) C7-C8.

120. D.S. Phillips and G.J. Vogt, *Mater. Res. Soc. Bulletin*, **12** [7] (1987) 54-58.

121. K.S. Mazdiyasni, C.T. Lynch, and J.S. Smith II, *J. Am. Ceram. Soc.*, **48** [7] (1965) 372-375.

122. J. Tanaka and A. Kato, *Yogyo Kyokai-Shi*, **81** [5] (1973) 179-183.

123. H. Takeuchi, M. Nagano, Y. Suyama, and A. Kato, *Yogyo Kyokai-Shi*, **83** [1] (1975) 23-27.

124. J.S. Haggerty in *Ultrastructure Processing of Ceramics, Glasses, and Composites*, eds. L.L. Hench and D.R. Ulrich (Wiley, New York, 1984), pp. 353-366.

125. J.D.F. Ramsay and R.G. Avery, *J. Mater. Sci.*, **9** (1974) 1681-1688, 1689-1695.

126. S.M.L. Hamblyn and B.G. Reuben, *Adv. Inorganic Chem. Radiochem.*, **17** (1975) 89-114.

127. P.C. Schultz, *Proc. IEEE*, **68** [10] (1980) 1187-1190.

128. Private communication, D.R. Powers, Corning Glass Works; see also D.R. Powers, *J. Am. Ceram. Soc.*, **61** [7-8] (1978) 295-297.

129. G.D. Ulrich, *Combustion Science Tech.*, **4** (1971) 47-57.

130. G.W. Scherer in *Better Ceramics Through Chemistry*, eds. C.J. Brinker, D.E. Clark, and D.R. Ulrich (North-Holland, New York, 1984), pp. 205-211.

131. D.W. Schaefer, *Mater. Res. Soc. Bulletin*, **13** [2] (1988) 22-27.

132. J.E. Martin, D.W. Schaefer, and A.J. Hurd, *Phys. Rev. A*, **33** [5] (1986) 3540-3543.

133. A.J. Hurd, D.W. Schaefer, and J.E. Martin, *Phys. Rev. A*, **35** (1987) 2361-2364.

134. E. Dickinson, *J. Colloid Interface Sci.*, **118** [1] (1987) 286-289.

135. N.Ya. Turova and M.I. Yanovskaya, *Sov. J. Inorganic Materials*, **19** [5] (1983) 625–638 (Eng. trans.).
136. *Advances in Ceramics, Vol. 21: Ceramic Powder Science*, eds. G.L. Messing, K.S. Mazdiyasni, J.W. McCauley, and R.A. Haber (Am. Ceram. Soc., Westerville, Ohio, 1987).
137. K.D. Budd and D.A. Payne in *Better Ceramics Through Chemistry*, eds. C.J. Brinker, D.E. Clark, and D.R. Ulrich (North-Holland, New York, 1984), pp. 239–244.
138. H. Salze, P. Odier, and B. Cales, *J. Non-Cryst. Solids*, **82** (1986) 314–320.
139. C. Marcilly, P. Courty, and B. Delmon, *J. Am. Ceram. Soc.*, **53** [1] (1970) 56–57.
140. C.E. Johnson, D.K. Hickey, and D.C. Harris in *Better Ceramics Through Chemistry II*, eds. C.J. Brinker, D.E. Clark, and D.R. Ulrich (Mater. Res. Soc., Pittsburgh, Pa., 1986), pp. 785–789.
141. J.J. Ritter and K.G. Frase in *Science of Ceramic Chemical Processing*, eds. L.L. Hench and D.R. Ulrich (Wiley, New York, 1986), pp. 497–503.
142. E.M. Rabinovich, *J. Mater. Sci.*, **20** (1985) 4259–4297.
143. F. Ehrburger, V. Guerin, and J. Lahaye, *Colloids and Surfaces*, **9** (1984) 371–383.
144. J.D.F. Ramsay and R.G. Avery, *Br. Ceram. Proc.*, **38** (1986) 275–283.
145. G.W. Scherer, *J. Am. Ceram. Soc.*, **60** [5–6] (1977) 236–239.
146. J.D.F. Ramsay, U.S. Patent 4,389,385 (June 21, 1983).
147. E.M. Rabinovich, D.W. Johnson, Jr., J.B. MacChesney, and E.M. Vogel., *J. Non-Cryst. Solids*, **47** (1982) 435–439.
148. E.M. Rabinovich, D.W. Johnson, Jr., J.B. MacChesney, and E.M. Vogel, *J. Am. Ceram. Soc.*, **66** [10] (1983) 683–688.
149. D.W. Johnson, Jr., E.M. Rabinovich, J.B. MacChesney, and E.M. Vogel, *J. Am. Ceram. Soc.*, **66** [10] (1983) 688–693.
150. D.L. Wood, E.M. Rabinovich, D.W. Johnson, Jr., J.B., MacChesney, and E.M. Vogel, *J. Am. Ceram. Soc.*, **66** [10] (1983) 693–699.
151. E.M. Rabinovich, D.L. Wood, D.W. Johnson, Jr., D.A. Fleming, S.M. Vincent, and J.B. MacChesney, *J. Non-Cryst. Solids*, **82** (1986) 42–49.
152. E.M. Rabinovich in *Sol-Gel Technology for Thin Films, Fibers, Preforms, Electronics, and Speciality Shapes*, ed. L.C. Klein (Noyes, Park Ridge, N.J., 1988), pp. 260–294.
153. G.W. Scherer and J.C. Luong, *J. Non-Cryst. Solids*, **63** (1984) 163–172.
154. G. Féat and S. Levine, *J. Colloid Interface Sci.*, **54** (1976) 34–44.
155. G.W. Scherer, U.S. Patent 4,574,063 (March 4, 1986).
156. R. Clasen, *J. Non-Cryst. Solids*, **89** (1987) 335–344.
157. R. Clasen, *Glastech. Ber.*, **60** [4] (1987) 125–132.
158. R. Clasen, *Glastech. Ber.*, **61** [5] (1988) 119–126.
159. M. Toki, S. Miyashita, T. Takeuchi, S. Kanbe, and A. Kochi, *J. Non-Cryst. Solids*, **100** (1988) 479–482.
160. C.P. Scherer and C.G. Pantano, *J. Non-Cryst. Solids*, **82** (1986) 246–255.
161. Z. Deng, E. Breval, and C.G. Pantano, *J. Non-Cryst. Solids*, **100** (1988) 364–370.

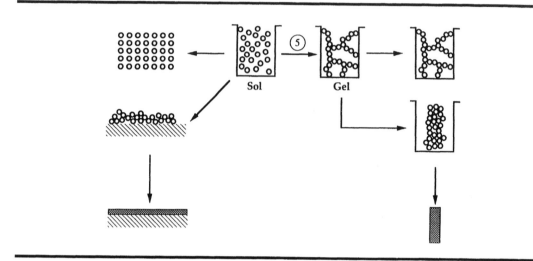

Sol

Gel

CHAPTER 5

Gelation

The hydrolysis and condensation reactions discussed in the preceding chapters lead to the growth of clusters that eventually collide and link together into a gel. Gels are sometimes defined as "strong" or "weak" according to whether the bonds connecting the solid phase are permanent or reversible. However, as noted by Flory [1], the difference between weak and strong is a matter of time scale. Even covalent siloxane bonds in silica gel can be cleaved, allowing the gels to exhibit slow irreversible (viscous) deformation. Thus the chemical reactions that bring about gelation continue long beyond the gel point, permitting flow and producing gradual changes in the structure and properties of the gel. This chapter describes the changes that occur in the vicinity of the gel point, and it reviews the theories that have been proposed to explain gelation. The following chapter describes the long-term aging of gels. The outline of this chapter's presentation is as follows:

1. *Phenomenology* describes the changes in properties that occur as a sol transforms into a gel.
2. *Classical Theory* explains the theory developed by Flory and Stockmayer to account for the gel point and the molecular-weight distribution in the sol. The most important deficiency of this model is that it neglects the formation of closed loops within the growing clusters, and this leads to unrealistic predictions about the geometry of the polymers.
3. *Percolation Theory* avoids the unrealistic assumptions of the classical theory and makes predictions about the properties of gelling systems that are in good accord with experimental observations. Unfortunately, only a few results can be obtained analytically, so these models must generally be

studied by computer simulations. Moreover, they are equilibrium models that teach nothing about the kinetics of gelation.

4. *Kinetic Models* are based on Smoluchowski's analysis of the growth and aggregation of clusters. Specific predictions are made about the size distribution of clusters and the conditions needed to produce gelation, but the theory says nothing about the geometry of the clusters. The evolution of the size distribution and shape of the clusters has been extensively studied by computer simulation and found to be in good agreement with experiment.

5. *Experimental Studies* shows that the predictions of percolation and kinetic models are valid for the structure and size distribution of growing polymers, but the static and dynamic properties are less well described.

1.

PHENOMENOLOGY

The simplest picture of gelation is that clusters grow by condensation of polymers or aggregation of particles until the clusters collide; then links form between the clusters to produce a single giant cluster that is called a gel. The giant *spanning cluster* reaches across the vessel that contains it, so the sol does not pour when the vessel is tipped. At the moment that the gel forms, many clusters will be present in the sol phase, entangled in but not attached to the spanning cluster; with time, they progressively become connected to the network and the stiffness of the gel will increase. According to this picture, the gel appears when the last link is formed between two large clusters to create the spanning cluster. This bond is no different from innumerable others that form before and after the gel point, except that it is responsible for the onset of elasticity by creating a continuous solid network. As one would expect from such a process, no latent heat is evolved at the gel point [2], but the viscosity rises abruptly and elastic response to stress appears, as shown in Fig.1.

The sudden change in rheological behavior is generally used to identify the gel point in a crude way. For example, the *time of gelation*, t_{gel}, is sometimes defined as corresponding to a certain value of viscosity, η; alternatively, it may be defined as the point where the gel shows so much elasticity that the probe (e.g., rotating spindle) tears the gel. The problem with this approach is that the rate of increase of the viscosity varies with the preparation conditions, so that a particular value of viscosity (say, 1000 Pa•s) might be observed seconds before t_{gel} in one system, but hours before t_{gel} in another. A more elegant and informative way to look at gelation is to measure the viscoelastic behavior of the gel as a function of shear rate, as has been

Fig. 1.

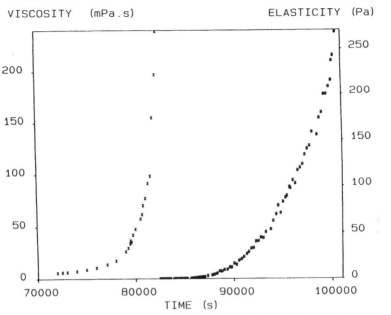

Evolution of viscosity (×) and elasticity (■) versus time for silica gel made from tetramethoxysilane. From Gauthier-Manuel *et al.* [3].

done by Sacks and Sheu [4,5]. They measured the complex shear modulus, G, using a narrow-gap viscometer to insure a well-defined shear rate; the cylinder oscillates with a small amplitude γ at a frequency ω. The modulus consists of a viscous contribution [the *loss modulus*, $G''(\omega)$] and an elastic contribution [the *storage modulus*, $G'(\omega)$], whose relative importance is indicated by the *loss tangent*,

$$\tan \delta = G''/G'. \tag{1}$$

The loss and storage moduli are sketched in Fig. 2 as functions of frequency, ω (at a given extent of reaction, p), or extent of reaction (at a fixed frequency). The energy, W, dissipated in viscous flow during harmonic oscillation is given by [6]

$$W = \pi G''(\omega)\gamma^2. \tag{2}$$

The energy is dissipated at the maximum rate when the sample is excited at the natural frequency for molecular motion,[†] ω_o (where G'' is greatest in

---

[†] This interpretation is essentially correct, but highly oversimplified. The relaxation behavior is characterized by a spectrum of relaxation frequencies, of which ω_o is the modal value. The following molecular interpretation of viscoelasticity is schematic.

Fig. 2.

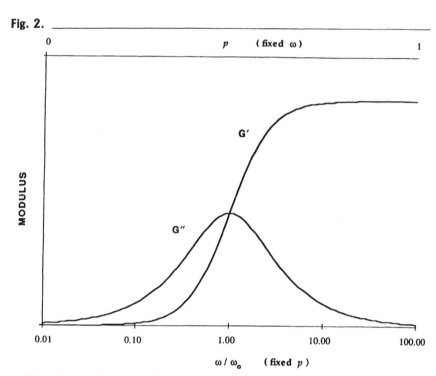

Schematic plot of storage modulus, G', and loss modulus, G'', versus frequency (ω) at fixed degree of reaction (p) on the lower abscissa and versus p at fixed ω on the upper abscissa.

Fig. 2). If the excitation frequency (ω) is much faster than ω_o, then the molecules do not have time to respond and there is no flow; if $\omega \ll \omega_o$, then the molecules slide easily past one another (i.e., the viscosity is low) and little energy is dissipated. Similarly, if the frequency is fixed, but the degree of reaction (extent of cross-linking) is low, the sol is fluid and W (or G'') is small; as the reaction proceeds (p increases), the viscosity increases and G'' passes through a maximum before the gel becomes purely elastic (when the network is too stiff to flow) and dissipation is arrested. The storage modulus, G', is the familiar elastic (static) shear modulus when $\omega \gg \omega_o$, but $G'(\omega)$ approaches zero when ω (or p) is small. The static shear modulus, $G'(0)$, is zero until a continuous gel network forms, but at high frequencies elasticity will appear even in the sol, because chain entanglements can briefly support a load.

For silica gels made from tetraethoxysilane (TEOS) at various water : alkoxide ratios and pH values, Sacks and Sheu [5] found that the properties evolved along a common path. Figure 3 shows typical plots of the shear

Fig. 3a. _____

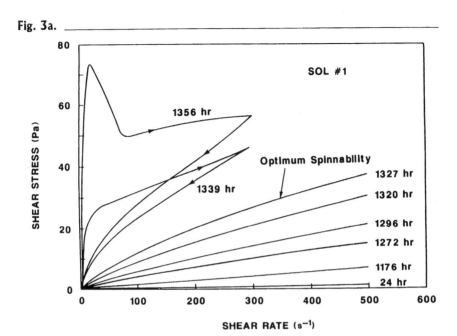

Fig. 3b. _____

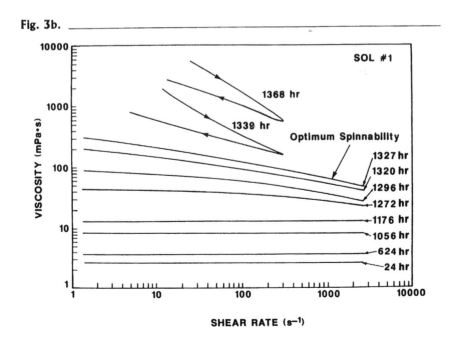

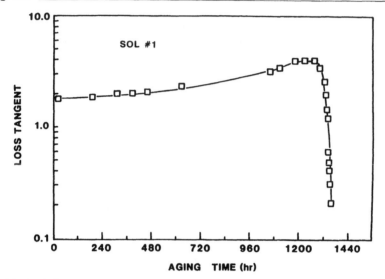

Rheological properties of silica sol made from TEOS/water/ethanol/HNO₃ = 1/2/5.8/0.1. (3a) Shear stress versus shear rate at indicated aging times. (3b) Viscosity versus shear rate. (3c) Loss tangent versus time. From Sacks and Sheu [5].

stress and strain (a) and the viscosity (b) for a silica sol; the data were taken while increasing and then decreasing the shear rate at a constant pace. First, the sol exhibits *Newtonian viscosity*, meaning that the shear rate, $\dot{\gamma}$, increases in proportion to the shear stress, τ_s, so that the viscosity, η, is constant:

$$\tau_s = \eta\dot{\gamma}. \qquad (3)$$

The viscosity increases as the polymers grow, because more work must be done to produce flow of the sol. The liquid within the clusters is not available for flow, and this increases the effective volume fraction of polymer. When the clusters become large (later than 1176 h in Fig. 3) they can be broken by the rotating cylinder, so the sol becomes *shear thinning*, which means that the viscosity decreases with increasing shear rate. When the network begins to form, the sol can support a static load without flowing; this is elastic behavior, and it can be seen in the data collected later than 1327 h in Fig. 3a. The sol behaves elastically until the stress reaches a certain level (the *yield stress*) after which it flows; the yield stress is ~25 Pa at 1339 h and ~75 Pa at 1356 h in Fig. 3a. The hysteresis seen in the curves for 1339 and 1356 hours in Fig. 3b indicates that the network is broken by the shear stress and does not recover immediately as τ_s is reduced below the yield stress. As the gel point approaches, the elastic character of the network increases faster than the viscosity, so the loss tangent goes through a maximum as shown in Fig. 3c. Sacks and Sheu noted that the time of gelation, t_{gel}, cannot be accurately defined by a certain value of viscosity, since the latter depends on shear rate. They suggest that t_{gel} would be best located by reference to the loss tangent (say, the location of the maximum or a particular value of tan δ).

Sacks and Sheu chose the reaction conditions to give very long gel times, so that the sol-gel transition could be easily observed, but gelation can be made to occur in minutes. Generally, t_{gel} is decreased by factors that increase the condensation rate. For gels made from silicon alkoxides, gelation is much faster in the presence of a base or HF than of other acids, as indicated in Table 1. Increases in the ratio of water to alkoxide [8,9], temperature [2,10], and concentration of alkoxide [9,11], and decrease in the size of the alkoxy group [7,12] all decrease t_{gel}. The influence of alkoxide concentration and water content are illustrated in Fig. 4, and the dramatic influence of pH is shown in Fig. 5 for silica gels made from TMOS. The temperature-dependence of gelation can be represented by the Arrhenius equation:

$$\ln(t_{gel}) = A + E/R_g T \tag{4}$$

where A is a constant, R_g is the ideal gas content, and T is temperature. The "activation energy", E, cannot be ascribed to any particular reaction, as gelation depends in a complicated way on the rates of hydrolysis, condensation, and diffusion of clusters. For silica gels, E is found to be ~10–20 kcal/mole [2,10], depending on the catalyst and alkoxy group.

Table 1.

Gelation time for Different Catalysts.

Catalyst[a]	Initial pH	Gelation Time (h)
HF	1.90	12
HCl	0.05[b]	92
HBr	0.20	285
HI	0.30	400
HNO$_3$	0.05[b]	100
H$_2$SO$_4$	0.05[b]	106
CH$_3$COOH	3.70	72
NH$_4$OH	9.95	107
None	5.00	1000

Source: Mackenzie [7].
[a] Concentration = (0.05 moles)/(mole TEOS).
[b] Between 0 and 0.05.

Fig. 4.

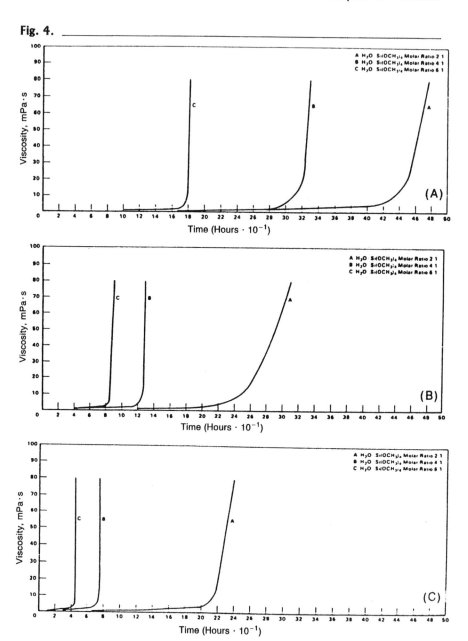

Effect of solution concentration on viscosity change during the sol-gel transition of
TMOS solutions with R = (moles H_2O)/(moles TMOS) = 2 (curve A), 4 (curve B), and
6 (curve C) in sols containing (A) 100, (B) 75, (C) 50 g/l SiO_2. From Debsikdar [11].
Reprinted by permission of the American Ceramic Society.

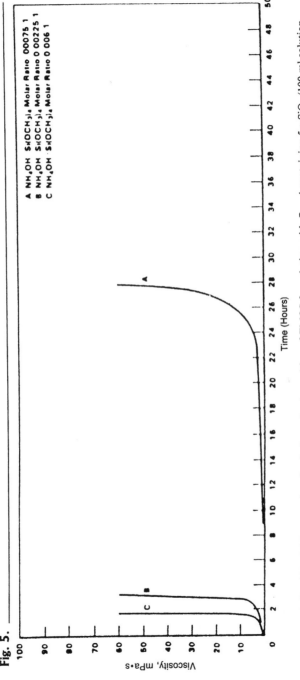

Fig. 5. Effect of NH$_4$OH additions on viscosity during sol-gel transition of TMOS for solution with $R = 4$ containing 5 g SiO$_2$/100-ml solution. From Debsikdar [11]. Reprinted by permission of the American Ceramic Society.

Fig. 6.

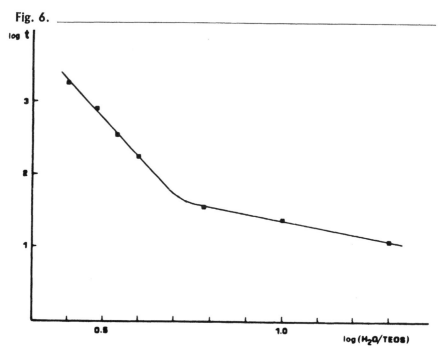

Logarithm of gelling time (hours) versus $\log_{10}$ ([H$_2$O]/[TEOS]). Solutions were kept in closed flasks at 60°C until no flow could be seen. From Gottardi *et al.* [8].

Figure 6 shows that the gel time decreases as the water content increases, even though the sol becomes more dilute. It can be shown, using *nuclear magnetic resonance* (NMR), that the silica network is more highly cross-linked when the water : alkoxide ratio is high. The data in Fig. 7 were obtained using ^{29}Si NMR, which gives the proportions of silicon atoms linked through oxygen to 0, 1, 2, 3, or 4 other silicon atoms; these species are accordingly labelled Q^n, where $n = 0, 1, 2, 3$, or 4. For example, TEOS and silicic acid are Q^0, a silicon in the middle of a linear chain is Q^2, and a fourfold coordinated silicon (such as is found in fused silica glass) is Q^4. Figure 7 shows that silica gels made with higher water concentrations have higher proportions of Q^3 and Q^4 at the gel point ($t_{gel} \leq 12$ h for all of these gels). This indicates that the condensation reaction is accelerated by excess water, and the faster polymerization is credited with reducing t_{gel}. Kelts *et al.* [14] used ^{1}H NMR and ^{29}Si NMR to study the hydrolysis and condensation of TEOS and TMOS (alkoxide/alcohol/water = 1/3.7/2.6; catalyst = HCl or NH$_4$OH). At the gel point, of the silicons with $n > 0$, 60% are Q^4 at pH 7.8, but only 26% are Q^4 at pH < 1; 70% of the TMOS remains unreacted at high pH. This Q^n distribution is readily understood in terms of the effects

Fig. 7.

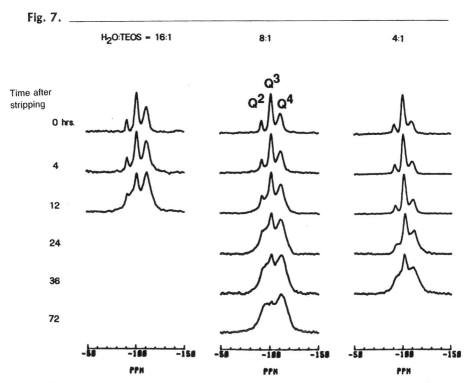

H_2O:TEOS = 16:1   8:1   4:1

Time after stripping

0 hrs.

4

12

24

36

72

^{29}Si NMR data for gels made from TEOS/ethanol/HCl = 1/4/0.01; molar ratio water/TEOS = 4, 8, 16. Each sol was refluxed for 2.5 h and then stripped under reduced pressure to a density of 1.1 g/cm^3. From Vega and Scherer [13].

of pH on the rate of condensation, as discussed in Chapter 3. The important point is that the structure at the gel point is highly variable.

The gel network contains a continuous liquid phase, which allows relatively fast transport. For example, Gits-Léon *et al.* [15] have shown that the diffusion coefficient for NaCl in a silica gel is only slightly slower than in pure water. (See Fig. 8.) Similarly, Dwivedi [16] has shown that the rate of evaporation of the liquid from the pores of an alumina gel is about the same as that from an open dish of the same liquid. Given our present understanding of the open structure of fractal clusters, these observations are not surprising, but it was not always so. The structure of gels was controversial in the 1930s, when it was thought that gels might be emulsions or foams. Hurd [2] used the circumstantial evidence available at that time to deduce a very accurate picture of the actual structure of gels. He cited their high ionic conductivity as evidence in favor of a polymeric model for the structure of silica gels made from silicic acid. The open structure makes it possible to

Fig. 8.

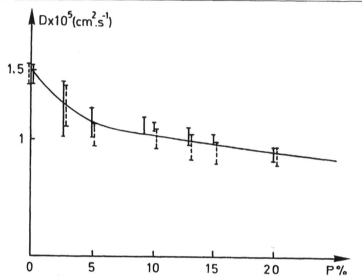

Diffusion coefficient (D) of NaCl in silica gel as a function of the tetramethoxysilane volume ratio (p). Bars indicate range of experimental results from two measurement techniques: extraction (solid) and holographic interferometry (dotted). From Gits-Léon *et al.* [15].

remove oligomers of the sol that are not connected to the spanning cluster. Hurd *et al.* [17], working with silicic acid sols made by mixing sulfuric acid with sodium silicate, found that 90% of the silica could be removed shortly after mixing; the removable fraction decreased with age after gelation, but was always greater than zero. It was noted that membranes with smaller pores passed less polysilicic acid, indicating that the molecular weight of the polymers was increasing up to the gel point.

The reactions that produce gelation do not stop at the gel point. There is a substantial fraction of oligomers that are free to diffuse and react, and even the spanning cluster retains enough internal mobility to allow further condensation rections. Therefore, the properties of a gel, such as elastic modulus, continue to change long after t_{gel}. This process, known as *aging*, may result in substantial structural reorganization of the network, including coarsening of the pores or precipitation of crystals, or simply a stiffening of the network through formation of additional cross-links. Aging is discussed in detail in the following chapter.

We now turn to an examination of several types of theories that have been proposed to explain the phenomenon of gelation and the time-dependence of the properties of a gelling system.

2.

CLASSICAL THEORY

The theory of gelation developed by Flory and Stockmayer is now generally known as the "classical" or "mean field" theory. It is clearly explained in Flory's book, *Principles of Polymer Chemistry* [18], and is related to percolation theories in the wonderfully lucid book by Zallen, *The Physics of Amorphous Solids* [19]. The theory seeks to answer the following question: of all the bonds that could form in a polymerizing system, what fraction (p_c) must form before an infinitely large molecule appears? Consider the condensation polymerization of a z-functional monomer (e.g., silicic acid, with $z = 4$) and suppose that a fraction p of all the possible bonds has already formed. The first assumption we make is that the reactivity of all the functional groups on a monomer is equal; that is, the probability of formation of a bond does not change when any of the other $z - 1$ bonds form. This means that the probability that any particular bond has formed is equal to p. The next assumption is that bonds form only between polymers, not within them, so that the polymers contain no closed loops. The growing polymer is illustrated in Fig. 9 for $z = 3$; this sort of structure, which branches without ever forming rings, is known as a *Cayley tree* or *Bethe lattice*. Travelling along a path of completed bonds, at each node we find $z - 1$ opportunities for continuing the journey; the probability that one of them is a completed bond is $(z - 1)p$. Since there are no closed loops, each completed bond leads into new territory. For the polymer to continue indefinitely (i.e., to form a gel), on average there must be at least one completed path from each node,

Fig. 9.

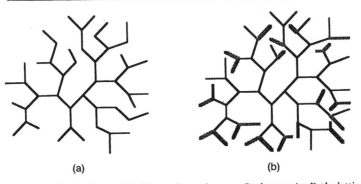

(a)   (b)

Disaster approaches for cluster (a) with $z = 3$ growing on a Cayley tree (or Bethe lattice), because lattice contains no loops, so its density increases without limit (b) as the radius grows. Additional bonds forming at the periphery indicated by thick lines.

so we require $(z - 1)p \geq 1$. Thus the critical condition for gel formation is

$$p_c = 1/(z - 1). \tag{5}$$

This condition is valid only when there are no closed loops; if the paths could double back on one another, a larger extent of reaction would be required to achieve gelation. This model predicts that $p_c = 1$ when $z = 2$, because such a system forms only linear chains that cannot cross-link into a network. For silica ($z = 4$), gelation is predicted to occur when one third of the possible siloxane bonds has formed.

Now let us find the probability that an unreacted bond on a given z-functional monomer is part of an *x-mer* (i.e., a polymer made up of x monomer units). An x-mer must contain exactly $x - 1$ completed bonds (each with probability p) and $(z - 2)x + 1$ unformed bonds (each with probability $1 - p$), so the probability of a given x-mer configuration is

$$p^{x-1}(1 - p)^{(z-2)x+1}.$$

Note that this derivation implicitly excludes infinite polymers (gels), which are bounded by the container, not by $(z - 2)x + 1$ unformed bonds. The probability p_x that an unreacted bond selected at random is part of an x-mer of any configuration is

$$p_x = c_x p^{x-1}(1 - p)^{(z-2)x+1} \tag{6}$$

where c_x is the number of configurations available to x-mers. Flory [18] showed that

$$c_x = \frac{(zx - x)!}{(zx - 2x + 1)! \, x!}. \tag{7}$$

Since the number of unreacted bonds in an x-mer is $(z - 2)x + 1$, the probability p_x is also given by

$$p_x = n_x \frac{(z - 2)x + 1}{n_0 z(1 - p)} \tag{8}$$

where n_x is the number of x-mers and n_0 is the original number of monomers. The weight fraction of x-mers, w_x, can be found from Eqs. 6 and 8:

$$w_x \equiv \frac{n_x x}{n_0} = \left[\frac{zx(zx - x)!}{(zx - 2x + 2)! \, x!} \right] p^{x-1}(1 - p)^{2+(z-2)x}. \tag{9}$$

The change in the weight fractions of x-mers as polymerization proceeds is shown in Fig. 10, which indicates that the monomer is the most prevalent species, even after gelation. To find the distribution of molecular weights at the gel point, the factorial terms in Eq. 9 can be approximated using

Fig. 10.

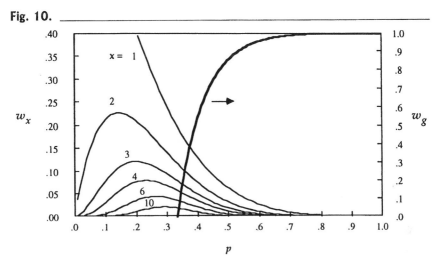

Weight fractions of various finite species (w_x, $x = 1$–10) and of gel (w_g) in tetrafunctional ($z = 4$) condensation as function of degree of reaction, p. Curves calculated from Eqs. 9 and 12.

Stirling's formula, with the result [20]

$$w_x \sim x^{-3/2} \tag{10}$$

so the number of x-mers at the gel point is

$$n_x \sim x^{-5/2} \tag{11}$$

where the tilde ($\sim$) means "varies as" and indicates that any constant factors have been dropped. The sum of w_x for all $1 \leq x < \infty$ is the weight fraction of the sol phase, which is 1 for $p < p_c$. Beyond p_c, the weight fraction of sol, $w_s < 1$, and the weight fraction of gel, $w_g = 1 - w_s$, can be obtained from Eq. 9:

$$w_s = \sum_x w_x = \frac{(1 - p)^2 \beta}{(1 - \beta)^2 p} \tag{12}$$

where the sum is taken over all finite-sized species and β is the smallest solution of

$$\beta(1 - \beta)^{z-2} - p(1 - p)^{z-2} = 0. \tag{13}$$

For $z = 3$, this leads to

$$w_g = 1 - w_s = 1 - \frac{(1 - p)^3}{p^3} \qquad (p > p_c). \tag{14}$$

The growth of the gel fraction in a tetrafunctional system ($z = 4$, as in silica), calculated from Eq. 12, is shown in Fig. 10. After the gel point, the largest oligomers disappear fastest and the most common species is the monomer. Thus, this theory predicts that the average molecular weight in the sol grows until the gel point and then shrinks again as the largest polymers attach themselves to the spanning cluster leaving the smaller polymers in the sol phase. At the gel point, there is a broad distribution of molecular weights, as indicated by Eq. 10. After the gel forms, the average molecular weight of the sol and the breadth of the x-mer distribution both regress over the same path that they followed on the way to the gel point. These predictions have been confirmed experimentally [20], but we shall see that other models make similar predictions.

The most appealing feature of the classical theory is that it provides formulas for the most important features of the gelling solution: the critical degree of reaction at the gel point and the distribution of molecular weights in the sol. The formulas just given refer to the simplest case, self-condensation of a monomer, but similar analyses are possible for systems involving several types of monomers and functional groups [18]. The theory requires the assumption that the reactivity of a functional group is independent of the degree of reaction; for example, this means that a hydroxl (OH) has the same reactivity on a silicic acid monomer and on a Q^3 group on the spanning cluster. As explained in the preceding chapters, this is generally not true, but is probably an adequate approximation. A more important assumption is that the growing polymers contain no closed loops (rings), which means that bonds form only between polymers, never within them. This situation is more probable at high concentrations of polymers, where it is easier for a polymer to link to another than to fold back onto itself. Recognizing this, Stockmayer and Weil [21] measured the gel point of pentaerythritol ($z = 4$) with adipic acid at various degrees of dilution with the solvent dimethoxytetraglycol. They found that the gel point was always at a greater degree of reaction than predicted by Eq. 5; Flory [18] had obtained similar results on a variety of systems. However, when p_c was plotted against the inverse concentration of monomer and extrapolated to infinite concentration ($1/c \rightarrow 0$), the theoretically predicted value for p_c was obtained. This elegant experiment reveals the importance of the neglect of closed loops.

A more important consequence of the neglect of rings can be seen in Fig. 9. The repeated branching results in crowding at the periphery of the polymer. In fact, it can be shown [22,23] that the mass, M, of such a polymer increases with the fourth power of the radius, R (i.e., $M \propto R^4$). Since the volume, V, increases as R^3, this leads to the conclusion that the density, ρ, increases in proportion to R (i.e., $\rho \propto R$). This result is physically unacceptable, because the density cannot increase indefinitely as the polymer grows. Thus, the classical model does not provide an entirely realistic picture of polymer growth.

3.

PERCOLATION THEORY

Percolation theory offers a description of gelation that does not exclude the formation of closed loops and so does not predict a divergent density for large clusters. The disadvantage of the theory is that it generally does not lead to analytical solutions for such properties as the percolation threshold or the size distribution of polymers. However, these features can be determined with great accuracy from computer simulations, and the results are often quite different from the predictions of the classical theory. Excellent reviews of percolation theory and its relation to gelation have been written by Zallen [19] and Stauffer *et al.* [24].

3.1. Percolation Threshold

Figure 11 illustrates percolation on a square lattice. Starting with an empty grid, circles are placed on sites (i.e., intersections of grid lines) at random; if two neighboring sites are filled, they are joined by a bond. This process, called *site percolation*, produces clusters of size s with frequency n_s; as the fraction of filled sites, p, increases, so does the average cluster size, s_{av}. When $p = 0.50$ (Fig. 11b) there is a broad range of cluster sizes, but none of them is a spanning cluster (equivalent to a gel) that reaches across the entire grid. The *percolation threshold*, p_c, is defined as the critical value of p at which the spanning cluster first appears ($s_{av} \rightarrow \infty$). It is evident that in Fig. 11c, $p > p_c$. A different version of percolation that is more appropriate as a model for gelation is *bond percolation*, in which the sites are initially filled (with monomers) and the bonds are filled in at random. The percolation threshold is lower for bond percolation than for site percolation, because a bond is attached to two sites, but a site is connected to several bonds.

Table 2 shows the values of p_c found for several lattices in 1, 2, 3, and higher dimensions. (Computer experiments are not limited to three dimensions as we mortals are!) In 1-d, the trivial result is that percolation requires every site to be filled. In higher dimensions, the value of p_c depends on the shape (e.g., simple cubic, face-centered cubic) of the lattice. This may seem unsatisfying, since real gelation does not occur on a lattice at all, but the situation is better than it appears. Taking the coordination number (number of bonds surrounding a site) of the lattice as equivalent to the functionality of the monomer, the threshold for bond percolation is seen to be close to that of the classical theory,

$$p_c \approx 1/(z - 1). \tag{15}$$

Fig. 11. _____

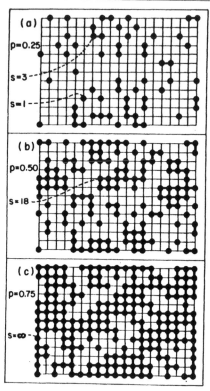

Site percolation on the square lattice, illustrating various cluster sizes (*s*) for three values
of *p*, the fraction of filled sites. For $p = 0.75$, an unbounded cluster is present. From
Zallen, *The Physics of Amorphous Solids* (Wiley, New York, 1983), chapter 4.

The prediction of percolation theory converges on that of the classical theory
as the dimensionality (*d*) increases, becoming the same for $d \geq 6$. A simple-
minded way of rationalizing this result is to note that the number of nearest
neighbors, z_c, in a close-packed lattice increases rapidly with dimension
($z_c = 2, 6, 12, 24, 40, 72$, respectively, for $d = 1, 2, 3, 4, 5, 6$). Thus, at high
d we approach the condition of infinite concentration of monomer, where
neglect of closed loops within the polymer is justified. In three dimensions
(where most of us have our laboratories), p_c is greater than the value
predicted by the classical theory, in keeping with experimental observations.

The threshold for site percolation in any particular dimension occurs at
approximately the same volume fraction, ϕ_c, as indicated in the last column
of Table 2. The filling factor, *v*, is defined as the fraction of (area or) volume
that would be occupied if every site were covered with a (circle or) sphere,

Table 2.

Percolation Threshold for Various Lattices.

Dimensionality d	Lattice[a]	Coordination z	$1/(z-1)$	p_c^{bond}	p_c^{site}	Filling Factor v	$\phi_c = v p_c^{site}$
1	chain	2	1	1	1	1	1
2	triangular	6	0.200	0.347	0.500	0.907	0.45
2	square	4	0.333	0.500	0.593	0.785	0.47
2	kagomé	4	0.333	0.45	0.653	0.680	0.44
2	honeycomb	3	0.500	0.653	0.698	0.605	0.42
3	fcc	12	0.091	0.119	0.198	0.741	0.147
3	bcc	8	0.143	0.179	0.245	0.680	0.167
3	sc	6	0.200	0.247	0.311	0.524	0.163
3	diamond	4	0.333	0.388	0.428	0.340	0.146
3	rcp[b]	~8	~0.143	—	~0.27	~0.637	~0.16
4	sc	8	0.143	0.160	0.197	0.308	0.061
4	fcc	24	0.043	—	0.098	0.617	0.060
5	sc	10	0.111	0.118	0.141	0.165	0.023
5	fcc	40	0.026	—	0.054	0.465	0.025
6	sc	12	0.091	0.094	0.107	0.081	0.009

Source: From Zallen [19].

[a] Note: fcc = face-centered cubic. bcc = body-centered cubic. sc = simple cubic. rcp = random close-packed.

[b] Less precise values, determined experimentally.

so v represents the accessible volume fraction. In 3-d, the percolation threshold for site percolation is at $\phi_c \sim 16$ vol%. This can be demonstrated [19] by mixing various volume fractions of equal-sized balls of glass and metal and measuring the electrical conductivity, σ. When the fraction of metal balls is $<16\%$, σ is zero, because the metal does not form a continuous path (i.e., there is no spanning cluster); above 16%, σ rises gradually with the fraction of metal. Note that there *is no lattice* in such an experiment—the balls are simply mixed randomly in a container. This is an example of *continuum percolation*, so-called because the objects are arranged randomly in space.

Figure 12 shows how the properties of the system change in the vicinity of the percolation threshold. Near p_c, the average cluster size (s_{av}) diverges,

Fig. 12.

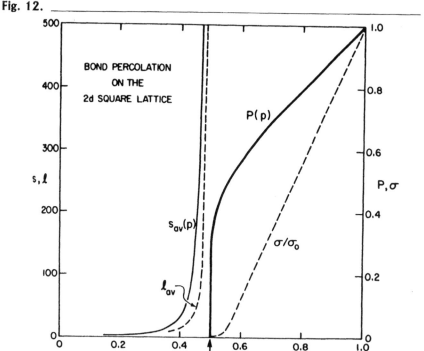

The behavior, as a function of the fraction (p) of filled bonds, of key properties that characterize bond percolation on the square lattice in two dimensions. The percolation probability $P(p)$, the average cluster size $s_{av}(p)$, and conductivity σ are results of computer studies; the spanning length l_{av} is schematic. From Zallen, *The Physics of Amorphous Solids* (Wiley, New York, 1983), chapter 4.

as does the *spanning length, l_{av}*. The spanning length is defined as the greatest distance between any two sites (or bond centers) in a cluster, so it gives a rough estimate of the cluster size. The percolation probability, $P(p)$, is the probability that a given site (or bond) is attached to the spanning cluster, so it is equivalent to the gel fraction, w_g. Below p_c, $P(p) = 0$, but it rises rapidly as a function of $p - p_c$ beyond the threshold. In contrast, the electrical conductivity rises gradually beyond p_c, because the current-carrying spine of the spanning cluster is sparse; most of the cluster is made up of "dead-ends" hanging off the spine. Similarly, the elastic modulus of the gel is predicted to increase slowly beyond p_c, because the dead ends support no load.

The best test of the validity of percolation theory as a description of gelation is its power to predict the behavior of the properties near the gel point (p_c), so we now examine those predictions in some detail.

3.2. Scaling Laws

It is always possible to represent a function, such as $s_{av}(p)$, in the vicinity of p_c by an expansion such as [24]

$$s_{av} = c(p_c - p)^{-\gamma} + c_1(p_c - p)^{-\gamma_1} + c_2(p_c - p)^{-\gamma_2} + \cdots \qquad (p \to p_c)$$

(16)

where p approaches p_c from below and $\gamma > \gamma_1 > \gamma_2 > \cdots$. (For a property that diverges as $\log(p_c - p)$, $\gamma = 0$ and for a property that diverges as $\exp(p_c - p)$, $\gamma \to \infty$.) The first term in this expansion is sufficiently accurate if we are close enough to the threshold, and such power law dependence has indeed been observed for the properties of percolating networks. The exponent of the first term (e.g., γ in Eq. 16) is called the *critical exponent*. The remarkable feature is that the critical exponent for a given property is *universal* in the sense that it is *independent of the lattice type*. For example, the average cluster size diverges with the exponent $\gamma = 1.7$ in three dimensions and 2.4 in two dimensions for any lattice. Exponents for the most important properties are summarized in Table 3; the last column, for $d \geq 6$, gives the exponents for the classical theory, as well as for percolation in those dimensions. The classical theory predicts that the critical exponent for the average cluster size is $\gamma = 1$, which is so different from the percolation prediction that it should be readily distinguishable experimentally. On the other hand, the distribution of cluster sizes at the gel point is characterized by a critical exponent τ that equals 2.2 according to percolation theory, close to the classical value of 2.5. (See Eq. 11.) The exponent d_f that relates the

Table 3

Critical Exponents for Near-Threshold Scaling Behavior in Percolation Theory.

Property	Scaling Near Threshold	Exponent	Value of Exponent in d Dimensions		
			$d = 2$	$d = 3$	$d = 6$
Gel fraction	$P(p) \sim (p - p_c)^\beta$	β	0.14	0.40	1
Conductivity	$\sigma(p) \sim (p - p_c)^t$	t	1.1	1.65	3
Mean cluster size	$S_{av}(p) \sim (p - p_c)^{-\gamma}$	γ	2.4	1.7	1
Spanning length	$l_{av} \sim (p - p_c)^{-\nu}$	ν	1.35	0.85	1/2
Viscosity	$\eta(p) \sim (p - p_c)^{-k}$	k	—	0, 0.7, 1.3[a]	—
Elastic modulus	$G(p) \sim (p - p_c)^{-T}$	T	—	$t \leq T \leq 4^a$	3
Cluster size distribution, $s \to \infty$	$n_s(s) \sim s^{-\tau}$	τ	2.06	2.2	5/2
Spanning length distribution[b] $s \to \infty$	$l(s) \sim s^{1/d_f}$	d_f	1.9	2.5	4

Source: After Zallen [19]. Some exponents have been obtained from computer simulations, so they are not exact.
[a] Note: Various models proposed. See text.
[b] d_f = fractal dimension.

scaling length to the size of the spanning cluster is the fractal dimension of the cluster. In three dimensions, $d_f = 2.5$, whereas the classical theory predicts $d_f = 4$ (so, as noted earlier, the mass increases as radius to the *fourth* power. Another important geometric characteristic of a cluster is the *connectivity correlation length*, or *z-averaged radius*,

$$R_z^2 \equiv \frac{\sum_{s=1}^{\infty} n_s s^2 R_g^2(s)}{\sum_{s=1}^{\infty} n_s s^2} \qquad (17)$$

where n_s is the number of s-mers (clusters containing s sites) and R_g, called the *radius of gyration*, is the typical radius of a s-mer. If $g_c(r)$ is the probability that two monomers separated by a distance r are part of the same cluster, then $g_c(r) \approx 0$ for $r > R_z$; thus, the correlation length is a measure of the spatial extent of the *connectivity function*, $g_c(r)$. The connectivity correlation length has the same critical exponent as the spanning length. Experimental measurements of critical exponents will be discussed in Section 5. The derivation of critical exponents and relations that are known to exist between them are discussed by Stauffer *et al.* [24].

Before continuing the discussion of percolation theory, we should note that Isaacson and Lubensky [25] and de Gennes [26] have extended the classical theory by taking account of excluded volume effects (i.e., the recognition that two polymers cannot occupy the same point in space). They showed that the fractal dimension of the polymer is 2.5 (in three dimensions) in an undiluted sol, and 2.0 in a sol diluted in a good solvent in which the polymers are swollen. For the undiluted sol, the critical exponents, as well as the fractal dimension, are in agreement with the percolation predictions. This observation offers some justification for the use of percolation theory as a model for gelation. Since the exponents are the same, gelation and percolation are said to be in the same *universality class*. The classical exponents were shown to be valid only in space with dimensionality ≥ 6.

The two properties that are most often studied in gelling systems are the viscosity and the elastic modulus. Unfortunately, the theoretical foundation for the critical exponents for such dynamic properties is much weaker than for the geometric properties, such as cluster size distribution. For example [24], to calculate the viscosity, one could start with the Einstein equation for the viscosity (η) of a sol,

$$\eta = \eta_0(1 + 2.5\phi + \cdots) \tag{18}$$

where η_0 is the viscosity of the solvent and ϕ is the volume fraction of polymer. We seek an expression of the form

$$\eta \propto (p_c - p)^{-k} \tag{19}$$

to represent the change in viscosity near the gel point. The volume fraction in Eq. 18 can be represented by

$$\phi = \sum_s n_s R^3(s) \tag{20}$$

where n_s and $R(s)$ are respectively the number and radius of clusters containing s monomers. Knowing how n_s and $R(s)$ diverge near the gel point, and neglecting higher-order terms in Eq. 18, we predict that the viscosity near p_c varies as

$$\eta \propto \log(p_c - p) \quad \text{(percolation)}$$
$$\eta \to \text{constant} \quad \text{(classical).} \tag{21}$$

According to this model, percolation theory predicts logarithmic divergence, corresponding to $k = 0$ in Eq. 19. On the other hand, assuming that the contribution to the viscosity of a cluster of size s is proportional to sR^2, instead of R^3, leads to $k = 1.3$ [27]. The classical theory surprisingly predicts that the viscosity is finite at the gel point [28]. In general, the classical theory predicts slower divergence of properties below the gel point, because of the slow increase in the volume fraction of the polymers (viz., proportional to the 1/4 power of the mass.) The derivation of Eq. 21 is crude, because it is not legitimate to neglect higher-order terms in Eq. 18 except in dilute solution; in fact the higher-order terms are much *larger* than the leading term in the vicinity of the gel point [24]. Another approach is to draw an analogy between viscosity and electrical conductivity through a network of superconductors and ordinary conductors. Above the percolation threshold, the conductivity through the spanning superconductor is infinite (as is the viscosity of the gel); below p_c, σ is finite (as is the viscosity of the sol). This reasoning leads to the prediction that $k = 0.7$ [29]. A theory developed by Martin *et al.* [30], based on the self-similarity of polymer dynamics (viz., the idea that a polymer of any size rotates through the same angle in the time it takes to diffuse a distance equal to its diameter), leads to several specific predictions that are supported by experimental evidence. They obtain the bounds $0 \leq k \leq 1.35$; the value for a given sol depends on the extent of the hydrodynamic interactions of the polymers.

Clearly, none of the derivations of the critical behavior of the viscosity could be called rigorous, and a similar problem exists for the behavior of the elastic modulus. De Gennes [31] proposes the following analogy between the modulus and the conductivity of a network. The elastic energy stored in a network can be written as

$$E_{el} = \tfrac{1}{2} \sum_{ij} K_{ij}(u_i - u_j)^2 \tag{22}$$

where u_i is the displacement of the ith site and K_{ij} is the spring constant connecting the ith and jth sites. This is known as *scalar elasticity*, because the energy does not depend on the direction of the displacement. Minimizing the stored energy leads to the force balance equation:

$$\sum_j K_{ij}(u_j - u_i) = 0 \qquad \text{(for all } i\text{).} \tag{23}$$

In an electrical network, Eq. 23 represents the Kirchoff equation (indicating that the net current flowing into a node is zero) with K_{ij} being the local conductance and u_i the voltage on node i. Just as the conductivity relates the current with the potential gradient, the modulus relates the stress to the displacement gradient (i.e., strain), so the modulus ought to scale as the conductivity. Thus, we expect

$$E \propto (p_c - p)^{-T} \tag{24}$$

where T, the elasticity exponent, is equal to the conductivity exponent, $t \approx 1.7$.[†] This prediction has been challenged by Kantor and Webman [32], who replace Eq. 22 with

$$E_{el} = G \sum g_{ij} g_{ik} \, \delta\phi_{ijk}^2 + K \sum g_{ij}(u_i - u_j)_{\parallel}^2 \tag{25}$$

[†] We retain this standard notation, since the context will preclude confusion with time, t, and temperature, T; for added clarity, the exponents are not set in italics.

where g_{ij} is a random variable that equals 0 or 1 with probability p and $(1 - p)$, respectively, $\delta\phi_{ijk}$ is the change in angle between bonds (i, j) and (i, k), and G and K are elastic constants. The second sum represents only the portion of the displacement that is parallel to the line between centers of the monomers on sites i and j, so it represents the hydrostatic component of strain, while the first sum represents the shear strain. This is called *tensor elasticity*, because the energy depends on the direction of the displacements. Based on Eq. 25 they estimate that the modulus scales with an exponent $2.85 < T < 3.55$. They explain the difference in the elastic and electrical behavior by noting that the conductance does not depend on the shape of the conductor, while the force constant of a chain depends on its shape and the direction of the applied force. Roux and Guyon [33] argue that *torques* propagate as electrical conduction, though forces do not, and obtain a bound for the elasticity exponent of $T \leq 3.78$ in three dimensions (and 3.96 in two dimensions). Martin *et al.* [30] predict $T = 2.67$ for a sol of branched polymers.

Feng and Sen [34] performed computer simulations of elastic deformation of a percolating fcc network using *only* central forces (similar to the second term in Eq. 25). They obtained an exponent of $t = 4.4$ above a percolation threshold of $p_{cen} = 0.42$ (whereas $p_c = 0.119$ on this lattice for scalar elasticity). Allowing for both shear and hydrostatic strains, they found a crossover from the scalar exponent (~ 1.6 in their simulation) to the tensor exponent as p increased from p_c to p_{cen}. The physical picture is that the shear term provides rigidity as soon as the percolation cluster forms, so the modulus rises with the scalar exponent immediately above p_c. Since central forces do not inhibit monomers from rotating around one another (at fixed separation), they cannot contribute to the rigidity until after the formation of a substantial number of load-bearing triangular or tetrahedral structures; this occurs beyond the higher threshold, p_{cen}. The experimental results discussed in Section 5 seem to indicate that such a transition does occur.

3.3. Other Percolation Models

The percolation models represent many features of a gelling system, but there are obviously many differences. For example, in bond percolation, the lattice is assumed to have a monomer on every site, whereas gelation generally occurs in dilute systems. This deficiency is addressed by *site-bond percolation*, in which the sites are randomly populated with monomers and solvent molecules. As the concentration of solvent rises from zero, it is found [35] that p_c increases continuously from the bond-percolation to the site-percolation threshold. (See Fig. 13.) The remarkable (and convenient) fact

Fig. 13.

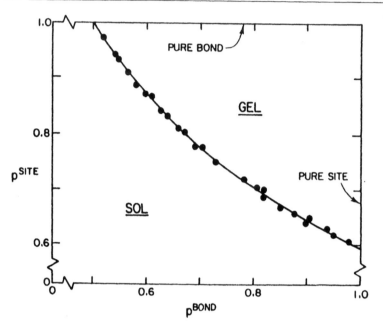

Phase diagram for site-bond percolation on the square lattice. The solid line separates the percolating regime (labelled "gel") from the nonpercolating regime (labelled "sol"). Left axis represents fraction of sites occupied by monomers. From Zallen, *The Physics of Amorphous Solids* (Wiley, New York, 1983), chapter 4.

is that the scaling exponents have the usual percolation values, regardless of the solvent concentration [19,24]. In *correlated site-bond percolation*, the distribution of solvent molecules and monomers on the lattice is influenced by an interaction energy. Consequently, at low "temperatures," there may be separation into phases rich in monomer (where random percolation can occur) and solvent (where the monomer is so dilute that gelation cannot occur even if $p = 1$). At higher temperatures there is a single phase (either sol or gel), and the random percolation exponents are found to apply along the entire sol-gel boundary [24].

It was noted earlier that the percolation threshold corresponds to a certain volume fraction of occupied sites, whether on a lattice or in a continuum (i.e., objects arranged at random without a lattice). For three-dimensional percolation, the threshold was found to be 16 vol% for isotropic objects, such as spheres. For *anisotropic* objects, such as plates or fibers, the volume fraction occupied at the percolation threshold depends on the aspect ratio of the object. For example, Boissonade *et al.* [37] studied site percolation on a simple cubic lattice where the percolating objects were fibers with various aspect ratios, $a \equiv L/w$, where w = diameter and L = length. They found that p_c decreased from 0.31 to 0.06 as a increased from 1 (sphere) to 15, but the critical exponent for the correlation length, R_z, was the same as for spheres. Carmona *et al.* [38] measured the threshold for electrical conductivity in an insulating matrix containing conducting fibers with high aspect ratios and found $\phi_c < 3$ vol% for fibers with $a > 100$.

For *permeable* objects (i.e., things such as polymer clusters that can interpenetrate), it has been shown [39,40] that percolation occurs at a critical value of the *excluded volume fraction*, V_{ex}. The excluded volume, $\langle V \rangle$, is the volume around an object into which the center of another similar object cannot enter without overlapping of the objects. This is illustrated in Fig. 14 for cylinders whose axes are parallel or skewed. For parallel objects (as in Fig. 14a), $\langle V \rangle$ is eight times the volume of the object, V, so a

Fig. 14. _____

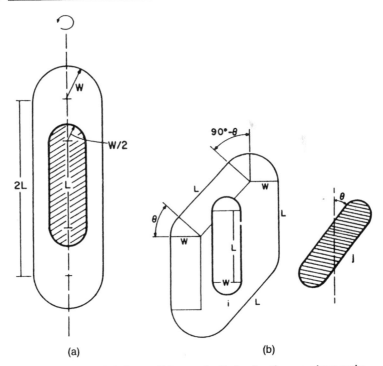

(a)   (b)

(a) Excluded volume (unshaded) for parallel capped cylinders has the same shape as the object, and volume eight times as large. (b) Excluded volume for capped cylinders with axes at angle θ, averaging θ for random distribution, leads to Eq. 26. From Balberg *et al.* [39].

condition of constant excluded volume fraction is equivalent to a condition
of constant volume fraction. However, when the objects are arranged at
random (as in Fig. 14b), $\langle V \rangle$ is no longer proportional to V. For the capped
cylinder,

$$\langle V \rangle = (4\pi/3)w^3 + 2\pi w^2 L + (\pi/2)wL^2 \tag{26}$$

whereas

$$V = (\pi/6)w^3 + (\pi/4)w^2 L. \tag{27}$$

Percolation occurs at a particular value of the excluded volume fraction,

$$V_{ex} = N_c\langle V \rangle \tag{28}$$

where N_c is the number density of objects (e.g., cylinders per unit volume).
Thus, Eqs. 26 and 28 indicate that the critical density varies as

$$N_c \propto 1/\langle V \rangle \propto (2/3 + a + a^2/4)^{-1}. \tag{29}$$

This agrees with computer simulations [37] that show $N_c \propto 1/a$ at intermediate values of a as well as experiments [38] that show $N_c \propto 1/a^2$ at
high aspect ratios ($a = L/w \gg 1$). The critical excluded volume fraction
differs for objects of different shapes (e.g., spheres versus cylinders), but
has the same value for objects of given shape differing in aspect ratio, and
it remains constant when a distribution of sizes is present [39]. It increases,
however, when the objects are oriented; for example, it is more difficult for
parallel cylinders to percolate than it is for cylinders with axes arranged at
random.

It is evident that percolation models have much in common with gelling
systems, and, as we shall see in Section 5, the predicted scaling exponents
are generally supported by experiments. However, these models also have
serious deficiencies. Both percolation and the classical theory (which is just
percolation on a Bethe lattice) are equilibrium theories that give no information about the kinetics of gelation. It is quite plausible that the reduced
mobility of large clusters near the gel point would prevent equilibrium from
being attained, so that the course of cluster growth would be altered. Another
difference between percolation and cluster growth is that bonds are randomly
(i.e., uniformly) distributed on the lattice, whereas growing clusters represent
concentrations of bonds. That is, bonds are *correlated* (grouped together) in
clusters, but *uncorrelated* in percolation. In the next section, we examine
kinetic models for cluster growth that account for bond correlation. It
appears that these models describe the early evolution of the sol, during
which clusters grow and fill space, then percolation describes the linking of
the clusters into a gel.

4.

KINETIC MODELS

An enormous amount of work has been done to model aggregation processes, leading to predictions of the kinetics of growth and the fractal structure of the resulting clusters. Work in this area has been comprehensively reviewed in several excellent articles by Meakin [41–44] and Martin [45].

4.1. Smoluchowski's Equation

The Smoluchowski equation describes the rate at which the number, n_s, of clusters of size s changes with time, t, during an aggregation process [46,47]:

$$\frac{dn_s}{dt} = \frac{1}{2} \sum_{i+j=s} K(i,j)n_i n_j - n_s \sum_{j=1}^{\infty} K(s,j)n_j. \tag{30}$$

The *coagulation kernel*, $K(i,j)$, is the rate coefficient for aggregation of a cluster of size i with another of size j. The first term in Eq. 30 gives the rate of creation of clusters of size s by aggregation of two smaller clusters, and the second term gives the rate at which clusters of size s are eliminated by further aggregation. For this equation to apply, the sol must be so dilute that collisions between more than two clusters can be neglected, and the clusters must be free to diffuse so that the collisions occur at random (which is not likely when a gel phase is present). Further, since K depends only on i and j, ignoring the range of structures that could be present in a cluster of a given size, this is a *mean-field* analysis that replaces structural details with averages. For realistic collision models the kernel is found to be a homogeneous function of i and j, which means that

$$K(ai, aj) = a^\lambda K(i,j) = a^\lambda K(j,i) \tag{31}$$

where λ is a constant. For collisions of large clusters, van Dongen and Ernst [48] investigated the form

$$K(i,j) \approx i^\mu j^\nu \qquad (j \gg i, \lambda = \mu + \nu). \tag{32}$$

The collision probability of two large clusters cannot grow faster than their volumes, so the upper bound for the kernel must be $K(j,j) \sim j^2$; thus, $\lambda \leq 2$ and $\nu \leq 1$. Three different classes of behavior can be identified, based on the value of μ [48]:

Class I   ($\mu > 0$) is dominated by collisions between large clusters.

Class II   ($\mu = 0$) has similar frequencies for collisions of large-with-large and large-with-small clusters.

Class III ($\mu < 0$) is dominated by collisions of large-with-small clusters.

The rate of change of the total mass in clusters up to size s is found from Eq. 30:

$$\frac{dM_s}{dt} = \sum_{j=1}^{s} j \frac{dn_j}{dt} = -\sum_{i=1}^{s} \sum_{j=s-i+1}^{\infty} iK(i,j)n_i n_j. \tag{33}$$

For gelation to occur, we must have $dM_\infty/dt > 0$ while the total mass, M,

$$M = \sum_{s=1}^{\infty} sn_s \tag{34}$$

is constant. It can be shown [48] that gelation will not occur if n_s decreases more rapidly than $n_s \sim s^{-\tau}$ as $s \to \infty$; given this dependence for n_s, the largest terms of Eq. 33 are finite only if $\tau > (\lambda + 3)/2$. The total mass, M, remains finite only if $\tau > 2$, leading to the requirement that $\lambda > 1$ for gelling systems and $\lambda \le 1$ for nongelling systems. For example [47], the classical theory assumes that all sites are equally reactive, so $K(i,j) \sim ij$; this means that $\mu = \nu = 1$ and $\lambda = 2$, so the system is expected to gel.

For systems with $\lambda \le 1$, the mean cluster size, $s_{av}(t)$, grows as [48]

$$s_{av} \sim t^z, \qquad z = 1/(1 - \lambda) \tag{35a}$$

$$\to e^{ct} \quad \text{as} \quad \lambda \to 1. \tag{35b}$$

Gelation cannot occur, because it requires an infinite time to produce a cluster of infinite size. For a gelling system ($1 < \lambda \le 2$), the average cluster size diverges according to

$$s_{av} \sim (t - t_{gel})^{-\gamma}, \qquad \gamma = (\lambda - 1)/2. \tag{36}$$

The distribution of cluster sizes in a nongelling system is given by

$$n_s = Ms_{av}^{-2} f(s/s_{av}) \tag{37}$$

where $f(x)$ is a function that indicates how the distribution scales with respect to the mean cluster size. For clusters smaller than the mean size, the distribution has the form of a power law,

$$f(x) \sim x^{-\tau} \qquad (x \to 0) \tag{38}$$

where $\tau = 1 + \lambda$ for class I and $\tau < 1 + \lambda$ for class II. At the large end of the distribution for all nongelling systems,

$$f(x) \sim x^{-\lambda} e^{-ax} \qquad (x \to \infty) \tag{39}$$

which means that the power law distribution has an exponential cutoff. For class III, the small clusters scale as

$$f(x) \sim \exp(-x^{-|\mu|}) \qquad (x \to 0) \tag{40}$$

so there is a bell-shaped distribution of sizes.

For gelling systems ($\lambda > 1$), the size distribution has the form

$$n_s = M s_{av}^{-\tau} f(s/s_{av}) \tag{41}$$

where s_{av} is given by Eq. 36. The distribution function has the form of Eq. 38 with $\tau = (\lambda + 3)/2$, both as $x \to 0$ in the pregel stage, and as $x \to \infty$ in the postgel stage. That is, the shape of the distribution regresses after gelation, just as it does in the classical theory.

These predictions for the kinetics and statistics for growing clusters are summarized in Table 4. Evidently the form of the reaction kernel is of paramount importance. We now examine computer simulations of cluster aggregation and see that the results compare well with the predictions of the kinetic theory. One aspect of growth that is not explained by Smoluchowski's theory is the geometry of the clusters: it is not possible to predict the fractal dimension of a cluster, given knowledge of the form of the reaction kernel. This information can only be obtained from the computer simulations.

Table 4.

Properties of Growing Clusters According to Smoluchowski's Theory Compared with Flory-Stockmayer and Percolation Theories.

$$K(i, j) \sim i^{\mu} j^{\nu} \qquad (j \gg i, \lambda = \mu + \nu)$$

	Gelling	λ	ν	$n_s \sim s^{-\tau}$	M_w
Smoluchowski:					
Class I[a]	No	$\nu < \lambda < 1$	—	$\tau = 1 + \lambda < 2$	$t^z, z = 1/(1 - \lambda)$
Class I[a]	Yes	$1 < \lambda \le 2$	$\nu < 1$	$\tau = (\lambda + 3)/2 < 2$	$\|t_{gel} - t\|^{-\gamma}, \gamma = 1/(\lambda - 1)$
Class II[a]	No	$\lambda = \nu \le 1$	—	$\tau < 1 + \lambda$	$t^z, z = 1/(1 - \lambda)$
Class III[a]	No	$\lambda < \nu$	$\nu < 1$	bell-shaped dist.	$t^z, z = 1/(1 - \lambda)$
Flory-Stockmayer	Yes	2	1	$\tau = 5/2$	$\|t_{gel} - t\|^{-\gamma}, \gamma = 1$
Percolation	Yes	—	—	$\tau = 2.2$	$\|t_{gel} - t\|^{-\gamma}, \gamma = 1.7$

[a] Following classification introduced by van Dongen and Ernst [48].

4.2. Computer Simulations of Cluster Growth

Computer models of aggregation use a "brute-force" approach, rather than making explicit use of a reaction kernel. In the simplest version, a monomer is placed at the center of a lattice and other monomers are launched toward it from some distance away. If the approaching monomers follow a linear path, the process is called *ballistic aggregation*. This would be most appropriate as a model of particle growth from the vapor phase [49], as in the processes described in Chapter 4, Section 3.2. Models that are more relevant to gelation allow the incoming monomer to follow a random walk (analogous to Brownian diffusion), where the direction and length of the diffusion step is random. Figure 15 shows a two-dimensional example [43] in which

Fig. 15.

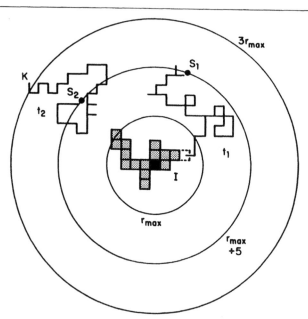

Simple two-dimensional lattice model for diffusion-limited monomer–cluster aggregation. Particles are created one at a time from randomly selected points (S_1 and S_2) on a "launching circle" with radius $r_{max} + 5$ (where r_{max} is the maximum radius of the cluster) that encloses the cluster. They are then moved to the nearest lattice site and undergo a random walk on the lattice. Two particle trajectories, t_1 and t_2, are shown: t_1 eventually brings the particle into an unoccupied lattice site on the surface of the cluster (dashed border), where the trajectory is stopped and the surface site is filled; t_2 moves the particle away from the cluster. When t_2 crosses the "killing" circle with radius of $3r_{max}$, the trajectory is terminated and a new one is begun from a random point on the lauching circle. From Meakin [44].

monomers are launched from random points on a circle whose radius is five lattice units larger than the growing cluster. Trajectory t_1 brings the monomer into contact with a site on the surface of the cluster, to which it becomes attached. Trajectory t_2 meanders outside of the "killing circle," whose radius is three times as great as that of the cluster, so the particle is eliminated and another is launched. Actual simulations use much larger launching and killing circles, and they often do not use a lattice; the techniques are described in detail by Meakin [41–44].

It is important to distinguish *monomer–cluster aggregation*, in which monomers are added to clusters, from *cluster–cluster aggregation*, in which whole clusters (as well as monomers) diffuse and collide. The latter process produces much less compact structures (i.e., fractal clusters) and is generally much more realistic. If the colliding clusters always stick together (sticking probability = 1), the rate of aggregation is determined by transport kinetics, and the process is known as *diffusion-limited aggregation*. The erratic path of a diffusing cluster makes it difficult for one cluster to penetrate another without colliding and sticking, so attachment tends to occur at the periphery. In many cases (for instance, when a repulsive electrostatic barrier is present), the sticking probability is much less than unity, so many collisions will occur before two clusters link together. This process, called *reaction-limited aggregation*, allows more opportunity for the clusters to interpenetrate (in the limiting case, all accessible attachment sites are equally probable), so the resulting clusters are more compact. As we shall see, most experimental studies of colloidal aggregation suggest that growth occurs by reaction-limited cluster–cluster aggregation (RLCCA); if no barrier to aggregation exists, diffusion-limited cluster–cluster aggregation (DLCCA) may predominate.

A computer simulation of cluster–cluster aggregation usually begins by distributing monomers at random on a lattice, then allowing them to diffuse, collide, and aggregate, as illustrated in Fig. 16. Suppose that at some time t there are $N(t)$ clusters of various sizes, s, and that the diffusion coefficient of a cluster of size s is $D(s)$. A cluster of size s is chosen at random and a random number, $0 < r < 1$, is generated; if $r < D(s)/D_{max}$, where D_{max} is the largest diffusion coefficient for any cluster in the system, then the cluster is moved by one lattice unit in a randomly selected direction. (Since r is uniformly distributed over the interval 0–1, this scheme insures that the probability of moving increases in proportion to D.) After moving, the cluster's perimeter is examined for contact with other clusters. In DLCCA, every collision results in the clusters being permanently joined, after which the new larger cluster diffuses as one entity. RLCCA is equivalent to DLCCA with a small sticking probability, so, as noted earlier, attachment is equally likely at every accessible site (on the periphery or the interior of the cluster).

Fig. 16A.

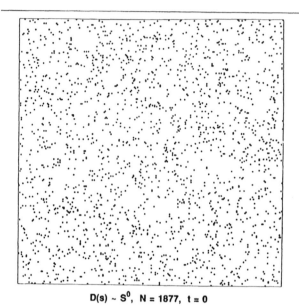

D(s) ~ S^0, N = 1877, t = 0

Fig. 16B.

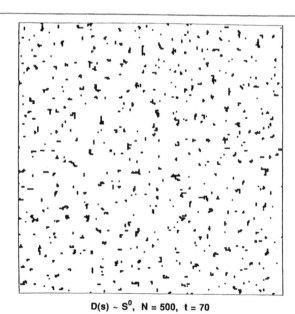

D(s) ~ S^0, N = 500, t = 70

Fig. 16C.

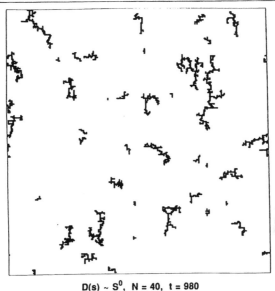

$D(s) \sim S^0$, N = 40, t = 980

Fig. 16D.

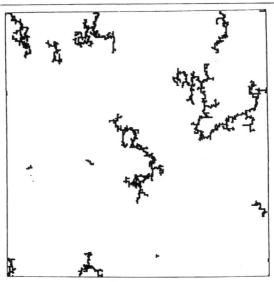

$D(s) \sim S^0$, N = 10, t = 2571

Four stages in a small-scale simulation of diffusion-limited cluster–cluster aggregation carried out on a square latice with a monomer concentration of 0.0305. Diffusion coefficient, D, independent of cluster size, s; i.e. $D(s) \sim s^0$. Reproduced with permission, from the *Annual Review of Physical Chemistry* **Vol. 39** © 1988 by Annual Review Inc.

Simulations implement this idea by picking particles at random (if they are found to be in the same cluster, a new choice is made) and bringing their respective clusters into contact with random orientations. If this does not cause overlap of any other particles in the clusters, then the chosen particles are linked; if there is overlap, another choice is made. This makes bonding equally likely at every accessible site. After each cluster has been selected, the time is incremented by $\Delta t = 1/[N(t)D_{max}]$, so that the rate of the aggregation process can be monitored. A more efficient, but less realistic, technique for simulating cluster–cluster aggregation is known as the *hierarchical method*. The calculation begins with a list of N monomers that are selected at random and paired to form $N/2$ dimers; these are selected randomly and combined into $N/4$ tetramers, and so on. Since the clusters are monodisperse, they do not readily interpenetrate, so this scheme tends to produce clusters with slightly smaller fractal dimensions than the *polydisperse method* illustrated in Fig. 16.

These procedures not only indicate the kinetics of aggregation, but also allow study of the size distribution and the fractal geometry of the clusters. Typical results for the fractal dimension of clusters "grown" in computer simulations are summarized in Table 5. Note that the monomer–cluster aggregation models produce dense nonfractal clusters, except for the diffusion-limited case, whereas cluster–cluster models yield fractal dimensions in good agreement with experiments on real colloids.

Table 5.

Fractal Dimensionalities from Simple Aggregation Models.

Dimensionality d	Path	Monomer–Cluster	Cluster–Cluster Polydisperse	Cluster–Cluster Hierarchical
2	Diffusion-limited	1.71	1.45	1.44
3		2.50	1.80	1.78
4		3.40	2.10	2.02
2	Ballistic	2.0	1.55	1.51
3		3.0	1.95	1.89
4		4.0	2.24	2.22
2	Reaction-limited	2.0	1.61	1.54
3		3.0	2.09	1.99
4		4.0	2.48	2.32

Source: From Meakin [44].

Usually, the clusters are forced to adhere with the relative orientation that they had at the moment of contact; in other cases, they are allowed to undergo restructuring. Meakin and Jullien [50] examined diffusion-limited, ballistic, and reaction-limited aggregation processes in which the colliding clusters were allowed to rearrange by rotation around points of contact. When a collision occurs, a radius is drawn from the center of each cluster to the particle on its surface that has made contact; the clusters are rotated around an axis perpendicular to the plane defined by those two radii, until another point of contact is established. This constitutes the first stage of internal rotation, IR = 1; further rotations may be permitted about the axes joining new points of contact (IR = 0 if no restructuring is allowed). Such restructuring leads to denser structures (i.e., higher fractal dimension), as indicated in Table 6. The first rotation has a large effect on the fractal dimension of the clusters, especially for those that are initially least dense (viz., DLCCA clusters). The most important result is that a single rotation raises the fractal dimension of a DLCCA cluster to that characteristic of an unmodified (IR = 0) RLCCA cluster. Therefore, it is not possible to infer the aggregation mechanism from the geometry of the clusters. An example of this restructuring process is presented in Fig. 8 of Chapter 4.

We now examine the connection between these computer simulations and the Smoluchowski equation. Consider the form of the kernel for aggregation of particles undergoing Brownian diffusion (see Fig. 17), where the diffusion coefficient of a particle of mass i is D_i. In the frame of reference of particle i, the diffusion coefficient of particle j is $(D_i + D_j)$. A collision occurs if the center of particle j enters a sphere of radius $(r_i + r_j)$ around particle i, so [46]

$$K(i, j) = (D_i + D_j)(r_i + r_j). \qquad (42)$$

For Brownian diffusion, the Stokes–Einstein equation predicts $D_i \propto 1/r_i$ and for an ordinary particle $r_i \propto i^{1/3}$, but for a cluster with mass fractal

Table 6.

Effect of Restructuring on Fractal Dimension.

Aggregation Path	Fractal Dimension			
	$IR^a = 0$	1	2	3
Diffusion-limited	1.80	2.09	2.17	2.18
Ballistic	1.95	2.13	2.18	2.19
Reaction-limited	2.09	2.18	2.24	2.25

Source: Based on results of Meakin and Jullien [50].
[a] Number of internal rotations allowed after collision.

Fig. 17. _____

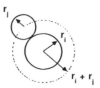

Illustration of Brownian collision between clusters of size i and j with diffusion coefficients D_i and D_j. In the frame of reference of i, the diffusion coefficient of j is $(D_i + D_j)$. Collision occurs when the center of j enters the circle with radius $r_i + r_j$.

dimension d_f, $r_i \propto i^{1/d_f}$, so [41,51]

$$K(i,j) \sim (i^{-1/d_f} + j^{-1/d_f})(i^{1/d_f} + j^{1/d_f}). \tag{43}$$

For $j \gg i$ this reduces to

$$K(i,j) \sim (j/i)^{1/d_f} \tag{44}$$

so $\mu = -1/d_f$, $v = 1/d_f$, and $\lambda = 0$, which means that Brownian coagulation is a *nongelling* process in class III. This category includes diffusion-limited aggregation processes, where the clusters stick immediately and irreversibly on contact. Therefore, we expect DLCCA to yield a peaked distribution of cluster sizes (according to Eqs. 39 and 40) and linear growth kinetics for the mean cluster size ($z = 1$ in Eq. 35a). The predicted distribution of sizes has indeed been found in computer simulations [52]. More important, experimental studies are in agreement with theory when aggregation of colloids is carried out with no repulsive interaction. When gold colloids are destabilized by large additions of pyridine (which adsorbs and reduces the electrostatic barrier), the fractal dimension of the clusters is found [53] by light scattering to be 1.75, as expected for DLCCA, and the cluster mass increases approximately in proportion to time. The cluster size distribution [54] is nearly flat, with depletion at large and small sizes, consistent with expectations.

For reaction-limited aggregation, Ball *et al.* [55] argue that the clusters *can* adhere if their surfaces approach within a distance w, and the probability that they *will* adhere is proportional to the volume V_c shown in Fig. 18. This is the volume traced out by the center of one cluster as it moves around the periphery of another cluster with the surfaces within the interaction range, w. The perimeter of V_c is the surface of the excluded volume for the clusters, as defined in Fig. 14. For clusters of sizes i and j with volume fractal dimension d_f, they estimate that

$$V_c = 4\pi(i^{1/d_f} + j^{1/d_f})^{d_s}w \tag{45}$$

Fig. 18. _____

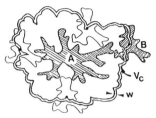

The excluded volume for clusters A and B. The volume V_c is a shell traced out by the center of cluster B with the thickness of the interaction range, w. From Ball *et al.* [55].

where d_s is the surface fractal dimension; in three dimensions, $2 \leq d_s \leq 3$. The reaction kernel is expected to be proportional to V_c, so for $i \approx j$,

$$\mathrm{K} \sim j^\lambda \qquad (i \approx j, \lambda = d_s/d_f). \tag{46}$$

For $j \gg i$, the large cluster can be regarded as consisting of j/i blobs of size i, so the kernel becomes

$$K(i, j) \sim \left(\frac{j}{i}\right) i^{d_s/d_f} \sim j i^{\lambda - 1} \qquad (j \gg i, \lambda = d_s/d_f). \tag{47}$$

In this case we expect $d_s \approx d_f \approx 2$ which leads to $\lambda \approx 1$, with $\nu \approx 1$ and $\mu \approx 0$; thus, RLCCA is predicted to be a nongelling process in class II. Therefore, Eq. 35b predicts exponential growth kinetics for the mean cluster size, and Eq. 38 predicts a power law distribution of cluster sizes with an exponent $\tau < 2$. Equations 46 and 47 are well approximated by $K(i, j) = i + j$ ($\sim i$ when $i = j$ or $i \gg j$) for which it is known [56,57] that $\tau = 1.5$ and s_{av} grows exponentially in time. Ball *et al.* [55] argue that the value of λ tends to be stabilized at unity. Increasing λ from that value favors the retention of small clusters, which more readily penetrate the large clusters; but this tends to raise d_f and thereby reduce λ. Similarly, reducing λ below 1 makes the cluster size distribution flatter, which reduces d_f, driving λ up.

Computer simulations of RLCCA support the form of Eq. 46 with $\lambda \approx 1$ [58,59], the prediction of exponential growth kinetics, and the power law form for the cluster size distribution [58]; however, the latter study found $\tau \approx 1.7$. RLCCA has been observed in many experimental studies of aggregation of colloids, based on the observed fractal dimension of ~ 2.1 [60–62]. Figure 19 compares the structures of clusters of gold particles grown under conditions of diffusion-limited and reaction-limited aggregation [61]. The size distribution is found to obey Eq. 38 with the value of τ in the expected range of ~ 1.5 [61] to 2 [63], and the growth kinetics are exponential [63]. Many other experimental studies are described by Meakin [41].

Fig. 19.

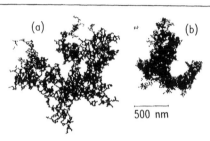

Transmission electron micrographs of clusters of gold particles (radius 7.5 nm) aggregated under diffusion-limited (a) and reaction-limited (b) conditions. Form Weitz *et al.* [61].

A particularly interesting study was performed by Aubert and Cannell [64], who examined the aggregation of Ludox® colloidal silica. They added 1 M NaCl to dilute (<1 wt%) suspensions and adjusted the pH over the range 4–12. Under conditions such that aggregation took days (pH 6.7), the fractal dimension of the clusters was $d_f = 2.1$, whereas at pH 8.6 the aggregation occurred within 30 seconds and $d_f = 1.75$. This suggests a transition from reaction-limited to diffusion-limited aggregation as the repulsive barrier is reduced. They also found that clusters with $d_f = 1.75$ were *unstable*, such that they rearranged with time until d_f reached 2.1. This is very much in keeping with the simulation results of Meakin and Jullien [50], which showed that a single internal rotation would convert a DLCCA cluster to the same fractal dimension as a RLCCA cluster. Surprisingly, at intermediate pH values Aubert and Cannell obtained fast aggregation with $d_f = 2.1$. They were able to demonstrate that under those experimental conditions the clusters were able to restructure within the time required to make the measurement; that is, the aggregation was diffusion-limited, but the rearrangement was too fast to be observed.

These studies show that the kinetic aggregation models provide a good description of the growth kinetics and the structure of clusters in colloids, but neither RLCCA nor DLCCA produces gelation. The reason is that the models are defined, and the supporting studies are performed, in very dilute suspensions. However, as the clusters grow, because of their fractal nature they occlude increasing amounts of solvent; that is, the density of the clusters decreases as $\rho \sim R^{d_f - 3} \sim 1/R$. Thus the amount of free solvent decreases and the effective volume fraction of polymer increases. Eventually the clusters becomes space-filling (when the volume fraction of monomer in the average cluster equals the initial volume fraction in the sol), as indicated in Fig. 20, by which time the reduced mobility will have caused the kinetic model, Eq. 30 to break down. Beyond that point, the interpenetrating

Fig. 20. _____

Space-filling fractal clusters.

clusters will begin to link together by a type of percolation process. In the next section, we survey experimental studies of the structure and properties of gelling systems, and see that the evidence generally supports this picture.

5.

EXPERIMENTAL STUDIES

We now explore two classes of experiments that test the predictions of the growth theories described in the preceding sections. Structural analyses provide the best test, because the predictions of the theories are unambiguous and the measurements can be made without damaging the sample. In contrast, theoretical predictions of properties, such as the viscosity of the sol or the modulus of the gel, are always based on secondary assumptions. As we have seen, within the context of percolation theory there are predictions of the critical exponent for the modulus ranging from 1.7 to 4, spanning the classical prediction of 3. Moreover, the experimental data are suspect on several counts [24]: (1) It is difficult to be sure that the measurement does not damage the fragile structure (most troublesome very near the gel point, where the data are most important). (2) The critical behavior is expected only very near p_c, but how near is "near"? (3) The calculation of the divergence of a property near p_c is very sensitive to the value of p_c, which is hard to

Fig. 21A.

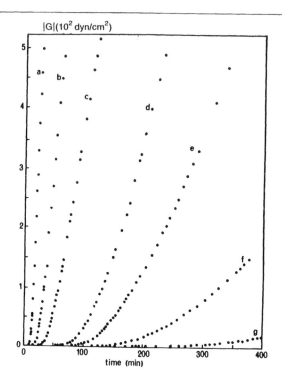

Fig. 21B.

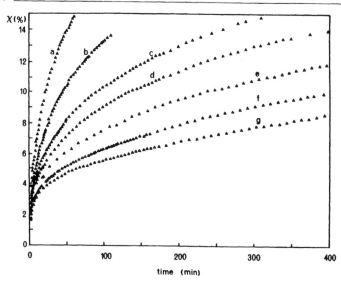

Fig. 21C.

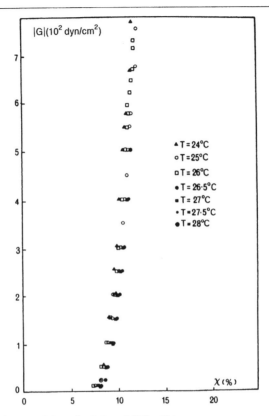

Increase of shear modulus of gelatin gel (4.7 wt%).
(A) Time-dependence of shear modulus at 0.15 Hz at temperatures (°C) 24 (a), 25 (b),
26 (c), 26.5 (d), 27 (e), 27.5 (f), 28 (g).
(B) Time-dependence of helix content at same temperatures.
(C) Relation between χ and G. From Djabourov *et al.* [65].

establish. (4) It is rarely possible to measure $p - p_c$ directly, so it is necessary
to assume that the departure from the critical point changes linearly with
time,

$$p - p_c \propto t - t_{gel} \qquad (48)$$

where t_{gel} is the time of gelation. The last of these assumptions is particularly
poor, but almost universally applied. In the vicinity of the gel point the
mobility of the polymer (including internal modes) is decreasing rapidly and
there is no reason to expect the rate of bond formation to be constant.
Djabourov *et al.* [65] measured the degree of formation, χ, of helical links
in gelatin gels (equivalent to p) and found, as shown in Fig. 21B, that χ is

not proportional to time, except over a very narrow interval near the critical point, χ_c. The modulus and viscosity were found to exhibit the critical exponents predicted by percolation when examined as a function of $\chi - \chi_c$. Zarzycki [66] closely examined the data of Yamane *et al.* [67] on reaction kinetics of TMOS and found serious discrepancy with the proportionality suggested in Eq. 48. Therefore, we should not be surprised to find that the critical exponents differ from theoretical predictions when they are based on that relation. If the error in t_{gel} is 1% and the critical interval, $\varepsilon \equiv (p - p_c)/p_c$, is 0.01, then the error in ε is 100%, even if the degree of reaction is exactly proportional to time.

Fortunately, there is another approach that avoids many of the problems just cited [24]: if more than one property is measured, the dependence on $p - p_c$ can be eliminated and a combined exponent can be determined. For example, suppose we measure both the average molecular weight,

$$M_w \sim (p - p_c)^{-\gamma} \tag{49}$$

and the z-averaged radius (or connectivity correlation length) R_z (which diverges as the spanning length, l_{av})

$$R_z \propto (p - p_c)^{-\nu}. \tag{50}$$

Then it is possible to solve for one in terms of the other:

$$M_w \sim R_z^{\gamma/\nu}. \tag{51}$$

Although this method removes the uncertainty about the location of the critical point, it does not necessarily help to distinguish between theories. For example, the exponent in Eq. 51 has the value $\gamma/\nu = 2$ for both percolation and the classical theory. (See values in Table 3.) Fortunately, this is not the case for every pair of properties, so the general approach can be successfully applied, as we shall see.

5.1. Structure

As noted in Chapters 3 and 4, small-angle X-ray scattering (SAXS) has shown that silicate polymers have a fractal dimension of $\sim$2.1 [68,69]. Similar results have been obtained for alumina aggregates grown in dilute solution [62]. These studies probe the clusters on a scale smaller than $\sim$10 nm, where the structure is found to agree with that expected for RLCCA, rather than percolation ($d_f = 2.5$) or the classical gelation theory ($d_f = 4$). SAXS also measures the spatial correlation length, ξ: if $g_s(r)$ is the probability that a monomer is located a distance r away from another monomer, then g_s is constant for $r > \xi$; for $r < \xi$, their exists *bond correlation*. This is the situation that one would expect during cluster growth: bond correlation would exist within clusters, because the clusters are denser than the surrounding solvent, so ξ would be a measure of the mean cluster size. However, remember that percolation theory requires that the bonds be placed (on the lattice or in the continuum) at random, so that they are uniformly distributed in space and there is no bond correlation. The fact that a spatial correlation length ($\xi \sim 10$ nm) is measured by SAXS implies that the sol is not exhibiting simple percolation. The process might be describable by correlated site-bond percolation, but the data indicate that the kinetics and geometry are consistent with the kinetic model (i.e., the Smoluchowski equation). As the gel point is approached, however, the spatial correlation length stops increasing [24,45], while the connectivity diverges, and *this* sounds like a percolation process. Thus, to study the connection between gelation and percolation it is necessary to explore length scales larger than ξ, which is most conveniently done using light scattering.

In the vicinity of the gel point, the scattered intensity, I, is given by [70]

$$I/c \sim q^{-\gamma/\nu} \tag{52}$$

where c is the concentration of the gelling species and q is the magnitude of the scattering vector, $q = 4\pi \sin(\theta/2)/\lambda$, θ is the scattering angle, and λ is the wavelength. The exponent in Eq. 52 is related to the fractal dimension and the polydispersity exponent[†] by [70]

$$\gamma/\nu = d_f(3 - \tau) \qquad (\tau > 2). \tag{53}$$

This creates a problem if one wants to distinguish between percolation and the classical gelation theory, because the former predicts $d_f \approx 2.5$ and $\tau \approx 2.2$ and the latter gives $d_f = 4$ and $\tau = 2.5$, and both lead to $\gamma/\nu \approx 2$. Fortunately, as noted earlier, when the sol is diluted, the fractal dimension of the cluster decreases from 2.5 to 2 [25,26], so the percolation prediction for a diluted sol becomes $\gamma/\nu \approx 1.6$.

A value near the percolation prediction was found by Martin and Keefer [72] for silica gels made from TMOS, using one-step hydrolysis with ammonium hydroxide ($\gamma/\nu \approx 1.60$) or ammonium fluoride ($\gamma/\nu \approx 1.59$), or using two-step hydrolysis with HCl, then NH_4OH ($\gamma/\nu \approx 1.69$). At times approaching the gel point, they diluted the sols and found overlapping clusters larger than a micron in radius; the scattering measurements were made on these larger swollen clusters. The q-dependence of the relaxation time, found from quasi-elastic light scattering, indicated that the clusters were flexible, rather than rigid structures. This large-scale flexibility is in

[†]Martin [71] has shown that the exponent in Eq. 53 is simply d_f, if $\tau < 2$. This is the case for RLCCA, so the SAXS data cited earlier do give d_f for the clusters.

contrast to the short-range structure revealed by SAXS [68,69] which, as discussed in Chapter 3, indicates that base-catalyzed silicates are stiff on a scale ≤ 10 nm, and only acid-catalyzed polymers are flexible.

This work was extended by Martin *et al.* [73], who also found that the value of γ/ν for swollen silica clusters was in reasonable agreement with percolation theory. For two-step acid–base catalysis, they found that Eq. 51 was obeyed with $\gamma/\nu \approx 1.79$, as shown in Fig. 22. However, when the critical behaviors of M_w (Fig. 23a) and R_z (Fig. 23b) were examined separately, the exponents were larger than expected. The departure from the critical point was assumed to follow Eq. 48 and t_{gel} was determined to within less than one minute. The cluster mass was found to diverge with an exponent of 2.7 ± 0.3, in contrast to the percolation value of 1.76 or the classical value of 1. The cluster radius diverged with an exponent of $\nu = 1.53 \pm 0.2$, and the dynamic light scattering showed that the hydrodynamic radius diverged with $\nu = 1.35$, whereas the percolation exponent for a swollen cluster is 1.1 and the classical theory predicts 0.5. These results indicate either a serious departure from theory or the inadequacy of Eq. 48. The error bounds on the exponents reflect the estimated uncertainty in t_{gel}, but the error in $p - p_c$ could be much greater, as previously noted. This may also explain the observations of Schmidt and Burchard [74] in a study of a divinylbenzene–styrene gel. Even though the internal structure of the gel included a large proportion of rings (which are forbidden by the classical theory), they found that the critical exponents for the polymer mass and radius and the gel fraction were all in agreement with the classical theory. However, when ε was excluded as in Eq. 51, then the results were in agreement with percolation theory.

Fig. 22.

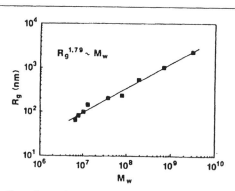

The z-average radius of gyration as a power of the weight-average cluster mass for an acid- and base-catalyzed system. The effective fractal dimension, $d_f(3 - \tau) = 1.79 \pm 0.12$, is in good agreement with the value 1.69 ± 0.05 found from the dependence of I/c on q (ref. [72]). From Martin *et al.* [73].

Fig. 23a.

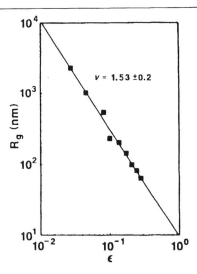

The z-averaged radius of gyration is shown to diverge with $\nu = 1.53 \pm 0.2$, where the error bound reflects the uncertainty in t_{gel}.

Fig. 23b.

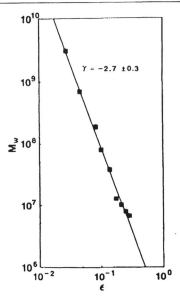

The critical exponent for the weight-average cluster mass is $\gamma = 2.7 \pm 0.3$; $\varepsilon \equiv p - p_c \approx t - t_{gel}$. From Martin *et al.* [73].

5.2. Properties

The properties that undergo the most dramatic changes at the gel point are the modulus, G, and viscosity, η, as illustrated in Fig. 1, and these are often measured as tests of the predictions of percolation theory. However, as noted by Stauffer et al. [24], it is particularly difficult to make accurate measurements of these properties in the critical region, because of the divergence of the relaxation time. That is, one needs to measure the static (i.e., zero-frequency) value, which requires that the experiment be continued for much longer than the average viscoelastic relaxation time of the system, $\tau_r \approx \eta/G$. Since τ_r diverges with η, the experiments require a long time, during which the properties of the system are changing rapidly. This problem was recognized by Gauthier-Manuel [75] in a study of an acrylamide-bisacrylamide gel. The data were obtained by magnetically levitating a metal sphere in the gel and measuring the resistance to its movement (revealed by the current required to keep the sphere fixed in position) as the gel was moved through the magnetic field at a constant velocity (shear rate $\sim 10^{-4}\,s^{-1}$). Assuming the validity of Eq. 48, the critical exponent for the viscosity was found to be $k = 2.7 \pm 0.2$, much larger than the predictions of any theory. (See Table 3.) However, the location of t_{gel} was found to depend on the frequency of measurement, and η was found to extrapolate to a finite value at the gel point as a result of the error introduced by the frequency-dependence. Similarly, Djabourov et al. [65] showed that long viscoelastic relaxation times prevented accurate measurements of η or G closer to the gel point than $|(\alpha - \chi_c)/\chi_c| \approx 0.1$.

Using the levitation apparatus [75], Gauthier-Manuel et al. [3,76] examined the critical behavior in silica gels made by hydrolyzing TMOS in pure water and in 0.5 M NaCl solution. The conditions were chosen to give gel times of ~ 1 day, so that measurements could be taken conveniently close to the gel point. They found $k = 1.0 \pm 0.1$ and 1.1 ± 0.3 for the water and salt solutions, respectively, values that are in the range expected from percolation theory [27]. Similar values ($k = 0.82$–0.93) were found for a polyurethane gel by Adam et al. [77], who used the same apparatus. On the other hand, Kozuka et al. [78] examined acid-catalyzed silica sols made from TEOS using a cone-and-plate rheometer. The viscosity was measured as a function of frequency and extrapolated to $\omega = 0$, and the critical exponent was found to be $k = 2.6$. Thus the data for a variety of systems give values of k ranging over more than a factor of 3.

Theoretical predictions for the evolution of the modulus in a percolating system range from critical exponents of $T \approx 1.7$ [31] to 3.55 [32], 3.78 [33], or possibly a crossover [34] from ~ 1.6 near t_{gel} to ~ 4.4 far above it. Most of the data seem to support the latter idea, while none agree with the classical prediction that t = 3 [24,79].

A direct test of the idea that the modulus scales as the electrical conductivity was made by Benguigui [80], who measured both σ and E on a metal foil as a function of the number of holes punched in it (on the sites of a square lattice). He found that the critical exponent for the conductivity was 1.2 ± 0.1, in agreement with the percolation prediction for two dimensions, while for the modulus a much higher value (T = 3.5 ± 0.4) was obtained. These measurements could not be made very near the critical threshold ($\varepsilon > 0.04$), because of the weakness of the film, but they clearly indicate that the scaling behavior of the conductivity and the modulus can be quite different. Deptuck et al. [81] measured E and σ as functions of the volume fraction, ϕ, on sintered silver compacts and found

$$E \sim (\phi - \phi_c)^a \tag{54}$$

and

$$\sigma \sim (\phi - \phi_c)^b \tag{55}$$

where $a = 3.8 \pm 0.5$, $b = 2.15 \pm 0.25$, and the critical volume fraction was $\phi_c \approx 0.062$. They identified the exponents a and b with T and t, respectively, and claimed that the results were in agreement with the predictions of Kantor and Webman [32]. However, it is not clear what it means to have a critical volume fraction in a powder compact. Woignier et al. [82] argue that the measured volume fraction corresponds to the gel fraction in a percolation model,

$$\phi = P(p) \sim (p - p_c)^\beta \tag{56}$$

so we expect

$$E \sim \phi^{T/\beta}. \tag{57}$$

Woignier et al. found $T/\beta = 3.8$ for a silica aerogel; given $\beta = 0.45$, this leads to T = 1.7, as expected for scalar elasticity. Replotting the data of Deptuck et al. with $\phi_c = 0$ one obtains $T/\beta \approx 7.0$–8.5, so T ≈ 3.2–3.8, and $t/\beta \approx 4.2$, so t ≈ 1.9. Thus the latter data are still in rough agreement with Kantor and Webman and show different scaling behavior for the modulus and conductivity.

Tokita et al. [83] examined the modulus of casein gel. This represents site rather than bond percolation, because the bonds are formed with 100% efficiency by addition of an enzyme, so the fraction of filled sites is equivalent to the volume fraction, ϕ, of monomer. In this way, they avoided reliance on Eq. (48). The critical exponent was found to be t = 2.06 near the critical volume fraction, ϕ_c, but at larger departures the shear modulus obeyed

$$G = c_1(\phi - \phi_c)^{2.06} + c_2(\phi - \phi_c)^{3.4}. \tag{58}$$

This indicates a crossover of the sort predicted by Feng and Sen [34]. A similar crossover is suggested by the data of Gauthier-Manuel *et al.* [3,76] on the modulus of silica gel. Far from the threshold they found t $\approx$ 3.7, but for small values of ε there were indications of a shift to a smaller exponent. They attributed the effect to the growth of necks between colloidal particles, which would create a resistance to bending of the "bonds" between particles and thereby satisfy the conditions for tensor elasticity. A similar effect was observed by Sonntag and Russell [84], who studied the effect of aging on the modulus of a gel made from a polystyrene latex. They found that t $\approx$ 2.5 for fresh gels, but the exponent increased to ~4.4 as the gels were aged. Many other studies of the critical behavior have been performed, with the resulting exponents typically being close to 1.9 [75] or 3.6 [76,77]. Thus, none of the data contradict the prediction of a crossover between those values, and some of the studies seem to show it directly.

The data for the critical behavior of the modulus are in general agreement with the prediction of percolation theory, but it is important to beware of believing data that suit one's preconceptions. After all, these measurements not only suffer the problems discussd previously for the geometrical parameters, they also have the inherent problem of the long relaxation time, just as for the viscosity. The clustering of values for the viscosity exponent, k, near 1 and 2.7 in different studies may result from a crossover phenomenon of the sort that apparently exists for the modulus. However, even if we accept this appealing explanation, it is still necessary to account for the unexpectedly rapid divergence of radius of gyration and molecular weight indicated in scattering measurements. It seems most prudent to accept experiments that use results for two properties to eliminate dependence on ε (most of which support percolation theory) and to be skeptical of direct measurements of the critical exponents, except in those rare cases (e.g., [65] and [83]) where the degree of reaction is accurately known.

6.

SUMMARY

Condensation reactions produce polymeric clusters whose growth kinetics and fractal structure follow the predictions of kinetic models based on Smoluchowski's equation. As the clusters grow, their density decreases (because $d_f < 3$), so the effective volume fraction of polymer increases. Eventually the clusters overlap and become nearly immobile, so that further bonding involves a percolative process with the "sites" being filled by large (~micron) polymeric clusters. The evolution of the properties in the vicinity of the gel point is generally in agreement with the critical behavior predicted by percolation theory, and in contradiction to the classical theory. This is particularly true when the critical behavior is established using pairs of properties (as in Eq. 51), presumably because these are the most reliable experiments. Much less consistent (or trustworthy) results have been found in direct measurements of the divergence as a function of the departure from the critical point $(t - t_{gel})$.

The reactions that bring about gelation continue long after the gel point. In the next chapter we examine the long-term aging of gels, during which they may stiffen, coarsen, and shrink.

REFERENCES

1. P.J. Flory, *Faraday Disc. Chem. Soc.*, **57** (1974) 7–18.
2. C.B. Hurd, *Chem. Rev.*, **22** (1938) 403–422.
3. B. Gauthier-Manuel, E. Guyon, S. Roux, S. Gits, and F. LeFaucheux, *J. Phys.* (Les Ulis, Fr) **48** [5] (1987) 869–875.
4. M.D. Sacks and R.-S. Sheu in *Science of Ceramic Chemical Processing*, eds. L.L. Hench and D.R. Ulrich (Wiley, New York, 1986), pp. 100–107.
5. M.D. Sacks and R.-S. Sheu, *J. Non-Cryst. Solids*, **92** (1987) 383–396.
6. R.M. Christiansen, *Theory of Viscoelasticity—An Introduction*, 2d ed. (Academic Press, New York, 1982), p. 119.
7. J.D. Mackenzie in *Science of Ceramic Chemical Processing*, eds. L.L. Hench and D.R. Ulrich (Wiley, New York, 1986), p. 113–122.
8. V. Gottardi, M. Guglielmi, A. Bertoluzza, C. Fagnano, and M.A. Morelli, *J. Non-Cryst. Solids*, **63** (1984) 71–80.
9. M.F. Bechtold, W. Mahler, and R.A. Schunn, *J. Polym. Sci.: Polym. Chem. Ed.*, **18** (1980) 2823–2855.
10. M.W. Colby, A. Osaka, and J.D. Mackenzie, *J. Non-Cryst. Solids*, **99** (1988) 129–139.
11. J.C. Debsikdar, *Adv. Ceram. Mater.*, **1** [1] (1986) 93–98.
12. K.C. Chen, T. Tsuchiya, and J.D. Mackenzie, *J. Non-Cryst. Solids*, **81** (1986) 227–237.
13. A.J. Vega and G.W. Scherer, *J. Non-Cryst. Solids*, **111** (1989) 153–166.
14. L.W. Kelts, N.J. Effinger, and S.M. Melpolder, *J. Non-Cryst. Solids*, **83** (1986) 353–374.
15. S. Gits-Léon, F. LeFaucheux, and M.C. Robert, *J. Crystal Growth*, **84** (1987) 155–162.
16. R.K. Dwivedi, *J. Mater. Sci.*, **5** (1986) 373–376.
17. C.B. Hurd, M.D. Smith, F. Witzel, and A.C. Glamm, Jr., *J. Phys. Chem.*, **57** (1953) 678–680.
18. P.J. Flory, *Principles of Polymer Chemistry* (Cornell Univ. Press, Ithaca, New York, 1953), chapter IX.
19. R. Zallen, *The Physics of Amorphous Solids* (Wiley, New York, 1983), chapter 4.
20. C.A.L. Peniche-Covas, S.B. Dev, M. Gordon, J. Judd, and K. Kajiwara, *Faraday Disc. Chem. Soc.*, **57** (1974) 165–180.
21. W.H. Stockmayer in *Advancing Fronts in Chemistry, Vol. I: High Polymers*, ed. S.B. Twiss (Reinhold, New York, 1945), pp. 61–73.
22. B.H. Zimm and W.H. Stockmayer, *J. Chem. Phys.*, **17** [12] (1949) 1301–1314.
23. P.G. de Gennes, *Biopolymers*, **6** (1968) 715–729.

24. D. Stauffer, A. Coniglio, and M. Adam, *Advances in Polymer Science*, **44** (1982) 103–158.
25. J. Isaacson and T.C. Lubensky, *J. Phys.* (Paris), **41** (1980) L469–L471.
26. P.G. de Gennes, *C.R. Acad. Sci.*, **291** (1980) 17–19.
27. P.G. de Gennes, *C.R. Acad. Sci.*, **286B** (1978) 131–133.
28. G.R. Dobson and M. Gordon, *J. Chem. Phys.*, **41** [8] (1964) 2389–2398.
29. P.G. de Gennes, *J. Physique (Paris) Lett.*, **40** (1979) 197–199.
30. J.E. Martin, D. Adolf, and J.P. Wilcoxon, *Phys. Rev. A*, **39** [3] (1989) 1325–1332.
31. P.G. de Gennes, *Scaling Concepts in Polymer Physics* (Cornell Univ. Press, Ithaca, New York, 1979), p. 142.
32. Y. Kantor and I. Webman, *Phys. Rev. Lett.*, **52** [21] (1984) 1891–1894.
33. S. Roux and E. Guyon in *On Growth and Form*, eds. H.E. Stanley and N. Ostrowski (Martinus Nijhoff, Boston, 1986), pp. 273–277.
34. S. Feng and P.N. Sen, *Phys. Rev. Lett.*, **52** [3] (1984) 216–219.
35. H. Nakanishi and P.J. Reynolds, *Phys. Lett.*, **71A** [2,3] (1979) 252–254.
36. P. Agrawal, S. Redner, P.J. Reynolds, and H.E. Stanley, *J. Phys. A.*, **A12** (1979) 2073–2085.
37. J. Boissonade, F. Barreau, and F. Carmona, *J. Phys. A: Math. Gen.*, **16** (1983) 2777–2787.
38. F. Carmona, R. Barreau, P. Delhaes, and R. Canet, *J. Physique—Lett.*, **41** (1980) L531–L534.
39. I. Balberg, C.H. Anderson, S. Alexander, and N. Wagner, *Phys. Rev. B*, **30** [7] (1984) 3933–3943.
40. I. Balberg, *Phil. Mag. B*, **56** [6] (1987) 991–1003.
41. P. Meakin in *Phase Transitions, 12* (Academic Press, New York, 1988), pp. 335–489.
42. P. Meakin, in *Advances in Colloid and Interface Science*, ed. A.C. Zettlemoyer (Elsevier, Amsterdam, 1988), pp. 1–83.
43. P. Meakin in *Time-Dependent Effects in Disordered Materials*, eds. R. Pynn and T. Riste (Plenum, New York, 1987), pp. 45–70.
44. P. Meakin, *Ann. Rev. Phys. Chem.*, **39** (1988) 237–267.
45. J.E. Martin in *Time-Dependent Effects in Disordered Materials*, eds. R. Pynn and T. Riste (Plenum, New York, 1987), pp. 425–449.
46. See discussion in sect. III.6 in S. Chandrasekhar, *Rev. Mod. Phys.*, **15** [1] (1943) 1–89.
47. M. Ernst in *Fractal Physics*, eds. L. Pietronero and E. Tosatti (North Holland, New York, 1986), pp. 289–302.
48. P.G.J. van Dongen and M.H. Ernst, *Phys. Rev. Lett.*, **54** [13] (1985) 1396–1399.
49. D.W. Schaefer, *Mater. Res. Bull.*, **13** [2] (1988) 22–27.
50. P. Meakin and R. Jullien, *J. Chem. Phys.*, **89** [1] (1988) 246–250.
51. R.M. Ziff, E.D. McGrady, and P. Meakin, *J. Chem. Phys.*, **82** [11] (1985) 5269–5274.
52. P. Meakin, T. Vicsek, and F. Family, *Phys. Rev. B*, **31** [1] (1985) 564–569.
53. D.A. Weitz, J.S. Huang, M.Y. Lin, and J. Sung, *Phys. Rev. Lett.*, **53** [17] (1984) 1657–1660.
54. D.A. Weitz and M.Y. Lin, *Phys. Rev. Lett.*, **57** (1986) 2037–2040.
55. R.C. Ball, D.A. Weitz, T.A. Witten, and F. Leyvraz, *Phys. Rev. Lett.*, **58** [3] (1987) 274–277.
56. F. Leyvraz in *On Growth and Form*, eds. H.E. Stanley and N. Ostrowski (Martinus Nijhoff, Boston, 1986), pp. 136–144.
57. A.M. Golvin, *Bull. Acad. Sci. USSR*, Geophys. Ser. No. 5 (1963) 482–487 (Eng. trans.).
58. P. Meakin and F. Family, *Phys. Rev.*, **A36** [11] (1987) 5498–5501.
59. W.D. Brown and R.C. Ball, *J. Phys.*, **A18** (1985) L517–L521.
60. D.W. Schaefer, J.E. Martin, P. Wiltzius, and D.S. Cannell, *Phys. Rev. Lett.*, **52** [26] (1984) 2371–2374.

61. D.A. Weitz, J.S. Huang, M.Y. Lin, and J. Sung, *Phys. Rev. Lett.*, **54** (1985) 1416–1419.
62. D.W. Schaefer, R.A. Shelleman, K.D. Keefer, and J.E. Martin, *Physica*, **140A** (1986) 105–113.
63. J.E. Martin, *Phys. Rev.*, **A36** (1987) 3415–3426.
64. C. Aubert and D.S. Cannell, *Phys. Rev. Lett.*, **56** [7] (1986) 738–741.
65. M. Djabourov, J. Leblond, and P. Papon, *J. Phys.* (France), **49** (1988) 333–343.
66. J. Zarzycki in *Science of Ceramic Chemical Processing*, eds. L.L. Hench and D.R. Ulrich (Wiley, New York, 1986), pp. 21–36.
67. M. Yamane, S. Inoue, and A. Yasumori, *J. Non-Cryst. Solids*, **63** (1984) 13–22.
68. D.W. Schaefer and K.D. Keefer in *Better Ceramics Through Chemistry*, eds. C.J. Brinker, D.E. Clark, and D.R. Ulrich (North-Holland, New York, 1984), pp. 1–14.
69. C.J. Brinker, K.D. Keefer, D.W. Schaefer, R.A. Assink, B.D. Kay, and C.S. Ashley, *J. Non-Cryst. Solids*, **63** (1984) 45–59.
70. J.E. Martin and B.J. Ackerson, *Phys. Rev. A*, **31** (1985) 1180–1182.
71. J.E. Martin, *J. Appl. Cryst.*, **19** (1986) 25–27.
72. J.E. Martin and K.D. Keefer, *Phys. Rev. A*, **34** [6] (1986) 4988–4992.
73. J.E. Martin, J. Wilcoxon, and D. Adolf, *Phys. Rev. A*, **36** [4] (1987) 1803–1810.
74. M. Schmidt and W. Burchard, *Macromolecules*, **14** (1981) 370–376.
75. B. Gauthier-Manuel, *Springer Proc. Phys.*, **5** (Phys. Finely Divided Matter) (1985) 140–147.
76. B. Gauthier-Manuel, E. Guyon, S. Roux, S. Gits, and F. LeFaucheux, *PhysicoChemical Hydrodynamics*, **9** [3/4] (1987) 605–609.
77. M. Adam, M. Delsanti, and D. Durand, *Macromolecules*, **18** (1985) 2285–2290.
78. H. Kozuka, H. Kuroki, and S. Sakka, *J. Non-Cryst. Solids*, **95 & 96** (1987) 1181–1188.
79. G.R. Dobson and M. Gordon, *J. Chem. Phys.*, **43** [2] (1965) 705–713.
80. L. Benguigui, *Phys. Rev. Lett.*, **53** [21] (1984) 2028–2030.
81. D. Deptuck, J.P. Harrison, and P. Zawadski, *Phys. Rev. Lett.*, **54** (1985) 913–916.
82. T. Woignier, J. Pelous, J. Phalippou, R. Vacher, and E. Courtens, *J. Non-Cryst. Solids*, **95 & 96** (1987) 1197–1202.
83. M. Tokita, R. Niki, and K. Hikichi, *J. Chem. Phys.*, **83** [5] (1985) 2583–2586.
84. R.C. Sonntag and W.B. Russell, *J. Colloid Interface Sci.*, **116** [2] (1987) 485–489.

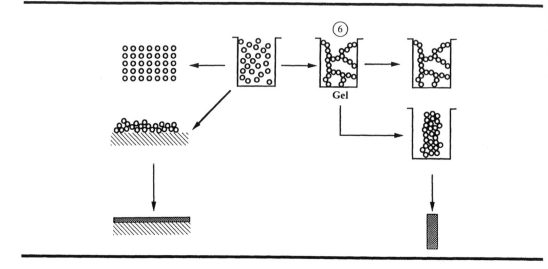
Gel

CHAPTER 6

Aging of Gels

This chapter describes changes in the structure and properties of gels that occur during aging in the pore liquid at room temperature or under hydrothermal conditions. The chemical reactions that cause gelation continue long after the gel point, producing strengthening, stiffening, and shrinkage of the network. Processes of dissolution and reprecipitation may coarsen the pore structure, and separation may occur into mixtures of solid–liquid, liquid–liquid, or solid–solid phases. These changes have a profound effect on the subsequent processes of drying and sintering. The outline of this chapter is as follows:

1. *Aging Processes* reviews the mechanisms by which changes in structure and properties occur.
2. *Structure* discusses changes in microstructure, chemical composition, and phase during aging.
  2.1. *Silicates* surveys the most thoroughly studied materials, for which the effects of pH, temperature, adsorbed organics, and other factors are well known.
  2.2. *Other Oxides* describes structural changes of alumina, titania, zirconia, chromia, and other gels during aging.
3. *Syneresis* explains the phenomenon of spontaneous shrinkage and shows how the kinetics of the process depend on pH, temperature, and other conditions, including sample size.
4. *Mechanical Properties* describes the dependence of the viscoelastic properties and strength of gels on time, shrinkage, and chemical environment.

1.

AGING PROCESSES

Gelation is a spectacular event, when a solution suddenly loses it fluidity and takes on the appearance of an elastic solid. The eye is deceived, however, by what appears to be a "freezing" process. The gel point represents the moment when the last link is formed in the chain of bonds that constitutes the spanning cluster. The network restrains the flow of the pore liquid, but there is no exotherm or endotherm and the chemical evolution of the system is virtually unaffected. The processes of change during aging after gelation are categorized as *polymerization, coarsening,* and *phase transformation.*

Polymerization is the increase in connectivity of the network produced by condensation reactions, such as

$$\equiv Si-OH + HO-Si\equiv \rightarrow \equiv Si-O-Si\equiv + H_2O. \tag{1}$$

Studies based on nuclear magnetic resonance (NMR) [1] and Raman spectroscopy [2,3] indicate that condensation in silica gels continues long after gelation, because of the large concentration of labile hydroxyl groups. For example, Fig. 7 of Chapter 5[†] shows substantial amounts of Q^2 at the gel point, but the proportions of Q^3 and Q^4 increase with time after gelation. These changes continue for months at room temperature; the rate of reaction is dependent on temperature and the concentration and pH of the solution. By creating new bridging bonds, such reactions stiffen and strengthen the network. (See Section 4.) In addition to condensation, aging can result in further hydrolysis,

$$\equiv Si-OR + H_2O \rightarrow \equiv Si-OH + ROH, \tag{2}$$

or in the reverse reaction, reesterification,

$$\equiv Si-OH + ROH \rightarrow \equiv Si-OR + H_2O. \tag{3}$$

As explained in Chapter 3, the latter reaction is suppressed by using excess water.

Syneresis is shrinkage of the gel network resulting in expulsion of liquid from the pores. This process, which is discussed in detail in Section 3, is believed to be caused by the same condensation reactions that produce gelation. A dramatic example of syneresis in a titania gel [4] is presented in Fig. 1: the gel (tipped over) initially filled the volume now occupied by liquid alone. The shrinkage is produced by the reactions

$$\equiv Ti-OH + HO-Ti\equiv \rightarrow \equiv Ti-O-Ti\equiv + H_2O \tag{4a}$$

[†] In the terminology of ^{29}Si NMR, Q^n represents a silicon atom bonded through oxygen to n other silicon atoms.

Fig. 1. _____

Shrinkage from syneresis in titania gel made from alkoxide. From Yoldas [4].

and

$$\equiv Ti-OR + HO-Ti\equiv \rightarrow \equiv Ti-O-Ti\equiv + ROH. \qquad (4b)$$

Equation (4b) is significant in this case, because only two moles of water per mole of alkoxide were initially provided; consequently the syneresis rate is accelerated by subsequent additions of water or peroxide (H_2O_2). A different mechanism of syneresis applies to particulate gels. In these systems, gelation results from collapse of the repulsive double layer that stabilizes the sol, and syneresis is believed to be driven by van der Waals forces [5]. Shrinkage is ultimately stopped by the remaining repulsive force; the more electrolyte that is added, the greater the final shrinkage [6].

The process exemplified in Fig. 1 is sometimes called *macrosyneresis* to distinguish it from *microsyneresis*, which is schematically illustrated in Fig. 2. This is a process of phase separation in which the polymers cluster together, creating regions of free liquid; the driving force is the greater affinity of the polymer for itslf than for the pore liquid. In organic polymers, microsyneresis produces turbidity (because the separate phases scatter light) that decreases as macrosyneresis compresses the structure [7].

Fig. 2.

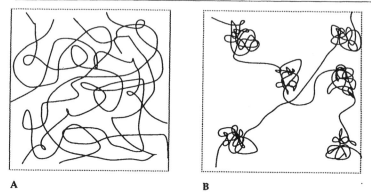

A   B

Schematic of microsyneresis: uniform network of polymers in solvent (A) and phase-separated network with polymers drawn together (B).

The porosity in dried organic gels is attributed to microsyneresis [7], but it is not clear whether the same process is important in inorganic gels, where porosity may reflect the structure of the clusters created in the sol. One apparent example of microsyneresis in a titania gel is provided by Quinson *et al.* [8], who used thermoporometry [9–11] to demonstrate that the pore size of the gel expanded while soaking in decane, but shrank while soaking in water. As shown in Fig. 3, the pore size triples in decane over a period of days, but returns to its original size when transferred into water. The swelling of the macroscopic sample during immersion in decane is only ~1–3 vol%.

Coarsening or *ripening* is a process of dissolution and reprecipitation driven by differences in solubility, s, between surfaces with different radii of curvature, r:

$$s = s_0 \exp(2\gamma_{SL} V_m / R_g T r) \qquad (5)$$

where s_0 is the solubility of a flat plate of the solid phase, γ_{SL} is the solid–liquid interfacial energy, V_m is the molar volume of the solid, R_g is the ideal gas constant, and T is the temperature. As shown in Fig. 4, particles have positive radii of curvature ($r > 0$), so they are more soluble than a flat plate of the same material; the smaller the particle, the greater its solubility. Therefore, smaller particles dissolve and the solute precipitates onto larger particles. Crevices and necks between particles have negative radii ($r < 0$), so their solubility is especially low and material accumulates there. The result of dissolution–reprecipitation is to reduce the net curvature of the solid phase: small particles disappear and small pores are filled in, so the interfacial area decreases and the average pore size increases. Note that this

Fig. 3.

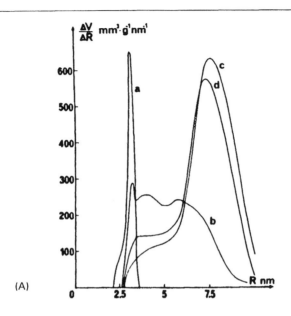

(A)

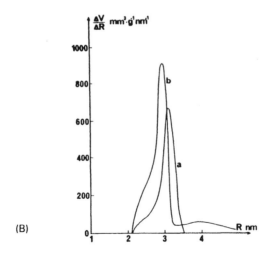

(B)

Pore radius distribution curves obtained by decane thermoporometry during aging.
(A) (a) Initial measurements; (b) after 1 day in decane; (c) after 9 days; (d) after 45 days.
(B) (a) Same curve as (a) in part A of figure; (b) after 1.5 months in decane and a subsequent treatment in water.
From Quinson *et al.* [8].

Fig. 4.

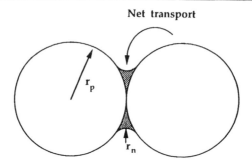

Net transport

Definition of positive and negative radii of curvature. The radius of the neck, r_n, is negative because the center of curvature is outside of the solid phase; the radius of the particle, r_p, is positive. Material tends to dissolve from surfaces with positive curvature and deposit in regions with negative curvature. The growing neck (shaded) strengthens the gel.

not produce shrinkage, because the centers of the particles do not move toward one another; to cause that, it is necessary to *remove* material from the neck region. (See discussion in Chapter 11.)

The growth of necks adds to the strength and stiffness of the network, as suggested by Fig. 5. This process is discussed at length by Iler [12] (pp. 222–230, 519–523). The rate of coarsening is influenced by the factors that affect solubility, such as temperature, pH, concentration, and type of solvent. The amount of shrinkage that occurs during drying is dependent on the stiffness

Fig. 5.

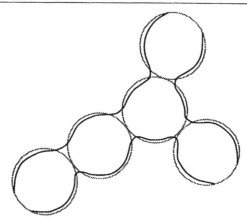

Dissolution and reprecipitation causes growth of necks between particles, increasing the strength and stiffness of the gel.

Fig. 6. _____

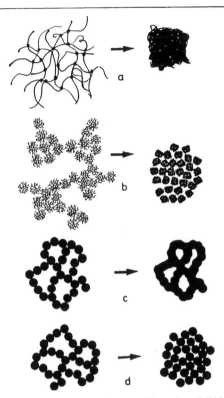

Schematic representation of gel desiccation for (a) acid-catalyzed, (b) base-catalyzed, and particulate silica gels aged under conditions of high (c) or low (d) solubility. From Brinker and Scherer [13].

of the network [13], as indicated schematically in Fig. 6. If the gel is aged under conditions of high solubility, the network may resist compression by capillary forces during drying; although this is easiest to visualize for particulate systems, it is equally true for polymeric gels.

Several types of *phase transformations* can occur during aging. We have already mentioned microsyneresis, in which solid phase separates from the liquid on a local scale. There may also be segregation of the liquid into two or more phases. For example, in base-catalyzed hydrolysis of silicon alkoxides, there may be isolated regions of unreacted alkoxide [14]. When a gel of that type is soaked in pure water, it turns white and opaque [15], apparently (as explained in Section 2) from segregation of droplets of partially reacted alkoxide. Aging may also lead to crystallization, as in the precipitation of nitrate crystals from alumina gel made from $Al(NO)_3$ [16], or zirconia crystals from a $Na_2O \cdot Zr_2O \cdot SiO_2 \cdot P_2O_5$ gel [17]. Many gels and precipitates of hydrous oxides are amorphous as formed [18], but aging of the solution allows reorganization of the structure by dissolution and reprecipitation, resulting in a crystalline product. For example, aging of hydrous alumina sols allows crystallization of aluminum hydroxide [19]. Similarly, amorphous zirconia gels can be made [20] by precipitation of $ZrCl_2$, but crystalline particles of ZrO_2 result [21] when the aqueous salt solution is aged at 90 to 116°C; the transformation is attributed to rearrangement of hydrous tetramers. Titania gels made by hydrolysis of $TiCl_4$ are also amorphous unless aged in solution [22]. These transformations are accelerated under hydrothermal conditions, as shown in Section 2.

The structural changes that occur during aging have an important effect on the drying process (discussed in detail in the following chapters). The capillary pressure that develops during drying is proportional to the interfacial area in the gel; if that area is reduced by coarsening, the maximum pressure generated during drying is smaller. The stiffer and stronger the gel network becomes, the better it can withstand the capillary pressure, so aged gels shrink and crack less. The structure of the gel also influences the sintering behavior, since the process of densification is driven by the interfacial energy. Moreover, as explained in Chapter 11, crystalline gels sinter much more slowly than amorphous gels with the same composition, so phase transformation during aging can be profoundly important.

2.

STRUCTURE

In this section we examine the effect of solution conditions (liquid composition, pH, temperature) on the evolution of microstructure of gels. Ongoing condensation reactions are most notable for their effects on syneresis and changes in the viscoelastic properties of the gel, effects that are discussed in the following sections. Here we concentrate on the processes of coarsening and phase transformation during aging.

2.1. Silicates

As shown in Fig. 7, the solubility and rate of dissolution of silica increase rapidly at high pH, so it is not surprising that the rate of coarsening of gels is similarly pH-dependent. A striking example of this effect is shown in Fig. 8. The microporous xerogel (LpH5) made under conditions of low

Fig. 7. _____

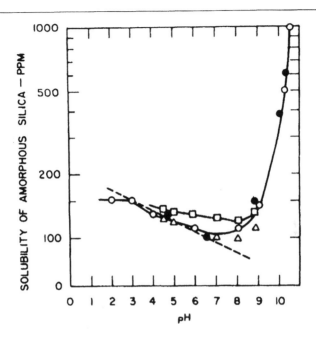

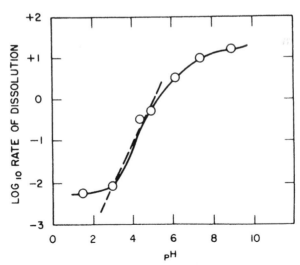

Data of several authors for (A) solubility and (B) log(rate of dissolution) versus pH for amorphous silica in water at 19 to 30°C. From Iler, *The Chemistry of Silica* (Wiley, New York, 1979).

Fig. 8.

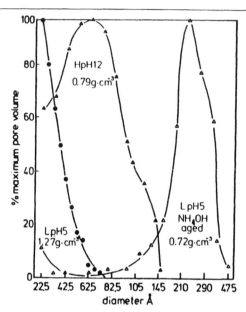

Pore-size distributions (normalized to 100% at the modal pore diameter) and bulk densities of xerogels LpH5 (made at pH 2.5 with 5 moles water/mole TEOS) and HpH12 (made at pH 9.2 with 12 moles water/mole TEOS), and LpH5 aged one week at 50°C in 3 M NH₄OH prior to drying. From Brinker and Scherer [23].

water and low pH is contrasted with the mesoporous specimen (HpH12) made using higher pH and more water during hydrolysis of the tetraethoxysilane (TEOS) precursor. If the wet gel LpH5 is aged in a strongly basic solution (3 M NH₄OH) at 50°C for one week, the derived xerogel has even larger pores than HpH12. Similar behavior has been reported for silica gels derived from alkali silicates by neutralization with mineral acids. Shamrikov *et al.* [24] made such gels which yielded xerogels with surface areas of $814 \, m^2/g$ and pore volumes of $0.45 \, g/cm^3$. After aging the wet gel at pH 9 for ≥ 10 h at 50°C, followed by drying, the pore volume of the xerogel was $0.7 \, g/cm^3$ and the surface area was $600 \, m^2/g$; aging at 95°C led to corresponding values of $0.9 \, g/cm^3$ and $400 \, m^2/g$. These differences reflect changes produced in solution *and* differences in drying behavior: the reduction in surface area is produced by dissolution–reprecipitation; the higher pore volume results because the stiffer gel produced by aging does not shrink as much under the influence of capillary pressure.

The effect of aging temperature on pore-size distribution was clearly illustrated by Yamane and Okano [25] for a gel made from tetramethoxysilane (TMOS). As shown in Fig. 9, the porosity and average pore size

Fig. 9. _____

Sample	Gelling temperature (°C)	Bulk density (g·cm⁻³)	True density (g·cm⁻³)	Porosity (%)
A	54	1.46	2.10	30.5
B	65	1.13	2.03	44.3
C	68	1.02	2.04	50.0
D	70	0.98	2.03	51.7

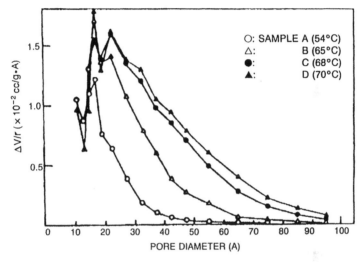

Pore-size distributions of xerogels made from silica gels aged at indicated temperatures. From Yamane and Okano [25].

increased rapidly as the temperature was increased from 54 to 70°C. This result reflects the influence of temperature on solubility of silica. Similarly, hydroxyl and fluorine ions both catalyze the dissolution of silica [12,26], so they enhance the rate of aging [26]. The solubility of silica rises again at very low pH (<0), so aging in strong solutions of mineral acids also leads to an increase in pore volume and loss of micropores [27]. At a given normality, Sheinfain *et al.* [27] found the effect to decrease in the order HCl $>$ $H_2SO_4 > H_3PO_4$; however, the rate of aging was independent of the type of acid if the activity of the proton was the same in each solution. Many other examples of the effects of temperature and pH on the structure of silica gels are presented by Iler [12] (pp. 528–533).

Organic liquids can retard aging by adsorbing on silanol groups and inhibiting condensation reactions. Some examples from the work of Sheinfain and Neimark [26] are given in Table 1. The gels were made from

Table 1.

Influence of Medium on Aging of Silica Gel.

Aging Medium	Duration of aging (days)			
	67		168	
	V_p (cm³/g)	S (m²/g)	V_p (cm³/g)	S (m²/g)
(No aging)	(0.71)	(700)	—	—
Water	0.99	550	1.03	420
Ethanol	0.72	700	0.67	680
Glycerine	0.66	720	0.66	740
Dioxane	0.66	720	0.68	700

Source: Sheinfain and Neimark [26].

solutions of alkali silicate neutralized with mineral acid, so the fresh gel is in an aggressive environment that causes a significant amount of aging (condensation) during drying. Therefore the drying shrinkage is greater (pore volume is smaller) for gels soaked in organic liquids before drying than for the control sample dried immediately after gelation. The lower surface tension of the organic liquids produces less capillary pressure during drying, but this is more than compensated by the lower modulus (resulting from less condensation). Table 1 shows that soaking in water leads to higher pore volume in the xerogel. Since water is the product of the condensation reaction, excess water might be expected to retard condensation; on the other hand, the solubility of silica and the concentration of catalyzing hydroxyl ions increase with the water content and these appear to be the dominant factors.

The rate of coarsening of silica increases with temperature and pressure. For example, van der Grift *et al.* [28] prepared silica gels by neutralization of alkali silicate solutions and aged them in an autoclave in water at various temperatures. When the temperature is above the boiling point of water, the sample is exposed to steam at high pressure, and the process is called *hydrothermal aging*. As shown in Fig. 10, the pore size increased by more than an order of magnitude as the temperature was raised ∼80°C (from ∼170 to 250°C). More detailed studies along this line were performed by Chertov *et al.* [29], who followed the effect of aging at pH of 1 and 11. At the high pH, the gel went back into solution after aging 50 h at 90°C or <1 h at $T > 130°C$; consequently, the highest pore volume (0.8 cm³/g) was obtained by drying immediately after adjusting the pH to 11, and smaller values (down to 0.2 cm³/g) resulted from aging at elevated temperatures. At pH 1, the surface area decreased and the pore volume increased at higher

Fig. 10.

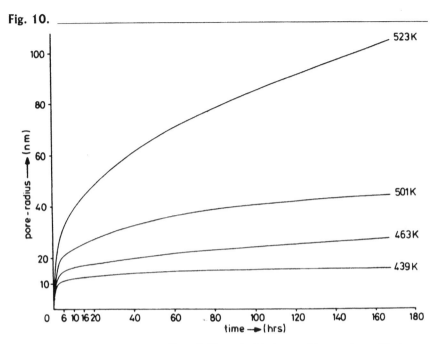

Change of the average pore radius of silica gel with time and temperature of hydrothermal treatment. From van der Grift *et al.* [28]

temperatures and longer times, but the rate of change was slow: on aging at 170°C, half of the total change occurred in 90 h at pH 1, 2 h at pH 6, and 0.25 h at pH 11. The rate of change of the surface area was characterized by an activation energy of 180 kJ/mole at pH 1 and 270 kJ/mole in base; these high energies are attributed to the need to break Si-O bonds for dissolution. Iler [12] (pp. 539–544) cites many studies showing that silica xerogels exposed to superheated steam (rather than water) show no loss in pore volume; since the pores are not full of liquid, no contraction is caused by drying. The surface area decreases in steam (more slowly than in water), presumably as surface diffusion carries material to regions of greater negative curvature.

For aqueous (particulate) silica gels, Chertov *et al.* [30] found that hydrothermal aging did not change the hydroxyl content, which was ~3 OH/100 Si within the solid phase and 4.8–5.8 OH/nm^2 on the surface of the particles. This type of gel is expected to consist of dense silica particles (see Chapter 4) with most of the hydroxyl groups on the surface; aging involves only dissolution–reprecipitation. A different result would be expected for alkoxide-derived gels, which can contain about as many hydroxyl groups within the solid phase as on the surface [1]. For such gels, aging involves condensation within the solid phase (which would tend to eliminate the internal hydroxyls) as well as coarsening.

Hydrothermal treatment at high temperatures (e.g., 375–450°C, steam pressure 21–30.5 MPa) can cause crystallization as well as coarsening [31]. The tendency to crystallize is related to the amount of alkali impurity in the gel. Lazarev *et al.* [32] showed that hydrothermal treatment at 400°C at a pressure of 20 MPa caused densification of alkoxide-derived silica gel with a decrease of surface area from ~200 m^2/g to ~1 m^2/g, but no crystallization occurred. If the original gel, with a total impurity content of 3 ppm, was treated with an alkali-containing mineralizer (procedure not described), hydrothermal treatment would produce cristobalite and then quartz. The kinetics of the transformation could be described by the Avrami equation [33],

$$\frac{v(0) - v(t)}{v(0) - v(\infty)} = 1 - \exp(-kt^n),\tag{6}$$

where v is the specific volume of the solid phase (reciprocal of skeletal density), k depends on the kinetics of nucleation and growth of crystals, and n is a parameter that was found to equal 4. This is the value expected for growth of spheriodal crystals with a constant growth rate and constant rate of nucleation; the activation energy was 290 ± 30 kJ/mole. It has been found that aging at room temperature allows the formation of nuclei that accelerate crystallization during subsequent treatment at elevated temperatures [34]. In sodium aluminosilicate gels, formation of nuclei during aging at room temperature is attributed to a dissolution–reprecipitation process [35], which has a low activation energy (21.5 kJ/mole) and accelerates with increasing pH.

Stability against crystallization of silica can be obtained, even when alkali impurities are present, by addition of B$_2$O$_3$ [36]. Stability against coarsening results from incorporation of a less soluble oxide, such as SnO$_2$. The SiO$_2$–SnO$_2$ system is phase-separated; at 10% SnO$_2$, cassiterite crystals appear on hydrothermal treatment above 300°C; above 25%, crystals are present at the time of gelation. Nevertheless, the insoluble oxide retards loss of surface area [37]. Titania also stabilizes silica against coarsening, as described in Section 2.2.2, and so does alumina [12] (pp. 532, 544).

Crystallization of gels may also result from reaction with carbon dioxide in the atmosphere. Prassas *et al.* [38] prepared Na$_2$O–SiO$_2$ gels from the methoxides and found that they reacted to form trona, Na$_3$H(CO$_3$)$_2 \cdot$2H$_2$O, when the soda content was <25 mole%, and Na$_2$CO$_3 \cdot$H$_2$O at higher soda levels. This can be a serious problem during subsequent processing, because sintering may occur before complete decomposition of the carbonates [39], leading to bloating and carbon retention. Gels containing formamide are resistant to attack by CO_2 for three months or more, because the formamide bonds to the surface of the gel and protects it [40]. Unfortunately, the formamide picks up moisture from the air, and this may lead to cracking; furthermore, it is hard to remove upon heating. If glycerol is present in the pore liquid, it may react to form carbonates when heated [40].

2.2. Other Oxides

2.2.1. ALUMINA

The most often cited process for making alumina gels, developed by Yoldas [41,42], is to hydrolyze aluminum *iso*-propoxide or *sec*-butoxide in a large excess of water, with an acid catalyst in the ratio 0.07 moles acid/mole Al. If the reaction is carried out at room temperature, the product is an amorphous gel that converts to bayerite [Al(OH)$_3$] over a period of ~24 h. At 80°C, the reaction produces boehmite [AlO(OH)]; if the process is begun at room temperature, the conversion to bayerite can be arrested by heating to 80°C. Bye *et al.* [43,44] found similar phase transformations in gels made by a somewhat different procedure. They dissolved aluminum *sec*-butoxide in benzene and hydrolyzed it by shaking the solution with water, aqueous ethanol, or aqueous glycerol (molar ratios unspecified). As the amorphous gel converts to pseudoboehmite at room temperature the surface area increases (~10–20%), apparently because the change in density of the phases causes cracking. The pseudoboehmite subsequently changes into bayerite by a dissolution–reprecipitation process, during which the surface area decreases (~30%). The initial crystallization is believed to involve either condensation between hydroxyls,

$$2(H_2O)_3(HO)_2Al\text{–}OH \rightarrow (H_2O)_3(HO)_2Al\text{–}O\text{–}Al(OH)_2(H_2O)_3 + H_2O, \quad (7)$$

or displacement of a water molecule and formation of a hydroxide link,

$$2(H_2O)_3(HO)_3Al \rightarrow (H_2O)_3(HO)_2Al\text{–}OH\text{–}Al(OH)_3(H_2O)_2 + H_2O. \quad (8)$$

These reactions require removal of a proton from –OH or H_2O, respectively, so the reaction is promoted by increasing pH; consequently, the surface area of the gel rises over the pH range from 4 to 6. Organic molecules, such as glycerol or ethanediol, inhibit the reaction by adsorbing on the hydroxyls. Conversion to bayerite is inhibited in dioxane, not because of adsorption, but because transfer of the proton is prevented by the aprotic solvent [45].

Nail *et al.* [46] neutralized aluminum chloride or sulfate solutions with ammonia and obtained plates (3.2 nm by 24.6 nm) of a poorly crystallized hydroxide. During aging at room temperature it was found that the OH : Al ratio increased as OH replaced Cl⁻ or SO₄⁼ ions adsorbed at the edges of the plates. The sulfate ions are more strongly bound, so aging was slower in that system. Sing *et al.* [44,47] made alumina gels by neutralizing aluminum chloride with NaOH; the residual Cl⁻ content corresponded to monolayer coverage. Wet samples of gel coarsened slowly with aging, the surface area decreasing from ~300 to ~60 m²/g. For reasons that are not explained, the process was greatly accelerated if the water content was reduced to the "critical concentration" corresponding to $Al_2O_3 \cdot 6H_2O$. Aging was negligible in vacuum, but became rapid if the humidity was raised so that the critical concentration of water was reached.

2.2.2. TITANIA

Ragai [48] neutralized an aqueous solution of titanium tetrachloride to pH 7 with ammonia and obtained a black powder that turned white when oxygen was bubbled through the sol. The precipitate was transferred into water or concentrated ammonia for aging ("hydrogels"), or was first dried and then placed into these liquids for aging ("xerogels"). After drying, the hydrogel aged in ammonia contained less than a tenth of the pore volume of the unaged powder, and the surface area had decreased from 240 to 43 m²/g; relatively little change occurred in the structure of the xerogel. The surprising loss of pore volume in the hydrogel suggests that the rigidity of the network was *reduced* by aging, possibly as a result of replacement of hydroxyls by ammonia ligands in the inner coordination sphere of the titania ions. Very little change occurred in either gel when aged in water. The exothermic transformation from the amorphous phase to anatase occurred at 395°C for unaged powders, but at temperatures up to 100°C higher after aging in ammonia. Chertov *et al.* [49] aged similar xerogels in water for 12 *years* at room temperature and found that the surface area decreased from 450 to 180 m²/g, but the pore volume changed very little.

Chertov and Sidorchuk [50] prepared TiO₂–SiO₂ gels from the respective chlorides and subjected the wet gels and xerogels to hydrothermal treatment. Samples containing <15 mole% TiO₂ were homogeneous when prepared, but exhibited phase separation and an increase in surface area after treatment at 100 to 150°C. Gels containing 15–50% TiO₂ were phase-separated as-made and crystallized to anatase after treatment at 100 to 150°C. For higher titania contents, anatase was fully formed in the fresh gel, so hydrothermal treatment merely decreased the surface area. Higher-temperature treatments caused decreases in surface area and microporosity for all of the samples. Mixed compositions were more resistant than either pure oxide to textural changes during aging.

2.2.3. OTHER SYSTEMS

There is relatively little information on the aging behavior of other oxides. Chromia gels [44] made by neutralizing the nitrate with ammonia (or by decomposition of urea) behave rather like alumina: the surface area increases with pH in the range from 4 to 6 on aging at room temperature. These gels are reported to have a large proportion of porosity that can be entered by water molecules, but not nitrogen. Halide-derived gels of ZrO_2 and SnO_2 experience neither phase change nor coarsening during 12 years of aging in water at room temperature [49]. Under the same conditions, in 3 years the surface area of silica gel decreases from 790 to 695 m^2/g and the pore volume increases from 0.33 to 0.51 cm^3/g. Iron oxide sols appear first as amorphous spheres 2–4 nm in diameter that link into chains and then coarsen by dissolution–reprecipitation into rods. Amorphous gels convert to goethite (α-FeOOH) at pH > 10 and to hematite at pH < 4 [51].

3.

SYNERESIS

This section discusses the process of macrosyneresis (hereafter, simply called syneresis), during which the network of the gel contracts and expels the pore liquid. This phenomenon has a superficial resemblance to the contraction of polyacrylamide gels described by Tanaka [52]. In that system, small changes in temperature, the composition of the pore liquid, or application of an electric field can cause the volume of the network to change by a factor of 100. The change is reversible and is produced by osmotic forces resulting from differences in the affinity of the polymers for each other and for the pore liquid. In most inorganic gels syneresis is irreversible, and swelling of the gel is not observed. (Local expansion of the pore size reported in ref. [8] is microsyneresis and does not involve significant dilatation of the sample.) Transferring the gel from one liquid to another or altering the temperature changes the rate of shrinkage, but does not reverse it. Syneresis in these systems is generally attributed to formation of new bonds through condensation reactions, as in Eqs. 1, 4a, or 7, or hydrogen bonding. In particulate gels, the structure is controlled by the balance between electrostatic repulsion and attractive van der Waals forces, so the extent of shrinking is controlled by additions of electrolyte. For example, Prakash and Dhar [6] made aqueous gels from salts in a wide variety of systems (including ferric arsenate, vanadium pentoxide, zirconium hydroxide, and manganese dioxide) and found that the final shrinkage increased continually with the concentration of electrolyte. Some other particulate systems, such as alumina gels and some clays, exhibit swelling, rather than syneresis [5] because water can invade the layer structure of the crystals and cause expansion.

How does bond formation produce shrinkage? It is easy to imagine that a bridging bond between two metal atoms, M–O–M, takes up less space than the two MOH groups from which it was formed, so the solid phase would contract as a result of condensation (Fig. 11a). Greater shrinkage could result from "diffusion" or flexure of the strands of solid phase (Fig. 11b): if a local movement of a strand brings surfaces close enough to form a bond, the deformation will be irreversible; this process will continue to cause a net contraction as long as the solid phase remains flexible. It has also been suggested [53] that contraction is driven by the tendency to reduce the huge solid–liquid interfacial area of the gel. This might be accomplished by flexure of the solid phase bringing surfaces into contact, in which case it would be indistinguishable from the reaction-driven mechanism just described. Alternatively, it might occur by flow or diffusion of the solid phase in a process analogous to sintering, which is also driven by interfacial energy.

Fig. 11.

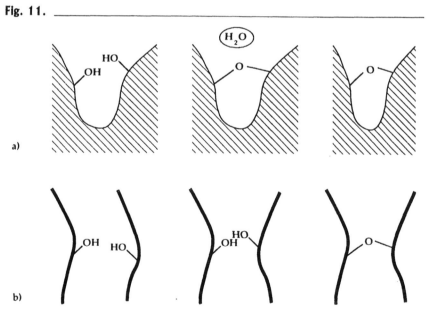

(a) Shrinkage results from condensation between neighboring groups on a surface as the strain in the new bond relaxes. (b) Movement of flexible chains may permit new bonds to form that prevent the chains from returning to their original position; this permits extensive shrinkage as long as the network remains flexible.

(See Chapter 11.). As we shall see, the latter suggestion is not supported by experimental results.

The system depicted in Fig. 1 showed enormous shrinkage after addition of water or peroxide, because the alkoxide had been incompletely hydrolyzed. Rapid shrinkage of silica gels was reported by Jones and Fischbach [54] in silica gels made by mixing TEOS directly into acidified water. The fully hydrolyzed sol (with a water : TEOS ratio ≥ 10) was then mixed with a solution of NH$_4$OH, which produced gelation and shrinkage of as much as 12 to 15% (linearly) in a matter of hours. This situation is analogous to Yoldas's titania gel [4] in that the addition of base promoted rapid condensation in the silicate solution, and this evidently was responsible for both gelation and syneresis. Most gels contract at a much more leisurely pace.

The most thorough studies of syneresis have been performed by Vysotskii and colleagues [55–57] using silica gel made by neutralization of sodium silicate with acid. The sol was cast into greased cylindrical tubes and the gel point (t_g) was roughly determined from the rate of fall of spheres dropped into the tube. Then the tube was sealed and shrinkage of the gel was observed with a cathetometer. The kinetics of shrinkage are shown in Fig. 12: there

Fig. 12. _____

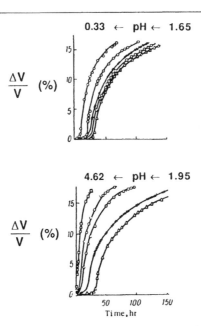

Kinetics of syneresis in silica gel as a function of time in solutions with various pH. Minimum contraction rate is at isoelectric point (pH ~2). From Klimentova *et al.* [57]. Note that these data are reproduced in ref. 55, but the curves are mislabelled.

is an initial "induction period" (t_i) during which no shrinkage occurs; then the shrinkage rate rises rapidly and eventually decreases again. A theory that accounts for the shape of these curves is presented in the next chapter. The shrinkage rate is characterized by the time (t_α) when the volume has contracted by 10%. As shown in Fig. 13, all of these characteristic times show the same dependence on pH: gelation takes longest and syneresis is slowest at pH $\approx$ 1.7. The shrinkage rate increases with the concentration of the solid phase in the gel, but the pH-dependence is unchanged. The importance of the data in Fig. 13 is that pH 1.7 is the *isoelectric point* of silica, where the surface charge is zero and the rate of condensation is least. (See discussion in Chapter 3.) (As shown in Table 2, the isoelectric point of silica depends on the acid used to make the gel. The samples represented in Fig. 13 were made with sulfuric acid.) Since the rates of gelation and syneresis were affected in the same way by the chemical conditions, it was concluded [26,55] that the same reaction (viz., condensation) was responsible for both phenomena.

Acker [58] examined gels made from sodium silicate and sulfuric acid, and found syneresis behavior (at pH 1.5) similar to that previously discussed:

Fig. 13.

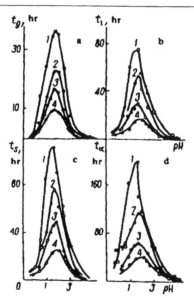

Dependence on pH of gel time (t_g), induction time before start of syneresis (t_i), time before beginning of syneresis ($t_s = t_i + t_g$), and time for volume contraction of 10% (t_α). Curves represent gels with concentrations (moles SiO_2/liter) of 1.09 (1), 1.33 (2), 1.56 (3), and 1.78 (4). From Klimentova *et al.* [57].

Table 2.

Dependence of Isoelectric Point on Acid.

Acid	IEP[a]	pKa[b]
HNO_3	1.45	−1.64
HCl	1.55	−7
H_2SO_4	1.70	−3
H_2CrO_4	1.80	−1
CCl_3COOH	2.10	0.66
H_3PO_4	2.15	2.1
HCOOH	3.15	3.75
CH_3COOH	3.50	4.76

Source: From Vysotskii and Strazhesko [55].
[a] IEP = isoelectric point determined from location of maximum in plot of gel time (t_g) versus pH.
[b] Equilibrium constant for dissociation of first proton from acid.

an induction period followed by shrinkage at a rate that increased with the solids content of the gel. He reported, as did Jones and Fischbach [54], that there was a minimum solids content (4–5 g SiO_2/100 ml of sol) below which no observable shrinkage occurred.

Snel [59] doped a silica gel with 1.7 mole% alumina and found that the rate of syneresis was still dependent on pH, increasing by a factor of 10 as the pH increased from 3.5 to 11.

Ponomareva *et al.* [60] found that the rate of syneresis increased with temperature, but the total amount of shrinkage decreased. When the logarithm of the shrinkage rate (defined as the slope of the rapidly rising portion of the shrinkage-versus-time curve, see Fig. 12) was plotted against reciprocal temperature, a line was obtained whose slope was interpreted as an "activation energy" of 25 kJ/mole. From this small value they concluded [60] that syneresis was produced by rearrangement of particles driven by hydrogen bonding. On the other hand, one could argue that each mole of silica represents four moles of hydroxyls, for which the condensation enthalpy is reported to be 4.7 kJ/mole per mole of OH [61], so the observed activation energy is actually similar to what might be expected ($4 \times 4.7 \approx$ 19 kJ/mole) from a process driven by condensation. Any such interpretation is dubious, because the rate of the process depends on factors other than the rate of reaction. To obtain shrinkage, the stress produced by the formation of a bond must be relaxed (see Fig. 11a), so the compliance of the network affects the shrinkage rate. Suppose, for example, that the rate of formation of bonds, dn/dt, has an activation energy of ΔE_b,

$$dn/dt \propto \exp(-\Delta E_b/R_g T), \qquad (9)$$

and that the viscosity of the solid phase, η, is characterized by activation energy ΔE_f (ignoring the time-dependence of η as new bonds form)

$$\eta \propto \exp(\Delta E_f / R_g T). \tag{10}$$

If the shrinkage rate, dV/dt, is given by

$$\frac{dV}{dt} \propto \frac{1}{\eta} \frac{dn}{dt} \propto \exp[-(\Delta E_b + \Delta E_f)/R_g T], \tag{11}$$

then the observed activation energy will be the sum of those for bonding and flow. Furthermore, shrinkage requires movement of the pore liquid through the pores. (Actually, it is the solid phase that passes through the liquid, but it is the relative movement that matters.) Thus, the kinetics of shrinkage and the temperature-dependence of the rate depend on the mechanical properties and the permeability of the network, and the viscosity of the pore liquid, as well as on the kinetics of the reaction. It is presumably this interplay of different processes that accounts for the observation that less shrinkage occurs at higher temperatures. Suppose that an increase in temperature raises the reaction rate by a factor of ten, but only reduces the viscosity of the pore liquid by a factor of two. The network might not be able to shrink ten times faster, because it cannot squeeze out the liquid fast enough; then the stress in the network will be relieved by shear deformation, rather than volume change, and the shrinkage produced by formation of a given number of bonds will be less than at a lower temperature.

It has long been known that the rate of shrinkage during syneresis is slower for larger gels [62]. Scherer [63] accounts for this by the following argument. As the gel network contracts, it subjects the pore liquid to a compressive load that forces it out of the body. If the gel is small enough, the liquid escapes easily; but if it is large, a steep pressure gradient is needed to drive the liquid to the surface. To test this idea, cylinders of radii between 0.3 and 0.67 cm were cast from silica gel (hereafter, gel A) made from TEOS (acid-catalyzed, molar ratio of water/TEOS = 16/1), and the diametral shrinkage during syneresis was measured [64]. Results for the largest and smallest cylinders are compared in Fig. 14. Clearly the smaller sample shrank more, as expected. The curves were calculated using a model discussed in detail in the following chapter. The calculation assumed that the inherent shrinkage rate (i.e., the syneresis rate in a very small gel, from which liquid can escape without resistance) was inversely proportional to the viscosity of the solid phase, as in Eq. 11. The decrease in shrinkage rate at long times was attributed to the increase in viscosity of the network as new bonds formed (which makes syneresis a self-retarding process). The shapes of the curves, which were based on the measured viscosity of the gel, match the data reasonably well. This supports the idea that the rate of shrinkage decreases

Fig. 14.

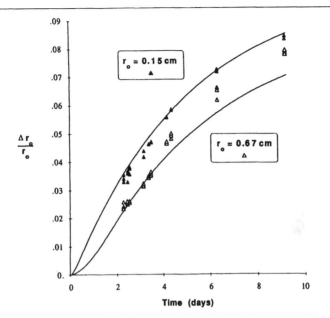

Radial shrinkage, $[r_o(0) - r_o(t)]/r_o(0)$, versus time for gel rods with initial radii of 0.15 (▲) and 0.67 (△) cm. Silica gel from acid-catalyzed hydrolysis of TEOS, with 16 moles of water/mole of TEOS. From Scherer [64].

because of stiffening of the network, rather then depletion of reactive groups. The time needed to establish a steady-state pressure gradient is related to the viscoelastic relaxation time of the network and the size of the body. This was suggested [63] as an explanation for the induction period preceding shrinkage in Fig. 12, and it can be seen that the shrinkage data in Fig. 14 indicate a longer lag for the larger sample. The absence of syneresis in gels with very small solids contents [54,58] may indicate that the low modulus of the network[†] is inadequate to squeeze liquid out of the pores at an appreciable rate.

The rate of syneresis depends on the composition of the liquid in the pores. Figure 15 shows the shrinkage of gel A when it was immersed in an ethanol–water solution [63,65]. The liquid in the pores at the gel point was 67 vol% water and 33% ethanol; 26 h after casting, the gel rods were removed from that liquid and transferred into a larger volume of solution with the indicated composition (vol%). The greater the water content of the solution, the faster the gel shrank; in pure ethanol there might be slight swelling, but the amount

[†] The modulus of the solid phase itself is probably independent of the solids content of the gel, but it is the modulus of the porous network as a whole that is relevant in this case.

Fig. 15.

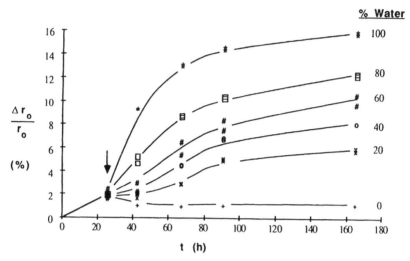

Radial strain of rods of gel immersed in ethanol–water solutions with indicated compositions (vol%). Samples were removed from pipettes ~26 h after casting (arrow), then placed in baths at 40°C; the pore liquid in the gel is initially 67% water. Shrinkage increases with water concentration. From Scherer [65].

was within the uncertainty in the measurement. If swelling does occur, it might be a result of re-esterification. Ohno *et al.* [66] found that TMOS gels aged in alcohol–water solutions richer in alcohol retained more organics; soaking in butanol led to more retention than soaking in methanol. The effect of water content on shrinkage rate is consistent with the observation [1] that the degree of condensation (i.e., the quantity of Q^4 measured by ^{29}Si NMR) increases with the water concentration, even when that quantity is well above the stoichiometrically required amount. The reason might be that the concentration of catalyzing hydroxyl or hydronium groups increases with the water : alcohol ratio. Thus, syneresis in these alkoxide-derived gels appears to be driven by condensation reactions, as in aqueous systems.

   Similar studies have been performed [15] using a base-catalyzed gel made by the two-step process (identical to gel B2 from ref. 14). No change in dimensions was observed when the sample was soaked in ethanol–acetone solutions ranging from 0 to 100% ethanol. When soaked in ethanol–water solutions, contraction was relatively rapid, and the rate increased with water concentration. (The initial pore liquid for this gel was ~9 ethanol/1 water by volume.) Gels soaked in solutions containing more than 80 vol% water turned white and opaque. This was not microsyneresis [7], which can cause the same appearance; as shown in Fig. 16, the SEM reveals spherical regions

Fig. 16.

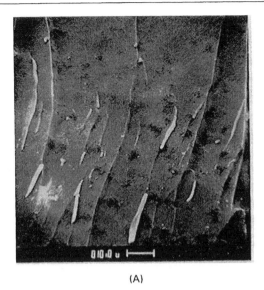

(A)

(B)

Scanning electron micrographs of fracture surface of B2 gel after aging in pure water and drying. Opacity of the gel is attributed to the circular regions $\sim 5\,\mu$m in diameter. (Bar = 10 μm). From Scherer [15].

of heterogeneity large enough to scatter light. These were attributed to phase separation of unhydrolyed TEOS [14]; in fact, the samples exuded a clear oil that appeared to be partially hydrolyzed and condensed TEOS [15]. Unpublished work by P.J. Davis, C.J. Brinker, and D.M. Smith indicates that this gel does not become opaque when immersed in water, if it is thoroughly washed first in ethanol. Presumably this treatment removes the unreacted TEOS.

The effect of dissolved salts on the shrinkage rate of gel A has been examined [67] by immersing rods of gel in distilled water containing 0.6 M or 3.0 M NaCl (10% and 50% of the solubility limit, respectively) or 0.67 M CaCl$_2$ (10% of the solubility limit). The pH of the solution was adjusted to 2 or 4 by addition of HCl, and was carefully maintained at that value throughout the study. The shrinkage was measured as a function of time using a hand gauge (as described in ref. [63]) with the results shown in Fig. 17. The shrinkage rate increased with the ion concentration of the solution (e.g., 0.6 M NaCl yields 1.2 moles/liter of ions and 0.67 M CaCl$_2$ provides 2 moles/liter). This is an osmotic effect: the water diffuses out of the gel to dilute the salt concentration of the surrounding liquid, causing the gel to shrink. Samples immersed in 3.36 M CaCl$_2$ shattered immediately, because the rapid exodus of water caused the surface of the gel to contract while the interior was still swollen with water; the differential shrinkage caused stress and fracture. This is analogous to the differential shrinkage (produced by capillary pressure) that causes stress and cracking during drying. As expected from Fig. 13, the shrinkage is more rapid at pH 4 than at pH 2. There are two other possible explanations for the accelerated shrinkage produced by the salt. First, it could indicate collapse of a repulsive double layer (electrostatic charge on the surface of the gel; see Chapter 4). That would not be important at the isoelectric point (pH 2), where the surface is uncharged; at pH 4 that mechanism might be significant, but then the Schulze-Hardy rule [68] would predict a much stronger effect from Ca^{2+} than from Na$^+$. Since Fig. 17 indicates that those ions are about equally effective at comparable molarity, electrostatic interactions do not seem to be important. In fact, comparing the rates of shrinkage at pH 2 and 4, the change in pH has as much effect on the sample in pure water as on the samples in salt solutions. Second, shrinkage could be retarded by strongly adsorbed molecular water on the surface of the gel, and their adsorption sites (silanol groups) could be removed by ion exchange in the salt solutions. However, neither sodium nor calcium is expected to cause significant ion exchange in this range of pH [69].

It has been suggested [53] that the contraction during syneresis is driven by the solid–liquid interfacial energy, γ_{SL}. To calculate the rate of shrinkage, the solid phase was represented by an array of cylinders intersecting at right angles [70,71]. (See discussion in Chapter 11.) Assuming that the cylinders

Fig. 17.

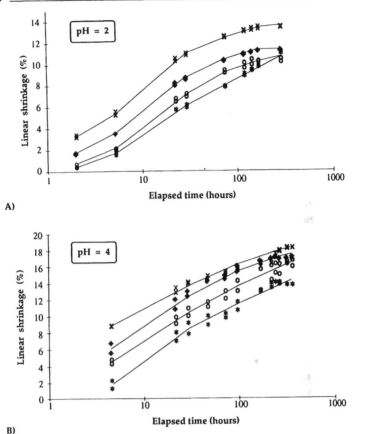

A)

B)

Shrinkage of gel A while immersed in water at controlled pH containing no salt (∗), 0.6 M NaCl (○), 0.67 M CaCl₂ (◆), or 3.0 M NaCl (×) at pH 2 (A) or pH 4 (B). From Scherer and Swiatek [67].

deform by viscous flow, the inherent rate of syneresis (i.e., in small samples not affected by fluid flow in the pores), $\dot{\varepsilon}_s$, is given by

$$\dot{\varepsilon}_s = -\frac{\gamma_{SL} S \rho_S}{6\eta} \tag{12}$$

where S is the interfacial area, ρ_S is the skeletal density, and η is the viscosity of the solid phase. The latter quantity is problematic, because one can only measure the viscosity of the porous network, and one must depend on a theoretical model to extract the viscosity of the solid phase itself. Using the cylinder model to obtain η from the measured viscosity of the gel, together with the shrinkage rate of the smaller sample in Fig. 14, Eq. 12 indicates that $\gamma_{SL} \approx 0.5$ ergs/cm^2, which is unrealistically small [64]. The interesting point is that this model says that the shrinkage should be *faster* than it actually is. If there is a mechanism by which faster shrinkage could occur, why doesn't it? Perhaps the solid phase cannot flow in such a way as to decrease the solid–liquid interfacial area. For example, Table 1 indicates that gels can shrink more (i.e., have smaller pore volumes), yet have the same surface area. Perhaps shrinkage is accompanied by "crumpling" that increases the roughness of the interface, and only a process of dissolution and reprecipitation can smooth it. Alternatively, it may be that Eq. 12 is valid, but that η was seriously underestimated. The cylinder model assumes that the viscosity of the network equals the viscosity of the solid phase multiplied by the load-bearing fraction of the cross section. However, the flow of the solid might depend on breaking of "weak links" in the network; a network of very stiff rods with occasional weak spots would be indistinguishable on a macroscopic scale from a network of uniform rods of lower viscosity. If the weak-link mechanism applied, the cylinder model would grossly underestimate η. Although a better estimate of η might improve the performance of Eq. 12, the dependence of shrinkage rate on pH and solvents is more consistent with a process driven by chemical reactions than interfacial energy.

4.

MECHANICAL PROPERTIES

Suppose we apply a stress σ_x to the ends of a bar of a viscoelastic material. The total strain along the axis (the x-direction) is [72]

$$\varepsilon_x = \varepsilon_E + \varepsilon_D + \varepsilon_V \tag{13a}$$

$$= \frac{\sigma_x}{E} + \frac{\sigma_x}{E_D}\phi(t) + \frac{\sigma_x t}{3\eta} \tag{13b}$$

where the three components of strain are (see Fig. 18) *instantaneous elastic* (ε_E), *delayed elastic* (ε_D), and *viscous* (ε_V). By definition, compressive stresses and strains are negative. The molecular motions responsible for these strains are complicated and not clearly understood, but they can be rationalized as resulting from stretching of bonds, disentangling of polymer chains, and sliding of chains, respectively. Stretching is instantaneous and reversible, and the strain is related to the applied stress by the Young's modulus, E. The rate of growth of the delayed elastic strain is governed by the *creep function*, ϕ, which rises from 0 to 1 as t increases from 0 to ∞; the maximum (final)

Fig. 18.

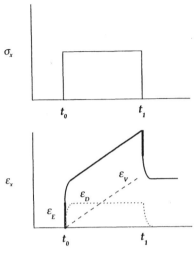

A) Viscoelastic

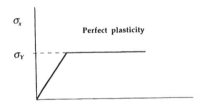

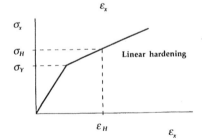

B) Plastic

(A) Schematic illustration of strain when a viscoelastic material is subjected to a constant stress, σ_x, at time t_0. Total strain, ε_x, is the sum of the instantaneous elastic (ε_E), delayed elastic (ε_D), and viscous (ε_V) strain. When the stress is removed at time t_1 the elastic (thick bar) and delayed elastic strains recover.

(B) Two idealized types of plastic behavior: perfectly plastic with a yield stress of σ_Y and linear hardening. When the stress is removed, the plastic strain does not recover. If the stress is sustained at σ_H, the strain remains constant at ε_H.

value is $\varepsilon_D = \sigma_x / E_D$. This strain is recovered after the stress is removed (one might say, as the chains resume their equilibrium entangled condition). The viscous strain increases continuously with time at a rate determined by the viscosity, η; if the material is *Newtonian*, the viscous strain rate is proportional to the applied stress. When the stress is removed, the viscous strain does not recover.

When a compressive load is applied on the end of a purely elastic bar (i.e., one that exhibits neither delayed nor viscous strain), it shortens in length and increases in width. The ratio between the strain in the direction of the applied stress, ε_x, and in the perpendicular direction (viz., along the radius of the bar), ε_y, is called *Poisson's ratio, v*:

$$\varepsilon_y = -v\varepsilon_x. \tag{14}$$

The volumetric strain of the bar, ε, in this case is given by [72]

$$\varepsilon = (1 - 2v)\varepsilon_x = \frac{(1 - 2v)\sigma_x}{E} \tag{15}$$

so a compressive load on the ends of the bar causes the volume of the bar to decrease ($\varepsilon < 0$). The more compressible the material, the smaller the value of v; when $v = 1/2$ the material is said to be *incompressible*, which means that $\varepsilon = 0$ regardless of the stress. In a viscoelastic material, Poisson's ratio relaxes from its instantaneous value toward $1/2$ as t approaches ∞; that is, the volume relaxes to its initial value ($\varepsilon = 0$) as the expansion in the radial direction compensates for the contraction in the axial direction.

If the bar is simply twisted, there is no volume change ($\varepsilon = 0$), only distortion. (Lines scratched along the sides of the bar would be converted to helices by the strain.) This type of deformation is called *shear*, and the constant of proportionality between the instantaneous elastic shear stress and strain is the shear modulus, G. The shear modulus is related to Young's modulus by

$$G = 2(1 + v)E. \tag{16}$$

If the bar is subjected to hydrostatic stress (for example, by placing it into a rubber membrane and compressing it with gas in a pressure vessel), the body is uniformly compressed without changing shape. That is, if a straight line were drawn between any two points in the body before the stress was applied, the line would remain straight (but would be shorter) after pressurizing. The constant relating the applied pressure, P, and the volumetric strain is the bulk modulus, K, is

$$\varepsilon = P/K. \tag{17}$$

It is related to Young's modulus by

$$K = 3(1 - 2v)E. \tag{18}$$

Particulate materials, such as clay or particulate gels of the type discussed in Chapter 4, may be *plastic*, rather than viscoelastic. Two simple types of plastic behavior are illustrated in Fig. 18b: a perfectly plastic material is elastic up to the *yield stress*, σ_Y, but it deforms without limit if a higher stress is applied; in a linearly hardening material there is a finite slope after the yield stress. In real plastic materials, the stress–strain relations are likely to be curved, rather than linear. If the stress is raised above σ_Y and then released, the elastic strain is recovered but the plastic strain is not. This differs from viscous behavior in its time-dependence: if the stress on a linearly hardening plastic material is raised to σ_H and held constant, the strain remains at ε_H; a viscoelastic material would continue to deform at a rate proportional to σ_H/η.

In a gel, it is necessary to distinguish between the properties of the network (the solid phase that would be revealed if the liquid could be drained away) and those of the fluid-filled gel (solid plus liquid). If the gel is subjected to a hydrostatic stress, negligible strain occurs, because the liquid phase is nearly incompressible and it supports the load. However, the bulk modulus of the network, if it were free of liquid, would be quite small because the solid phase typically occupies a small fraction of the volume of the gel. A good analogy is offered by a soft sponge. When saturated, the liquid in the pores makes it incompressible (as long as the liquid can't escape); if it is squeezed along one direction, it expands correspondingly in the perpendicular direction ($v \approx 1/2$, $\varepsilon = 0$). When dry, it can be compressed without expanding laterally at all ($v \approx 0$). When a load is suddenly applied to the saturated sponge, it will deform without changing in volume; but as the liquid drains out, the volume will contract. Thus, the sponge (or gel) mimics viscoelastic behavior, even if the solid network is elastic, because of the time-dependence of fluid flow in the pores; of course, the solid phase may be viscoelastic in its own right. The important point is that a sudden deformation of a gel causes shape change, but not volume change, so it provides a measure of the *shear* modulus of the network. Young's modulus of a gel can be measured only if the load is sustained for so long that the pressure in the liquid drops to zero and all the load is supported by the solid network. This has been overlooked in most studies of gels: when the modulus is measured by beam-bending it is usually reported as E, but it can be shown [73] that the measured quantity is actually G.

It is important to recognize that the mechanical properties of gels are influenced by the capillary pressure that exists in the pore liquid. When evaporation exposes the solid phase of the gel, the liquid goes into tension as it stretches to cover the solid–vapor interface. The tension is balanced by compression in the solid phase, and this leads to an increase in the strength and stiffness of the body. This phenomenon is discussed in detail by Rumpf [74] and Schubert [75]. In a particularly nice study, Schubert [76] measured the strength and capillary pressure simultaneously in agglomerates of limestone particles. As shown in Fig. 19, the strength is equal to the capillary pressure, which rises as menisci form at the surface of the body; a maximum is passed as air begins to invade the pores, and the strength decreases in proportion to the saturation (i.e., the volume fraction of the pores containing liquid). Similar results were reported by Newitt and Conway-Jones [77]. This effect is expected to be important for gels, which have such large interfacial areas that they can develop enormous capillary pressures. For example, it is sometimes found that particulate gels fall apart when immersed, but the wet

Fig. 19. _____

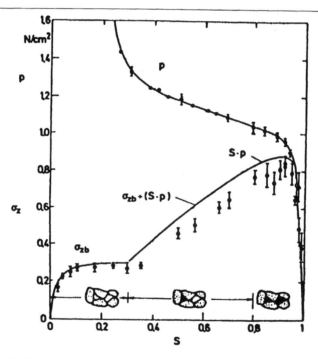

Calculated (solid curve) and measured tensile strengths (σ_z) and measured capillary pressure (p_p) in moist agglomerates of limestone particles (71 μm) as function of saturation, S (fraction of the pore space filled with liquid). In the saturated body, $\sigma_z = p_p$; when air enters the pores ($S < 0.95$), the strength decreases with saturation, $\sigma_z = Sp_p + \sigma_{z,b}$, where $\sigma_{z,b}$ is the strength of a bridge of liquid at a point of contact between two particles. From Schubert [76]. Shape of p versus S curve discussed on pp. 418-419.

gels have measurable strength and stiffness when measured under ambient conditions; this can be attributed to elimination of the meniscus upon immersion, although dilution of electrolytes may also be a factor.

4.1. Elastic Modulus

Several studies cited in Chapter 5 showed that the shear modulus rises from zero at the gel point and increases as new bonds are formed. Keesman *et al.* [78] showed that the *rate* of increase of the shear modulus after gelation was influenced by the degree of hydrolysis of the alkoxide precursor. They used a two-step process (acid-catalyzed hydrolysis followed by base-catalyzed condensation) to prepare silica gel from TEOS and found that G increased faster when more time was allowed in the initial hydrolysis. This is understandable, since the stiffening of the network results from the formation of new bonds by condensation of hydroxyls; the greater the concentration of Si–OH, the faster the reaction. These reactions can continue at a slow rate for a very long time. Dumas *et al.* [79] prepared acid-catalyzed silica gels with various water contents (r = moles water/moles TEOS) and found that G increased for ~3 weeks when $r \geq 8$ and for 4 to 6 months when $r \leq 6$. (See Fig. 20.) Again, the more completely hydrolyzed gels (higher r) stiffen faster; in this case, the plateau values decrease as r increases, because the gels are more dilute.

West *et al.* [80] examined the modulus of silica gels made by acid-catalyzed hydrolysis of TMOS. Bars of wet gel were aged in their pore liquid at temperatures from 25 to 105°C for periods up to 32 days, then their moduli and strength were measured by four-point bending (with the samples exposed to the atmosphere). At temperatures $\leq 60°C$, those properties increased by more than an order of magnitude over the course of 2 to 4 weeks; at $\geq 90°C$, plateau values (~0.2–0.25 MPa modulus and ~0.35–0.45 MPa strength) were reached in a few days, and little change occurred thereafter. At the lower temperatures, the modulus increased roughly in proportion to log (time).

Klimentova *et al.* [81] soaked aqueous silica gels for nine months in solutions with pH values between 0 and 5.5. The moduli were measured (in air) by loading the bars of gel in a cantilever configuration or in three-point bending. Surprisingly, they found the greatest modulus in the gel aged at the isolectric point (pH 1.7), where the rate of condensation is smallest [55]. Since the samples were not immersed, it is possible that the results reflect the difference in capillary pressure developed during the measurement: the samples aged at pH 1.7 would have the finest pores and might therefore develop more pressure on exposure to the atmosphere. A more likely explanation is offered by Iler [12] (p. 523). He notes that the number (N)

Fig. 20.

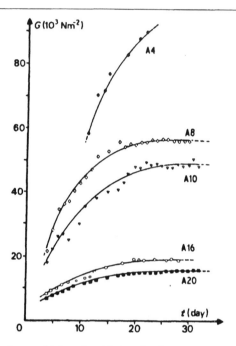

Variation of the shear moldulus, G, with time for five concentrations of silica (wt%): 10.5 (A4), 8.4 (A8), 7.7 (A10), 6 (A16), 5.2 (A20); gels identified by Ar, where r = moles water/moles TEOS. From Dumas *et al.* [79]

of chains of particles of radius R in a gel with concentration C of solid phase is $N \propto C/R^3$, and the number of chains intersecting any cross section of the gel is proportional to $N^{2/3}$. As indicated in Fig. 21, the area of the neck joining a pair of particles is proportional to $(x/R)^2$, so the load-bearing fraction of the cross-sectional area is $A \propto N^{2/3}(x/R)^2$. The radius of curvature of the neck, r, is related to x and R by the Pythagorean theorem, so we can write [12] (p. 521)

$$A \propto C^{2/3}\left[\frac{r}{R} + \left(\frac{r}{R}\right)^2\left(1 - \sqrt{1 + \frac{R}{r}}\right)\right].\tag{19}$$

Both the modulus and the strength of the gel are expected to be proportional to A, so they increase with r/R. The radius of curvature depends on the concentration of solid phase dissolved in the surrounding liquid, according to Eq. 5. Therefore, as shown in Fig. 22, at a given concentration r/R is greater for smaller particles. This accounts for the results of Klimentova *et al.*, because the smallest particles are formed near the isoelectric point.

Fig. 21.

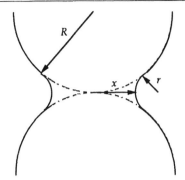

Two particles of radius R joined by a neck with radius x. The radius of curvature of the neck, r, is determined by the solubility of the solid, so it varies with temperature and pH.

Fig. 22.

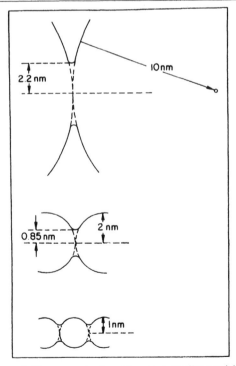

These gels have equal silica concentration. However, smaller particles develop greater coalescence and strength and the structure becomes more fibrillar. From Iler, *The Chemistry of Silica* (Wiley, New York, 1979).

Pardenek *et al.* [82] measured the moduli of gels made from TEOS, as well as particulate gels made from fumed silica. (See Rabinovich [83] for a discussion of the later process.) For both systems they found that the load–displacement curve was linear up to the point of failure; i.e., there was no yield point before the sample fractured. The modulus increased with age in both systems; for the particulate gel, aging presumably results from growth of interparticle necks by a process of dissolution–reprecipitation. It is significant that the particulate gels could not be immersed for measurement, because they become too soft; obviously the capillary pressure is contributing to the strength of these gels.

Scherer *et al.* [73] measured the effect of aging on the properties of silica gel made by acid-catalyzed hydrolysis of TEOS with $r = 16$ (gel A); the solution was refluxed for 2.5 h, then the excess solvent was removed under vacuum until the density of the sol reached 1.100 g/cm^3. The shear modulus was measured by beam bending with the sample immersed in a bath with the same composition as the pore liquid, as illustrated in Fig. 23. This method is ordinarily used to measure Young's modulus, but (as previously explained) because of the short duration of the experiment and the slow flow of liquid out of the pores, deformation of a wet gel gives a measure of the *shear* modulus. As shown in Fig. 24, the elastic shear modulus increased with approximately parabolic kinetics for the entire four weeks of observation. Each point in the figure represents a separate sample; to achieve that high degree of reproducibility it was found necessary to control the starting density of the sol precisely. Studies of this system using ^{29}Si NMR [1] showed that the rate of condensation was faster when the sol was more concentrated.

Fig. 23.

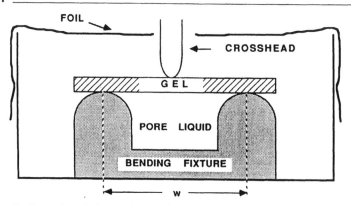

Schematic illustration of three-point bending apparatus for measurement of modulus and MOR; samples immersed in liquid with the same composition (67 vol% water/33% ethanol) as pore liquid. From Scherer *et al.* [73].

Fig. 24.

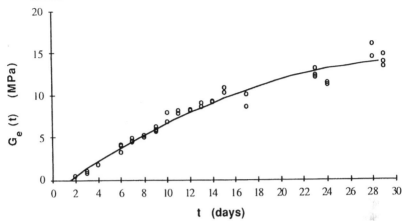

Shear modulus of wet gel as function of time. Silica gel from acid-catalyzed hydrolysis of TEOS, with 16 moles of water/mole of TEOS. Measurements made by beam bending while sample immersed in bath with same composition (67 vol% water and 33% ethanol) as pore liquid. From Scherer *et al.* [73].

That study was extended [15] by examining the effect of soaking gel A in various ethanol–water solutions. Figure 15 shows that the shrinkage of gel A was faster in solutions containing more water, and it was found that the increase in the modulus was also faster. Other samples of gel were allowed to dry partially in air. The moduli were measured with the samples immersed if they had shrunk <30% linearly; gels that had dried beyond that point would crack if immersed, so they were sprayed with an aerosol of Telflon® (to prevent further evaporation) and measured dry. When the modulus was plotted against the shrinkage of the gel, all of the data fell on a single curve (Fig. 25). Thus, for gel A there seems to be a unique relationship between shrinkage and shear modulus, regardless of whether the shrinkage is produced by aging in an aqueous solution or by drying in air. The driest samples had moduli (~3 GPa) close to those measured by Murtagh *et al.* [84] for similar dried gels.

The same sort of experiments were performed [15] using a two-step base-catalyzed gel called B2 [14] (made from TEOS with $r = 3.7$, pH 7.9). When soaked in the pore liquid (89 vol% ethanol and 11% water) the shear modulus increased for as much as one year, as shown in Fig. 26; the rate of stiffening was greater at 35°C than at room temperature. Other samples were soaked in various ethanol–water solutions or allowed to dry partially before measurement. Samples aged in pure ethanol or in ethanol–acetone solutions showed neither shrinkage nor increase in modulus over a period of

Fig. 25.

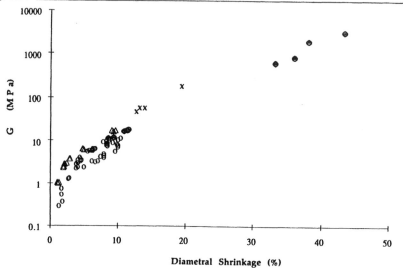

Shear modulus (MPa) of gel A versus linear shrinkage (%) for samples aged in (○) pore liquid (67 vol% water and 33% ethanol), (△) other water–alcohol solutions, and (×) water at pH 2, plus (●) samples allowed to dry in air. From Scherer [15].

Fig. 26.

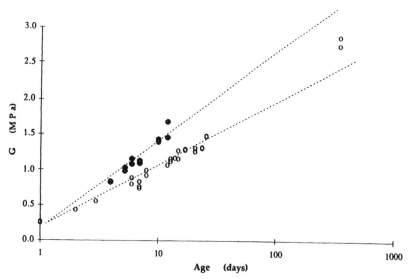

Shear modulus (MPa) of gel B2 versus age (days) for samples aged in pore liquid (11 vol% water and 89% ethanol) at room temperature (○) or at 35°C (●). From Scherer [15].

10–12 days. When the modulus was plotted against shrinkage (Fig. 27), most of the data fell near the line given by

$$G(\text{MPa}) = 0.13 \exp(0.17\varepsilon_d) \qquad (20)$$

where ε_d is the linear strain, $\varepsilon_d \equiv 100[d(0) - d(t)]/d(0)$, and d is the diameter of the rod of gel. The samples aged in the pore liquid show a somewhat higher slope, so the modulus depends on more than ε_d alone. The numbers next to the open circles in Fig. 27 indicate the age of the gel (in days) and reveal that the drying rate is inconsequential; the dominant factor controlling the modulus is the amount of shrinkage. The interpretation offered for these data was that aging results from condensation of silanols buried within the solid phase or crowded together on the surfaces of pores. This causes stiffening, but relatively little shrinkage, and becomes slow as the network stiffens and the supply of vicinal (closely proximate) silanols is depleted. When drying is allowed, the capillary pressure produces shrinkage that brings more silanols together and allows further condensation to occur. These two processes (aging and shrinkage by drying) cause different amounts of change in volume per bond formed, so the modulus is not uniquely related to

Fig. 27.

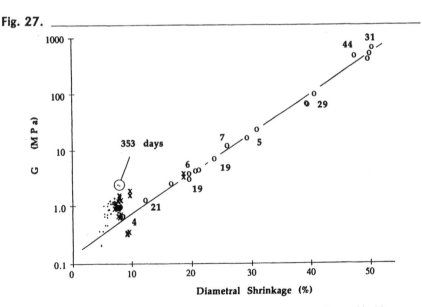

Shear modulus (MPa) of gel B2 versus linear shrinkage (%) for samples aged in (•) pore liquid (11 vol% water and 89% ethanol) or (×) other water–alcohol solutions, or (O) allowed to dry. Numbers indicate age (in days) of adjacent partially dried samples, except for 353-day-old samples, which were aged in pore liquid without evaporation. From Scherer [15].

shrinkage. The increase in modulus is too great to be explained by standard models (such as the Hashin–Shtrikman equation [85] or the cylinder model [71]) for the dependence of modulus on porosity [15]. Both the change in modulus of the solid phase and the shrinkage contribute to the increase.

4.2. Viscosity

A cross-linked network is purely elastic if the links remain intact, but viscous deformation is possible otherwise. For example, a cross-linked organic polymer exhibits stress relaxation if it is subjected to stress in an environment where it can be oxidized or hydrolyzed [86]. This is illustrated in Fig. 28 for sulfur-cured natural rubber. The samples were stretched to a fixed elongation and the force, $f(t)$, required to sustain that elongation was measured as a function of time. For strains up to ~100% the stress relaxation is linear (i.e., the relaxation rate is independent of the applied strain). The relaxation is much faster in air, in which the cross-links can be oxidized, than in nitrogen. The stress eventually decays even in a nitrogen atmosphere, because of oxygen impurities in the rubber.

Inorganic gels are obviously immersed in a chemically aggressive environment, so it is not surprising that they exhibit irreversible flow under a constant load. The viscosity of gel A (described previously) was measured by beam bending [73] with results of the type shown in Fig. 29. The kink in the curve near 100 minutes was attributed to escape of the pore liquid from the deformed bar; after 10 to 15 h, fluid flow and delayed elastic effects were essentially complete and the bar exhibited a constant strain rate corresponding to a viscosity of the network (not the solid phase itself) of $F = 2.4 \times 10^{12}$ Pa·s. Figure 30 shows that the logarithm of the viscosity increased linearly for weeks, but at a somewhat slower rate than the shear modulus.

Stress-relaxation measurements were attempted in the same study, but it was found that the results were confounded by fluid flow effects. When the bar is deflected by a certain amount and held in that position, the gel is compressed on the upper surface and stretched on the lower, so fluid tends to flow through the bar. At the same time, the solid phase is trying to flow to relieve the stress. The stress required to hold the bar in the deflected position decreases with time as the gel relaxes and the liquid escapes. A characteristic bend was found [73] in the relaxation curves that appeared sooner after application of the load in thinner rods, but at the same time in rods of a given size, regardless of aging time. That bend was attributed to escape of fluid from the pores (which takes longer with a thicker rod). At much shorter times, the relaxation of the load was assumed to result

Fig. 28.

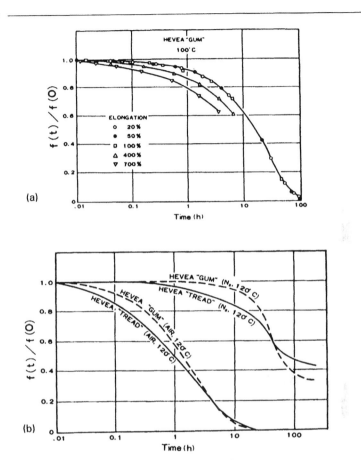

Chemorheological behavior of sulfur-cured natural rubber.
(a) Effect of elongation on chemical stress relaxation, where $f(t)$ is force at time t required to sustain fixed elongational strain.
(b) Chemical stress-decay curves in air and nitrogen; "gum" contains no filler, "tread" contains 33.3 wt% carbon black.
From Murakami and Ono [86].

from relaxation of the network at a constant content of pore liquid (i.e., pure shear relaxation). The relaxation curve was found to indicate a broad distribution of relaxation times, with an average relaxation (τ) time similar in magnitude to the Maxwell relaxation time [72], $\tau \approx F/[2(1 + \nu)G]$. This sort of study should be repeated using a pure shear deformation (such as torsion) instead of beam bending, so that fluid flow effects can be avoided.

Fig. 29.

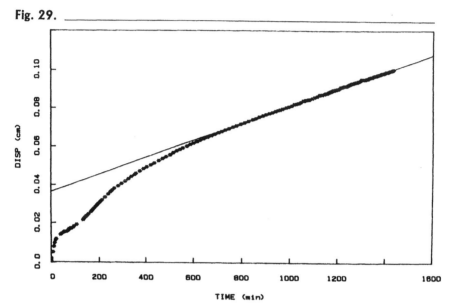

Typical beam-bending curve for wet gel A. Gel cast 12 days before start of measurement; rod diameter 0.58 cm; load 19.9 g; equilibrium viscosity 2.4×10^{12} Pa·s. From Scherer *et al.* [73].

4.3. Strength

The strength depends on the load-bearing fraction of the cross-sectional area of the gel, just as the modulus does. For particulate gels, a relation such as Eq. 19 is expected to apply. The problem with using such an equation is that it is difficult to predict or measure the radius of the neck; if r is predicted by assuming equilibrium between the neck and the solution, it is likely to be overestimated. Woignier and Phalippou [87] encountered the latter problem in an attempt to account for the strength of aerogels (using an equation related to Eq. 19 derived by Rumpf [88]). Regardless of the problems of explaining the data quantitatively, we can expect the strength to be increased by the factors that raise the solubility of the solid phase; therefore the important variables are pH, temperature, and type of solvent.

The *modulus of rupture* (MOR) is determined by continuing the beam-bending deflection to the point that the rod breaks; the maximum tensile stress on the rod (which occurs on the bottom surface) is the MOR. This measure of strength has been examined in several studies [15,82,73] and it generally increases in parallel to the shear modulus; for example, compare

Fig. 30.

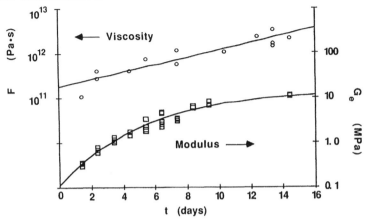

Uniaxial viscosity of network (not viscosity, η, of solid phase itself), F, and elastic shear modulus, G, versus time after gelation. From Scherer *et al.* [73].

Fig. 31 with Fig. 27. There is much more scatter in the data for MOR than for G, because the strength is dominated by flaws on the surface of the gel. Assuming that the flaw population is independent of age, the data suggest that the same factors (new bonds, growth of necks, decrease in porosity) that increase the stiffness of the gel lead to strengthening.

Zarzycki [89] measured the *critical stress intensity factor*, K_{IC}, in particulate silica gels made from Ludox® with and without additions of Aerosil® fumed silica, and in *sonogels* made by hydrolyzing TEOS in the presence of ultrasonic agitation [90]. The K_{IC} is a measure of the stress at the tip of a crack that grows spontaneously. (See Chapter 8, Section 3.) The experimental arrangement is shown in Fig. 32; by introducing a crack of known size, the influence of random flaws is eliminated. The particulate gels were measured in air, but the sonogels were measured in a solvent-saturated atmosphere to prevent cracking. For this crack geometry, K_{IC} is related to the energy of the fracture surface, Γ, by

$$2\Gamma = K_{IC}^2(1 - v^2)/E. \tag{21}$$

The quantity $E/(1 - v^2)$ was measured directly by an indentation technique. The results for Ludox® are shown in Fig. 33: the increase in K_{IC} was largely due to the increase in elastic modulus, as the energy of the fracture surface was constant after the first day. Assuming that $\Gamma = \rho\gamma_{SL}$, where ρ is the volume fraction of solid phase and γ_{SL} is the solid–liquid interfacial energy, the final value $\Gamma = 0.025 \pm 0.005$ J/m^2 corresponds to $\gamma_{SL} \approx 0.11$ J/m^2; this is near the range (0.13–0.14 J/m^2 [12]) reported for the surface energy

Fig. 31.

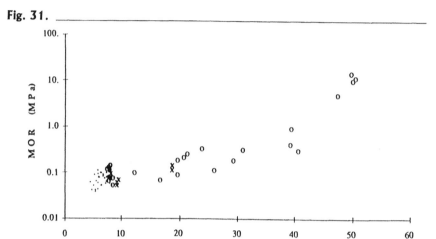

Modulus of rupture of gel B2 versus linear shrinkage (%) for same samples as in Fig. 27. From Scherer [15].

of fully hydrated silica. Figure 33d indicates that the fracture energy was independent of the quantity of fumed silica, so the links in the network were apparently all provided by the Ludox®. For the sonogels, K_{IC} rose from ~160 to ~280 Pa•m$^{1/2}$ between 24 and 40 h of aging, but the fracture energy remained roughly constant at 0.05 J/m^2. Given the volume fraction of solids of 0.377 in these gels, this value of Γ is in good agreement with the known value of γ_{SL}. It should be noted that this correspondence of Γ with γ_{SL} is not typically observed in fracture of ordinary glasses, where the fracture energy is often an order of magnitude greater than the interfacial energy [91]. This calculation of Γ is not very accurate because the load-bearing fraction of the cross-sectional area depends on necksize; therefore, it changes with time and is smaller than p.

Fig. 32.

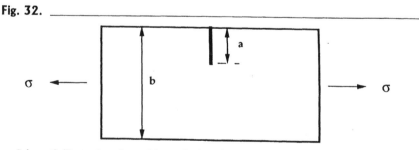

Schematic illustration of one-side notched tension bar used for K_{IC} determination. From Zarzycki [89].

Fig. 33a.

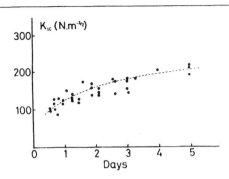

Fig. 33b.

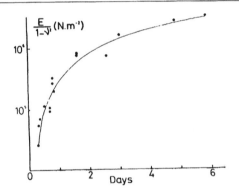

Fig. 33c.

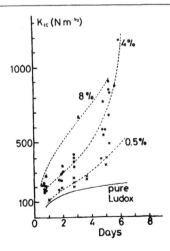

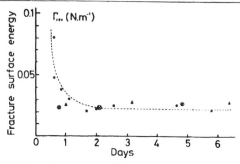

Effect of aging on (a) K_{IC} and (b) $E/(1 - \nu^2)$ for pure Ludox® gel, and (c) K_{IC} for gel containing Ludox® plus indicated volume fraction of Aerosil OX-50®, and (d) fracture energy, Γ, for pure Ludox® (●) and composites containing (×) 4 vol% and (△) 8% Aerosil®. From Zarzycki [89].

In measurements of strength, it is particularly important to consider the effect of capillary pressure since, as shown in Fig. 19, the strength of particulate systems is proportional to that pressure. The goal is usually to understand the strength that governs cracking during drying, so it may be appropriate to allow evaporation during the measurement. However, it should be recognized that this may make the strength dependent on time and ambient humidity.

5.

SUMMARY

The condensation reactions that bring about gelation also result in syneresis and progressive strengthening and stiffening of the gel during aging. Since the same mechanism is responsible for all of these phenomena, they depend in the same way on such factors as pH, temperature, and composition of the solution. In liquids in which the solid phase is soluble, dissolution and reprecipitation processes cause coarsening as material is transported to surfaces of lower curvature. This may result in loss of micropores, decrease in surface area, and stiffening through growth of interparticle necks.

The limited data available indicate that silica gels are linearly viscoelastic materials, and that the modulus and viscosity rise by orders of magnitude during aging. There is a close correlation between the shrinkage and the modulus during drying. The increase in modulus is too large to be explained simply by loss of porosity; the difference comes from the increased stiffness of the solid phase of the gel.

References

1. A.J. Vega and G.W. Scherer, *J. Non-Cryst. Solids*, **111** [2,3] (1989) 153–166.
2. T.W. Zerda, I. Artaki, and J.J. Jonas, *J. Non-Cryst. Solids*, **81** (1986) 365–379.
3. G. Orcel, L.L. Hench, I. Artaki, J. Jonas, and T.W. Zerda, *J. Non-Cryst. Solids*, **105** (1988) 223–231.
4. B.E. Yoldas, *J. Mater. Sci.*, **21** (1986) 1087–1092.
5. A.C. Pierre and D.R. Uhlmann in *Better Ceramics Through Chemistry III*, eds. C.J. Brinker, D.E. Clark, D.R. Ulrich (Mat. Res. Soc., Pittsburgh, Pa., 1988), pp. 207–212.
6. S.P. Prakash and N.R. Dhar, *J. Ind. Chem. Soc.*, **7** (1930) 417–434.
7. K. Dusek in *Polymer Networks, Structure and Mechanical Properties*, ed. A.J. Chompff and S. Newman (Plenum, New York, 1971), pp. 245–260.
8. J.F. Quinson, N. Tchipkam, J. Dumas, C. Bovier, J. Serughetti, C. Guizard, A. Larbot, and L. Cot, *J. Non-Cryst. Solids*, **99** (1988) 151–159.
9. M. Brun, A. Lallemand, J.F. Quinson, and C. Eyraud, *Thermochimica Acta*, **21** (1977) 59–88.
10. C. Eyraud, J.F. Quinson, and M. Brun in *Characterization of Porous Solids* (Elsevier, Amsterdam, 1988), pp. 295–305.
11. J.F. Quinson and M. Brun in *Characterization of Porous Solids* (Elsevier, Amsterdam, 1988), pp. 307–315.
12. R.K. Iler, *The Chemistry of Silica* (Wiley, New York, 1979).
13. C.J. Brinker and G.W. Scherer, *J. Non-Cryst. Solids*, **70** (1985) 301–322.
14. C.J. Brinker, K.D. Keefer, D.W. Schaefer, R.A. Assink, B.D. Kay, and C.S. Ashley, *J. Non-Cryst. Solids*, **63** (1982) 45–59.
15. G.W. Scherer, *J. Non-Cryst. Solids*, **109** (1989) 183–190.
16. A.C. Pierre and D.R. Uhlmann in *Ultrastructure Processing of Advanced Ceramics*, eds. J.D. Mackenzie and D.R. Ulrich (Wiley, New York, 1988), pp. 865–871.
17. A. Dauger, F. Chaput, J.C. Pouxviel, and J.F. Boilot, *J. de Physique*, **46** [12] (1985) C8-455–C8-459.
18. E. Matijević in *Science of Ceramic Chemical Processing*, eds. L.L. Hench and D.R. Ulrich (Wiley, New York, 1986), pp. 463–481.
19. J.M. Fletcher and C.J. Hardy, *Chem. and Ind.*, **18** (1968) 48–51.
20. E. Crucean and B. Rand, *Trans. J. Brit. Ceram. Soc.*, **78** [3] (1979) 58–64.
21. A. Bleier and R.M. Cannon in *Better Ceramics Through Chemistry II*, Mater. Res. Soc. Symp. Proc., **73**, eds. C.J. Brinker, D.E. Clark, and D.R. Ulrich (Mater. Res. Soc., Pittsburgh, Pa., 1986), pp. 71–78.
22. J.L. Woodhead, *Sci. Ceram.*, **9** (1977) 29–37.
23. C.J. Brinker and G.W. Scherer in *Ultrastructure Processing of Ceramics, Glasses, and Composites*, eds. L.L. Hench and D.R. Ulrich (Wiley, New York, 1984), pp. 43–59.
24. V.M. Shamrikov, V.L. Struzhko, V.I. Malkiman, I.E. Neimark, G.M. Kesareva, T.A. Loseva, and L.A. Bondar, *Sov. J. Colloids*, **46** [3] (1984) 544–546 (Eng. trans.).
25. M. Yamane and S. Okano, *Yogyo Kyokai-shi*, **87** [8] (1979) 56–60.
26. R.Yu. Sheinfain and I.E. Neimark in *Adsorption and Adsorbents*, no. 1, ed. D.N. Strazhesko (Wiley, New York, 1973), pp. 87–95.
27. R.Yu. Sheinfain, O.P. Stas, and T.F. Makovskaya, *Sov. J. Colloids*, **34** [6] (1972) 869–871 (Eng. trans.).
28. C.J.G. van der Grift, J.W. Geus, H. Barten, R.G.I. Leferink, J.C. van Miltenburg, and A.T. den Ouden in *Characterization of Porous Solids* (Elsevier, Amsterdam, 1988), pp. 619–628.

29. V.M. Chertov, D.B. Dzhambaeva, and I.E. Neimark, *Ukr. Khim. Zh.*, **31** [12] (1965) 1253–1258.

30. V.M. Chertov, D.B. Dzhambaeva, A.S. Plachinda, and I.E. Neimark, *Russ. J. Phys. Chem.*, **40** [3] (1966) 282–285.

31. V.B. Lazarev, I.L. Voroshilov, G.P. Panasyuk, G.P. Budova, N.I. Ivanov, and B.M. Zhigarnovskii, *Sov. J. Inorganic Materials*, **23** [3] (1987) 381–387 (Eng. trans.).

32. V.B. Lazarev, G.P. Panasyuk, G.P. Budova, and I.L. Voroshilov, *Sov. J. Inorg. Mater.*, **20** [10] (1984) 1445–1449 (Eng. trans.).

33. J.W. Christian, *Theory of Transformation in Metals and Alloys, Part I*, 2d ed. (Pergamon, New York, 1975).

34. J.D. Cook and R.W. Thompson, *Zeolites*, **8** [4] (1988) 322–326.

35. S.P. Zhdanov and N.N. Samulevich, *Proc. 5th Int. Conf. Zeolites*, ed. L.V.C. Rees (Heyden, London, 1980), pp. 75–84.

36. S.I. Molodchikova, N.A. Shabanova, E.S. Lukin, and Yu.G. Frolov, *Sov. J. Glass and Ceramics*, **9** (1981) 478–480 (Eng. trans.).

37. V.V. Sidorchuk and V.M. Chertov, *Sov. J. Inorg. Mater.*, **22** [11] (1986) 1692–1695 (Eng. trans.).

38. M. Prassas, J. Phalippou, L.L. Hench, and J. Zarzycki, *J. Non-Cryst. Solids*, **48** (1982) 79–95.

39. M. Prassas and L.L. Hench in *Ultrastructure Processing of Ceramics, Glasses, and Composites*, eds. L.L. Hench and D.R. Ulrich (Wiley, New York, 1984), pp. 100–125.

40. L.L. Hench in *Better Ceramics Through Chemistry*, eds. C.J. Brinker, D.E. Clark, and D.R. Ulrich (North-Holland, New York, 1984), pp. 101–110.

41. B.E. Yoldas, *J. Appl. Chem. Biotechnol.*, **23** (1973) 803–809.

42. B.E. Yoldas, *J. Mater. Sci.*, **10** (1975) 1856–1860.

43. G.C. Bye and J.G. Robinson, *Koll. Z. und Z. Polym.*, **198** [1–2] (1964) 53–60.

44. G.C. Bye and K.S.W. Sing in *Particle Growth in Suspensions*, ed. A.L. Smith (Academic Press, New York, 1973), pp. 29–65.

45. J. Pawlaczyk, Z. Kokot, and A. Borkowski, *Pol. J. Pharmacol. Pharm.*, **32** (1980) 409–417.

46. S.L. Nail, J.L. White, and S.L. Hem, *J. Pharm. Sci.*, **65** [8] (1976) 1192–1195.

47. M.R. Harris and K.S.W. Sing, *J. Appl. Chem.*, **7** (1957) 397–401.

48. J. Ragai, *Chem. Tech. Biotechnol.*, **35A** (1985) 263–269.

49. V.M. Chertov, V.V. Tsyrina, and N.T. Okopnaya, *Ukr. Khim. Zh.*, **51** [6] (1985) 613–614.

50. V.M. Chertov and V.V. Sidorchuk, *Sov. J. Inorganic Mater.*, **22** [11] (1986) 1689–1692.

51. C.M. Flynn, Jr., *Chem. Rev.*, **84** (1984) 31–41.

52. T. Tanaka, *Scientific American*, **244** (1981) 124–138.

53. G.W. Scherer, *J. Non-Cryst. Solids*, **87** (1986) 199–225.

54. W.M. Jones and D.B. Fischbach, *J. Non-Cryst. Solids*, **101** (1988) 123–126.

55. Z.Z. Vysotskii and D.N. Strazhesko in *Adsorption and Adsorbents*, no. 1, ed. D.N. Strazhesko (Wiley, New York, 1973), pp. 55–71.

56. Z.Z. Vysotskii, V.I. Galinskaya, V.I. Kolychev, V.V. Strelko, and D.N. Strazhesko in *Adsorption and Adsorbents*, no. 1, ed. D.N. Strazhesko (Wiley, New York, 1973), pp. 72–86.

57. Yu.P. Klimentova, L.F. Kirichenko, and Z.Z. Vysotskii, *Ukr. Khim. Zh.*, **36** [1] (1970) 56–58 (Eng. trans.).

58. E.G. Acker, *J. Colloid Interface Sci.*, **32** [1] (1970) 41–54.

59. R. Snel, *Appl. Catalysis*, **11** (1984) 271–280.

60. T.P. Ponomareva, S.I. Kontorovich, M.I. Chekanov, and E.D. Shchukin, *Sov. J. Colloids*, **46** [1] (1984) 118–120 (Eng. trans.).

61. H.-J. Tiller, R. Göbel, and U. Hartung, *J. Non-Cryst. Solids*, **105** (1988) 162–164.
62. H.N. Holmes, W.E. Kaufmann, and H.O. Nicholas, *J. Am. Chem. Soc.*, **41** (1919) 1329–1336.
63. G.W. Scherer, *J. Non-Cryst. Solids*, **108** (1989) 18–27.
64. G.W. Scherer, *J. Non-Cryst. Solids*, **108** (1989) 28–36.
65. G.W. Scherer in *Better Ceramics Through Chemistry III*, eds. C.J. Brinker, D.E. Clark, and D.R. Ulrich (Mat. Res. Soc., Pittsburgh, Pa., 1988), pp. 179–186.
66. M. Ohno, T. Yamada, and N. Takato, *Yogyo Kyokai-shi*, **94** [7] (1986) 644–650.
67. G.W. Scherer and R.M. Swiatek, unpublished work.
68. R.H. Ottewill, *J. Colloid Interface Sci.*, **58** [2] (1977) 357–373.
69. L.H. Allen and E. Matijević, *J. Colloid Interface Sci.*, **31** [3] (1969) 287–296.
70. G.W. Scherer, *J. Am. Ceram. Soc.*, **60** [5–6] (1977) 236–239.
71. G.W. Scherer, *J. Non-Cryst. Solids*, **34** (1979) 239–256.
72. G.W. Scherer, *Relaxation in Glass and Composites* (Wiley, New York, 1986).
73. G.W. Scherer, S.A. Pardenek, and R.M. Swiatek, *J. Non-Cryst. Solids*, **107** [1] (1988) 14–22.
74. H. Rumpf in *Agglomeration*, ed. W. Knepper (Interscience, New York, 1962), pp. 379–417.
75. H. Schubert, *Powder Technol.*, **37** (1984) 105–116.
76. H. Schubert, *Chem. Ing. Techn.*, **51** (1979) 277–282.
77. D.M. Newitt and J.M. Conway-Jones, *Trans. Instn. Chem. Engrs.*, **36** (1958) 422–442.
78. M.J. Keesman, P.H.G. Offermans, and E.P. Honig, *Mater. Lett.*, **5** [4] (1987) 140–142.
79. J. Dumas, S. Baza, and J.Serughetti, *J. Mater. Sci. Lett.*, **5** (1986) 478–480.
80. J.K. West, R. Nickles, and G. Latorre in *Better Ceramics Through Chemistry III*, eds. C.J. Brinker, D.E. Clark, and D.R. Ulrich (Mat. Res. Soc., Pittsburgh, Pa., 1988), pp. 219–224.
81. Yu.P. Klimentova, L.F. Kirichenko, and Z.Z. Vysotskii, *Ukr. Khim. Zh.*, **37** [3] (1971) 20–23 (Eng. trans.).
82. S.A. Pardenek, J.W. Fleming, and L.C. Klein in *Ultrastructure Processing of Advanced Ceramics*, eds. J.D. Mackenzie and D.R. Ulrich (Wiley, New York, 1988), pp. 379–389.
83. E.M. Rabinovich in *Sol-Gel Technology for Thin Films, Fibers, Preforms, Electronics, and Speciality Shapes*, ed. L.C. Klein (Noyes, Park Ridge, N.J., 1988), pp. 260–294.
84. M.J. Murtagh, E.K. Graham, and C.G. Pantano, *J. Am. Ceram. Soc.*, **69** [11] (1986) 775–779.
85. Z. Hashin and S. Shtrikman, *J. Mech. Phys. Solids*, **11** (1963) 127–140.
86. K. Murakami and K. Ono, *Chemorheology of Polymers* (Elsevier, New York, 1979).
87. T. Woignier and J. Phalippou, *J. Non-Cryst. Solids*, **100** (1988) 404–408.
88. H. Rumpf, *Chem. Ing. Tech.*, **30** (1958) 144–158.
89. J. Zarzycki, *J. Non-Cryst. Solids*, **100** (1988) 359–363.
90. L. Esquivas and J. Zarzycki in *Ultrastructure Processing of Advanced Ceramics*, eds. J.D. Mackenzie and D.R. Ulrich (Wiley, New York, 1988), pp. 255–270.
91. S.W. Freiman in *Glass Science and Technology, Vol. 5: Elasticity and Strength in Glasses* (Academic Press, New York, 1980), pp. 21–78.

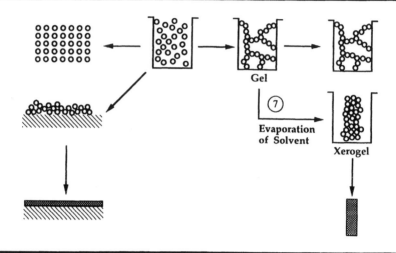

Gel

⑦
Evaporation
of Solvent

Xerogel

CHAPTER 7

Theory of Deformation and Flow in Gels

This chapter deals with the flow of liquid in the pores and the deformation of the skeleton of a gel during syneresis and drying. The driving forces for shrinkage include chemical effects, such as condensation reactions, and physical effects, such as capillary pressure. Fluid transport can occur by flow down a pressure gradient or diffusion down a chemical potential gradient, and deformation of the network may involve elastic, plastic, or viscoelastic strains. The interaction of the pressure in the pore liquid and the shrinkage of the network has been thoroughly examined theoretically, and the results are reviewed. Readers who are not theoretically inclined may choose to skip Sections 3 and 4; the discussion of drying in the following chapter does not require understanding of the details of the theory presented here. However, we do recommend reading Sections 1 and 2 of this chapter. The outline is as follows.

1. *Driving Forces for Shrinkage* explains the role of chemical reactions and the pressure exerted by osmotic, disjoining, and capillary phenomena. The concept of moisture stress, widely used in soil science, is a phenomenological measure of all of these factors.
2. *Liquid transport* discusses the processes of fluid flow and diffusion in porous media, with particular reference to the unusual behavior to be expected when the pore diameter approaches atomic dimensions.
3. *Rheology of the Porous Network* formulates the relations between stress and strain in a fluid-filled network when the solid phase is viscoelastic.
4. *Theory of Deformation* presents a theory that relates the pressure in the liquid to the shrinkage of the network. The pressure distributions and stresses are calculated for gels undergoing syneresis or drying.

1.

DRIVING FORCES FOR SHRINKAGE

1.1. Chemical Reactions

In the previous chapter it was shown that the shrinkage of gels during aging is attributable to ongoing condensation reactions between M–OH groups. The total contraction resulting from condensation can be enormous, as illustrated in Fig. 1 of Chapter 6, but in more typical cases is on the order of 10% (linear). For example, after a year of syneresis, the B2 gel in Fig. 27 of Chapter 6 has contracted ~8% in diameter, but it shrinks several times as much in a few days if evaporation is permitted. Thus, the driving force for shrinkage provided by chemical reactions is small compared to the other factors operating during evaporation. Moreover, organic groups that adsorb or chemisorb on the M–OH groups inhibit condensation and thereby further reduce the influence of chemical reactions on shrinkage.

It is important to recognize that the mechanical properties of gels are profoundly affected by the formation of new bridging bonds. The modulus and viscosity of the gel increase during aging, and they rise even faster during drying-induced shrinkage. It seems likely that the contraction brings reactive M–OH groups into proximity so that further condensation is possible, and the shrinkage is irreversible. This process is also influenced by adsorbed organics: in alcohol–acetone solutions, where condensation is inhibited by adsorption, the modulus of silica gel B2 does not increase during aging [1]. Thus chemical reactions are less important for their contribution to the driving force for shrinkage than for their impact on the rheology of the gel.

1.2. Osmosis

Osmosis is a process of diffusion driven by a chemical potential gradient. For example, if pure liquid is separated from a salt solution by a membrane that is permeable to liquid, but not to salt, pure liquid will diffuse into the salt-rich side. This type of transport occurs in general when liquid A and B are separated by a membrane permeable only to B. A dramatic example is shown in Fig. 1: Yoldas [2] prepared a sol of alumina in water, placed it into a bag made from a dialysis membrane, and immersed the bag in alcohol; as the water diffused out of the sol into the alcohol bag, the sol contracted and finally gelled. Of course, it is not necessary to have a membrane to have

Fig. 1.

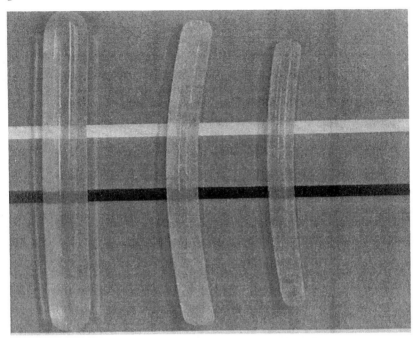

Sample on left is dialysis bag containing fresh sol (~5% solids in water) of alkoxide-derived alumina; bag was suspended in an alcohol bath, allowing the water to diffuse out of the bag. Loss of water caused the sample to gel (middle sample). Further loss caused more shrinkage (right sample). After removal from the bath, the sample was completely dried by exposure to air, but no further shrinkage occurred. The shrinkage in the bath took place in a matter of hours. From Yoldas [2].

osmotic flow. The gel network acts as its own membrane, permitting diffusion while offering high frictional resistance to counterflow of liquid. For example, it was shown in Chapter 6 that silica gels shrink faster when immersed in salt solutions than they do in pure water. A more typical situation that arises during drying of a gel with a solution of alcohol and water in its pores is that the alcohol evaporates preferentially. This creates a concentration gradient of alcohol in water that induces interdiffusion of the liquids within the gel. It is shown in Section 4 that this phenomenon may be exploited to reduce drying stresses.

The presence of a solute reduces the vapor pressure of the solvent from p_o to p_V, so the chemical potential, μ, of the solvent is lowered by [3]

$$\Delta\mu = R_g T \ln(p_V/p_o) \equiv -\Pi V_m \tag{1}$$

where R_g is the ideal gas constant, T is the temperature, and V_m is the molar volume of the solvent. The *osmotic pressure*, Π, is defined in Eq. 1 by equating the change in chemical potential to mechanical work. If a pressure of that magnitude were applied to the solution, the vapor pressure would become equal to that of the pure solvent (because the molecules, being crowded together by the pressure, would be more inclined to evaporate). In dilute solutions (where Raoult's law is obeyed), the osmotic pressure can be expressed in terms of the mole fraction of solvent, x, by [3]

$$\Pi = -\left(\frac{R_g T}{V_m}\right) \ln(x). \qquad (2)$$

In organic polymers it is recognized that there is a difference in the affinity of polymer chains for one another and of the chains for the solvent; if the concentration of chains changes (e.g., by evaporation of solvent), or if the quality of the solvent changes, the change in chemical potential of the chains will induce diffusion of the solvent. This osmotic effect contributes to the enormous volume changes seen in polyacrylamide gels upon a change in the ratio of acetone to water in the pore liquid [4]. In contrast, the microsyneresis reported in a titania gel by Quinson *et al.* [5] resulted in a 1–3% increase in volume when the gel was transferred into pure decane; the volume returned to the original value when the sample was immersed in pure water. This small effect was produced by liquids that represent the extremes of polar and nonpolar behavior, so it is not surprising that osmotic swelling has not been observed in polymeric inorganic gels in alcohol–water systems. For silicate gels there is no report of swelling upon immersion in any liquid.[†] The data in the preceding chapter show that the rate of *shrinkage* may change when a gel is transferred from one liquid to another, but the change is always in the direction expected from the effect of the new liquid on the rate of condensation reactions. Osmotic forces are important in particulate systems such as clays [8] and transition metal gels [9], which consist of covalently bonded sheets held together by ions. Water tends to invade the space between the sheets and push them apart, so such materials may swell and even disperse when immersed in liquid. The swelling pressure (i.e., the pressure that the swelling body can exert against a confining force) in clays can be as great as a few megapascals. Spitzer [10] has developed a model that relates the pressure to the electrostatic repulsion between the particles and he finds excellent agreement with experimental data for a variety of clays.

[†] Kistler [6] and Kawaguchi *et al.* [7] observed swelling when partially dried silica gels were immersed in liquid, but this was presumably from release of elastic strains imposed by drying stresses, rather than an osmotic effect.

1.3. Disjoining Forces

Disjoining forces are a class of short-range forces resulting from the presence of a solid–liquid interface. The most important examples are double-layer repulsion between charged surfaces (see discussion in Chapter 4) and interactions caused by structure created in the liquid by dispersion forces. Liquid molecules, especially water [11,12], tend to adopt a special structure in the vicinity of a solid surface, as illustrated schematically in Fig. 2. The interaction with the surface is so strong that the adsorbed layers resist freezing. Brun *et al.* [13], using a calorimetric method, found a noncrystal-lizing layer with a thickness of 0.8 nm for water and 1.3 nm for benzene. Water in the pores of silica gel is also found to have a higher thermal expansion coefficient and density than bulk water [14]. These *structured layers* inhibit close approach of surfaces, as they resist overlapping. The effect of adsorption forces on the mobility of molecules near solid surfaces is discussed in Section 2.2.

Fig. 2. _____

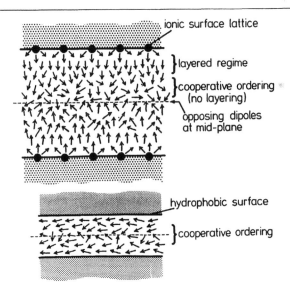

Orientation of water molecules around a charged hydrophilic surface. The molecules closest to each surface exhibit a diffuse layered structure (oscillatory hydration force regime); beyond that there may be some cooperative ordering (monotonic hydration force regime). Because the mean orientations of the water molecules are antiparallel on either side of the midplane, the net hydration force is repulsive. Bottom: cooperative alignment of water molecules between two hydrophobic surfaces, resulting in a net attractive solvation force. From Israelachvili [12].

Fig. 3.

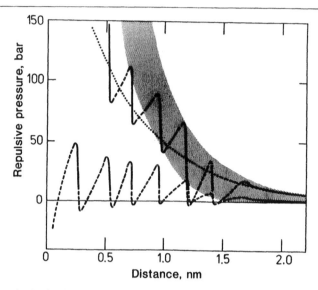

Short-range hydration forces between mica surfaces plotted as pressure against distance. Force measured in 1 M KCl, where there is 1 K$^+$ ion adsorbed per 0.5 nm^2 (surfaces 95% saturated with K$^+$) (upper curve) and 0.001 M KCl, where there is 1 K$^+$ per nm^2 (40% coverage). From Israelachvili [12].

At low concentrations of electrolyte, the force between surfaces is found to obey the Derjaguin–Landau–Vervey–Overbeck (DLVO) theory (see Chapter 4) until the separation is ~1.5 nm; then the force oscillates, as shown in Fig. 3, as individual layers of molecules are displaced. The same behavior is observed in a wide range of nonaqueous liquids. In water containing high concentrations of electrolytes, ions adsorbed on the solid surface develop a layer of hydration that creates an additional component of repulsion that decays exponentially over the range of ~1.5 to 4 nm (decay length ~0.6–1.1 nm). This monotonic component results from hydration forces, not ordering of molecules, and is not observed in liquids other than water [12].

As evaporation occurs and solid surfaces are brought together, repulsive forces arising from electrostatic repulsion, hydration forces, and solvent structure resist contraction of the gel. The pore liquid will diffuse or flow from the swollen interior of the gel toward the exterior to allow the surfaces to move further apart. The disjoining forces thus produce an osmotic flow. Since these forces become important when the separation between surfaces is small, they are most likely to be important near the end of drying, when the pore diameter may approach 2 nm.

1.4. Capillary Forces

Atoms on a surface have a higher energy than those within the bulk, so there is an energy associated with the existence of any interface. When the specific energy (J/m^2) of a solid–vapor interface, γ_{SV}, is greater than that of a solid–liquid interface, γ_{SL}, liquid tends to flow over an exposed solid surface. A layer of liquid on a plane solid surface has two interfaces, solid–liquid and liquid–vapor, so the change in energy, ΔE, produced by spreading of the liquid film is, as shown in Fig. 4a,

$$\Delta E = \gamma_{SL} + \gamma_{LV} - \gamma_{SV}. \tag{3}$$

If $\Delta E < 0$, the energy of the system is reduced so the liquid will spread spontaneously; otherwise, the solid–liquid–vapor will be characterized by a *contact angle*, θ, defined in Fig. 4b. The balance of tensions at the point of intersection leads to a relationship between the surface tensions that is known as *Young's equation*:

$$\gamma_{SV} = \gamma_{SL} + \gamma_{LV} \cos(\theta). \tag{4}$$

If $\theta = 0$ (as in Fig. 4a), the liquid is said to be a *spreading liquid*, and the solid will be covered with a liquid film. There is often hysteresis in the contact angle [15] as a solid is wetted and then drained, because of surface roughness and impurity effects.

Fig. 4.

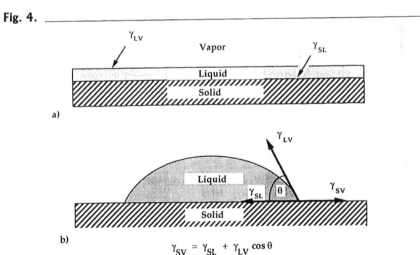

(a) Liquid film on solid surface has two interfaces with energies γ_{SL} and γ_{LV}.
(b) Contact angle θ is determined by balance of forces at intersection of solid–vapor and liquid–vapor interfaces.

Liquid rises in a capillary tube because in so doing it replaces a solid–vapor interface with a solid–liquid interface, gaining energy equal to $2\pi a h(\gamma_{SV} - \gamma_{SL})$, where a is the radius of the tube and h the height to which the liquid rises. The liquid does work against gravity equal to $P_c\,\Delta V$, where P_c is called the *capillary pressure* and the volume of liquid moved is $\Delta V = \pi a^2 h$. Equating the work done to the energy gained, the capillary pressure is found to be

$$P_c = -\frac{2(\gamma_{SV} - \gamma_{SL})}{a} = -\frac{2\gamma_{LV}\cos(\theta)}{a}. \tag{5}$$

The negative sign indicates that the liquid is in tension. (The sign convention for pressure is opposite to that for stresses, which are positive when tensile.) Figure 5 shows that when the radius of the tube, a, is very small the pressure is large. Laplace proved [15,16] that there is a pressure difference across a curved liquid–vapor interface equal to

$$\Delta P = \gamma_{LV}\left(\frac{1}{R_1} + \frac{1}{R_2}\right) \equiv \gamma_{LV}\kappa \tag{6}$$

where R_1 and R_2 are the principal radii of curvature of the interface and κ is called the *curvature* of the interface. In the cylindrical capillary, the pressure difference is simply P_c and symmetry requires that $R_1 = R_2$, so comparison of Eqs. 5 and 6 indicates that the *meniscus* (liquid–vapor interface) is hemispherical with a radius of curvature of $-a/\cos(\theta)$. (The radius of curvature of the interface is negative, by definition, because the center of curvature is outside of the condensed phase. If the radius in Eq. 5 is understood to be that of the interface, rather than that of the tube, the negative sign is unnecessary. In the course of discussion when we refer to the size of a radius of curvature, we use the absolute value.) The exact shape of the meniscus is different in large tubes, where gravitational effects must be considered, and in submicron tubes, where disjoining forces can affect liquid structure near the wall.

The capillary pressure in an unsaturated porous body can be predicted by a simple argument [17,18], if the pore size is uniform. Suppose that the liquid in the saturated portion of the body is suddenly subjected to a negative pressure of P_c (as if the interfacial energies were just now "turned on"). It will cause the solid network to contract so that the pore space in the saturated region decreases by ΔV, forcing that volume of liquid to redistribute into the previously unsaturated region. The liquid will cover an area equal to $\Delta V S_P/V_P$, where S_P/V_P is the surface-to-volume ratio of the empty pores, so the redistribution process will yield energy equal to $(\gamma_{SV} - \gamma_{SL})\,\Delta V S_P/V_P$. The work done by the liquid to cause that contraction is simply $P_c\,\Delta V$,

Fig. 5.

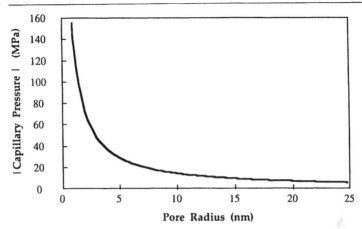

Effect of radius of curvature of meniscus on magnitude of capillary pressure in liquid, $|P_c|$, calculated from Eq. 5 using $\gamma_{LV} = 0.072$ J/m^2 and $\theta = 0$, so $\kappa = -2/a$.

so balancing these energies we find that the capillary pressure is

$$P_c = -\frac{(\gamma_{SV} - \gamma_{SL})S_P}{V_P} = -\frac{\gamma_{LV}\cos(\theta)S_P}{V_P}. \tag{7}$$

For a simple cylinder, $S_P = 2\pi ah$ and $V_P = \pi a^2 h$, so Eqs. 5 and 7 are equivalent. The specific surface area of a porous body (interfacial area per gram of solid phase), S, is related to the surface-to-volume ratio by [18]

$$\frac{S_P}{V_P} = \frac{S\rho_s\rho}{1 - \rho} \tag{8}$$

where ρ is the *relative density*, $\rho = \rho_b/\rho_s$, ρ_b is the *bulk density* of the solid network (not counting the mass of the liquid), and ρ_s is the density of the solid skeleton (the *skeletal density*).

The preceding analysis is strictly valid only if the pore size of the body is uniform. Otherwise, the liquid will preferentially invade the smaller pores in which the capillary suction is greatest; that is, the liquid will invade the pores with the greatest surface-to-volume ratio, because that route will yield the greatest reduction in the energy of the system. One might think that the pores would fill in order, from smallest to largest, so that the pressure could be predicted as a function of volume fraction of liquid, if the pore size distribution were known. That may not be true in general, however, because the pores of a given size might not be connected with one another. For example, the smallest pores might be separated by much larger ones that the liquid would not tend to enter, so some intermediate pores would fill before all of the smallest ones could. This is a problem of percolation [19] (see Chapter 5 for a brief discussion), because it depends on the connectivity of randomly distributed pores.

Comparison of Eqs. 6 and 7 indicates that the curvature of the mensicus in a porous body is

$$\kappa = -\frac{S_P \cos(\theta)}{V_P} = -\frac{S\rho_s \rho \cos(\theta)}{1 - \rho}. \tag{9}$$

The vapor pressure of the liquid, p_V, is related to the curvature (or, equivalently, to the capillary pressure) by the *Gibbs–Thompson* (or *Kelvin*) *equation*,

$$p_V = p_o \exp\left(\frac{V_m P_c}{R_g T}\right) = p_o \exp\left(\frac{V_m \kappa \gamma_{LV}}{R_g T}\right), \tag{10}$$

where p_o is the vapor pressure over a flat liquid surface. Recalling that P_c and κ are negative, we see that the vapor pressure is reduced over a concave interface. Physically, it is easy to understand that molecules are less likely to leave when tension (suction) already exists in the liquid phase. The effect of curvature on the vapor pressure of water is illustrated in Fig. 6. Since the pore diameters in gels are typically in the range of 2 to 10 nm, it is clear that significant reductions in p_V can occur.

It is obvious that Eq. 10 must cease to be valid when the radius of the pore approaches atomic dimensions, because the concepts of interfacial energy and contact angle lose their meaning. This equation has been subjected to direct tests [20–22] and it has been shown that the relation between liquid pressure and vapor pressure is valid for radii as small as 1 nm for organic liquids such as cyclohexane, n-hexane, and benzene. For water, one study [22] concludes that Eq. 10 is valid for vapor pressures as low as $p_V/p_o \geq 0.7$, corresponding to a radius of curvature ≥ 1.5 nm; another [21] finds departures exceeding 10% when $p_V/p_o < 0.9$, when the radius of curvature drops below ~ 5 nm. The difference between the behavior of water and the organic liquids is attributed to the long-range hydration forces in water (see Fig. 2), which could influence the surface tension in films less than 5 nm thick [21]. These studies thus indicate that the concepts of bulk thermodynamics can be applied without serious error even on the scale of sizes of interest in inorganic gels.

The importance of capillary pressure to the drying of gels is not its effect on the vapor pressure, but its effect on the solid phase. The tension in the liquid is supported by the solid phase. (For example, the walls of a simple capillary tube are subjected to a tensile stress equal to $-P_c$ that pulls them radially inward.) The gel contracts as the liquid evaporates from its pores,

Fig. 6.

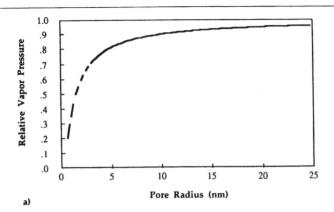

a)

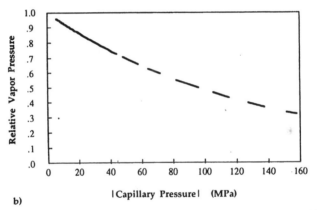

b)

Illustration of relative vapor pressure, p_V/p_o, of water over concave liquid–vapor interface, calculated from Eq. 10 using $V_m = 18$ cm³/g, $\gamma_{LV} = 0.072$ J/m² and $\theta = 0$, so $\kappa = -2/a$: relative vapor pressure versus radius of pore (a) and magnitude of capillary pressure, $|P_c|$ (b). Dashed curves for $p_V/p_o < 0.7$ indicate probable limit of applicability of bulk thermodynamic properties.

because of the huge stresses (see Fig. 6) generated by the menisci in the tiny pores. In ordinary ceramic materials, where the pore diameters are on the order of microns, these stresses are relatively small, but in gels they can exceed the strength of the network by a considerable margin. If the pressure is uniformly applied on the network (i.e., if it is placed under hydrostatic compression by the liquid), it will shrink uniformly and have no tendency to crack. However, as we shall see, large pressure gradients can develop through the thickness of the gel, such that the network is compressed more at the exterior surface than in the interior of the body, and this differential strain causes cracking. Such pressure gradients are quickly relaxed by fluid flow in materials with large pores, but flow is difficult within gels, so the gradients persist.

1.5. Moisture Stress

Moisture stress or *moisture potential*, ψ, is the partial specific Gibbs free energy (J/g) of liquid in a porous medium, and is given by [23]

$$\psi = \left(\frac{R_g T}{\rho_L V_m}\right) \ln\left(\frac{p_v}{p_o}\right) \tag{11}$$

where ρ_L is the density of the liquid. In soil science [24] it is conventional to define the moisture potential in terms of the height h to which it would draw a column of water, so a factor of g (the gravitational acceleration) would be included in the denominator on the right side of Eq. 11. Comparison of Eqs. 10 and 11 reveals that ψ is equivalent to P_c/ρ_L when only capillary forces are acting (or, according to the alternative definition, $h = P_c/\rho_L g$). However, the moisture potential is quite general, because the vapor pressure is depressed by other factors, including osmotic pressure, hydration forces, and adsorption forces. Thus, moisture potential subsumes all of the driving forces previously discussed and can be obtained by measuring the vapor pressure of the liquid in the system. For that reason, Zarzycki [25] recommends it as the appropriate potential driving shrinkage of gels during drying. The difficulty in implementing that suggestion is that capillary pressure gradients produce bulk flow, while concentration gradients (that produce osmotic pressure) cause diffusion, so it is necessary to apply portions of the total potential to different transport processes. In soil science, it is customary to assume that bulk flow is driven by the gradient in moisture potential, but it is done with the understanding that factors other than capillary pressure and gravitation are negligible [24].

When the moisture potential is defined in units of height (length), it is convenient to use a logarithmic scale defined by

$$pF = \log_{10} h(\text{cm}) = \log_{10}\left[\left(\frac{R_g T}{\rho_L g V_m}\right) \ln\left(\frac{p_v}{p_o}\right)\right]. \tag{12}$$

A typical plot of pF versus moisture content is illustrated in Fig. 7. The body is saturated with liquid between points a and b, and pF rises almost linearly as liquid evaporates. Shrinkage stops near point b, and air invades the pores from b to c; the more uniform the distribution of pore sizes, the flatter

Fig. 7.

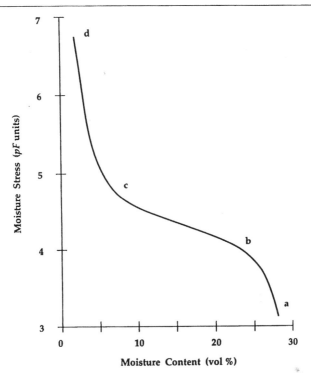

Moisture potential curve (schematic): body shrinks but remains saturated from a to b; shrinkage stops and air enters the pores between b and c; adsorbed liquid is removed between c and d. Compare with plot of capillary pressure (p) versus saturation (S) in Fig. 19 of Chapter 6.

the curve. Between c and d, the adsorbed layers of water are removed, and the potential increases in steps [26] (not shown) corresponding to the individual layers. These are tightly bound, so very high potentials (i.e., very low partial pressures of vapor) are required to remove them. There is substantial hysteresis in such curves [23], because liquid in large pores may be unable to drain through small openings, whereas it easily advances into such openings.

The relation between ψ and moisture content in Fig. 7 is intended to represent equilibrium, but no such relationship exists in gels that exhibit syneresis and viscoelastic relaxation. The liquid content of such gels (e.g., the silica gels discussed in the previous chapter) decreases spontaneously during syneresis, and the process can continue for many months. Even if there were no syneresis (as might occur if organic molecules adsorb onto the M–OH groups), the pressure in the liquid would be relieved by relaxation of the solid phase. These phenomena make it difficult to measure the moisture potential, because the vapor pressure changes with time. In contrast, soils [27] and clay [28] are plastic materials with a yield stress that depends on moisture content. They contract under a constant load, but stiffen as liquid drains out; shrinkage stops when the yield stress becomes equal to the applied stress. Therefore it is possible to establish an equilibrium curve relating liquid content and vapor pressure (corresponding to the point where the capillary pressure is balanced by the yield stress of the solid phase). The same is probably true of particulate gels in general.

2.

LIQUID TRANSPORT

2.1. Darcy's Law

Fluid flow through porous media obeys *Darcy's law* [29,30], which states that the flux of liquid, J, is proportional to the gradient in pressure in the liquid, ∇P_L:

$$J = -\frac{D}{\eta_L} \nabla P_L.$$ (13)

The flux is in units of volume per area of the porous body (*not* the area occupied by the liquid) per time (m/s), P_L is the force per unit area of the liquid (Pa), η_L is the viscosity of the liquid (Pa•s), and D is called the *permeability* and has units of area (m^2). Positive flux moves in the direction of increasingly negative pressure (i.e., the flow is toward regions of greater tension in the liquid). Equation 13 is an empirical equation derived from observation of flow of water through soil [31], but it is analogous to *Poiseuille's law* for flow of liquid through a straight circular pipe:

$$Q = -\frac{\pi}{8} \frac{\Delta P_L}{h} \frac{a^4}{\eta_L}$$ (14)

where Q is the rate of flow (m^3/s), a is the radius of the pipe, and h is its length; ΔP_L is the pressure difference between the ends of the pipe, so the pressure gradient is $\nabla P_L = \Delta P_L/h$. Since the flux through the pipe is $J = Q/\pi a^2$, comparison of Eqs. 13 and 14 indicates that the "permeability" of the single pipe is $a^2/8$.

Darcy's law is based implicitly on the principle that the flow is uniform on a scale large compared to the diameters of the pores, but small compared to the whole body. The permeability averages away the geometric details of the pore structure, taking the form

$$D = (1 - \rho)r_h^2/f_S f_T \qquad (15)$$

where $1 - \rho$ is the porosity and f_S and f_T are two finagle factors accounting for the noncircular cross section and the nonlinear path, respectively, of the actual pores; f_T is called the *tortuosity*. The *hydraulic radius*, r_h, is defined as the ratio of pore volume to surface area, $r_h = V_P/S_P$; for a circular capillary, $r_h = a/2$. The analogy to Poiseuille's law has naturally led to many models for the permeability of porous media based on representations of the pores by arrays of tubes. A number of these are discussed in the excellent texts by Scheidegger [29] and Dullien [30]; van Brakel [32] offers a critical review of over three hundred such models. Every conceivable shape and distribution of pores has been considered, including tubes with various cross-sectional shapes, with and without constrictions and intersections along their length; some represent the solid phase by isolated solid objects and analyze the flow around them. The most popular model, because of its simplicity and accuracy is the *Carman-Kozeny equation*, which gives the permeability in terms of the relative density and specific surface area:

$$D = \frac{(1 - \rho)^3}{5(\rho S \rho_s)^2}. \qquad (16)$$

The factor of 5 is an estimate of the shape and tortuosity factors based on experimental results. This equation is reasonably successful for many types of granular materials, but it often fails, and should be applied with caution. Happel and Brenner [33] illustrate the calculation (by solution of the Navier–Stokes equations) of the permeability of various arrays of cylinders, spheres, and other obstacles. They find that the flux is proportional to the pressure gradient, as required by Darcy's law, but the Kozeny constant, $f_S f_T$, agrees only with the order of magnitude of the empirical value of ~ 5. Adler [34–36] makes the interesting observation that such solutions for flow around *regular* arrays of objects do not agree with Eq. 16; however, when the submerged objects are *fractal*, the permeability does obey the Carman-Kozeny equation. He infers that it is the randomness in the structure of real systems that accounts for the empirical success of that equation. For our purposes it is sufficient to note that the large porosity of gels favors high permeability, but the very high surface area (or, equivalently, very small pores) drives D down.

Very little is known about the permeability of inorganic gels, but the importance of microstructure is revealed by studies of polyacrylamide (PA) gels. Phase separation occurs when PA gels are cooled to $-17°C$, whereupon the polymer chains cluster together (as in Fig. 2 of Chapter 6) leaving relatively large pores. Using a light-scattering technique, Tanaka *et al.* [37] found that phase separation resulted in an increase in permeability of more than an order of magnitude. Weiss *et al.* [38] found that the permeability of PA gels increased by orders of magnitude when the cross-link density was increased at a fixed concentration of monomer. This was attributed to the heterogeneous distribution of crosslink, resulting in segregation into regions of dense polymer and open pores. By analogy, we would expect base-catalyzed silica gels to be more permeable than acid-catalyzed gels with the same concentration of silica, because the "granular" nature of the former systems implies the existence of larger pores. This idea is supported by data presented in Section 2.2.

2.2.   Flow in Small Pores

The pores of inorganic gels are so small that solvent structure of the sort illustrated in Fig. 2 can extend throughout the volume of the liquid, so the permeability of the gel can be controlled by liquid whose viscosity is different from that of the bulk liquid. There is abundant evidence that the mobility of liquid molecules is lowered in the vicinity of a solid surface. For example, nuclear magnetic resonance reveals that the reorientation time of water molecules is longer than for bulk liquid [11]; this phenomenon is used to estimate the surface area of porous materials by determining the relative amounts of rapidly and slowly relaxing solvent molecules [39]. Warnock *et al.* [40,41] examined the mobility of nitrobenzene (a planar, polar molecule) and carbon disulfide (linear, nonpolar) within the pores of two types of silica gel, one derived from alkoxides and the other by Shoup's method [42] from potassium silicate. Using an optical spectroscopic technique, they found that nitrobenzene in a gel with 4.4-nm pores exhibited two reorientation times: one characteristic of the bulk liquid and one three times slower that was attributed to the molecules adjacent to the solid surface. If the gel was esterified by boiling in alcohol, the slow reorientation time was not observed; carbon disulfide showed only bulk behavior. These results imply that the viscosity of the polar molecule is increased by a factor of ~3 in the vicinity of the hydroxylated surface of the gel, but the mobility is unaffected if the surface or the molecule is nonpolar.

Chan and Horn [43] determined the viscosity of thin films by measuring the force required to push together two sheets of mica with a layer of intervening liquid. For large molecules (n-$C_{14}H_{30}$, n-$C_{16}H_{34}$, and octa-methylcyclotetrasiloxane) they found bulk behavior for separations

>50 nm, and increasing viscosity on closer approach. Their results were interpreted to indicate the presence of a layer of immobile liquid two molecules thick adjoining the solid surface. Contradictory results were obtained by Israelachvili [44], who examined the viscosity of liquid between two oscillating mica sheets at very small separations. In dilute aqueous NaCl (0.002 M) there was no deviation from the bulk viscosity when the surfaces were ≥5 nm apart; in more concentrated solutions (0.15 M) no deviation was found at separations ≥ 2 nm, in spite of the presence of a much thicker layer of solvation. The difference between these results and those of Chan and Horn may result from the higher shear rate used by the latter workers (~1800 s^{-1} versus 300 s^{-1} for Israelachvili). Israelachvili concludes from his study that the adsorbed layers are not more viscous than bulk liquid. However, it was subsequently recognized [45] that the technique used in both of those studies gives misleading results. Since the mica surfaces are curved, the measurement examines liquid in a gap of varying width; in fact, the measured viscosity is dominated by liquid in regions where separation of the surfaces is large (compared to the point of closest approach). Thus there are no reliable direct measurements of viscosity very close to solid surfaces.

There are very few direct measurements of flow through inorganic gels. Debye and Cleland [46] studied the permeability of a related material, VYCOR® porous glass, which has 5–10-nm pores and porosity of ~35%. In accordance with Darcy's law, Eq. 13, the flux was found to be proportional to the pressure drop through the sample. The permeability was in the range $D \approx 0.02$–0.04 nm^2 (which means that a pressure drop of one atmosphere through a plate 1 mm thick would allow water to flow at a rate of 10^{-7} cm^3/cm^2·s!); the Carman–Kozeny equation predicts $D \approx 0.09$ nm^2. When the temperature was changed, the flux did not change at the same rate as the viscosity of the bulk liquid. This indicates that the temperature-dependence of the liquid in the pores is different from that of the bulk liquid, and/or that the thickness of the adsorbed (structured) layer changes with temperature. (Note than an underestimate of η_L would produce a corresponding underestimate of D; if the pore liquid is more viscous than the bulk liquid, then D is smaller than previously indicated, and the Carman–Kozeny equation is in error by more than the indicated factor of ~3.)

Scherer and Swiatek [47] measured the permeability of silica gels made from tetraethoxysilane (TEOS) using acid catalysis and $r =$ (moles water)/(moles TEOS) $= 16$ (gel A), or two-step acid–base catalysis with $r = 3.7$ (gel B2). The sol was cast onto a Teflon® filter over the porous bottom of a stainless steel housing, and allowed to gel overnight. The housing and an attached microcapillary tube were then filled with liquid having the same composition as the pore liquid, and pressure was applied to the liquid. As shown in Fig. 8, the flux was found to be roughly proportional to the

Fig. 8.

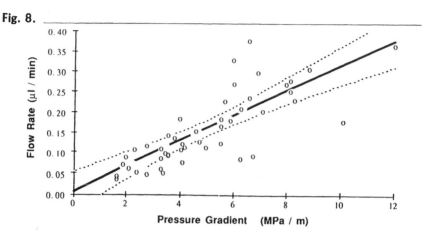

Flow rate = flux (J) × area (A) versus pressure gradient (P/L) for plate of gel A with thickness $L \approx 2$ mm. The slope of the dark line indicates that the permeability is $D \approx 1$ nm^2. From Scherer and Swiatek [47].

applied pressure, in keeping with Darcy's law. The scatter in the data (each representing a separate sample) reflects a number of experimental difficulties, including changes in the volume of the gel caused by the applied pressure and syneresis, and changes in liquid volume from temperature drift. Given the permeability of ~1 nm^2 for gel A and the relative density of $\rho \approx 0.1$, the surface area found from Eq. 16 is ~1400 m^2/g, which is ~2.5 times as large as the BET area of the corresponding xerogel. The base-catalyzed gel B2 had a permeability ~10 times greater than that of gel A, as expected (qualitatively) from its coarser pore structure.

These studies indicate that flow through gels obeys Darcy's law to the extent that the flux is proportional to the pressure gradient in the liquid. Further, it appears that D/η_L is within a factor of ~3 of the value predicted by using the Carman–Kozeny equation and the bulk viscosity. However, the flow process is more complicated than in materials with larger pores, and it is probably not appropriate to assume that η_L is the same as for the bulk liquid (differences being likely in absolute value and temperature dependence). The errors are likely to increase as the gel shrinks during drying, as a larger proportion of the liquid in the pores is in structured layers.

2.3. Unsaturated Media

An *unsaturated* body contains both gas and liquid in its pores. Gels become unsaturated near the end of drying, when shrinkage stops and the meniscus recedes into the pores. The liquid is said to be in the *pendular* state when

Fig. 9.

Saturated   Funicular   Pendular

Illustration of *pendular state*, where liquid (light gray) exists in isolated pockets in regions of negative curvature between particles (dark gray); *funicular state*, where liquid is contiguous but does not fill the pore space; and *saturated state*, where pore space is filled with liquid.

it is trapped in isolated pockets, as illustrated in Fig. 9; at higher liquid contents, the pockets become contiguous and are said to be in the *funicular* state. In the pendular state, liquid is transported only by diffusion of the vapor, but funicular liquid can still be transported by flow, and the flux can be related to the gradient in capillary pressure by Darcy's law [29,48]. The permeability is a strong function of liquid content, as shown in Fig. 10, and shows considerable hysteresis as the liquid content is raised and lowered.

Fig. 10.

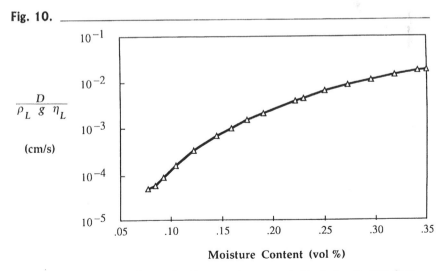

Dependence of permeability on liquid content in unsaturated body (sand). Data from Watson [48].

2.4. Diffusive Transport

Diffusion is a random-walk process by which material is transported down a gradient in concentration. According to *Fick's law*, the diffusive flux, J_D (moles/cm$^2 \cdot$s), is proportional to the gradient, ∇C [49]:

$$J_D = -D_c \, \nabla C \tag{17}$$

where D_c is the chemical diffusion coefficient (cm^2/s) and C is the concentration (moles/cm^3). There are several ways of defining the diffusion coefficient. If the interdiffusion of species *1* and *2* results in a change in volume, then the net velocity of a molecule contains contributions from both diffusion and flow. The *intrinsic diffusion coefficients*, D_1 and D_2, are defined with respect to a (moving) plane across which there is no net transfer of volume. If interdiffusion causes no change in volume, or if the diffusion coefficients are defined with respect to a plane across which there is no net transfer of mass, then the flux can be defined by a single diffusion coefficient called the *mutual diffusion coefficient*, defined by [ref. 49, p. 210]

$$D_c = V_{m2} C_2 D_1 + V_{m1} C_1 D_2 \tag{18}$$

where V_{mi} and C_i are the partial molar volume and concentration of component *i*, respectively ($i = 1$ or *2*), and $V_{mi} C_i$ is the corresponding volume fraction.

It was shown in Section 1 that diffusion contributes to the shrinkage of gels in special cases (e.g., when the gel is immersed in a salt solution), and may be important during evaporative drying, if a concentration gradient develops in the pores by preferential evaporation of one component of the pore liquid. In organic gels, a gradient in polymer concentration can produce osmotic flow. For example, Tanaka and Fillmore [50] found that spheres of polyacrylamide gel swelled by 10 to 14% linearly when transferred from paraffin oil into water; the infinite quantity of water in the bath then floods into the water-starved (concentrated) polymer. An analogous process in inorganic systems is the swelling of clay in water [8].

Rather than assuming that fluid transport occurred by diffusion, Tanaka and Fillmore [50] analyzed the rate of swelling by assuming that the network was effectively stretched by the osmotic pressure, Π, which created suction in the pore liquid and induced fluid flow (according to Darcy's law) into the gel. The kinetics of swelling were shown to depend on the elastic modulus and permeability of the gel network. Given that a pressure gradient exists,[†]

[†] Actually, this is a controversial idea. Tanaka and Fillmore [50] imagine that the osmotic pressure "inflates" the network as if there were a gas pressure in the pores, but it is generally argued [3] that there is no actual pressure. However, it is hard to argue with the fact that the swelling kinetics were in agreement with independently measured values of the elastic modulus and permeability of the gel.

it is interesting to see under what circumstances bulk flow will dominate over diffusive transport. The chemical potential, μ, can be written in terms of the concentration as

$$\Delta\mu = R_g T \ln(\gamma C) \tag{19}$$

where γ is the activity coefficient, so the chemical potential gradient is

$$\nabla\mu = \frac{R_g T}{\gamma C} \nabla(\gamma C) \approx \frac{R_g T}{C} \nabla(C) \tag{20}$$

(the approximation is exact if Henry's law [3] applies). Fick's law can be written as

$$J_D = -\frac{D_c C}{R_g T} \nabla\mu = \frac{D_c C V_m}{R_g T} \nabla\Pi \tag{21}$$

where the second equality follows from Eq. 1. Note that $CV_m = 1 - \rho$ is the volume fraction of liquid in the gel. We want to know when the volume flux from Darcy's law, J, exceeds that from diffusion, $J_D V_m$. From Eqs. 13 and 21 we find that fluid flow dominates when

$$\frac{D}{\eta_L} > \frac{D_c(1 - \rho)V_m}{R_g T}. \tag{22}$$

The chemical diffusion coefficient can be related to the viscosity of the liquid by the Stokes–Einstein equation [51]

$$D_c = \frac{R_g T}{6\pi N_A r_m \eta_L} \tag{23}$$

where N_A is Avogadro's number and r_m is the radius of the diffusing molecule. With Eqs. 15 and 23, 22 becomes

$$r_h > 2r_m. \tag{24}$$

This means that flow dominates whenever the pore size is larger than a few times the molecular dimension. Of course, Darcy's law and the Carman–Kozeny equation cannot be expected to apply to such small pores, so the result must be taken as an order-of-magnitude estimate. Nevertheless, this is a physically appealing conclusion: diffusion dominates in dense objects (as in the initial swelling of a polymer by a solvent), but flow takes over whenever channels are present that are much larger than the diffusing molecules (as in osmotic swelling of gels).

Note that flow can relieve a gradient in polymer concentration by swelling the network, but cannot reduce a concentration gradient in the liquid phase. For example, if a gel is immersed in a salt solution, flow of the solution into the pores does not affect the difference in salt concentration between the bath and the original pore liquid; that can be achieved only by diffusion. Similarly, if evaporation creates a concentration gradient in the pore liquid, flow from the interior cannot eliminate it, only interdiffusion within the pores can do so. Therefore, the preceding analysis applies only to situations such as the swelling experiment by Tanaka *et al.*, and to flow within a clay body (where tension in the liquid is produced by disjoining forces), where there is a gradient in concentration of solid phase.

Fluid flow is the principal transport mechanism in both saturated and unsaturated media (as long as the liquid phase is continuous), but diffusion of vapor becomes increasingly important as the liquid content of the pores decreases. After the pendular state is reached, liquid can be removed from the gel only by diffusion of its vapor. Isolated pockets of liquid actually participate in the transport process, as shown by Philip and de Vries [52], because the gradient in vapor concentration causes liquid to condense on one side of the pocket and evaporate from the other; as a result, the flux through the liquid is the same as that through the vapor in the open pores. Whitaker [53] has analyzed the coupled flows of mass and heat during drying of unsaturated bodies, using the same principle of local averaging that is implicit in Darcy's law. That is, the transport parameters are assumed to be uniform on a scale that is large compared to the pores, but small compared to the drying body. Applications of his model are discussed in the following chapter.

3.

RHEOLOGY OF THE POROUS NETWORK

In this section we discuss a theoretical model for the properties of a drained gel network (i.e., the solid phase with the liquid removed). It is convenient to imagine that the properties of the liquid and solid are independent, but it should be kept in mind that this is an approximation, because (as Johnson [54] points out) disjoining forces may contribute to the rigidity of the gel. The *network* may be elastic, viscoelastic, or plastic,[†] but we assume that the liquid and solid phases are incompressible; that is, stresses can deform the network, but cannot change the volume of the constituents. (For example, there might be 10 cm^3 of polymeric material in 100 cm^3 of dry sponge; when you crush it in your hand, the volume of the sponge might decrease to 20 cm^3, but there would still be very nearly 10 cm^3 of polymer. The compressibility

---

[†] These terms were defined in the preceding chapter.

of the network is orders of magnitude greater than that of the solid phase.) In the preceding chapter we saw that alkoxide-derived silicate gels are viscoelastic, and that is the behavior that we shall concentrate upon in the following discussion. It should be noted, however, that other types of behavior are seen in related materials. For example, particulate gels may be plastic [55] and may even redisperse when agitated [56]; other plastic materials include clay [8], soil [27], and wood [57].

In a classic series of papers, Biot developed a theory for the deformation of saturated and unsaturated porous media, assuming the solid skeleton to be homogeneous and elastic [58], or homogeneous but anisotropic and viscoelastic [59,60], or heterogeneous, anisotropic, and viscoelastic [61]. The model was tailored to describe propagation of acoustic waves through soil [62], but could also be applied to consolidation and drying of soil. It allowed for compressibility of the solid and liquid phases, and for dissipation of energy within both phases, as well as by their relative movement. A number of other workers (e.g., [63–65]) have developed related models, but Johnson and Chandler [54,66] have shown that the correct ones are special cases of, or approximations to, Biot's. The model[†] discussed in this chapter [67,68] is also closely related, but is particularly adapted to the case of drying of gels in that it allows for syneresis and for simple viscoelastic behavior. In the remainder of this chapter we follow the development in the latter papers.

3.1. Constitutive Equation for the Drained Network

3.1.1. ELASTIC NETWORK

Consider the cube of porous material shown in Fig. 11, which has a relative density of ρ and a volume fraction of porosity of $1 - \rho$. The area of each face is $A = A_S + A_L$ where $A_S = \rho A$ is the fraction of that area occupied by the solid phase and $A_L = (1 - \rho)A$ is the fraction occupied by liquid.[††] We are interested in the *constitutive equation* for this material, which is to say, the relationship between the stresses applied and the strains that result. For the moment we assume that the liquid has been drained away, and consider the behavior of the solid phase alone. If the solid phase is elastic,

---

[†] Several mistakes in the original treatment [67] are discussed and corrected in ref. 68.

[††] When isotropic inclusions (in this case, the solid network) are randomly distributed in a matrix (the liquid), the area fraction occupied by inclusions on any arbitrary plane through the body is equal to the volume fraction of inclusions [69].

Fig. 11.

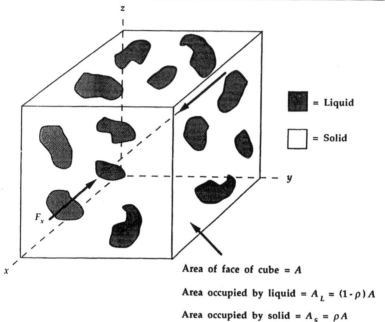

= Liquid

= Solid

Area of face of cube = A

Area occupied by liquid = A_L = $(1 - \rho) A$

Area occupied by solid = A_s = ρA

Illustration of porous network. Shaded areas represent intersection of liquid-filled pores with surface of cube of gel; white areas represent solid phase. Volume fraction of solid phase is ρ, so the average area fraction occupied by liquid on surface of cube is $1 - \rho$. Force on x-face = F_x; force on liquid on x-face = F_{xL}; force on solid on x-face = F_{xS}; total stress on x-face = F_x/A; stress on liquid on x-face = F_{xL}/A; pressure in liquid on x-face = $-F_{xL}/A$; stress on solid on x-face = F_{xS}/A.

the strains in the x, y, and z directions are given by [70]

$$\varepsilon_x = \frac{1}{E_P} [\tilde{\sigma}_x - v_P(\tilde{\sigma}_y + \tilde{\sigma}_z)] \tag{25a}$$

$$\varepsilon_y = \frac{1}{E_P} [\tilde{\sigma}_y - v_P(\tilde{\sigma}_x + \tilde{\sigma}_z)] \tag{25b}$$

$$\varepsilon_z = \frac{1}{E_P} [\tilde{\sigma}_z - v_P(\tilde{\sigma}_x + \tilde{\sigma}_y)]. \tag{25c}$$

For example, ε_x is the fractional change in length of the cube when the network is subjected to stress. Suppose that force F_x is applied perpendicular to the x-face of the cube; then the stress $\tilde{\sigma}_x$ is the force on the solid phase divided by the area of that face, $\tilde{\sigma}_x = F_x/A$; the stress is negative if the force compresses the cube. If that is the only stress applied, the strain in the x-direction is $\varepsilon_x = \tilde{\sigma}_x / E_P$, and the strains in the perpendicular directions are $\varepsilon_y = \varepsilon_z = -v_P \varepsilon_x$, where E_P is *Young's modulus* and v_P is *Poisson's ratio* for the porous network. The *volumetric strain* (the fractional change in volume), ε, is the sum of the linear strains:

$$\varepsilon = \varepsilon_x + \varepsilon_y + \varepsilon_z. \tag{26}$$

In the case of a uniaxial stress (e.g., only $\tilde{\sigma}_x$ applied), the volumetric strain is

$$\varepsilon = (1 - 2v_P)\varepsilon_x \tag{27}$$

so there is no change in volume if $v_P = 1/2$, in which case the material is said to be *incompressible*. We assume that the solid phase (not the network) is incompressible. Adding the three equations 25, we find

$$\varepsilon = \tilde{\sigma}/3K_P \tag{28}$$

where

$$\tilde{\sigma} = \tilde{\sigma}_x + \tilde{\sigma}_y + \tilde{\sigma}_z \tag{29}$$

is the *hydrostatic stress* and K_P is the *bulk modulus* of the porous network, defined by

$$K_P = \frac{E_P}{3(1 - 2v_P)}. \tag{30}$$

The bulk modulus relates the volume change to the applied stress; note that no volume change is possible when $v_P = 1/2$, because the bulk modulus becomes infinitely large. *Shear strains* produce a change in shape of the body, but no change in volume. Shear strains and stresses are related by the *shear modulus*, G_P, defined by

$$G_P = \frac{E_P}{2(1 + v_P)}. \tag{31}$$

For an isotropic elastic material, only two of the four elastic constants, E_P, v_P, G_P, and K_P are needed to define the response of the material to any state of stress, since they are related by Eqs. 30 and 31.

3.1.2. VISCOUS NETWORK

A Newtonian viscous material exhibits a constant *rate* of strain when subjected to a constant load. The constitutive equations for a porous material of this type can be written as [7]

$$\dot{\varepsilon}_x = \frac{1}{F}[\tilde{\sigma}_x - N(\tilde{\sigma}_y + \tilde{\sigma}_z)] \tag{32a}$$

$$\dot{\varepsilon}_y = \frac{1}{F}[\tilde{\sigma}_y - N(\tilde{\sigma}_x + \tilde{\sigma}_z)] \tag{32b}$$

$$\dot{\varepsilon}_z = \frac{1}{F}[\tilde{\sigma}_z - N(\tilde{\sigma}_x + \tilde{\sigma}_y)] \tag{32c}$$

where F is the *uniaxial viscosity* and N is Poisson's ratio for the viscous porous network. The difference between N and v_P is discussed in the next subsection. The superscript dot on the strains indicates the partial derivative with respect to time (i.e., $\dot{\varepsilon}_x = \partial\varepsilon_x/\partial t$). A nonporous viscous material is incompressible, but the network is not, because the pores can be compressed; as the porosity decreases, N approaches 1/2. The strain rate of an *incompressible* viscous material subjected to a uniaxial stress is

$$\dot{\varepsilon}_x = \tilde{\sigma}_x/3\eta \tag{33}$$

where η is the shear viscosity of the material (e.g., the solid phase of the gel); thus, as the porosity of the network approaches zero, F must approach 3η. The volumetric strain rate

$$\dot{\varepsilon} = \dot{\varepsilon}_x + \dot{\varepsilon}_y + \dot{\varepsilon}_z \tag{34}$$

is related to the hydrostatic stress by

$$\dot{\varepsilon} = \tilde{\sigma}/3K_G \tag{35}$$

where K_G is the *bulk viscosity* of the porous gel network, defined by

$$K_G = \frac{F}{3(1 - 2N)}. \tag{36}$$

This is the viscosity that controls the rate of change of the volume of the porous network (with no liquid in the pores) when it is subject to a hydrostatic stress. As the porosity decreases, N approaches 1/2 and the network becomes incompressible. Shear stresses and strains are related by the *shear viscosity*, G_G,

$$G_G = \frac{F}{2(1 + N)}. \tag{37}$$

When the porosity vanishes, since $F \to 3\eta$ and $N \to 1/2$, the shear viscosity of the gel network approaches that of the skeleton, $G_G \to \eta$.

3.1.3. VISCOELASTIC NETWORK

A viscoelastic material behaves like an elastic material immediately after application of a load, and then gradually approaches the behavior of a viscous material [72]. Therefore, it is not surprising that (in the simplest possible case) the constitutive equation consists of two parts resembling the purely elastic and purely viscous extremes [73]:

$$\dot{\varepsilon}_x = \frac{1}{F}[\tilde{\sigma}_x - N(\tilde{\sigma}_y + \tilde{\sigma}_z)] + \frac{1}{E_P}\frac{\partial}{\partial t}[\tilde{\sigma}_x - \nu_P(\tilde{\sigma}_y + \tilde{\sigma}_z)]. \tag{38}$$

The transition from elastic to viscous behavior occurs after a period of time roughly equal to the *relaxation time*,

$$\tau_R = F/E_P. \tag{39}$$

That is, if a constant load $\tilde{\sigma}_x$ is imposed at time $t = 0$, the elastic strain $\varepsilon_x^E = \tilde{\sigma}_x/E_P$ appears immediately, and the viscous strain $\varepsilon_x^V = \tilde{\sigma}_x t/F$ becomes larger than ε_x^E when $t > \tau_R$. At the same time, the strain in the y direction instantaneously jumps to $\varepsilon_y^E = -\nu_P\varepsilon_x^E$, then gradually approaches $\varepsilon_y^V = -N\varepsilon_x^V$. Thus, Poisson's ratio ranges from ν_P at $t = 0$ to N as $t \to \infty$. The behavior of actual silica gels is more complicated than is indicated by Eq. 38, as experiments suggest that a broad distribution of relaxation times is needed to describe the relaxation behavior [74]. However, the essential features of viscoelastic behavior are contained in the simpler equation; further complication of the analysis is not justified, given the present lack of data for the constitutive parameters (viz., viscosity and Poisson's ratio).

3.2. Constitutive Equation for Gel

Two more features are needed to complete the constitutive equation for a gel: allowance for the syneresis strain rate (which occurs in the absence of applied stress) and for the presence of the liquid phase. When a force F_x is applied to the face of the sample cube (see Fig. 11), a portion F_{xS} of that force is borne by the solid phase and a portion F_{xL} is borne by the liquid (and $F_x = F_{xS} + F_{xL}$). The pressure in the liquid phase, P_L, is equal to the force on the liquid divided by the area of the liquid, $P_L = -F_{xL}/A_L$. The negative sign is present because the pressure in the liquid is defined as positive when compressive (so a *negative pressure* means a tensile stress in the liquid), whereas a stress is defined as being negative when it is compressive. It is convenient to define a *stress* in the liquid, $P = F_{xL}/A_L$, which obeys the same sign convention as a stress in the solid phase. The *total stress* on the face of the cube is defined as $\sigma_x = (F_{xS} + F_{xL})/A$. Thus,

$$\sigma_k = \tilde{\sigma}_k + (1 - p)P = \tilde{\sigma}_k - (1 - p)P_L \qquad (k = x, y, z). \qquad (40)$$

Suppose we use impermeable pistons (so the liquid cannot escape) to apply stresses σ_x, σ_y, σ_z (or forces F_x, F_y, F_z) to the faces of the sample cube. To find the resulting strains, we must adapt Eq. 32 to allow for the presence of the liquid. Since the liquid has no rigidity, it can exert only hydrostatic stress; that is, it exerts the same force on each face of the cube, so it must affect each component of strain equally. Therefore we can write

$$\dot{\varepsilon}_x = \frac{1}{F} [\tilde{\sigma}_x - N(\tilde{\sigma}_y + \tilde{\sigma}_z)] + cP \qquad (41a)$$

$$\dot{\varepsilon}_y = \frac{1}{F} [\tilde{\sigma}_y - N(\tilde{\sigma}_x + \tilde{\sigma}_z)] + cP \qquad (41b)$$

$$\dot{\varepsilon}_z = \frac{1}{F} [\tilde{\sigma}_z - N(\tilde{\sigma}_x + \tilde{\sigma}_y)] + cP \qquad (41c)$$

where c is a constant to be determined. Since the pistons are impermeable and the components of the gel are incompressible, the volumetric strain rate must be zero, so Eqs. 34 and 41 lead to

$$\dot{\varepsilon} = \tilde{\sigma}/3K_G + 3cP = 0. \qquad (42)$$

From Eqs. 29 and 40,

$$\tilde{\sigma} = \sigma - 3(1 - p)P = -9K_G cP \qquad (43)$$

where the second equality follows from Eq. 42 and the *total hydrostatic stress* is

$$\sigma = \sigma_x + \sigma_y + \sigma_z. \qquad (44)$$

The average force on the liquid is $F_L = (1 - p)(F_x + F_y + F_z)/3$, and the stress in the liquid is $P = F_L/(1 - p)A$, so $\sigma = (F_x + F_y + F_z)/A = 3P$, and Eq. 43 leads to

$$cP = -pP/3K_G \qquad (45)$$

so Eq. 41 becomes

$$\dot{\varepsilon}_x = \frac{1}{F} [\sigma_x - N(\sigma_y + \sigma_z)] - \frac{P}{3K_G} \qquad (46a)$$

$$\dot{\varepsilon}_y = \frac{1}{F} [\sigma_y - N(\sigma_x + \sigma_z)] - \frac{P}{3K_G} \qquad (46b)$$

$$\dot{\varepsilon}_z = \frac{1}{F} [\sigma_z - N(\sigma_x + \sigma_y)] - \frac{P}{3K_G}. \qquad (46c)$$

From Eq. 43 we see that $\tilde{\sigma} = 3\rho P = \rho\sigma$, so the total force on the solid phase is $\rho(F_x + F_y + F_z)$; that is, the load is borne by the solid phase in proportion to the volume fraction that it occupies.

Finally, we identify the linear *syneresis strain rate* as $\dot{\varepsilon}_s$ and modify the constitutive equation by adding that quantity to the right side of each line in Eq. 46. That is, we assume that the stress-induced strain rate and the inherent syneresis strain rate are linearly additive. Thus the constitutive equations for an isotropic viscous gel are

$$\dot{\varepsilon}_x = \dot{\varepsilon}_f + \frac{1}{F}[\sigma_x - N(\sigma_y + \sigma_z)] \tag{47a}$$

$$\dot{\varepsilon}_y = \dot{\varepsilon}_f + \frac{1}{F}[\sigma_y - N(\sigma_x + \sigma_z)] \tag{47b}$$

$$\dot{\varepsilon}_z = \dot{\varepsilon}_f + \frac{1}{F}[\sigma_z - N(\sigma_x + \sigma_y)] \tag{47c}$$

where $\dot{\varepsilon}_f$ is called the *free strain rate* and is defined by

$$\dot{\varepsilon}_f = \dot{\varepsilon}_s - P/3K_G = \dot{\varepsilon}_s + P_L/3K_G \tag{48}$$

and the volumetric strain is

$$\dot{\varepsilon} = \dot{\varepsilon}_x + \dot{\varepsilon}_y + \dot{\varepsilon}_z = 3\dot{\varepsilon}_f + \sigma/3K_G. \tag{49}$$

To understand the meaning of the free strain rate, consider a small cube of gel in isolation. If there is stress P in the liquid, then the force on the liquid on the x-face of the cube is $F_{xL} = PA_L$, and Newton's first law requires that it be balanced by an equal and opposite force on the solid phase, $F_{xS} = \tilde{\sigma}_x A = -F_{xL}$. If there is tension in the liquid ($P > 0$), then there is a corresponding compression on the network $[\tilde{\sigma}_x = -(1 - \rho)P]$, and Eqs. 40 and 44 indicate that $\sigma = 0$. In that case, according to Eq. 49, the volumetric contraction rate of the gel is given by the free strain rate. That is, $\dot{\varepsilon}_f$ is the linear contraction rate of the network when the forces are locally in balance. In general, however, the cube of gel is not free, but is incorporated in a larger body in which the stress in the liquid varies from place to place; each cube of gel tries to shrink at its own free strain rate (corresponding to the pressure in its pores), but they must all accommodate one another and shrink at some average rate (corresponding to force balance across the whole body). Therefore, each cube is subjected to an additional stress imposed by neighboring regions that have different free strain rates, and the local strain rate differs from $\dot{\varepsilon}_f$ because of this constraint.

In the form of Eq. 47 the constitutive equations take account of the pressure in the liquid, as well as the stress on the solid phase. Moreover, they are in the familiar form used in analysis of thermal stresses, where the free strain is equal to $\varepsilon_f = X\,\Delta T$, X is the linear thermal expansion coefficient, and ΔT is the change in temperature. The analogy between drying stresses and thermal stresses, which has been noted by several authors [57,63,75,76], is as follows. Suppose we blow cold air on the surface of a hot plate, so that the surface cools by ΔT. The surface tries to contract by $\varepsilon_f = X\,\Delta T$ as it cools, but its shrinkage is inhibited by the hot interior region to which it is attached. The difference between the free strain (i.e., the amount the surface would contract if it were not connected to the rest of the body) and the actual strain gives rise to a tensile stress in the surface. Similarly, when a body dries by evaporation from the surface, capillary pressure (negative pressure, tensile stress) develops in the liquid, and the suction makes the solid network contract. As in the case of the cooling plate, the surface is not free to shrink (in response to the tension in the liquid), so stresses develop in the network and cracking may result.

As a simple example of the analysis of stresses in a gel, consider a plate with thickness $2L$ with its faces at $z = L$ and $z = -L$ parallel to the x–y plane. There is no external force exerted on the faces of the plate (i.e., $\tilde{\sigma}_z = -(1-\rho)P$), so $\sigma_z = 0$. However, the plate has begun to dry and a capillary pressure gradient exists in the pores; the pressure (and consequently the free strain rate) varies with z, so stresses develop in the plane of the plate. Symmetry requires that $\sigma_x = \sigma_y$, so Eq. 47 becomes

$$\dot{\varepsilon}_x = \dot{\varepsilon}_f + \sigma_x(1-N)/F. \tag{50}$$

Suppose that lines were scratched on the edge of the plate along the z-direction at the start of drying; if the plate dries from both sides and shrinks uniformly, we expect those lines to remain straight and parallel as the plate shrinks ([72], p. 95). That means that the strain in the x-direction does not vary with z. Averaging Eq. 50 over the thickness of the plate (or over half of it, since it is symmetrical), we obtain

$$\int_0^L \dot{\varepsilon}_x\,dz = L\dot{\varepsilon}_x = \int_0^L \dot{\varepsilon}_f\,dz + \frac{(1-N)}{F}\int_0^L \sigma_x\,dz. \tag{51}$$

The integral over σ_x represents the average force over the edge of the plate, which must be zero (unless some external force, such as a clamp, is applied). Therefore, the strain rate in the plane of the plate is

$$\dot{\varepsilon}_x = \frac{1}{L}\int_0^L \dot{\varepsilon}_f\,dz = \dot{\varepsilon}_s - \langle P \rangle / 3K_G \tag{52}$$

where the second equality follows from Eq. 48, and the average stress in the liquid is defined by

$$\langle P \rangle = \frac{1}{L} \int_0^L P(z, t) \, dz. \tag{53}$$

Rearranging Eq. 50 and using Eq. 52 we find that the stress in the plane of the plate is related to the stress in the liquid by

$$\sigma_x = \left(\frac{F}{1 - N} \right)(\dot{\varepsilon}_x - \dot{\varepsilon}_f) = C_N(P - \langle P \rangle) \tag{54}$$

where $C_N = (1 - 2N)/(1 - N)$. Thus, the stress results from the difference between the average pressure and the local pressure in the pores (or, equivalently, the difference between the average and local free strain rates). When the pressure is uniform, the network is hydrostatically compressed (assuming $P > 0$); that is $\tilde{\sigma}_x = -(1 - \rho)P$ throughout the thickness of the plate, so the plate contracts uniformly. Stress and fracture are likely only if different regions of the plate try to shrink at different rates (i.e., have different free strain rates, because of variation in P). Then regions with higher free strain rates (more negative values of $\dot{\varepsilon}_f$) are inhibited from shrinking at their natural rate; they are held in tension ($\sigma_x > 0$) by the slower-shrinking surrounding material, and are likely to crack. In Section 4.5 we shall see how to find the pressure distribution that develops during drying, which will allow the stress to be calculated from Eq. 54.

4.

THEORY OF DEFORMATION

4.1. Continuity

The volume of the cube in Fig. 11 is $V = V_S + V_L$, where $V_S = \rho V$ is the volume occupied by the solid network and $V_L = (1 - \rho)V$ is the volume of liquid.[†] In a saturated body (i.e., one containing no gas bubbles), the liquid volume must be equal to the volume of pore space, V_P, in the drained body. If liquid is flowing only in the x-direction, the rate of change of the liquid content of the cube depends on the difference between the flux into the face at x, $J(x, t)$, and the flux out of the face at $x + \Delta x$, $J(x + \Delta x, t)$:

$$V_L(t + \Delta t) = V_L(t) + J(x, t)A \, \Delta t - J(x + \Delta x, t)A \, \Delta t \tag{55}$$

[†] As Johnson [54] points out, the density of the solid phase in the gel might be different from that of the same material with the liquid removed, as the solvent molecules might invade the solid phase; the liquid density in the gel might be different from that of the free liquid, because of structure developed at the solid–liquid interface.

since J is the flux of liquid per unit area of the body (not per unit area of liquid). The relative rate of accumulation of liquid in the cube is found by rearranging Eq. 55 and taking the limits as Δx and Δt go to zero. Recognizing that the volume of liquid is given by $V_L = (1 - \rho)A\,\Delta x$, we find

$$\frac{V_L(t + \Delta t) - V_L(t)}{V_L\,\Delta t} \rightarrow \frac{\dot{V}_L}{V_L} = -\frac{1}{1 - \rho}\frac{\partial J(x, t)}{\partial x}. \tag{56}$$

If liquid is flowing in all directions, Eq. 56 is generalized to

$$\frac{\dot{V}_L}{V_L} = -\frac{1}{1 - \rho}\nabla \cdot J = -\frac{1}{1 - \rho}\left(\frac{\partial J}{\partial x} + \frac{\partial J}{\partial y} + \frac{\partial J}{\partial z}\right). \tag{57}$$

Some authors have written J in terms of chemical diffusion [75,77] and others in terms of Darcy's law [58,64,67,78]. For now we assume that fluid flow prevails; concurrent diffusion is considered in the next subsection. From Eqs. 13 and 57 we find

$$\frac{\dot{V}_L}{V_L} = -\frac{1}{1 - \rho}\nabla \cdot \left(\frac{D}{\eta_L}\nabla P\right). \tag{58}$$

Since liquid is not created or destroyed within the cube, *continuity* (i.e., conservation of volume) of the incompressible liquid requires that the liquid content be equal to the pore space (of the drained network), $V_L = V_P$, so

$$\frac{\dot{V}_L}{V_L} = \frac{\dot{V}_P}{V_P}. \tag{59}$$

The volumetric strain rate of the network is defined as

$$\dot{\varepsilon} = \frac{\dot{V}}{V} = \frac{\dot{V}_P}{V} = \frac{\dot{V}_P}{V_P}(1 - \rho) \tag{60}$$

where the second equality reflects the incompressibility of the solid phase ($\dot{V}_S = 0$), and the third equality follows from the definitions of the relative density of the network, $\rho = V_S/V$, and the porosity, $1 - \rho = V_P/V$. From Eqs. 58–60, the *equation of continuity* is found to be

$$\dot{\varepsilon} = -\nabla \cdot J = -\nabla \cdot \left(\frac{D}{\eta_L}\nabla P\right). \tag{61}$$

Now we need to express $\dot{\varepsilon}$ in terms of the stress in the liquid using a constitutive equation. Various authors have does this by using empirical (nonlinear elastic) equations [24,78], or by assuming elastic behavior with the solid and liquid phases compressible [63,58] or incompressible [79,64], or by allowing the network to be purely viscous [67] or viscoelastic [73,59,60,61,80]; Biot [61] allows for anisotropy of the mechanical properties as well as for viscous dissipation within both the solid and liquid phases.

When the network is assumed to be elastic, Eq. 61 has the mathematical form of the diffusion equation. Philip [24] discusses at length the methods for solving the nonlinear version of that equation that results when the permeability and elastic properties vary with the porosity (and therefore with position in the body). In the following, we use the viscous constitutive equation developed in Section 3.

For an explicit example of the use of Eq. 61, we return to the case of a flat plate of gel with a purely viscous network drying from both faces. In that case (as shown in Section 3.2), $\sigma = 2\sigma_x$, so Eqs. 48, 49, and 54 lead to

$$\dot{\varepsilon} = 3\dot{\varepsilon}_s - \frac{\beta P + (1 - \beta)\langle P \rangle}{K_G} \tag{62}$$

where the constant $\beta \ [=1 - 2C_N/3]$ is defined by

$$\beta = \frac{1 + N}{3(1 - N)}. \tag{63}$$

Since N is expected to be small when the volume fraction of solid is small, $\beta \approx 1/3$. Assuming that D and η_L are not functions of z, Eqs. 61 and 62 lead to the following differential equation describing the pressure distribution in the plate [68]:

$$\frac{D}{\eta_L} \frac{d^2 P}{dz^2} + \frac{\beta P + (1 - \beta)\langle P \rangle}{K_G} = -3\dot{\varepsilon}_s. \tag{64}$$

This is solved in Sections 4.3 and 4.4 using boundary conditions appropriate for syneresis and evaporation, respectively.

4.2. Flow and Diffusion

In many cases, the liquid in the pores of a gel is a solution of liquids that differ in volatility, so that a change in composition occurs upon evaporation; such a solution is called a *zeotrope*. An extreme example is a solution of acetone and ethanol: the liquid that evaporates is almost pure acetone. A gel containing a zeotropic solution will develop a concentration gradient in its pores as one of the liquids evaporates preferentially, and that gradient will result in chemical diffusion within the pores. Some solutions form an *azeotrope*, which means that the composition of the vapor is the same as that of the liquid. For example, a solution of 4.5 wt% water in ethanol will evaporate with no change of composition, and a gel containing that liquid will not develop a diffusive flux. The compositions of azeotropes are sometimes surprising, in that the species that evaporates preferentially is the one with the *lower* vapor pressure in the pure state. For example, the azeotrope in the butanol–water systems contains 62 wt% butanol, so a solution containing less than that amount of alcohol can be converted to pure water by evaporation, even though the vapor pressure of the pure alcohol (0.9 kPa at 25°C) is much lower than that of pure water (3.1 kPa). The azeotropic composition might be affected by the presence of the gel network (through preferential adsorption of one component of the liquid, for example), but this phenomenon does not seem to have been studied. Diffusion is also important if the gel is immersed in a liquid with a composition different from that in the pores. In such cases, the continuity equation, Eq. 61, must have an additional term allowing for the volumetric flux produced by diffusion [81].

If the pores of the gel contain a solution of liquids 1 and 2, the diffusion flux of each component (moles/cm^2·s) is given by

$$J_{D_i} = -D_i \nabla C_i \qquad (i = 1, 2) \tag{65}$$

where D_i is the intrinsic diffusion coefficient of liquid i (discussed in Section 2.4). The volume flux per unit area of liquid is $V_{mi}J_{D_i}$, so the flux per unit area of gel is $(1 - \rho) V_{mi}J_{D_i}$. By an analysis equivalent to that leading to Eq. 57, it can be shown [81] that there is a contribution to the liquid content of the sample cube from diffusion given by

$$\frac{\dot{V}_L}{V_L} = -\nabla \cdot (V_{m1} J_{D1} + V_{m2} J_{D2}) = \nabla \cdot [V_{m1}(D_1 - D_2) \nabla C_1] \tag{66}$$

where the second equality follows from the identity

$$V_{m1} C_1 + V_{m2} C_2 = 1. \tag{67}$$

The diffusion term must be added to the flow term, so Eq. 61 becomes

$$\dot{\varepsilon} = -\nabla \cdot \left(\frac{D}{\eta_L} \nabla P\right) + (1 - \rho)\nabla \cdot [V_{m1}(D_1 - D_2) \nabla C_1]. \tag{68}$$

Note that diffusion has no influence if the intrinsic diffusion coefficients of the two liquids are equal, because the diffusive volume fluxes are then equal and opposite (i.e., diffusion produces no volume flow). In the following chapter it is shown that drying stresses can be reduced considerably when the diffusion term is significant.

4.3. Finite Deformation

During drying of a gel there is a huge change in volume (often one or two orders of magnitude), which is not allowed for in the analysis given in Section 4.1. There are two ways of handling the contraction of the network: the

Eulerian [82] and the *Lagrangian* methods [83–85]. The Eulerian approach uses a coordinate system fixed in the frame of reference of the laboratory, and it takes account of the velocity of the body relative to that frame as the volume of the body changes. The Lagrangian method uses a coordinate system fixed in the gel, such that a fixed volume of solid phase is contained in any volume element. The solution is obtained in terms of the *material coordinate*, *m*, where

$$dm = \rho \, dx \tag{69}$$

when the shrinkage is uniaxial (in the *x*-direction). This approach is simpler than the Eulerian, but still leads to a nonlinear differential equation that must be solved by numerical methods, except in special cases.

Note that a Lagrangian coordinate system was used in Section 4.1, where it was assumed that $V_S = 0$ in the cube (Fig. 11). If the shrinkage varies strongly with position in the gel, then $\rho = \rho(x)$ and Eq. 69 indicates that the Lagrangian coordinate is not linearly proportional to a coordinate fixed in the frame of the laboratory. However, if the volumetric contraction is nearly uniform throughout the gel, solutions obtained from Eq. 61 are valid regardless of the amount of shrinkage that has occurred. For example, if ρ is approximately uniform, Eq. 54 gives the stress in a plate at any time during drying as a function of z/L, where $L(t)$ is the current thickness of the shrinking plate.

4.4. Syneresis

As explained in the preceding chapter, syneresis is a process of spontaneous contraction of the network of the gel that results from continuing condensation reactions. As the network shrinks it must move through the liquid. (One usually says that the liquid flows out of the gel during syneresis, but it is actually the gel that collapses into the stationary liquid.) Relative movement of the solid and liquid is difficult because of friction generated in the small pores of the network, so a substantial pressure gradient is required to drive the flow. This is simply a restatement of Darcy's law, Eq. 13, for a material with a low permeability. (This law pertains to *relative movement* of solid and liquid, regardless of which phase is stationary.) As the network contracts it compresses the liquid and squeezes it out of the gel. The liquid near the exterior surface escapes freely, so its pressure is zero, but the liquid deep inside the body may be strongly compressed ($P < 0$). If the intrinsic syneresis strain rate of a small slice of gel (from which the liquid flows freely) is $\dot{\varepsilon}_s$, then Eq. 48 shows that shrinkage is completely suppressed ($\dot{\varepsilon}_f = 0$) if the stress in the liquid equals P_s, where

$$P_s = 3K_G \dot{\varepsilon}_s. \tag{70}$$

This negative stress (positive pressure) pushes the solid phase, preventing its contraction. In a thin slice of gel (in which $P = 0$), Eq. 32 shows that a viscous network subjected to a hydrostatic stress equal to P_s (i.e., $\tilde{\sigma}_x = \tilde{\sigma}_y = \tilde{\sigma}_z = P_s$) will exhibit a linear strain rate of

$$\dot{\varepsilon}_x = \frac{P_s(1 - 2N)}{F} = \frac{P_s}{3K_G} = \dot{\varepsilon}_s. \tag{71}$$

Thus the syneresis strain rate can be regarded as the result of an "effective stress" on the network equal to P_s.

To find the pressure distribution in the liquid during syneresis, we need to solve Eq. 61 with the boundary condition that P equals zero at the surfaces of the body (since that liquid freely escapes as the network shrinks). Let us consider a wide plate, large enough so that relatively little liquid escapes from the edges. Then the flow of the fluid is virtually all in the direction perpendicular to the faces of the plate, and Eq. 61 reduces to Eq. 64. Assuming that the properties of the materials are independent of z, the solution is [86]

$$P(z) = P_s \left[\frac{1 - \cosh(\alpha z/L)/\cosh(\alpha)}{1 - (1 - \beta)\tanh(\alpha)/\alpha} \right] \tag{72}$$

where

$$\alpha \equiv \sqrt{\frac{\eta_L L^2}{DH_G}} \tag{73}$$

and

$$H_G = K_G + \tfrac{4}{3}G_G = K_G/\beta. \tag{74}$$

The viscosity H_G is related to the "uniaxial modulus" that appears in models using an elastic constitutive equation [54]: if a sample is placed in a cylinder (so that no lateral strain is possible) and compressed with a piston (assumed to be porous so that the liquid can escape), the strain rate produced by the applied stress, σ_z, is $\dot{\varepsilon}_z = \sigma_z/H_G$. (See ref. 64, p. 86, for an analysis of this problem for an elastic or viscoelastic material.) The pressure distribution given by Eq. 72 is illustrated in Fig. 12 for several values of the parameter α; as α increases, a greater portion of the interior of the gel experiences high pressure and correspondingly slow contraction. As indicated by Eq. 73, this situation is favored by large samples with low permeability.

The stress in the network in the plane of the plate is given by Eqs. 54 and 72. The rate of contraction of the plate is given by [86]

$$\dot{L} = \int_0^L \dot{\varepsilon}_z \, dz. \tag{75}$$

Fig. 12.

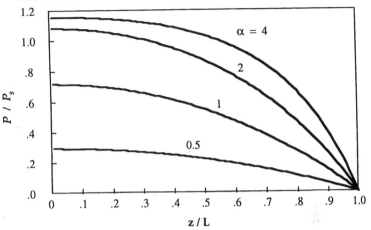

Distribution of stress, P/P_s, in liquid during syneresis in plate, based on Eq. 72; surface of plate is at $z = L$, and midplane is at $z = 0$.

In this case the constitutive equation reduces to (see Eq. 47c)

$$\dot{\varepsilon}_z = \dot{\varepsilon}_f - 2N\sigma_x/F = \dot{\varepsilon}_s - [3\beta P + (1 - 3\beta)\langle P \rangle]/3K_G \qquad (76)$$

where the second equality follows from Eq. 54. Equations 72, 75, and 76 lead to

$$\frac{\dot{L}}{L} = \dot{\varepsilon}_s - \frac{\langle P \rangle}{3K_G} = \dot{\varepsilon}_s\left[\frac{\beta\tanh(\alpha)/\alpha}{1 - (1 - \beta)\tanh(\alpha)/\alpha}\right], \qquad (77)$$

which is illustrated in Fig. 13. This indicates that the shrinkage rate of a sample of gel depends on its size, and is equal to the inherent syneresis strain rate only if the permeability is high or the sample is small. The syneresis rate for a long cylinder can be found by a similar analysis [86], and the result is shown in Fig. 13. The shrinkage rate of the cylinder is less sensitive to its thickness than a plate, because liquid flowing out of the center of the cylinder passes through a volume of gel that increases as r^2, so its meets less resistance than liquid flowing out of a plate.

A viscoelastic gel is predicted [86] to show a similar dependence of shrinkage rate on size, but it also exhibits a "lag time" before shrinkage starts. When the network starts to contract and is resisted by pressure in the liquid, that resistance is initially accommodated by elastic strain of the network. After the elastic strain is fully developed, which occurs over a period similar to the stress relaxation time ($\sim F/E_P$), the viscous strain rate is observed. As shown in Fig. 14 of Chapter 6, the kinetics predicted by

Fig. 13.

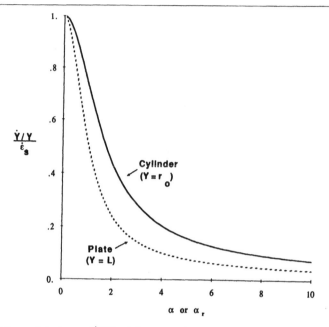

Ratio of observed strain rate ($\dot{Y}/Y$) to inherent strain rate ($\dot{\varepsilon}_s$) versus α parameter for plate ($Y = L$, see Eq. 77) and cylinder ($Y = r_0$). From Scherer [86].

this model, using experimental data for F and E_P, are in reasonable agreement with experiment. At long times, the shrinkage rate decreases again as the network ages and stiffens. It was suggested by van Dijk *et al.* [85] that syneresis is driven by an effective pressure that gradually decreases, becoming zero when the observed rate of shrinkage is negligible. That would be reasonable if the driving force (viz., condensable hydroxyl groups) disappeared over time. However, the data mentioned previously are consistent with curves calculated on the assumption that the driving force is constant and the syneresis rate decreases as the network stiffens (i.e., $\dot{\varepsilon}_s \propto 1/K_G$).

4.5. Drying Stress

As liquid evaporates from the gel, the network becomes exposed and a solid–vapor interface appears where a solid–liquid interface had been. This raises the energy of the system, because $\gamma_{SV} > \gamma_{SL}$, so liquid tends to flow from the interior of the gel to cover the exposed solid. As it stretches toward the exterior, the liquid goes into tension, and this has two consequences: (1) liquid tends to flow from the interior along the pressure gradient, according to Darcy's law and (2) the tension is balanced by compressive stress in the network that causes shrinkage. The lower the permeability, the more difficult it is to draw liquid from the inside of the gel and, therefore, the greater the pressure gradient that develops. As the pressure gradient increases, so does the variation in free strain rate, with the surface tending to contract faster than the interior. As explained in Section 3.2, it is the *differential strain* (i.e., the spatial variation in strain (for an elastic material) or strain rate (for a viscous material) that produces stress.

As the stress in the liquid increases, the vapor pressure decreases according to Eq. 10 with $P_c = -P$. As we shall see in the next chapter, however, during most of the drying process the pressure is too small to have a significant effect on the evaporation rate. In that case, we can find the pressure and stress distributions in a drying plate by solving Eq. 64 subject to the boundary condition that the flux at the exterior surface is constant:

$$\frac{D}{\eta_L} \frac{dP}{dz}\bigg|_{z=L} = \dot{V}_E \tag{78}$$

where $\dot{V}_E$ is the constant evaporation rate (cm^3 per cm^2 of gel per second). Assuming that the properties of the materials are independent of z, the result is [68]

$$P = P_s - H_G\left(\frac{\dot{V}_E}{L}\right)\left[1 - \beta - \frac{\alpha \cosh(\alpha z/L)}{\sinh(\alpha)}\right] \tag{79}$$

which is shown in Fig. 14 for several values of α. The tensile stress in the liquid at the surface of the plate causes shrinkage of the network; if α is large (D is small), the suction near the surface cannot easily draw liquid from the interior, so the contraction of the plate drives that internal liquid into compression. (See curve for $\alpha = 4$ in Fig. 14.) Remember that the solution of Eq. 64 becomes increasingly inaccurate as α increases, because large differences in local shrinkage rates cause the physical properties to vary with z. In addition, the "material coordinate" ceases to be proportional to the frame of reference of the laboratory when $\rho = \rho(z)$.

The stress in the plane of the plate is found from Eqs. 54 and 79 to be

$$\sigma_x = C_N H_G\left(\frac{\dot{V}_E}{L}\right)\left[\frac{\alpha \cosh(\alpha z/L)}{\sinh(\alpha)} - 1\right] \tag{80}$$

or, when α is small,

$$\sigma_x \approx C_N\left(\frac{L\eta_L \dot{V}_E}{2D}\right)\left(\frac{z^2}{L^2} - \frac{1}{3}\right) \qquad (\alpha \leq 1). \tag{81}$$

Fig. 14.

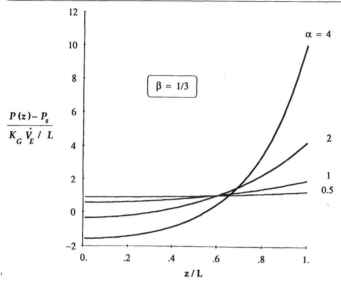

Normalized pressure, $(P - P_s)/(H_G \dot{V}_E/L)$, in liquid phase of a drying plate for several values of α; calculated from Eq. 79 assuming $\beta = 1/3$. From Scherer [68].

Thus there is a parabolic distribution of stress, with the greatest tension appearing at the drying surface, $z = L$. The stress increases in proportion to the thickness of the plate and the rate of evaporation, and in inverse proportion to the permeability; that is, the stress is increased by those factors that steepen the pressure gradient. Comparing the stress at the surface of a drying plate, cylinder, and sphere, it is found [68,87] that the tension decreases in the ratio plate : cylinder : sphere $= \frac{1}{3}/\frac{1}{4}/\frac{1}{5}$. The lower stress reflects the shallower pressure gradients in the cylinder and sphere, where the liquid flowing from the interior passes through a volume that increases as r^2 and r^3, respectively. Since these results are derived from Eq. 61 they are only valid as long as the pores remain filled with liquid. At some point the network will stop shrinking and the meniscus will retreat into the gel; then Eq. 61 will apply only within the saturated pores inside the gel.

If the network is assumed to be viscoelastic, rather than purely viscous, the stress distribution is given by [73,68]

$$\sigma_x = C_N H_G \left(\frac{\dot{V}_E}{L}\right) \left\{ \frac{\alpha \cosh(\alpha z/L)}{\sinh(\alpha)} - 1 - 2\alpha^2 e^{-\phi} \right.$$

$$\left. \times \sum_{n=1}^{\infty} \left[\frac{(-1)^n \cos(n\pi z/L)}{\alpha^2 + n^2\pi^2} \right] \exp\left(-\frac{n^2\pi^2\phi}{\alpha^2}\right) \right\} \tag{82}$$

Fig. 15.

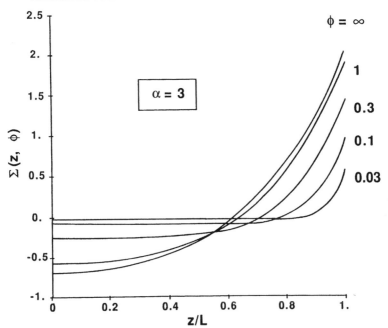

The function $\Sigma \equiv \sigma_x/[C_N H_G(\dot{V}_E/L)]$, the quantity in braces in Eq. 82, as a function of reduced time, $\phi = K_P t/K_G$, when $\alpha = 3$. The viscoelastic solution approaches the viscous solution given by Eq. 80 (labelled ∞) when $\phi > 1$. From Scherer [73,68].

where ϕ is the reduced time defined by $\phi = K_P t/K_G \approx E_P t/F$. As shown in Fig. 15, this stress profile develops toward that predicted by Eq. 80 over a period of time about equal to the stress-relaxation time. For silica gels [74], $\phi = 1$ corresponds to about one day, so the viscous solution is expected to apply through most of the drying period.

Whereas the stress in the plate comes from the internal pressure distribution, the stress in a film results principally from the external constraint of the substrate. If the substrate lies in the x–y plane, then no strain can occur in that direction (the film is too thin to deform the substrate), so $\varepsilon_x = \varepsilon_y = 0$. Then the stress in the film is [68,76]

$$\sigma_x = \sigma_y = -F\dot{\varepsilon}_f/(1 - N) = -C_N(3K_G\dot{\varepsilon}_f). \tag{83}$$

The volumetric strain comes entirely from contraction in the direction (z) perpendicular to the plate:

$$\dot{\varepsilon} = \dot{\varepsilon}_z = 3\beta\dot{\varepsilon}_f = 3\beta(\dot{\varepsilon}_s - P/3K_G). \tag{84}$$

In this case, the strain does not depend on the average pressure. From Eqs. 61 and 84, the equation governing the pressure is found to be

$$L^2 \frac{d^2P}{dz^2} - \alpha^2 P = -\alpha^2 P_s. \tag{85}$$

The solution subject to the boundary condition, Eq. 78, is

$$P = P_s + H_G\left(\frac{\dot{V}_E}{L}\right)\left[\frac{\alpha \cosh(\alpha z/L)}{\sinh(\alpha)}\right] \tag{86}$$

so, according to Eqs. 48, 83, and 86,

$$\sigma_x(z) = C_N H_G\left(\frac{\dot{V}_E}{L}\right)\left[\frac{\alpha \cosh(\alpha z/L)}{\sinh(\alpha)}\right] \approx P(z). \tag{87}$$

The approximation $\sigma_x \approx P$ recognizes that the syneresis stress is small and the constant $C_N \approx 1$. This indicates that the stress in the film can be as large as the capillary stress in the liquid. The remarkable fact, discussed in the next chapter, is that this enormous stress does not cause fracture of thin films of gel.

5.

SUMMARY

The driving forces for shrinkage during drying of a gel include stresses produced by chemical reactions, as well as osmotic, disjoining, and capillary forces. Usually the most important of these is the capillary force arising from the difference between the solid–vapor and solid–liquid interfacial energies. The huge interfacial area of gels ($\sim$300–1000 m²/g) can result in capillary pressures as large as $\sim$100 MPa. These driving forces produce compressive stresses that draw the solid network into the liquid (so there is a flux of solid through the liquid). When the pores are larger than a few times the diameter of the liquid molecules, transport is expected to obey Darcy's law. Experimental evidence supports the prediction that the flux is proportional to the pressure gradient in the liquid in a gel, but it is expected that the liquid in the very small pores of the gel has a higher viscosity than the bulk liquid. Diffusive transport will predominate only in special cases, such as when the gel is suspended in a dialysis membrane (see Fig. 1) or when the pores contain a mixture of liquids differing greatly in volatility. To model the stresses and strains developed during drying, one needs to know the constitutive behavior of the network. Assuming that the network is elastic, viscous, or viscoelastic, one can derive a differential equation for the distribution of pressure in the liquid phase of a saturated gel. The pressure gradient drives fluid flow and also gives rise to differential strain that causes drying stresses. The magnitude of the tension in the network is found to increase in proportion to the size of the body and the rate of drying.

In the next chapter, we examine the phenomenology of drying and show that the theory presented here is in qualitative accord with observations. Quantitative tests of the model are not possible, because of lack of data for properties such as the permeability and viscosity of gels.

REFERENCES

1. G.W. Scherer, *J. Non-Cryst. Solids*, **109** (1989) 183–190.
2. B.E. Yoldas, Westinghouse Co., Pittsburgh, Pa. Private communication.
3. K. Denbigh, *The Principles of Chemical Equilibrium* (Cambridge Univ. Press, Cambridge, England, 1968).
4. T. Tanaka, *Scientific American*, **244** (1981) 124–138.
5. J.F. Quinson, N. Tchipkam, J. Dumas, C. Bovier, J. Serughetti, C. Guizard, A. Larbot, and L. Cot, *J. Non-Cryst. Solids*, **99** (1988) 151–159.
6. S.S. Kistler, *J. Phys. Chem.*, **36** (1932) 52–64.
7. T. Kawaguchi, H. Hishikura, and J. Iura, *J. Non-Cryst. Solids*, **100** (1988) 220–225.
8. H. van Olphen, *Introduction to Clay Colloid Chemistry* (Interscience, London, 1977).
9. J. Livage, *J. Solid State Chem.*, **64** (1986) 322–330.
10. J.J. Spitzer, *Langmuir*, **5** (1989) 199–205.
11. K.J. Packer, *Phil. Trans. R. Soc. Lond. B.*, **278** (1977) 59–87.
12. J.N. Israelachvili, *Chem. Scr.*, **25** (1985) 7–14.
13. M. Brun, A. Lallemand, J.F. Quinson, and C. Eyraud, *Thermochimica Acta*, **21** (1977) 59–88.
14. B.V. Derjaguin, V.V. Karasev, and E.N. Khromova, *J. Colloid Interface Sci.*, **109** [2] (1986) 586–587.
15. A.W. Adamson, *Physical Chemistry of Surfaces*, 2nd ed. (Wiley Interscience, New York, 1967).
16. P.C. Hiemenz, *Principles of Colloid and Surface Chemistry* (Marcel Dekker, New York, 1977).
17. L.R. White, *J. Colloid Interface Sci.*, **90** [2] (1982) 536–538.
18. G.W. Scherer, *J. Non-Cryst. Solids*, **92** (1987) 375–382.
19. R. Zallen, *The Physics of Amorphous Solids* (Wiley, New York, 1983), chapter 4.
20. L.R. Fisher and J.N. Israelachvili, *J. Colloid Interface Sci.*, **80** [2] (1981) 528–541.
21. L.R. Fisher and J.N. Israelachvili, *Colloids and Surfaces*, **3** (1981) 303–319.
22. H.K. Christenson, *J. Colloid Interface Sci.*, **121** [1] (1988) 170–178.
23. R.Q. Packard, *J. Am. Ceram. Soc.*, **50** [5] (1967) 223–229.
24. J.R. Philip, *Adv. Hydrosci.*, **5** (1969) 215–296.
25. J. Zarzycki in *Ultrastructure Processing of Ceramics, Glasses, and Composites*, eds. L.L. Hench and D.R. Ulrich (Wiley, New York, 1984), pp. 27–42.

26. D. Croney and J.D. Coleman in *Proc. Conf. Pore Pressure and Suction in Soils*, Brit. Nat'l. Soc. of Int. Soc. Soil Mechanics and Foundation Eng. (Butterworth, Washington, D.C., 1961), pp. 31–37.

27. L.D. Baver, W.H. Gardner, and W.R. Gardner, *Soil Physics*, 4th ed. (Wiley, New York, 1972).

28. W.D. Kingery and J. Francl, *J. Am. Ceram. Soc.*, **37** [12] (1954) 596–602.

29. A.E. Scheidegger, *The Physics of Flow through Porous Media*, 3d ed. (Univ. Toronto Press, Toronto, Canada, 1974).

30. F.A.L. Dullien, *Fluid Transport and Pore Structure* (Academic Press, New York, 1979).

31. H. Darcy, *Les Fontaines Publiques de la Ville de Dijon* (Librairie des Corps Impériaux des Ponts et Chaussées et des Mines, Paris, 1856).

32. J. van Brakel, *Powder Tech.*, **11** (1975) 205–236.

33. J. Happel and H. Brenner, *Low Reynolds Number Hydrodynamics* (Martinus Nijhoff, Boston, 1986).

34. P.M. Adler, *Phys. Fluids*, **29** [1] (1986) 15–22.

35. P.M. Adler, *C.R. Acad. Sc. Paris*, **302** [1] (1986) 691–693.

36. P.M. Adler in *Characterization of Porous Solids* (Elsevier, Amsterdam, 1988), pp. 433–439.

37. T.Tanaka, S. Ishiwata, and C. Ishimoto, *Phys. Rev. Lett.*, **38** [14] (1977) 771–774.

38. N. Weiss, T. van Vliet, and A. Silberberg, *J. Polym Sci: Polym. Phys. Ed.*, **17** (1979) 2229–2240.

39. C.L. Glaves, C.J. Brinker, D.M. Smith, and P.J. Davis, *Chem. Mater.*, **1** (1989) 34–40.

40. J. Warnock, D.D. Awschalom, and M.W. Shafer, *Phys. Rev. B*, **34** [1] (1986) 475–478.

41. M.W. Shafer, D.D. Awschalom, J. Warnock, and G. Ruben, *J. Appl. Phys.*, **61** [12] (1987) 5438–5446.

42. R.D. Shoup, *Colloid and Interface Science*, **III** (Academic Press, New York, 1976), pp. 63–69.

43. D.Y.C. Chan and R.G. Horn, *J. Chem. Phys.*, **83** [10] (1985) 5311–5324.

44. J.N. Israelachvili, *J. Colloid Interface Sci.*, **110** [1] (1986) 263–271.

45. J. van Alsten, S. Granick, and J.N. Israelachvili, *J. Colloid Interface Sci.*, **125** [2] (1988) 739–740.

46. P. Debye and R.L. Cleland, *J. Appl. Phys.*, **30** [6] (1959) 843–849.

47. G.W. Scherer and R.M. Swiatek, *J. Non-Cryst. Solids*, **113** (1989).

48. K.K. Watson, *Water Resources Res.*, **2** [4] (1966) 709–715.

49. J. Crank, *Mathematics of Diffusion* (Clarendon Press, Oxford, England, 1975).

50. T. Tanaka and D.J. Fillmore, *J. Chem. Phys.*, **70** [3] (1979) 1214–1218.

51. R.B. Bird, W.E. Stewart, and E.N. Lightfoot, *Transport Phenomena* (Wiley, New York, 1960).

52. J.R. Philip and D.A. de Vries, *Trans. Am. Geophys. Union*, **38** [2] (1957) 222–232, erratum p. 594.

53. S. Whitaker, *Adv. Heat Transfer*, **13** (1977) 119–203.

54. D.L. Johnson, *J. Chem. Phys.*, **77** [3] (1982) 1531–1539.

55. J. Zarzycki, *J. Non-Cryst. Solids*, **100** (1988) 359–363.

56. E.M. Rabinovich in *Sol-Gel Technology for Thin Films, Fibers, Preforms, Electronics, and Speciality Shapes*, ed. L.C. Klein (Noyes, Park Ridge, N.J., 1988), pp. 260–294.

57. P.F. Lesse, *Wood Sci. Tech.*, **6** (1972) 204–214, 272–283.

58. M.A. Biot, *J. Appl. Phys.*, **12** (1941) 155–164.

59. M.A. Biot, *J. Appl. Phys.*, **25** [11] (1954) 1385–1391.

60. M.A. Biot, *J. Acoustical Soc. Am.*, **34** [9] (1962) 1254–1264.

61. M.A. Biot, *J. Appl. Phys.*, **33** [4] (1962) 1482–1498.

62. M.A. Biot, *J. Acoust. Soc. Amer.*, **28** [2] (1956) 168–178, 179–191.
63. J. Geertsma, *J. Mech. Phys. Solids*, **6** (1957) 13–16.
64. T. Tanaka, L.O. Hocker, and G.B. Benedek, *J. Chem. Phys.*, **59** [9] (1973) 5151–5159.
65. J.R. Rice and M.P. Cleary, *Rev. Geophys. Space Phys.*, **14** [2] (1976) 227–241.
66. R.N. Chandler and D.L. Johnson, *J. Appl. Phys.*, **52** [5] (1981) 3391–3395.
67. G.W. Scherer, *J. Non-Cryst. Solids*, **87** (1986) 199–225.
68. G.W. Scherer, *J. Non-Cryst. Solids*, **109** (1989) 171–182.
69. W.D. Kingery, H.K. Bowen, and D.R. Uhlmann, *Introduction to Ceramics*, 2d ed. (Wiley, New York, 1976), p. 526.
70. S.P. Timoshenko and J.N. Goodier, *Theory of Elasticity* 3d ed. (McGraw-Hill, New York, 1970), p. 8.
71. G.W. Scherer, *J. Non-Cryst. Solids*, **34** (1979) 239–256.
72. G.W. Scherer, *Relaxation in Glass and Composites* (Wiley, New York, 1986).
73. G.W. Scherer, *J. Non-Cryst. Solids*, **99** (1988) 324–358.
74. G.W. Scherer, S.A. Pardenek, and R.M. Swiatek, *J. Non-Cryst. Solids*, **107** [1] (1988) 14–22.
75. A.R. Cooper in *Ceramics Processing before Firing*, eds. G.Y. Onoda, Jr., and L.L. Hench (Wiley, New York, 1978), pp. 261–276.
76. G.W. Scherer, *J. Non-Cryst. Solids*, **89** (1987) 217–238.
77. T.K. Sherwood, *Ind. Eng. Chem.*, **21** [1] (1929) 12–16.
78. H.H. Macey, *Trans. Brit. Ceram. Soc.*, **41** [4] (1942) 73–121.
79. G.W. Scherer, *J. Non-Cryst. Solids*, **92** (1987) 122–144.
80. P.J. Banks in *Drying '85*, eds. R. Toei and A.S. Mujumdar (Hemisphere, New York, 1985), pp. 102–108.
81. G.W. Scherer, *J. Non-Cryst. Solids*, **107** (1989) 135–148.
82. J.R. Philip, *Aust. J. Soil Res.*, **6** (1968) 249–267.
83. J.R. Philip and D.E. Smiles, *Aust. J. Soil Res.*, **7** (1969) 1–19.
84. J.R. Philip in *Fundamentals of Transport Phenomena in Porous Media* (Elsevier, New York, 1972), pp. 341–355.
85. H.J.M. van Dijk, P. Walstra, and J. Schenk, *Chem. Eng. J.*, **28** (1984) B43–B50.
86. G.W. Scherer, *J. Non-Cryst. Solids*, **108** (1989) 18–27.
87. G.W. Scherer, *J. Non-Cryst. Solids*, **91** (1987) 101–121.

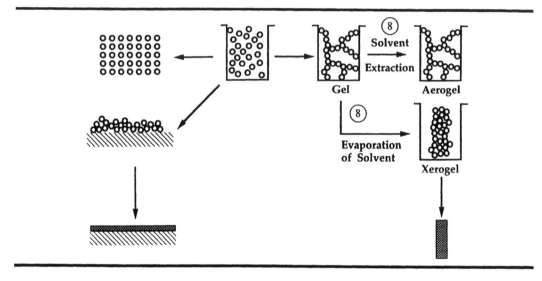

Gel

Aerogel

⑧ Solvent Extraction

⑧ Evaporation of Solvent

Xerogel

CHAPTER 8

Drying

The process of drying of a porous material can be divided into several stages: At first the body shrinks by an amount equal to the volume of liquid that evaporates, and the liquid–vapor interface remains at the exterior surface of the body. The second stage begins when the body becomes too stiff to shrink and the liquid recedes into the interior, leaving air-filled pores near the surface. Even as air invades the pores, a continuous liquid film supports flow to the exterior, so evaporation continues to occur from the surface of the body. Eventually, the liquid becomes isolated into pockets and drying can proceed only by evaporation of the liquid within the body and diffusion of the vapor to the outside. In this chapter we discuss the transport processes acting during these stages. The factors affecting stress development are discussed and various strategies for avoiding warping and cracking are described. The outline is as follows.

1. *Phenomenology* describes the stages of drying in detail.

    1.1. *Constant Rate Period* discusses the first stage, when the decrease in volume of the gel is equal to the volume of liquid lost by evaporation. The compliant gel network is drawn into the liquid by capillary and/or osmotic forces.

    1.2. *Critical Point* examines the end of the first stage, when shrinkage stops and cracking is most likely to occur.

    1.3. *First Falling Rate Period* explains the process of liquid flow through partially empty pores.

    1.4. *Second Falling Rate Period* discusses the final stage of drying, when liquid can escape only by diffusion of its vapor to the surface.

2.   *Drying Stress* shows how stresses result from the pressure gradient in the
pores of the gel. Using results of the theory presented in the preceding
chapter, the stresses are related to the drying rate, size, and properties of the
gel.
3.   *Avoiding Fracture* discusses the connection between stress and fracture,
and reviews several strategies that reduce or eliminate cracking. These
include aging, chemical additives, and supercritical drying.
4.   *Films* briefly discusses the peculiar behavior of films, which crack on
drying only if they exceed a critical thickness.

1.

PHENOMENOLOGY

The stages of drying were clearly discussed in the classic work of Sherwood
[1-3] in 1929 and 1930. Several texts provide qualitative descriptions of the
phenomenology and detailed discussion of the technology of drying [4-6].
The scientific aspects are discussed in several very good reviews (e.g., [7,8])
and in the series of books called *Advances in Drying* [9,10].

1.1.   Constant Rate Period

The first stage of drying is called the *constant rate period* (CRP), because the
rate of evaporation per unit area of the drying surface is independent of time
[7,11]. The evaporation rate is close to that from an open dish of liquid, as
indicated by data for drying of alumina gel [12], shown in Fig. 1. The rate
may differ slighly, depending on the texture of the surface. For example,
as sand beds dry, the water conforms to the shapes of the particles, so the
wetted area is larger than the planar area of the surface of the body, and
the rate of evaporation is correspondingly higher [13]. The distribution of
a spreading liquid is illustrated schematically in Fig. 2a. The chemical
potential, μ, of the liquid in the adsorbed film is equal to that under the
concave mensicus—otherwise liquid would flow from one to the other to
balance the potential. As explained in the preceding chapter, μ is lower than
for bulk liquid because of disjoining and capillary forces, so the vapor
pressure (p_v) is lowered according to

$$p_v/p_0 = \exp(\Delta\mu/R_g T) \tag{1}$$

where p_0 is the vapor pressure of bulk liquid (i.e., over a planar liquid
surface), R_g is the ideal gas constant, and T is the temperature. The rate

Fig. 1. ──

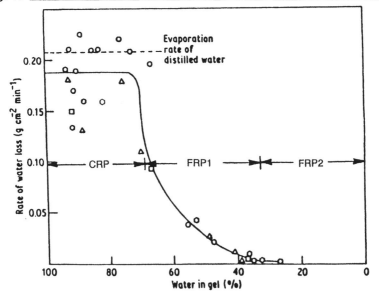

Rate of water loss from alumina gel versus water content of gel for various initial thicknesses (○ 7.5; ○ 3.0; □ 1.8; △ 0.8 mm). During the first stage (CRP) the evaporation rate from the gel is about the same as from a dish of water. From Dwivedi [12].

of evaporation, $\dot{V}_E$, is proportional to the difference between p_V and the ambient vapor pressure, p_A:

$$\dot{V}_E = k(p_V - p_A) \tag{2}$$

where k is a constant that depends on the design of the drying chamber, draft rate, etc. The fact that the evaporation rate is similar to that of bulk liquid indicates that the vapor-pressure reduction is insignificant during the CRP.

It seems reasonable to conclude that the surface of the body must be covered with a film of liquid (as in Fig. 2a), because the rate would decrease as the body shrank if evaporation occurred only from the menisci (Fig. 2b). However, Suzuki and Maeda [14] proved that the evaporation rate can remain constant even when dry patches form on the surface of the body. As indicated in Fig. 3, there is a stagnant (or slowly flowing) boundary layer of vapor over the drying surface. If the breadth of the dry patches is small compared to the thickness of the layer, diffusion parallel to the surface homogenizes the boundary layer at the equilibrium concentration of vapor. This would certainly be expected in the situation depicted in Fig. 2b, where the expanse of dry gel between menisci would be on the order of nanometers.

Fig. 2.

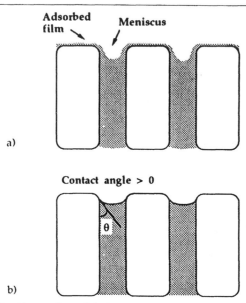

a)

b)

Distribution of liquid at surface of drying porous body, when liquid is (a) spreading (contact angle $\theta = 0°$) or (b) wetting, but nonspreading ($90° > \theta > 0°$). The chemical potential of the liquid in the adsorbed film is equal to that under the meniscus.

Fig. 3.

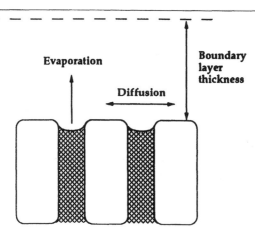

The area occupied by menisci decreases faster than the total area of the body as it shrinks, and the rate of evaporation might be expected to decrease correspondingly. However, lateral diffusion within a boundary layer allows the vapor pressure to equilibrate, so that evaporation continues at a constant rate per unit area of surface even when the surface is partially dry. After Suzuki and Maeda [14].

Therefore, transport of vapor across the boundary layer obeys Eq. 2, and the rate of evaporation per unit area of surface is constant,[†] whether or not there are small dry patches.

Evaporation causes cooling of a body of liquid, but the reduced temperature leads to a lower rate of evaporation and less cooling. When equilibrium is established, the temperature of the liquid is called the *wet-bulb temperature*, T_w. As indicated by Eq. 2, V_E increases as p_A decreases, so T_w decreases with the ambient humidity. The exterior surface of a drying body is at the wet-bulb temperature during the CRP [2]. As illustrated by the data of Simpkins *et al.* [15], shown in Fig. 4, the surface temperature rises only after the rate of evaporation decreases (in the falling rate period discussed in Section 1.3). Sherwood [2] points out that a body drying from one side is warmer than one drying from both sides, so the former dries at much more than half the rate of the latter. For alkoxide-derived gels the vapor pressure

Fig. 4.

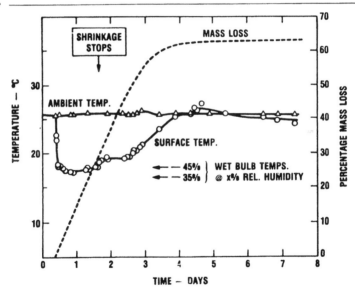

The surface of a particulate gel remains at the wet-bulb temperature during the CRP and then rises as the drying rate decreases. In this case, because of the relatively large pores, the CRP continues beyond the critical point, when shrinkage stops. From Simpkins *et al.* [15].

[†] Note added in proof: Recent work by Wilson (M.S. thesis, "Drying Kinetics of Pure Silica Xerogels," Univ. Florida, Gainesville, 1989) shows that there is no CRP when evaporation causes a change in the composition of the pore liquid. Nevertheless, the meniscus remains at the exterior surface until shrinkage stops, and cracking is most likely near the critical point.

Fig. 5.

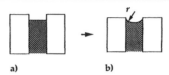

a)   b)

Formation of concave menisci visualized in two steps. (a) Evaporation removes liquid from thin layer near surface. (b) Liquid covers exposed solid surface. The radius of curvature of the meniscus, r, is defined as negative when the center of curvature is outside the liquid as in (b). Adsorption forces determine the thickness of the adsorbed film on the surface, and affect the contact angle. (See Fig. 2.)

must be kept high to avoid rapid drying, so the temperature of the sample remains near ambient.

If there were no interaction between the liquid and solid components of the gel, the liquid would evaporate from the pores leaving the network exposed, but otherwise unchanged. In reality, adsorption and capillary forces oppose exposure of the solid phase, so liquid flows from the interior to replace that which evaporates. Imagine that all of the liquid evaporates from the pores in a thin slice near the surface of a plate of gel. The remaining liquid wants to cover the newly exposed solid–vapor interface, but the reduced volume of liquid cannot do so without developing concave menisci. (See Fig. 5.) As explained in Chapter 7, the tension (P) in the liquid is related to the radius of curvature (r) of the meniscus by[†]

$$P = -2\gamma_{LV}/r \tag{3}$$

where γ_{LV} is the liquid–vapor interfacial energy. When the center of curvature is in the vapor phase, as in Fig. 5, the radius of curvature is negative and the liquid is in tension ($P > 0$). For liquid in a cylindrical pore of radius a, the radius of the meniscus is

$$r = -a/\cos(\theta) \tag{4}$$

where θ is the contact angle (defined in Fig. 4 of Chapter 7). If θ is 90°, then the liquid does not wet the solid and the liquid–vapor interface is flat ($r \to -\infty$, $P = 0$). If $\theta = 0°$ the solid surface is covered with a liquid film, as in Fig. 2a.

The tension in the liquid is supported by the solid phase, which therefore goes into compression. If the network is compliant, as it is in alkoxide-derived gels, the compressive forces cause it to contract into the liquid. As indicated in Fig. 6, it doesn't take much force to submerge the solid phase, so the capillary pressure is low and the radius of the meniscus is much larger

[†]The stress in the liquid, P, is positive when the liquid is in tension. The pressure, P_L, follows the opposite sign convention ($P_L = -P$), so tension is "negative pressure."

Fig. 6.

STAGES OF DRYING

a) Initial condition

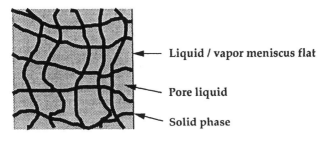

← Liquid / vapor meniscus flat

Pore liquid

Solid phase

b) Constant rate period

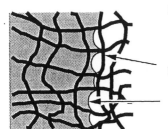

Evaporation

Shrinkage

c) Falling rate period

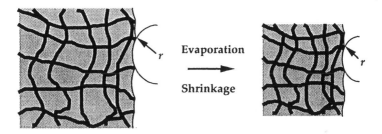

Maximum capillary pressure:

$$P_R = \frac{(\gamma_{SV} - \gamma_{SL}) S_p}{V_p}$$

Empty pores

Minimum radius of curvature

Schematic illustration of drying process. Capillary tension develops in liquid flows to prevent exposure of the solid phase by evaporation, and the network is drawn back into liquid. The network is initially so compliant that little stress is needed to keep it submerged, so the tension in the liquid is low and the radius of the meniscus (r) is large (b). As the network stiffens, the tension rises and r decreases. At the critical point, the radius of the meniscus becomes equal to the pore radius; then the CRP ends and the liquid recedes into the gel (c).

than the pore radius. As drying proceeds, the network becomes increasingly stiff, because new bonds are forming and the porosity is decreasing, and the tension in the liquid rises correspondingly. Once the radius of the meniscus becomes equal to the radius of the pores in the gel (assuming, for the moment, that the pores are uniform in size), the liquid exerts the maximum possible force. That marks the end of the CRP: beyond that point, the tension in the liquid cannot overcome further stiffening of the network, so the meniscus recedes into the pores, leaving air-filled pores near the outside of the gel. Thus, during the CRP, the shrinkage of the gel is equal to the volume of liquid evaporated; the meniscus remains at the exterior surface, but r decreases continuously.

This behavior is illustrated by the data of Kawaguchi *et al.* [16], presented in Fig. 7; equivalent results have been reported for particulate gels made from fumed silica [15]. The shrinkage behavior of the gel can be divided into three regions. In region I, syneresis is occurring, so free liquid is accumulating on the balance pan and the recorded weight loss is less than the amount of liquid that actually escapes from the sample. In region II, the loss by evaporation is greater than the rate of liquid expulsion during syneresis, so the gel enters the CRP, where the change in volume is proportional to the loss of weight. The weight of the gel is

$$W(t) = \rho_s V_s + \rho_L V_L(t) \tag{5}$$

where ρ_s and ρ_L are the densities and V_S and V_L are the volumes of the solid and liquid phases, respectively. As long as the pores are full of liquid (i.e., during the CRP), the volume of the gel is $V(t) = V_s + V_L(t)$, so Eq. 5 leads to

$$\frac{W(t)}{W(0)} = \frac{V(t)/V(0) + b}{1 + b} \tag{6}$$

where

$$b = \frac{(\rho_s - \rho_L)V_s}{\rho_L[V_s + V_L(0)]}. \tag{7}$$

Since $V_S/[V_S + V_L(0)]$ is the initial volume fraction of solid phase in the gel and $(\rho_s - \rho_L)/\rho_L \approx 1$ for silica gel, we find that $b \ll 1$ and $W(t)/W(0) \approx V(t)/V(0)$. Therefore, during the CRP the plot in Fig. 7 should have a slope of ~ 1, as it does in region II. The break between regions II and III, where the gel stops shrinking, is discussed in the next section.

Factors other than capillary pressure may operate during drying of gels, and some authors (e.g., [17,18]) consider them to be of primary importance. *Adsorption forces* bind solvent into an ordered layer a few molecules thick near a solid surface; the attractive force may drive flow of a film over an exposed solid surface [19] and influence the curvature of the meniscus (Fig. 8a). However, if there were no capillary force ($\gamma_{LV} = 0$), no compressive

Fig. 7. _____

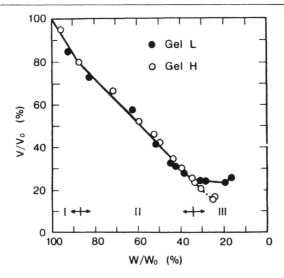

Volume shrinkage versus weight loss (both normalized by initial values) for base-catalyzed (L) and acid-catalyzed (H) silica gels. The CRP extends over region II, $80\% \geq V/V_0 \geq 25\%$; very little shrinkage occurs after the critical point ($V/V_0 \approx 25\%$ for L and 15% for H). From Kawaguchi et al. [16].

force would be exerted on the network as evaporation emptied the pores and no shrinkage would result from adsorption forces unless the initial pore radius were on the order of the film thickness (Fig. 8b). Adsorption forces draw water into the sheet structure of certain clays (e.g., montmorillonite), causing the volume to increase by about a factor of two. Removal of this water is difficult and would occur in the last stage of drying, after the inter-particle space was drained. There is also an *osmotic pressure* that induces water to enter the interlayer space in clay particles to equalize the concentration of ions in that space and in the surrounding bulk liquid [20]. This could be important during the CRP as evaporation raises the concentration of ions near the outer surface, inducing diffusion of water from the interior.[†] Macey [11] argues that *electrostatic repulsion* between particles of clay produces tension in the liquid that draws flow from the interior of a drying body. Spitzer [22] has shown that the electrostatic repulsion between clay particles is great enough to account for swelling pressures of several megapascals. Even for clays, in which these phenomena are most evident, it

[†] This is analogous to the swelling of polyacrylamide gels when immersed in water [21], where the lower chemical potential of the pore liquid drives flow into the gel. During drying, the source of liquid is the interior of the body, so flow is toward the surface.

Fig. 8.

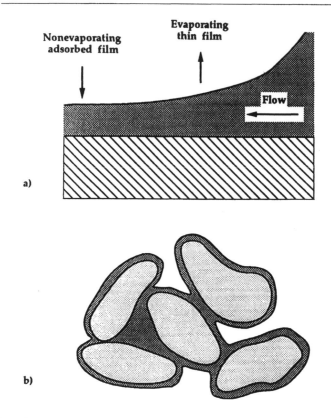

a)

b)

(a) Adsorption forces cause retention of a film of liquid on the surface, and draw flow to compensate for evaporation. (b) The range of adsorption forces is small, so they do not resist the emptying of pores larger than ~1 nm; if γ_{LV} were zero, there would be no compressive force exerted on this aggregate and drying would proceed without shrinkage until the films evaporated.

has been argued [7] that osmotic forces must be less important than capillary pressure, because moisture gradients persist in clays for long periods when evaporation is prevented. In addition, it has been shown (albeit for a non-swelling kaolinite clay) that the final shrinkage of clay during drying is directly related to the surface tension of the pore liquid [23]. The swelling pressure of clays in water is $<10\,MPa$ [20,22], which is comparable to the capillary pressure in pores with radii $>14\,nm$ (according to Eq. 3, assuming $\gamma_{LV} = 0.072\,J/m^2$ for water). Since the pores of gels are generally smaller than that, capillary forces are expected to dominate during drying, even in swelling materials. Electrostatic repulsion and other disjoining forces may be

Fig. 9.

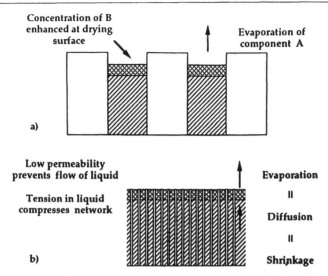

Evaporation of volatile component A from pores increases concentration of B near the drying surface, so A diffuses from the interior; the flux of A creates tension in the interior that is relieved by counterflow if the pores are large (a). If the pores are small enough (b), the low permeability inhibits flow, so tension develops in the liquid and the network is compressed. In the limiting case, the rate of contraction of the network is equal to the rate of evaporation and the meniscus remains at the surface. Capillary forces are neglected in this illustration.

more significant for their effects on the rheology of the network (e.g., increasing the resistance of the network to compression) and of the liquid (e.g., increasing the viscosity near a solid–liquid interface).

Even when capillary forces predominate, osmosis (i.e., diffusion) can be a significant factor in liquid transport within the gel. Evaporation will generally change the composition of the pore liquid (for example, increasing the concentration of electrolytes near the exterior surface, or changing the water–alcohol ratio), producing an osmotic pressure that draws liquid from the interior. First consider the case of a body with large pores. (See Fig. 9a.) Again we imagine that evaporation has emptied the pores near the surface, this time creating a higher concentration of solute in the remaining liquid, which induces diffusion of solvent from the interior. Suppose that the solvent has a larger diffusion coefficient than the solute, so that diffusion produces a net flux of volume from the interior. The resulting tension in the liquid causes bulk flow toward the interior that tends to balance the diffusive flux. If there is no resistance to flow through the network, the counterflow instantly compensates for the volume of liquid transported by diffusion, so no pressure develops in the pores and there is no contraction of the network. However, if the pores of the network are small enough, the network may offer so much resistance (friction) to flow that the gel acts as a dialysis membrane, allowing diffusion, but inhibiting bulk flow. This creates tension in the wake of the faster diffusing species. Then, as indicated in Fig. 9b, the rate of evaporation of the volatile component may be balanced by contraction of the network. This situation is analyzed in Section 2.4.

Transport of liquid in drying bodies has been attributed to diffusion along gradients in concentration of liquid [1,24]. This approach has been strongly disputed [13,25] on the grounds that flows goes in the direction of higher capillary tension, even if that opposes the gradient in moisture content. For example, Ceaglske and Hougen [13] placed a layer of fine sand particles on top of coarser particles, and showed that the pores between the fine particles remained saturated as the pores between the larger particles emptied. The capillary pressure in the small pores drew the water from the underlying larger pores, so that layer dried as water flowed *into* the saturated surface layer of fine particles. In fact, they were able to show quantitatively that the moisture distribution was controlled by the capillary pressure in the pores. Similarly, Moore [7] points out that gradients in moisture content in clay can persist for a very long time under nondrying conditions; that is, if the pressure in the liquid is uniform (because no evaporation is occurring), there is no transport of liquid. It is generally accepted (e.g., [26,27]) that when the pore liquid is in the saturated or funicular state (see Fig. 9 of Chapter 7) the dominant mechanism of transport is fluid flow obeying Darcy's law. However, in the presence of concentration gradients *within* the liquid phase (as discussed in Section 2.4), diffusion may also be important. Temperature gradients in drying bodies have an important influence on vapor transport in unsaturated pores, but they do not produce fluid flow in saturated pores [28].

During the CRP a gel will typically shrink in volume by a factor of 5 to 10. There are obviously problems associated with transporting that quantity of liquid through the gel and avoiding adhesion to the mold during such huge shrinkage. In addition, the concentration of any solute in the pore liquid increases dramatically and destructive precipitation of crystals can result. This process is called *efflorescence*, and it can result in the precipitation of crystals (such as nitrates [29]) within the gel or growth of fibers out of the surface [30]. One solution is to use a more soluble salt (e.g., acetates instead of nitrates [31]). The term efflorescence is also used to describe a related phenomenon, involving chemical reaction with the atmosphere, that occurs in bricks when CaOH reacts with ambient carbon dioxide to produce carbonate crystals that grow out of the surface of the brick [32]; similar reactions occur in alkali silicate gels [33].

1.2. Critical Point

As the gel shrinks, the tension in the pores increases and the vapor pressure of the liquid in the pores decreases according to

$$p_V = p_0 \exp\left(-\frac{PV_m}{R_g T}\right) \tag{8}$$

where V_m is the molar volume of the liquid and P is the tensile stress in the liquid (the *pressure*, $P_L = -P$, has the opposite sign). From Eqs. 2–4 and 8 we see that evaporation will continue as long as

$$p_A < p_0 \exp\left(-\frac{2V_m \gamma_{LV} \cos(\theta)}{R_g T a}\right) \tag{9}$$

where p_A is the ambient vapor pressure. If the network remains compliant and $p_A = 0$, shrinkage could continue until the pores collapse completely ($a \rightarrow 0$) [34]. In practice, however, the network stiffens as it shrinks and at some point becomes able to withstand the capillary pressure. Shrinkage stops at the *critical point* (or, in clay technology, the *leatherhard point*), which can be defined in terms of the volume fraction of solid phase, ρ_c, or porosity (both liquid- and air-filled), $1 - \rho_c$. Since further shrinkage is negligible, the relative density of the dried gel, ρ_0, is approximately equal to ρ_c. Evaporation beyond the critical point necessarily creates unsaturated pores. In clay or coarse particulate gels, one often sees haze at the surface at the critical point as air enters the pores. The saturated body is translucent or transparent because of the similarity in refractive index of the liquid and solid, but the lower index of air causes scattering of light. This is usually not visible in alkoxide-derived gels, because the pores are too small, but opacity can occur for other reasons that are discussed in Section 1.3.

At the critical point, the radius of curvature (r) of the meniscus is small enough to enter the pores. For an ordinary porous material that does not have cylindrical pores, r can be defined in terms of V_P and S_P, the volume and surface area of the pore space, respectively; then r obeys Eq. 4 with the pore radius given by $a = 2V_P/S_P$. Alternatively, from Eqs. 7 and 8 of Chapter 7, the capillary stress (tension in the liquid) at the critical point can be written as

$$P_R = \gamma_{LV} \cos(\theta)\left(\frac{S\rho_b}{1 - \rho}\right) \tag{10}$$

where S is the specific surface area of the drained network, ρ_b is its bulk density, and ρ is its relative density. For an alkoxide-derived gel with $S \sim 300\text{–}800 \text{ m}^2/\text{g}$, $\rho_b \sim 0.4\text{–}1.6 \text{ g/cm}^3$, $\rho \sim 0.2\text{–}0.6$, and $\gamma_{LV} \cos\theta \sim 20\text{–}70 \text{ ergs/cm}^2$, this is an enormous pressure: $P_R \sim 3\text{–}200 \text{ MPa}$!

Some authors interpret the critical point as the moment when the particles comprising the gel first come into contact, assuming that a liquid film separates the particles initially and that film becomes thinner as drying proceeds. That picture is clearly inapplicable to network structures, such as alkoxide-derived silicate gels. However, it does account for the fact that some particulate systems can be redispersed by gentle agitation in excess liquid, which would not be possible if primary bonds were formed between the particles. Even for particulate gels and clays, the particles are often in contact at the gel point. (For example, some clays form an edge-to-face "house of cards" structure.) During drying the rising capillary pressure forces the particles to rearrange into closer packing, which is initially easy because of the flimsy structure of the network. As shrinking proceeds, the particles ultimately become too crowded to rearrange further, and shrinkage stops. Thus the critical point could occur long after the particles make contact, or it could occur in spite of the presence of a liquid film between particles if they were packed into an arrangement stiff enough to withstand the capillary pressure.

The apparent abruptness of the critical point and the amount of weight loss thereafter depend on the structure of the gel. For example, the data of Kawaguchi et al. [16] (shown in Fig. 7) indicate that the base-catalyzed silica gel (L) exhibits a critical point (i.e., the break between regions II and III) when ~35% of the initial weight remains, and continues to dry until ~15% remains. The resulting porosity of the dried gel is ~70%. If no shrinkage occurred while those pores emptied, 100 cm³ of gel would lose 70 cm³ ≈ 70 g of liquid, leaving 30 cm³ ≈ 60 g of silica, for a weight loss of ~54%. This corresponds very well to the observed loss (from 35 to 15% of the initial weight) following the critical point. In contrast, the acid-catalyzed gel (H) seems to shrink continually. Actually, the final porosity of the sample is ~20%, so there would necessarily be a period of weight loss at constant volume as those pores empty; however, the loss would amount to ~3% of the initial mass, and would barely show up on the scale of Fig. 7. Similar results were obtained by Yamane et al. [35]. This behavior is in accord with the structure of silica gels, as described in Chapter 3: the acid-catalyzed gels are less heavily cross-linked and therefore less robust, so they collapse more readily under the capillary stress. In addition, that stress is higher in the acid-catalyzed systems, because of their smaller pore network; the denser clusters in base-catalyzed gels are separated by larger pores that generate less pressure.

The amount of shrinkage that precedes the critical point depends on the magnitude of the maximum capillary stress, P_R. As indicated by Eq. 10, P_R increases with the interfacial energy (γ_{LV}) and with decreasing pore size. It is not surprising, therefore, to find that the porosity of a dried body is greater (because less shrinkage has occurred) when surfactants are added to the liquid. For example, Kingery and Francl [23] found a linear proportionality between γ_{LV} and ρ_o for clay bodies mixed with surfactants. It is important to recognize, however, that the pressure depends on the contact angle, and the surfactant could increase θ while reducing γ_{LV}. The importance of contact angle is nicely illustrated by the work of Mitsyuk et al. [36]. They prepared aqueous silica gels from sodium silicate, then soaked them in various alcohols (methanol, ethanol, n-propanol, n-butanol) to replace the pore liquid. When the gels were dried, the final porosity was found to be linearly related to the heat of wetting, as shown in Fig. 10. The heat of wetting is related [37] to the quantity $\gamma_{SV} - \gamma_{SL} = \gamma_{LV} \cos \theta$; in this case, γ_{LV} is nearly the same for all the alcohols, so the variation in capillary stress is caused by θ.

Beyond the critical point the gel *expands* slightly as drying continues, because the compressive forces exerted on the network are released as the liquid evaporates [3]. This is clearly illustrated in Fig. 11, where the base-catalyzed silica gel expands $\sim 1.2\%$ linearly after the critical point. Using the known [38] properties of this gel, $S = 1380 \text{ m}^2/\text{g}$, $\rho_b \approx 0.6 \text{ g/cm}^3$, $\rho \approx 0.3$, and assuming that the liquid in the pores is water ($\gamma_{LV} = 0.072 \text{ J/m}^2$, $\theta = 0°$), we estimate from Eq. 10 that $P_R \approx 0.08 \text{ GPa}$. From the approximation $P_R \approx 2\gamma_{LV}/a$, this pressure corresponds to $a \approx 1.7 \text{ nm}$, whereas the measured [38] pore size distribution ranges over $1.2 < r < 4 \text{ nm}$. Thus the pressure is an upper bound, since the liquid in the pores is actually a

Fig. 10.

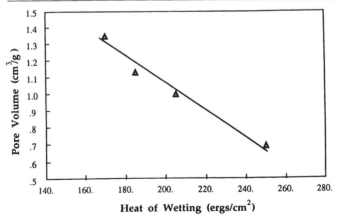

Pore volume versus heat of wetting for silica gels with pure alcohols (methanol, n-propanol, n-butanol) in the pores. Heat of wetting decreases with polarity of molecule (MeOH > BuOH). Data from Mitsyuk et al. [36].

Fig. 11.

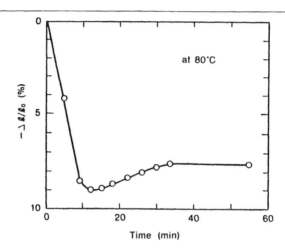

Isothermal linear shrinkage versus drying time at 80°C for gel L; initial state corresponds to $W/W_0 = 45\%$ in Fig. 7. Expansion occurs after critical point as capillary pressure is released. From Kawaguchi *et al.* [16].

water–alcohol solution with a surface tension less than that of water, and the larger pores in the distribution will moderate the pressure. The volumetric strain is given by

$$\varepsilon = P/K_\mathrm{P} \tag{11}$$

where K_P is the bulk modulus of the porous gel. Estimating $K_\mathrm{P} \approx 0.4$ GPa, based on the measured modulus of a similar gel [39], Eq. 11 predicts a linear strain of $\varepsilon_x = \varepsilon/3 \approx 6.7\%$ upon release of P_R. Thus the (overestimated) capillary pressure easily accounts for the observed strain. Expansion after the critical point has also been observed in particulate gels [15] and in wood [3].

In coarse materials, such as sand beds [13], the CRP may continue long after the critical point, to the point that unsaturated pores extend throughout the body. Even in particulate gels made from fumed silica [15] the CRP extends somewhat beyond the critical point, probably because these gels contained a significant fraction of relatively large interagglomerate pores. In fine materials, however, such as clay [7] and alkoxide-derived gels [16], the critical point coincides with the end of the CRP. After shrinkage stops, further evaporation drives the meniscus into the body and the rate of evaporation decreases, as discussed in the next subsection. Since this is the moment when the tension in the liquid reaches its maximum value, it is not surprising that cracking is most likely at the end of the CRP in systems as diverse as clays [6], particulate gels [15,40], and alkoxide-derived

Fig. 12. _____

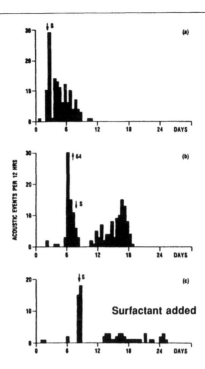

Histograms of acoustic activity in gels made from fumed silica, dried at relative humidities of ~36% (a) and 70% (b), and with surfactant added to solution (c). Shrinkage stopped at point indicated by arrow and letter S. From Simpkins *et al.* [15].

gels [12,41]. Simpkins *et al.* [15] used an acoustic detector to establish that cracking was most likely to occur at about the time that shrinkage stopped. (See Fig. 12.) Note that the amount of acoustic activity (cracking) was drastically reduced by adding surfactants to the pore liquid. Surfactants have been found [42] to reduce cracking in alkoxide-derived gels, as well.

1.3. First Falling Rate Period

When shrinkage stops, further evaporation drives the meniscus into the body, as illustrated in Fig. 13; as air enters the pores, the surface may become less translucent [15]. In the *first falling rate period* (FRP1), the rate of evaporation decreases and the temperature of the surface rises above the wet-bulb temperature. The liquid in the pores near the surface remains in the funicular condition, so there are contiguous pathways along which flow

Fig. 13. _____

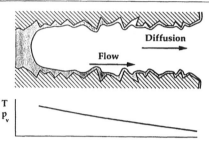

After the critical point, the liquid–vapor meniscus retreats into the pores of the body. In the first falling rate period, liquid is in the funicular state, so transport by fluid flow is possible. There is also diffusion toward the exterior in the vapor phase, since the temperature (T) and vapor pressure (p_V) both decrease in that direction.

can occur. *Most* of the evaporation is still occurring at the exterior surface, and the surface remains below the ambient temperature, and the rate of evaporation is sensitive to the ambient temperture and vapor pressure [2,7]. Inhomogeneity can result as the flow carries solutes (e.g., nitrates) toward the surface where they may precipitate (and even effloresce). At the same time, some liquid evaporates within the unsaturated pores and the vapor is transported by diffusion. Analysis of this situation involves coupled equations for flow of heat and liquid and diffusion of vapor, with transport coefficients that are generally dependent on temperature and concentration. There are several good reviews [8,43,44] of the many theories that have been proposed to described the FRP1. The most complete and rigorous treatment is by Whitaker [26,45].

Fluid flow in the FRP1 was originally attributed to diffusion of liquid [1], but it is now understood [8,13] that flow is driven by the gradient in capillary stress, P. Evaporation occurs at the outer surface because the vapor pressure is lower there than inside the relatively "humid" pores, and Eq. 8 indicates that the gradient in p_V results in a gradient in P within the pores. Therefore the funicular liquid flows toward the exterior surface, where the tension in the liquid is greatest, in accordance with Darcy's law. The permeability of the unsaturated body decreases very rapidly with the liquid content, as shown in Fig. 10 of Chapter 7; nevertheless, the transport is dominated by flow (rather than diffusion of vapor) as long as the liquid has a continuous path to the outside. The gradient in capillary pressure is affected by the temperature gradient, because the vapor pressure and surface tension are functions of temperature [26]. Both of these factors favor flow toward the outer surface by raising the tension in the liquid there, because the cooler surface will have lower p_V (giving a smaller r) and higher γ_{LV}. The flow rate may also be influenced by adsorption forces [19]: as indicated in Fig. 8a, the surface

Fig. 14.

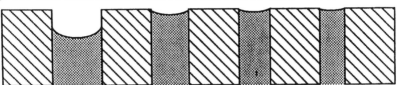

At a given vapor pressure, all menisci have the same curvature; therefore larger pores empty first.

tries to preserve a thick adsorbed layer of liquid, and this causes flow of the film. This mechanism (which is operative only when $\theta = 0°$) is most likely to be important when the vapor pressure at the surface is very low.

As indicated in Fig. 14, when there is a range of pore sizes in the body, the larger pores will empty first. If the large pores are interconnected, pockets of smaller saturated pores may become surrounded by regions of drained larger pores. The drained regions may be big enough to scatter light, even if the pores themselves are much too small to do so. Therefore, some materials (e.g., porous VYCOR® glass or some base-catalyzed silica gels) turn opaque during drying,[†] but are clear when fully saturated or fully dried.

The isolation of clusters of saturated pores occurs during *immiscible displacement*, wherein a liquid under pressure drives another liquid out of a saturated pore space (as when water is pumped into an oil field to drive oil out of the pores in rock formations). If the entering liquid is much less viscous than the one present in the pores, the interface between the liquids is unstable [46]: in a process called *viscous fingering*, portions of the interface surge ahead of the advancing front. If isolated pockets of liquid can never escape (e.g., oil pockets surrounded by water-filled pores), the interface becomes fractal with the same fractal dimension as a cluster grown by diffusion-limited aggregation (DLA) [47]. If isolated pockets are not trapped (as when the invading fluid is air, and the surrounded liquid can evaporate or flow away), the fractal dimension is the same as for a percolation cluster [48]. During drying, the invading fluid is air, which is much less viscous than the liquid in the pores, so the interface would be expected to be unstable. However, drying differs from immiscible displacement in that the invaded fluid (pore liquid) escapes by flowing along a pressure gradient through the invading fluid (air), and this has important consequences for the shape of the advancing liquid–vapor front.

[†] Opacity could also result from phase separation in the pore liquid, or exsolution of gas from the liquid. The haze in nominally dried gels is usually attributed to scattering from the pores, but is sometimes caused by pockets of residual liquid; drying under high vacuum may restore clarity to the gel.

Fig. 15.

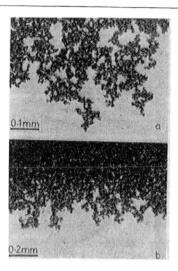

The drying front (i.e., the advancing liquid–vapor interface) is rough on the scale of the particles (a), but relatively smooth on a larger scale (b). In this photo, saturated pores are white and empty pores are black. From Shaw [49].

Shaw [49,50] performed an elegant series of experiments, illustrated in Fig. 15, showing that the drying front (i.e., the liquid–vapor interface) is fractally rough on the scale of the pores, but stable on a much larger scale. Using the process of Stöber et al. [51] (discussed in Chapters 3 and 4), he prepared silica spheres ~0.5 μm in diameter, and deposited a layer 20–30 particles deep in a cell formed between two microscope slides. The water filling the interstices between the particles was allowed to evaporate from one edge of the cell; the air-filled pores scattered light, so the empty pores appear dark in Fig. 15. The deposit was rigid, so there was no CRP; the liquid–vapor front advanced into the deposit at a rate inversely proportional to the square root of time. The vapor pressure gradient in the drained pores results in a gradient in capillary pressure, and that drives flow of the liquid to the edge of the cell according to Darcy's law. The importance of capillary transport of the funicular liquid is illustrated by the observation that the deposit dried an order of magnitude faster than a cell containing liquid, but no particles (which empties only by diffusion of vapor). The roughness of the front results from the distribution of pore sizes, because large pores empty at a higher value of p_v than small pores.

It often happens that relatively small pores form the entrances to large groups of pores, so liquid can be removed from those groups only when p_v drops low enough for the meniscus to pass through the entrance. At that point p_V is well below the equilibriium vapor pressure for the inside of the group, so it drains instantly; some of the liquid evaporates, but much of it is sucked into unsaturated pores in the advanced part of the front, where p_V is higher. Shaw [49] found that thousands of pores would drain between frames of his videotape (~0.05 s).

It is the pressure gradient in the unsaturated region that is responsible for the large-scale stability of the drying front: the capillary pressure is so low in advanced regions of the front that the radius of the meniscus is too large to pass through the pores. Shaw found that the scale of roughness of the front (defined as the distance from the most advanced to the least advanced portion of the front at a given instant), w, decreased as its velocity, v, increased:

$$w \propto v^{-m} \tag{12}$$

where $m = 0.48 \pm 0.1$. Since Darcy's law says that v is proportional to the pressure gradient in the liquid, Eq. 12 can be written as

$$w \propto (\nabla P)^{-m}. \tag{13}$$

This quantifies the phenomenon illustrated in Fig. 16: as the pressure gradient increases, the scale of roughness decreases parabolically ($\nabla P \sim 1/w^2$). An important consequence of this relation is that the drying front roughens as it advances into the body, because the pressure gradient in the drained region decreases.

Fig. 16. _____

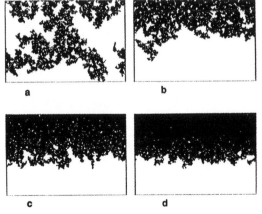

This computer simulation shows that the drying front becomes smoother as the pressure gradient in the unsaturated (black) region increases. In (a) there is no gradient; in (b-d) tension in the funicular liquid increases from the top to the bottom of the figure (gradient in b < c < d). From Shaw [49].

Since the scale of irregularity of the drying front is a few hundred times the size of the pores, it is very small compared to the dimensions of the body. Even in a particulate gel with 60 nm pores [52], if a partially dried gel is broken in half, the drying front is visible as a sharp line between the translucent saturated region and the opaque dry region. No doubt this line would be rough if observed in the SEM, but it is quite smooth on a macroscopic scale. This is likely to be true even for materials with a broad pore size distribution, although they will exhibit rougher interfaces. Sometimes isolated regions drain ahead of the interface (dark spots ahead of the front in Fig. 15b), presumably because the tension in the liquid causes expansion of pre-existing bubbles. (The tension is much too small to *nucleate* a bubble.) This was found to be a serious problem in particulate gels [52] if the solvent was not de-aired with a partial vacuum before gelation.

1.4. Second Falling Rate Period

As the meniscus recedes into the body, the exterior does not become completely dry right away, because liquid continues to flow to the outside; as long as the flux of liquid is comparable to the evaporation rate, the funicular condition is preserved. However, as the distance from the exterior to the drying front increases, the capillary pressure gradient decreases and therefore so does the flux. Eventually (if the body is thick enough) it becomes so slow that the liquid near the outside of the body is isolated in pockets (i.e., enters the pendular condition), so flow to the surface stops and liquid is removed from the body only by diffusion of its vapor. At this stage, drying is said to enter the *second falling rate period* (FRP2), where evaporation occurs inside the body [2] (see Fig. 17). The temperature of the surface approaches the ambient temperature and the rate of evaporation becomes less sensitive to external conditions (temperature, humidity, draft rate, etc.).

As indicated in Fig. 17, the drying front is drained by flow of funicular liquid which evaporates at the boundary of the funicular–pendular regions. In the pendular region, vapor is in equilibrium with isolated pockets of liquid and adsorbed films, and the vapor pressure is significantly influenced by adsorption forces. Theories for the prediction of water activity in such situations are reviewed by Bruin and Luyben [43]. As noted earlier, the gradient in p_v may induce flow in adsorbed layers [19], but the principal transport process is expected to be diffusion of vapor. Since the surface is warmer than the interior, there is a driving force for diffusion *into* the body; thus there may be diffusion toward the outside in the very dry exterior region, and diffusion toward the drying front from points within the partially

Fig. 17.

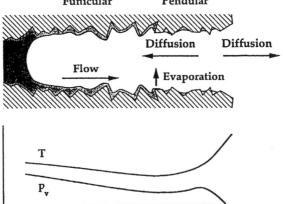

Schematic illustration of transport during the second falling rate period, when liquid near the exterior is in the pendular state (isolated in pockets). Since evaporation occurs inside the body (at the pendular–funicular boundary), the surface rises above the wet-bulb temperature. The vapor pressure increases with the temperature, so the concentration gradient may cause some diffusion toward the interior; but near the outer surface the ambient vapor pressure is low and the flux is toward the exterior. Deeper inside the body the principle transport mechanism is flow of funicular liquid toward the surface.

drained pores [53]. This phenomenon was observed in studies of convective drying of porous sandstone by Wei *et al.* [54]; when microwave heating was used, the temperature was higher inside, so the flux of vapor was always toward the surface [55]. Philip and DeVries [56] made the interesting observation that the "diffusion flux" passes through isolated pockets of liquid as well as through the drained pores. As shown in Fig. 18, the gradient in p_v causes a difference in curvature on opposite sides of the pocket, so there is a corresponding gradient in P that produces flow. At equilibrium, the flux through the pocket must equal the flux through the vapor phase.

The pores of a gel are so small that the vapor moves by *Knudsen diffusion*, which means that molecules collide more frequently with the walls of the pores than with each other [57]. In a study of alumina gel (pore size 2.5–3.5 nm), Suzuki *et al.* [58] found that the rate of permeation of small molecules decreased with the inverse square root of the molecular weight, as predicted by theory. However, the permeation rate was anomalously low for polar molecules, because of interaction with the solid surface. The slow diffusion of vapor through the gel results in an abrupt drop in the drying rate at the start of FRP2, as indicated by the data in Fig. 1. As noted earlier, slower drying encourages instability of the drying front [50] which may lead to transient opacity during drying.

Fig. 18. _____

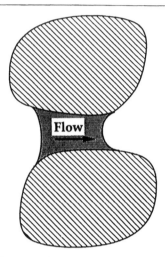

Flux through an isolated pocket of liquid driven by difference in curvature: the vapor pressure decreases from left to right, so the right meniscus has greater curvature; the tension in the liquid induces flow. At equilibrium, the flux through the liquid is equal to the flux in the surrounding vapor. After Philips and de Vries [56].

The capillary pressure is high in the isolated pockets, but they occupy such a small volume fraction that they do not exert much force on the solid network. This is reflected in Fig. 19 of Chapter 6, which shows that the strength of an aggregate decreases with the saturation, because the force that holds the aggregate together comes principally from the capillary pressure in the saturated pores. As the saturated region recedes into the body, the body expands slightly as the total stress on the network is relieved [3,16,15]. At the same time, differential strain builds up because the solid network is being compressed more in the saturated region than near the drying surface. This can cause warping in a plate dried from one side, as faster contraction of the wet side makes the plate convex toward the drying side [59]. The fact that the warping is permanent (i.e., does not spring back when drying is complete) indicates that the unsaturated region of the gel retains some viscosity or plasticity during FRP2. As the saturated region becomes thinner, its contraction is more effectively prevented by the larger unsaturated region; as explained in the next section, this raises the tension in the network in the saturated region. This phenomenon accounts for the observation by Simpkins *et al.* [15] that cracks in drying gels often originated near the nondrying surface.

Whitaker [26,45] developed an analysis of heat and mass transfer during drying of rigid materials that offers the most complete description of the falling rate periods. He uses transport coefficients that are local averages for regions that are large compared to the pore size, but small compared to the sample. This is analogous to the averaging implicit in Darcy's law, where the permeability, D, "smears out" the geometrical details of the microstructure. Use of Whitaker's model requires knowledge of a large number of physical properties (permeability, thermal conductivity, diffusivity of vapor), and the analysis must be performed numerically. An excellent test of the model was performed by Wei *et al.* [54,55] who studied the drying of porous sandstone. They measured the permeability of the stone, determined the capillary pressure–saturation relation (as in Fig. 7 of Chapter 7), and used measured values or reliable estimates for the remaining physical properties. The evaporation rate and distribution of temperature within the stone were found to be in good agreement with calculations based on Whitaker's model, for both convective and microwave heating of the sample. As noted earlier, with convective heating the surface of the sample was warmer than the interior, so vapor diffused *into* the sample while liquid flowed out; microwave heating made the interior warmer, so both fluxes moved toward the surface. In this material the calculations indicate that unsaturated pores extend throughout the sample, with no sharp front between the saturated and unsaturated regions.

2.

DRYING STRESS

We now review the phenomena described in Section 1 in terms of the theory developed in the preceding chapter. The equations presented here are based on the assumption that the mechanical properties of the network and the rate of contraction (and, consequently, such properties as the porosity and permeability) are uniform in the gel. In actuality, the shrinkage will initially be greater at the drying surface, so that region will be stiffer and less permeable than the interior. However, the difference in shrinkage rate from surface to center is not large (in our lab, for example, we have never been able to detect density gradients in dried gels), so the assumption is probably not unreasonable. The result is therefore semiquantitative, but is believed to incorporate the essential features of a correct theory. More precise calculations will be possible when the relevant physical properties of gels have been measured. The effects of nonuniform shrinkage are discussed in Section 2.5.

2.1. Pressure Distribution

The course of drying during the CRP is illustrated schematically in Fig. 6. For the moment we assume that the pore size is uniform and that diffusion in the pore liquid is insignificant. Liquid flows toward the outside to prevent exposure of the solid network. As the liquid stretches to cover the solid phase it goes into tension (concave menisci form). The tension is balanced by compressive stresses on the solid phase that tend to suck the network under the surface of the liquid. The more compliant the network, the less effort has to be expended to keep it submerged; the tension in the liquid, P, rises only as high as necessary to achieve that. The radius of curvature of the menicus, related to P by Eq. 3, is initially much larger than the pore radius. The faster the evaporation and the stiffer the network, the greater the tension in the liquid must be to pull the network under. As long as the network is sufficiently compliant, the liquid–vapor interface remains at the exterior surface of the body. However, the maximum pressure that the liquid can exert is related to the pore size of the network, as indicated by Eq. 10. The critical point is reached when that negative pressure (P_R) cannot compress the network fast enough for the volumetric contraction rate to equal the rate of evaporation.

During the CRP the tension in the liquid does two things: it compresses the network and induces flow from the interior. For the meniscus to remain at the surface of the network, the rate of evaporation ($\dot{V}_E$) must equal the flux of liquid (J) to the surface,[†] which is given by Darcy's law:

$$J_{\text{surface}} = \frac{D}{\eta_L} \nabla P|_{\text{surface}} = \dot{V}_E \qquad (14)$$

where D is the permeability of the network and η_L is the viscosity of the liquid. The lower the permeability, the greater the pressure gradient must be to support a given evaporation rate. The reason that gels are more difficult to dry than ordinary ceramics is that the permeability of gels is very low (because of the small pore size), so modest drying rates produce very steep pressure gradients. The higher tension in the liquid near the outer surface makes that portion of the network shrink faster than the body as a whole, so it tends to crack. The steeper ∇P, the greater the difference in shrinkage rate between the exterior and interior, and the more likely the gel is to fracture.

Consider a slice in the interior of the gel with liquid flowing through its pores. The net loss of liquid from the slice is given by the difference between

[†] Actually it is a flux of solid—the contracting network—through the stationary liquid, but it is their relative movement that is described by Darcy's law.

the rate of flow out of one face and the rate of flow into the other face, and that depends on the difference in ∇P at those faces. Assuming that gas bubbles are not nucleated within the gel, the rate of removal of liquid from the slice must be equal to the rate of contraction of the pore space in the network, which depends on the viscosity (or modulus) of the network. As shown in Chapter 7, equating the changes in pore volume and liquid volume in the slice leads to a differential equation for the pressure as a function of time and position. We now examine the solution of that equation, with the boundary condition given by Eq. 14. In this section we assume that the network is a viscous material. The implications of viscoelastic behavior are examined in Section 2.3.

Syneresis is contraction of the network driven by condensation reactions. There is a limit, P_s, to the stress that this mechanism can generate; if the liquid were prevented from leaving, the network would squeeze the liquid until the pressure rose to that level ($P = P_s$), then contraction of the network would stop. (P is the *stress* in the liquid, so it is negative when—as in this case—the liquid is compressed.) This is a small stress (typically <1 atm) compared to the capillary pressure that develops during drying. For a plate of thickness $2L$ drying by evaporation from both faces (each with area A), the rate of loss of liquid is $2A\dot{V}_E$, so the total volumetric strain rate of the drying body is

$$\langle \dot{\varepsilon} \rangle = \frac{1}{V}\frac{dV}{dt} = -\frac{2A\dot{V}_E}{2LA} = -\frac{\dot{V}_E}{L}. \tag{15}$$

For a gel exhibiting Newtonian viscosity, the strain rate is proportional to the stress:

$$\langle \dot{\varepsilon} \rangle = -\frac{P_E}{K_G} = -\frac{\dot{V}_E}{L} \tag{16}$$

where K_G is the bulk viscosity of the network and P_E is the average tension produced in the liquid by evaporation. The average tension in the liquid during the CRP is the sum of P_s and P_E,

$$\langle P \rangle = P_s + P_E = P_s + K_G\dot{V}_E/L. \tag{17}$$

The average pressure in the liquid is zero (since P_s and P_E differ in sign) when the rate of evaporation is equal to the rate of syneresis; higher rates of evaporation create tension in the liquid ($P > 0$).

Of course, the pressure in the pores is not uniform, because of the low permeability of the network. As shown in Chapter 7, the pressure distribution in the plate during the CRP is given by [60]

$$P = \langle P \rangle + \frac{H_G\dot{V}_E}{L}\left[\frac{\alpha\cosh(\alpha z/L)}{\sinh(\alpha)} - 1\right] \tag{18}$$

where the faces of the plate are at $z = \pm L$ and the midplane is at $z = 0$. The parameter α is defined by

$$\alpha = \sqrt{\frac{L^2 \eta_L}{D H_G}} \tag{19}$$

where H_G is the viscosity of the network ($H_G = K_G + \frac{4}{3} G_G \approx 3 K_G$, $G_G =$ shear viscosity of network). The quantity in brackets in Eq. 18, which represents the pressure variation through the body, is shown in Fig. 19 for several values of α. The difference in pressure between the middle and the surface of the plate increases with α, so it is large when the permeability is small (as in gels) or the plate is thick.

If the drying gel is in the form of a cylinder or sphere, the quantity in brackets in Eq. 18 is different [60]. The pressure gradients are shallower than in a plate, so (as shown in Fig. 20) the tensile stresses are smaller than in a plate. This reflects the fact that liquid flowing out of a cylinder passes through a volume of gel that increases with (radius)^2 (or (radius)^3, for a sphere), so it meets less resistance than liquid flowing in a straight line out of a plate.

As the gel shrinks, the viscosity of the network increases and this causes an increase in P, as indicated by Eq. 18. The syneresis pressure is negligible

Fig. 19.

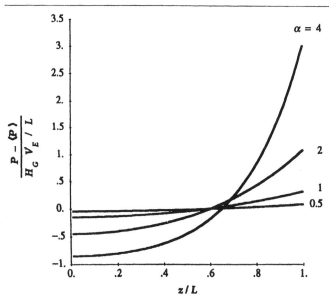

Pressure distribution in liquid in pores of plate drying by evaporation from both faces, calculated from Eq. 18 for several values of the parameter α.

Fig. 20.

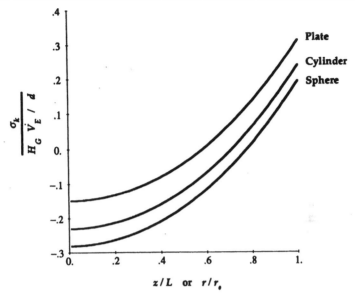

Comparison of pressure distributions in plate, cylinder, and sphere. σ_k and d are σ_x and L for the plate and σ_θ (circumferential stress) and r_o (outer radius) for the cylinder and sphere. Stresses are tensile when $\sigma_k > 0$. Calculated from formulas given by Scherer [60].

compared to the capillary stress near the end of the CRP, so Eqs. 17 and 18 indicate that the tension in the liquid at the surface of the plate is

$$P(L) \approx \frac{K_G \dot{V}_E}{L} \left[\frac{3\alpha}{\tanh(\alpha)} - 2 \right] \qquad (20)$$

(where we have assumed $H_G \approx 3K_G$ [60]). The CRP ends when the increasing viscosity causes $P(L)$ to rise to P_R, the maximum value given by Eq. (10); then the radius of the meniscus becomes equal to that of the pore, and the liquid–vapor interface recedes into the gel. During FRP1, there is a region of funicular liquid extending from the outside to the main drying front. The boundary condition is no longer given by Eq. 15, because the rate of shrinkage of the gel is now less than the rate of evaporation. However, Eq. 14 still applies, because evaporation still occurs at the exterior surface, and transport to the surface is dominated by flow through the unsaturated pores. The analysis of the falling rate period is complicated by the strong dependence of D on the liquid content of the unsaturated pores [26,27]. Qualitatively, we know that the unsaturated region is less compressed than the saturated region, as explained in Section 1.4.

2.2. Warping

If a plate dries from only one side, the pressure distribution is asymmetrical [60], as shown in Fig. 21. Since the network is compressed more on the drying face, the plate becomes concave toward the drying side [59,61], as shown in Fig. 22a. The shape will be affected by gravity, if the plate is big, and by adhesion to the mold or substrate; if those effects are not serious, the radius of curvature, r_c, of the plate is given by [60]

$$\frac{d}{dt}\left(\frac{1}{r_c}\right) = -\frac{\dot{V}_E}{2L^2}\left\{\frac{\alpha^2[1 - \tanh(\alpha)/\alpha]}{\alpha^2 - 2[1 - \tanh(\alpha)/\alpha]}\right\} \tag{21a}$$

$$\approx -\frac{\dot{V}_E}{2L^2}\alpha^2 \approx -\frac{\dot{V}_E\eta_L}{2DH_G} \quad (\alpha \leq 1) \tag{21b}$$

$$\approx -\frac{\dot{V}_E}{2L^2} \quad (\alpha \gg 1). \tag{21c}$$

Thus the curvature is greater (r_c is smaller in magnitude) when the evaporation rate is great and the permeability and viscosity of the network are small. When α is small, $P(z)$ varies linearly through the plate (see the curve for $\alpha = 0.5$ in Fig. 21); consequently, the curvature is small and is independent of the thickness of the plate. When α is larger, Eq. 21c leads to $r_c \approx -2L^2/\dot{V}_E t$, so the curvature is less (r_c is greater) in thicker plates. The

Fig. 21.

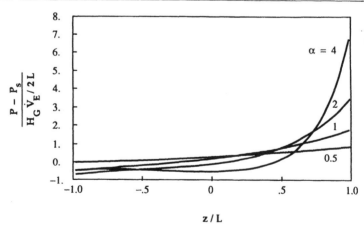

z/L

Pressure distribution in a plate of thickness $2L$ drying from the face at $z = L$, but not from the face at $z = -L$; calculated for several values of the parameter α from formulas given by Scherer [60].

Fig. 22. _____

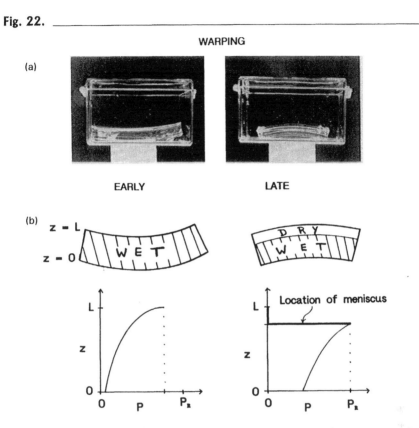

Photos of warping plate of silica gel. (a) During the CRP the plate is concave toward the drying surface. (b) After the critical point, the saturated side is compressed more than the drying surface and the curvature reverses. From Scherer [61].

curvature of thicker plates would be even less if gravity were included in the analysis. After the CRP, the unsaturated suface is less compressed than the saturated (nondrying) side, so the curvature reverses, as shown in Fig. 22b. The warping is permanent, which means that viscous (or plastic) deformation of the gel is still possible after the critical point.

2.3. Stress

If the pressure in the liquid were uniform, the network would be uniformly compressed and there would be no tendency to crack. However, the low permeability of the gel gives rise to a pressure gradient, so the tension in

Fig. 23.

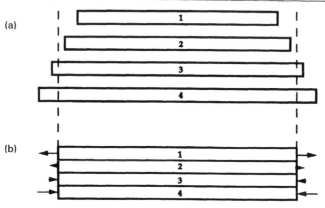

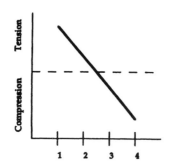

Schematic illustration of origin of drying stresses. If the drying plate were cut into slices parallel to the drying surface, the slices closer to the surface (smaller numbers) would shrink more because of the higher tension in the liquid. However, because they are bound together they must shrink the same amount, so the surface region is stretched and the interior is compressed. The lengths of the slices in (a) represent the free strains, ε_f, and the uniform lengths in (b) represent the true strain, ε_x; the stress in the plate results from the difference between ε_x and ε_f.

the liquid is greater near the drying surface, and the contraction of the network is consequently greater. The difference in shrinkage rate between the inside and outside of the body is the cause of drying stress. As shown in Fig. 23, the gel near the drying surface ($z = L$) would shrink with a strain rate of

$$\dot{\varepsilon} = P(L)/K_G \qquad (22)$$

but it is attached to a plate with a slower average contraction rate given by

$$\langle \dot{\varepsilon} \rangle = \langle P \rangle / K_G \qquad (23)$$

(when the network is viscous). The portion of the network at $z = L$ is stretched (i.e., prevented from shrinking at its natural rate) by the bulk of the plate, which has a slower contraction rate. The result, as shown in Chapter 7, is that the network experiences a tensile stress, σ_x, equal to

$$\sigma_x \approx P - \langle P \rangle. \tag{24}$$

The greatest tension occurs at the drying surface where P is greatest. From Eqs. 18 and 24 the stress at $z = L$ is

$$\sigma_x = \frac{H_G \dot{V}_E}{L} \left[\frac{\alpha \cosh(\alpha z / L)}{\sinh(\alpha)} - 1 \right] \tag{25}$$

or, when $\alpha \leq 1$,

$$\sigma_x(L) \approx L \eta_L \dot{V}_E / 3D. \tag{26}$$

This result accounts for the observation that cracking is most likely for thick gels (large L) and high drying rates. It also emphasizes the role of permeability, which is large enough in ordinary ceramics to avoid large drying stresses. It was shown in Chapter 6 that a typical value for the strength of a fresh gel is ~ 0.1 MPa; given $\eta_L = 0.001$ Pa·s, $L = 1$ cm, and $D = 10^{-14}$ cm^2 [62], Eq. 26 indicates that the strength will be exceeded for any drying rate $\dot{V}_E > 0.03\ \mu$m/s, corresponding to a minimum drying time of $L/\dot{V}_E \approx 4$ days. It is clear, therefore, why a gel breaks into pieces if simply exposed to the atmosphere. If Eq. 26 applies, the stress rises during the CRP because D decreases faster than L. (See Fig. 24.) For a typical ceramic with a

Fig. 24.

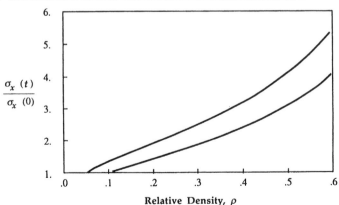

$\dfrac{\sigma_x(t)}{\sigma_x(0)}$

Relative Density, ρ

Increase in stress with time, t, as gel shrinks, assuming that the permeability, D, is given by the Carman–Kozeny equation, Eq. 36, and the stress by Eq. 26. Initial relative density (volume fraction of solid) assumed to be 0.05 or 0.1.

porosity of 50% and pore size of $1 \mu m$, the Carman–Kozeny equation (Eq. 15 of Chapter 7) predicts $D \approx 10^{-10} \, cm^2$, so the stress is 10,000 times smaller than in a gel drying at the same rate, and fracture is highly unlikely.

The capillary pressure does not appear in Eq. 26. The stress results from the *difference* in pressure through the thickness of the body, as indicated in Eq. 24, not on the absolute value of the pressure. That difference is usually much smaller than P_R—otherwise fracture would be unavoidable, since the capillary stress is about 1000 times greater than the strength of a fresh gel! The total shrinkage during drying depends on P_R, because the critical point (time t_R) occurs when the viscosity of the network is so high that $P(L) = P_R$; then, according to Eq. 20,

$$P_R = \frac{K_G \dot{V}_E}{L} \left[\frac{3\alpha}{\tanh(\alpha)} - 2 \right] \qquad (t = t_R). \qquad (27)$$

Equations 25 and 27 indicate that the stress at the critical point is

$$\frac{\sigma_x(L)}{P_R} = \frac{3[\alpha - \tanh(\alpha)]}{3\alpha - 2\tanh(\alpha)} \approx \begin{cases} \dfrac{3\alpha - 2}{3\alpha - 3}, & \alpha \gg 1 \\[3mm] \dfrac{\alpha^2}{1 + \frac{2}{3}\alpha^2}, & \alpha \le 1 \end{cases} \qquad (t = t_R). \qquad (28)$$

Generally, α is expected to be small, so the second approximation applies:

$$\sigma_x(L) \approx \alpha^2 P_R \qquad (t = t_R). \qquad (29)$$

The greater the capillary stress, the greater the stress at the critical point. The lower P_R, the less the shrinkage (and therefore the greater the permeability) at the end of the CRP. Thus, factors (such as changes in γ_{LV}) that reduce P_R also reduce the stress indirectly through their effect on the permeability.

When $\alpha \gg 1$, Eq. 25 reduces to

$$\sigma_x(L) \approx \dot{V}_E \sqrt{\frac{H_G \eta_L}{D}}. \qquad (30)$$

In this case, the stress does not depend on the thickness of the plate, because the permeability is so low that the pressure drops to zero between the surface and the midplane. During drying the network stiffens, the effective viscosity of the pore liquid rises (because a larger proportion of it is near a solid surface), and the permeability decreases (because of the decrease in pore size and porosity), and all these factors contribute to a continual increase in stress. However, unless the gel is quite large, it seems unlikely that α will be large enough for Eq. 30 to apply.

A gel is a viscoelastic material [63], so the stress builds up gradually to the value given by Eq. 26, and may not reach that value before the critical point.

The initial response of a viscoelastic material is the same as that of one that is purely elastic. For a gel with an elastic network, as shown in Fig. 25, the stress rises over a period of time of the order of [60,64]

$$\tau = L^2 \eta_L / D H_p \tag{31}$$

where $H_p = K_p + \frac{4}{3} G_p$ is an elastic modulus for the network. (G_P and K_P are the shear and bulk elastic moduli of the network, respectively.) Note that τ is rather like a viscoelastic relaxation time, but it arises from the friction between the pore liquid and the purely elastic network. According to Fig. 25, the stress in the liquid rises until $P(L, \theta)$ becomes equal to P_R at time $\theta_R = t_R / \tau$; this is the critical point when shrinkage essentially stops, and the liquid/vapor meniscus enters the gel. The maximum stress at the critical point approaches P_R when the evaporation rate is very high or the permeability is very low, as shown in Fig. 26; the maximum stress is approximately the same for an elastic or viscous network. The pressure distribution becomes approximately parabolic by the time $t/\tau > 1/3$ (see Fig. 25c) and retains that form until t_R. After that time, the tension in the liquid at the surface cannot rise further, but the tension in the interior of the body rises until the pressure becomes hydrostatic (Fig. 25e). Since the stress is produced by the pressure gradient, it reaches a "steady state" value, given by Eq. (26), during the time interval $\theta_R > \theta > 1/3$ and therefore decreases (Fig. 25f). If the evaporation rate is high, it can happen that $\theta_R < 1/3$, so the parabolic distribution is never achieved. According to Fig. 1 of [64], this happens when the parameter μ (proportional to the evaporation rate) is greater than ~1.5. However, in that case, Fig. 26 indicates that the stress at the surface of the plate is $\sigma_x > P_R/3$, and that is so high that the gel will inevitably break.

If the gel network is viscoelastic, rather than elastic, the viscoelastic relaxation time is defined by [65]:

$$\tau_{VE} = H_G / H_p = \tau / \alpha^2 \tag{32}$$

so when $t > \tau/3$,

$$t/\tau_{VE} > \alpha^2/3. \tag{33}$$

Viscoelastic relaxation is complete when $t \geq \tau_{VE}$, as indicated in Fig. 15 of Chapter 7, so the value of α is very important. If $\alpha < \sqrt{3}$, then relaxation is completed *after* the pressure distribution becomes parabolic, and there is no effect on the stress (since Eq. (26) applies for either a purely viscous or a purely elastic network when α or μ is small). If $\alpha > \sqrt{3}$, viscoelastic relaxation is complete before the pressure distribution becomes parabolic and the viscous solution applies before t_R. In that case, the stress at the surface of the plate at time t_R is given by Eq. (28), but when $\alpha > \sqrt{3}$, $\sigma_x > 0.7 P_R$ and the gel will certainly break. Thus, under conditions such that disastrous

Fig. 25.

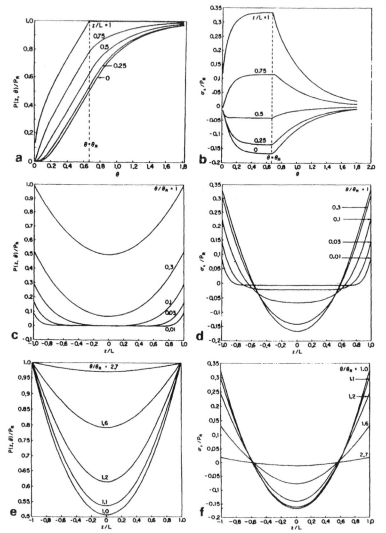

Drying behavior of flat plate that does not shrink. (a) Normalized pressure (P/P_R) in liquid versus reduced time ($\theta = t/\tau$) at several locations in the plate (exterior surface at $z/L = 1$ and midplane at $z/L = 0$). (b) Normalized stress (σ_x/P_R) versus reduced time at same locations as in (a). (c) Pressure distribution at several times during CRP ($\theta < \theta_R$). The CRP is short, because the plate is assumed not to shrink. (d) Normalized stress distribution at same times as in (c). (e) Normalized pressure distribution at several times during FRP ($\theta > \theta_R$). If shrinkage were permitted, the CRP would last longer, and P would equilibrate in $<2.7\theta_R$. (f) Normalized stress distribution at same times as in (e). From Scherer [64].

Fig. 26.

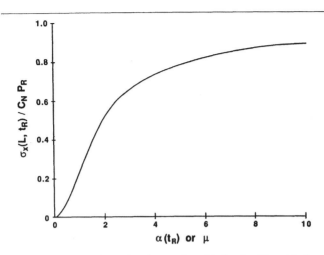

Normalized stress at the surface of a plate at the critical point (time t_R); the constant $C_N \approx 1$. The same curve applies to a viscous gel (characterized by the parameter α defined in Eq. 19) and to an elastic gel (characterized by the parameter $\mu = \dot{V}_E L \eta_L / D P_R$). The larger α or μ, the greater the stress, which asymptotically approaches a maximum value of P_R. From Scherer [65].

drying stresses do not arise, the parabolic pressure distribution is expected to develop before the shinkage of the gel stops (viz., at time t_R), and the stress will be given by Eq. (26). The stress will increase throughout the CRP because of the decrease in permeability, as shown in Fig. 24.

2.4. Diffusion during Drying

In some cases, the pore liquid is transported to the exterior by diffusion, rather than flow. One example is shown in Fig. 1 of Chapter 7, where an aqueous sol was placed in a dialysis bag and immersed in alcohol. The water rapidly diffused through the bag, causing shrinkage and gelation. Beyond the gel point the dialysis membrane is not really needed; as noted by Flory [66], the gel network acts as a membrane inhibiting flow. This is illustrated by the shrinkage of gels immersed in salt solutions (see Fig. 17 of Chapter 6), where the gel shrinks as water diffuses out of the pores of the gel. If the salt concentration is high enough, the diffusion of water causes such rapid shrinkage that the gel cracks almost immediately. During drying, a concentration gradient will develop if one component of the pore liquid is more volatile than the other(s), as is likely to be the case in the usual water–alcohol solution.

For example, if ethanol evaporates faster than water, a gradient of alcohol will develop and ethanol will diffuse from the interior of the sample; at the same time, water will diffuse into the interior. The interdiffusion will affect shrinkage only if the intrinsic diffusion coefficients of the components are different *and* the permeability of the network is low. (See Fig. 9.) If the diffusion coefficients are equal, the fluxes of the components will be equal and opposite, so no net volume change will occur in the saturated pores. If the diffusion coefficients are different, flow will tend to compensate for the volume change (say, by the rapid diffusion of alcohol to the drying surface and counterflow of the pore liquid into the interior). If the permeability is low, that flow will be inhibited, and the diffusion of the volatile component from the interior will cause the tension to rise in the liquid within the gel. This will compensate for the rising tension in the liquid at the exterior, reducing the differential strain and stress.

The diffusion coefficients of water and alcohol may not differ enough for this effect to be of general importance in alkoxide-derived gels. However, incorporation of much more viscous (and consequently slower-diffusing) liquids, such as glycerol, might make diffusion of water and/or alcohol the principal transport process. The importance of this is that interdiffusion of liquids is fast, so that a substantial flux can be produced by a shallow concentration gradient [67]. That is, the more volatile component can diffuse from the interior so readily that its concentration changes almost as rapidly in the middle of the gel as it does at the surface. In that case, the volumetric contraction of the gel would be relatively uniform, so there would be less differential shrinkage and stress during drying. This may be one of the functions of chemical additives, such as formamide and glycerol (discussed in Section 3.3), that are reported [68] to permit faster drying without cracking.

Figure 27a shows the predicted influence of diffusion on the stress in the liquid. The *mutual diffusion coefficient*, D_c, is the volume-weighted average of the intrinsic diffusion coefficients (D_1 and D_2) of the two components of the pore liquid; in the calculation it is assumed that only component 1 evaporates and $D_1 \gg D_2$. (The volatile component diffuses much faster.) When D_c is large, the transport is dominated by diffusion, rather than flow, so the pressure gradient is flattened as diffusion extracts liquid from the interior. As shown in Fig. 27b, the stress in the gel is dramatically reduced at a given value of $\mu = \dot{V}_E L\eta_L/DP_R$ when diffusion is the dominant mechanism of transport. This phenomenon could be exploited by adding a nonvolatile liquid to the pores; then, everything else being equal, the gel could safely be dried much faster. For example, in the absence of diffusion, Fig. 27b shows that when $\mu = 1$ the stress at the surface of a drying plate is $\sigma_x \approx 0.33P_R$ at the critical point (time t_R). If a similar gel is prepared with an additive of lower volatility and diffusivity in the pore liquid, such that

Fig. 27. _____

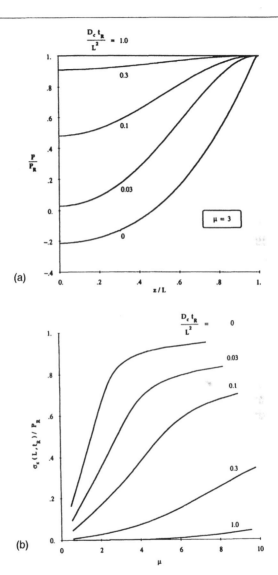

(a)

(b)

(a) Stress distribution $P(z)$ in pore liquid in plate of gel at the critical point (time t_R) when $\mu = \dot{V}_E L\eta_L/DP_R = 3$, calculated assuming that the volatile component diffuses much faster than the nonvolatile component. The greater the mutual diffusion coefficient, D_c, the flatter the pressure distribution. (b) Predicted stress in network at drying surface ($z = L$) as function of parameter μ (proportional to evaporation rate). When evaporation is rapid and diffusion is negligible ($D_c t/L^2 = 0$), stress approaches limiting value of $\sigma_x = P_R$; stress drops rapidly as extent of diffusion increases. From Scherer [67].

$D_c t_R / L^2 = 0.1$, then the maximum stress will be lower as long as $\mu \le 3.5$; that is, the gel can be dried 3.5 times faster with the same stress if the viscosity of the pore liquid (η_L) is the same. However, an additive with high viscosity (such as glycerol) is not helpful, because the increases in η_L offsets the potential increase in $\dot{V}_E$; if η_L is 10 times greater when the additive is present, then the original evaporation rate gives $\mu = 10$, so the stress is $\sim 0.75 P_R$, and the situation has been made worse by the additive. It is suggested [67] that the most favorable approach would be to use an additive (such as acetone) that is *more* volatile and *less* viscous than the original pore liquid, so that the additive would evaporate preferentially. The remaining liquid would be removed after the critical point had been reached, when the danger of cracking is small. This concept remains to be demonstrated experimentally.

2.5. Nonuniform Contraction

The simple analytical results given in this section neglect the variation of properties with location in the body. For example, if the shrinkage near the drying surface is much greater than deep inside the gel, then the permeability will be lower and the viscosity of the network will be greater at the surface. The distribution of pressure and stress could be calculated from the general equations given in Chapter 7, but no simple analytical results would be obtained; the problem would have to be done numerically. An extreme example of such a situation is the drying of a paint film, where a dense skin forms while the interior is still fluid. If a stiff skin formed on a gel while the interior was still soft and wet, the subsequent shrinkage of the interior would be inhibited, and the dried gel would have a density gradient (with lower density inside). The opposite result was obtained by Kawaguchi *et al.* [16], who found that the density was lower at the outer surface of a large piece of xerogel. That region also contained a higher amount of retained organics, so it is possible that reesterification during drying influenced the shrinkage by inhibiting condensation; it is likely that the alcohol content of the pore liquid decreases continually during drying, so the last region to dry (viz., the interior) should be more hydroxylated than the outer surface. Lower density at the surface could also result simply because the interior is subjected to compression for a longer period of time; during the falling rate period, the unsaturated region should shrink more slowly then the saturated region. The existence of such an effect is supported by the warping behavior illustrated in Fig. 22: if the saturated region did not continue to shrink faster than the unsaturated region, there would be no residual convexity.

The stresses discussed in the preceding sections result from differences in strain rate, so they are transient; once shrinkage stops, the stress disappears.

That will not be the case, however, if the viscosity rises so much during drying that relaxation is arrested in part of the body while the rest continues to shrink. For example, if the surface of the plate becomes rigid (elastic) at the end of the CRP, the continued shrinkage of the interior will impose a compressive elastic strain on the surface. As the rest of the body becomes rigid, the elastic strains will be partially retained, with net compression near the original drying surface and tension in the interior. This has been observed to occur in the drying of wood [69]. Kawaguchi *et al.* [16] found optical retardation indicating residual *tension* in the surface of their gel that reversed when the samples were heated above ~300°C; since that is the temperature where the organics burn out, it seems likely that the apparent tension was an artifact produced by the organics, and that the true stress distribution was revealed when they were removed.

Very few studies have looked for evidence of residual stress or inhomogeneity in microstructure. As long as the variations are not great (say, if α does not vary by more than 10–20% through the sample), the equations given previously should provide useful estimates of the stress and strain in the drying gel. Large gradients should not develop unless the initial value of α is large (>1). To analyze residual stresses, it is necessary to use the full viscoelastic analysis, allowing for time-dependent modulus and viscosity. While this could be done in principle, the necessary data are not yet available.

3.

AVOIDING FRACTURE

We have seen that drying produces a pressure gradient in the liquid phase of a gel, which leads to differential shrinkage of the network. When the exterior of the gel tries to shrink faster than the interior, tensile stresses arise that tend to fracture the network at the exterior. As shown in Fig. 28, the material on either side of the crack can contract more freely, so it is favorable for the crack to grow into the drying surface. It may seem odd that *compression* of the network by the liquid causes fracture. In fact, if the pressure in the liquid were uniform, the whole network would be isotropically compressed and the gel would shrink without risk of cracking. However, as illustrated in Fig. 23, the higher tension in the liquid at the exterior causes greater contraction of the network in that region. Since that contraction is inhibited by the slower-contracting interior (where the tension in the liquid is less), the network at the exterior is effectively *stretched*

Fig. 28. _____

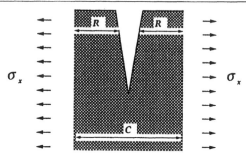

Schematic illustration of stress relief by cracking at the drying surface of a gel. The network in region R is free to relax (contract) in response to the compression applied by the liquid, but the network ahead of the crack in region C is constrained. The stress, σ_x, results from the gradient in P, as indicated in Fig. 23, and is relatively uniform far from the crack.

(prevented from contracting), and this promotes cracking. Thus, it is the differential contraction that produces macroscopic tension in the network and causes cracking.[†]

Cracking occurs when the stress in the network exceeds its strength. Since the classic work of Griffith [70], it has been understood that fracture of brittle materials depends on the presence of flaws that amplify the stress applied to the body. That is, if a uniform stress σ_x is applied to a body containing a crack with a length of c, the stress at the tip of the crack is proportional to $\sigma_x\sqrt{c}$, and failure occurs when that stress exceeds the strength of the material. The theory of linear elastic fracture mechanics (LEFM), which is discussed in several excellent textbooks [71–73], indicates that catastrophic crack propagation occurs when

$$\sigma_x\sqrt{\pi c} \geq K_{IC} \tag{34}$$

where K_{IC} is a material property called the *critical stress intensity factor* and σ_x represents the applied stress. (See Fig. 29.) Crack growth is called "catastrophic," because the stress intensity (the left side of Eq. 34) increases with the size of the crack, so the bigger the crack gets, the faster it goes until it reaches the speed of sound.

[†] An analogous argument can be offered in the case of thermal stress. One could say that a body contracts on cooling because of an effective compressive force that drives atom centers toward one another as the temperature decreases. If a temperature gradient exists in a body such that the surface is cooler, the atoms at the surface are being "compressed" more strongly, but they cannot freely move toward one another because they are bound to the warmer interior of the plate. The interior effectively stretches the surface beyond its equilibrium interatomic spacing, and cracking may result. Then the atoms on either side of the crack can move closer together.

Fig. 29. _____

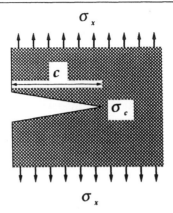

Illustration of applied stress on sample containing crack. The stress at the tip of the crack, σ_c, is greater than the stress applied to the body, σ_x, and increases with the crack size according to $\sigma_c \propto \sigma_x \sqrt{c}$.

Unfortunately, there is only one report of measurement of K_{IC} for an alkoxide-derived gel. Zarzycki [74] determined K_{IC} as a function of aging for two particulate gels and a sonogel (prepared by ultrasonic mixing of TEOS and water [75]); for the latter, K_{IC} was in the range of 160 to 270 Pa$\sqrt{m}$, increasing slightly with age. If the full capillary pressure ($P_R \approx 10$ MPa) were applied to that material, Eq. 34 indicates that any flaw with $c > 0.1$ nm would grow uncontrollably. This unrealistically small size for the critical flaw size accounts for the instantaneous failure of gels when suddenly exposed to the atmosphere. In fact, the stress developed during slow drying, given by Eq. 25, is much smaller than P_R. As shown in Chapter 6, the strength (i.e., the stress that causes fracture in a bending test) of similar gels is found to be ~0.1 MPa [64]; with Zarzycki's data, this corresponds to a critical flaw size of $c \approx 1$–$2\,\mu$m. Flaws of that size are inevitable, unless the mold is highly polished and the sample is handled with the greatest care, so the drying stresses must be kept well below the maximum (viz., $\sigma_x = P_R$).

The theory of LEFM applies to brittle elastic materials, whereas gels are viscoelastic, so the theory would be expected to apply only when the strain rate is too fast for significant relaxation to occur. It can be shown [73] that there is an elastic region near the tip of a moving crack whose dimension, d, is given by

$$d \sim v_c \tau_{VE} \tag{35}$$

where v_c is the velocity of the crack and τ_{VE} is the viscoelastic relaxation time.

There is a zone of plastic deformation near the crack tip with a characteristic dimension of

$$d_{\mathrm{p}} \sim (K_{\mathrm{IC}}/\sigma_{\mathrm{Y}})^2 \tag{36}$$

where σ_{Y} is the yield stress. The concept of critical stress intensity is meaningful as long as $d_{\mathrm{p}} \ll d$ [73]. No yield was seen in the strength measurements [63], so it is safe to say that $\sigma_{\mathrm{Y}} > 0.1$ MPa for that gel, which means that $d_{\mathrm{p}} \leq 1\,\mu$m; since τ_{VE} is on the order of days for a fresh silica gel [63], the elastic theory is expected to apply when $v_{\mathrm{c}} > d/\tau_{\mathrm{VE}} \sim$ microns/day. It appears, therefore, that LEFM provides a reasonable approximation, but there is clearly a need for more experimental studies of the fracture mechanics of wet gels.

Cracking is sometimes attributed to the existence of a pore size distribution in the gel [42,68]. As indicated in Fig. 30, when larger pores are emptied by evaporation, the wall between adjoining pores is subjected to uneven stress that can cause cracking. This provides a simple explanation for the observation that cracking often occurs at the critical point, as the pores begin to empty. However, it does not explain why cracking is prevented by slower evaporation. If this were the principal mechanism of failure, cracking would depend only on the pore size distribution and, for a sample with a wide distribution, cracking would be inevitable when the critical point was reached. In fact, slower drying makes the drying front more irregular on the scale of the pore size [49], so fracture would seem to be more probable at

Fig. 30. ─────────────────────────────────────

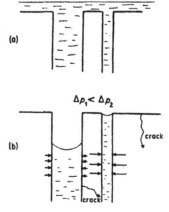

Schematic illustration of cracking resulting from draining of nonuniform pores. (a) Liquid covers surface before drying starts; (b) larger pores empty first, after critical point. The higher tension in the smaller pore creates stress that cracks the "wall" between the pores. From Zarzycki *et al.* [42].

slower drying rates if uneven draining were the problem. Another difficulty with the model is that the resulting flaw would be a "point defect" similar in size to the pore diameter, which is too small to produce catastrophic failure. However, one could argue that these small cracks could link together as the drying front advanced until a critical flaw size (on the order of microns) was produced; that flaw would propagate catastrophically if Eq. 34 were satisfied. The portion of the flaw under tension would be equal to the width of the drying front, as indicated in Fig. 31, so c in Eq. 34 should be replaced with w. Since Eqs. 25 and 12 indicate that $\sigma_x\sqrt{w}$ increases as

Fig. 31.

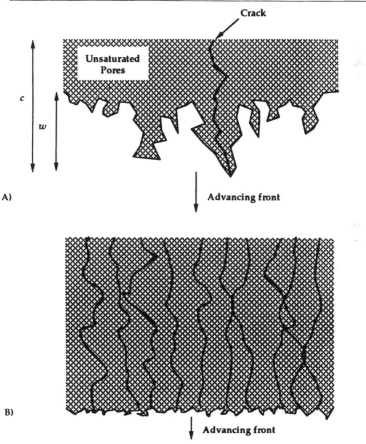

(A) Growing flaw is under stress only over the width of the drying front, w, not the whole length of the crack, c. (B) If w is smaller than the critical flaw size, the dried body could be traversed by cracks that are generated by the drying front, but are too small to propagate catastrophically.

(drying rate)$^{3/4}$, this version of the model accounts for the success of slow drying; however, it suggests that the slowly dried body could be full of subcritical cracks (Fig. 31b). If cracks actually open up during nominally successful drying (i.e., when macroscopic cracks do not appear), they must heal when drying is complete, because large pores (corresponding to open cracks) are not revealed by nitrogen adsorption.

Thus we have two models for the origin of drying cracks. According to the theory presented in Section 2.3, the stress rises continually until the critical point (time t_R). If drying is slow enough that $t_R > \tau/3$, the purely viscous solution (Eq. 25, for a plate of gel) applies near the end of the CRP. It is unlikely that the α parameter is large at t_R, so the stress is probably given by Eq. 29, $\sigma_x \approx P_R L^2 \eta_L / D H_G$. The tension in the network is macroscopic (e.g., the tension described by Eq. 25 extends over a third of the thickness of the plate) and will cause catastrophic growth of pre-existing flaws of size c if Eq. 34 is satisfied. The magnitude of the stress on the network is proportional to the maximum capillary stress, P_R. If this theory is correct, Eqs. 26 and 29 indicate that cracking will be reduced by slower drying, larger pores (which increase D and reduce P_R), smaller $\gamma_{LV} \cos(\theta)$ (which reduces P_R), and less viscous pore liquid (lower η_L). The other model of fracture depends on microscopic processes (i.e., damage accumulating on the scale of the pore diameter); if the microcracks percolate into a macroscopic flaw, catastrophic failure will occur. The formation of microcracks results from the application of the pressure P_R across the pore wall and (in its simplest form) is not influenced by drying rate. If this model is correct, successful drying requires minimizing the capillary stress and/or increasing the strength of the gel. We favor a "hybrid" model: Flaws are generated by severe local stresses at the irregular drying front; these flaws (length $\sim w$) are propagated by the macroscopic stress, σ_x. A variety of strategies for avoiding fracture during drying are described next.

3.1. Larger Pores

Since Eq. 3 indicates that the capillary pressure is inversely proportional to the pore size, the most obvious way to avoid drying stress is to prepare bodies with larger pores. In addition, the Carman–Kozeny equation (see Eq. 15 of Chapter 7) indicates that the permeability is related to the microstructure by

$$D \propto (1 - \rho)a^2 \tag{36}$$

where ρ is the relative density of the network and a is the pore size. In general, Eq. 25 predicts that the drying stress is proportional to $1/D$ and is much less than P_R. The scale of the microstructure is, of course, one of the principle differences between conventional ceramics and gels. Larger pores do allowfaster drying, but the resulting bodies must be sintered at higher temperatures (as explained in Chapter 11). Favorable compromises between drying and firing rates are often obtained with particulate gels or by combination of particles with alkoxides.

Shoup [76] developed a method for making silica gels with extremely uniform pores. (See Chapter 4.) Samples several centimeters in thickness can be dried in a matter of hours or days without risk of fracture, as long as the pore size is greater than ~ 60 nm. Similarly, studies of particulate gels made from fumed silica have found that Aerosil OX-50® (particle size ~ 50 nm) allows drying of large pieces without cracks, but smaller particles (~ 20 nm) lead to cracking [77–79]. Rabinovich [80] used the smaller particles and allowed them to crack during drying, and then redispersed and gelled aggregates of the particles. The larger interaggregate pores enabled the bodies to dry the second time without cracking, but those pores drive the sintering temperature up. Another variation on this theme was performed by Toki *et al.* [81], who mixed OX-50® with TEOS. The resulting gels had larger pores (they were opaque white, rather than clear, as alkoxide-derived silica gels usually are) and large samples could be dried much faster than gels not containing the particles. The advantage of larger pores is generally ascribed to the lower capillary pressure they produce, but it also results in part from the much higher permeability of the gel. In addition, the particles contribute to the strength of the gel; Zarzycki [74] found that additions of Aerosil® to gels made from Ludox® colloidal silica raised K_{IC} substantially. All of these relatively coarse silica gels must be sintered at temperatures of 1300–1500°C, whereas alkoxide-derived gels densify at comparable rates at ~ 1000°C.

3.2. Aging

Aging a gel before drying helps to strengthen the network and thereby reduce the risk of fracture [42]. This has been demonstrated by direct measurements of the modulus of rupture [63,82] and stress intensity factor [74] of gels subjected to various periods of aging in their own pore liquid. The process is accelerated in an aggressive chemical environment, as explained in Chapter 6, where the rate of condensation of M–OH groups is accelerated. For example, Mizuno *et al.* [83] found that they could dry silica gels up to five times faster after soaking for 24 h in 4N HCl or 2N NH₃. The lower bulk density of the aged gels indicates that they were coarser, so the capillary pressure was lower and the permeability higher, and they were probably stiffer and stronger.

3.3. Chemical Additives

Surfactants can be added to the pore liquid to reduce the interfacial energy and thereby decrease the capillary stress. It has been demonstrated for particulate [15] and alkoxide-derived [42] gels that cracking is reduced by surfactants, though it is not necessarily eliminated. The shrinkage at the critical point is reduced by surfactants [23,42], and this will have a beneficial effect on the permeability of the gel, contributing to reduction of the stress during the CRP.

Another group of chemicals, known as *drying control chemical additives* (DCCA) [68], is reported to allow faster drying. One example is formamide (NH_2CHO), which is used to replace ~half of the solvent ordinarily used when making silica gels from alkoxide. The resulting gel is found to be harder (as determined by indentation tests [84]) and to have a larger and more uniform pore size, and all of these features help to reduce cracking. The coarser structure may be a result of the higher pH produced by hydrolysis of formamide [85]. Unfortunately, the additive is difficult to remove upon heating, so bloating and cracking result. Oxalic acid is another DCCA that produces a narrower and larger pore size distribution (though the pores are smaller than when formamide is used) [68]. No data have been presented describing the effect of oxalic acid on the maximum drying rate or the strength of the gel.

The original claims [86] of rapid processing (~48 h) for centimeter-thick pieces of gel processed with formamide have not been repeated nor reproduced, but promising results have been reported for dimethyl formamide (DMF) [87,88]. That additive yields gels with larger pores, and they are even larger after aging at elevated temperatures (~150°C). Gels made with DMF do not crack at drying rates that destroy gels made with formamide, or without any DCCA. Interestingly, the dried gel cracks when exposed to vapors of water ($\gamma_{LV} \approx 0.072$ J/m^2) or formamide (0.058 J/m^2), but not vapors of methanol (0.023 J/m^2) or DMF (0.036 J/m^2), so the lower surface tension of the additive may be important. The gel is readily sintered to clear glass at 1050°C.

To the extent that these additives are effective, their success can be attributed to coarsening of the microstructure (which increases D and decreases P_R) and strengthening of the network. They may also provide a medium through which the more volatile components (water and alcohol) can diffuse, thereby allowing diffusion to reduce the pressure differential within the body [67]. A different type of DCCA is glycerol, which has been found to reduce cracking in alumina gels [89] and composites with alumina gel matrices [90], and is used as a ''plasticizer'' in freeze-drying [43]. Since this molecule has three hydroxyl groups, it adsorbs strongly on the gel

surface [91,92] and is likely to reduce the capillary pressure in two ways:
(1) by forming a film on the surface, it reduces the contact angle; (2) because
of its low vapor pressure it will not evaporate, but will be retained in the
smallest pores. The fact that it provides a plasticizing effect in the "dried"
gel indicates that it is retained; indeed, it is not removed until heating to high
temperatures, and even then it tends to decompose into carbonates rather
than evaporating [93].

3.4. Supercritical Drying

Since shrinkage and cracking are produced by capillary forces, Kistler [94]
reasoned that those problems could be avoided by removing the liquid from
the pores above the *critical temperature* (T_c) and *critical pressure* (P_c) of the
liquid. As indicated in the phase diagram in Fig. 32, there is no longer any
distinction between the liquid and vapor phases: the densities become equal,
there is no liquid–vapor interface and no capillary pressure. In the process
of *supercritical* (or *hypercritical*) *drying*, a sol or wet gel is placed into an
autoclave and heated along a path such as the one indicated in Fig. 32. The
pressure and temperature are increased in such a way that the phase boundary
is not crossed; once the critical point[†] is passed, the solvent is vented at a

Fig. 32.

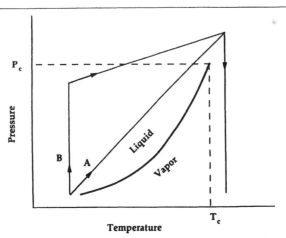

Schematic diagram of liquid–vapor phase boundary (dark curve). At the critical point
(T_c, P_c) the density of the liquid and vapor phases are the same. The conventional
temperature–pressure path followed during supercritical drying is represented by A,
while B corresponds to the use of nitrogen overpressure.

[†] Not to be confused with the "critical" point that marks the end of the CRP.

Table 1.

Critical Points of Selected Solvents.[†]

Substance	Formula	T_c (°C)	P_c (MPa)
Carbon dioxide	CO_2	31.1	7.36
Freon® 116	CF_3CF_3	19.7	2.97
Methanol	CH_3OH	240	7.93
Ethanol	C_2H_5OH	243	6.36
Water	H_2O	374	22.0

[†] Data collected by Tewari, Hunt, and Lofftus, ref. 117.

constant temperature ($>T_c$). The resulting gel, called an *aerogel*, has a volume similar to that of the original sol. This process makes it possible to produce monolithic gels as large as the volume of the autoclave. Table 1 contains values of T_c and P_c for some relevant liquids.

Kistler [94] found that silica gel would dissolve if subjected to supercritical conditions when the pores contained water, so he exchanged the liquid by washing the gel in ethanol. In a survey of many types of gels he found that alcohol would attack some, so further exchanges were required (e.g., ether for tungstic and stannic oxides, pentane for ferric oxide). Teichner *et al.* [95,96] have studied many oxide systems (e.g., Al_2O_3, TiO_2, NiO/Al_2O_3, Fe_2O_3/Al_2O_3, MgO/ZrO_2) prepared from alkoxides and found that they could be supercritically dried directly (i.e., with a water–alcohol solution in the pores) without time-consuming washing. The resulting powders were found to be useful as catalysts because of their high surface area and exceptional chemical activity.

The structure of silica aerogels made from alkoxides has been extensively studied [97]. The skeletal density is found by helium pycnometry [98,99] to be $\sim 2 \text{ g/cm}^3$, as it is for xerogels. Small-angle X-ray scattering (SAXS) [100] indicates that primary particles with diameters of ~ 3 nm are linked into structures with a fractal dimension of ~ 2; the density becomes uniform above a scale of ~ 8.5 nm. There is another level of microstructure beyond that resolved by SAXS, as illustrated by the beautiful transmission electron micrographs by Mulder *et al.* [101] shown in Fig. 33. It is estimated [102] that 90% of the pore volume is in pores with diameters between 60 and 200 nm. Characterization of the structure of these materials is difficult, because much of the porosity is too large to be detected by nitrogen adsorption–desorption techniques [102]. Large pores are customarily measured using mercury penetration porosimetry, but that causes compression of the very fragile aerogels [98] and gives the false impression that very large pores are present.

Fig. 33. _____

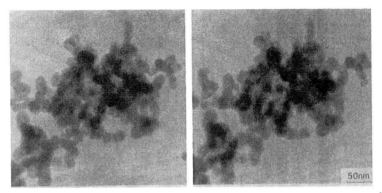

Stereo pair of electron micrographs of silica aerogel with bulk density of 0.15 g/cm^3. From Mulder _et al._ [101].

Although there is little shrinkage of the gel during drying (0–25% under acidic conditions and <5% under basic conditions [99]), the structure is clearly not the same as that of the network of the wet gel. Thermoporometry provides a direct indication that pore coarsening occurs during supercritical treatment [103]. Moreover, if the aerogel is carefully rewet by allowing vapors to condense in its pores, the shrinkage on subsequent drying is smaller than in a xerogel formed directly from the sol. This reflects the strengthening of the network by the high-temperature treatment. In the case of alumina, the aerogel contains boehmite, whereas a xerogel formed from the same sol is amorphous [96]. These observations are not surprising in view of the effect of temperature and on the pressure aging process, as discussed in Chapters 3 and 6.

If an aerogel is immersed in liquid, it collapses immediately [94]. When it is subsequently dried, the resulting xerogel has the same surface area as the aerogel [102,104]. When vapors are allowed to condense to a small extent in the pores of an aerogel, the redried gel is found to have lost a portion of its larger pores, as shown in Fig. 34. These observations reveal that the large-scale structure is weak and easily collapsed by capillary pressure; the micropores and mesopores that produce the high surface area are present in a more rigid network.

Two groups succeeded at about the same time in making large monolithic gels by supercritical drying. In one case [105] the aerogel itself was the objective: the low density of the silica gel was required for a Cherenkov radiation detector [106]. Large samples were prepared in an autoclave with a capacity of 3000 liters (which subsequently exploded! [107]). The other group [108,109] wanted to make monolithic gels to be sintered into dense glasses

Fig. 34.

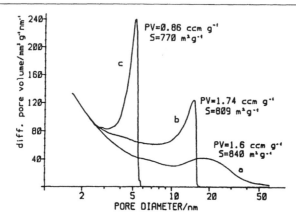

Pore size distribution for silica aerogel after treatments with water. (a) Original sample.
(b) After storage in water vapor. (c) After storage in liquid water. These data obtained
by nitrogen adsorption do not reveal the larger pores that constitute most of the pore
volume. From Schuck *et al.* [102].

or ceramics. Large crack-free bodies could be made within wide ranges of
concentration of reactants. The most important factor was found to be the
presence of excess alcohol in the autoclave, so that the vapor pressure would
be high enough to prevent evaporation from the gel during heating. Since the
pores contain a mixture of alcohol and water, the temperature and pressure
must exceed the critical point for the mixture. Some workers have found that
it is difficult to avoid cracking unless the autoclave is initially 100% full of
alcohol, and T and P exceed T_c and P_c for water [110]. A much more
convenient method was introduced by van Lierop *et al.* [111]. They found
that premature evaporation of alcohol could be suppressed by pressurizing
the autoclave with 80 bars (8 MPa) of nitrogen. If the vapors behaved ideally,
the vapor pressure of alcohol would be unaffected by the presence of
nitrogen, but at these pressures there is evidently a strong interaction. The
result is faster processing and no shrinkage of the gel during drying. Silica
gels with bulk densities as low as 0.03 g/cm^3 were made in this way.

Supercritical drying has now been used to make monolithic gels from a
wide range of oxides. Woignier *et al.* [112] prepared borosilicates, phosphosilicates, and ternary compositions. It was necessary to age the gels at
temperatures of 25–60°C for weeks to strengthen them before supercritical drying. The phosphosilicates showed traces of crystallinity, and the
ternaries contained crystals of BPO$_4$. Complex glasses such as cordierite
(2MgO•2Al$_2$O$_3$•5SiO$_2$) [113] and mullite (3Al$_2$O$_3$•5SiO$_2$) [114] have also
been successfully prepared.

Although supercritical drying gives very good results for silica, the high temperatures and pressures make the process expensive and dangerous. A convenient alternative is to exchange the pore liquid for a substance with a much lower critical point. As shown in Table 1, carbon dioxide has $T_c = 31°C$ and $P_c = 7.4$ MPa, so the process can be performed near ambient temperatures. Kistler [94] tried this approach in an unsuccessful attempt to make an aerogel from swollen rubber. Now, however, supercritical drying following CO_2 exchange has become a standard technique for preparing biological samples for TEM examination [115]. It was apparently first applied for producing monolithic silica gels by Woignier [116] and was independently developed by Tewari *et al.* [117] for making large windows. For some materials, supercritical treatment in alcohol causes dissolution, so a milder process is essential. Brinker *et al.* [118] used CO_2 exchange to make aerogels of lithium borate compositions that would dissolve in alcohol. This would seem to be an ideal way of making aerogels, but it does have some disadvantages. Long times can be required to achieve complete solvent exchange, especially because CO_2 is not miscible with water. (Kistler [94] notes that liquid–liquid interfaces formed by immiscible liquids could produce capillary compression of the gel.) It may be necessary to exchange first with a mutual solvent such as amyl acetate [118] and then to flush for hours with liquid CO_2. The aerogels produced by this method are hydrophilic, whereas aerogels made directly from alcohol are hydrophobic, because of reesterification at high temperatures [109]. The hydrophilic gels can pick up ambient moisture and crack long after drying [119].

3.5. Freeze-Drying

Another way of avoiding the presence of the liquid–vapor interface is to freeze the pore liquid and sublime the resulting solid under vacuum. This process of *freeze-drying* is widely used in the preparation of foods [43], but does not permit the preparation of monolithic gels. The reason is that the growing crystals reject the gel network, pushing it out of the way until it is stretched to the breaking point. It is this phenomenon that allows gels to be used as hosts for crystal growth [120,121]: the gel is so effectively excluded that crystals nucleated in the pore liquid are not contaminated with the gel phase; the crystals can grow up to a size of a few millimeters before the strain is so great that macroscopic fractures appear in the gel. If a silica sol is frozen, flakes of silica gel (sometimes called *lepidoidal silica*) are produced [122]; if freezing is done unidirectionally, fibers of gel are obtained [123,124]. Attempts to freeze-dry gels typically result in flakes (e.g., [125]) or in translucent bodies with large pores that are the "fossils" of the crystals.

This approach might work if the pore liquid vitrified, rather than crystallized; however, the glassy solvent might disproportionate during sublimation, and that could lead to reappearance of a liquid that would boil under the reduced pressure and destroy the gel.

4.

FILMS

The process of deposition of films, of which drying is an integral part, is discussed in detail in Chapter 13. The only aspect of the problem that we need to consider here is the phenomenon of fracture of films. Actually, the *phenomenal* part is that films don't *always* crack! According to Eq. 87 of Chapter 7, the stress in a film is approximately equal to the tension in the liquid, $\sigma_x \approx P$. Therefore, as the film become rigid, the tension in the film becomes equal to the capillary stress, $\sigma_x \approx P_R$, which would shatter a macroscopic gel. There is no indication that the capillary stress is absent in a film; on the contrary, it seems sufficient to cause complete collapse of the porosity in some cases [126], as predicted [34] for a material with a very compliant matrix. The fascinating observation is that inorganic films thinner than $\sim 0.5\,\mu$m do not crack, regardless of the drying rate, whereas films thicker than $\sim 1\,\mu$m are virtually impossible to dry without cracks. (See, e.g., [127–131].)

This behavior may be analogous [132] to crack growth in composites containing inclusions, where thermal expansion mismatch causes cracks to appear near large inclusions, but not near small ones. The explanation [133] is that the energy that drives crack growth comes from relief of stress in a volume proportional to the size of the inclusion, while the energy invested to grow the crack is proportional to the surface area created. When the inclusion is small, this surface-to-volume ratio becomes a barrier to growth of the crack; that is, the energy invested to extend the crack is greater than the energy gained from relief of stresses near the inclusion. This line of argument is supported by the work of Thouless [134], who investigated the conditions for growth of flaws of the type shown in Fig. 35. It was shown that the energy release rate, G, for such flaws is approximately proportional to the thickness of the film, $G \sim L$ (when $R/a \approx 1$); there is also a very weak dependence on flaw size ($G \sim R^{0.1}$). The stress intensity, K_I, at the tip of the crack is given by

$$K_I \propto \sqrt{G} \propto \sqrt{L} \tag{37}$$

so as $L \to 0$, it inevitably falls below the critical stress intensity of the material, K_{IC}. Unfortunately, we cannot apply this result to see whether it

Fig. 35. _____

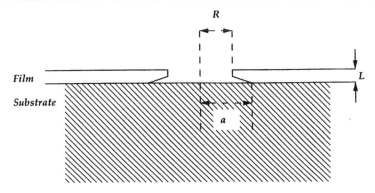

Pinhole of radius R in film extends to radius a along interface. Film thickness L is much less than the thickness of the substrate. The flaw propagates by peeling (a grows, R fixed). From Thouless [134].

accounts quantitatively for the observed limit of $\sim 1\,\mu m$ for films, because the physical properties of the film are not known.

Two exceptions to the "micron rule" should be noted. Schmidt *et al.* [135] were able to dry films $20\,\mu m$ thick by using a silicon precursor with two nonhydrolyzing phenyl substituents. This evidently provided sufficient compliance (rubberiness) to prevent fracture. Cracking did occur when the organics were burned out, but large (compared to L) areas remained intact. This appears to be a promising route to preparation of thicker films. Another remarkable exception to the rule was produced by Schlichting [136], who made dense films $\sim 10\,\mu m$ thick from alkoxides. It appears that he dipped the substrates in *unhydrolyzed* alkoxides for 10–15 minutes and allowed hydrolysis to occur in the air during withdrawal of the substrate, then heated the samples slowly (37°C/h) to 700 to 900°C. Perhaps these mild conditions left a substantial fraction of ligands unhydrolyzed during drying (as contrived deliberately by Schmidt *et al.*), allowing the rubbery film to remain intact. This remarkably simple approach deserves to be re-examined.

5. _____

SUMMARY

Silica and most other inorganic gels shrink spontaneously while immersed in their pore liquid. In these materials, electrostatic and osmotic forces are relatively unimportant, and shrinkage during evaporation is caused by capillary stresses. Some systems, including alumina and some clays, expand when immersed, because of the strong affinity of the solid phase for water (adsorption forces) and electrostatic repulsion between the particles. In such gels, osmotic pressure may play an important role in shrinkage during drying; electrostatic repulsion is expected to produce flow of the pore liquid, but concentration gradients (e.g., of electrolytes) will cause diffusion.

As liquid evaporates, the solid network is drawn into the remaining liquid to prevent exposure of the energetic solid–vapor interface. The network is sucked under the surface of the liquid by the capillary tension, and as long as the network remains sufficiently compliant, the liquid–vapor meniscus remains at the exterior surface and the evaporation rate is constant. Once the network becomes stiff enough to withstand the pressure, evaporation drives the meniscus into the body and the evaporation rate begins to drop. For a time, liquid is able to flow along contiguous channels of liquid and evaporate at the exterior; later, evaporation occurs within the body and transport to the surface requires diffusion of the vapor.

If the pressure in the liquid were uniform, the network would be hydrostatically compressed and there would be no risk of fracture. However, because of the low permeability of the gel network, the tension is greater in the liquid near the drying surface, and this produces differential strain leading to cracking. The greatest stress and highest probability of fracture occurs when shrinkage stops and the meniscus begins to recede into the body. The tension in the network, given approximately by $\sigma_x \approx \alpha^2 P_R$, is reduced by factors that increase the permeability of the network or reduce the capillary pressure (e.g., lower surface tension, larger pore size). The same factors affect the degree of warping of bodies that are dried nonuniformly, such as a plate with evaporation from only one side. Different rules apply to the drying of films, which seem to be immune to fracture when they are thinner than $\sim 0.5\,\mu\text{m}$. Films thicker than $\sim 1\,\mu\text{m}$ have only been dried successfully by incorporating organics into the network to provide extra compliance.

REFERENCES

1. T.K. Sherwood, *Ind. Eng. Chem.*, **21** [1] (1929) 12–16.
2. T.K. Sherwood, *Ind. Eng. Chem.*, **21** [10] (1929) 976–980.
3. T.K. Sherwood, *Ind. Eng. Chem.*, **22** [2] (1930) 132–136.
4. R.W. Ford, *Ceramics Drying* (Pergamon, New York, 1986).
5. R.B. Keey, *Drying, Principles and Practice* (Pergamon, New York, 1972).
6. F.H. Clews, *Heavy Clay Technology* (Academic Press, New York, 1969).
7. F. Moore, *Trans. Brit. Ceram. Soc.*, **60** (1961) 517–539.
8. M. Fortes and M.R. Okos in *Advances in Drying*, vol. 1, ed. A.S. Mujumdar (Hemisphere, New York, 1980), pp. 119–154.
9. *Advances in Drying*, vol. 1, ed. A.S. Mujumdar (Hemisphere, New York, 1980).
10. *Advances in Drying*, vol. 2, ed. A.S. Mujumdar (Hemisphere, New York, 1983).
11. H.H. Macey, *Trans. Brit. Ceram. Soc.*, **41** [4] (1942) 73–121.
12. R.K. Dwivedi, *J. Mater. Sci. Lett.*, **5** (1986) 373–376.
13. N.H. Ceaglske and O.A. Hougen, *Ind. Eng. Chem.*, **29** (1937) 805–813.
14. M. Suzuki and S. Maeda, *J. Chem. Eng. Japan*, **1** [1] (1968) 26–31.
15. P.G. Simpkins, D.W. Johnson, Jr., and D.A. Fleming, *J. Am. Ceram. Soc.*, **72** [10] (1989) 1816–1821.
16. T. Kawaguchi, J. Iura, N. Taneda, H. Hishikura, and Y. Kokubu, *J. Non-Cryst. Solids*, **82** (1986) 50–56.
17. A.C. Pierre and D.R. Uhlmann in *Better Ceramics Through Chemistry III*, Mater. Res. Soc. Symp. Proc., **vol. 121**, eds. C.J. Brinker, D.E. Clark, and D.R. Ulrich (Mater. Res. Soc., Pittsburgh, Pa., 1988).
18. R. Toei in *Advances in Drying*, vol. 2, ed. A.S. Mujumdar (Hemisphere, New York, 1983), pp. 269–297.
19. P.C. Wayne, Jr., Y.K. Kao, and L.V. LaCroix, *Int. J. Heat Mass Trans.*, **19** (1976) 487–492.
20. H. van Olphen, *An Introduction to Clay Colloid Chemistry*, 2d ed. (Wiley, New York, 1977).
21. T. Tanaka and D.J. Fillmore, *J. Chem. Phys.*, **70** [3] (1979) 1214–1218.
22. J.J. Spitzer, *Langmuir*, **5** (1989) 199–205.
23. W.D. Kingery and J. Francl, *J. Am. Ceram. Soc.*, **37** [12] (1954) 596–602.
24. A.R. Cooper in *Ceramics Processing before Firing*, eds. G.Y. Onoda, Jr., and L.L. Hench (Wiley, New York, 1978), pp. 261–276.
25. O.A. Hougen, H.J. McCauley, and W.R. Marshall, Jr., *Trans. Am. Inst. Chem. Eng.*, **36** (1940) 183–209.
26. S. Whitaker, *Adv. Heat Transfer*, **13** (1977) 119–203.
27. J.R. Philip, *Adv. Hydrosci.*, **5** (1969) 215–296.
28. J.M. Kuzmak and P.J. Sereda, *Soil Sci.*, **84** (1957) 291–299, 419–422.
29. M. Yamane and T. Kojima, *J. Non-Cryst. Solids*, **44** (1981) 181–190.
30. M. Atik and J. Zarzycki, *J. Mater. Sci. Lett.*, **8** (1989) 32–36.
31. I.M. Thomas in *Sol-Gel Technology for Thin Films, Fibers, Preforms, Electronics, and Speciality Shapes*, ed. L.C. Klein (Noyes, Park Ridge, N.J., 1988), pp. 2–15.
32. T. Deichsel, *Betonwerk & Fertigteil-Technik.*, **48** [10] (1982) 590–597.
33. M. Prassas, J. Phalippou, L.L. Hench, and J. Zarzycki, *J. Non-Cryst. Solids*, **48** (1982) 79–95.
34. W.H. Banks and W.W. Barkas, *Nature*, **158** (1946) 341–342.
35. M. Yamane, A. Shinji, and T. Sakaino, *J. Mater. Sci.*, **13** (1978) 865–870.

36. B.M. Mitsyuk, Z.Z. Vysotskii, and M.V. Polyakov, *Kokl. Akad. Nauk SSSR*, **155** [6](1964) 416–418 (Eng. trans.).

37. A.W. Adamson, *Physical Chemistry of Surfaces* (Interscience, New York, 1967), p. 541.

38. T. Kawaguchi, H. Hishikura, J. Iura, and Y. Kokubu, *J. Non-Cryst. Solids*, **63** (1984) 61–69.

39. G.W. Scherer, *J. Non-Cryst. Solids*, **109** (1989) 183–190.

40. R. Clasen, *J. Non-Cryst. Solids*, **89** (1987) 335–344.

41. P. Anderson and L.C. Klein, *J. Non-Cryst. Solids*, **93** (1987) 415–422.

42. J. Zarzycki, M. Prassas, and J. Phalippou, *J. Mater. Sci.*, **17** (1982) 3371–3379.

43. S. Bruin and K.Ch.A.M. Luyben in *Advances in Drying*, vol. **1**, ed. A.S. Mujumdar (Hemisphere, New York, 1980), pp. 155–215.

44. J. van Brakel in *Advances in Drying*, vol **1**, ed. A.S. Mujumdar (Hemisphere, New York, 1980), pp. 217–267.

45. S. Whitaker in *Advances in Drying*, vol. **1**, ed. A.S. Mujumdar (Hemisphere, New York, 1980), pp. 23–61.

46. R. Lenormand, *C.R. Acad. Sc. Paris*, **301** [5] (1985) 247–250.

47. K.J. Måløy, J. Feder, and T. Jøssang, *Phys. Rev. Lett.*, **55** [24] (1985) 2688–2691.

48. D. Wilkinson and J.F. Willemsen, *J. Phys. A*, **16** (1983) 3365–3376.

49. T.M. Shaw in *Better Ceramics Through Chemistry II*, Mater. Res. Soc. Symp., vol. **73**, eds. C.J. Brinker, D.E. Clark, and D.R. Ulrich (Mater. Res. Soc., Pittsburgh, Pa., 1986), pp. 215–223.

50. T.M. Shaw, *Phys. Rev. Lett.*, **59** [15] (1987) 1671–1674.

51. W. Stöber, A. Fink, and E. Bohn, *J. Colloid Interface Sci.*, **26** (1968) 62–69.

52. G.W. Scherer, unreported work involving gels of the type described in ref. 77.

53. S. Whitaker in *Drying '85*, ed. R. Toei and A.S. Mujumdar (Hemisphere, New York, 1985), pp. 21–32.

54. C.K. Wei, H.T. Davis, E.A. Davis, and J. Gordon, *AIChE J.*, **31** [8] (1985) 1338–1348.

55. C.K. Wei, H.T. Davis, E.A. Davis, and J. Gordon, *AIChE J.*, **31** [5] (1985) 842–848.

56. J.R. Philip and D.A. de Vries, *Trans. Am. Geophys. Union*, **38** [2] (1957) 222–232, erratum p. 594.

57. W.A. Kauzman, *Kinetic Theory of Gases* (W.A. Benjamin, New York, 1966).

58. F. Suzuki, K. Onozato, and Y. Kurokawa, *J. Non-Cryst. Solids*, **94** (1987) 160–162.

59. G.W. Scherer, *J. Non-Cryst. Solids*, **91** (1987) 83–100.

60. G.W. Scherer, *J. Non-Cryst. Solids*, **109** (1989) 171–182.

61. G.W. Scherer, *J. Non-Cryst. Solids*, **100** (1988) 77–92.

62. G.W. Scherer and R.M. Swiatek, *J. Non-Cryst. Solids*, **113** (1989).

63. G.W. Scherer, S.A. Pardenek, and R.M. Swiatek, *J. Non-Cryst. Solids*, **107** [1] (1988) 14–22.

64. G.W. Scherer, *J. Non-Cryst. Solids*, **92** (1987) 122–144.

65. G.W. Scherer, *J. Non-Cryst. Solids*, **99** (1988) 324–358.

66. P.J. Flory, *Principles of Polymer Chemistry* (Cornell Univ. Press, Ithaca, N.Y., 1953), p. 577.

67. G.W. Scherer, *J. Non-Cryst. Solids*, **107** (1989) 135–148.

68. L.L. Hench in *Science of Ceramic Chemical Processing*, eds., L.L. Hench and D.R. Ulrich (Wiley, New York, 1986), pp. 52–64.

69. P.F. Lesse and R.S.T. Kingston, *Wood Sci. Tech.*, **6** (1972) 272–283.

70. A.A. Griffith, *Phil. Trans. Roy. Soc.*, **A221** (1920) 163–198.

71. B.R. Lawn and T.R. Wilshaw, *Fracture of Brittle Solids* (Cambridge Univ. Press, Cambridge, England, 1975).

72. H.L. Ewalds and R.J.H. Wanhill, *Fracture Mechanics* (Edward Arnold, Victoria, Australia, 1984).
73. G.P. Cherepanov, *Mechanics of Brittle Fracture* (McGraw-Hill, New York, 1979).
74. J. Zarzycki, *J. Non-Cryst. Solids*, **100** (1988) 359–363.
75. L. Esquivias and J. Zarzycki in *Ultrastructure Processing of Advanced Ceramics*, eds. J.D. Mackenzie and D.R. Ulrich (Wiley, New York, 1988), pp. 255–270.
76. R.D. Shoup, *Colloid and Interface Science*, **vol. III** (Academic Press, New York, 1976), pp. 63–69.
77. G.W. Scherer and J.C. Luong, *J. Non-Cryst. Solids*, **63** (1984) 163–172.
78. G.W. Scherer in *Better Ceramics Through Chemistry*, eds. C.J. Brinker, D.E. Clark, and D.R. Ulrich (North-Holland, New York, 1984), pp. 205–211.
79. R. Clasen, *Glastech. Ber.*, **61** [5] (1988) 119–126.
80. E. Rabinovich in *Sol-Gel Technology for Thin Films, Fibers, Preforms, Electronics, and Speciality Shapes*, ed. L.C. Klein (Noyes, Park Ridge, N.J., 1988), pp. 260–294.
81. M. Toki, S. Miyashita, T. Takeuchi, S. Kanbe, and A. Kochi, *J. Non-Cryst. Solids*, **100** (1988) 479–482.
82. J.K. West, R. Nikles, and G. Latorre in *Better Ceramics Through Chemistry III*, Mater. Res. Soc. Symp. Proc., **vol. 121**, eds. C.J. Brinker, D.E. Clark, and D.R. Ulrich (Mater. Res. Soc., Pittsburgh, Pa., 1988), pp. 219–224.
83. T. Mizuno, H. Nagata, and S. Manabe, *J. Non-Cryst. Solids*, **100** (1988) 236–240.
84. S.H. Wang and L.L. Hench in *Better Ceramics Through Chemistry*, Mater. Res. Soc. Symp. Proc., **vol. 32**, eds. C.J. Brinker, D.E. Clark, and D.R. Ulrich (North-Holland, New York, 1984), pp. 71–77.
85. G. Orcel and L.L. Hench in *Better Ceramics Through Chemistry*, Mater. Res. Soc. Symp. Proc., **vol. 32**, eds. C.J. Brinker, D.E. Clark, and D.R. Ulrich (North-Holland, New York, 1984), pp. 79–84.
86. S. Wallace and L.L. Hench in *Better Ceramics Through Chemistry*, Mater. Res. Soc. Symp. Proc., **vol. 32**, eds. C.J. Brinker, D.E. Clark, and D.R. Ulrich (North-Holland, New York, 1984), pp. 47–52.
87. T. Adachi and S. Sakka, *J. Mater. Sci.*, **22** (1987) 4407–4410.
88. T. Adachi and S. Sakka, *J. Non-Cryst. Solids*, **99** (1988) 118–128.
89. S.M. Wolfrum, *J. Mater. Sci. Lett.*, **6** (1987) 706–708.
90. J.J. Lannuti and D.E. Clark in *Better Ceramics Through Chemistry*, eds. C.J. Brinker, D.E. Clark, and D.R. Ulrich (North-Holland, New York, 1984), pp. 375–381.
91. G.C. Bye and K.S.W. Sing in *Particle Growth in Suspensions*, ed. A.L. Smith (Academic Press, New York, 1973), pp. 29–65.
92. R. Yu. Sheinfain and I.E. Neimark in *Adsorption and Adsorbents*, **vol. 1**, ed. D.N. Strazhesko (Wiley, New York, 1973), pp. 87–95.
93. L.L. Hench in *Better Ceramics Through Chemistry*, Mater. Res. Soc. Symp. Proc., **vol. 32**, eds. C.J. Brinker, D.E. Clark, and D.R. Ulrich (North-Holland, New York, 1984), pp. 101–110.
94. S.S. Kistler, *J. Phys. Chem.*, **36** (1932) 52–64.
95. S.J. Teichner, G.A. Nicolaon, M.A. Vicarini, and G.E.E. Gardès, *Adv. Colloid Interface Sci.*, **5** (1976) 245–273.
96. S.J. Teichner in *Aerogels*, ed. J. Fricke (Springer-Verlag, New York, 1986), pp. 22–30.
97. J. Fricke and G. Reichenauer, *J. Non-Cryst. Solids*, **95 & 96** (1987) 1135–1142.
98. F.J. Broecker, W. Heckmann, F. Fischer, M. Mielke, J. Schroeder, and A. Strange in *Aerogels*, ed. J. Fricke (Springer-Verlag, New York, 1986), pp. 160–166.
99. T. Woignier and J. Phalippou, *J. Non-Cryst. Solids*, **93** (1987) 17–21.

100. D.W. Schaefer, K.D. Keefer, J.H. Aubert, and P.B. Rand in *Science of Ceramic Chemical Processing*, eds. L.L. Hench and D.R. Ulrich (Wiley, New York, 1986), pp. 140-147.

101. C.A.M. Mulder, G. van Leeuwen-Stienstra, J.G. van Lierop, and J.P. Woerdman, *J. Non-Cryst. Solids*, **82** (1986) 148-153.

102. G. Schuck, W. Dietrich, and J. Fricke in *Aerogels*, ed. J. Fricke (Springer-Verlag, New York, 1986), pp. 148-153.

103. C. Eyraud, J.F. Quinson, and M. Brun in *Characterization of Porous Solids* (Elsevier, Amsterdam, 1988), pp. 295-305.

104. M.F.L. Johnson and H.E. Ries, Jr., *J. Am. Chem. Soc.*, **72** (1950) 4289.

105. S. Henning and L. Svensson, *Physica Scripta*, **23** (1981) 697-702.

106. G. Poelz in *Aerogels*, ed. J. Fricke (Springer-Verlag, New York, 1986), pp. 176-187.

107. S. Henning in *Aerogels*, ed. J. Fricke (Springer-Verlag, New York, 1986), pp. 38-41.

108. M. Prassas, thesis, Univ. des Sciences et Techniques du Languedoc, Montpellier, France, 1981.

109. M. Prassas, J. Phalippou, and J. Zarzycki, *J. Mater. Sci.*, **19** (1984) 1656-1665.

110. R.A. Laudise and D.W. Johnson, Jr., *J. Non-Cryst. Solids*, **79** (1986) 155-164.

111. J.G. van Lierop, A. Huizing, W.C.P.M. Meerman, and C.A.M. Mulder, *J. Non-Cryst. Solids*, **82** (1986) 265-270.

112. T. Woignier, J. Phalippou, and J. Zarzycki, *J. Non-Cryst. Solids*, **63** (1984) 117-130.

113. H. Vesteghem, A.R. di Giampaolo, and A. Dauger, *J. Mater. Sci. Lett.*, **6** (1987) 1187-1189.

114. M.N. Rahaman, L.C. de Jonghe, S.L. Shinde, and P.H. Tewari, *J. Am. Ceram. Soc.*, **71** [7] (1988) C338-C341.

115. A.A. Bartlett and H.P. Burstyn in *Scanning Electron Microscopy 1975, Part I* (ITT Research Inst., Chicago, 1975), pp. 305-316.

116. T. Woignier, thesis, Univ. des Sciences et Techniques du Languedoc, Montpellier, France, 1984.

117. P.H. Tewari, A.J. Hunt, and K.D. Lofftus, *Mater. Lett.*, **3** [9,10] (1985) 363-367.

118. C.J. Brinker, K.J. Ward, K.D. Keefer, E. Holupka, P.J. Bray, and P.K. Pearson in *Aerogels*, ed. J. Fricke (Springer-Verlag, New York, 1986), pp. 57-67.

119. J. Phalippou, T. Woignier, and M. Prassas, *Revue de Physique Appliquée*, **24** [4] (1989) c4-47-c4-52.

120. H.K. Henisch, *Crystal Growth in Gels* (Penn. State Univ. Press, University Park, Pa., 1970).

121. M.C. Robert and F. Lefaucheux, *J. Cryst. Growth*, **90** (1988) 358-367.

122. R.K. Iler, *The Chemistry of Silica* (Wiley, New York, 1979), pp. 21-23.

123. W. Mahler and M. Bechtold, *Nature*, **285** (1980) 27-28.

124. T. Maki and S. Sakka, *J. Non-Cryst. Solids*, **82** (1986) 239-245.

125. E. Degn Egeberg and J. Engell, *Revue de Physique Appliquée*, **24** [4] (1989) c4-23-c4-28.

126. G.C. Frye, A.J. Ricco, S.J. Martin, and C.J. Brinker in *Better Ceramics Through Chemistry III*, eds. C.J. Brinker, D.E. Clark, and D.R. Ulrich (Mat. Res. Soc., Pittsburgh, Pa., 1988), pp. 349-354.

127. H. Schroeder in *Physics of Thin Films*, ed. G. Hass (Academic Press, New York, 1969), pp. 87-141.

128. H. Dislich in *Sol-Gel Technology for Thin Films, Fibers, Preforms, Electronics, and Specialty Shapes*, ed. L.C. Klein (Noyes, Park Ridge, N.J., 1988), pp. 50-79.

129. S. Sakka, K. Kamiya, K. Makita, and Y. Yamamoto, *J. Non-Cryst. Solids*, **63** (1984) 223-235.

130. I. Strawbridge and P.F. James, *J. Non-Cryst. Solids*, **82** (1986) 366-372.

131. I. Strawbridge and P.F. James, *J. Non-Cryst. Solids*, **86** (1986) 381–393.
132. G.W. Scherer, *J. Non-Cryst. Solids*, **89** (1987) 217–238.
133. F.F. Lange in *Fracture Mechanics of Ceramics*, **vol. 2**, *Microstructure, Materials, and Applications*, eds. R.C. Bradt, D.P.H. Hasselman, and F.F. Lange (Plenum, New York, 1974), pp. 599–609.
134. M.D. Thouless, *Acta Metall.*, **36** [12] (1988) 3131–3135.
135. H. Schmidt, G. Rinn, R. Nass, and D. Sporn in *Better Ceramics Through Chemistry III*, eds. C.J. Brinker, D.E. Clark, and D.R. Ulrich (Mat. Res. Soc., Pittsburgh, Pa., 1988), pp. 743–754.
136. J. Schlichting, *J. Non-Cryst. Solids*, **63** (1984) 173–181.

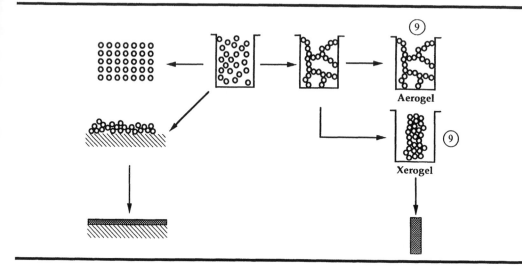

CHAPTER 9

Structural Evolution during Consolidation

For all types of gels, so-called "polymeric" or "particulate," fibers, films, or bulk, the receding liquid during the final stages of drying exposes an interconnected porous network within the gel that completely surrounds the solid or *skeletal* phase. The average dimensions of the pores and the thickness of the skeleton depend upon the structure that exists at the gel point and the extent of the collapse or distortion of this structure that results from drying [1]. (See, for example, the schematic of drying, Fig. 6, Chapter 6.) Dried gels, either *xerogels* or *aerogels*[†], are distinguished from other porous, ceramic green bodies by their enormous surface areas and correspondingly small pore sizes. Quite often the dimensions of the pores and skeletal phase comprising the dried gel are sufficiently small (<1/10 the wavelength of visible light) that despite very high volume percent porosites (V_p) xerogels and aerogels are transparent or translucent. (See Figs. 1 and 2.) Typical surface areas for xerogels range from 500 to 900 m^2/g, while aerogel surface areas may exceed 1000 m^2/g. Such high surface areas and small pore sizes are, of course, a result of the molecular- to colloidal-scale structures that form as products of hydrolysis and condensation during gelation, aging, and drying. Structural evolution during gel consolidation is, in turn, largely a consequence of this surface area and porosity. Thus the relationship between the sol-to-gel and gel-to-glass (or ceramic) conversions is in essence one of cause and effect.

The structure of dried gels has been compared to that of a porous glass such as VYCOR® (a leached, phase-separated borosilicate glass); however we must remember that for gels the skeletal phase has evolved through

---

[†] Xerogels are dried by evaporation, whereas aerogels are dried by supercritical extraction of solvent.

Fig. 1.

Transparent xerogels prepared by acid-catalyzed hydrolysis of TMOS in methanol–formamide followed by drying at 60°C [2].

continued condensation reactions without melting. As we shall see, the skeletal structure is rarely equivalent to that of melted glass. By comparison it is less highly condensed, may be more or less homogeneous, and often contains excess free volume (i.e., has a greater molar volume). Only as the gel is heated to higher temperatures does the skeletal structure approach that of conventional glass made by melting. For example, in amorphous systems, this requires heat treatment in the vicinity of the glass transition temperature (T_g) of the corresponding melted glass.

This chapter examines the structural changes that occur during the initial stages of gel consolidation. Final consolidation by sintering to form dense glasses or polycrystalline ceramics is discussed in Chapter 11. Because of the emphasis placed on silicate systems by the gel research community, this chapter is largely devoted to structural studies of silicate gels. However multicomponent systems and nonsilicate systems are discussed where appropriate structural information exists. The chapter is organized in two major sections that discuss gel structure after drying at low temperature and structural evolution during heating. An outline of the chapter is as follows:

1.   *Structures of Porous Gels: Xerogels and Aerogels* addresses the physical and chemical structures of porous gels prepared by either evaporation or supercritical drying.

Fig. 2.

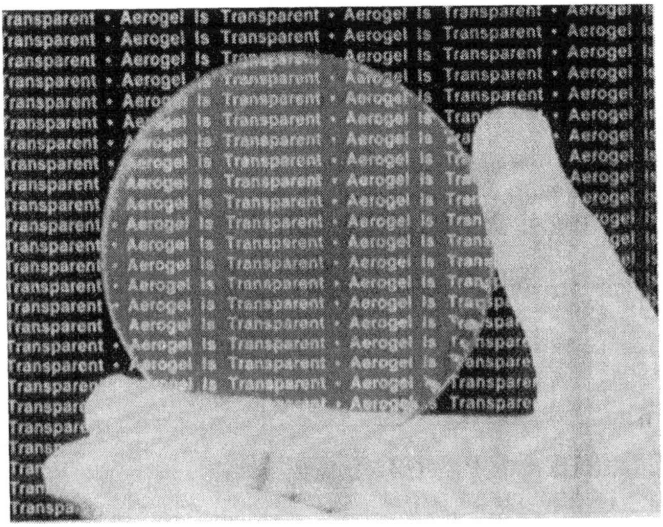

Transparent silica aerogels prepared by base-catalyzed hydrolysis of TMOS in methanol followed by liquid CO_2 solvent exchange and CO_2 supercritical extraction [3].

    1.1. *Porosity* examines structural models (utilizing fractal or Euclidian geometry) for porous gels based primarily on the results of gas adsorption-desorption studies.

    1.2. *Small-Angle-Scattering Investigations* provides information concerning porosity on 1 to 100 nm length scales derived from SAXS and SANS studies.

    1.3. *Mechanical Properties* relate modulus and density measurements to structural models of porous gels.

    1.4. *Chemical Structure of Dried Gels* examines structure on short length scales based on the results of NMR and vibrational spectroscopy.

2. *Structural Changes during Heating: Amorphous Systems* discusses the evolution of structure during gel consolidation. For amorphous systems, three regions are defined depending on the relationship between shrinkage and weight loss.

    2.1. *Region I* describes structural changes that occur at low temperatures as water and alcohol are desorbed.

    2.2. *Region II* reviews numerous studies of physical changes that accompany gel shrinkage at intermediate temperatures.

    2.3. *Thermodynamic and Kinetic Aspects of Gel Consolidation* points out that dry gels have high free energies and that consolidation kinetics depend on the competition between condensation and structural relaxation.

    2.4.   Structural Studies of Silicate Gels discusses molecular-scale structural changes that occur during silicate gel consolidation based on the results of NMR, electron paramagnetic resonance (EPR), and vibrational spectroscopy measurements.

    2.5.   Structural Studies of Multicomponent Systems presents illustrative examples of structural changes that occur in multicomponent borosilicate and titania-silicate systems.

    2.6.   Crystalline Systems discusses the structure of aluminate systems to illustrate structural changes dominated by the effects of phase transformations.

3.   *Summary* provides concluding remarks concerning structural evolution during heating.

1.

STRUCTURES OF POROUS GELS: XEROGELS AND AEROGELS

1.1.   Porosity

According to Iler [4], surface tension forces created in a gel during solvent removal cause the network to fold or crumple as the coordination of the particles is increased. Porosity develops when, due to additional crosslinking or neck formation, the gel network becomes sufficiently strengthened to resist the compressive forces of surface tension. Thus the dried xerogel structure (which comprises both the skeletal and porous phases) will be a contracted and distorted version of the structure originally formed in solution.

    The relationship between the structure of the gel and xerogel is shown schematically in Fig. 3 along with a TEM micrograph of the corresponding xerogel or aerogel for four characteristic gel types [5]. The most weakly branched systems (e.g., the two-step acid-catalyzed silica gel discussed in Chapter 3), which are formed under conditions where the condensation rate is low, tend to be overlapped (interwoven) at the gel point (Fig. 3a). Because there are weak excluded volume effects (i.e., the structures can freely interpenetrate) and the condensation rate remains low, the compliant structure can rather freely shrink in response to solvent removal. As discussed in the preceding chapters, the enormous capillary pressure (up to 200 MPa) attained at the final stage of drying due to the tiny pores causes a further compaction of the structure. The resulting xerogel is characterized by an extremely fine "texture" when imaged by TEM. No evidence of the individual polysilicate precursor species remains.

Fig. 3a.

(A) (B) (C)

Schematic representation of weakly cross-linked acid-catalyzed gel (A), desiccated xerogel (B), and TEM micrograph of xerogel prepared by two-step acid-catalyzed hydrolysis of TEOS ($r = 5$) followed by drying at 50°C, bar = 25 nm (C) [5].

More highly branched structures (e.g., the two-step acid–base-catalyzed silica gel discussed in Chapter 3) are prevented from interpenetrating due to strong intercluster, steric screening effects. As solvent is removed, individual clusters undergo shrinkage and rearrangement to achieve higher coordination numbers. Shrinkage stops at an earlier stage of drying due to the stiffness of the impinging clusters. This results in larger pores, so that the maximum capillary pressure is reduced compared to the previous example. The xerogel is characterized by a globular structure (most likely a compacted version of the original gel clusters) that is evident in the TEM micrographs (see Fig. 3c). Often porosity exists on two length scales: microporosity within clusters and mesoporosity between clusters. Scanning electron microscopy (SEM) of fractured surfaces often reveals a globular structure on a larger length scale suggestive of a hierarchical morphology in which small globules are organized into larger globules. However, there is no evidence of porosity created by the packing of these larger globules, which suggests that they are space-filling.

Fig. 3b.

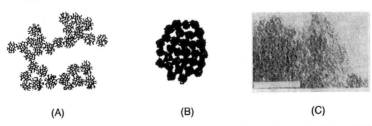

(A) (B) (C)

Schematic representation of base-catalyzed gel (A), desiccated xerogel (B), and TEM micrograph of xerogel prepared by two-step acid–base-catalyzed hydrolysis of TEOS ($r = 3.8$) followed by drying at 50°C, bar = 100 nm (C) [5].

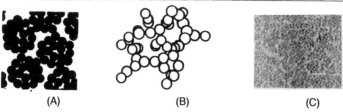

| (A) | (B) | (C) |

Hierarchical packing of colloidal particles (A), random packing of colloidal particles, coordination number = 3 (B), and TEM micrograph of colloidal xerogel prepared by single-step base-catalyzed hydrolysis of TEOS ($r = 2.25$) followed by drying at 50°C (C). Bar = 100 nm.

Particulate xerogels (Fig. 3c) are composed of uniform particles readily identified by TEM. Because particle interpenetration is not possible, drying can serve only to rearrange the particle assemblage to achieve higher coordination numbers. However because the maximum capillary pressure is inversely related to the particle size and extent of aggregation, it is much lower for particulate systems than for the previous two polymeric systems. Thus there is a lower driving force for compaction and rearrangement. For nonaggregated monodisperse systems, the pore volume of the xerogels depends only on the particle-packing geometry (see Table 1) [6], and the pore size increases with the particle size. For aggregated systems, both the pore volume and pore size depend on the aggregate structure.

The removal of a solvent above its critical point occurs with no capillary pressure because there are no liquid–vapor interfaces. Thus in the aerogel process there is a greatly reduced driving force for shrinkage. Compared to xerogels, aerogels are expanded structures (Fig. 3d) that are often more closely related to the structure of the gel that existed at the gel point.

Fig. 3d.

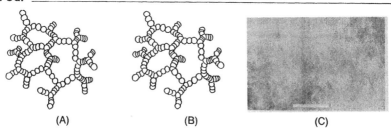

| (A) | (B) | (C) |

Schematic representation of gel (A), desiccated aerogel (B), and TEM micrograph of aerogel prepared by two-step acid–base-catalyzed hydrolysis of TEOS ($r = 3.8$) followed by aging at 50°C, liquid CO_2 solvent exchange and CO_2 supercritical extraction (C). Bar = 50 nm.

Table 1.

Porosities and Coordination Numbers of Regular and Random
Packings of Spheres.

Packing Type	Porosity (V_p)	Coordination Number (z)
Hexagonal	0.260	12
Body-centered cubic	0.320	8
Simple cubic	0.426	6
Tetrahedral	0.660	4
Trihedral	0.815	3
Random dense	ca. 0.36	ca. 7.5

TEM micrographs of aerogels often reveal a tenuous assemblage of clusters that bound larger interstitial cavities as shown in Fig. 3d. Although super-critical drying eliminates the capillary pressure because there are no liquid-vapor interfaces, it does not completely eliminate shrinkage. In fact, solid-vapor interfaces created at the final stage of supercritical drying can cause weakly condensed gels to shrink up to 50%.

For all categories of gels, aging prior to solvent extraction can alter the porosity of the dried product [1]. As described in Chapter 6, aging gels under conditions in which there is appreciable solubility of the skeletal phase allows the structure to reorganize via dissolution–reprecipitation reactions. Typically this has the effect of strengthening the gel so that it stops shrinking earlier in the drying process, causing the resulting xerogels or aerogels to have a greater pore volume.

The preceding, rather qualitative discussion is largely based on gel texture observed by TEM. Quantitative determinations of surface area, pore size distribution, and pore volume may be derived using gas adsorption-condensation and mercury porosimetry. In the following subsections quantitative data obtained using these methods are used to elucidate xerogel porosity.

1.1.1. GAS ADSORPTION-CONDENSATION

Figure 4 shows *nitrogen adsorption–desorption isotherms* for the same four dried gels examined by TEM in Figs. 3a–d. According to the original BDDT classification [8], the isotherm obtained for the two-step acid-catalyzed xerogel, which is shown to have a very fine texture by TEM (Fig. 3a), is of type I characteristic of a microporous solid (pore radii ≤ 1.5 nm). The volume adsorbed at the lowest relative pressure represents about 65% of the total pore volume, which indicates a large volume of extremely small pores.

Fig. 4.

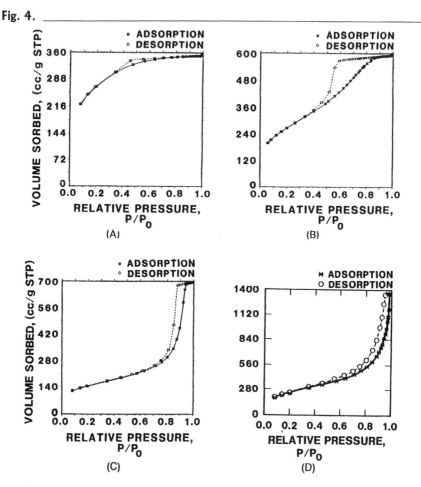

Nitrogen adsorption–desorption isotherms at 77 K for the desiccated silica gels examined by TEM in Fig. 3. (A) Two-step acid-catalyzed xerogel. (B) Two-step acid–base-catalyzed xerogel. (C) Particulate xerogel. (D) Two-step acid–base-catalyzed aerogel. ∗: absorption; O: desorption [7].

The virtual lack of hysteresis in the desorption branch is generally interpreted to mean that the pores are smooth and cylindrical. The pore volume, surface area, and average pore size determined from the isotherm are presented in Table 2 along with the bulk density.[†]

[†] It should be noted that due to capillary condensation in microporous solids, surface area measurements (determined by the BET method) and pore size distributions (determined on the basis of the Kelvin equation) may be inaccurate [9].

Table 2.

Summary of Porosity of Silicate Xerogels and Aerogels.

Sample	Pore Volume (cm³/g N₂ STP)	$V_p{}^a$	Surface Area (m²/g)	Pore Diameter (Å) (adsorption)	Pore Diameter (Å) (desorption)	Bulk Density^b (g/cm³)
Two-step acid-catalyzed xerogel (A2 in Table 8)	345	0.54	740	10–50	18	1.54
Two-step acid–base-catalyzed xerogel (B2 in Table 8)	588	0.67	910	10–100	46	0.99
Particulate (one-step base-catalyzed xerogel)	686	0.70	515	10–200	125	~0.6
Two-step acid–base-catalyzed aerogel	1368	0.82^c	858	$10–500^c$	186^c	0.30

[a] Volume fraction porosity based on the theoretical SiO_2 skeletal density of 2.2 g/cm³.
[b] Measured at ~25% RH.
[c] Because most of the adsorption occurs near P/P_0 near 1, pore volumes and pore size distributions may be inaccurate for aerogels.

The isotherm for the two-step acid–base-catalyzed xerogel (Fig. 4b) is a type IV isotherm. A characteristic feature of type IV isotherms is the hysteresis that is normally attributed to the existence of pore cavities larger in diameter than the openings (throats) leading into them (so-called ink-bottle pores) [9]. In order to empty the large cavities, the relative pressure must be reduced sufficiently to empty the throats. The steepness of the desorption branch indicates the uniformity in diameter of the throats. Ink-bottle pores are expected to occur from compaction of a globular or particulate gel structure during desiccation. Table 3 lists the relative size of the throats and cavities for several types of uniform packing geometries.

Compared to the type I isotherm discussed previously, which exhibited a large adsorbed volume at low relative pressure and a plateau in the volume adsorbed at intermediate pressures, the type IV isotherm (Fig. 4b) shows lower adsortion at low pressure and a monotonic increase in adsorption with increasing relative pressure. This indicates that this xerogel has less micro-porosity and a broader distribution of larger pores (*mesopores*, radii 1.5 to 50 nm). The microporosity results from the compaction of the individual globular structures. The mesopores result in turn from the packing of globules as observed in TEM. The surface area, pore volume, and average pore sizes determined from the adsorption and desorption branches are listed in Table 2 along with the bulk density determined geometrically. The pore size distributions calculated by differentiation of the pore volume adsorbed with respect to pore size (related to the relative pressure through the Kelvin equation; see Eq. 10, Ch. 7) are shown in Figs. 5a and b. The broad pore size distribution determined from the adsorption branch corresponds to the cavity size distribution. The narrow pore size distribution determined from the desorption branch corresponds to the throat size distribution.

The adsorption–desorption isotherm for a particulate silica xerogel prepared by a single-step base-catalyzed process is shown in Fig. 4c. It is

Table 3.

Cavity and Throat Sizes for Different Packings of Uniform Particles.

Packing	z	V_p	Radius of Sphere Inscribed in Cavity (Fraction of Particle Radius)	Radius of Sphere Inscribed in Thoat (Fraction of Particle Radius)
Rhombohedral	12	0.24	0.225/0.412	0.155
Tetragonal	10	0.30	0.291	0.155/0.265
Body-centered cubic	8	0.32	0.291	0.226
Orthorhombic	8	0.39	0.528	0.155/0.414
Cubic	6	0.48	0.732	0.414

Fig. 5.

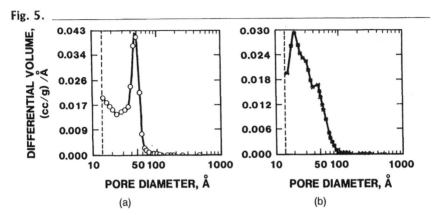

Pore size distributions for the two-step acid–base-catalyzed xerogel determined from (a) the adsorption isotherm and (b) the desorption isotherm in Fig. 4b.

also a type IV isotherm, but compared to the previous example, less volume is adsorbed at low relative pressures and a sudden increase in adsorption occurs at high pressure. The desorption branch is also steeper. These features are consistent with a xerogel composed of nonporous, rather monodisperse particles. The surface area, pore volume, and pore sizes determined from the adsorption and desorption branches are listed in Table 2 along with the bulk density.

An adsorption–desorption isotherm for a particulate gel prepared from LUDOX® (an aqueous colloidal silica) and potassium silicate is shown in Fig. 6. During gel formation the potassium silicate forms necks between particles resulting in a vermicular structure (Fig. 16, Ch. 4). Shafer *et al.* [11] interpret the adsorption commencing at point A as resulting from an instability in the adsorbate film. Above a critical thickness the film becomes unstable and capillary condensation occurs. All the pores are just filled at point B. During desorption the menisci curvature increases with decreasing pressure until the liquid minimizes its energy by changing from a metastable state at point C to a thick adsorbed film at point D. This interpretation uses adsorbate film instability rather than pore throats to explain the hysteresis but requires there to be smooth cylindrical porosity. Because the formation of necks between spherical particles is expected to result in a solid phase composed of cylindrical elements [12], it is difficult to understand how this LUDOX®-derived gel could acquire nonintersecting, cylindrical pores. Most likely the steep desorption branch results from a very narrow throat size distribution.

The adsorption–desorption isotherm of an aerogel prepared by two-step acid–base-catalyzed hydrolysis of TEOS ($r = 3.8$) in ethanol followed by

Fig. 6.

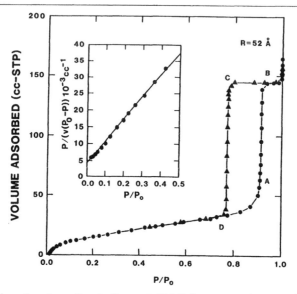

Oxygen adsorption–desorption isotherm at 89.5 K for a xerogel prepared from an aqueous dispersion of colloidal silica and potassium silicate by the "Shoup process" discussed in Ch. 4. Inset is BET plot [11].

CO_2 solvent exchange and supercritical drying is shown in Fig. 4d. Compared to the corresponding xerogel (Fig. 4b), the aerogel contains over twice the pore volume and the pore size is considerably greater as is evident from the larger amount of adsorption that occurs at high relative pressures (>0.9). At low relative pressures, the aerogel isotherm is similar to the xerogel isotherm which indicates that both samples contain considerable microporosity. The surface area, pore volume, and pore size of the aerogel are listed in Table 2.

1.1.2.   GEOMETRICAL MODELS

Information derived from gas adsorption–desorption analyses has been interpreted on the basis of several geometrical models. For models based on the packing of spheres (potentially appropriate for describing globular or colloidal structures), the type of packing, extent of aggregation, and the sphere density determine the bulk xerogel density and porosity. (See Table 1.) As shown in Fig. 7 for xerogels composed of either aggregated or non-aggregated colloidal particles, porosity is nearly independent of the sphere size [6]. The effect of aggregation is an increase in the pore volume due to a reduction in the coordination number, Z, from about 8, corresponding to

Fig. 7.

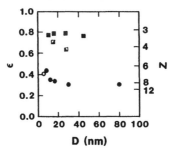

Porosity, ε, and coordination number, Z, of oxide gels derived from colloidal sols having different primary particle sizes, D. ● and ○ refer to nonaggregated sols of silica and ceria, respectively. ■, ◪, and ◩ refer to aggregated sols of silica, alumina, and titania, respectively. Z is the average coordination number for regular sphere packing having the corresponding porosity [6].

random close packing, to about 3, which on average is equivalent to trihedral packing. Figures 8 and 9 show that surface area decreases with an increase in sphere size according to the inverse relationship:

$$S = 6000/D\rho_{skeleton} \qquad (1)$$

(where S is the surface area in m²/g and D equals the particle diameter in nanometers), whereas the pore size increases linearly with an increase in sphere size. (See Fig. 8.). The effect of aggregation (Fig. 9) is primarily an increase in the pore size that corresponds to a given sphere size, although the surface areas are slightly reduced, presumably due to neck formation. Sphere density or, more generally, *skeletal density*, $\rho_{skeleton}$, can be determined from the pore volume V_p by the following relationship:

$$V_p = 1/\rho_{gel} - 1/\rho_{skeleton} \qquad (2)$$

Fig. 8. _____

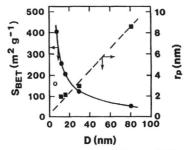

Specific surface areas, S_{BET}, and mean pore radii, r_p, of silica xerogels derived from nonaggregated colloidal sols having particle diameters, D [6].

Fig. 9.   _____

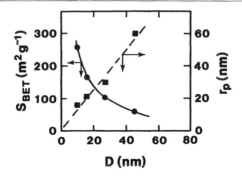

Specific surface areas, S_{BET}, and mean pore radii, r_p, of silica xerogels derived from aggregated colloidal sols having particle diameters, D [6].

where ρ_{gel} is the bulk density and $\rho_{skeleton}$ is the density of the solid phase not accessible to the adsorbate.

In a study of a multicomponent borosilicate silicate xerogel prepared by a multistep hydrolysis procedure involving mainly alkoxide precursors, Brinker and Scherer [13] used the surface area, $496 \, m^2/g$, in Eq. 1 and the values of the pore volume, $0.31 \, cm^3/g$, and bulk density, $1.10 g/cm^3$, in Eq. 2 to arrive at values for the cluster size and skeletal density, 7.3 nm and $1.8 \, g/cm^3$, respectively. Based on a theoretical gel density of $2.48 g/cm^3$, which corresponds to the melted glass density,[†] the bulk and skeletal densities correspond to relative densities of 0.48 and 0.73, respectively. The packing efficiency, $\rho_{bulk}/\rho_{skeleton} \times 100$, is therefore 67%. This corresponds quite closely to random close packing (see Fig. 10a) in which the average coordination number of each particle is about 9.

This physical picture is reasonably consistent with the TEM micrograph shown in Fig. 11, although the estimate of the particle size is somewhat high. An overestimate of the particle size, based on Eq. 1, means that the surface area is less than that expected for particles of the equivalent spherical diameter observed by TEM. This indicates that the interparticle contacts are necked or deformed. According to Meissner *et al.* [14] the ratio of cavity radius to throat radius should equal about 1.4 for spheres compacted to a coordination number of about 8. The ratio obtained from values of the average pore sizes derived from the adsorption and desorption branches of the nitrogen isotherm was closer to 1.3, consistent with some interparticle neck formation.

[†] Due to the greater hydroxyl (OH) content of the gel, the melted glass density represents an upper limit of the theoretical gel density.

Fig. 10.

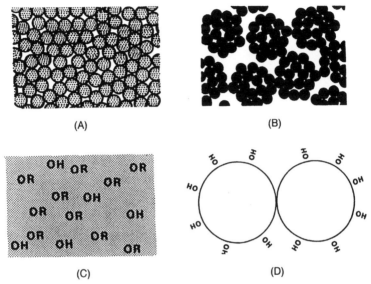

Schematic representations of two mesoporous xerogels. (A) Random-close packing of low-density spheres. (B) Hierarchical random-close packing of dense spheres. (C) Schematic representation of microporous xerogel illustrating uniform distribution of terminal sites. (D) Schematic representation of the particulate xerogel illustrating the dense oxide core and nonuniform distribution of terminal sites.

Fig. 11.

(A) TEM micrograph of the multicomponent borosilicate xerogel. Gel fragment dried at 150°C is suspended on a holey carbon grid. Bar = 50 nm [13]. (B) SEM micrograph of a fractured surface after drying at 50°C. Bar = 500 nm. There is no evidence for any porosity formed by the packing of the larger clusters observed in the SEM micrograph. Thus these clusters must be deformed in such a way as to completely fill space [13].

Fig. 12.

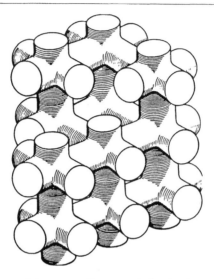

Cylindrical cubic array model corresponding to a relative density, $\rho/\rho_{solid} = 0.48$ [13].

An alternative geometrical model that is pertinent to necked globular structures is the cylindrical array shown in Fig. 12. This model was developed to describe silica "soots" formed by oxygen pyrolysis of $SiCl_4$ during fiber optic preform processing [12]. In the cylinder model, the cylinder diameter, $2a$, represents the average diameter of the globules that comprise the gel. The bulk density of the array depends on the skeletal density and the ratio of the cylinder radius to its length, a/l. The pore diameter is related to the cube dimension, l, by equating the cross-sectional area of the opening in the side of the cell to the equivalent circular area of the pore. Using values of bulk density and surface area measured for the multicomponent borosilicate xerogel, and assuming the relative skeletal density to be 0.73, the cylinder model predicts a pore diameter of 1.5 nm and a cylinder diameter of 3.1 nm, which, based on TEM and nitrogen adsorption analyses, are low estimates.

Thus, the analysis of the globular gel structure according to a particle packing model results in an overestimate of the particle diameter, whereas the cylindrical model results in an underestimate of the cylinder diameter. The appropriate geometry, therefore, appears to be a necked, random close-packed assemblage of particles whose relative density is 0.73. The necks are not large enough, however, to constitute smooth cylinders.

Both the skeletal density and the pore size distributions determined for the multicomponent borosilicate xerogel dispel the notion that the low bulk densities generally observed in alkoxide-derived xerogels are a consequence of hierarchical microstructures (Fig. 10b) in which fully dense particles are randomly close-packed into agglomerates that are in turn randomly close-packed. For the borosilicate xerogel, this microstructural model would exhibit approximately the correct bulk density, $1.02 \, g/cm^3$; however, the measured skeletal density is not that of the corresponding melted glass, nor is there evidence for the larger pores created between agglomerates (>27 nm diameters for random close packing).

Particle-packing models are not appropriate for describing microporous xerogels in which no evidence of primary particles exists. For microporous systems both the "thickness" of the skeletal phase bounded by the porosity and the pore size itself are typically only a few angstroms. Thus inhomogeneities such as terminal alkoxy and hydroxyl groups are distributed on a molecular scale. The xerogel structure is similar to a single-phase system of low skeletal density. (See Fig. 10c.) By comparison, particulate gels are composed of large, highly condensed regions in which terminal groups exist mainly on the surface of particles. (See Fig. 10d.)

1.1.3. FRACTAL MODELS

Avnir and Pfiefer [15,16] pioneered the analysis of adsorption isotherms according to fractal geometry. As discussed in Chapter 3, for *mass fractals* the *fractal dimension*, d_f, equals the *surface fractal dimension*, d_s, with $1 \leq d_s \leq 3$. For *surface fractals* the mass fractal dimension equals the embedding dimension (normally 3) and $2 \leq d_s \leq 3$. If adsorbent molecules of different cross-sectional areas are adsorbed on the xerogel surface, then the number of adsorbed molecules, n, adsorbed on an adsorbent volume, V_0, decreases with increasing molecule size, σ, and increasing adsorbent particle diameter, D, according to [15]:

$$n = V_0 \sigma^{-d_f/2} D^{d_f - 3}. \tag{3}$$

This relationship is valid for both mass and surface fractals. It assumes that the shape of the molecules and of the adsorbent particles remains constant as either σ or D are varied, causing all shape or packing factors to coalesce into the dimensionless prefactor. The mass or surface fractal dimension, can in theory be determined by varying the size of either the adsorbate molecule or adsorbent particle.[†] From the slope of a log–log plot of n versus σ for

---

[†] In practice, the latter method is generally invalid because the range of self-similarity does not extend to sufficiently large length scales, i.e., particles are Euclidian (uniform) on long length scales corresponding to manageable particle sizes.

nitrogen and various normal, secondary, and tertiary alcohols adsorbed on silica xerogel, Pfeifer *et al.* [15] determined d_f according to Eq. 3. They found $d_f = 2.97 \pm 0.02$ indicative of a mass fractal or extremely rough surface fractal ($d_f = d_s \approx 3$). A similar result was obtained from adsorption of polystyrene of varying molecular weights on porous alumina: $d_f = 2.79 \pm 0.03$ [16]. Unfortunately, these fractal investigations of xerogels have not addressed gel-synthesis conditions in order to understand the origins of the fractal structure.

Self-similarity requires pores of all sizes (over the range of self-similarity) increasing in number with decreasing size. The pore size distribution $V_{pore}(r)$, defined as the volume of pores smaller than r, has been expressed in terms of d_s by Pfeifer *et al.* [15]:

$$\frac{d}{dr} V_{pore}(r) \propto r^{2-d_s}. \tag{4}$$

Values of d_s calculated according to Eq. 4 for porous aluminas equal 3.04 ± 0.02 over the range $r = 2\text{-}195$ Å. Pfeifer *et al.* interpreted this as evidence of fractally rough surfaces. Avnir [16] has compiled a table that summarizes a large number of fractal analyses of adsorption data obtained for natural and synthetic oxides. A portion of this table is listed in Table 4. It appears that many porous silicates and aluminates can be described by a surface or mass fractal dimension.

1.2.   Small-Angle–Scattering Investigations

Small-angle–scattering employing X-rays (SAXS) or neutrons (SANS) has been used by several groups to investigate the structure of desiccated gels, both xerogels and aerogels. Schaefer *et al.* [17–20] examined silica aerogels

Table 4.

Surface Fractal Dimensions Determined by Surface-Area Probe-Size Relationships.[†]

Material	d_s	Molecular Probe
Silica gel	2.97 ± 0.02	Spherical alcohols ($C_1\text{-}C_7$)
Silicic acid	2.94 ± 0.04	N_2, 5 alcohols
Activated alumina	2.97 ± 0.03	Polystyrene fractions
Porous AlOOH (isostructural with FeOOH)	2.57 ± 0.04	N_2, *n*-alkanes

Source: From Avnir [16].

[†]Note added in proof: Based on results of TEM, SAXS, molecular size dependent adsorption, and direct energy transfer, Drake, Levitz, and Klafter dispute the claims of Avnir that these materials have fractal surfaces. (*New J. of Chem.*, in press).

Fig. 13.

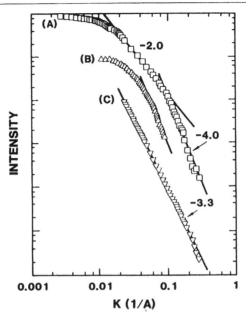

SAXS Porod plots of porous silicates. (A): Aerogel prepared from a particulate precursor, $\rho_{\text{bulk}} = 0.09\,\text{g/cm}^3$. (B): Aerogel prepared from a particulate precursor, $\rho_{\text{bulk}} = 0.21\,\text{g/gm}^3$. (C): Xerogel prepared from a single-step based-catalyzed hydrolysis of TEOS ($r = 2$) followed by drying at room temperature [18].

and xerogels and observed that, depending on the details of the preparation procedures, three distinct structures are produced: nonfractal, mass fractal, and surface fractal. Figure 13C shows a *Porod plot* for a xerogel prepared from fractally rough silicate precursors synthesized by a one-step base-catalyzed hydrolysis procedure. (See Fig. 49 in Chapter 3.) The slope of the line, -3.3, is identical to the slope observed for the silicate precursor prior to gelation and extends over about the same range of self-similarity (3–50 Å). The value of the surface fractal dimension derived from the slope, $d_s = 6 - 3.3 = 2.7$, implies that the fractal surface is preserved during drying, although the gel volume is reduced by over 50%. Schaefer *et al.* [18] interpret the xerogel structure as consisting of uniform porosity with fractally rough surfaces. However, from the scattering curve alone, a power law size distribution of uniform pores [21] cannot be excluded as a possible structure.

Himmel *et al.* [22] investigated silicate xerogels prepared by slowly drying acid-catalyzed TEOS-derived gels ($r = 2$). For samples prepared with acid concentrations ranging from 0.02 to 0.1 M, they observed in all cases a limiting value of the Porod slope equal to -3 (slit-smeared). For slit-smeared conditions this corresponds to Porod's law [23] indicating uniform structure (nonfractal) on the 0.1–1.0-nm length scale. Because silicate species prepared under similar conditions are fractal before gelation on this length scale [24,25], this result indicates that the fractal structure is collapsed by the high capillary pressure created by drying. Comparison with the results of Schaefer *et al.* [18] suggests that for silicate systems fractal features are better preserved by aging and drying under basic conditions. As discussed in Chapter 6, basic conditions promote ripening mechanisms that strengthen the gel network and minimize gel shrinkage during drying. Apparently reduced shrinkage helps to prevent the collapse of delicate, fractal structures. As we shall see, this is evident from SAXS studies of aerogel structures.

Small-angle scattering investigations of silicate aerogels have been performed by a large number of groups. (See, for example, refs. [18,26,27].) The aerogel process avoids liquid–vapor menisci during drying, thus eliminating the capillary pressure and reducing gel shrinkage. The most common aerogel procedure involves removing methanol or ethanol above their respective critical points [28] ($T_c = 240$ and $243°C$, $P_c = 78.5$ and 63 bars). An alternate, low-temperature method [3] involves replacing alcohol with liquid CO_2 and removing CO_2 above its critical point ($T_c = 31°C$, $P_c = 73$ bars).

Figure 14 shows Porod scattering curves for a series of aerogels investigated by Woignier *et al.* using SANS [26]. The precursor gels were prepared from TMOS ($r = 4$) under acidic (10^{-4} N HNO_3), neutral, or basic (5×10^{-2} N NH_4OH) conditions. Supercritical extraction of methanol at $300°C$ and 160 bars was used to produce the aerogels. Several trends are apparent in Fig. 14. One-step base hydrolysis followed by high-temperature solvent extraction (sample B (0.19) in Fig. 14) yields smooth, uniform particles as indicated by the Porod slope of -4 at high $q^\dagger$ (length scales <1.2 nm). On length scales greater than 1.2 nm, a crossover to a fractal structure is observed (mass fractal dimension $d_f = 1.9$), which indicates that the uniform particles are assembled into fractal networks. This behavior is essentially identical to that observed by Schaefer and Keefer [18] for a low-density aerogel (0.09 g/cm^3) prepared from a particulate precursor. (See sample A in Fig. 13.) Schaefer and Keefer interpreted this behavior as evidence of particle aggregation. (See Fig. 13 in Chapter 4.) Basically the aerogel looks like a random, reaction-limited aggregate ($D \simeq 2$) of uniform particles (Porod slope $= -4$).

The aerogels prepared from neutral and acid-catalyzed silica gels have mass fractal dimensions, $d_f \cong 2.4$. (See, e.g., the N series of SANS results

---

†The scattering vector q is equivalent to K used by Schaefer and coworkers.

Fig. 14.

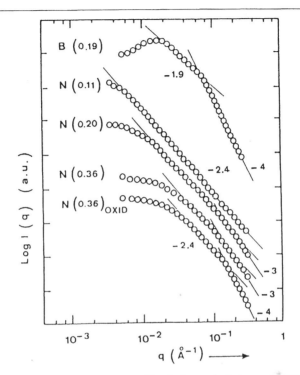

SANS Porod plot of silica aerogels prepared by single-step hydrolyses of TMOS ($r = 4$) under neutral (N) or basic ((B) 5×10^{-2} M NH_4OH) conditions followed by supercritical extraction of methanol. Numbers in parentheses are the corresponding bulk densities. The oxidation (heat) treatment for the 0.36-g/cm³ sample was performed at 500°C [26].

in Fig. 14.) The range of self-similarity increases with decreasing density. For higher-density aerogels (0.36 g/cm³), prepared from solutions with a higher TMOS concentration, the limiting value of the slope at high q approaches −3, indicating that the particles are very rough. Oxygen annealing at 500°C smooths out the particles (Porod slope = −4) but does not affect the mass fractal structure on longer length scales (lower q) [26].

Because considerable shrinkage accompanies methanol supercritical drying for acid and neutral aerogels compared to the base aerogel (linear shrinkage = 21, 16, and 4%, respectively), it is not clear how the aerogel structures compare to the original silicate structures developed at the gel point for acid and neutral conditions. It might be expected that the capillary strain arising from solid-vapor interfaces at the final stages of drying, causes a compaction or collapse of the smallest fractal features (increased D). (Although if the range of self-similarity is sufficiently large, no change of

D will be observed with a scale reduction.) Also, as discussed in Chapter 3, the pentacoordinate reaction intermediates involved in the hydrolysis and condensation reactions and the reverse reactions, esterification and siloxane bond hydrolysis, lead to a negative activation volume. With respect to aerogel formation, this means that pressure will accelerate the reaction kinetics in either the forward or reverse directions. According to Le Chatelier's principle, because there is limited water and an excess of alcohol, reesterification and alcohol-producing condensation are the most probable reactions to occur during supercritical drying. It is also well known that these reactions are thermally activated. Thus the increased pressure and temperature experienced during supercritical drying will increase the condensation rate when reactive terminal groups come into close proximity and, due to the excess of alcohol, promote reesterification. It is also likely that pressure-temperature-enhanced dissolution–reprecipitation may lead to coarsened structures compared to the original gel.

Schaefer *et al.* [29] have used SAXS to compare the structure of CO_2 extracted aerogels to the precursor silicate structures attained at the gel point for two-step acid- and two-step acid–base-catalyzed preparation procedures. Although the mass fractal dimensions are similar for the acid and acid–base-catalyzed gels, 1.9 and 2.0, respectively, values of the Porod slope, $P = -2d_f + d_s$, are quite different for the corresponding aerogels. $P > -3$ obtained for the acid–base aerogel indicates a polymer-like network; however, compared to the gel, the structure is more compact (greater d_f) on the 1–10-nm length scale. This compaction of structure occurs with little shrinkage of the gel upon solvent extraction (<10%). By comparison the acid-catalyzed gel suffers considerable shrinkage during CO_2 extraction (>50%). $P = -3.3$ obtained for the acid-catalyzed gel requires $d = 3$ and $d_s = P + 6 = 2.7$. This implies that the acid-catalyzed gel structure collapses to form a uniform structure with fractally rough surfaces. Schaefer *et al.* [29] interpret this result as evidence that the acid-catalyzed gel structure is not sufficiently robust to support itself against the solid–vapor surface tension forces present during the final stage of supercritical drying. Although structure on the 1.0–10-nm scale structure in both gel systems is essentially identical (as a result of reaction-limited cluster–cluster aggregation), Schaefer's result implies that weaker branching on shorter length scales under acid-catalyzed conditions (determined by ^{29}Si NMR, see Chapter 3) leads to significant gel shrinkage accompanied by collapse of the fractal structure.[†]

[†] Aerogel shrinkage can be reduced to zero by aging under basic conditions prior to alcohol exchange and CO_2 extraction [30]. As shown by ^{29}Si NMR investigations, aging strengthens the gel structure on short length scales (presumably by dissolution–reprecipitation) causing the relative concentration of Q^4 silicon species to increase [30]. (See Chapter 6.)

1.3. Mechanical Properties

The elastic moduli of silica xerogels and aerogels have been measured using pulse-superposition interferometry [31] and conventional three-point flexural techniques [26,32]. Murtagh *et al.* analyzed the elastic properties of xerogels prepared from acid-catalyzed silicate gels (r = 4, 12, 16, or 24) using a composite model that accounted for pore volume, pore shape, and skeletal density. For this series of xerogels, an inverse relationship was observed between bulk density and skeletal density as a function of r. (See Fig. 15.) Larger r values increase the skeletal density, increasing the resistance of the gel network toward collapse during solvent evaporation. Figure 16 shows that the compressional wave velocities, proportional to the elastic moduli (see Table 5), were less than predicted by the composite model assuming either spherical or needle-shaped pores. The 6x xerogels (r = 24), which had the greatest skeletal density, exhibited elastic moduli closest to the predicted values. A progressively poorer agreement with the model predictions was observed as the bulk gel density increased and the skeletal density decreased (decreasing r). This indicates that the stiffness of the skeletal phase was reduced with decreasing r, presumably due to the retention of more terminal OH and OR groups. This hypothesis could be confirmed by solid-state ^{29}Si NMR but has not been tested at this writing.

Fig. 15.

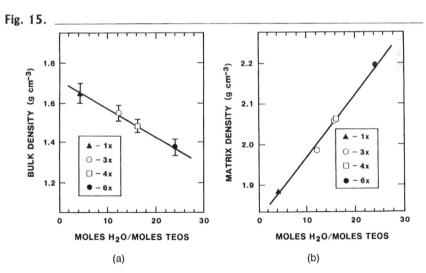

(a)   (b)

(a) Variation in bulk density (0% RH) with the H_2O : Si ratio, r, for xerogels prepared by single-step acid-catalyzed hydrolysis of TEOS [31].
(b) Variation in matrix density (skeletal density) with r for the acid-catalyzed TEOS xerogels in Fig. 15a [31].

Fig. 16.

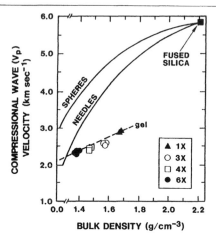

Compressional wave velocities versus bulk density for the series of acid-catalyzed TEOS xerogels investigated by Murtagh *et al.* [31]. The solid lines correspond to the relationship predicted from a fused silica–porosity composite assuming either spherical- or needle-shaped pores. Gel values fall beneath the predicted values, indicating that the skeletal phase is not as stiff as conventional fused silica [31].

Woignier *et al.* [26,32] have investigated the elastic properties of silica aerogels prepared from TMOS hydrolyzed under acidic (10^{-4} M HNO_3), neutral, or basic (5×10^{-2} M NH_4OH) conditions ($r = 4$). Aerogel density depended on the initial TMOS concentration and the shrinkage that occurred during supercritical extraction of methanol. As discussed in Section 1.2, considerable shrinkage was observed for the acid-catalyzed and neutral systems. Figure 17 shows that a log–log plot of Young's modulus, measured by a three-point flexural method, versus aerogel density yields a straight

Table 5.

Physical Properties versus H_2O/Si Ratio, r.

r	Bulk Density (g/cm³)	Skeletal Density (g/cm³)	Young's Modulus (GPa)	Shear Modulus (GPa)	Bulk Modulus (GPa)	Poisson's Ratio
4	1.65	1.883	9.95	3.95	6.53	0.245
12	1.55	1.986	8.72	3.54	5.28	0.228
16	1.494	2.069	7.71	3.21	4.42	0.201
24	1.388	2.195	6.55	2.83	3.51	0.192

Source: Murtagh *et al.* [31].

Fig. 17.

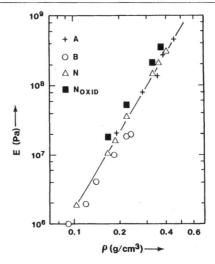

Log–log plot of Young's modulus, E, versus bulk density for aerogels prepared from TMOS hydrolyzed under neutral (N), acid-catalyzed (A), or base-catalyzed (B) conditions ($r = 4$). Oxidation treatment (N_{OXID}) was at 500°C [26].

line defined by the power law relationship:

$$E \propto \rho^{3.7 \pm 0.3}. \qquad (5)$$

This scaling behavior is similar to that expected in a percolating network where the elasticity scales in the same way as the electrical conductivity (see Chapter 5):

$$E \propto \rho^{T/\beta} \qquad (6)$$

where ρ is the density of the infinite cluster and T and β are percolation exponents related to the elasticity and gel fraction, respectively. However because the shrinkage associated with syneresis and supercritical drying is not taken into account, percolation theory, which may adequately describe structure at the gel point, only qualitatively describes the elasticity (see p. 351) [32].

Whereas in most mechanical property models of porous materials, elastic moduli depends only on the bulk density (and in some cases pore shape), it is apparent from Fig. 17 that for any particular bulk density a range of Young's moduli, E, are obtained that depend on the synthesis conditions employed. For the same values of bulk density, base-catalyzed synthesis conditions resulted in lower values of E than acid or neutral conditions. Heat treatments at 500°C increased the modulus of the neutral-synthesized samples. Based in part on SAXS investigations, Woignier et al. [32] attribute the lower E of the base-catalyzed samples to larger primary particles with lower particle–particle connectivity. Neutral and acid-catalyzed conditions result in polymeric precursors composed of smaller particles that for similar values of bulk density have many more particle–particle contacts, increasing the aerogel stiffness. Heat treatments were shown to remove microporosity associated with fractally rough surfaces. (See Fig. 14.) As evident from the Young's modulus results, this sintering process must also increase the network connectivity.

1.4.   Chemical Structure of Dried Gels

1.4.1.   NMR STUDIES

NMR studies of dried gels, both xerogels and aerogels, have been limited primarily to silicate and aluminate systems. Klemperer and coworkers [33] conducted a solid-state study of silicate gels prepared under neutral conditions from TMOS employing r values of 4 or 10. Figures 18a and b show ^{29}Si Fourier transform (FT) magic angle spinning (MAS) spectra for TMOS gels hydrolyzed with four equivalents of water ($r = 4$) gelled at 40°C, and dried at 60°C for eight hours (Fig. 18a) or gelled at 25°C and dried open for 11 months at 25°C (Fig. 18b). Figure 18c shows a ^{29}Si FT MAS spectrum for a two-step acid–base-catalyzed gel prepared from TEOS ($r = 3.8$), gelled at 50°C for one week, and dried at 50°C for more than one year [34]. All these xerogels are weakly condensed compared to v-SiO$_2$,[†] which shows only a Q^4 resonance in the ^{29}Si FT MAS spectrum (Fig. 18d). However differences exist depending on the drying and synthesis procedures. More extensive drying increases the extent of condensation as is evident from the reduced intensity of the Q^1 resonance and increased intensity of the Q^3 resonance in Fig. 18b compared to Fig. 18a. The acid–base-synthesis procedure results in a more highly condensed structure as shown by the greater relative intensity ratio, Q^4/Q^3, in Fig. 18c compared to Q^4/Q^3 in both Figs. 18a and 18b.

The weaker crosslinking of alkoxide-derived xerogels compared to v-SiO$_2$ is consistent with their lower skeletal densities (see Table 6), lower elastic moduli, and larger coefficients of linear thermal expansion (to be discussed in Section 3). The weaker cross-linking of the xerogels prepared without added catalyst, compared to that observed for acid–base-catalyzed conditions employing an equivalent value of r, is consistent with the more particle-like character observed in TEM (see Fig. 3b) when base catalysts are employed.

[†] Fully dense, vitreous silica.

Fig. 18.

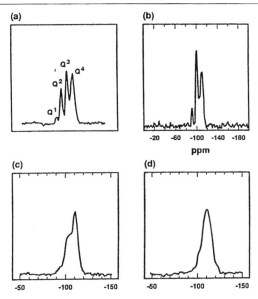

^{29}Si FTMAS spectra of (a) xerogel prepared by single-step hydrolysis of TMOS ($r = 4$) with no added catalyst followed gelation at 40°C and drying at 60°C for eight hours [33]; (b) xerogel as in 18a after drying open for 11 months at 25°C [33]; (c) xerogel prepared by a two-step acid–base-catalyzed hydrolysis of TEOS ($r = 3.8$) followed by drying at 50°C for over one year [34]; (d) v-SiO$_2$ prepared by sintering the xerogel in 18c at 1100°C for one hour [34].

1.4.2. RAMAN AND INFRARED INVESTIGATIONS

Due to the selection rules, Raman and infrared (IR) investigations are complementary in nature, providing molecular-scale information about the xerogel framework. Raman and infrared spectra of silicate xerogels prepared from TEOS using an acid-catalyzed process with a large excess of water ($r = 50$) are shown in Figs. 19a and 19b, respectively, where they are compared to the corresponding spectra of the silicate precursor, TEOS. Bertoluzza *et al.* [35] assigned the framework vibrations on the basis of the structure and vibrational spectra of conventional vitreous silica (v-SiO$_2$) [36,37]. The major features of the Raman spectrum associated with network vibrational modes are the 430, 800, 1070, and 1180 cm^{-1} bands. Corresponding IR bands are observed at 460, 800, 1080, and 1220 cm^{-1}. The 1180 (Raman) and 1220 (IR) cm^{-1} bands and the 1070 (Raman) and 1080 (IR) cm^{-1} bands are assigned to LO and TO Si–O–Si asymmetric stretching modes, respectively. The ~800 cm^{-1} vibration is associated with

Table 6.

Skeletal Densities of Various Xerogels after Drying and the
Values Approached ($\rightarrow$) during Heating.

Commercial aqueous silica sols and gels [4]	$2.2 \rightarrow 2.3 \, \text{g/cm}^3$
Base TMOS [121]	$2.08 \rightarrow 2.2 \, \text{g/cm}^3$
Base NaBSi [121]	$1.45 \rightarrow 2.4 \, \text{g/cm}^3$
Acid TEOS [147]	$1.7 \rightarrow 2.2 \, \text{g/cm}^3$
NaAlBSi [145]	$1.65 \rightarrow 2.27 \, \text{g/cm}^3$

Fig. 19a.

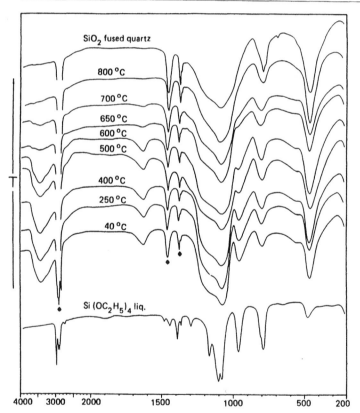

IR spectra of xerogels prepared by a single-step acid-catalyzed hydrolysis of TEOS
($r = 50$) followed by drying at 40°C or heat treatments in air at temperatures ranging
from 250 to 800°C. Shown for comparison are the spectra of the TEOS precursor and
conventional v-SiO$_2$. The samples are diluted in a nujol emulsion, and bands due to the
nujol are indicated by dots [35].

Fig. 19b.

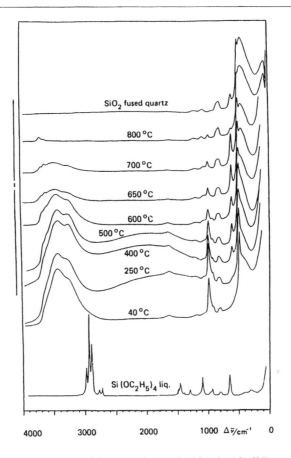

Raman spectra of the xerogels described in Fig. 19a [35].

symmetric Si–O–Si stretching or vibrational modes of ring structures. Bertoluzza *et al.* [35] assign the 430 (Raman) and 460 (IR) cm^{-1} vibrations to Si–O–Si bending modes. Although the ~490 cm^{-1} band, which is observed to be very prominent in the Raman spectrum, was associated with a defect such as a broken bond [35], more recent investigations have assigned this band to either an oxygen ring breathing mode involving four-membered rings [34] (cyclic tetrasiloxanes) or a symmetric oxygen stretching vibration of Q^3 surface silanol sites [38]. In fact, both types of vibrations may contribute to the large intensity of the 490 cm^{-1} band observed in the xerogel [34]. The structural origin of the 490 cm^{-1} band will be discussed further in Section 2.4.1 and in the following chapter.

The absence of any distinct peaks in the 2800–3000 cm^{-1} region of the Raman spectrum (assigned to C–H stretching as evident from the TEOS spectrum, Fig. 19b) suggests that, due to the large excess of water, the ethoxy groups are completely hydrolyzed and no reesterification occurs during drying. Lower r values during gel synthesis, however, do lead to the retention of alkoxy groups in the xerogel. This is demonstrated in Fig. 20 which shows the Raman spectra of two-step acid- and two-step acid–base-catalyzed xerogels prepared with $r = 5$ and 3.8, respectively, and the Raman spectrum of a base-catalyzed xerogel prepared with $r = 2.3$ [39]. Although complete hydrolysis was confirmed by ^{1}H NMR prior to gelation for the two-step acid-catalyzed system (see Chapter 3), reesterification during drying results in significant OR retention as indicated, for example, by the C–H scissors and deformation modes at 1453 cm^{-1} that correlate with the intense C–H stretching bands in the 2800–3000 cm^{-1} region of the spectrum.

Hydroxyl groups are evident in both the Raman and IR spectra of silica xerogels (Figs. 19 and 20) from bands at ~980 cm^{-1} (assigned to Si–OH stretching) and broad bands centered near 3400 cm^{-1} (assigned to various

Fig. 20.

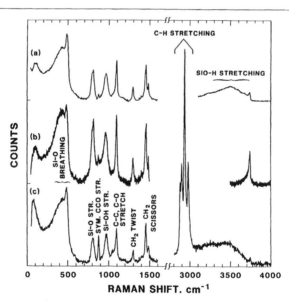

(a) Raman spectrum of xerogel prepared by two-step acid-catalyzed hydrolysis of TEOS ($r = 5$) followed by drying at 50°C for one month [39], (b) Raman spectrum of xerogel prepared by two-step acid–base-catalyzed hydrolysis of TEOS ($r = 3.8$) followed by drying at 50°C for one month [39], (c) Raman spectrum of xerogel prepared by single-step base-catalyzed hydrolysis of TEOS ($r = 2.3$) followed by drying at 50°C for one month [39].

isolated and hydrogen-bonded SiO–H stretching vibrations and hydrogen-bonded water). Superimposed on the broad bands are shoulders at 3220, 3595, and 3650 cm^{-1} (Raman) and 3210 and 3630 cm^{-1} (IR). The contributions of the various silanol species to the envelope of vibrations near 3400 cm^{-1} may be determined from near IR absorption spectra. Figure 21 shows near IR spectra of xerogels prepared from acid-catalyzed hydrolysis of TMOS. Orgaz and Rawson [40] have assigned the sharp band at about 7326 cm^{-1} to the first overtone of the O–H stretching vibration of isolated (nonhydrogen-bonded) SiOH on the surface of the silicate skeleton. Wood and Rabinovich [41] associate both this band and the ~4560 cm^{-1} bands with isolated surface silanols. The bands at about 7142 and 4544 cm^{-1} are

Fig. 21.

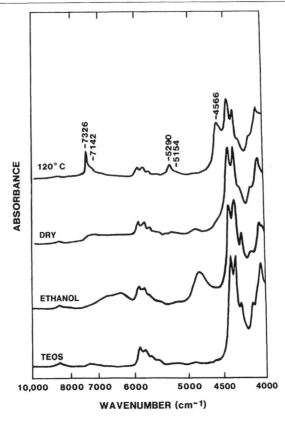

Near IR spectra of xerogels prepared from single-step acid-catalyzed hydrolysis of TEOS followed by drying at 70 (DRY) or 120°C. Shown for comparison are the near IR spectra of the TEOS precursor and ethanol [40].

associated with hydrogen-bonded silanols [41] and the band at 5290 cm^{-1} is assigned to hydrogen bonded water. These various configurations of surface silanols are represented schematically by Iler [4] in Fig. 22, where neighboring Q^3 silanol sites are defined as *vicinal silanols*, and Q^2 silanol sites are defined as *germinal silanols*.

Combined IR and Raman investigations show that when very large excesses of water are used in the gel synthesis procedure (e.g., $r = 50$), the spectral features begin to approach those of conventional v-SiO$_2$, although many hydroxyl groups are required to terminate the silicate surface. Lower r values result in significant OR retention and, apart from the envelope of vibrations near 430 cm^{-1} (Raman) and 1080 cm^{-1} (IR), the spectra look quite dissimilar to those of v-SiO$_2$. Thus although the tetrahedral silicate building blocks are the same, the assemblage of tetrahedra in the silicate framework is sufficiently disrupted in the gel that the average structure is quite different from v-SiO$_2$. Therefore it should come as no surprise that in general the xerogel structure cannot be well approximated by that of porous, melted glass.

Fig. 22.

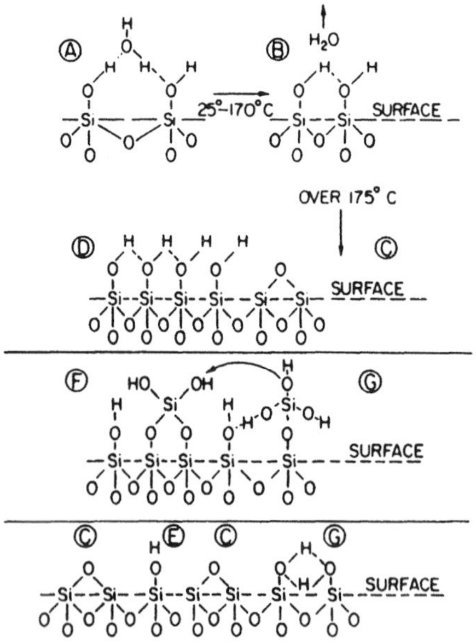

Types of hydroxyl groups on the surface of silica postulated by Iler [4]. A: vicinal hydrated. B: vicinal anhydrous. C: siloxane-dehydrated. D: hydroxylated surface. E: isolated. F: geminal. G: vicinal, hydrogen-bonded. According to Iler, the Q^1 and Q^2 species probably do not exist on a dried surface [4].

2.

STRUCTURAL CHANGES DURING HEATING: AMORPHOUS SYSTEMS

The preceding section on dried gel structure is by no means exhaustive. Further information concerning the structure of dried gels will be apparent as we discuss the structural changes that occur during heating. This topic has been addressed by numerous authors in a variety of binary and multi-component systems as evident from Table 7. However pure silica systems are by far the most carefully studied amorphous systems and will be emphasized here. Crystalline Al_2O_3 systems will be discussed in a following section.

The most obvious physical change that occurs when an amorphous gel is heated above room temperature is shrinkage. Figure 23 shows a typical shrinkage curve for a multicomponent borosilicate gel, during heating at 0.5°C/min [145]. The shrinkage curve is divided into three regions defined by the accompanying weight loss. In region I, weight loss occurs with little shrinkage. In region II, both shrinkage and weight loss are substantial. In region III, which commences in the vicinity of T_g for the corresponding melt-prepared glass, shrinkage occurs with little weight loss. Physical, chemical, and structural changes that occur in these regions are discussed in the following three sections.

2.1. Region I

Structural changes responsible for these three regions may be partially understood by consideration of the differential thermal analysis (DTA) trace (inset) and sequence of Fourier transform infrared (FTIR) spectra shown in Fig. 24. According to the DTA trace, the weight loss in region I corresponds to the endotherm attributed to desorption of physically adsorbed water (or perhaps residual solvent). This effect is not as obvious in the FTIR spectra, because the samples were re-exposed to ambient conditions ($\sim 25\%$ RH H_2O) prior to analysis. However, DTA studies by Orgaz and Rawson [40] have confirmed that the $\sim 100°C$ endotherm results from the desorption of physically adsorbed water and that the adsorption–desorption process is reversible. Figure 25 shows DTA traces of silica xerogels heated to 120°C causing initial desorption, exposed to various relative humidities at room temperature, and reheated in the DTA. Exposures to progressively higher relative humidities caused an increase in the endotherm at $\sim 100°C$ and a decrease in the exotherm at $\sim 275°C$. This latter effect may be attributed to more complete hydrolysis of residual alkoxide species with exposures to

Table 7.

Partial List of Binary, Ternary, and Multicomponent Compositions
Produced by Sol-Gel Processing.

Binary glasses prepared by gel route
  SiO_2-Al_2O_3 [42–50]
  SiO_2-B_2O_3 [45,51–63]
  SiO_2-CaO [64]
  SiO_2-Fe_2O_3 [65]
  SiO_2-GeO_2 [56,66–73]
  SiO_2-Li_2O [74–77]
  SiO_2-Na_2O [35,45,59,62,74,78–83]
  SiO_2-PbO [84,85]
  SiO_2-P_2O_5 [53,55,56,69,86]
  SiO_2-R_mO_n (R is Cr, Mn, Fe, Co, Ni, Cu, or V.) [87–91]
  SiO_2-SrO [92]
  SiO_2-TiO_2 [42,43,48,55,56,62,69,90,93–101]
  SiO_2-Y_2O_3 [69]
  SiO_2-ZrO_2 [42,48,56,60,61,90,101–107]
  B_2O_3-Li_2O [108,109]
  GeO_2-PbO [110]
  P_2O_5-Na_2O [111,112]

Ternary glasses prepared by gel route
  SiO_2-Al_2O_3-B_2O_3 [45]
  SiO_2-Al_2O_3-CaO [113,114]
  SiO_2-Al_2O_3-Li_2O [59,115]
  SiO_2-Al_2O_3-MgO [114,115]
  SiO_2-Al_2O_3-Na_2O [117–119]
  SiO_2-B_2O_3-Na_2O [59,85,120–124]
  SiO_2-B_2O_3-PbO [84,85]
  SiO_2-B_2O_3-P_2O_5 [53]
  SiO_2-B_2O_3-TiO_2 [125]
  SiO_2-B_2O_3-ZnO [84,85]
  SiO_2-CaO-Na_2O [126]
  SiO_2-TiO_2-ZrO_2 [42,48,127]
  SiO_2-ZnO-K_2O [128]
  SiO_2-ZrO_2-Na_2O [42,104]

Multicomponent glasses prepared by gel route
  General [84,120,126,128–134]
  SiO_2-Al_2O_3-B_2O_3-K_2O-Na_2O [84,135]
  SiO_2-Al_2O_3-TiO_2-Li_2O [136]
  SiO_2-Al_2O_3-ZrO_2-P_2O_5 [133]
  SiO_2-Al_2O_3-Li_2O-Na_2O [137]
  SiO_2-Al_2O_3-(Tl_2O, Cs_2O, Ag_2O) [138]
  SiO_2-Al_2O_3-B_2O_3-Na_2O-BaO [15,129,130,139]
  SiO_2-B_2O_3-Na_2O-Al_2O_3 [140,141]
  SiO_2-B_2O_3-Na_2O-V_2O_5 [120]
  SiO_2-La_2O_3-Al_2O_3-ZrO_2 [52,142]
  SiO_2-TiO_2-BaO-ZrO_2 [84]

Source: James [144].
Note: Reference numbers are in brackets.

Fig. 23.

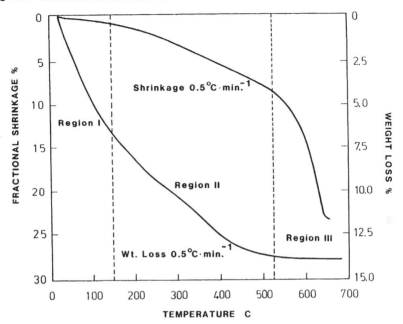

Fractional linear shrinkage and percent weight loss for a multicomponent borosilicate xerogel during heating in air at 0.5°C/min. Three regions are indicated corresponding to (I) primarily weight loss, (II) shrinkage proportional to weight loss, and (III) primarily shrinkage [145].

progressively higher RH. Combined near IR and TGA investigations by Orgaz and Rawson [40] showed that the amount of molecular water physically adsorbed upon exposure to 100% RH correlates with the concentration of surface silanol groups that remain after heat treatments at 120 or 800°C. (See, e.g., Fig. 26.) This strongly indicates that molecular water preferentially adsorbs on surface silanol sites.

The slight shrinkage accompanying the region I weight loss results from the increase in surface energy caused by the desorption of water or alcohol. For example, increasing the surface energy, γ, from 0.03 J/m^{-2} (corresponding to a surface saturated with physically adsorbed water and alcohol) to 0.15 J/m^{-2} (corresponding to a pure silanol surface [4]) causes an increase in the capillary pressure, P. The resulting linear strain, ε, is [145]:

$$\varepsilon = (1 - v)S\rho_s \, \Delta\gamma/E \qquad (7)$$

where S = surface area, ρ_s = skeletal density, $\Delta\gamma$ = change in specific

Fig. 24.

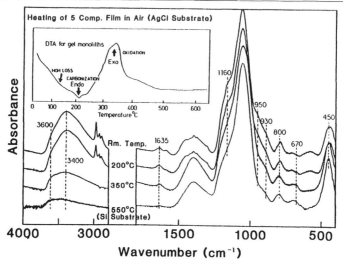

FTIR spectra of a multicomponent borosilicate xerogel after drying at room temperature or heating to temperatures ranging from 200 to 550°C in air. Samples were diluted in KBr except the 550°C sample, which was prepared as a thin film on an intrinsic Si substrate. Inset is DTA trace during heating in air at 10°C/min [146].

surface energy, and v and E are Poisson's ratio and Young's modulus of the skeleton, respectively. Using the measured values of surface area and the density of the corresponding melted glass (496 m^2/g and 2.27 g/cm^3) and the values $v = 0.2$ and $E = 15$–20 GPa [31], Eq. 7 predicts $\varepsilon = 0.5$–0.7%, in excellent agreement with the measured value of 0.6% (Fig. 23).

For more weakly branched xerogels, a measurable dilation is observed in region I [5,147]. (See, e.g., Fig. 27a.) The calculated coefficient of linear thermal expansion (α) for the two-step acid-catalyzed xerogel (A2, $r = 5$) is 46.9×10^{-6} K over the temperature range 20–100°C [145]. (See Fig. 28.) This value is nearly 100 times greater than α for conventional v-SiO$_2$ and comparable to values measured for common linear, organic polymers. Similar results were obtained by Kawaguchi [148] for silicate xerogels prepared by a single-step acid-catalyzed process. This indicates that on a molecular scale the skeletal structure is so disrupted that it contains no macroscopic regions of fully condensed silica. As shown by the results of the incremental heating and cooling experiment (Fig. 28), the thermal expansion coefficient of the skeleton decreases substantially as the skeletal structure becomes more highly condensed during heating above 300°C. From comparisons of the region I expansion coefficients of A2, B2 (two-step acid–base-catalyzed, $r = 3.8$) and the particulate gel (prepared from fumed

Fig. 25.

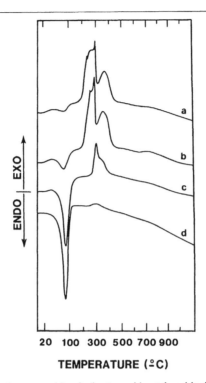

DTA traces of xerogels prepared by single-step acid-catalyzed hydrolysis of TEOS and dried at 70°C (a) or heated for two hours at 120°C and exposed to 32% RH (b), 62% RH (c), or 87% RH (d) prior to the DTA experiment [40].

silica) (Fig. 27a), it is evident that α reflects the extent of condensation of the skeleton. For weakly condensed systems, the capillary strain, ε (Eq. 7), resulting from desorption of solvent, may be overwhelmed by dilation arising from the thermal expansion.

2.2. Region II

Concurrent weight loss and shrinkage in region II are attributed primarily to removal of organics (principally weight loss), polymerization (shrinkage $\propto$ weight loss), and structural relaxation (shrinkage only). Continued desorption and dehydroxylation that increase surface energy, γ, by about a factor of 2, should result in an additional capillary strain (Eq. 7) of $\ll 1\%$, which represents a minor contribution to the region II shrinkage.

Fig. 26.

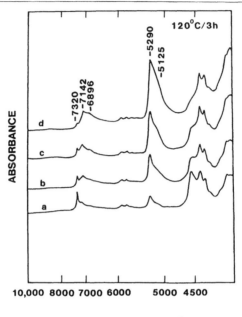

WAVENUMBER (cm⁻¹)

Near IR spectra of xerogels prepared by single-step acid-catalyzed hydrolysis of TEOS after drying at 120°C for 3 hours and exposure to 3% (a), 19% (b), 32% (c), or 47% RH water vapor [40]. Increasing RH causes an increase in molecular water H-bonded to silanol (5290, 5125 cm⁻¹), a decrease in vicinal free silanol (7320 cm⁻¹), and an increase of silanol H-bonded to water (7142, 6896 cm⁻¹).

2.2.1. REMOVAL OF ORGANICS

Removal of organic substituents from the multicomponent borosilicate xerogel (in this case resulting from acetate precursors of Na and Ba as well as unhydrolyzed alkoxide groups) is evident in the 350°C FTIR spectrum (Fig. 24). Note the reduced relative intensities of the sharp bands near 2900 and 1400 cm⁻¹ (assigned to C–H and C–O stretching vibrations, respectively) and the shoulder at 1160 cm⁻¹ (assigned to various M–OR stretching vibrations). In the DTA trace (Fig. 24 inset), the removal of organics is evident from endothermic carbonization near 200°C followed by exothermic oxidation between 300 and 400°C. The gel turns amber in color at 200°C and transparent again by 400°C. For a silicate system, Carturan *et al.* [118] observed that the 300–400°C exotherm disappeared when the gel was heated under inert conditions, where oxidation would be suppressed.

Fig. 27a.

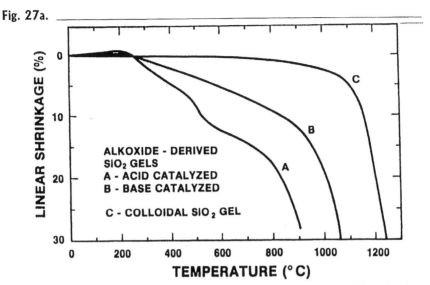

Percent linear shrinkage for silica xerogels prepared by two-step acid-catalyzed hydrolysis of TEOS ($r = 5$) (A); two-step acid–base-catalyzed hydrolysis of TEOS ($r = 3.8$) (B); or destabilization of a particulate sol prepared from Aerosil® (C). Heating treatments were performed at 2°C/min in air [5].

Fig. 27b.

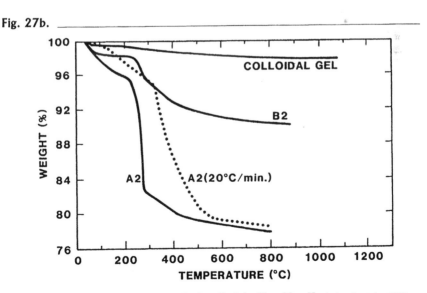

Percent weight loss for the xerogels described in Fig. 27a. Heat treatments were performed at 2°C/min or 20°C/min (dotted curve) in air [5].

Fig. 28.

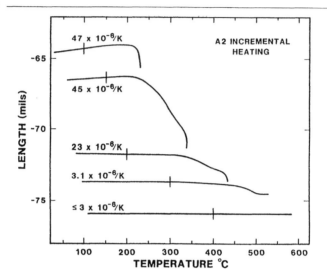

Changes in length of a single xerogel prepared by two-step acid-catalyzed hydrolysis of TEOS ($r = 5$) during incremental heating and cooling cycles performed between room temperature and 500°C. Coefficients of linear thermal expansion (25–250°C) are indicated for each heating interval [149]. Total length change is indicated on ordinate. From top to bottom each heating sequence is separated by a cooling step (cooling curves not shown).

Several studies have illustrated the effect of the gel synthesis conditions on the retention of organics in the resulting xerogel. Figure 29 shows the percent weight loss in the temperature intervals 150–600°C and 600–1100°C versus log r (single-step acid-catalyzed TEOS gels $r = 2$–20) [150]. The weight loss attributed primarily to organics (150–600°C) decreases substantially between $r = 2$ and $r = 6$, whereas the weight loss attributed to dehydroxylation at higher temperatures increases. For silica xerogels prepared by two-step acid or two-step acid–base-catalyzed procedures, the retained organic content was a sensitive function of r and the catalyst added in the second stage (acid or base) [151]. Table 8 and Fig. 30 show that for the two-step acid-catalyzed series, the ethoxide content determined from combustion analysis decreases with increasing r. Comparison with the two-step acid-base-catalyzed series (Table 8) shows that for the same r value less organics are retained under basic conditions. Because under acid-catalyzed conditions it was confirmed by ^{1}H NMR that hydrolysis was complete at the gel point for all r values, whereas unhydrolyzed monomer persisted in the B2 sample, these results confirm that reesterification during drying occurs more readily under acidic conditions than under basic conditions as discussed in Chapter 3.

Fig. 29.

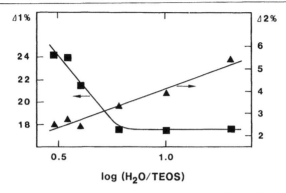

Percent weight loss in the temperature range 150–600°C ($\Delta 1\%$) or 600–1100°C ($\Delta 2\%$) versus log (H_2O/TEOS) for single-step acid-catalyzed xerogels [150].

Although the removal of organics contributes to the weight loss, it occurs with little associated shrinkage. This is evident in Figs. 31a and 31b [151] where the shrinkage and weight loss of the two-step acid- and acid-base-catalyzed gels may be compared. Xerogel A2 loses ~14 wt% in the temperature interval 220–260°C; however the dilatometer curve shows no enhanced shrinkage in this temperature range. In fact BET analyses [152] indicate that the surface area of silica gels often increases as organics are removed. Pyrolysis creates tiny "holes" in the skeletal network that contribute to the pore volume and significantly increase the surface area because there is no associated shrinkage.

Table 8.

Summary of Properties of Silicate Xerogels Prepared by Two-Step Hydrolyses of TEOS.

Sample	H_2O/Si	Apparent[a] pH	EtO/Si[b]	OH/Si[b]	O/Si[b]	BET Surface Area (m^2/g)
A1	4.2	2.7	0.35	0.33	1.66	734
A2	5.1	0.8	0.22	0.39	1.69	740
A3	12.4	3.1	0.19	0.53	1.64	806
A4	12.7	2.1	0.02	0.47	1.75	—
A5	15.3	1.4	0.01	0.42	1.79	—
B2	3.8	7.9	0.21	0.33	1.73	910
B3	8.1	8.2	0.10	0.23	1.83	—
B5	3.6	8.6	0.12	0.30	1.79	—

[a] Measured with a nonaqueous pH electrode near the gel point prior to drying.
[b] Assume HO/Si + EtO/Si + 2(O/Si) = 4.

Fig. 30.

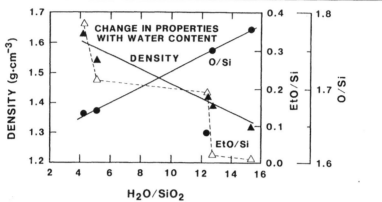

Bulk density (▲), average number of ethoxide groups per Si (EtO/Si) (△) and average number of bridging oxygens per Si (O/Si) (●) for a series of two-step acid-catalyzed xerogels prepared with varying H_2O/SiO_2 ratios, r, assuming EtO/Si + OH/Si + 2(O/Si) = 4 [151].

Fig. 31a.

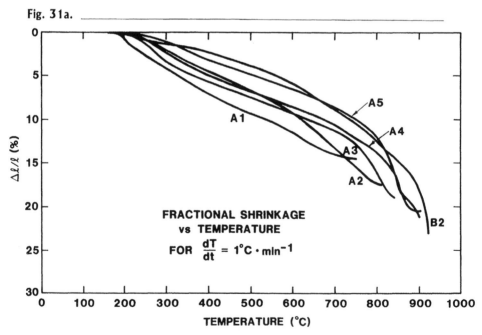

Percent shrinkage versus temperature for a series of xerogels prepared by two-step acid-catalyzed hydrolysis of TEOS, A1–A4 (see Table 8 for compositions) or two-step acid–base-catalyzed hydrolysis of TEOS (sample B2 in Table 8). Heat treatments were performed in air at 1°C/min [151].

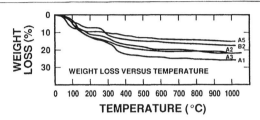

Percent weight loss versus temperature for the same series of xerogels as in Fig. 31a. Heat treatments were performed in air at 1°C/min [151].

2.2.2. CONDENSATION REACTIONS

Shrinkage in region II is attributed to continued condensation reactions, for example:

$$M\text{–OH} + HO\text{–M} \rightarrow M\text{–O–M} + H_2O \tag{8}$$

which occur both within and on the surface of the inorganic skeleton. The network is further densified through *structural relaxation*, an irreversible process in which the free energy decreases through bond restructuring or rearrangements with no associated weight loss [153].

There is considerable evidence that condensation reactions are responsible for shrinkage. Because shrinkage should be proportional to weight loss, we can compare shrinkage and weight loss above ~400°C where the pyrolysis of organics is complete. Comparing Figs. 27a and 27b, we observe that above about 500°C weight loss and shrinkage for gels A2 and B2 are nearly identical (suggestive of shrinkage ~ weight loss), and that the particulate gel, which loses little weight, exhibits a correspondingly small shrinkage. Increasing the heating rate reduces the region II weight loss observed at any particular temperature for the A2 xerogel and consequently causes a reduction in the associated shrinkage.

A second line of evidence relating shrinkage to continued condensation reactions in region II is the increasing skeletal density. Table 6 shows that the skeletal density of a variety of alkoxide-derived xerogels prepared below 100°C is substantially less than that of the corresponding melt-prepared glasses. Comparing Fig. 23, which shows shrinkage and weight loss for the multicomponent borosilicate gel, to Fig. 32, which plots shrinkage and the corresponding skeletal density (determined according to Eq. 2), shows that the skeletal density increases progressively in region II over the temperature range where considerable weight loss occurs. A similar trend of increasing skeletal density in region II is observed for all gels listed in Table 6. The

Fig. 32.

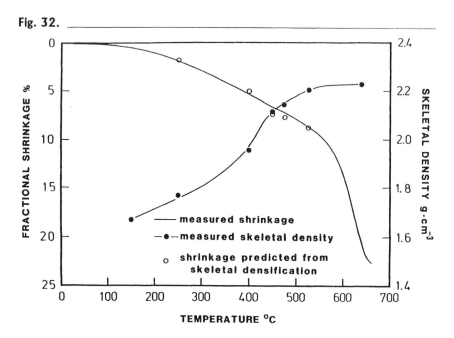

Percent linear shrinkage (solid line) and skeletal density (filled circles) for the multi-component borosilicate xerogel during heating at 2°C/min. The open circles represent the shrinkage above 150°C that is accounted for by the increasing skeletal density [145].

accompanying weight loss and evolution of water (a by-product of condensation) [141] indicate that at least a portion of the region II shrinkage is attributable to continued condensation reactions.

It should be re-emphasized at this point that in region II condensation reactions occur both within the skeleton (contributing to skeletal densification and associated shrinkage) and on the gel surface (increasing the surface energy, contributing to capillary strain, <1%; see Eq. 7). The relative proportion of surface versus bulk hydroxyls depends on the synthesis method employed. For example, from an FTIR investigation of single-step acid- and base-catalyzed silica xerogels ($r = 4$), Orcel *et al.* [154] showed that acid-catalyzed conditions resulted in a larger concentration of internal hydroxyl groups. This causes acid-catalyzed xerogels in general to show a greater region II shrinkage than base-catalyzed gels prepared with the same r value [151]. In particulate gels the hydroxyls exist primarily on the surfaces of fully condensed particles. (See Fig. 10.) As shown in Fig. 27a, particulate xerogels exhibit little region II shrinkage.

Infrared and Raman spectroscopy provide further evidence that region II shrinkage is related to condensation reactions. The sequence of infrared spectra obtained for the multicomponent borosilicate xerogel (Fig. 24) shows a progressive reduction in the $\sim 3660 \, \text{cm}^{-1}$ band assigned to internal and mutually hydrogen-bonded surface hydroxyls, over the 200–550°C temperature interval associated with region II. The series of Raman spectra obtained by Krol and van Lierop [44] (Fig. 33) show a reduction both in the $3750 \, \text{cm}^{-1}$ band assigned to surface silanols and the $\sim 3660 \, \text{cm}^{-1}$ shoulder associated

Fig. 33.

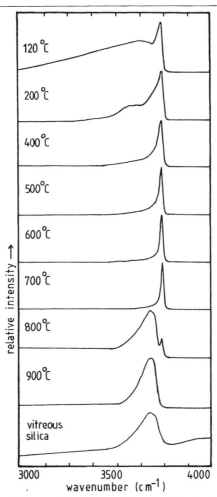

Raman bands due to O–H stretching for xerogels prepared by single-step acid-catalyzed hydrolysis of TEOS followed by heat treatments between 120° and 900°C for several weeks. Also shown is the Raman spectrum of conventional v-SiO$_2$ containing 500 ppm OH [44].

with hydrogen-bonded or internal silanols over the 200–700°C temperature interval corresponding to region II in silica xerogels. Similar behavior has been documented in numerous silicate and multicomponent silicate systems, confirming the relationship between shrinkage and condensation reactions in region II.

2.2.3. STRUCTURAL RELAXATION

A second mechanism that can contribute to shrinkage in region II is structural relaxation, a process by which excess free volume is removed, allowing the structure to approach the configuration characteristic of the metastable liquid [153]. Structural relaxation occurs by diffusive motions of the network. The resulting shrinkage occurs irreversibly with no associated weight loss. Evidence for structural relaxation is shown in Figs. 34a and 34b for the two-step acid-catalyzed silica xerogel [61,140,147]. The abrupt shrinkage at ∼450–550°C occurs without a corresponding increase in weight loss. (See Fig. 34a.) DSC measurements (Fig. 34b) made in this temperature interval show a strong exotherm (DSC 1) superimposed on an endothermic background (to be discussed in Section 2.4.1). The repeat DSC scan (DSC 2) shows that this exothermic process is irreversible. Skeletal density measurements (Fig. 34b) indicate that these phenomena are accompanied by a corresponding densification of the skeleton. ($\rho_{skeleton}$ increases from ∼1.8 to ∼2.2 g/cm^3.) Irreversible, exothermic skeletal densification occurring

Fig. 34a.

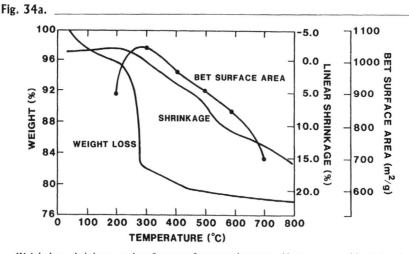

Weight loss, shrinkage, and surface area for xerogels prepared by two-step acid-catalyzed hydrolysis of TEOS during heating at 2°C/min in air [147].

Fig. 34b.

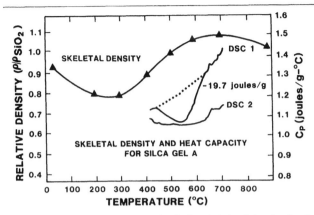

Heat capacity (C_p, DSC-1, and DSC-2) and calculated skeletal density for the xerogel described in Fig. 34a. Heat capacities were measured in air at 10°C/min. The repeat scan, DSC-2 was performed after cooling from 700°C at 80°C/min [147].

with no associated weight loss is circumstantial evidence for structural relaxation. It might be argued that this densification process is simply viscous sintering. However the net energy of this process (-4.7 cal/g, from DSC 1) is more negative than the sum of energies associated with the reduction in surface energy (due to sintering), the heat of dehydration, and the skeletal heat capacity. Also, because the rate of viscous sintering is proportional to the surface energy, γ, divided by the product of viscosity, η, and pore diameter, d, the pores responsible for sintering at low temperatures where η is high must be considerably smaller than the <20 Å pores responsible for sintering at $T > 800°C$, where η is considerably reduced. Such small "pores" would be of atomic dimensions and thus indistinguishable from free volume. Finally, the reduction in surface area over the temperature range 450–550°C is accounted for by the contraction of the skeleton resulting from skeletal densification:

$$(\rho_{\text{skeleton(initial)}}/\rho_{\text{skeleton(final)}})^{2/3} = S_{\text{(final)}}/S_{\text{(initial)}}. \tag{9}$$

Thus there is also no evidence for a dramatic reduction in surface area associated with the sintering of microporosity.[†]

We expect that structural relaxation will be most important for acid-catalyzed systems prepared with low r values. As discussed in Chapter 3, condensation occurs irreversibly under these conditions, "freezing-in" a metastable structure far from equilibrium. These conditions also favor the

---

[†] The removal of free volume does not constitute a change in surface area, since free volume is inaccessible even to helium sorbate molecules.

retention of organic groups in the final dried gels due both to incomplete hydrolysis and reesterification. Both factors promote more open structures, containing excess free volume. Because shrinkage in the two-step acid-catalyzed sample occurs just after the removal of the organic substituents, it is conceivable that in this case the excess free volume removed during structural relaxation is primarily that created by pyrolysis.

Structural relaxation may also explain the DTA results of Puyane et al. [78], who compared the thermal behavior of alkoxide-derived sodium silicate gels to the corresponding glasses rapidly quenched from the melt. Both systems showed an exotherm in the vicinity of T_g, the upper bound of region II (for the sodium silicate composition). Puyane et al. therefore concluded that the gel structure can be described by a high *fictive temperature* (large excess free volume). Presumably the exotherm was accompanied by shrinkage resulting from structural relaxation.

Although to date the investigation by Brinker et al. [140,147] is the only thorough documentation of structural relaxation occurring in region II, similar inflections are observed in shrinkage curves of other acid-catalyzed silica systems, strengthening the evidence for structural relaxation contributing to the shrinkage in region II. Further evidence for structural relaxation is derived from studies of sintering kinetics discussed in Chapter 11.

2.2.4. SKELETAL DENSIFICATION

Both continued condensation reactions and structural relaxation contribute to skeletal densification. A comparison of region II skeletal densification and shrinkage is shown in Fig. 32 for a multicomponent borosilicate xerogel [145]. The calculated values of the shrinkage expected from the measured increase in skeletal density (open circles) exactly account for all the region II shrinkage. Thus none of the observed region II shrinkage results from particle rearrangement within the skeletal phase to higher coordination sites. Instead the matrix shrinks isotropically by continued condensation reactions and structural relaxation without changing the coordination number or spatial relationships of the units that comprise it. (See Fig. 35.)

2.3. Thermodynamic and Kinetic Aspects of Gel Consolidation

As pointed out by Roy [155], xerogels are high–free-energy materials compared to their dense oxide counterparts [13]. (See Fig. 36.) Surface areas of xerogels range from 100 to 1000 m^2/g and surface energies increase from

Fig. 35.

SKELETAL DENSIFICATION

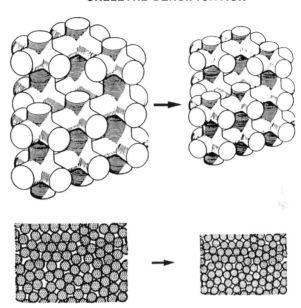

Schematic representation of structural changes resulting from skeletal densification. Shrinkage occurs isotropically causing no change in the surroundings of the structural elements in either the cylindrical array (top) or the random-close-packed array (bottom) [145].

about 0.15 to 0.28 J/m² with the extent of dehydroxylation. Thus the surface free energy provides a significant driving force (30–300 J/g) for shrinkage in region II. Reduced crosslinking and increased excess free volume further contribute to the excess free energy. The molar ratio of non-bridging oxygens (OH + OR) to Si in silica xerogels has been estimated to range from 1.48/Si to 0.33/Si compared to 0.003/Si for conventional vitreous silica (500-ppm OH) [13]. Because silicate condensation forming unstrained siloxane bonds is weakly exothermic,[†]

$$Si(OH)_4 \rightarrow SiO_2 + 2H_2O \ \Delta G_{f(298 \text{ K})} = -14.9 \text{ kJ/mol}, \qquad (10)$$

there exists a thermodynamic driving force for continued condensation reactions contributing to region II shrinkage. For silica xerogels containing 0.33 to 1.48(OH + OR)/Si, the corresponding energy ranges from ~20 to 100 J/g. A further contribution to the free energy of xerogels is the excess

---

[†] Condensation reactions forming strained siloxane bonds, defined by the Si–O–Si bond angle, $\phi < \sim 140°$, may be significantly endothermic as discussed in Section 2.4.1.

Fig. 36.

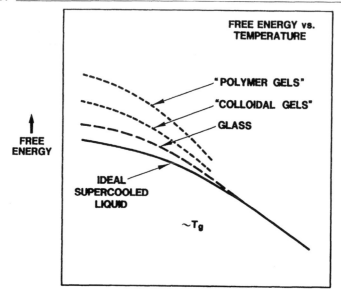

Schematic representation of free energy versus temperature for polymer gels, colloidal
(particulate) gels, and glass compared to an ideal supercooled liquid [13].

free volume that results ultimately from the metastable structure frozen-in
at the gel point or by modification of this structure during aging and drying.
As shown in Fig. 34b, the removal of excess free volume by structural relaxation is an exothermic process. Roy and coworkers [156] have discussed the
additional free energy in multicomponent diphasic or multiphasic gels. The
heat of reaction of the different phases or presumably the heat of mixing in
an amorphous system provides extra energy that may be available to drive
the densification process.

Region II shrinkage occurs by condensation reactions of internal silanol
groups, structural relaxation, and to a lesser extent capillary strain. As
previously discussed, these processes are all exothermic. In general there
exists no thermodynamic barrier to region II shrinkage (see Section 1.2 in
Ch.12). However condensation reactions, structural relaxation, and capillary
strain all depend on material transport and thus are kinetically limited
processes [13]. For example, it is reasonable to expect that condensation
reactions within the skeleton are limited by the diffusion of the individual
reactant species (viz. hydroxyls) to locations where a second reactant species
is encountered, or perhaps by the diffusion of the by-product, (water) to

a free surface [61]. If the condensation reaction occurs rapidly compared to the diffusional processes, the rate of reaction is diffusion-controlled, so shrinkage is proportional to $t^{1/2}$. Both isothermal-shrinkage experiments conducted by Prassas *et al.* [62] and constant–heating-rate experiments performed by Nogami and Moriya [60] show $t^{1/2}$ shrinkage kinetics in region II. For example, replotting the constant–heating-rate shrinkage results of Nogami and Moriya for SiO_2, B_2O_3–SiO_2, and ZrO_2–SiO_2 gels according to $L = (Dt)^{1/2}$ (where L is the shrinkage; D, the diffusion coefficient, equals $D_0 \exp(-Q/RT)$; and Q is the activation energy for diffusion) yields straight lines (correlation coefficients >0.996) suggestive of a diffusion-limited process [62]. (See Fig. 37.)[†]

Fig. 37.

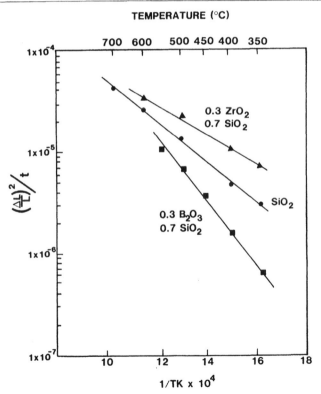

Xerogel shrinkage data of Nogami and Moriya (5°C/min, [60]) plotted assuming (shrinkage)$^2 \propto Dt$ [61].

[†] Linearity of such plots, however, is a rather insensitive criterion for determining reaction kinetics.

Fig. 38.

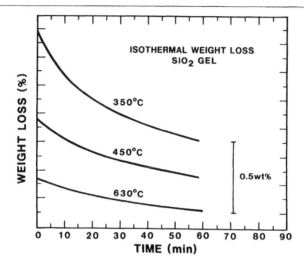

Isothermal weight loss for xerogels prepared by two-step acid–base-catalyzed hydrolysis of TEOS ($r = 3.8$) [61].

Gallo *et al.* [141] have shown that weight loss above $\sim400°C$ is completely accounted for by water, the by-product of condensation. Therefore the kinetics of the condensation reaction may be monitored by isothermal weight loss experiments. Isothermal weight loss for the two-step acid–base-catalyzed silicate system is shown in Fig. 38. At $350°C$ weight loss varies approximately as $t^{1/2}$ (weight loss (%) $= 0.13t^{0.45}$; correlation coefficient $= 0.98$), whereas after about 20 min at the higher temperatures it varies approximately as e^{kt}. This suggests that the condensation process is initially diffusion-limited and at higher temperatures eventually exhibits first-order kinetics.

To explain this behavior, Brinker and Scherer [61] proposed that at lower temperatures condensation occurs primarily within the skeleton and is diffusion-limited. At higher temperatures, where the skeleton is more fully densified, hydroxyls are located mainly on surfaces. Condensation of isolated surface hydroxyls obeys first-order kinetics. Raman results of Krol and van Lierop [44] are consistent with this explanation (see Fig. 33). At temperatures below $\sim500°C$ a low-frequency shoulder is observed near 3660 cm^{-1}, the portion of the spectrum associated with SiO–H stretching. This shoulder is assigned to hydrogen-bonded silanols residing in the bulk or on the surface. Above $500°C$, primarily isolated surface silanols remain as indicated by the narrow 3750 cm^{-1} band. Thus the transition from diffusion-limited to first-order kinetics coincides closely with the removal of hydrogen-bonded silanols.

Isothermal shrinkage experiments performed at higher temperatures (800–900°C) also show an apparent reduction in the shrinkage kinetics. Orgaz argues [157] that, as the skeleton becomes more condensed, the diffusion coefficient of water is reduced, causing a reduction in the condensation kinetics. However, at such high temperatures, these results could also be explained by the effect of increasing viscosity (due to continued condensation or structural relaxation) on the kinetics of viscous sintering. This subject is discussed at length in Chapter 11.

2.3.1. CONSTANT-HEATING-RATE EXPERIMENTS

In order to derive kinetic data pertinent to the proposed shrinkage mechanisms, *constant heating rate* (CHR) experiments have been performed. By assuming that at any combination of shrinkage and temperature the isothermal and nonisothermal shrinkage rates are identical and that at any temperature only one rate-controlling mechanism predominates, CHR shrinkage data have been analyzed according to the equation [158]:

$$(\Delta l/l_0)^{n+1} = [K_0 R T^2 (n + 1)/aQ] \exp(-Q/RT) \qquad (11)$$

where the activation energy, Q, and the parameter n are characteristic of the rate-controlling mechanism, and a is the heating rate. From Eq. 11, a plot of $\ln(\Delta l/l_0)$ versus $1/T$ gives a straight line of slope $-Q/(n + 1)R$. The parameter, n, may be evaluated by performing CHR experiments at different heating rates. A plot of $\ln(\Delta l/l_0)$ versus $\ln(a)$ at constant temperature results in a line of slope $-1/(n + 1)$. This simple kinetic analysis is appealing for gels, because changes in slope, $-Q/(n + 1)R$, imply a change in the activation energy, indicating a change in the rate-controlling mechanism.

A plot of $\ln(\Delta l/l_0)$ versus $1/T$ is shown in Fig. 39 for a five-component borosilicate system [145]. Two linear regions are observed that correspond quite closely to regions II and III defined in Fig. 23. The transition from region II to region III occurs over a temperature range (~400–525°C) corresponding to a third linear region. As shown by the weight loss and skeletal density curves (Figs. 23 and 32), the temperature interval 400–525°C represents a region where considerable skeletal densification occurs with little associated weight loss. This suggests that structural relaxation, by which the skeleton densifies without expulsion of water, is the predominant shrinkage mechanism in this temperature interval. Structural relaxation is characterized by a different activation energy and therefore may account for the observed change in slope near the region II–III transition (400–525°C).

Fig. 39.

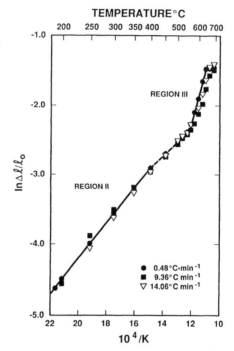

Shrinkage versus temperature for the multicomponent borosilicate xerogel plotted according to the CHR expression (Eq. 11): $\log(\Delta l/l_0)$ versus $1/TK$ for three different heating rates [145].

However, despite the apparent success of the CHR plot in defining regions that correspond to different shrinkage mechanisms, there is no heating-rate–dependence of the shrinkage in region II (contrary to the CHR model), whereas the amount of weight loss observed at any temperature is reduced as the heating rate, a, increases. (See Fig. 40.)

Two hypotheses were proposed to explain this unusual behavior [145].

1. A single mechanism that is thermodynamically favorable but not kinetically limited is responsible for region II shrinkage.
2. Region II shrinkage is kinetically limited, and the rate increases with increased excess free volume and decreases with increased condensation. These effects closely compensate for each other, causing no heating-rate–dependence.

In the first-case, shrinkage occurs spontaneously by condensation reactions, as reactive species come within close proximity as a result of diffusive

Fig. 40a. _____

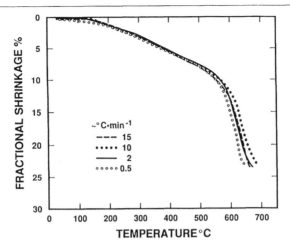

Percent shrinkage for the multicomponent borosilicate xerogel in Fig. 39 during heating in air at rates ranging from 0.5 to 15°C/min [145].

Fig. 40b. _____

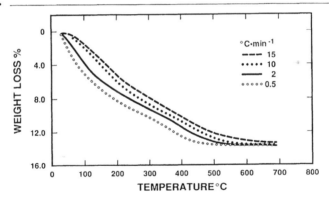

Percent weight loss for the multicomponent borosilicate xerogel in Fig. 39 during heating in air at rates ranging from 0.5 to 15°C/min [145].

motion. Samples that have undergone identical amounts of shrinkage should therefore be chemically and physically identical, requiring the heating-rate-dependence of the weight loss (Fig. 40b) to result strictly from the diffusion-limited evolution of water from the low permeability, porous network.

In the second case, shrinkage occurs both by condensation and structural relaxation, which are both kinetically limited processes. The reduced time spent at each increment of temperature as the heating rate is increased causes less condensation to occur (as reflected in the weight-loss data). Less highly condensed structures in turn exhibit a greater rate of structural relaxation. No heating-rate–dependence is observed if the greater extent of structural relaxation just compensates for the reduced extent of condensation.

To distinguish between these two cases, a series of DSC and TGA experiments were performed [145]. Gel samples were heated to 525°C (the upper bound of region II for the multicomponent composition) at heating rates between 2 and 30°C/min and quenched to room temperature. The quenched samples exhibited nearly identical shrinkages and their surface areas differed by less than 3%. These samples were then reheated in the DSC or TGA at 40°C/min to 717°C (region III), causing complete densification. The series of DSC and TGA curves in Figs. 41a and 41b show that, although the samples initially heated at different rates exhibited identical shrinkages at 525°C, the samples are not chemically or physically identical. Higher initial heating rates caused progressively greater endothermic behavior in the temperature interval 550–675°C that corresponded to progressively greater weight loss. Based on the OH concentration of the fully densified gels (determined by FTIR) and the weight loss during reheating, the OH concentration of the samples quenched from 525°C was calculated. Using an extinction coefficient of 56 l/mol-cm [159], OH concentrations of 5.5×10^{-4} and 1.0×10^{-3} moles OH/g_{gel} were determined for initial heating rates of 2 and 30°C/min, respectively. Thus the differences in total weight loss measured at 525°C (Fig. 40b) reflect not only the possible diffusion-limited evolution of water, but are at least partially accounted for by actual

Fig. 41a.

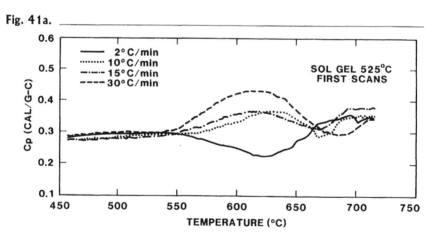

Heat-capacity curves (460–717°C) determined at 40°C/min for multicomponent borosilicate xerogels after an initial heat treatment to 525°C at rates ranging from 2 to 30°C/min [145].

Fig. 41b.

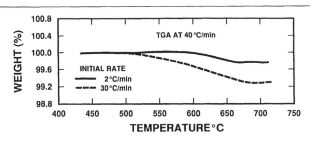

Weight loss measured at 40°C/min for the samples in Fig. 41a previously heated to 525°C at either 2 or 30°C/min [145].

differences in the extent of condensation, depending on the initial heating rate. For there to be equal amounts of shrinkage yet different amounts of condensation, condensation alone cannot account for the region II shrinkage. Therefore it was concluded [145] that at least two shrinkage mechanisms operated concurrently: reduced condensation, resulting from the reduction in time spent at each increment of temperature, was compensated for by increased structural relaxation, causing no heating-rate-dependence of the measured shrinkage.

Because region II shrinkage occurs by concurrent condensation reactions and structural relaxation in the multicomponent gel, the underlying assumption in Eq. 11, that only one shrinkage mechanism predominates, is invalid. A plot of $\ln(\Delta l/l_0)$ versus $\ln(a)$ yields a line of slope 0, requiring n to be infinite. Thus although CHR plots are often composed of essentially linear segments [157] and the changes in slope closely coincide with transitions from region II to region III (suggestive of a change in the predominant shrinkage mechanism), kinetic analyses of CHR data according to Eq. 11 generally do not provide meaningful kinetic parameters describing gel densification. The preceding discussion pertains to region II data. In region III, the concurrence of structural relaxation, continued condensation, and viscous sintering also preclude the use of CHR analyses. This subject is discussed further in Chapter 11.

The DSC curves shown in Fig. 41a provide information concerning the thermodynamics of gel consolidation from which we may infer structural information. These curves result from the exothermic contribution of region III sintering ($-\gamma \Delta S$ where γ is the average surface energy and ΔS is the change in surface area) plus the endothermic contributions of (1) the vaporization and subsequent heating of water formed as the by-product of condensation and (2) the heat required to raise the temperature of the skeleton (defined by the heat capacity curves of the fully densified gels). (See Fig. 42.)

Fig. 42.

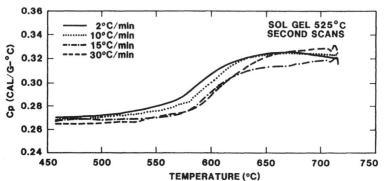

Repeat heat-capacity curves measured at 40°C/min for the samples in Fig. 41a after cooling from 717°C at 80°C/min [145].

It is also necessary to account for the heat of formation of the condensed inorganic product. Estimates of the heats of formation of silicates from silicic acid show the polymerization process to be weakly exothermic as indicated in Eq. 10. However, the change in surface area, heat of vaporation of water, and the heat capacity of the skeleton do not account for the magnitude of the endothermic DSC peaks unless the net contribution of the heat of formation of siloxane bonds, Si–O–Si, is positive [145]. A positive heat of formation associated with condensation is evidence for the formation of strained, cyclic structures as discussed in the following section. Apparently, the progressively larger endotherms that occurred with higher initial heating rates result principally from the larger endothermic contributions of siloxane bond strain (and to a lesser extent, vaporation). Consequently the exotherm expected from sintering is completely obscured at initial heating rates greater than 2°C/min. The endothermic background (400–700°C) observed in Fig. 34b for the two-step acid-catalyzed silicate gel (on which the DSC exotherm associated with structural relaxation is superimposed) is also attributed to the heat of formation of strained, cyclic species.

2.4. Structural Studies of Silicate Gels

The preceding sections have documented that shrinkage in region II results primarily from continued condensation reactions and structural relaxation. The accompanying structural changes have been investigated by vibrational spectroscopy (Raman and IR), solid-state magic angle spinning (MAS) NMR, and electron paramagnetic resonance spectroscopy (EPR).

2.4.1 NMR AND RAMAN SPECTROSCOPIC INVESTIGATIONS OF SILICATE FRAMEWORK STRUCTURES

Brinker and coworkers combined Raman and ^{29}Si MAS NMR spectroscopy with DSC–TGA investigations to elucidate the silicate structures formed by continued condensation reactions in region II [34,160–162]. Figure 43 compares a sequence of Raman spectra of two-step acid–base-catalyzed silica gels ($r = 3.8$) after drying at 50°C and after heat treatments between 200 and 1100°C. Also shown is the Raman spectrum of conventional vitreous silica (v-SiO$_2$). Corresponding ^{29}Si MAS NMR and ^{1}H cross-polarization (CP) ^{29}Si MAS NMR spectra are shown in Fig. 44 along with ^{29}Si MAS NMR spectra of the dried or heated gels after exposure to 100% RH water vapor for 24 h at room temperature.

Fig. 43.

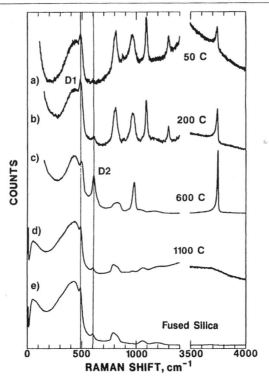

Raman spectra of the two-step acid–base-catalyzed gel ($r = 3.8$) after drying at 50°C or heating to 200, 600, or 1100°C compared to the spectrum of conventional v-SiO$_2$. Xerogel samples treated at 50 to 600°C possess high surface areas (>800 m^2/g), whereas the 1100°C sample and v-SiO$_2$ are fully dense [34].

Fig. 44.

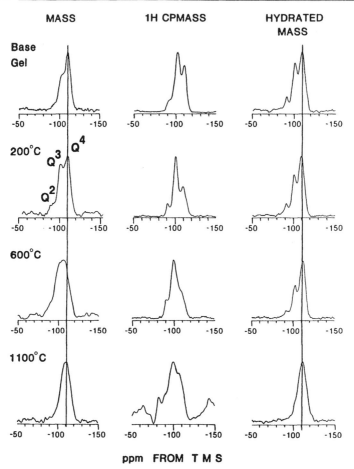

^{29}Si solid-state MAS NMR and ^{1}H cross-polarization MAS NMR spectra of the silica xerogels described in Fig. 43. (Base gel corresponds to the sample dried at 50°C.) The hydrated MAS spectra were collected after exposure of the original samples to 100% RH for 24 hours at 25°C. The ^{1}H CP MAS spectrum of the 1100°C sample is greatly scale-expanded in order to reveal the Q^2 and Q^3 resonances [34].

The SiO_2 framework vibrations that occur at about 430, 800, 1070, and 1180 cm^{-1} in the fused silica spectrum (discussed in Section 1.4.2) can be explained by a vibrational calculation on a *continuous random network* (CRN) [36,37]. The 1070 and 1180 cm^{-1} bands are assigned to the TO and LO modes of the Si–O asymmetric stretching vibration, respectively. Due to the selection rules, Raman-active modes involve symmetric vibrations, which explains the low relative intensity of these two asymmetric stretching bands. The 800 cm^{-1} band has been assigned to a symmetric Si–O–Si stretching mode. The 430 cm^{-1} band is assigned to a symmetric ring-breathing mode involving mainly oxygen motion. In the following discussion we associate the narrow bands labelled D1 and D2[†] with small cyclic structures not accounted for in the CRN model: cyclotetrasiloxane (D1) and cyclotri-siloxane (D2). Bands at ca. 3740 and 980 cm^{-1} in the 50, 200, and 600°C samples are assigned to SiO–H and Si–OH stretching vibrations, respectively. The narrow bands at ca. 1100 and 1300 cm^{-1} in the 50 and 200°C samples result from C–H and C–O stretching vibrations of unhydrolyzed ethoxide species often retained under base-catalyzed hydrolysis conditions. (For assignments, see Fig. 20.)

With regard to the framework vibrations, the ca. 430 and 800 cm^{-1} bands are broadened during heating, corresponding to a more varied distribution of Si–O bond lengths and Si–O–Si and O–Si–O bond angles, ϕ and θ, respectively. The band labelled D1 is slightly broadened and exhibits a reduction in relative intensity over the temperature interval 200–1100°C. The band labelled D2 exhibits the most remarkable changes in relative intensity: it is absent after drying at 50°C (Fig. 43a), appears near 200°C (Fig. 43b), becomes quite intense at intermediate temperatures (600–900°C, Fig. 43c), and is reduced in the fully densified gel (1100°C, Fig. 43d) to a level comparable to that in conventional v-SiO$_2$ (Fig. 43e).

The increase in relative intensity of the D2 band at intermediate temperatures is correlated with a reduction in relative intensity of the ca. 3740 cm^{-1} band assigned to surface SiO–H stretching vibrations [147]. (See Fig. 45.) This suggests that the species responsible for the D2 band forms by condensation reactions involving isolated vicinal silanols on the silica gel surface as confirmed by ^{18}O isotopic enrichment studies [160,161].

$$\begin{array}{c} \text{H} \\ ^{18}\text{O} \\ | \\ \text{Si} \\ \diagup | \diagdown \end{array} \quad + \quad \begin{array}{c} \text{H} \\ \text{O} \\ | \\ \text{Si} \\ \diagup | \diagdown \end{array} \quad \rightleftharpoons \quad \begin{array}{c} \quad ^{18}\text{O} \\ \text{Si}\diagup \quad \diagdown\text{Si} \\ \diagup | \diagdown \quad \diagup | \diagdown \\ \textbf{PORTION OF} \\ \textbf{D2 "DEFECT"} \end{array} \quad + \quad \text{H}_2\text{O}. \quad\quad (12)$$

Further information regarding the structure of the silicate species formed as products of condensation reactions in region II is obtained from the series of ^{29}Si NMR spectra (Fig. 44). The three prominent peaks at chemical shifts of ca. -91, -101, and -110 ppm, corresponding to Q^2, Q^3, and Q^4 sites [163], appear in the MAS and ^{1}H CPMAS spectra of the gels heated to 50

[†] The D designation originates from the association of these bands with defects, e.g., broken bonds, in the v-SiO$_2$ network [65].

Fig. 45.

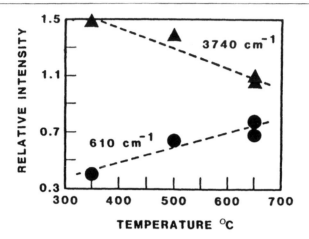

Relative intensities of the 3740 and 610 cm^{-1} Raman bands for the xerogel prepared by two-step acid–base-catalyzed hydrolysis of TEOS ($r = 3.8$) during heating between 350 and 650°C in air [147].

and 200°C and those rehydrated after heating between 50 and 600°C. Terminal oxygens associated with the Q^2 and Q^3 silicons are bonded to hydrogen (OH groups) as evidenced by the ^{1}H CPMAS spectra, in which the intensities of the Q^2 and Q^3 sites are enhanced relative to those associated only with bridging oxygens (Q^4 sites) [163,164]. The MAS spectrum of the gel is quite different after heating to 600°C: the Q^3 and Q^4 peaks are not resolved, and there is a broad band of intensity centered at about -107 ppm. A linear relationship exists between the ^{29}Si chemical shift (δ) of Q^4 resonances and the average of the four Si–O–Si angles per Q^4 site (ϕ) [165]:

$$\delta \text{ (ppm)} = -0.59(\phi) - 23.21. \tag{13}$$

Accordingly a Q^4 resonance at -107 ppm corresponds to an average ϕ of 142°. Statistically acceptable decomposition of this spectrum, using peak positions of -91 and -101 pm for Q^2 and Q^3 sites requires a peak at about -105 ppm [34]. Because Q^2 and Q^3 sites containing OH groups are not known to resonate in this chemical shift range, the additional peak must be due to a second Q^4 site with a small ϕ value, 138°. The lack of preferential enhancement in this chemical shift range in the ^{1}H CPMAS spectrum of the 600°C sample indicates that OH is not associated with this site, confirming the assignment to a Q^4 species. Exposure of the 600°C sample to water vapor causes a narrowing and shift of the Q^4 peak position back to its original value in the 50 and 200°C samples.

Heating to 1100°C (region III) completely densifies the gel, reducing its surface area and causing loss of most of the silanol groups as indicated in the Raman spectrum (Fig. 43d). The ^{29}Si MASS NMR spectrum of this sample consists of a Q^4 peak at about -111 ppm corresponding to $\phi = 149°$. ^1H CPMAS yields very little signal for Q^2 and Q^3 sites, consistent with a very low silanol content. The spectrum of the 1100°C sample exposed to 100% RH water vapor is unchanged, indicating the lack of exposed surface area available for rehydration.

The increase in relative intensity of the D2 Raman band correlates well with a reduction in the average Si–O–Si bond angle of Q^4 sites, ϕ, determined from the Q^4 peak positions in the NMR data. D2 is practically absent at 50 and 200°C ($\phi = 148°$), intense at 600°C ($\phi = 138$ and $148°$), and quite low in relative intensity after heating to 1100°C ($\phi = 149°$). The relative intensity of D2 is also quite low for the sample exposed to water vapor after heating to 600°C ($\phi = 148°$). By comparison, although the D1 species can form by condensation reactions on the silica gel surface, its formation is not correlated with a reduction in the average value of ϕ.

The association of the species responsible for D2 with the presence of Q^4 sites with reduced values of ϕ and conversely the elimination of this species with an increase in ϕ is consistent with Galeener's [166] assignment of the D2 vibration in conventional v-SiO_2 to an oxygen ring breathing mode of cyclic trisiloxanes (three-membered rings). Galeener made this assignment partly on the basis of the close agreement with the respective oxygen ring breathing frequency of the cyclic molecule, hexamethylcyclotrisiloxane (587 cm^{-1}) [167,168]. Molecular orbital (MO) calculations performed on the model, planar, cyclic molecule, $H_6Si_3O_3$, have established 136.5° and 103° as the optimum values of ϕ and θ, respectively [169]. For all larger rings MO calculations predict optimum angles of $\phi \cong 148°$ (attained by ring puckering) and the tetrahedral angle, $\theta = 109.5°$.

The NMR–Raman data indicate that in gels three-membered rings are absent at low temperatures. They form at intermediate temperatures primarily on the silica gel surface by condensation reactions involving isolated vicinal silanol groups located on unstrained precursors [32]:

$$d(Si\text{-}O) = 1.626 \text{ Å} \qquad d(Si\text{-}O) = 1.646 \text{ Å}$$

(14)

The cyclic trisiloxanes are composed exclusively of Q^4 silicons as shown by the MAS and ^{1}H CPMAS spectra of the 600°C sample in which shifts are observed for the Q^4 peak position, whereas the ^{1}H CPMAS and hydrated MAS spectra show that the Q^2 and Q^3 peak positions are unaffected by the formation or elimination of the D2 species. The H–O distances in the neighboring silanol groups of the precursor (left side Eq. 14) exceeds 5.5 Å, causing them to exhibit SiO–H vibrations characteristic of isolated species [164].

MO calculations [163,165] indicate that a reduction in the value of ϕ below about 140° costs energy, causing the heats of formation of small cyclic molecules to be positive: $\Delta H_f = 55.4$ and 24.4 kcal/(mole of rings) for two-membered ($\phi = 91°$) and three-membered rings, respectively. During ring formation according to Eq. 14, all of this energy is associated with ring closure; i.e., the precursor structure is unstrained. Therefore according to theory we expect to measure 24.4 kcal/(mole by-product H_2O) as three-membered rings are formed via dehydroxylation of the silica gel surface. This is close to the value (~ 23 kcal/mole) found in a careful DSC study [160,161]. This positive heat of formation explains the DSC endotherms (550–650°C) observed in Fig. 41a and the endothermic background (400–700°C) observed in Fig. 34b.

Based on decomposition of the MAS ^{29}Si NMR results (Fig. 44) [170] and estimates of the extent of dehydroxylation according to Eq. 14, the maximum percentage of silicons incorporated in three-membered rings in the two-step acid–base-catalyzed gels studied by Brinker and coworkers exceeds 20%. The NMR results indicate that after heating at intermediate temperatures (e.g., 600°C), the silica gel surface is composed primarily of OH terminated Q^3 sites and two different Q^4 sites with quite different bond angles and bond lengths (Q^4 silicons contained in three-membered rings and Q^4 silicons contained in four-membered and higher-order rings). As discussed in the following chapter, the variations in bond lengths and angles may impart different Lewis acid-base character to the two different Q^4 sites, causing the siloxane bonds contained in the three-membered rings to be very susceptible to hydrolysis. Since the vibrational spectra of dried gels approximates the species distribution present near the gel point, the susceptibility of three-membered rings to hydrolysis also explains the virtual absence of the D2 band in the Raman spectra of xerogels dried at low temperaures. Only in water-starved solutions or very basic conditions (pH > 13 [171]) do cyclic trimers constitute a significant portion of the species population.

The behavior of the D1 vibration in gels appears consistent with its assignment to a four-membered ring: the relative intensity of the D1 band is relatively unchanged by exposure to 100% RH water vapor, and the Q^2-Q^4 peak positions are unaffected by the changing concentrations of the D1 species (Fig. 44). In addition, because cyclic tetrasiloxanes are stable toward hydrolysis [167], they can constitute a large fraction of the oligomeric species in TEOS-derived systems [172]. Thus it is reasonable that cyclic tetrasiloxanes also contribute to the intensity of the D1 band in the low-temperature xerogels. (See, e.g., the 50 and 200°C samples in Fig. 43.)

Mulder and Damen [38] have proposed localized vibrations of O atoms bonded to network terminating Q^3 sites as an alternate structural model for the D1 band in "wet" gels and in high–surface-area gels prior to consolidation. This model is consistent with ^{29}Si NMR and Raman investigations of silicate solutions during the latter stages of polymerization [173]. (See Fig. 38 in Chapter 3.) Although Q^3 silanols most likely contribute to the intensity at 490 cm^{-1} (especially in gels that have not been heated) [34], the relative intensity of the D1 vibration is unaffected by the silanol concentration at 600°C [174] (see Fig. 46) and Q^3 silanols cannot account for the D1 vibration in conventional v-SiO$_2$ (which may be substantially OH-free). Therefore it is likely that four-membered rings are responsible for the D1 vibration at intermediate and higher temperatures.

Fig. 46.

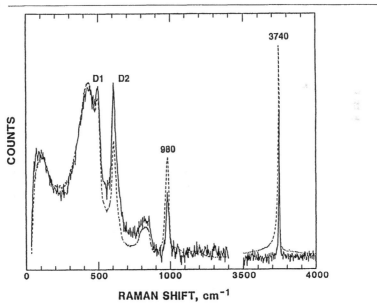

RAMAN SHIFT, cm^{-1}

Raman spectra of a two-step acid–base-catalyzed silicate xerogel after heating to 600°C in oxygen for 1 hour (dashed line) or after heating to 600°C in oxygen followed by vacuum (3×10^{-7} torr) for 24 hours (solid line) [174]. The 490 cm^{-1} band (D1) is relatively unchanged although the 980 and 3740 cm^{-1} bands indicate that the vacuum treatment reduces the SiOH concentration by more than 2×.

Heat treatments in the vicinity of the glass-transition temperature (T_g) of v-SiO$_2$ (about 1000°C) reduce the concentrations of the D1 and D2 species to levels comparable to those in conventional v-SiO$_2$ and cause the ~430 cm^{-1} band to broaden. (See, e.g., the 1100°C spectrum in Fig.43.) From the drastic reduction in the relative intensity of the D2 band (and to a lesser extent the relative intensity of the D1 band), we infer that as the temperature is increased, the reduced viscosity allows the surface to reconstruct by a disproportionation process in which stable five-membered and higher-order rings are formed at the expense of less stable, smaller rings. The assignment of the ~430 cm^{-1} band to O ring-breathing modes of five-membered and higher-order rings is consistent with the broadening observed at 1100° and the trend of decreasing vibrational frequency with increasing ring size (see Table 9) [162].

The effect of viscosity on the surface reconstruction process is evident from Raman investigations of xNa$_2$O–(100–x)SiO$_2$ gels where $0.1 \leq x \leq 2.0$ [35]. The temperature at which the relative intensity of the D2 vibration (and to a lesser extent the D1 vibration) is suddenly reduced decreases with x in the same manner as T_g. A relationship between viscosity and the relative intensity of the D1 and D2 vibrations was also observed for silica gels by Krol and van Lierop [44]. After heating at 800°C one portion of the sample contained rather large concentrations of the D1 and D2 species characteristic of the surface prior to reconstruction, whereas a second portion of the sample at nominally the same temperature exhibited concentrations of the D1 and D2 species comparable to those in v-SiO$_2$. Thus within a very narrow temperature or compositional (OH) range the surface suddenly reconstructs forming five-membered and higher-order rings at the expense of three- and four-membered rings. From these observations, we infer that, once formed, small cyclic species are kinetically stabilized by the exceedingly high matrix viscosity at low and intermediate temperatures.

Table 9.

Si–O Stretching Frequencies of Siloxane Rings in Model Compounds and v-SiO$_2$.

Functional Group	Source Material	Stretching Frequency (cm^{-1})
$\geq$ 5-fold rings	v-SiO$_2$/SiO$_2$ gel	430
4-fold rings	Octamethyl cyclotetrasiloxane	480
3-fold rings	Hexamethyl cyclotrisiloxane	587
2-fold rings	Tetramesityl cyclodisiloxane	873
D1	SiO$_2$ gel	490
D2	SiO$_2$ gel	608

Source: Brinker *et al.* [162].

2.4.2. INFRARED INVESTIGATIONS OF FRAMEWORK SILICATE STRUCTURES

Infrared (400–$4000\,cm^{-1}$) spectroscopic investigations of silica gel consolidation have been performed by numerous research groups. (See, for example, refs. [35,146,154].) Due to the selection rules, asymmetric vibrations are infrared allowed, therefore the infrared spectra of silica gels complement the Raman spectra discussed in the previous section. One drawback of IR spectroscopy is that bulk silicates are totally absorbing for wave numbers below about $2400\,cm^{-1}$, so infrared investigations of the fundamental framework vibrations often require dilution in infrared transparent media such as KBr or Nujol.[†]

The complementary nature of IR and Raman spectra is apparent from the comparison of the IR and Raman spectra obtained during silica gel consolidation in region II. (For example, compare Fig. 19a and Fig. 47 with Figs. 19b or Fig. 43.) The most intense 1080 and $1220\,cm^{-1}$ bands in the IR spectrum are assigned to the TO and LO modes of the asymmetric Si–O$^-$ stretching vibration, respectively [90]. The corresponding Raman bands are very weak. The $800\,cm^{-1}$ band is assigned to symmetric Si–O stretching and the $460\,cm^{-1}$ band to a Si–O–Si bending mode. Si–OH and SiO–H stretching of terminal silanol groups occurs at $960\,cm^{-1}$ and in the envelope of vibrations centered near $3400\,cm^{-1}$, respectively. There is no evidence in the IR spectra for symmetric ring breathing modes corresponding to the D1 and D2 vibrations in the Raman spectra.

There is comparatively little information obtainable from the IR spectra concerning the evolution of the silicate network in region II. Above about $200°C$ the major features of the IR spectra change little with temperature. (See Figs. 19a and 47.) Apart from the sharp spectral features associated with C–H and C–O bending vibrations in the $110°C$ spectrum of the $r = 1$ gel, there are only subtle spectral changes with the value of r (1–20) used in the gel synthesis procedure [90] (Fig. 47). Continued condensation with increasing temperature is evident from the reduction in the relative intensity of the $\sim 970\,cm^{-1}$ band assigned to Si–OH stretching along with a progressive increase in the frequency of the Si–O stretching vibration for the $r = 1$ gel. This latter trend has been observed in many silicate gels and is interpreted to result from a strengthening of the network through crosslinking. As shown in Fig. 48, the increasing Si–O$^-$ stretching frequency is, in general, strongly correlated with increasing density and physical properties such as microhardness and refractive index. The absence of this

[†] These procedures may introduce additional H_2O. Therefore studies of hydroxyl content or adsorbed water in KBr- or Nujol-prepared samples may be unreliable.

Fig. 47.

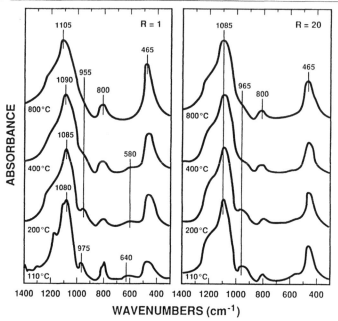

IR spectra of xerogels prepared by single-step hydrolysis of TEOS ($r = 1$ or $r = 20$) after drying at 120°C or after heating between 200 and 800°C [90].

trend at temperatures below 800°C in the $r = 20$ gel (Fig. 47) is evidence that this synthesis condition results in a gel structure composed of extensively crosslinked regions that undergo little further internal condensation in region II. Presumably the stretching frequency of the $r = 20$ gel does increase as the gel densifies by viscous sintering above 800°C in region III.

2.4.3. IR–RAMAN INVESTIGATIONS OF GEL DEHYDROXYLATION

IR (400–4000 cm^{-1}), near-IR (4000–10000 cm^{-1}), and Raman spectroscopy have been used extensively to investigate the dehydroxylation of silica gel surfaces in region II. (See, e.g., refs. [40,41,154].) Figure 49a–c shows infrared spectra of TMOS-derived gels (prepared using a one-step hydrolysis procedure ($r = 4$) under neutral, acidic, or basic conditions) during low-temperature dehydration (25–170°C) [154]. The broad adsorption band between about 3000 and 3800 cm^{-1} corresponds to the fundamental stretching vibrations of different hydroxyl groups. It is composed of a

Fig. 48.

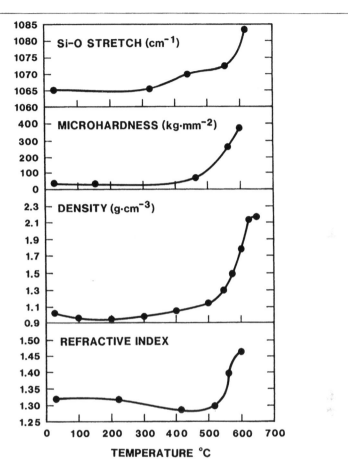

Si–O⁻ asymmetric stretching vibration, microhardness, density, and refractive index of multicomponent borosilicate gels after heat treatments between room temperature and 600°C (heating rates = 0.5–2.0°C/min) [175].

superposition of SiO–H stretching vibrations:

$3750\ cm^{-1}$	isolated vicinal SiO–H stretching
$3660\ cm^{-1}$	mutually H-bonded SiO–H stretching or internal SiO–H stretching
$3540\ cm^{-1}$	SiO–H stretching of surface silanols hydrogen-bonded to molecular water
$3500–3400\ cm^{-1}$	O–H stretching of hydrogen bonded molecular water.

Fig. 49.

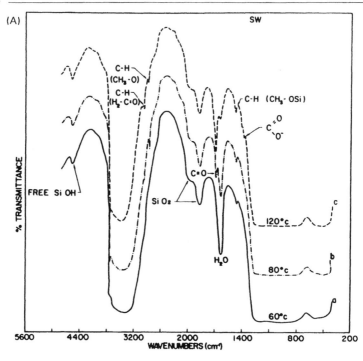

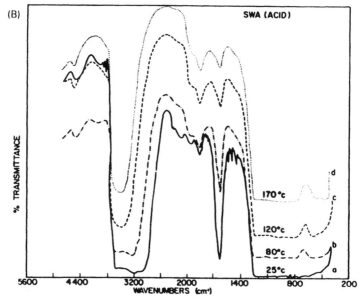

Fig. 49.

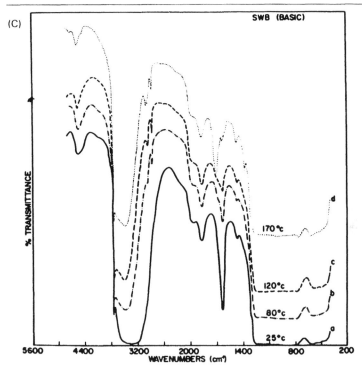

FTIR spectra of xerogels prepared by single-step hydrolysis of TMOS ($r = 4$) followed by low-temperature heat treatments (60–120°C). (a) SW gel prepared with no added catalyst. (b) SWA gel prepared with 0.1 N HCl. (c) SWB gel prepared with 0.01 N NH$_4$OH [154].

In addition to the envelope of vibrations corresponding to O–H stretching, the band at 1620 cm^{-1} is assigned to a deformation mode of adsorbed molecular water.

The principal change on heating is a decrease in the concentrations of silanols and adsorbed water. However, the spectra reveal several important differences that depend on the synthesis conditions. The acid-catalyzed gel exhibits the simplest spectrum (Fig. 49b). Above 25°C there is no evidence for C–H stretching vibrations associated with unhydrolyzed methoxy groups (sharp features near 2950 and 1470 cm^{-1} in the neutral and base-catalyzed samples), an indication of complete hydrolysis and no reesterification.[†]

[†] This result is quite different from those obtained for one- and two-step acid-catalyzed samples prepared from TEOS [39,40] in which significant reesterification occurred during drying. (See, e.g., Fig. 23.) This may reflect a reduced alcohol–water ratio in the original synthesis procedure.

By comparison, the neutral and base-catalyzed samples (Figs. 49a and 49c) contain unhydrolyzed methoxy groups, because unhydrolyzed monomers and dimers persist long past the gel point under these conditions. (See Chapter 3.) For the neutral and base-catalyzed gels methoxy ligands oxidize to produce formaldehyde and then formic acid and formates that remain strongly bound to the silica surface at temperatures in excess of 300°C. Acid-catalyzed gels contain more internal silanols and, due presumably to their microporous nature, lose water less easily than base-catalyzed or neutral gels. Therefore acid-catalyzed gels generally exhibit lower viscosities than base-catalyzed gels at equivalent temperatures, causing the transition to region III to occur at lower temperatures. (See, e.g., Fig. 27.)

Near-IR spectra are useful in distinguishing between physically adsorbed water and various hydrogen-bonded silanol species. This spectral region is composed of overtones and combination bands associated with various silanol species, adsorbed water, and residual organics. Figure 50 shows a sequence of near-IR spectra obtained during dehydroxylation of an acid-catalyzed TEOS-derived gel [40]. The major spectral features are assigned as follows [40]:

• Triply split bands near $4350 \, cm^{-1}$	combination C–H stretching-deformation
• Triply split bands near $5780 \, cm^{-1}$	first overtone of C–H stretching
• $4566 \, cm^{-1}$	combination stretching–bending of vicinal free SiO H
• $7326 \, cm^{-1}$	first overtone of vicinal free SiOH
• $5290, 5154 \, cm^{-1}$	combination stretching–bending of H-bonded water
• $7142, 6896 \, cm^{-1}$	first harmonic of mutually H-bonded SiO H.

The triply split bands assigned to C–H stretching weaken and disappear near 315°C, the temperature at which the bands assigned to silanols and molecular water exhibit their maximum intensities. The dry gel shows no band at $5290 \, cm^{-1}$, attributable to hydrogen bonded molecular water. This behavior indicates that the dried gel is hydrophobic due to essentially complete esterification of the terminal sites. Heating in air removes the organic groups creating a hydrophilic silanol surface. Further heating causes the intensities of the silanol-related bands to decrease. The reduction in the $7142 \, cm^{-1}$ band and the sharpening of the $7326 \, cm^{-1}$ band indicate that as

Fig. 50.

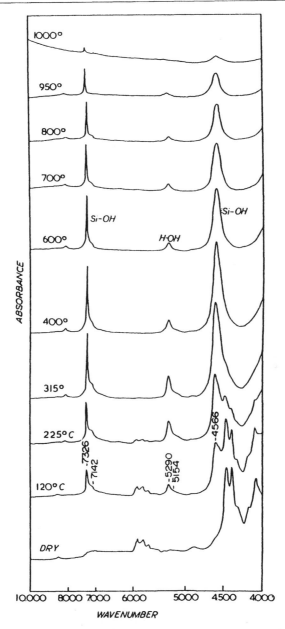

Near IR spectra of xerogels prepared by single-step acid-catalyzed hydrolysis of TEOS and dried at 70°C or heated to temperatures between 120 and 1000°C [40].

temperature is increased, the hydrogen bonded silanols are more easily removed, creating a surface composed primarily of isolated vicinal silanol species.

A sequence of Raman spectra is shown in Fig. 33 for an acid-catalyzed TEOS-derived gel ($r = 10$). The sharp 3750-cm^{-1} band is assigned to SiO–H stretching of isolated vicinal silanols. The broad, low-frequency shoulder (3680 cm^{-1}) is assigned to mutually hydrogen-bonded surface silanols, internal silanols, and/or hydrogen-bonded molecular water. As observed in the near-IR spectra, heating first causes the progressive removal of the low-frequency shoulder, then the removal of isolated silanols. At 800°C some portions of the gel have completely densified causing the formation of a broad band at 3680 cm^{-1} assigned to internal (bulk) silanols. By 900°C the gel is completely densified leaving only internal silanols. By comparison with the v-SiO$_2$ spectrum (500 ppm OH), the densified gel is estimated to contain 5000 ppm OH.

2.4.4. ELECTRON SPIN RESONANCE STUDIES

Electron spin resonance (ESR) has been used to investigate radiation-induced *paramagnetic defects* in silicate gels after various stages of gel consolidation [176–180]. Paramagnetic defects are not present in unirradiated gels; however their formation during *radiolysis* provides structural information concerning precursor species present in low concentrations (<1%), undetectable by vibrational spectroscopy.

Wolfe *et al.* [178] evaluated the effects of the processing temperature on the types and concentrations of paramagnetic defects in TEOS-derived silica gels. Prior to drying or thermal treatment, the ESR spectra were dominated by a variety of organic radical species (e.g., the ethanol radical, CH$_3$CH•OH, and the methyl radical •CH$_3$). Anneals at ~600°C prior to irradiation resulted in substantially lower yields of organic radicals and revealed notable concentrations of *peroxy radicals* (O$_2^-$-type defects). Kordas and coworkers [176,177] observed that vacuum-dried gels prepared from TEOS using a range of r values displayed mainly peroxy radical spectra, regardless of preradiation heat treatments to temperatures up to ~500°C. The seeming ubiquity of the peroxy radical as a radiolysis product has led to the hypothesis that peroxy linkages, Si–O–O–Si, may form as condensation products on the silica gel surface.

Griscom and coworkers [179,180] prepared ^{17}O-enriched and unenriched silica gels from TEOS using a two-step acid–base hydrolysis procedure with $r = 3.8$. After heating, the gels were sealed under vacuum and thereafter not subjected to atmospheric contamination during or after irradiation.

Fig. 51.

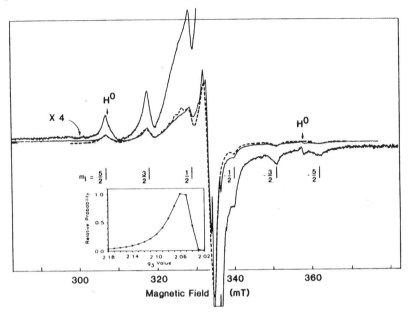

ESR spectrum of two-step acid–base-catalyzed xerogel prepared with ^{17}O enrichment. Spectrum was recorded at 105 K following 15.5 Mrad x-irradiation at 77 K. Dashed curve is a computer simulation based on a NBOHC. Six ^{17}O hyperfine lines are indicated by their nuclear magnetic quantum numbers, m_I [179–180].

Figure 51 shows the ESR spectrum of an ^{17}O-enriched, 600°C sample after 15.5 Mrad x-irradiation at 77 K. The computer-simulated spectrum, which utilizes the g-values and ^{17}O hyperfine coupling constants typical of *non-bridging oxygen hole centers* (NBOHC), unambiguously identifies NBOHC that presumably result from radiolysis of silanol groups according to

$$\equiv Si\text{-}OH \rightarrow \equiv Si\text{-}O\bullet + H^0. \tag{15}$$

Figure 52 shows the spectrum of an unenriched sample that was annealed at 600°C in air and irradiated (5.5 Mrad x-irradiation) at 77 K. Based on the computer simulation, the many "bumps" seen to the low-field side of the derivative zero-crossing are assigned to a sequence of O_2^- ions or peroxy radicals, *ROO•* (where *R* represents an unspecified radical or part of the glass structure), in at least five distinguishable chemical environments. An isochronal anneal experiment showed that radicals in different environments decayed at distinctly different rates [179] consistent with their different environments.

Fig. 52.

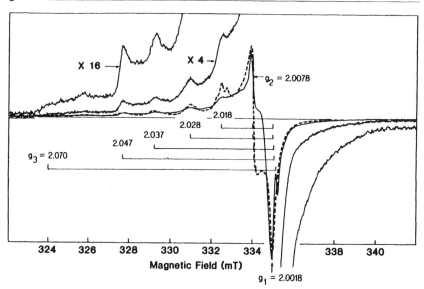

ESR spectrum of two-step acid–base-catalyzed silicate sample after heating to 600°C
followed by 5.5 Mrad x-irradiation at 77 K. The dashed curve is a computer simulation
based on an O_2^- model: five individual component spectra corresponding to the five
g_3-value "combs" were added together in the ratios 1:1:1:1:1 [179].

The E′ center $\equiv$Si•, which is a prominent paramagnetic defect in conventional v-SiO$_2$, was detected in both the ^{17}O-enriched and unenriched
samples [179,180]; however, its concentration was found to be 10 to 500 times
weaker than those of the O-related or organic species. Griscom *et al.* [180]
attribute the higher concentration of defects in gels compared to conventional v-SiO$_2$ to the radiolysis of molecular species associated with terminal
hydroxyl or alkoxy groups, organics (e.g., ethanol), or species adsorbed from
the atmosphere (e.g., O_2 or H_2O). Thus the "excess" defect concentration
in gels is related primarily to the high surface area and the complex surface
chemistry that occurs within the pores during gel consolidation.

2.5.   Structural Studies of Multicomponent Systems

Numerous multicomponent silicate and some nonsilicate gels have been
consolidated to glass as is evident from the extensive list compiled by James
in Table 7. It is certainly not the purpose of this chapter to present an account
of structural changes that occur for all these systems; however, examples of

structural evolution occurring in several representative systems are in order. In the following subsections we present information concerning alterations in the region II–III transition temperature and evidence for mixed oxide bonding, changes in coordination number, and crystallization occurring in region II.

2.5.1. BOROSILICATE GELS

The region II–III transition is characterized by a substantial increase in the shrinkage rate without a corresponding increase in the weight loss. Figure 53 shows shrinkage at 1°C/min for five multicomponent borosilicate gels prepared under similar conditions of hydrolysis, order of addition, etc. It is evident that the transition temperature depends on composition: it increases with decreasing alkali content. Alkali additions are known to lower the viscosities of conventional silicate glasses due to depolymerization:

$$Na_2O + \equiv Si-O-Si\equiv \rightarrow 2(\equiv Si-O^-Na^+). \tag{16}$$

Thus, from Fig. 53 we infer that the transition temperature depends on the gel viscosity and that alkali additions reduce the viscosity of silicate gels in a manner similar to conventional silicate glasses.

Fig. 53.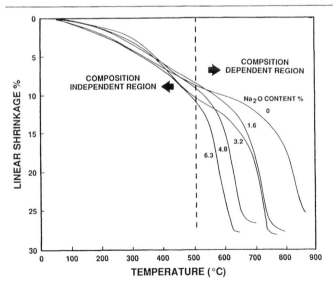

Percent linear shrinkage measured during heating at 1°C/min for a series of borosilicate gels prepared with varying Na_2O contents [130]. The region II–III transition increases from about 525 to 750°C as the Na_2O content is reduced from 6.3 to 0 wt%.

The correlation of the transition temperature with the viscosity is strong evidence for viscous sintering being the predominant shrinkage mechanism in region III.[†] In viscous sintering the shrinkage rate is proportional to the surface energy divided by the product of the pore size and viscosity. Thus for gels of equivalent surface area and pore size, the transition temperature will correspond to the minimum temperature at which the viscosity is sufficiently low to observe measureable shrinkage. This temperature should increase with decreasing alkali content as observed in Fig. 53. As a general rule the transition temperature is in the vicinity of the glass transition temperature, T_g (defined as the temperature at which $\eta \approx 10^{13.5}$ poises), for the corresponding melt-prepared glass.

Figure 53 shows that for multicomponent borosilicate gels prepared under similar conditions, the shrinkage in region II is not substantially affected by the alkali content. This implies that for this series of compositions, region II shrinkage occurs primarily by continued condensation reactions involving the borosilicate network and therefore depends on the extent of polymerization that results from the particular conditions of gel synthesis. The effect of synthesis conditions on region II densification is apparent in Fig. 54. As for pure silicates, larger values of r and more basic conditions lead to highly condensed gels that undergo little further densification in region II. Conversely, lower r values and acidic conditions result in weaker branching, promoting densification in region II.

The general trend observed in amorphous multicomponent gels is that shrinkage in region II depends on the connectivity of the tri- and tetra-functional network-forming species. Apparently mono- and difunctional network modifiers such as Na^+ and Ba^{2+} initially occupy network-terminating sites as does H^+ and therefore do not significantly influence shrinkage. However, unlike protons, modifiers are not removed by condensation reactions so that they influence (reduce) the temperature of the transition to region III. Obviously, as the transition temperature is lowered, the extent of dehydroxylation that can occur in region II is reduced. The residual hydroxyl content strongly influences subsequent region III sintering kinetics as discussed in Chapter 11.

FTIR investigations of condensation in region II in binary borosilicates [54] (Fig. 55) and multicomponent borosilicates (Fig. 24) reveal that borosiloxane bonding occurs almost exclusively during heating at temperatures below 500°C. Figure 55 shows that $\equiv$B-O-Si$\equiv$ bonding is quite rare in the "wet" gel as indicated by the low relative intensities of the 670 and 915 cm^{-1} vibrations assigned to Si-O-B bending and stretching modes,

---

[†] Although shrinkage occurs primarily by viscous sintering, continued polymerization and structural relaxation also contribute to region III shrinkage, significantly affecting the sintering kinetics. This subject is discussed in detail in Chapter 11.

Fig. 54.

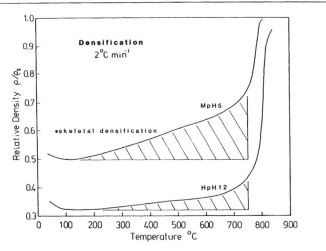

Relative bulk density during heating at 2°C/min for two four-component borosilicate samples prepared under different conditions. Sample MpH5 was prepared at pH 6.8 with $r = 5$. Sample HpH12 was prepared at pH 9.2 with $r = 12$ [130]. Skeletal densification is indicated by the cross-hatched areas.

respectively. Boron is mainly present as boric acid $(B(OH)_3)$ as is evident from the bands at $560 \, cm^{-1}$ (Figs. 24 and 55) and $1415 \, cm^{-1}$ (Fig. 24) assigned to B–O deformation and stretching modes, respectively. As discussed in Chapter 2, the absence of $=B{-}O{-}Si\equiv$ bonding in the "as-dried" gel is due to $=B{-}O{-}Si\equiv$ bond hydrolysis or alcoholysis reactions. The absence of water or alcohol during heating allows $=B{-}O{-}Si\equiv$ bonds to form at the expense of boric acid. Heating to about 450°C converts the boric acid mainly to B–O–Si, although a small amount of B–O–B bonding is indicated by the shoulder at $\sim720 \, cm^{-1}$ [54] (Fig. 55).

Borosilicates are representative of systems that, due to the hydrolytic instability of mixed oxide bonding, are chemically inhomogeneous on a molecular scale in the wet gel and desiccated xerogel. Homogeneity is achieved only during subsequent heating. Many other multicomponent systems may initially be inhomogeneous due to different rates of hydrolysis of the alkoxide precursors or a preference for homo- rather than heterocondensation during gel formation. In both cases, depending on the spatial scale of the inhomogeneity, complete homogenization may occur during heating in region II. By comparison, multicomponent diphasic gels (see, e.g., refs. [49, 56,181]), which are intentionally prepared under conditions that enhance compositional or structural inhomogeneities, remain inhomogeneous in region II, altering sintering and/or crystallization kinetics in region III.

Fig. 55.

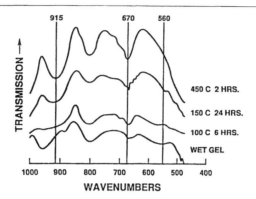

FTIR spectra of binary borosilicate gels prepared by acid-catalyzed hydrolysis of TMOS ($r = 2$) plus trimethyl borate during various stages of thermal treatment [54].

2.5.2. TITANIA–SILICA SYSTEMS

Sol-gel processing of TiO_2–SiO_2 compositions has been studied extensively as a low-temperature route to ultralow expansion (ULE®) glasses [94,95, 182–184] and as a method of obtaining optical thin films with refractive indices ranging from about 1.4 to 2.3 [185]. According to Evans [186] the addition of TiO_2 to a v-SiO_2 network occurs by the continuous substitution of tetrahedrally coordinated Ti^{4+} for Si^{4+} up to 11.5 wt% TiO_2. However Sandstrom et al. [187] find some octahedrally coordinated titanium at 3.4–9.5 wt% TiO_2. The structures of titania–silicate gels during heating have been investigated by infrared and Raman spectroscopy [94,182,183], X-ray diffraction (XRD) [183,184], extended X-ray absorption fine structure (EXAFS) [184], X-ray absorption near-edge structure (XANES) [184] and ESR [183]. Changes in physical properties [94] have been investigated by dilatometry, DTA, and microhardness measurements.

Emili et al. [184] combined XRD, XANES, and EXAFS to study structural evolution in TiO_2–SiO_2 gels containing 3.4, 7.7, or 15 mol% TiO_2 at temperatures ranging from 430 to 1200°C. XRD showed that all compositions were amorphous up to 700°C. Anatase was detected at 1200 and 1000°C for the compositions containing 7.7 and 15 mol% TiO_2, respectively, whereas the 3.4 mol% TiO_2 sample remained amorphous. The XANES and EXAFS results for all compositions prior to crystallization showed that the coordination number of oxygens in the first coordination shell and the Ti–O bond lengths (0.182 nm) are consistent with tetracoordinated Ti, corroborating the structural model of Evans [186]. With increasing temperature, the distortion of the tetrahedral environment of Ti was reduced in the amorphous samples. At 1200°C all the Ti–O bond lengths were identical in the 3.4 mol% TiO_2 sample, indicating perfect tetrahedral symmetry. Analysis of the XANES and EXAFS data for the 15 mol% TiO_2 sample after heating at 1200°C indicated a coordination number of 5.9 and a Ti–O bond length of 1.95 Å. The coordination number of ~6 and the increased Ti–O bond length are consistent with octahedrally coordinated Ti. Corresponding XRD studies indicated that the coordination number and bond length changes coincided with the crystallization of anatase.

Similar XANES and EXAFS experiments performed by Greegor *et al.* [188] on flame-pyrolyzed SiO_2–TiO_2 glasses were consistent with the model of Sandstrom *et al.* [187] in which Ti exists in both tetrahedral and octahedral sites for all investigated Si–Ti ratios. The results on sol-gel–prepared glasses therefore indicate that Ti is kinetically stabilized in tetrahedral coordination. Only at high temperatures (as in flame pyrolysis), where there is increased mobility, is the thermodynamically favored structure established. Kinetic, rather than thermodynamic, control of structure is a characteristic common to sol-gel processing.

Best and Condrate [182] used Raman spectroscopy to investigate sol-gel–derived SiO_2–TiO_2 glasses containing up to 65 mol% TiO_2. Raman spectra obtained after 500 and 700°C heat treatments are shown in Fig. 56a and 56b. Compared to the spectra obtained on pure silica gels, the major spectral differences observed for the 3.3 and 6.3 mol% TiO_2 gels are a broadening of the ~430 cm^{-1} band and new bands at ~952 and 1100 cm^{-1} associated with vibrational modes involving Si–O–Ti linkages. Heating increases the relative intensity of the latter two bands in most of the samples indicative of further Ti–O–Si bonding during heating. The Raman results are suggestive of homogenization occurring during heating as for the borosilicate system previously discussed. This reflects differences in the hydrolysis rates and tendency for homo- versus heterocondensation for the TEOS and titanium tetra-isopropoxide precursors. Best and Condrate [182] attribute the broadening of the 430 cm^{-1} band to a distortion of the SiO_4 tetrahedra that results from the incorporation of tetrahedrally coordinated Ti^{4+} into the SiO_2 network. Thus the ordering of the tetrahedral environment surrounding Ti [184] is apparently accomplished by a distortion of the Si environment.

The spectrum of the sample containing 11.2 mol% TiO_2 exhibits further broadening of the 430 cm^{-1} band, consistent with greater incorporation of Ti into tetrahedral sites in the SiO_2 network. However, a new band appears at 665 cm^{-1} associated with octahedrally coordinated Ti. Octahedrally coordinated Ti was observed in the XANES and EXAFS investigations described previously [184] for 15 mol% TiO_2 samples only after heating to 1200°C. This discrepancy may be explained either by a greater sensitivity of

Fig. 56.

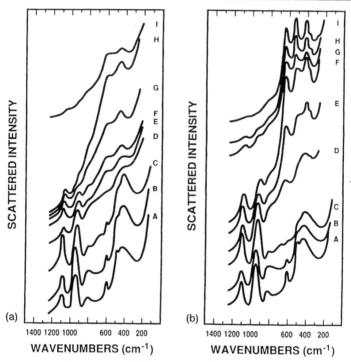

(a) ... (b)

Raman spectra of a series of binary gels of composition xTiO$_2$-(1-x)SiO$_2$: (A) $x = 3.3$; (B) $x = 6.3$; (C) $x = 11.2$; (D) $x = 20.1$; (E) $x = 29.6$; (F) $x = 36.8$; (G) $x = 43.1$; (H) $x = 50.2$; (I) $x = 65.4$. Figure 56a: after heating to 500°C. Fig. 56b: after heating to 700°C [182].

the Raman experiments or actual differences that exist due to the original gel synthesis conditions. No XRD are available to determine whether or not the appearance of octahedrally coordinated Ti is accompanied by crystallization of anatase.

Increasing the TiO$_2$ concentration above 11.2 mol% causes a progressive reduction in the 952 and 1100 cm^{-1} bands associated with Si-O-Ti bonding. This result indicates a preferential association of titanate species with themselves rather than with silicate species. Best and Condrate [182] interpret this as evidence for phase separation which precedes cystallization of anatase at 700°C. It is interesting to observe evidence for phase separation occurring at such low temperatures in these very viscous compositions. There are several documented cases of metastable, homogeneous glasses being prepared within the immiscibility region by sol-gel processing at temperatures less than $\sim T_g$ (e.g., SrO–SiO$_2$ glasses [152]). Therefore the results of Best and Condrate may reflect preferential homocondensation of titanate species in solution. Unfortunately Raman spectra of the original wet gels and as-dried xerogels are not presented to confirm this hypothesis.

Hayashi *et al.* [94] correlated several physical property changes (shrinkage, weight loss, hardness, and expansion coefficient) with structural changes determined by XRD and IR spectroscopy for TiO_2–SiO_2 compositions made under different conditions of hydrolysis. Weight loss and shrinkage are shown in Fig. 57 for a 7.4 wt% TiO_2 sample prepared with $r = 16$ or 50. It is apparent that lower r values promote shrinkage in region II (ca. 300–600°C) which is accompanied by substantial weight loss. When $r = 50$ there is less shrinkage in region II and a correspondingly lower weight loss in the temperature interval 200–700°C. Density measurements and XRD investigations show that the sample hydrolyzed with $r = 16$ is fully dense and completely amorphous at 800°C. By comparison, at 800°C the sample with $r = 50$ is porous and partially crystalline.

These results demonstrate a remarkable effect of the extent and manner of hydrolysis on the densification and crystallization behavior of TiO_2–SiO_2 gels. Lower concentrations of water added very slowly result in quite dense, homogeneous, transparent gels. Upon heating these gels densify almost

Fig. 57.

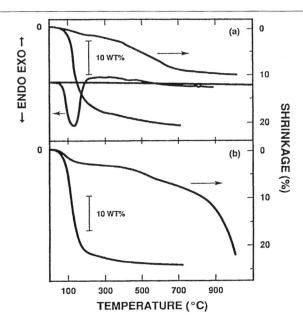

DTA, TGA, and shrinkage curves of TiO_2–SiO_2 gels containing 7.4 wt% TiO_2: (a) $r = 16$; and (b) $r = 50$ [94].

Fig. 58.

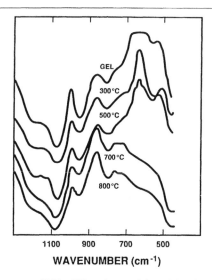

WAVENUMBER (cm⁻¹)

Infrared absorption spectra of TiO_2–SiO_2 gels containing 12.9 wt% TiO_2 after drying and after heating at various temperatures [94].

completely in region II (producing a transparent, amorphous glass). Densification occurs by continued condensation reactions as evidenced from the weight loss (Fig. 57) and the IR spectra (Fig. 58), which indicate a progressive increase in both the relative intensity of the ~735 cm^{-1} band, attributed to Ti–O–Si bonding, and the frequency of the Si–O asymmetric stretching vibration at ~1080 cm^{-1}. Higher water concentrations or more rapid additions of water result in quite porous, opaque gels. These gels exhibit little region II shrinkage or weight loss, suggestive of a highly condensed skeletal phase, and they crystallize to form anatase as low as 500°C. Apparently, these synthesis conditions produce quite inhomogeneous gels in which there is a strong preference for Ti–O–Ti bonding, facilitating crystallization at low temperature. Unfortunately, detailed structural analyses of gels resulting from the different synthesis conditions are not available.

Refractive index and thickness measurements of TiO_2–SiO_2 thin films [185] indicate that for high Ti–Si ratios, apparent densification occurs at remarkably low temperatures. For a series of binary compositions ranging from 70–95 mol% TiO_2, the thickness decreases continuously between 25 and 400°C, while the refractive index increases and appears to plateau at about 400°C. (See, e.g., results for the 90% TiO_2 film in Fig. 59.) TEM of the 400°C films showed that the films were partially crystalline, containing

Fig. 59.

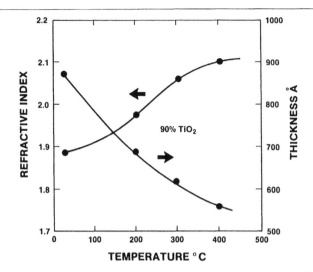

Refractive index and thickness measured by ellipsometry for 90 mol% TiO_2–SiO_2 films after five-minute heat treatments in air at the indicated temperatures [185].

extremely small anatase crystals. Therefore the refractive index increase may reflect the extent of crystallization rather than complete densification. Yoldas [189], however, used TiO_2-rich films heated at only 450 to 500°C to protect silvered mirror surfaces from corrosion. This suggests that in fact complete densification occurs at very low temperatures.

Both the preceding thin film results and the dilatometric results by Hayashi *et al.* [94] (see Fig. 57) provide evidence for considerable shrinkage occurring at low temperatures in TiO_2–SiO_2 systems by a mechanism other than viscous sintering (i.e., prior to the onset of region III). The corresponding physical and structural investigations have not been performed to understand these low-temperature densification processes. This area remains a promising one for future research.

2.6. Crystalline Systems

Compared to amorphous systems, structural evolution in crystalline systems may be dominated by the effects of phase transformations [190] which often occur in conjunction with dehydration or when metastable transitional phases are involved. The most thoroughly studied crystalline gel system is alumina prepared by the Yoldas process [47,191,192]. (See Chapter 2.) Xerogels, which are often completely transparent, are typically composed of

Fig. 60.

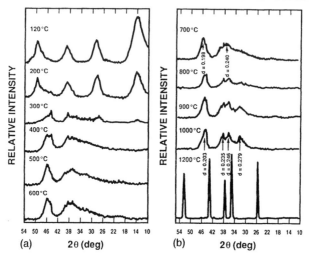

XRD patterns of boehmite xerogels after heat treatments at temperatures between 120 and 1200°C for one hour. Below 300°C, the patterns correspond to boehmite. The 300–700°C patterns correspond primarily to γ-Al$_2$O$_3$. The 800–1000°C patterns correspond to δ-Al$_2$O$_3$, and the 1200°C pattern indicates well-crystallized α-Al$_2$O$_3$ with a much larger grain size as indicated by the substantially narrower diffraction peaks [193].

boehmite (γ-AlOOH) or *pseudo-boehmite* (a less well crystallized boehmite containing 1.7 H$_2$O/Al). Heating causes dehydration and rearrangement leading to a series of *transitional aluminas* and finally α-Al_2O_3. A typical sequence of crystallization deduced from XRD [193–195] (see Fig. 60) is the following:

$$\text{AlOOH} \xrightarrow{300°C} \gamma\text{-Al}_2\text{O}_3 \xrightarrow{850°C} \delta\text{-Al}_2\text{O}3$$
$$\xrightarrow{1100°C} \theta\text{-Al}_2\text{O}_3 \xrightarrow{\geq 1200°C} \alpha\text{-Al}_2\text{O}_3. \qquad (17)$$

Room-temperature aging of pseudo-boehmite gels prior to drying produces well-crystallized bayerite, Al(OH)$_3$. A typical crystallization sequence for bayerite conversion to α-Al$_2$O$_3$ is [190]:

$$\text{Al(OH)}_3 \xrightarrow{400°C} \eta\text{-Al}_2\text{O}_3 \xrightarrow{1150°C} \theta\text{- and } \alpha\text{-Al}_2\text{O}_3. \qquad (18)$$

Structural and morphological changes accompanying these phase transformations have been investigated by numerous researchers employing such methods as TEM [190,194–197], XRD (see, e.g., refs. [193–195]), ^{27}Al MAS NMR [198], DTA/TGA [192,193] and gas adsorption–condensation [47,199]. Several TEM and XRD investigations have shown that boehmite gels are composed of 5–10 nm by 50–100 nm fibrillar or needle-like crystals that show preferred orientation when the gels are cast as thin plates. (See, e.g., [195].) By comparison, bayerite gels are composed of larger, more equiaxed crystals. The surface areas and pores sizes of several boehmite and bayerite gels investigated by Wolfrum [199] are listed in Table 10. The larger crystallite sizes resulting from the bayerite synthesis are evident from the reduced surface areas as well as narrower diffraction lines in XRD [195]. Lombardi and Klein [200] have shown that the surface areas of boehmite gels are quite sensitive to the H_2O : Al ratio, r, used in the gel-synthesis procedure. Figure 61 shows that the surface area is increased from about 200 to over 600 m^2/g by reducing the value of r from 100 to 7. Yang et al. [197], using isobutanol impregnation, determined that the density of the boehmite phase was only 2.2 g/cm^3, compared to 3.014 for crystalline boehmite. This means that the boehmite phase contains 27 vol% porosity not accessible to isobutanol.

In boehmite the oxygens are arranged in a distorted octahedral configuration around aluminum and are organized in parallel layers linked by hydrogen bonds, each layer of octahedra comprising two sublayers as shown in Fig. 62 [201]. Pierre and Uhlmann [201] have shown that XRD patterns of acid-peptized boehmite gels prepared at 25°C are essentially featureless for $2\Theta < 10°$ except for a gradually increasing intensity with decreasing angle (see Fig. 63). This is due to stacking faults, which lead to folding of the layers, and to intercalation of water, which causes swelling. These gels have been called "superamorphous."

Table 10.

Characteristics of Alumina Gels Prepared by Controlled Evaporation.

Gel	S-BET $m^2 g^{-1}$	d-BET nm	Porosity[a] (%)
AlOOH[b]	275	7.2	40
AlOOH[c]	106	19	24
AlOOH[d]	163	12	30
AlOOH[e]	243	8.2	43
Al(OH)₃	6	330	9
Al(OH)₃	25	80	21
Amorphous	71	28	31

Source: Wolfrum [199].
[a] Detection limit = 1.3 nm.
[b] Prepared from aluminum sec-butoxide.
[c] Prepared from aluminum iso-propoxide.
[d] 1% glycerol added.
[e] Prepared using 0.05 g l⁻¹ submicrometer α-alumina seeds.

Fig. 61.

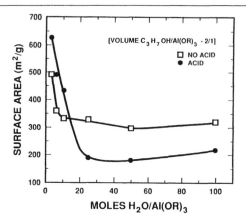

Surface areas of boehmite xerogels prepared with or without acid peptization as a function of r $(H_2O/Al(OR)_3)$ [200].

Upon heating, boehmite gels exhibit a weight loss over the range 50–600°C [193] due to loss of structural and intercalated water and gradually transform to γ-Al_2O_3. This *topotactic transformation*[†] is accomplished by internal condensation of protons and hydroxyls between boehmite layers that removes half the oxygens from the layers, causing a collapse and rearrangement of the oxygens into cubic close-packing [202]. ^{27}Al MAS NMR has been used to investigate the environment of Al during the boehmite $\rightarrow$ γ-Al_2O_3 conversion [198]. Figure 64 shows the ^{27}Al MAS NMR spectra of gels prepared from commercial boehmite (Dispural®) and dried at 60°C or heated to 300 or 500°C. The 7.2 ppm resonance in the dried gel and 300°C gel is evidence that Al is in octahedral coordination in boehmite and remains octahedrally coordinated as structural water is lost from within the layers during the initial stages of the boehmite $\rightarrow$ γ-Al_2O_3 conversion. A similar spectrum, indicating exclusively octahedrally coordinated Al, was obtained for a boehmite gel prepared from Al sec-butoxide after drying at 80°C [198]. Heating at 500°C causes a portion of the Al to adopt tetrahedral coordination as indicated by the new resonances at 67.8 and 69.5 ppm in Fig. 64. XRD indicates that the coordination-number change coincides with the transformation to γ-Al_2O_3. Other XRD investigations of boehmite gels indicate a partial conversion to γ-Al_2O_3 as low as 300°C [195]. Deconvolution of the NMR spectra yields tetrahedral octahedral Al ratios of 0.49 and 0.63 for gels prepared from Dispural® and Al sec-butoxide, respectively. The breadth of the 68–69 ppm resonances is attributed to short-range disorder in

[†] The topography of the boehmite gel is maintained in the transformed product.

Fig. 62.

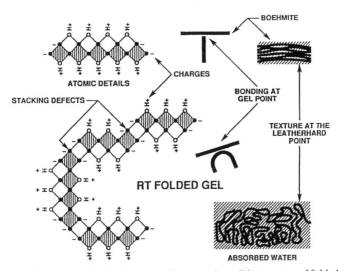

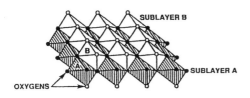

Ideal boehmite layer (top) indicating two sublayers and possible structure of folded layers accomplished by introduction of stacking faults (bottom). Also shown are schematic illustrations of the corresponding textures at the "leatherhard" point (see p. 465) [201].

the tetrahedral Al environments. Based on structural studies of boehmite macrocrystals, Wilson and Stacey [202] concluded that γ-Al_2O_3 formed by boehmite conversion is a distorted defect spinel structure with vacancies on the tetrahedral sites.

The transformation of boehmite to γ-Al_2O_3 and subsequent transformations to δ- and θ-Al_2O_3 at about 800 and 1000°C, respectively (see Fig. 60), are accompanied by a coarsening of the microstructure as evidenced by the decreasing values of the BET surface area (Fig. 65) and the increasing values of the average pore size shown in Fig. 66. However, Yang *et al.* [197] find that, despite this apparent coarsening of the microstructure, the density of the skeletal phase not penetrated by isobutanol remains near 2.2 g/cm^3 throughout the progressive transformation sequence. Thus the transition

Fig. 63.

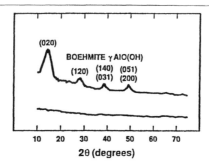

XRD patterns of two boehmite xerogels. The bottom pattern indicates a "super-amorphous" structure achieved by the folded and swollen boehmite layers depicted in Fig. 62 [201].

aluminas retain a significant volume fraction porosity that is inaccessible to isobutanol.

The $\gamma \rightarrow \delta \rightarrow \theta$-$Al_2O_3$ transformation sequence does not involve loss of water, and TGA shows no weight loss above about 600°C [193]. According to Wilson and Stacey [202] the $\gamma \rightarrow \delta$ transition represents the first step toward the ideal, random distribution of vacancies on the tetrahedral sites.

Fig. 64.

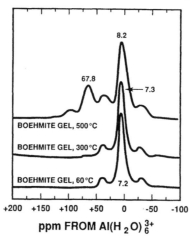

^{27}Al solid-state MAS NMR spectra of a boehmite gel prepared from a commercial sol (Dispural®) after drying at 60°C and during heating between 300 and 500°C. The sharp resonance located at about 8 ppm is due to octahedrally coordinated Al. The broader resonance at about 68 ppm is due to tetrahedrally coordinated Al. The smaller peaks surrounding the main resonances are spinning side bands [198].

Fig. 65.

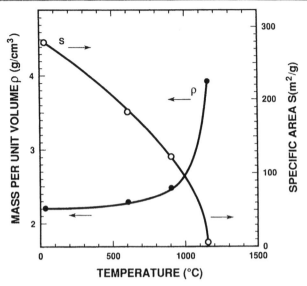

Density and specific surface area of boehmite xerogels after 24-hour heat treatments at the indicated temperatures. Density measurements were determined by impregnation with iso-butyl alcohol and hence measure the density of the skeletal phase not accessible to the alcohol [197].

The $c:a$ ratio adjusts toward 1, the cubic value, but concurrent ordering on the octahedral sites prevents the attainment of cubic symmetry. Their studies on macrocrystalline samples ($>1 \, \mu m$) indicate that the $\gamma \rightarrow \theta$ conversion causes a decrease in the accessibility of planar, hexagonally shaped pores that are formed parallel to the former $\langle 010 \rangle$ boehmite layers, It is not clear that this accounts for the density measurements obtained by Yang *et al.* [197] for the much finer grained gel-derived δ-Al_2O_3.

The topotactic $\delta \rightarrow \theta$-$Al_2O_3$ transformation occurs by a rearrangement of the Al cations within the cubic close-packed oxygen array. Wilson and Stacey [202] observed that θ-Al_2O_3 retains the same pore morphology as δ-Al_2O_3, but the pore size increases from about 6 to 10 nm. The pore size distributions shown in Fig. 66 for gel-derived aluminas suggest an increase in pore radius from about 6.5 to 7.2 nm accompanying the $\delta \rightarrow \theta$-$Al_2O_3$ transformation.

The $\theta \rightarrow \alpha$-$Al_2O_3$ transformation occurs above 1100°C (see XRD results in Fig. 60) and involves a reorganization of the oxygens into a denser, hexagonal close-packed configuration. This is accomplished by a nucleation and growth process that results in a drastic coarsening of the microstructure as shown by the BET surface areas and TEM micrographs in Figs. 65 and 67, respectively. According to Yoldas [47] this transformation is accompanied

Fig. 66.

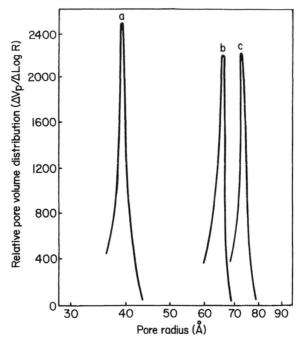

Pore size distributions ($\log \Delta V_p / \Delta \log R$) determined from the desorption branches of nitrogen isotherms after heat treatments for 24 hours at (a) 500°C; (b) 900°C; (c) 1000°C [191].

by a 18.7% linear shrinkage, of which only 2.9% can be accounted for by the higher-density α-Al$_2$O$_3$ phase. The remaining shrinkage involves the loss of porosity. The density results of Yang *et al.* [197] (Fig. 65) indicate that at 1150°C the skeletal density approaches that of α-Al$_2$O$_3$, 3.96 g/cm^3. Thus the nucleation and growth process and/or the shrinkage accompanying the conversion apparently eliminate all inaccessible porosity. Wilson and Stacey [202] also found this to be the case in the macrocrystalline samples.

Isothermal heating at 1150°C causes the number of pores per unit area to decrease but also causes the pore size to increase dramatically: from 15 nm after 1 hour to 85 nm after 20 hours, essentially arresting the sintering process. Densification of such coarsened microstructures is accomplished only at much higher temperatures, e.g., 1600°C. For this reason the potential of gel-processed alumina as a low-temperature route to dense, polycrystalline α-Al$_2$O$_3$ was not realized until the $\theta \rightarrow \alpha$-Al$_2O_3$ transformation was controlled by the introduction of α-Al$_2$O$_3$ *seeds* that serve as heterogeneous nucleation sites. Messing and coworkers (see, e.g. [203–205]) seeded

Fig. 67. _____

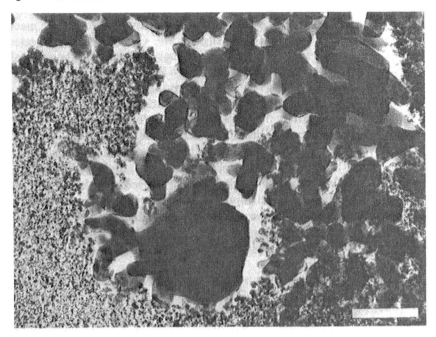

TEM micrograph of a boehmite-derived xerogel during the θ- to α-Al$_2$O$_3$ transformation. The θ-Al$_2$O$_3$ matrix on the lower left is being consumed by the coarser, vermicular α-Al$_2$O$_3$ colony on the upper right [196]. Bar = $0.4 \,\mu$m.

boehmite gels with small α-Al$_2$O$_3$ particles. During the $\theta \rightarrow \alpha$-Al$_2O_3$ transformation, TEM showed that each seed acted as a multiple nucleation site for the transformation [205]. This seeding process maintained high surface area by preventing the formation of a coarsened vermicular microstructure. As discussed in Chapter 11, the seeding process results in a fine, uniform microstructure that sinters at significantly lower temperatures.

3. _____

SUMMARY

In the introduction to this chapter we stated that the structure of noncrystalline, porous gels had been compared to that of porous glasses. This chapter has documented several important distinctions between the structure of gels (products of condensation) and the structure of glass (a product of melting).

Although drying may cause significant compaction, xerogels and especially aerogels often retain tenuous, fractal structures created during the gelation process. Aerogels and xerogels can generally be described as mass or surface fractals on the 1–10 nm length scale. On shorter length scales uniform structures characteristic of the primary particles or the skeletal framework are observed. On sufficiently large length scales, a crossover to uniform structure is also observed. By comparison, during melting of a glass, structure is governed by thermodynamics, producing uniform structure on all length scales greater than the "molecular" scale for single-phase systems.

In mass fractals there is no distinction between surface and interior, and in surface fractals the terminating surface is so convoluted that it begins to occupy volume. Both cases represent situations where terminal hydroxyl and alkoxide groups are uniformly distributed on a molecular scale throughout the porous gel. This random disruption of the structure causes the skeletal density of many xerogels to be significantly less than that of the corresponding melted glass. By comparison, in porous glasses and particulate xerogels, the hydroxyl groups exist primarily on the surfaces of much larger, fully condensed regions.

The amount and distribution of terminal groups are the greatest distinguishing factors between so-called polymeric and particulate xerogels or porous glasses with regard to structural changes during heating. Whereas porous glasses and particulate gels densify only by viscous sintering at $T \geq T_g$, polymeric gels may shrink dramatically at low temperatures ($\ll T_g$) by continued condensation reactions and structural relaxation, which cause the skeletal density to increase toward that of the melted glass. We have shown examples of systems in which all the shrinkage occurring below T_g is accounted for by this skeletal densification.

The extremely high surface area of gels causes the molecular-scale structure to be dominated by surface structure. Condensation reactions occurring on the surface of silicate gels at temperatures below T_g produce quite large concentrations of strained, cyclic trimers that are kinetically stabilized on the gel surface by the exceedingly high matrix viscosity. Because they are produced at low temperatures as products of condensation, the concentrations of these strained cyclic species in gels vastly exceeds that in melted glasses where thermodynamics largely dictates structure.

In multicomponent systems, the elemental distribution reflects that of the preceding hydrolysis and condensation steps. The gel may be homogeneous on a molecular scale or essentially phase separated due to a preference for homo- versus heterocondensation occurring in solution. In the latter case the structure may become homogeneous during heating ($\geq T_g$) if the scale of the inhomogeneities is not too great. If the structure is made intentionally multiphasic, the inhomogeneities will persist at T_g, influencing the sintering and crystallization kinetics (to be discussed in Chapter 11). Alternatively, because kinetics rather than thermodynamics dictates structure below T_g, glasses within the immiscibility region may be prepared as homogeneous single-phase materials (see Chapter 12). This depends primarily on achieving homogeneity in solution or during low-temperature heat treatments.

In crystalline systems, we have shown that microstructural evolution is dominated by phase transformations that accompany gel dehydration or that result from phase metastability. Both phase transformations and crystallization of amorphous systems may involve changes in the coordination numbers of the network-forming species (e.g., Al or Ti). In transformations that occur by nucleation and growth, the addition of "seeds" that serve as multiple nucleation sites appears to be a viable approach to microstructural control.

REFERENCES

1. G.W. Scherer, *J. Non-Cryst. Solids*, **100** (1988) 77–92.
2. S. Wallace and L.L. Hench in *Better Ceramics Through Chemistry*, eds. C.J. Brinker, D.E. Clark, and D.R. Ulrich (Elsevier, North-Holland, New York, 1984), pp. 47–52.
3. R.E. Russo and A.J. Hunt, *J. Non-Cryst. Solids*, **86** (1986) 219–230.
4. R.K. Iler in *The Chemistry of Silica* (Wiley, New York, 1979).
5. C.J. Brinker, W.D. Drotning, and G.W. Scherer in *Better Ceramics Through Chemistry*, eds. C.J. Brinker, D.E. Clark, and D.R. Ulrich (Elsevier, North-Holland, New York, 1984), pp. 25–32.
6. J.D.F. Ramsay and R.G. Avery, *Br. Ceram. Proc.*, **38** (1986) 275–283.
7. C.J. Brinker, "The Structure of Sol-Gel Silica," in *Glass: Science and Technology*, Vol. 4, eds. D.R. Uhlmann and N.J. Kreidl (Academic Press, Boston, to be published 1990).
8. S. Brunauer, L.S. Deming, W.S. Deming, and E. Teller, *J. Am. Ceram. Soc.*, **62** (1940) 1723.
9. S.J. Gregg and K.S.W. Sing, *Adsorption, Surface Area, and Porosity* (Academic Press, London, 1967).
10. R.D. Shoup, *J. Colloid Int. Sci.*, **8** (1976) 63–69.
11. M.W. Shafer, D.D. Awschalom, J. Warnock, and G. Ruben, *J. Appl. Phys.*, **61** (1987) 5438–5446.
12. G.W. Scherer, *J. Am. Ceram. Soc.*, **60** (1977) 236–239.
13. C.J. Brinker and G.W. Scherer, *J. Non-Cryst. Solids*, **70** (1985) 301–322.
14. H.P. Meissner, A.S. Michaels, and R. Kaiser, *Ind. Eng. Chem. Process Des. Div.*, **3** (1964) 202.
15. P. Pfeifer, D. Avnir, and D. Farin, *J. Statistical Physics*, **36** (1984) 699–716.
16. D. Avnir in *Better Ceramics Through Chemistry II*, eds. C.J. Brinker, D.E. Clark, and D.R. Ulrich (Mat. Res. Soc., Pittsburgh, Pa., 1986), pp. 321–329.
17. D.W. Schaefer and A.J. Hurd, in *Chemistry and Physics of Composite Media*, eds. M. Tomkiewicz and P.N. Sen (Electrochemical Society, Pennington, NJ), **85-8** (1985) 54–62.

18. D.W. Schaefer and K.D. Keefer in *Fractals in Physics*, ed. L. Pietronero and E. Tosatti (North-Holland, Amsterdam, 1986), pp. 39-45.

19. D.W. Schaefer, J.E. Martin, A.J. Hurd, and K.D. Keefer in *Physics of Finely Divided Matter*, eds. N. Boccara and M. Daoud (Springer-Verlag, Berlin, 1985), p. 31.

20. D.W. Schaefer, K.D. Keefer, J.H. Aubert, and P.B. Rand in *Science of Ceramic Chemical Processing*, eds. L.L. Hench and D.R. Ulrich (Wiley, New York, 1986), pp. 140-147.

21. P.W. Schmidt, *J. Non-Cryst. Solids*, **15** (1982) 567-569.

22. B. Himmel, Th. Gerber, and H. Bürger, *J. Non-Cryst. Solids*, **91** (1987) 122-136.

23. G. Porod, *Kolbid. Z*, **124** (1951) 83.

24. D.W. Schaefer, K.D. Keefer, and C.J. Brinker, *Polym. Preprints. Am. Chem. Soc.*, **24** (1983) 239.

25. C.J. Brinker, K.D. Keefer, D.W. Schaefer, R.A. Assink, B.D. Kay, and C.S. Ashley, *J. Non-Cryst. Solids*, **63** (1984) 45.

26. T. Woignier, J. Phalippou, and R. Vacher in *Better Ceramics Through Chemistry III*, eds. C.J. Brinker, D.E. Clark, and D.R. Ulrich (Mat. Res. Soc., Pittsburgh, Pa., 1988), pp. 697-702.

27. A. Craievich, M.A. Aegerter, D.I. dos Santos, T. Woignier, and J. Zarzycki, *J. Non-Cryst. Solids*, **86** (1986) 394-406.

28. S.S. Kistler, *J. Phys. Chem.*, **36** (1932) 52.

29. D.W. Schaefer, C.J. Brinker, J.P. Wilcoxon, D-Q. Wu, J. Phillips, and B. Chu in *Better Ceramics Through Chemistry III*, eds. C.J. Brinker, D.E. Clark, and D.R. Ulrich (Mat. Res. Soc., Pittsburgh, Pa., 1988), pp. 691-696.

30. C.S. Ashley, C.J. Brinker, and R.J. Kirkpatrick, unpublished results.

31. M.J. Murtagh, E.K. Graham, and C.G. Pantano, *J. Am. Ceram. Soc.*, **69** (1986) 775-79.

32. T. Woignier, J. Phalippou, R. Sempere, and J. Pelous, *J. Phys. France*, **49** (1988) 289-293.

33. W.G. Klemperer, V.V. Mainz, and D.M. Millar in *Better Ceramics Through Chemistry II*, eds. C.J. Brinker, D.E. Clark, and D.R. Ulrich (Mat. Res. Soc., Pittsburgh, Pa., 1986), pp. 15-26.

34. C.J. Brinker, R.J. Kirkpatrick, D.R. Tallant, B.C. Bunker, and B. Montez, *J. Non-Cryst. Solids*, **99** (1988) 418-428.

35. A. Bertoluzza, C. Fagnano, M.A. Morelli, V. Gottardi, and M. Guglielmi, *J. Non-Cryst. Solids*, **48** (1982) 117-128.

36. P.N. Sen and M.F. Thorpe, *Phys. Rev.*, **B15** (1977) 4030.

37. F.L. Galeener, *Phys. Rev.*, **B19** (1979) 4292.

38. C.A.M. Mulder and A.A.J.M. Damen, *J. Non-Cryst. Solids*, **93** (1987) 387.

39. D.R. Tallant and C.J. Brinker, unpublished results.

40. F. Orgaz and H. Rawson, *J. Non-Cryst. Solids*, **82** (1986) 57-68.

41. D.L. Wood and E.M. Rabinovich, *J. Non-Cryst. Solids*, **82** (1986) 171-176.

42. S. Sakka in *Treatise on Materials Science and Technology*, **22**, eds. M. Tomozawa and R.H. Doremus (Academic Press, New York, 1982), p. 129.

43. K. Kamiya, S. Sakka, and I. Yamanaka, *Tenth International Congress on Glass* (The Ceramic Society of Japan, 1974), **13**, pp. 44-48.

44. D.M. Krol and J.G. Van Lierop, *J. Non-Cryst. Solids*, **63** (1987) 131-144.

45. B.E. Yoldas, *J. Non-Cryst. Solids*, **63** [1,2] (1984) 145.

46. B.E. Yoldas, *J. Mater. Sci.*, **12** (1977) 1203.

47. B.E. Yoldas, *Amer. Ceram. Soc. Bull.*, **59** (1980) 479.

48. M. Nogami and Y. Moriya, *Yogyo Kyokai Shi*, **85** (1977) 59, 448.

49. D.W. Hoffman, R. Roy, and S. Komarneni, *J. Am. Ceram. Soc.*, **67** (1984) 468-471.

50. J.C. Pouxviel, J.P. Boilot, A. Dauger, and L. Huber in *Better Ceramics Through Chemistry II*, eds. C.J. Brinker, D.E. Clark, and D.R. Ulrich (Mat. Res. Soc., Pittsburgh, Pa., 1986), pp. 269–274.
51. B.E. Yoldas, *J. Mater. Sci.*, **14** (1979) 1843–1849.
52. M. Decottignies, J. Phalippou, and J. Zarzycki, *J. Mater. Sci.*, **13** (1978) 2605.
53. T. Woignier, J. Phalippou, and J. Zarzycki, *J. Non-Cryst. Solids*, **63** (1984) 81.
54. A.D. Irwin, J.S. Holmgren, T.W. Zerda, and J. Jones, *J. Non-Cryst. Solids*, **89** (1987) 191–205.
55. R. Jabra, J. Phalippou, and J. Zarzycki, *J. Non-Cryst. Solids*, **42** (1980) 489.
56. I. Matsuyama, K. Susa, and T. Suganuma, U.S. Patent No. 4323381 (April 6, 1982).
57. S.L. Antonova and V.V. D'Yakova, *Sov. J. Glass Phys. Chem.*, **5** (1979) 607.
58. R. Jabra, J. Phalippou, and J. Zarzycki, *Rev. Chim. Min.*, **16** (1979) 245.
59. J. Phalippou, M. Prassas, and J. Zarzycki, *J. Non-Cryst. Solids*, **48** (1982) 17.
60. M. Nogami and Y. Moriya, *J. Non-Cryst. Solids*, **48** (1982) 359.
61. C.J. Brinker and G.W. Scherer, *J. Am. Ceram. Soc.*, **69** (1986) C12–14.
62. M. Prassas and L.L. Hench in *Ultrastructure Processing of Ceramics, Glasses, and Composites*, eds. L.L. Hench and D.R. Ulrich (Wiley, New York, 1984), pp. 100–125.
63. M. Tohge, A. Matsuda, and T. Minami, *Yogyo Kyokai Shi*, **95** (1987) 182.
64. T. Hayashi and H. Saito, *J. Mater. Sci.*, **15** (1980) 1971.
65. M. Guglielmi and G. Principi, *J. Non-Cryst. Solids*, **48** (1982) 161.
66. E.M. Rabinovich, *J. Mater. Sci.*, **20** (1985) 4259.
67. S.P. Mukherjee and S.K. Sharma, *J. Non-Cryst. Solids*, **71** (1985) 317.
68. G. Kordas and S.P. Mukherjee, *J. Non-Cryst. Solids*, **82** (1986) 160.
69. R. Puyane, A.L. Harmes, and C.J.R. Gonzales-Oliver, Proc. 8th Eu. Conf. Opt. Commun. (Conference Européenne sur les Communications Optiques, Paris, France, 1982), p. 623.
70. S.P. Mukherjee and S.K. Sharma, *J. Am. Ceram. Soc.*, **69** (1986) 806.
71. S. Shibata, T. Kitagawa, F. Hanawa, and M. Horiguchi, *J. Non-Cryst. Solids*, **88** (1986) 345.
72. S.P. Mukherjee in *Better Ceramics Through Chemistry*, eds. C.J. Brinker, D.E. Clark, and D.R. Ulrich (Elsevier, North-Holland, New York, 1984), p. 111.
73. J. Schlichting and S. Neumann., *J. Non-Cryst. Solids*, **48** (1982) 185.
74. L.L. Hench in *Science of Ceramic Chemical Processing*, eds. L.L. Hench and D.R. Ulrich (Wiley, New York, 1986), p. 52.
75. S. Wallace and L.L. Hench, *Ceram. Eng. and Sci. Proc.* **5** [7–8] (1984) 568–573.
76. I. Schwartz, P. Anderson, H. de Lambilly, and L.C. Klein, *J. Non-Cryst. Solids*, **83** (1986) 391.
77. F. Branda, A. Aronne, A. Marotta, and A. Buri, *J. Mat. Sci. Lett.*, **6** (1987) 203–206.
78. R. Puyane, P.F. James, and H. Rawson, *J. Non-Cryst. Solids*, **41** (1980) 105–115.
79. M. Prassas, J. Phalippou, L.L. Hench, and J. Zarzycki, *J. Non-Cryst. Solids*, **48** (1982) 79.
80. M. Prassas, J. Phalippou, and L.L. Hench, *J. Non-Cryst. Solids*, **63** [1,2] (1984) 375.
81. A. Yasumori, S. Inoue, and M. Yamane, *J. Non-Cryst. Solids*, **82** (1986) 177–182.
82. L.L. Hench, M. Prassas, and J. Phalippou, *J. Non-Cryst. Solids*, **53** (1982) 183–193.
83. D. Ravaine, J. Traore, L.C. Klein, and I. Schwartz in *Better Ceramics Through Chemistry*, eds. C.J. Brinker, D.E. Clark, and D.R. Ulrich (Elsevier, North-Holland, New York, 1984), p. 139.
84. I. Strawbridge, Ph.D. thesis, University of Sheffield, England (1984).
85. I. Strawbridge and P.F. James in *Novel Ceramic Fabrication Processes and Applications*, **38** (1986) 251.

86. I.M. Thomas, U.S. Patents 3767432 and 3767434 (1973).
87. S. Sakka, K. Kamiya, K. Makita, and Y. Yamamoto, *J. Non-Cryst. Solids*, **63** (1984) 223.
88. F. Hutter, H. Schmidt, and H. Scholze, *J. Non-Cryst. Solids*, **82** (1986) 373.
89. F. Orgaz and H. Rawson, *J. Non-Cryst. Solids*, **82** (1986) 378.
90. A. Duran, J.M. Fernandez-Navarro, P. Casariego, and A. Joglar, *J. Non-Cryst. Solids*, **82** (1986) 69.
91. S. Sakka, S. Ito, and K. Kamiya, *J. Non-Cryst. Solids*, **71** (1985) 311.
92. M. Yamane and T. Kojima, *J. Non-Cryst. Solids*, **44** (1981) 181.
93. C.J.R. Gonzalez-Oliver, P.F. James, and H. Rawson, *J. Non-Cryst. Solids*, **48** (1982) 129.
94. T. Hayashi, T. Yamada, and H. Saito, *J. Mat. Sci.*, **18** (1983) 3137–3142.
95. H. Morikawa, T. Osuka, F. Marumo, A. Yasumori, M. Yamane, and M. Momura, *J. Non-Cryst. Solids*, **82** (1986) 97–102.
96. K. Kamiya and S. Sakka, *J. Mater. Sci.*, **15** (1980) 2937.
97. B.E. Yoldas, *J. Non-Cryst. Solids*, **38, 39** (1980) 81.
98. M. Yamane, S. Inoue, and K. Nakagawa, *J. Non-Cryst. Solids*, **48** (1982) 153.
99. C.P. Scherer and C.G. Pantano, *J. Non-Cryst. Solids*, **82** (1986) 246.
100. S. Sakka and K. Kamiya, *J. Non-Cryst. Solids*, **42** (1980) 403.
101. A. Smithson, Ph.D. thesis, University of Sheffield, England (1985).
102. M. Nogami, *J. Non-Cryst. Solids*, **69** (1985) 415–423.
103. T. Osuka, H. Morikawa, F. Marumo, K. Tohji, Y. Udagawa, A. Yasumori, and M. Yamane, *J. Non-Cryst. Solids*, **82** (1986) 154–159.
104. K. Kamiya, S. Sakka, and Y. Tatemichi, *J. Mater. Sci.*, **15** (1980) 1765.
105. A. Maddalena, M. Guglielmi, V. Gottardi, and A. Raccanelli, *J. Non-Cryst. Solids*, **82** (1986) 356.
106. S.H. Wang and L.L. Hench in *Better Ceramics Through Chemistry*, eds. C.J. Brinker, D.E. Clark, and D.R. Ulrich (Elsevier, North-Holland, New York, 1984), pp. 71–78.
107. M. Nogami, *Yogyo Kyokai Shi*, **95** (1987) 145.
108. M.C. Weinberg, G.F. Neilson, G.L. Smith, B. Dunn, J.S. Moore, and J.D. Mackenzie, *J. Mater. Sci.*, **20** (1985) 1501.
109. C.J. Brinker, K.J. Ward, K.D. Keefer, E. Holupka, and P.J. Bray in *Better Ceramics Through Chemistry II*, eds. C.J. Brinker, D.E. Clark, and D.R. Ulrich (Mat. Res. Soc., Pittsburgh, Pa., 1986), p. 57.
110. S.P. Mukherjee, *J. Non-Cryst. Solids*, **82** (1986) 293–300.
111. G. Carturan, B. Ancora, V. Gottardi, and M. Guglielmi, *J. Non-Cryst. Solids*, **82** (1986) 110.
112. P.F. James, *J. Non-Cryst. Solids*, **100** (1988) 93–114.
113. V. Gottardi, *J. Non-Cryst. Solids*, **49** (1982) 461.
114. F. Pancrazi, J. Phalippou, F. Sorrentino, and J. Zarzycki, *J. Non-Cryst. Solids*, **63** (1984) 81–93.
115. B.J.J. Zelinski, B.D. Fabes, and D.R. Uhlmann, *J. Non-Cryst. Solids*, **82** (1986) 307.
116. J. Covino, F.G.A. De Laat, and R.A. Welsbie, *J. Non-Cryst. Solids*, **82** (1986) 329.
117. S. Höland, E.R. Plumat, and Ph. Duvigneaud, *J. Non-Cryst. Solids*, **48** (1982) 205.
118. G. Carturan, V. Gottardi, and M. Graziani, *J. Non-Cryst. Solids*, **29** (1978) 41–48.
119. G. Carturan, G. Facchin, V. Gottardi, M. Guglielmi, and G. Navazio, *J. Non-Cryst. Solids*, **48** (1982) 219.
120. G. Carturan, G. Facchin, V. Gottardi, and G. Navazio, *J. Non-Cryst. Solids*, **63** (1984) 273.
121. N. Tohge, G.S. Moore, and J.D. Mackenzie, *J. Non-Cryst. Solids*, **63** (1984) 95–104.
122. H. Schmidt, H. Scholze, and A. Kaiser, *J. Non-Cryst. Solids*, **48** (1982) 65.

123. S.P. Mukherjee, *J. Non-Cryst. Solids*, **63** (1984) 35.

124. S.P. Mukherjee and W.H. Lowdermilk, *J. Non-Cryst. Solids*, **48** (1982) 117.

125. G. Orcel and L.L. Hench in *Better Ceramics Through Chemistry*, eds. C.J. Brinker, D.E. Clark, and D.R. Ulrich (Elsevier, North-Holland, New York, 1984), pp. 79–84.

126. S. Dave and R.K. MacCrone, *J. Non-Cryst. Solids*, **71** (1985) 303.

127. B.E. Yoldas, *J. Non-Cryst. Solids*, **51** (1982) 105.

128. Z. Chongshen, H. Lisong, G. Fuxi, and J. Zhonjhong, *J. Non-Cryst. Solids*, **63** [1,2] (1984) 105.

129. L.V. Nikolaeva and A.I. Borisenko, *J. Non-Cryst. Solids*, **82** (1986) 343.

130. C.J. Brinker and G.W. Scherer in *Ultrastructure Processing of Ceramics, Glasses, and Composites*, eds. L.L. Hench and D.R. Ulrich (Wiley, New York, 1984), pp. 43–59.

131. C.J. Brinker and S.P. Mukherjee, *J. Mater. Sci.*, **16** (1981) 1980–1988.

132. H. Dislich in *Glass: Science and Technology* 2, eds. N.J. Kriedl and D.R. Uhlmann (Academic Press, Boston, 1984), chapter 8.

133. S.P. Mukherjee, *J. Non-Cryst. Solids*, **42** (1980) 477.

134. N. Blanchard, J.P. Boilot, Ph. Colomban, J.C. Pouxviel, *J. Non-Cryst. Solids*, **82** (1986) 205.

135. C.J.R. Gonzalez-Oliver and P.F. James, *Glass*, **58** [8] (1981) 304.

136. I. Strawbridge, J. Phalippou, and P.F. James, *Phys. Chem. Glasses*, **25** (1984) 134.

137. G. Orcel and L.L. Hench in *Science of Ceramic Chemical Processing*, eds. L.L. Hench and D.R. Ulrich (Wiley, New York, 1986), p. 224.

138. G. Orcel, J. Phalippou, and L.L. Hench, *J. Non-Cryst. Solids*, **82** (1986) 301.

139. C.J.R. Gonzalez-Oliver and J. Kume, *J. Non-Cryst. Solids*, **82** (1986) 256.

140. C.J. Brinker, E.P. Roth, G.W. Scherer and D.R. Tallant, *J. Non-Cryst. Solids*, **71** (1985) 171–185.

141. T.A. Gallo, C.J. Brinker, L.C. Klein, and G.W. Scherer in *Better Ceramics Through Chemistry*, eds. C.J. Brinker, D.E. Clark, and D.R. Ulrich (Elsevier, North-Holland, New York, 1984), pp. 85–90.

142. G.W. Scherer, C.J. Brinker, and E.P. Roth, *J. Non-Cryst. Solids*, **82** (1986) 191.

143. S.P. Mukherjee, J. Zarzycki, and J.P. Traverse, *J. Mater. Sci.*, **11** (1976) 341.

144. P.J. James., *J. Non-Cryst. Solids*, **100** (1988) 93–114.

145. C.J. Brinker, G.W. Scherer, and E.P. Roth, *J. Non-Cryst. Solids*, **72** (1985) 345–368.

146. C.J. Brinker and D.M. Haaland, *J. Am. Ceram. Soc.*, **66** (1983) 758–765.

147. C.J. Brinker, E.P. Roth, D.R. Tallant, and G.W. Scherer in *Science of Ceramic Chemical Processing*, eds. L.L. Hench and D.R. Ulrich (Wiley, New York, 1986), 37–51.

148. T. Kawaguchi, J. Iura, N. Taneda, H. Hishikura, and Y. Kokubu, *J. Non-Cryst. Solids*, **82** (1986) 50–56.

149. W.D. Drotning and C.J. Brinker, unpublished results.

150. V. Gottardi, M. Guglielmi, A. Bertoluzza, C. Fagnano, and M.A. Morelli, *J. Non-Cryst. Solids*, **63** (1984) 71–80.

151. C.J. Brinker, K.D. Keefer, D.W. Schaefer, and C.S. Ashley, *J. Non-Cryst. Solids*, **48** (1982) 47–64.

152. M. Yamane, S. Aso, S. Okano, and T. Sakaino, *J. Mat. Sci.*, **14** (1979) 607–611.

153. G.W. Scherer, *Relaxation in Glasses and Composites* (Wiley, New York, 1986).

154. G. Orcel, J. Phalippou, and L.L. Hench, *J. Non-Cryst. Solids*, **88** (1986) 114–130.

155. R. Roy, *J. Am. Ceram. Soc.*, **52** (1969) 344.

156. R. Roy, S. Komarneni, and D.M. Roy in *Better Ceramics Through Chemistry*, eds. C.J. Brinker, D.E. Clark, and D.R. Ulrich (Elsevier, North-Holland, New York, 1984), pp. 347–360.

157. F. Orgaz-Orgaz, *J. Non-Cryst. Solids*, **100** (1988) 115–141.

158. J.L. Woolfrey and M.J. Bannister, *J. Am. Ceram. Soc.*, **53** (1972) 390.

159. J.P. Williams, *Am Ceram. Soc. Bull.*, **55** (1976) 524.

160. C.J. Brinker, D.R. Tallant, E.P. Roth, and C.S. Ashley in *Defects in Glasses*, eds. F.L. Galeener, D.L. Griscom, and M.J. Weber (Mat. Res. Soc., Pittsburgh, Pa., 1986), pp. 387–411.

161. C.J. Brinker, D.R. Tallant, E.P. Roth, and C.S. Ashley, *J. Non-Cryst. Solids*, **82** (1986) 117–126.

162. C.J. Brinker, B.C. Bunker, D.R. Tallant, and K.J. Ward, *J. Chimie Phys.*, **11–12** (1986) 851–858.

163. D.W. Sindorf and G.E. Maciel, *J. Am. Chem. Soc.*, **105** (1983) 1487.

164. C.A. Fyfe, *Solid State NMR for Chemists* (CRC Press, Guelph, Ontario, 1983).

165. R. Oestrieke, W.H. Yang, R.J. Kirkpatrick, R.L. Herrig, A. Navrotsky, and B. Montez, *Geochim. Cosmochim. Acta*, **51** (1987) 2199.

166. F.L. Galeener in *The Structure of Non-Crystalline Materials*, ed. P.H. Gaskell, J.M. Parker, and E.K. Davis (Taylor & Francis, London, 1982), pp. 337–359.

167. C.A. Balfe, K.J. Ward, D.R. Tallant, and S.L. Martinez in *Better Ceramics Through Chemistry II*, eds. C.J. Brinker, D.E. Clark, and D.R. Ulrich (Mat. Res. Soc., Pittsburgh, Pa., 1986), pp. 619–626.

168. A.F. Smith and D.R. Anderson, *Appl. Spectr.*, **38** (1984) 822.

169. M. O'Keeffe and G.V. Gibbs, *J. Chem. Phys.*, **81** (1984) 876–879.

170. R.J. Kirkpatrick, unpublished.

171. R.K. Harris and C.T.G. Knight, *J. Chem. Soc. Faraday Trans. 2*, **79** (1983) 1525.

172. R.W. Kelts and N.J. Armstrong in *Better Ceramics Through Chemistry III*, eds. C.J. Brinker, D.E. Clark, and D.R. Ulrich (Mat. Res. Soc., Pittsburgh, Pa., 1988), pp. 519–522.

173. J.L. Lippert, S.B. Melpolder, and L.M. Kelts, *J. Non-Cryst. Solids*, **104** (1988) 139–147.

174. C.J. Brinker and D.R. Tallant, unpublished.

175. C.J. Brinker, unpublished.

176. G. Kordas, R.A. Weeks, and L.C. Klein, *J. Non-Cryst. Solids*, **71** (1985) 327–333.

177. G. Kordas and L.C. Klein in *Science of Ceramic Chemical Processing*, eds. L.J. Hench and D.R. Ulrich (Wiley, New York, 1986), p. 108.

178. A.A. Wolfe, E.J. Friebele, and D.C. Tran. *J. Non-Cryst. Solids*, **71** (1985) 345–350.

179. D.L. Griscom and C.J. Brinker, *Diff. and Defect Data*, **53-34** (1987) 213–226.

180. D.L. Griscom, C.J. Brinker, and C.S. Ashley, *J. Non-Cryst. Solids*, **92** (1987) 295–301.

181. R.A. Roy and R. Roy, *Mat. Res. Bull.*, **19** (1984) 169–177.

182. M.F. Best and R.A. Condrate, Sr., *J. Mat. Sci. Lett.*, **4** (1985) 994–998.

183. J. Cheng and D. Wang, *J. Non-Cryst. Solids*, **100** (1988) 288–291.

184. M. Emili, L. Incoccia, S. Mobilio, G. Fagherazzi, and M. Guglielmi, *J. Non-Cryst. Solids*, **74** (1985) 129–146.

185. C.J. Brinker and M.S. Harrington, *Solar Energy Materials*, **5** (1981) 159–172.

186. D.L. Evans, *J. Non-Cryst. Solids*, **52** (1982) 115.

187. D.R. Sandstrom, F.W. Lytle, P.S.P. Wei, R.G. Greegor, J. Wong, and P. Schultz, *J. Non-Cryst. Solids*, **41** (1980) 201.

188. R.G. Greegor, F.W. Lytle, D.R. Sandstrom, J. Wong, and P. Schultz, *J. Non-Cryst. Solids*, **55** (1983) 27.

189. B.E. Yoldas, private communication.

190. F.W. Dynys, M. Ljungberg, and J.W. Halloran in *Better Ceramics Through Chemistry*, eds. C.J. Brinker, D.E. Clark, and D.R. Ulrich (Elsevier, North-Holland, New York, 1984), pp. 321–326.

191. B.E. Yoldas, *Am. Ceram. Soc. Bull.*, **54** (1975) 286–288.

192. B.E. Yoldas, *J. Mater. Sci.*, **10** (1975) 1856-1860.
193. R.K. Dwivedi and G. Gowda, *J. Mat. Sci. Lett.*, **4** (1985) 331-334.
194. V. Saraswati, G.V.N. Rao, and G.V.R. Rao, *J. Mater. Sci.*, **22** (1987) 2529-2534.
195. A.C. Pierre and D.R. Uhlmann in *Better Ceramics Through Chemistry*, eds. C.J. Brinker, D.E. Clark, and D.R. Ulrich (Elsevier, North-Holland, New York, 1984), pp. 119-124.
196. F.W. Dynys and J.W. Halloran in *Ultrastructure Processing of Ceramics, Glasses, and Composites*, eds. L.L. Hench and D.R. Ulrich (Wiley, New York, 1984), pp. 142-151.
197. X. Yang, A.C. Pierre, and D.R. Uhlmann, *J. Non-Cryst. Solids*, **100** (1988) 371-377.
198. S. Komarneni, R. Roy, C.A. Fyfe, and G.J. Kennedy, *J. Am. Ceram. Soc.*, **68** (1985) 243-245.
199. S.M. Wolfrum, *J. Mat. Sci. Lett.*, **6** (1987) 706-708.
200. T. Lombardi and L.C. Klein, *Adv. Ceram. Mat.*, **3** (1988) 167-170.
201. A.C. Pierre and D.R. Uhlmann, *J. Non-Cryst. Solids*, **82** (1986) 271-276.
202. S.J. Wilson and M.H. Stacey, *J. Coll. and Int. Sci.*, **82** (1981) 507.
203. G.L. Messing, J.L. McArdle, and R.A. Shelleman in *Science of Ceramic Chemical Processing*, eds. L.L. Hench and D.R. Ulrich (Wiley, New York, 1986), pp. 471-482.
204. G.L. Messing, M. Kumagai, R.A. Shelleman, and J.L. McArdle in *Science of Ceramic Chemical Processing* eds. L.J. Hench and D.R. Ulrich (Wiley, New York, 1986), pp. 259-271.
205. R.A. Shelleman, G.L. Messing, and M. Kumagai, *J. Non-Cryst. Solids*, **82** (1986) 277-285.

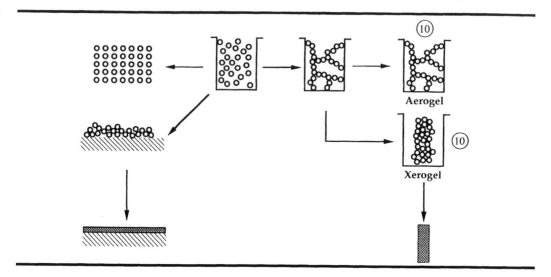

CHAPTER 10

Surface Chemistry and Chemical Modification

The preceding chapter has shown that dried gels, either xerogels or aerogels, are generally characterized by very high surface areas over wide ranges of processing temperatures. According to Iler [1], the properties of high-surface-area amorphous silicates, from the smallest colloidal particles to macroscopic gels, are dominated largely by the surface chemistry of the solid phase. Surface chemistry is important in such diverse topical areas as catalysis, chromatography, corrosion, and environmentally induced fracture.

This chapter discusses the surface chemistry of porous silicate-based gels and the consequences of high surface area with regard to sol-gel processing of silicate glasses. From the standpoint of preparing ultrahigh-purity glasses for optical communications, we shall show that high surface area is problematic because the surface is generally terminated by hydroxyl groups that must be removed prior to sintering. From the standpoint of chemical modification, however, high surface area is advantageous because high surface area necessarily implies accessibility of reactant species to the solid phase.

The chemistry of other oxide surfaces, e.g., Al_2O_3 or TiO_2, are not discussed here due to the extensive coverage of these topics in the catalyst literature. Interested readers are referred to references [2–11]. It should be noted however that many of the ideas developed in this chapter are generally applicable to systems other than silica. An outline of the contents of this chapter is as follows:

1. *Definition of Surface* briefly discusses the topic of what constitutes a surface based on several procedures for measuring surface areas.

2.   *Surface Coverage of OH and OR* examines the concentrations and types of hydroxyl and alkoxide species existing on gel surfaces as a function of processing conditions.
3.   *Dehydroxylation.*
    3.1.   *Thermal Dehydroxylation* describes methods to dehydroxylate surfaces based on high-temperature and/or vacuum treatments.
    3.2.   *Chemical Dehydroxylation* describes methods based on chemical reactions such as halogen treatments.
4.   *Chemistry of Dehydroxylated Surfaces* discusses the reactivity of strained surface sites created by the dehydroxylation process.
5.   *Compositional Modification* describes methods by which the surface reactivity can be used to modify the chemical composition, for example, ammonolysis reactions that produce oxynitride glasses.
6.   *Consequences of Surface Morphology* examines surface structure from the viewpoints of Euclidean or fractal geometry with regard to effects on reaction kinetics and diffusion.

1.

DEFINITION OF SURFACE

A surface is defined in many respects by the method of surface area measurement. The classical method is the determination of the *monolayer capacity* of an adsorbent molecule of known cross-sectional area. The relevant *surface area* is then

$$A = N_0 m\sigma \qquad (1)$$

where A is the apparent surface area (m^2/g), m is the experimental monolayer value (mol/g$_{solid}$) for an adsorbent of cross-sectional area, σ (m^2), and N_0 is Avagodro's number. Normally A is determined from nitrogen adsorption on the basis of the BET equation [12]. For example Iler [1] equated the surface area of silicate particles and gels to the BET area of nitrogen at 77 K. However, because the area of a nitrogen molecule is 16.2 Å^2, this definition of the surface excludes some microporosity that is accessible, for example, to water molecules. Iler considered these microporous regions as part of the underlying solid phase. According to this definition of surface, Iler concluded that all silicate surfaces are the same provided there is no microporosity, i.e., the topology of the underlying surface makes little difference to the adsorption behavior.
    The definition of surface according to Eq. 1 is valid only for a flat surface [13] where, e.g., if σ is reduced by a factor of 2, twice as many

moles, m, are required to obtain a monolayer, and A does not change. The preceding chapter, however, established that for many conditions of gel synthesis the resulting dried gels can be described as *mass* or *surface fractals*. For mass fractals there is no distinction between surface and interior and the *surface fractal dimension*, d_s, equals the *mass fractal dimension*, d_f. For uniform objects bounded by a surface, the surface fractal dimension, d_s, varies from 2 for a smooth pristine surface (e.g., a soap bubble) to 3 for a surface that is so convoluted that it completely fills space (e.g., a tightly crumpled napkin). For such an irregular, *fractally rough* surface the accessibility of adsorbent molecules to the surface is diminished. The appropriate scaling relationship describing physical adsorption (physisorption) is [13]:

$$m \propto \sigma^{-d_{s,a}/2} \tag{2}$$

where $d_{s,a}$ is the *surface fractal dimension for adsorption*, $2 \le d_{s,a} \le 3$. Equation 2 provides a quantitative measure of the surface accessibility, which decreases as $d_{s,a}$ increases from the classical case where $d_s = 2$ [13]. The apparent surface area (normalized to the nitrogen surface area) then decreases with increasing σ for $d_{s,a} > 2$ according to [13]:

$$A_{ad}/A_{N_2} = (\sigma_{ad}/\sigma_{N_2})^{-d_{s,a}/2} \tag{3}$$

where A_{ad} is the apparent surface area for an adsorbate molecule of cross-sectional area, σ_{ad}. The consequences of Eq. 3 with regard to estimating the surface coverage of hydroxyl and alkoxy groups are discussed in the following section. The effects of irregular surfaces on chemical reactivity and diffusion are discussed in Section 6.

Because water molecules have a smaller cross-sectional area than N_2, the monolayer capacity of water should be a more accurate definition of the surface of porous gels according to the preceding discussion. However, it is experimentally difficult to measure the monolayer capacity of H_2O, since the surfaces must be fully hydroxylated and the cross-sectional area of H_2O is poorly defined [14]. The adsorption behavior of water on porous silicates as revealed by water-adsorption isotherms was reviewed by Belyakova *et al.* [15] in 1959. A second method of surface-area measurement that uses water as a structural probe is based on the difference in spin-lattice relaxation times of protons in close proximity to a surface and "bulk" protons existing in a liquid. Determination of the weighted-average ^1H spin-lattice relaxation times of bulk and surface protons introduced by water condensation provides a measure of the surface-to-volume ratio of the pore space. The application of this NMR method to porous gels is discussed by Gallegos and coworkers [16].

2.

SURFACE COVERAGE OF OH AND OR

Kiselev [17] proposed that the surface of active silica gel was covered by OH groups as early as 1936, and by 1940 Carmen [18] recognized that water can hydrolyze the silicate surface to create silanol groups. Since that time the nature of the silica surface has been reviewed by numerous authors, most notably Iler [1], Kiselev [19], Barby [20], Okkerse [21], and Hair [22]. In 1966 Boehm [23] published two extensive reviews on surface groups of different solids including silica. One topic all these reviews have in common is the emphasis on surface coverage of OH groups, presumably because surface coverage largely determines the adsorption behavior and consequently the surface reactivity.

The types of hydroxyl species that may exist on the silicate surface are shown schematically in Fig. 25 of the preceding chapter. Excluding for the moment the effects of physically adsorbed water, the main types of silanol species are postulated to be vicinal, geminal or isolated:

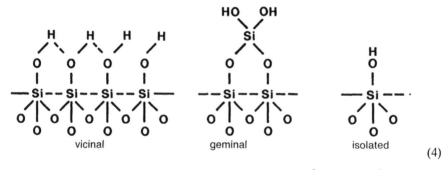

$$(4)$$

An *isolated silanol* is an OH group located on a Q^3 *silicon* site[†] that is not hydrogen-bonded; that is the minimum O–H---O distance between neighboring silanols exceeds about 0.33 nm equivalent to the normal van der Waals O--O contacts for nonbonded oxygen atoms [24]. Their existence is normally established by vibrational spectroscopy (IR or Raman) as a narrow band at *ca.* 3750 cm^{-1}. *Vicinal silanols* are hydroxyl groups located on neighboring Q^3 sites in which the O–H---O distance is sufficiently small (less than the van der Waals contact by *ca.* 0.02 nm) that hydrogen bonding may occur. Hydrogen bonding causes a reduction in the O–H stretching frequency, the magnitude of which depends on the strength of the hydrogen bond, and thus on the O–H---O distance.

[†] In Q^n terminology, n equals the number of bridging oxygens (–OSi) bonded to the central silicon ($n = 0$-4).

Geminal silanols are defined as two OH groups located on a Q^2 silicon site. Although their existence on a dried silica surface was disputed by Okkerse [21] on the basis of vibrational spectroscopy, solid-state MAS ^{29}Si NMR has clearly established the presence of Q^2 silanol species in a variety of porous silicates as shown, for example, in Fig. 1 for a commercial silica gel [25]. (For other examples, see refs. [26–28].) The location of at least a portion of these geminal silanol groups on surfaces (as opposed to within the bulk) has been confirmed by *trimethylsilylation* studies [25] in which reaction of the silica gel with chlorotrimethylsilane causes a reduction in the intensities of both the Q^2 and Q^3 resonances. (See Fig. 2.) It should be noted that complete elimination of Q^2 and Q^3 species by silylation reactions is not anticipated on the basis of accessibility considerations as discussed in the previous section. It is also likely that many gel synthesis conditions result in internal silanol groups that serve to reduce the skeletal density (see Section 2.2.4 in the preceding chapter) but that are inaccessible to adsorbent or reactant molecules. Based on ^{29}Si MAS NMR studies, Fyfe *et al.* [25] determined that the ratio of $Q^4 : Q^3 : Q^2$ silicon sites for commercial, Fisher S-157 silica gel is $8.8 : 5.7 : 1$.

The amorphous silicate surface may be visualized as the truncation of a random network composed of siloxane rings containing on average six silicons per ring. (See Fig. 3.) Open rings created by the surface are terminated with hydroxyl groups. Michalske and Bunker evaluated which open-ring configurations corresponded to isolated or hydrogen-bonded silanol groups on the basis of the ring size [29] and number of hydroxyl groups per terminating silicon site [30]. They concluded that OH groups contained on two neighboring Q^3 silicons that share a bridging oxygen are sufficiently separated that hydrogen bonding does not occur. By definition these hydroxyls are isolated. Neighboring Q^3 silanols located on larger open rings containing three or more silicons may be isolated or hydrogen bonded (vicinal) depending on the extent of ring opening. A geminal (Q^2) silanol site

Fig. 1.

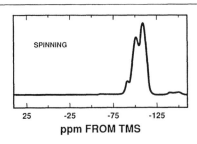

Solid-state ^{29}Si MAS NMR spectrum of commercial Fisher S-157® silica gel [25]. Peaks at about -90, -100, and -100 ppm correspond to Q^2, Q^3, and Q^4 silicons, respectively.

Fig. 2.

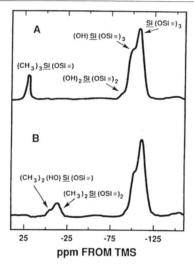

Solid-state ^{29}Si MAS NMR spectrum of commercial Fisher S-157 silica gel (A) after reaction with trimethylchlorosilane and (B) after reaction with dimethyldichlorosilane [25].

bonded to a neighboring Q^3 silicon through a single siloxane bridge results in a hydrogen bonded pair. The remaining OH experiences very weak hydrogen bonding as reflected by its O–H stretching vibration, 3748 cm^{-1}, compared to 3749 cm^{-1} for isolated silanols [30,31]. Identification of these various configurations of surface silanols species by high-resolution IR spectroscopy is discussed by Hoffman and Knozinger [31].

In addition to the ring sizes, degree of ring opening, and number of OH groups per surface silicon site, the extent of hydrogen bonding of surface

Fig. 3.

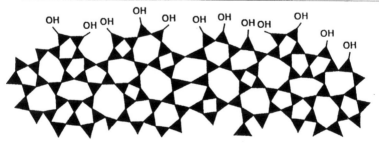

Schematic representation of an amorphous silica surface composed of open and closed rings of various sizes. Open rings are terminated by hydroxyl groups.

Fig. 4.

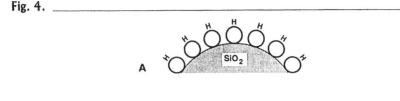
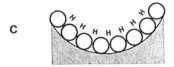

Effects of surface curvature on hydrogen bonding. (A) Small positive radius of curvature (small particles) has fewer hydrogen bonds facilitating dehydroxylation. (B) Large radius of curvature (large particles flat surface) allows more hydrogen bonds, inhibiting dehydroxylation. (C) Small negative radius of curvature (small cylindrical pores or necks) has the most hydrogen bonding and is the most difficult to dehydroxylate [1].

OH groups is influenced by the surface curvature. (See Fig. 4.) A small positive radius of curvature increases the O–H---O distance of neighboring hydroxyl sites, reducing the extent of hydrogen bonding on a particle compared to a flat surface with the same surface coverage of OH groups. Conversely, a small negative radius of curvature, as in cylindrical pores or "necks" between particles, reduces the O–H---O separation between neighboring hydroxyl groups, increasing the extent of hydrogen bonding over that of a flat surface.

Iler [1] considered the types of silanols on all the various forms of crystalline and amorphous silicas and concluded that it would be surprising if they all behaved exactly the same, i.e. if they all had the same surface coverage and speciation of OH groups (hydrogen bonded versus isolated, etc.). Indeed, early studies [32–36] of OH coverage on silicate particles and gels (see Table 1) resulted in surface coverage values ranging from 4 to 10 OH/nm^2. However the estimates were unreliable (often high), because the contribution to the surface area by microporosity was not accounted for, and, in measurements based on loss on ignition, it was not always possible to distinguish between physisorbed water, structural (bulk) OH, and surface OH groups.

More recently, Zhuravlev [37] determined the OH coverage of amorphous silicates using a deuterio-exchange method [38,39] that distinguished between surface and bulk OH. He found that the average hydroxyl coverage

Table 1.

Hydroxyl Coverage of a Variety of High–Surface-Area Silicates Determined by Different methods.

Type of SiO$_2$	Drying Method °C	Atm.	Time	Specific Surface Area (m^2 g^{-1})	Method of Determination	OH nm^{-2}
Gel	155	—	—	300	Diborane	7.9
Deionized colloid	110	Air	16 hr	182	Ignition, 1100°C	10.0
Deionized colloid	120	Air	4 days	160	Ignition, 1100°C	8.0
Deionized colloid	120	Air	4 days	169	Ignition, 1100°C	8.1
	150	Air	2 hr	160	Ignition, 1100°C	6.3
Mallinckrodt	120	—	17 hr	591	Ignition, 1200°C	9.6
					Infrared	8.6
Mallinckrodt, heated to 800°C, soaked in water, 5×	120	—	17 hr	166	Ignition, 1200°C	5.2
					Infrared	4.0
Rehydrated gels, aerogels, Aerosils	150	—	—	39–750	Combination methods	5.0–5.7

Source: Iler [1].

for one-hundred different samples of silicate particles and gels (Kr BET surface areas, S, varied from 9.5 to 950 m^2/g) equalled 4.9 OH/nm^2 with all values falling between 4 and 6 OH/nm^2. (See Fig. 5.) The single measurement made on a gel prepared by hydrolysis of TEOS resulted in a coverage value of 4.2 OH/nm^2. Using the deuterio-exchange method, high coverage values may still result due to microporosity unaccounted for by the BET measurement, and low values may result if all surface silanols are not exchanged due to inaccessible porosity. Therefore the rather narrow distribution of OH coverage values suggests that in fact the surface coverage of OH groups is independent of the form or synthesis conditions of the amorphous silica; i.e., there is no apparent effect of the surface area, type or size distribution of the pores, particle packing density, or structure of the underlying SiO$_2$ skeleton. Presumably on such small length scales, there is no effect of surface irregularity (fractal surfaces) so that all forms of silica exhibit the same hydroxyl coverage.

The coverage value of 4.9 OH/nm^2 corresponds quite closely to the value expected if on average the amorphous silica surface was represented by the ⟨111⟩ face of cristobalite which has a density and refractive index similar to that of amorphous silica [1]. Figure 6 shows the atomic arrangement of

Fig. 5.

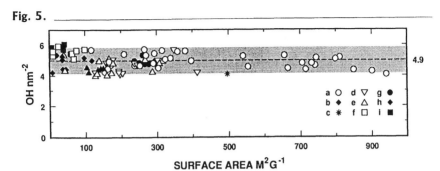

Surface coverage of hydroxyl groups as a function of surface area for one hundred different samples of amorphous silica: (a) silica gel obtained by acid neutralization of sodium silicate; (b) silica gels obtained by the acidic method followed by hydrothermal treatment; (c) silica gel obtained by hydrolysis of TEOS; (d) silica gels obtained by the Bard ion-exchange method from alkaline or acidic sols; (e) particulate gels prepared from aerosil; (f) porous glasses obtained by leaching borosilicate glasses; (g) rehydroxylated commercial silica gels; (h) rehydroxylated particulate silica gels; (i) rehydroxylated porous glasses [37].

Fig. 6. _____

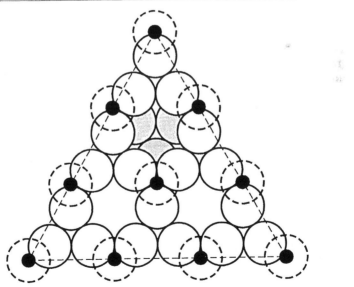

Atomic arrangement of the ⟨111⟩ octahedral face of cristobalite. Large circles: oxygen atoms. Small shaded-in circles: silicon atoms at surface. Dashed circles: position of hydroxyl groups on surface attached to underlying silicon atoms. Hydroxyl surface coverage equals 4.55 OH/nm². Atomic sizes not to scale [1].

the ⟨111⟩ octahedral face of cristobalite which is composed of six-membered rings of silicate tetrahedra. Termination of the Q^3 Si sites with hydroxyl groups results in a coverage value of $4.6 \, OH/nm^2$.

The speciation of the OH groups on the amorphous silica surface has been investigated by several groups. Armistead et al. [40] showed that out of $4.6 \, OH/nm^2$, $1.4 \, OH/nm^2$ are isolated and $3.2 \, OH/nm^2$ mutually hydrogen bonded. Armistead et al. also determined that their samples contained an additional[†] $1.6 \, OH/nm^2$ corresponding to internal silanols irreversibly removed upon annealing. Peri and Hensley [41] concluded that on a dry fully hydroxylated surface, one half of the hydroxyls are present as isolated OH groups due to the random structure of the silicate surface. Boehm [23] found that there were at least two different types of hydrogen bonded sites identified by IR bands at 3520 and $3660 \, cm^{-1}$ that correspond to stronger and weaker hydrogen bonding, respectively. Presumably the percentage and distribution of hydrogen bonded hydroxyl groups on gel surfaces depend largely on surface curvature effects. (See Fig. 4.) The curvature in turn depends on the gel synthesis and aging conditions. In general, particulate gels characterized by small positive radius of curvature surfaces (small particles) should contain less hydrogen bonded silanol groups than "polymeric" gels that are microporous, with extremely small negative curvatures. (See Section 1.1 in Chapter 9.)

Surface hydroxyl groups are sites where physical adsorption of water (and other polar molecules) occurs. On a completely hydroxylated surface $(4.9 \, OH/nm^2)$ H_2O will first cover all accessible OH sites forming a multiply hydrogen bonded layer. In a near IR study of water adsorption on air-equilibrated silica gels (Fig. 7), Wood and Rabinovich [42] observed that the removal of water under vacuum at room temperature essentially eliminated the $5292 \, cm^{-1}$ band attributed to hydrogen bonded water and caused the appearance of a relatively narrow band at $7342 \, cm^{-1}$ assigned to isolated OH. Bands at 7194 and $4545 \, cm^{-1}$ attributed to hydrogen bonded surface silanols were also reduced in intensity by the vacuum treatment. From this study we can conlude that both isolated and hydrogen bonded surface silanols serve as adsorption sites for molecular water on the fully hydroxylated silica surface.

Hertl and Hair [43] studied water adsorption on separated, hydrogen bonded vicinal silanol pairs created by partial trimethylsilylation of the silica surface. The water adsorption isotherm on this partially hydroxylated surface showed steps at 9.5 and 16 torr that were explained by distinct stages of water adsorption. First, the OH pair adsorbed a single water molecule. In the second stage two additional water molecules were adsorbed, and in an

---

[†] Determined by weight loss per unit area.

Fig. 7.

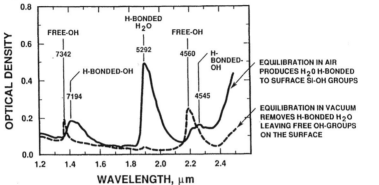

Near-infrared spectrum of a silica gel equilibrated with room air (full curve) and with vacuum (dashed curve) [42].

indistinct third stage three more water molecules were adsorbed, resulting in the following cluster:

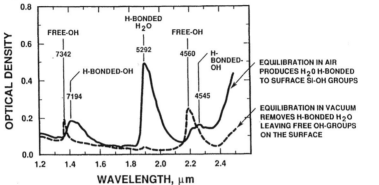

(5)

Klier and Zettlemoyer [44] found that on a surface composed primarily of isolated silanols (created at 700°C) water adsorbs preferentially on sites where water has already adsorbed rather than on isolated silanol sites. This is explained on the basis of heats of adsorption:

$$Si_sOH + H_2O = Si_sOH:OH_2 \qquad\qquad 6.0\,kcal/mol \qquad (6)$$

$$Si_sOH:OH_2 + xH_2O = Si_sOH:OH_2(OH_2)_x \qquad 10.5\,kcal/mol, \qquad (7)$$

where Si_s refers to a surface silicon site. Thus on partially hydroxylated surfaces it is likely that water clusters form before all the available OH sites are covered.

Complete coverage of alkoxide (OR) groups on the silica surface should correspond to that of OH ($4.9/nm^2$). However, due to steric limitations imposed by the adsorbate or limited accessibility to the surface, values less than $4.9 \, OR/nm^2$ are observed. For example, Peri and Hensley [41] found the surface coverage of methoxide groups (OMe) to be 4–5 OMe/nm^2. Stöber *et al.* [45] autoclaved silica with anhydrous alcohol for six hours at $200°C$. For normal alcohols, OR surface coverage decreased with the alkyl chain length: 4.7 OMe/nm^2, 3.7 OEt/nm^2, 3.5 OPr/nm^2, and 3.2 OBu/nm^2. Of course for branched alcohols the surface coverage is strongly dependent on steric factors. Iler [1] states that for branched alcohols it has been established experimentally that each alcohol molecule will cover $0.14n \, nm^2$ where n is the "branch" number defined as the width of the hydrocarbon groups at their widest point if "spread out flat." This expression holds for $n > 2$; however it is doubtful that this relation pertains to irregular surfaces characterized by $2 \le d_s \le 3$.

3.

DEHYDROXYLATION

The high surface areas of gels, combined with surface hydroxyl coverages of $4.9 \, OH/nm^2$ and additional physically adsorbed water, result in quite large quantities of water ($OH + H_2O$) bound to air-equilibrated porous gels. For example a silica gel with a BET surface area of $900 \, m^2/g$ contains $6.7 \, wt\% \, H_2O$ (calculated as $2SiOH \rightarrow Si\text{-}O\text{-}Si + H_2O$) that would be lost upon dehydroxylation due to chemically bound OH alone. Such large OH contents are problematic when trying to prepare glasses by a sol-gel process that are identical to the corresponding melt-prepared composition (typical OH content $\approx 500 \, ppm$).

Although OH is removed through *condensation reactions*,

$$
\begin{matrix}
\text{H} & & \text{H} & & & & \\
\text{O} & & \text{O} & & & & \\
| & + & | & \rightleftharpoons & \text{Si}\diagdown\text{O}\diagup\text{Si} & + & \text{H}_2\text{O}, \quad (8)\\
\text{Si} & & \text{Si} & & & & \\
\end{matrix}
$$

viscous sintering often commences before dehydroxylation is complete, leading to *bloating* or foaming of the gel in the final stages of sintering (see Fig. 8) [46]. Even if sufficient hydroxyl is removed so that sintering can occur without bloating, subsequent heating of the consolidated gel to its softening point, for example, during fiber drawing or sealing, may cause bloating. The hydroxyl content of consolidated gels is also detrimental to applications

Fig. 8.

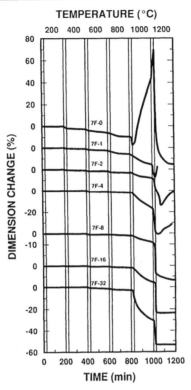

Dimensional change versus time for a series of silicate gels containing 0–32 wt% F during an incremental heating procedure consisting of a sequence of isothermal holds at temperatures between 200 and 1200°C. Bloating is denoted in the 7F–O sample (0 wt% F) by the dramatic expansion above 800°C [46].

related to optical communications where transparency in the IR portion of the spectrum is of utmost importance. This set of problems involving OH contents of gels has been the motivation to investigate means of gel dehydroxylation.

3.1. Thermal Dehydroxylation

Thermal dehydroxylation relies on condensation reactions (Eq. 8) occurring on the gel surface. Thermal dehydroxylation of silicates has been investigated extensively, and summaries of numerous studies are reported by Iler [1]. Hydroxyl coverage versus temperature is shown in Fig. 9 for a variety of

Fig. 9.

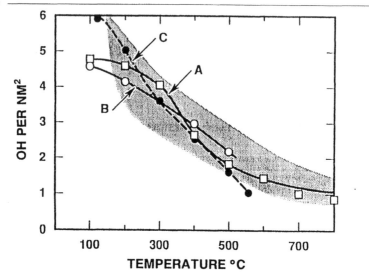

Hydroxyl coverage of the silica surface versus dehydroxylation temperature [1]. Shaded area: range of data on a variety of silicas investigated by Davydov *et al.* [36]. Dehydration of annealed (700°C) and rehydrated silica during heating in air: A; and in vacuum: B [47]. Broken line, C: data on unannealed silica [48–51].

silicates heated in either air or vacuum. In general, OH groups are gradually lost with increasing temperature, but at 800°C where the extent of viscous sintering may be substantial (see, e.g., the surface area versus temperature data in Fig. 10), the OH coverage according to Fig. 9 is still about 1 OH/nm^2. Thus thermal dehydroxylation is often not sufficient to avoid bloating in silica gels. For multicomponent systems that sinter at lower temperatures, the problem is even more severe, as discussed shortly.

The sequence of surface dehydration is the initial removal of physically adsorbed water at low temperatures followed by the progressive removal of weakly hydrogen bonded hydroxyls, strongly hydrogen bonded hydroxyls, and finally isolated hydroxyls. (See the schematic in Fig. 11.) It is postulated that the removal of isolated silanols occurs in part by diffusion of protons along strained siloxane bridges followed by condensation after an adjacent pair of hydroxyls is formed [1]. As previously discussed, two adjacent silanols that share a single bridging oxygen also appear "isolated" in vibrational spectroscopy. In this case condensation results in a highly strained, cyclic disiloxane species (Fig. 11).

Surface curvature may increase the extent of hydrogen bonding causing greater OH retention than anticipated from Fig. 9. For example, silica gels

Fig. 10.

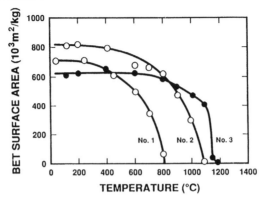

BET surface area versus temperature for a series of silica gels with varying bulk densities (ρ_b, g/cm^3) and pore diameters (d_p, nm) illustrating the effect of pore size on sintering temperature: No. 1, $\rho_b = 1.34$ and $d_p = 17$; no. 2 $\rho_b = 1.10$ and $d_p = 22$; no. 3 $\rho_b = 0.60$ and $d_p = 81$ [52].

Fig. 11.

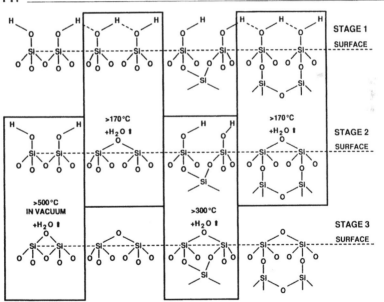

Schematic illustration of various stages of surface dehydroxylation. Stage 1: removal of hydrogen bonded silanols resulting in four-membered and larger rings. Stage 2: condensation reactions between isolated silanols resulting in strained two- and three-membered rings. Stage 3: fully dehydroxylated surface.

with pore diameters of 1.0, 2.0, and 2.7 nm were dehydrated at a series of temperatures and their water contents measured [53]. The results compiled in Table 2 show that for small pore gels the OH coverages are considerably greater than indicated in Fig. 9, although values greater than 4.9 OH/nm^2 presumably result from microporosity not accounted for in the BET surface-area measurement.

The strategy for achieving extensive thermal dehydroxylation involves optimization of both the gel microstructure and the thermal processing conditions. Pertinent microstructural variables are pore size, surface area, and curvature. For any system the most beneficial microstructural improvement is increased pore size. Larger pores enhance the diffusion of the by-product, H_2O, and retard the sintering rate (proportional to surface area divided by pore size) allowing more complete dehydroxylation to occur prior to pore closure at the final stages of sintering. Pore size may be increased by any of several approaches: (1) the formation of large particles (pore size scales with particle size [54]); (2) particle aggregation [54]; (3) supercritical drying [55]; (4) stiffening of the network by increased condensation rates or aging [1], causing less shrinkage upon drying; and (5) double-dispersion procedures that produce hierarchical microstructures [56].

The effects of pore size on dehydroxylation are demonstrated by comparing the hydroxyl contents of silica aerogels and xerogels during heating and following complete consolidation by viscous sintering in a dry nitrogen ambient [55]. (See the Raman spectra in Fig. 12.) The S series of samples are xerogels prepared by a single-step acid-catalyzed hydrolysis of TEOS. The A series of samples are aerogels prepared by a two-step acid–base-catalyzed hydrolysis of TEOS followed by supercritical drying at 300°C and

Table 2.

Hydroxyl Contents of Silica gels as a Function of Pore Diameter and Temperature.

Temperature of Dehydration (°C)	Average Pore Diameter								
	10 Å			20 Å			27 Å		
	Area (m^2 g^{-1})	Bound H$_2$O (%)	OH nm^{-2}	Area (m^2 g^{-1})	Bound H$_2$O (%)	OH nm^{-2}	Area (m^2 g^{-1})	Bound H$_2$O (%)	OH nm^{-2}
115	400	6.5	10.8	540	6.1	7.5	450	3.8	5.6
300	480	4.4	6.1	500	4.3	5.7	500	4.0	5.3
600	375	3.4	6.0	400	2.9	4.8	420	2.3	3.6
700	280	1.5	3.6	340	1.7	3.3	210	1.1	3.5

Source: Iler [1], Dzis'ko *et al.* [53].

Fig. 12.

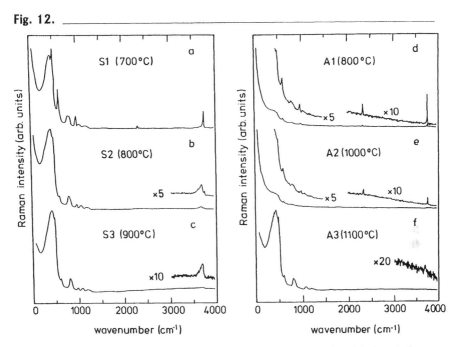

Raman spectra of silica xerogels (S series) and aerogels (A series) after dehydroxylation at temperatures between 700 and 1100°C [55].

250 atmospheres. Raman bands at 3750 and 3680 cm^{-1} are assigned to SiO–H stretching of isolated surface silanols and hydrogen bonded or bulk silanols,[†] respectively. At 800°C the xerogel is nearly completely dense as evidenced by the greatly reduced relative intensity of the surface 3750 cm^{-1} band and the appearance of the bulk 3680 cm^{-1} band. The fully dense glass processed at 900°C contains a significant concentration of bulk silanols resulting from pore closure prior to complete dehydroxylation. By comprison the aerogel sample remains porous at 1000°C (there is no evidence of the bulk silanol band) allowing more complete dehydroxylation prior to densification at 1100°C. The hydroxyl content of the final densified glass is much less than that of the densified glass prepared from the xerogel.

Surface curvature is optimized by choosing synthesis conditions that result in particulate as opposed to polymeric microstructures. Particulate micro-structures provide positive curvature surfaces (except in regions of neck

[†] Due to their varied environments and the close proximity of other species capable of hydrogen bonding, bulk silanols located within an amorphous network are characterized by broad bands at lower vibrational frequencies than isolated silanols.

formation) that are more easily dehydroxylated due to reduced hydrogen bonding, and minimize the surface to volume ratio, reducing the OH concentration on a per-gram basis. Smaller particles are beneficial from the standpoint of curvature, but they increase the surface area (and hence the OH content) and provide many more necks between particles, increasing the portion of the surface area exhibiting negative radii of curvature. Presumably an optimum (intermediate) particle size exists, but apparently no systematic investigation of this topic has been conducted. In any case, in order to facilitate dehydroxylation, it is wise to avoid synthesis conditions that produce microporous gels, characterized by small cylindrical pores.

Heat treatment procedures are designed to maximize the time spent at the highest possible temperatures prior to the onset of sintering and to minimize the water content of the processing ambient. The effects of heat treatment procedures on dehydroxylation are demonstrated by two related studies conducted by Gallo *et al.* [57] and Brinker and Haaland [58] on a borosilicate xerogel of nominal composition (wt%): $83SiO_2$, $15B_2O_3$, $1.2Na_2O$, $0.8Al_2O_3$. Extensive dehydroxylation of such multicomponent systems is difficult in practice due to the low sintering temperature. (In this case, rapid sintering begins at about 500°C.)

Figure 13 shows the BET surface area and hydroxyl coverage as a function of temperature for one series of samples heated at 2°C/min in ultrahigh-purity oxygen to temperatures between 175 and 650°C followed by quenching to room temperature (filled symbols) and a second series of samples heated to temperatures between 400 and 650°C followed by an 18-hour isothermal hold and quenching to room temperature (open symbols). OH/nm^2 values were determined by combined IR and TGA techniques: (1) the OH content of the fully densified gel was measured by FTIR in transmission using an extinction coefficient of 56 l/mol-cm [59];[†] (2) OH contents of the porous, quenched samples were then back-calculated for each sample from its weight loss, assuming that above 300°C all the weight loss was attributable to water. (The validity of this assumption was confirmed by Karl Fischer titration of the evolved gas.) The effect of the isothermal treatments at 440 and 490°C was to decrease the hydroxyl coverage while maintaining high surface areas. The surface area decreased at 540°C, but sufficient accessibility to the surface remained to allow a further reduction in the OH coverage. By 595°C, the accessible surface area for both series was reduced to about zero. Therefore 540°C appears to be the optimum processing temperature, although after an 18-hour hold a significant OH coverage remains.

[†] The low values of OH/nm^2 obtained for this multicomponent borosilicate gel compared to pure silica gels may result from an inappropriate choice of extinction coefficient or may reflect a different surface structure.

Fig. 13.

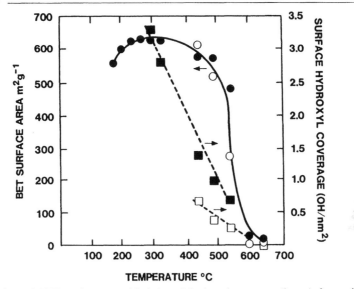

Variation of BET surface area (circles) and hydroxyl coverage (boxes) for multi-component borosilicate samples during heating at 2°C/min (filled symbols: no isothermal hold; open symbols: 18-hour isothermal hold) [57].

The effects of processing ambient on the residual hydroxyl content and shrinkage behavior are shown in Figs. 14 and 15 for the same multi-component borosilicate composition investigated by Gallo *et al.* [57]. The hydroxyl contents determined by FTIR on fully dense samples (Fig. 14) exhibit a trend that reflects the water contents of the processing gases. Except for the vacuum treatment, the hydroxyl contents of the densified gels are considerably greater than for the corresponding melt-prepared glass. A corresponding trend is seen in the shrinkage behavior (Fig. 15): the onset of viscous sintering occurs at progressively higher temperatures for processing ambients that result in lower residual hydroxyl contents. As discussed in Chapter 11, this behavior is explained by an increase in gel viscosity that results from more extensive dehydroxylation when gels are processed in drier ambients.

The effect of dehydroxylation on the sintering temperature may be exploited to design an incremental heating procedure in which the gel is first held isothermally just below the sintering temperature. Dehydroxylation that occurs during the isothermal hold increases the sintering temperature so that a second higher isothermal heat treatment can be employed. Repeating this procedure several times allows more complete dehydroxylation to be achieved.

Fig. 14.

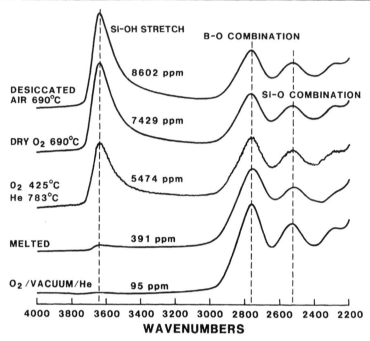

IR spectra of multicomponent borosilicate gels completely densified in various processing ambients compared to the IR spectrum of the corresponding melt-prepared composition. OH contents in ppm are determined using an extinction coefficient of 56 l/mol-cm. Further details concerning the heat treatments are given in Fig. 15 and in the text [58].

Such an incremental heat-treatment procedure was combined with a vacuum ambient ($\sim 1 \times 10^{-7}$ torr using a cryopump) to effect quite low residual OH contents, as shown in Figure 14 (95 ppm, determined using an extinction coefficient of 56 l/mol-cm [59]). In addition to incremental heating, the effectiveness of the vacuum procedure relied on the quality of the vacuum (at such low pressure a monolayer of water is removed faster than it can form) and the capture characteristics of cryopumps which are very efficient in water removal compared to oil-diffusion pumps. A residual OH content of 95 ppm was sufficiently low to avoid bloating of the gel during forming procedures at temperatures near 1000°C. However, even with lower OH contents than melted glass, heat treatments in the vicinity of the softening point would occasionally cause the formation of a few small bubbles (~ 1 or $2/cm^3$). This surprising appearance of bubbles may be due to the expansion of tiny, pre-existing, water-containing bubbles (invisible

Fig. 15.

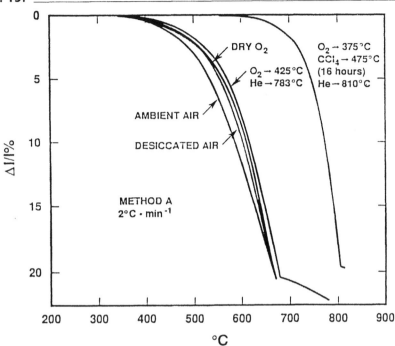

Linear shrinkage for a multicomponent borosilicate xerogel during heating at 2°C/min in various processing ambients. Dehydration and dealkalization (CCl_4 treatment) increase sintering temperature [58].

to the naked eye) that were sealed off during the initial stages of dehydroxylation, although near IR spectroscopy showed no evidence of molecular water [58]. Perhaps bubble formation occurs at lower OH contents in sintered gels compared to melts because a few isolated pores present on a statistical basis[†] serve to efficiently nucleate bubble formation.

3.2. Chemical Dehydroxylation

The preceding section illustrates that thermal dehydroxylation normally does not reduce the OH content to ppb levels required, for example, for fiber optic preform manufacturing. For this reason, chemical dehydroxylation procedures, primarily employing halogens, have been widely investigated.

[†] Unusually large pores that are not completely removed by sintering near T_g.

Fig. 16.

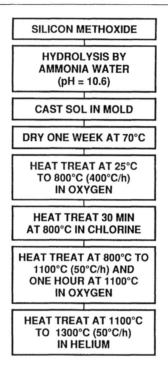

Scheme for preparing optically clear, bubble-free, and OH-free silica glass employing a halogen treatment followed by annealing in oxygen and sintering in He [52].

Figure 16 outlines a heat-treatment schedule devised by Matsuyama *et al.* [52] to dehydroxylate TMOS-derived gels intended for use as fiber optic preforms. After an initial heat treatment to 800°C in oxygen to remove any residual organics and promote thermal dehydroxylation, the gel samples were exposed to pure chlorine gas at 800°C for 30 min followed by a second oxygen treatment at 1100°C and finally sintering in helium at 1300°C. Figure 17 shows the residual OH content of fully dense gels as a function of the temperature of the chlorine treatment. We see from Fig. 17 that the residual OH content is lowered only by processing temperatures ≥700°C, which corresponds to the decomposition temperature of molecular Cl_2 [60].

Hair [22] proposed that Cl_2 dehydroxylates the silica surface according to:

$$2Si_sOH + 2Cl_2 \rightarrow 2Si_sCl + 2HCl + O_2 \qquad (9)$$

The data in Fig. 17 indicate that the decomposition of Cl_2 is a necessary first step so that Eq. 9 is correctly expressed as the sum of two simplified

Fig. 17. _____

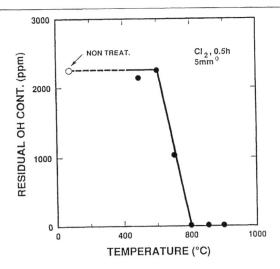

Dependence of the residual OH content of silica gels on the temperature of the Cl_2 treatment (30 min) prior to sintering in He at 1300°C. Each sample was treated at 800°C in oxygen prior to the Cl_2 anneal [52].

reactions:

$$2Cl_2 \xrightarrow{>700°C} 4Cl \tag{10}$$

and

$$2Si_sOH + 4Cl \longrightarrow 2Si_sCl + 2HCl + O_2 \tag{11}$$

where Eq. 10 is rate-limiting, at least at low temperatures. This is demonstrated by dehydroxylation of the silica surface using CCl_4 as the halogen source [22,61]:

$$Si_sOH + CCl_4 \xrightarrow{>350°C} Si_sCl + COCl_2 + HCl. \tag{12}$$

Dehydroxylation according to Eq. 12 is effective above 350°C, corresponding to the decomposition temperature of CCl_4. A more complete list of dehydroxylation reactions involving halogens appears in Table 3.

The effectiveness of halogen treatments compared to dry gases in removing hydroxyl groups prior to consolidation is evident from the IR spectra shown in Fig. 18 for SiO_2 aerogels densified at 1300°C in either dry oxygen (<5 ppm H_2O) or dry oxygen to 800°C followed by a chlorine containing ambient (O_2/Cl_2 or O_2 bubbled through $SOCl_2$) [55]. The comparison is more striking upon recognizing that the Cl-treated samples are ten to thirteen times thicker than the O_2-treated samples.

Table 3.

Reactions Forming Si-Halogen Bonds.

$$Si_sOH + NH_4F = Si_sF + NH_3 + H_2O$$
$$Si_sOH + HF = Si_sF + H_2O$$
$$2Si_sOH + 2SOCl_2 = 2Si_sCl + 2HCl + 2SO$$
$$2Si_sOH + 2Cl_2 = 2Si_sCl + 2HCl + O_2$$
$$2Si_sOH + CCl_4 = 2Si_sCl + CO_2 + 2HCl$$
$$Si_sOH + SiCl_4 = Si_sCl + Si\ oxychlorides$$
$$Si_sOH + SiCl_4 = Si_sOSiCl_3 + HCl$$

$$2Si_sOH + SiCl_4 = \begin{array}{c} Si_s{-}O \\ \\ Si_s{-}O \end{array}\!\!\Big\backslash\ \underset{Si}{\overset{/}{\underset{\backslash}{}}}\ \begin{array}{c} Cl \\ \\ Cl \end{array} + 2HCl$$

$$Si_sOH + CCl_4 \overset{>350°C}{=} SiCl + COCl_2 + HCl$$

Source: Iler [1], Tertykh *et al.* [61], Elmer *et al.* [62], Chukin [63], Uytterhoevan and Naveau [64], Hair and Hertl [65].

Despite virtually complete dehydroxylation of the silica surface with chlorine treatments, foaming of the consolidated gels can occur during heating in the vicinity of the glass softening point [66]. This problem results from the evolution of chlorine used to replace OH groups on the silica surface according to the preceding reaction schemes (Eqs. 11 and 12). The chlorine-foaming problem is mitigated by a second heat treatment in dry oxygen below the sintering temperature (1000–1100°C) following the chlorine treatment. (See Fig. 16.)

Susa *et al.* [66] have investigated the kinetics of chlorine removal and the effects of sample size and structure on the extent of removal. They conclude that the residual chlorine content decreases exponentially with time, indicating first-order kinetics. There are apparently two oxidation reactions involved with activation energies of 113 and 138 kJ/mol, compared to 473 kJ/mol for gas-phase oxidation of $SiCl_4$ at 1000 to 1230°C. In regions of incomplete chlorination, Cl may also be removed according to the condensation reaction:[†]

$$SiCl + HOSi \rightarrow Si{-}O{-}Si + HCl. \tag{13}$$

From the sample-size dependence of chlorine removal, they claim that diffusion of either oxygen or the chlorine containing oxidation products is rate controlling. The diffusion process is retarded by repeated adsorption-desorption from the pore walls, therefore Susa *et al.* recommend gel-synthesis conditions that result in dried gels containing large pores and small

---

[†]This reaction could also contribute to bloating in densified gels.

Fig. 18.

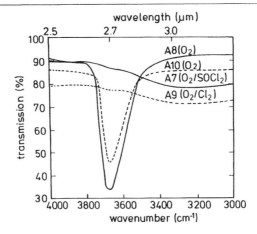

Transmission IR spectra of silica aerogels densified at 1300°C in O_2 or a Cl containing ambient. Thicknesses of samples in mm: A7 = 10.5, A8 = 0.8, A9 = 9.54, A10 = 0.31 [55].

surface areas (e.g., base-catalyzed hydrolysis of TMOS as indicated in Fig. 16). Reduction of the Cl content by high-temperature heat treatments prior to chlorination (which reduce both the surface area and hydroxyl coverage) are also recommended [66].

A second problem associated with chlorine dehydroxylation procedures is dealkalization [58]. Figure 15 shows that heat treatment of an alkali borosilicate gel in CCl_4 (375 to 475°C followed by a 16 hour isothermal hold, oxygen anneal, and consolidation in He) causes the shrinkage curve to be displaced to considerably higher temperatures (>700°C) than after treatments in dry gases. Although a portion of this behavior may be due to more complete dehydroxylation (<10 ppm residual OH [58]), 700°C greatly exceeds the T_g of the corresponding melt-prepared glass, ~575°C, and thus dehydroxylation alone cannot explain the observed sintering behavior.

Microprobe elemental mapping of a cross section of the fully densified gel (Fig. 19) shows both that considerable Cl is retained in the densified gel and that chlorine removal causes essentially complete dealkalization of the exterior gel surface. Quantitative elemental analysis (Fig. 20) indicates that the interior of the sample contains 1.5 wt% Cl and 0.9 wt% Na_2O (compared to 1.2 wt% Na_2O in the original gel), whereas the exterior surface (~300 μm thick) contains 0.6 wt% Cl and 0 wt% Na_2O. A Cl content of 1.5 wt% corresponds to replacement of 38% of the surface hydroxyls with Cl, based on a hydroxyl coverage at 450°C of 1.2 OH/nm^2 [56] and a BET specific

Fig. 19.

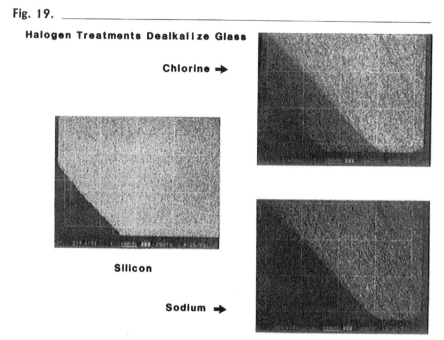

Halogen Treatments Dealkalize Glass

Chlorine →

Silicon

Sodium →

Elemental maps of Cl, Si, and Na on a cross section of a multicomponent borosilicate gel processed in CCl_4 followed by annealing in oxygen and sintering in vacuum. Edge on lower left is the exterior glass surface. Bar = 100 μm [58].

surface area of 560 m^2/g. Since some Cl may have been removed from the interior region as well, 38% OH replacement represents a lower limit for the extent of the chlorine dehydroxylation reaction. Because the densified gel contains a low OH content (<10 ppm), it is likely that dehydroxylation also occurred in part by condensation reactions involving surface OH groups. The dehydroxylation scheme may be schematically represented as follows:

$$
\begin{array}{ll}
\text{M—OH} & \text{M—Cl} \\[4pt]
\text{M—OH} & \text{M—Cl} \\[4pt]
\quad\quad +CCl_4 \longrightarrow & \quad +2\,HCl + H_2O + CO_2. \\[4pt]
\text{M—OH} & \text{M} \\[-2pt]
 & \quad\;\diagdown\!O \\[2pt]
\text{M—OH} & \text{M}
\end{array}
\tag{14}
$$

Fig. 20.

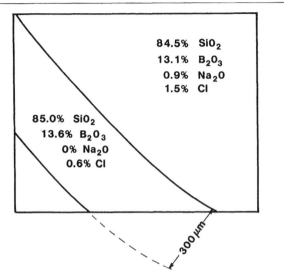

84.5% SiO_2
13.1% B_2O_3
0.9% Na_2O
1.5% Cl

85.0% SiO_2
13.6% B_2O_3
0% Na_2O
0.6% Cl

300 μm

Analyzed composition of the densified multicomponent borosilicate gel (wt%) in the two regions defined by the Cl and Na contents in Fig. 19. Edge on lower left is exterior glass surface[58].

The mole ratio of Na to Cl removed from the exterior surface is ~1.2:1. This suggests that the mechanism of dealkalization is the volatilization of sodium chloride[†] or sodium oxychloride. Regardless of the specific mechanism, it is apparent that dealkalization with Cl can be quite complete (Fig. 20) and that dealkalization can significantly alter the sintering temperature (Fig. 15).

The foaming problem due to chlorine evolution during softening of Cl dehydroxylated gels can be virtually eliminated by using fluorine instead of chlorine as the dehydroxylation reagent (see the shrinkage results in Fig. 8) [42,46,67]. Fluorine introduced as HF, NH_4F, or SiF_4 vapors replaces surface OH groups in a manner similar to Cl. (See, for example, reactions listed in Table 3.) Fluorine may also be introduced in solution using HF, HNH_4F_2, NH_4F, etc. [67] or in the vapor phase by replacement of Cl-terminated surface sites [67]. According to the fluorine dehydroxylation reactions listed in Table 3, F should be retained in the final gel at the same concentration as Cl. However foaming is eliminated in the case of fluorine due to stronger Si–F bonds: the heats of formation of compounds containing Si–F bonds are much greater than their Cl containing homologues [67].

[†] The heat treatment employed is close to the melting point of NaCl (~800°C) where the vapor pressure is appreciable.

Rabinovich *et al.* [67] claim that whereas 0.3 wt% Cl results in the evolution of 0.76 cm^3 Cl$_2$/cm^3 SiO$_2$, up to 2 wt% F can be accommodated in pure SiO$_2$ glass without foaming.

Wood and Rabinovich [42] and Nassau *et al.* [46] report several other advantages of fluorine treatments with regard to facilitating dehydroxylation. When added during the solution stage as a catalyst, F generally results in a reduction in the surface area and an increase in the pore size of the resulting xerogel compared, e.g., to HCl or NH$_4$OH catalysts [42]. As discussed in Chapter 3, this behavior is explained by the ability of F to promote ripening mechanisms. A reduction in surface area causes a reduction in the OH content. An increase in pore size causes an increase in the diffusion rate of H$_2$O. Both cause an increase in sintering temperature.

Fluorine treatments also make the surface less hydrophilic [46]. By analogy with the strength of HOH---OH$_2$ versus HF---HOH hydrogen bonds, the strength of SiF---HOH is a factor of 2 less than SiOH---OH$_2$. Therefore the replacement of OH with F reduces the extent of hydrogen-bonding with molecular water and facilitates the removal of physisorbed water during heating. Figure 21 shows near-IR spectra of a series of silica xerogels

Fig. 21.

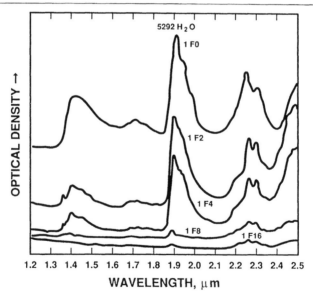

Near-infrared spectra of a series of silica gels prepared with added F contents ranging from 0-16 g/(100 g SiO$_2$) (as indicated by the sample numbers) after equilibration in room air. The analyzed F contents after heating to 150°C ranged from 0 to 6.7 g/(100 g SiO$_2$) [42].

prepared with either HCl or HF and equilibrated with room air for several days [42]. The reduction in the $5292 \, cm^{-1}$ band assigned to physisorbed molecular water as the fluorine content is increased from 0 (sample 1F0) to $6.7 \, g/(100 \, g \, SiO_2)$ (sample 1F16) is clearly evident. This effect is not only due to a reduction in surface area, since samples 1F2 and 1F4 have surface areas of 912 and $757 \, m^2/g$, respectively, compared to $549 \, m^2/g$ for sample 1F0. Thus F must increase the surface hydrophobicity; the surface of sample 1F16 is apparently quite hydrophobic.

Mulder [68] and Krol and Rabinovich [69] have used Raman spectroscopy to investigate the structural changes resulting from F incorporation in PCVD and gel-derived silica glass, respectively. New bands due to F are observed in gels at $935 \, cm^{-1}$ (Si–F stretching), 3300, 3395, and $3500 \, cm^{-1}$ (H–F stretching of various H-bonded HF species) and $240 \, cm^{-1}$ (unassigned). In addition to the new bands, F causes a reduction in the relative intensities of OH-related bands at 980 and $3750 \, cm^{-1}$. This is caused by both a reduction in surface area and replacement of terminal OH groups with F. In contrast to the results on gels, Mulder finds that F incorporation results in a band at $478 \, cm^{-1}$ in dense PCVD SiO_2 that he assigns to the symmetric O–Si stretching of Q^3 silicon sites terminated by an F ligand. This band is not evident in high surface area F-containing gels. Further structural studies of this subject are warranted.

4.
CHEMISTRY OF DEHYDROXYLATED SURFACES

As shown in Fig. 11, surface dehydroxylation progressively removes physisorbed water, hydrogen-bonded vicinal (or geminal) hydroxyls, and isolated hydroxyls. Because surface hydroxyls have been shown to be the principal sites for physisorption of water (see, e.g., Fig. 7), dehydroxylation makes the surface progressively more hydrophobic. The extent of physical adsorption of other molecular species capable of hydrogen bonding, such as ammonia and alcohol, is reduced as well.

In addition to altering the surface character with respect to physisorption, condensation reactions involving neighboring surface OH groups produce a distribution of cyclic species (rings) on the dehydroxylated surface. (See Fig. 3.) Investigations of the reactivity of such surfaces by Morrow and Cody [70–72] and more recently by Bunker and coworkers [29,30,73] and Brinker and coworkers [28,74] have shown that the kinetics of dissociative chemisorption reactions such as:

$$(15)$$

are significantly affected by the ring size, i.e., the number of Si–O bonds comprising the ring. Thus both physical and chemical adsorption processes are affected by the extent of surface dehydroxylation.

This section examines reactions of several molecular species with dehydroxylated surfaces. These reactions are important for gels, because of their intrinsically high surface areas and because dissociative chemisorption reactions can be used to alter the composition of gel-derived glasses as discussed in the following section. Reactions of molecular species with dehydroxylated surfaces are generally interesting in the areas of chromatography, separations, and catalysis [1].

4.1. Ring Statistics

The OH coverage of the silica surface is quite close to the expected value based on a $\langle 111 \rangle$ β-cristobalite structural model (six-membered rings, see Fig. 6) [1,75–77]. However, studies combining ^{29}Si MAS NMR [28], CP MAS NMR [28,77], and vibrational spectroscopy [28–30,70–72] suggest that the hydroxylated silica surface is terminated by a distribution of open and closed rings, where the number of silicons per ring varies from two to more than six. (See Fig. 3.) For example, combined ^{29}Si MAS NMR and Raman spectroscopy investigations [28] reported in the preceding chapter (Figs. 43 and 44 in Chapter 9) show that surface dehydroxylation above ~200°C produces a surprisingly large concentration of cyclic trisiloxanes (three-membered rings) according to

$$(16)$$

where the left side of Eq. 16 represents an "open ring." "Closed" three-membered rings (right side Eq. 16) are identified by a narrow Raman band at ~605 cm^{-1} [28 and references therein] and by the less negative value of the ^{29}Si chemical shift (δ) of Q^4 resonances, resulting from the reduction in the Si–O–Si bond angle (ϕ) according to the relationship

$$\delta(\text{ppm}) = -0.59(\phi) - 23.21 \qquad (17)$$

where ϕ is the average of the four Si–O–Si angles per Q^4 site [78,79]. The maximum surface coverage of three-membered rings is estimated to be 2.2–4.5/nm^2 based on combined NMR and BET results [80].†

Comparisons of the vibrational spectra of model cyclodisiloxanes and high surface area silicates [73] show that dehydroxylation above 650°C in vacuum produces two-membered rings on the silica surface:

$$\begin{array}{c}\text{(structure)}\end{array} \qquad (18)$$

with $\phi = 91.5$, $\theta = 88.5$, $d(\text{Si-O}) = 1.67$ Å, $+\,H_2O$

identified by strong IR bands at 908 and 888 cm^{-1} [70–72]. Although molecular orbital calculations have established that the optimum geometry of the model compound $(H_2SiO)_2$ is planar with D_{2h} symmetry and $\phi = 91.5°$ [81], the concentration of two-membered rings is probably too low (~0.1 nm^2 [73]) to be identified by ^{29}Si NMR according to Eq. 17.

Studies using isotopic enrichment [73,82,83]:

$$\begin{array}{c}\text{(structure)}\end{array} \rightleftharpoons \begin{array}{c}\text{(structure)}\end{array} + H_2O \qquad (19)$$

have proven that four-membered rings as well as two- and three-membered rings can be formed on the silica surface by dehydroxylation reactions. Four-membered rings are identified by a narrow Raman band at ~490 cm^{-1} [84]. (See Fig. 43 in Chapter 9.)

† Based on a surface coverage of silicon atoms of 7.8 Si$_s$/nm^2 [1] and a BET surface area of 840 m^2/g for a vacuum dehydroxylated gel exhibiting a very high concentration of three-membered rings (see Fig. 46 in Chapter 9), 65% of the silicons are calculated to be surface silicons (Si$_s$). From decomposition of the corresponding ^{29}Si MAS NMR spectrum, 19–39% of the total silicons are contained in three-membered rings (Si$_3$). The surface coverage of three-membered rings is Si$_3$/Si$_s \times$ Si$_s$/nm^2 and is estimated to range from 2.2 to 4.5 Si$_3$/nm^2 depending on the value of Si$_3$.

Since two-, three-, and four-membered rings have been shown to form by dehydroxylation, it seems likely that five-membered and larger rings may also form on the silica surface. However, these larger rings are not uniquely identified by vibrational spectroscopy; moreover, puckering allows the Si–O–Si angle to approach the equilibrium angle ($\phi \sim 150°$) for rings containing four or more silicons, so four-membered and larger rings cannot be identified on the basis of the ^{29}Si chemical shift. Thus a distribution of two-membered and larger rings seems probable, but, as of this writing, it is not possible to determine accurately the distribution of ring sizes on the dehydroxylated silica surface.

Since dehydroxylation of the $\langle 111 \rangle$ β-cristobalite surface produces equal numbers of three-, five-, and six-membered rings according to the following scheme:

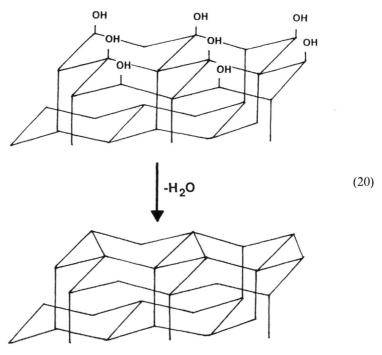

$$\downarrow \text{-H}_2\text{O} \hspace{4cm} (20)$$

it could be argued that a crystalline-like surface is required to accommodate the unusually large concentrations of three-membered rings observed in Raman experiments. However, since two OH groups are consumed per three-membered ring, the maximum surface coverage expected according to the preceding scheme is 2.3 (three-membered rings)/nm^2 = 6.9 Si$_3$/nm^2, exceeding the surface coverage previously estimated, viz. 2.2–4.5 Si$_3$/nm^2. Thus, although dehydroxylation of six-membered rings, according to

Eq. 20, may account for the formation of three-membered rings on the silica surface, it is not necessary to invoke a cristobalite surface model to explain their large concentrations. Instead we expect the observed concentrations of three-membered rings to be accounted for by dehydroxylation of a random surface. (See, e.g., Fig. 6.)

Based on the results of molecular orbital (MO) calculations of model cyclic molecules, the heats of formation of two- and three-membered rings are estimated to be 55.4 and 24.4 Kcal/(mole of rings), respectively [81,85]. These positive values reflect the strain energies required to reduce the Si–O–Si and O–Si–O bond angles (ϕ and θ in Eqs. 16 and 18) and increase the Si–O bond distance. Ring strain causes the optimum configuration of two- and three-membered rings to be planar. Larger rings are able to pucker allowing ϕ, θ, and the Si–O bond length to approach the equilibrium values. Thus four-membered and larger rings are essentially unstrained.

Ring strain and associated steric factors influence chemical reactivity as discussed in the following section. Ring strain also influences the observed temperature dependence of the ring concentrations. Three-membered rings appear at about 200°C and their concentration is maximized at or above 600°C. Two-membered rings result from dehydroxylation at or above 650°C. These temperatures presumably reflect the thermal energies required to form the rings. Once formed, the rings are well preserved in the stiff silicate matrix until sufficiently high temperatures (near T_g) are reached that the surface can undergo a reconstructive process via bond breakage and reformation. At these temperatures the concentration of rings rapidly approaches the equilibrium concentration observed for conventional fused silica. (See Figs. 43 and 44 in Chapter 9.)

4.2. Kinetics and Mechanism of Surface Rehydration

The rehydration of dehydroxylated surfaces has been investigated extensively. (See, e.g. [1,75,76,86,87].) The rehydration process involves water adsorption followed by dissociative chemisorption (Eq. 15). Either the adsorption step or the subsequent dissociative chemisorption step could be rate-determining [30]. Hair [22] states that rehydration is sluggish for surfaces dehydroxylated above about 450°C due to surface hydrophobicity (physisorption rate determining). Below this temperature there is sufficient hydroxyl coverage that the adsorption step is facile.

Data by Volkov et al. [87] indicate that on highly dehydroxylated surfaces water first adsorbs on strained surface sites. Subsequent adsorption occurs on areas adjacent to these sites so that rehydration takes place in "patches" nucleated by the strained sites. Studies by Morrow and Cody [70–72], Bunker *et al.* [30], and Brinker *et al.* [28,74] indicate that the extent of strain (ring size) in turn dictates the kinetics of the initial physisorption step.

In a study of dissociative chemisorption of water, ammonia, alcohol, or amines on highly strained two-membered rings, Bunker and coworkers [30] concluded that two types of rings with different acid–base sites for adsorption exist on surfaces dehydroxylated above 650°C:

$$a = \text{ACID}$$
$$b = \text{BASE}$$

The concentrations of these two different rings depend on the extent of surface dehydroxylation. For the ring designated d_4, the strained silicon site is a Lewis acid since it possesses an unoccupied d orbital that is available as an electron acceptor [30]. The lone pair electrons on the bridging oxygen can function as a Lewis base (electron donor) or Brønsted base (proton acceptor) [30]. The ring designated d_3 contains in addition a silanol group that could function as a Lewis or Brønsted base or a Brønsted acid (proton donor) [30].

Bunker and coworkers [30] performed an *in situ* FTIR investigation of dissociative chemisorption reactions of a series of molecules containing basic lone-pair electrons and acidic protons: H_2O, CH_3OH, NH_3, CH_3NH_2 (in order of increasing basicity) with the two types of two-membered rings. They concluded that physisorption is rate-limiting and is controlled by the molecular basicity [30]. Thus the relative reaction rates in general follow the trend $H_2O \leq CH_3OH \leq NH_3 \leq CH_3NH_2$.

Brinker and coworkers [74] used *in situ* Raman spectroscopy to compare the hydrolysis kinetics of three- and four-membered rings embedded in the silica surafce to the hydrolysis rates of model cyclic siloxane molecules in solution [88]. Three-membered rings slowly hydrolyze when exposed to water vapor. (See Fig. 22.) Figure 23 shows a pseudo–first-order kinetic plot (excess water) for the hydrolysis of three-membered rings formed on the silica surface by dehydroxylation at 650°C; at 25°C and 100% RH, the rate constant is $k = 5.2(\pm 0.5) \times 10^{-3}\,\text{min}^{-1}$ ($t_{1/2} = 2.2$ hours). This value is comparable to the rate constant of hexamethylcyclotrisiloxane hydrolyzed in solution, viz. $k = 3.8(\pm 0.4) \times 10^{-3}\,\text{min}^{-1}$ ($t_{1/2} = 3.0$ hours) [88]. By comparison, under the same conditions, octamethylcyclotetrasiloxane and

Fig. 22.

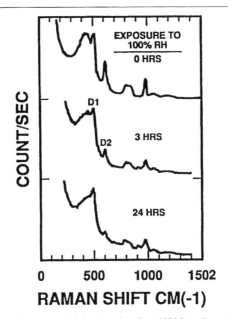

Raman spectra of a silica xerogel dehydroxylated at 600°C and exposed to 100% RH water vapor for the indicated times [74].

four-membered rings located on the silica surface are stable [88]. (See Figs. 43 and 44 in Chapter 9.) This indicates that the stability of siloxane bonds toward dissociative chemisorption of water depends on bond strain, and that the strain (and hence geometry) in surface rings is similar to that in isolated molecules. It is also apparent that surface irregularity does not significantly lower the accessibility of water to the three-membered rings.

The dramatic effect of ring strain on siloxane bond hydrolysis is summarized by Bunker and coworkers [30] in Fig. 24 where the hydrolysis rate constants [1,74,89,90] are plotted versus ring strain energies obtained from molecular orbital calculations [85]. The ring strains for four-membered and larger rings are assumed to be nearly zero and the corresponding hydrolysis rates have been equated to the hydrolysis rate of conventional fused silica which is assumed to be unstrained. Figure 24 shows that two-membered rings react with water over 5 million times faster than does fused silica! The less reactive three-membered rings which are more prevalent on the dehydroxylated surface react about 100 times faster than fused silica.

The foregoing discussion indicates that ring strain increases the acidity and accessibility of the silicon site promoting rate-limiting physisorption of molecular bases. The subsequent dissociative chemisorption step is very

Fig. 23.

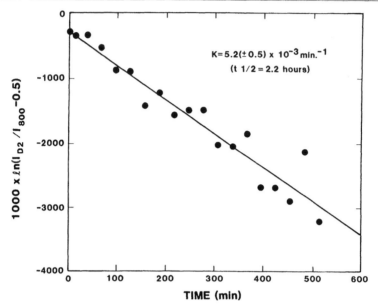

Pseudo–first-order kinetic plot for the hydrolysis of three-membered rings at 25°C and 100% RH water[74].

rapid, presumably due to the ring strain. The basicity of the oxygen in the strained ring appears to be less important in governing the reactivity. If physisorption were controlled by the molecular acidity of the adsorbate and basicity of the ring, the predicted reaction rates should follow the trend $H_2O \geq CH_3OH \geq NH_3$ which is opposite of the observed trend for the two-membered rings [30].

The effect of ring strain on the basicity of the oxygen appears in fact to be a highly disputed topic. Molecular orbital calculations performed by Gibbs [91] on $(HO)_6Si_2O$ at the 6-31G* level indicate that the Mulliken charge on the bridging oxygen increases from -0.88 to -0.945 as the Si–O–Si angle ϕ decreases from 170 to 120°. Table 4 shows that for a constant angle, $\phi = 150°$, the charge also increases with increasing bond length, d. Because ϕ decreases and d increases with increasing ring strain (decreasing ring size), we expect from the MO calculations that the Lewis basicity of the bridging oxygen should increase with decreasing ring size. However calculations based on a tight-bonding formalism (see, e.g. [92,93]) show the opposite dependence of charge on ϕ as shown by the results of Hollinger *et al.* [93] listed in Table 5. Although the ionicity increases with d, it decreases (rather than increases) with decreasing angle. Since the bond ionicity calculated

Fig. 24.

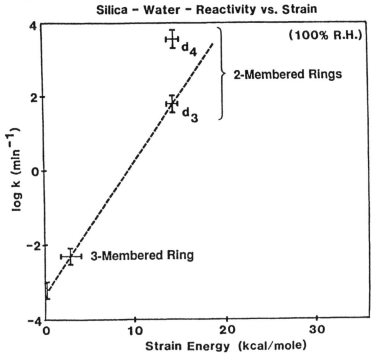

Silica – Water – Reactivity vs. Strain

bond strain normalized to an activity of water corresponding to 100% RH [30]. Data at zero strain represent dissolution data for fused silica. Strain energies of two- and three-membered rings are values obtained from MO calculations [85]. (d_3 and d_4 are the two types of two-membered rings described on page 650.)

Table 4.

Mulliken Charges Obtained from MO Calculations versus Si–O
Bond Lengths and Si–O–Si Bond Angles.

	$d(Si-O)$ Å			
ϕ	1.592	1.602	1.622	1.642
120°	—	—	−0.945	—
130°	—	—	−0.930	—
140°	—	—	−0.920	—
150°	−0.880	−0.890	−0.909	−0.926
170°	—	—	−0.880	—

Source: Gibbs [91]. Mulliken charge is defined as the partial charge on the oxygen nucleus where a charge of zero would indicate neutrality.

Table 5.

Si–O Bond Ionicities versus Si–O Bond Lengths and Si–O–Si
Bond Angles Derived from a Tight-Binding Formalism.

	d(Si–O) Å			
ϕ	1.58	1.61	1.65	1.70
132°	0.44	0.449	0.461	0.471
144°	0.461	0.467	0.488	0.494
156°	0.498	0.508	0.504	—
168°	0.529	0.536	—	—

Source: Hollinger *et al.* [93].

by Hollinger *et al.* is more a sensitive function of bond angle than bond length, the net effect of reducing the ring size would be a decrease in the charge on the bridging oxygen corresponding to a decrease in Lewis basicity. Unfortunately X-ray photoelectron spectroscopy (XPS) is not sensitive enough to resolve this question [79].

Studies of molecular photoemission on amorphous silica surfaces provide additional information concerning the effects of surface dehydroxylation on adsorption behavior. DeMayo [94] concluded that the most chemically inhomogeneous silica surface, as judged by multiple exponential decay of adsorbed pyrene probe molecules, was the surface created by vacuum dehydroxylation at 700°C. Subsequent exposure to water vapor rendered the surface much more homogeneous. Based on the preceding discussion, it is likely that the inhomogeneous surface at 700°C results from the altered acid–base properties of the strained, two- and three-membered rings, combined with Q^3 silanol sites capable of hydrogen-bonding with the π-electron system of pyrene and the hydrophobic nature of the remaining (unstrained) siloxane surface. Exposure to water vapor hydrolyzes the strained sites, proving additional sites for water adsorption, leading finally to a more homogeneous (hydroxylated) surface.

Based on the preceding discussions of water adsorption on dehydroxylated surfaces, the most likely mechanism of rehydration of silicate surfaces dehydroxylated above 450°C is adsorption on acidic silicon sites contained in strained two- and three-membered rings, followed by dissociative chemisorption. Since two-membered rings comprise a small fraction of the silica surface, the rehydration kinetics will initially reflect the rate of dissociative chemisorption of three-membered rings which cover approximately one quarter of the dehydroxylated surface. Subsequent water adsorption occurs preferentially on silanols formed by hydrolysis of three-membered rings.

The hydrolysis of the remaining surface occurs adjacent to these patches of rehydrated surface and exhibits kinetics determined by the hydrolysis rate of unstrained four-membered and larger rings ($\sim 10^2$ times slower).

5.

COMPOSITIONAL MODIFICATION

The intrinsically high surface areas of gels permit access of gases to a substantial portion of the solid phase. For example, in a silica gel with a BET surface area of 850 m^2/g, 65% of the silicon atoms are on a surface. The accessibility of the surface makes gas- or liquid-phase reactions a viable means of "bulk" compositional modification. This section first examines nitridation via ammonolysis, which is the most thoroughly studied example of compositional modification of gels. Other classes of reactions leading to compositional modification are briefly summarized in Section 5.2.

5.1. Nitridation

The motivation for nitridation of oxide glasses is the improvement in physical properties that result when bridging oxygen atoms, M–O–M, are replaced by
$$\begin{array}{c} M \\ | \end{array}$$
trigonally coordinated N atoms, M–N–M [95]. The strategy for surface nitridation is first to devise reaction schemes that replace M–O bonds with M–N bonds, and then to create additional M–N bonds to establish N in trigonal coordination with its surrounding metal atoms.

Reactions A–E which follow have been proposed by Brinker and Haaland [96] as possible schemes for surface nitridation via ammonolysis. Lewis acid adsorption (A) is a possible scheme for electrophilic metals capable of formally increasing their coordination numbers, e.g., trigonally coordinated boron or tetrahedrally coordinated aluminum. Lewis acid adsorption may be followed by dissociative chemisorption as in B. This scheme depends on the Lewis acidity of the metal site but does not necessarily involve a stable intermediate with a formal increase in coordination number. As discussed in the previous section, dehydroxylation of the silica surface at temperatures above $\sim 250°C$ progressively creates strained surface silicon species with enhanced Lewis acidity. The importance of scheme B for silica is therefore expected to increase with the extent of suface dehydroxylation. Reaction C was proposed by Mulfinger [97] to account for dehydroxylation of silica glass melts. Morrow *et al.* [98] proposed this reaction involving isolated silanols to account for the reaction of NH_3 with silica at 650°C. Reaction D is known to occur for organosilicon compounds [99] and to occur in general for reactions of metal alkoxides with amino compounds (e.g., dibutyltin diethoxide with ethanolamine [100]). In the presence of reducing agents such as hydrogen or carbon, introduced for example as alkoxide or acetate groups or by dissociation of NH_3, reduction of a dehydroxylated surface can generally occur according to scheme E. This reaction is favored at higher temperatures where ammonia is likely to be dissociated and where M-O-M bonds are more easily broken.

$$\rangle M \Big| \ + \ :\!\overset{H}{\underset{H}{N}}\!:\!H \quad \blacktriangleright \quad \rangle M\Big|\overset{H}{\underset{H}{N}}\!:\!H \qquad \text{(Scheme A)}$$

$$\begin{array}{c} \overset{H}{\underset{M{:}O\text{-}M}{H{:}N{:}H}} \end{array} \longrightarrow \begin{array}{c} \overset{H\diagdown\,H}{N} \quad \overset{H}{O} \\ | \qquad | \\ M \ + \ M \end{array} \qquad \text{(Scheme B)}$$

$$\text{M{-}OH} \ + \ NH_3 \quad \blacktriangleright \quad \text{M{-}NH}_2 \ + \ H_2O \qquad \text{(Scheme C)}$$

$$\text{M{-}OR} \ + \ NH_3 \quad \blacktriangleright \quad \text{M{-}NH}_2 \ + \ ROH \qquad \text{(Scheme D)}$$

$$10)\quad \begin{array}{c} M\diagdown \\ M\!\!\prec^{O}_{O} \\ M\diagup \end{array} \ + \ HC \ + \ 2NH_3 \quad \blacktriangleright \quad \begin{array}{c} M\diagdown \\ M\!\!\prec^{NH} \\ M\text{-}NH_2 \end{array} \ + CO + H_2O + H_2 \quad \text{(Scheme E)}$$

Reactions A–E lead to the formation of amines, $-NH_2$, or imines, $=NH$. Repeated reactions are required to obtain nitrogen trigonally coordinated with M. For gas-surface reactions it is likely that amines are formed at low and intermediate temperatures, but trigonal coordination occurs only during pore collapse with a corresponding loss of NH_3:

$$\text{M{-}NH}_2 \quad \text{H}_2\text{N{-}M} \qquad \overset{H}{\underset{}{\text{M{-}N{-}M}}} \ + \ NH_3 \qquad \text{(Scheme F)}$$

$$\begin{array}{c} M\diagdown \\ M\diagup^{NH} \end{array} \quad \text{H}_2\text{N{-}M} \qquad \begin{array}{c} M\diagdown \\ M\diagup^{N\text{-}M} \end{array} \ + \ NH_3 \qquad \text{(Scheme G)}$$

Figure 25 shows the analyzed N and C contents as a function of temperature for multicomponent 66.6 SiO_2 17.1 B_2O_3 6.6 Al_2O_3 6.3 Na_2O 3.4 BaO (wt%) gels processed in flowing NH_3 at standard pressure, NH_3 at high pressure (0.7–1.3 MPa), or flowing air. Several trends are apparent in the data. We see no nitrogen incorporation in air ($\sim 80\%$ N_2), whereas the N content increases sharply and then gradually decreases with temperature in flowing NH_3. High-pressure ammonia treatments increase the N contents at all temperatures. Pure silica gels, however, processed in flowing nitrogen under similar conditions contained only low N contents (0.07–0.49 wt%) for heat treatments up to 1050°C.

In flowing air, the residual carbon content is sharply reduced at 350°C due to oxidation, whereas in flowing NH_3 it is reduced gradually over the temperature interval 25–750°C, finally reaching a level comparable to that achieved in air. It should be emphasized that heating in pure nitrogen did not lead to carbon removal, and an opaque black glass resulted, while the NH_3 treatment followed by sintering in vacuum at 900°C yielded a colorless transparent glass [96].

In situ FTIR spectroscopy experiments [101] performed on binary borosilicate compositions indicate that boron is primarily responsible for nitridation in the multicomponent gel referred to in Fig. 25. Figure 26 shows a series of spectra obtained for a 20 B_2O_3 – 80 SiO_2 (mol%) film heated in flowing ammonia between 20 and 700°C. Dissociative chemisorption

Fig. 25.

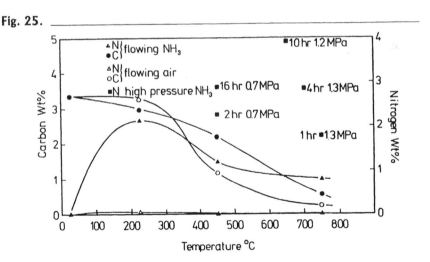

Measured carbon and nitrogen contents for a multicomponent borosilicate gel processed in flowing air, flowing NH_3, or high-pressure NH_3 [96]. Times and pressures refer to the duration and pressure of the high-pressure NH_3 treatments.

Fig. 26.

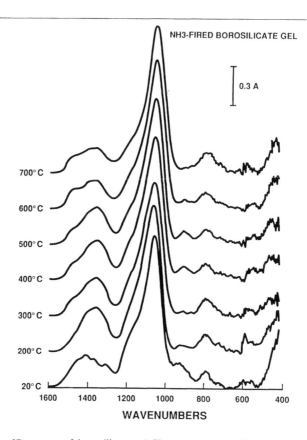

Transmission IR spectra of borosilicate gel films processed in flowing NH_3 (15-min holds) at temperatures between 20 and 700°C [101].

reactions of NH_3 with $=B-O-Si\equiv$ bonds according to

Scheme (H)

are evident above about 400°C from the reduction in relative intensity of the ~915 cm^{-1} band assigned to B–O–Si stretching and the appearance of a new band (shoulder) at ~1490 cm^{-1} assigned to B–N stretching.

The role of boron in the dissociative chemisorption process (H) may be either to promote adsorption or dissociation (or both). Trigonally coordinated boron is a Lewis acid: it can accept the lone pair electrons of nitrogen, promoting NH_3 adsorption (left-side of scheme H). Low and Ramasubramanian [102] have pointed out that B-O-Si bonds are more easily ruptured than Si-O-Si bonds, thus promoting dissociation (right side of scheme H). The combined effects result in increased nitridation compared to pure silica gels.

The creation of B-N bonds at 450°C, which is evident in the FTIR spectra of both binary borosilicates (Fig. 26) and the multicomponent gels [96], is not apparent in Fig. 25, which shows a $2\times$ reduction in the analyzed nitrogen content over the temperature interval 220-750°C. This suggests that some of the physical or chemical adsorption processes responsible for nitridation at low temperatures are not stable with respect to the reverse reactions as the temperature is increased. Alternatively the reduction in N content at higher temperatures may be explained by condensation of amines (schemes F and G) which, if complete, could account for a $3\times$ reduction in nitrogen. FTIR studies of the 3000-4000 cm^{-1} portion of the spectrum of the multicomponent gel show that some N-H bonds are retained even after heating to 900°C. Thus schemes F and G may account for the reduction in N content, but they do not go to completion.

The gradual reduction in the carbon content with temperature (Fig. 26) may be explained by the combined effects of reactions D and E. The reduction reaction, E, is certainly the primary mechanism for carbon removal above 400°C where residual alkoxy groups are decomposed. As discussed for reaction H, the susceptibility of B-O-Si bonds to rupture may promote the reduction reaction.

Extensive nitridation of pure silica systems has been accomplished by several groups. Brow and Pantano [103-105] succeeded in nitriding silica gel films ($<0.1\ \mu$m thick) by ammonia treatments at temperatures between 600 and 1200°C. Although FTIR investigations showed evidence of Si-NH_x bonding at temperatures below about 600°C, less than 5 mole% nitrogen was incorporated at these temperatures, as analyzed by XPS. Between 700 and 1000°C, the nitrogen content increased to over 40 mole% (Fig. 27a), corresponding closely to silicon oxynitride, Si_2N_2O. This increase was accompanied by a decrease in the Si $2p$ binding energy (BE), consistent with a progressive formation of nitrogen trigonally coordinated with Si (Fig. 27b):

$$\equiv Si - N \overset{H}{\underset{H}{<}} \rightarrow \overset{\equiv Si}{\underset{\equiv Si}{>}} N - H \rightarrow \overset{\equiv Si}{\underset{\equiv Si}{>}} N - Si \equiv \qquad \text{(Scheme I)}$$

increasing treatment temperature
decreasing N $1s$ binding energy

Fig. 27.

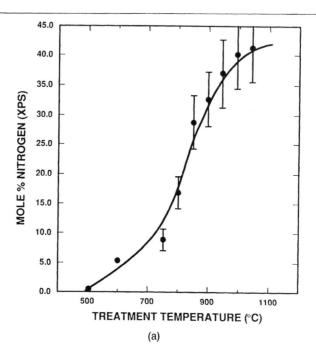

(a)

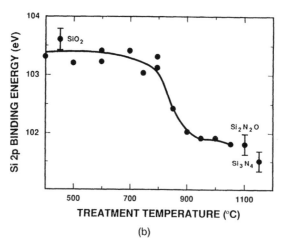

(b)

(a) Nitrogen concentrations of silica gel films treated in NH_3 for 2 h at temperatures between 500 and 1025°C [103] measured by XPS. (b) Effect of processing temperature on the Si $2p$ binding energy for silica gel films treated in NH_3 at various temperatures. Values for SiO_2, Si_2N_2O, and Si_3N_4 are shown for reference [104].

The Si XPS results are consistent with Si–NH$_2$ bonding at or below 800°C

$$\overset{\text{Si}}{\underset{|}{}}$$

and primarily Si–N–Si bonding at or above 900°C. Since extensive nitridation occurs above 800°C, these data suggest that nitridation occurs predominantly by a mechanism whereby N can immediately acquire trigonal coordination with Si. Perhaps at these high temperatures reduction of the network (reaction E) is accompanied by pore collapse (reactions F and G) or

$$\overset{\text{Si}}{\underset{|}{}}$$

surface reconstruction, facilitating the formation of Si–N–Si species. The fact that bulk silica gels processed in flowing NH$_3$ exhibit much lower levels of nitridation implies that the nitridation process is in some way transport-limited.

In a study of ammonolysis of silica gel fibers, Kamiya and coworkers [106] observed that little nitridation occurred in fibers prepared from TEOS, but substantial nitridation occurred in fibers prepared from methyltriethoxysilane (MTES). (See Fig. 28.) Since both TEOS and MTES provide a source of carbon that serves as a reducing agent, Kamiya *et al.* postulated that the reduction reaction, E, could not account for the dramatic differences shown in Fig. 28. Based on FTIR results that showed bands attributable to Si–H (2200 cm^{-1}) and Si–NH$_2$ (3300 cm^{-1}) for the MTES-prepared fibers at

Fig. 28.

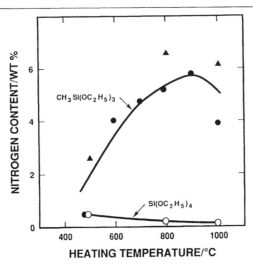

Analyzed nitrogen contents of gel-derived silica fibers prepared from TEOS (O) or methyltriethoxysilane (●) following 1 h isothermal holds in NH$_3$. ▲ denote 5 h isothermal holds for the MTES-derived fibers [106].

temperatures above about 600°C, they proposed the following reactions to explain the role of the methyl group:

$$\text{Si-CH}_3 \xrightarrow{600°C} \text{Si-H} + (\text{C} + \text{H}_2) \tag{21}$$

$$\text{Si-H} + \text{NH}_3 \longrightarrow \text{Si-NH}_2 + \text{H}_2. \tag{22}$$

Subsequent condensation reactions involving the Si-NH_2 groups during pore collapse (F and G) lead to N trigonally coordinated with Si and a corresponding loss of N above $\sim 900°C$. (See Fig. 28.)

5.2.   Other Schemes for Surface Modification

Other important schemes of surface modification include silylation reactions and, in general, reactions with MX_m where M is a metal and X is a halide, alkoxide, alkyl, allyl, carbonyl, etc. Silylation is the displacement of an active hydrogen (usually in a hydroxyl group) by an organosilyl group [107]. A major application for silylated surfaces has been in modifying glass fibers and fillers for preparation of organic polymer composites. Silylation is also used to impart hydrophobicity or organic nature to inorganic surfaces. The silylating materials may act as a means of bonding organic materials to oxide surfaces to obtain the advantages of reactive groups on a stable, insoluble substrate. Silylation methods and applications of silylated surfaces are discussed by Iler [1, Chapter 6].

Commercial coupling agents are, in general, of the type RSiX_3, i.e., they contain three groups that can potentially react with surface hydroxyl groups according to

$$3\text{Si}_s\text{OH} + \text{X}_3\text{SiR} \rightarrow -(\text{Si}_s\text{-O})_3\text{-SiR} + 3\text{HX}. \tag{23}$$

Representative commercial coupling agents are listed in Table 6 where X (Eq. 23) is an methoxy group. Monofunctional (R_3SiX) and difunctional (R_2SiX_2) silanes can react with surface hydroxyl groups according to [108]:

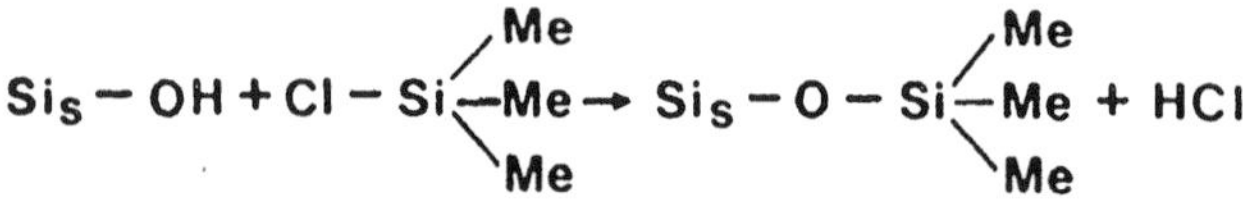

$$\tag{24}$$

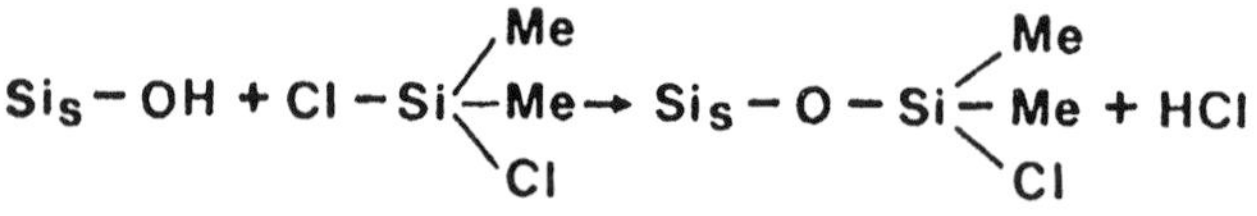

$$\tag{25}$$

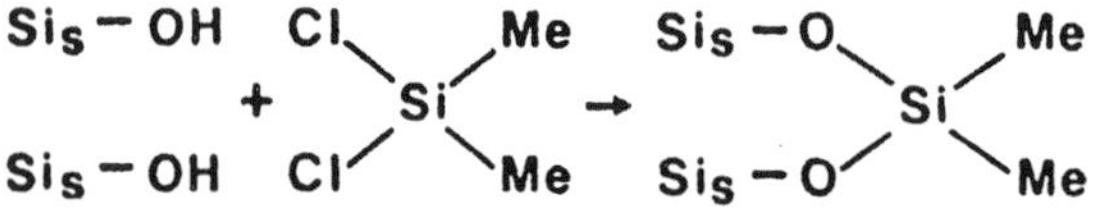

$$\tag{26}$$

Table 6.

Representative Commercial Coupling Agents.

Organofunctional Group	Chemical Structure
A. Vinyl	$CH_2\!=\!CHSi(OCH_3)_3$
B. Chloropropyl	$ClCH_2CH_2CH_2Si(OCH_3)_3$
C. Epoxy	$\overset{\displaystyle O}{\overset{\diagup\ \diagdown}{CH_2CHCH_2OCH_2CH_2CH_2Si(OCH_3)_3}}$
D. Methacrylate	$\overset{\displaystyle CH_3}{\overset{\mid}{CH_2\!=\!C\text{-}COOCH_2CH_2CH_2Si(OCH_3)_3}}$
E. Primary amine	$H_2NCH_2CH_2CH_2Si(OC_2H_5)_3$
F. Diamine	$H_2NCH_2CH_2NHCH_2CH_2CH_2Si(OCH_3)_3$
G. Methyl	$CH_3Si(OCH_3)_3$
H. Cationic styryl	$CH_2\!=\!CHC_6H_4CH_2NHCH_2CH_2NH(CH_2)_3Si(OCH_3)_3\!\cdot\!HCl$
I. Phenyl	$C_6H_5Si(OCH_3)_3$

Source: Sindorf and Maciel [26].

Hair and Hertl [109] showed that isolated hydroxyls, not hydrogen bonded hydroxyls, participate in Reactions 24–26. Reaction 24 obeys first-order kinetics [108], whereas Reactions 25 and 26 are expected to obey first-and second-order kinetics, respectively. A mixed, 1.6-order reaction is observed [108] suggesting that Reactions 25 and 26 both occur and that 60% of the isolated silanols must be sufficiently close to each other (neighboring, isolated silanols or geminal silanols) that they can react in a bifunctional manner.

The silylation literature associated with organic polymer composites and chromatography is far too extensive to review here. As yet, silylation reactions are not widely employed in sol-gel processing of glasses and ceramics; however, porous sol-gel–derived oxides are expected to be ideal low-temperature hosts, so the introduction of organic functionality is of much potential interest in the areas of sensors, catalysts, and optics. General reviews of silylation chemistry are found in refs. [110–112].

Reactions forming Si–O–M bonds have been extensively explored in the catalyst literature. The simplest scheme to anchor metals on the oxide surface is by reaction with the surface hydroxyls [113]:

$$n\text{Si-OH} + MX_m \rightarrow (\text{SiO})_n\text{-}MX_{m-n} + n\text{HX}. \qquad (27)$$

Organometallics, metal alkoxides, metal halogenides, salts of organic acids, etc. can be attached to oxide surfaces according to this scheme. Table 7 shows several representative reactions with aluminum compounds that form Si–O–Al bonds useful for introducing Lewis acidity to the surface.

Table 7.

Several Reactions with Aluminum Compounds Leading
to Si–O–Al Bonding.

$$Si_sOH + AlR_3 = Si_sOAlR_2 + RH$$
$$Si_sOH + Al(i\text{-Bu})_3 = Si_sOAl(i\text{-Bu})_2 + C_4H_9$$
$$Si_sOH + AlBr_3 = Si_sOAlBr_2 + HBr$$
$$Si_sOH + AlCl_3 = Si_sOAlCl_2 + HCl$$
$$2Si_sOH + AlCl_3 = (Si_sO)_2AlCl + 2HCl$$

Source: Iler [1], Kochlschütter and Bögel [115], Peglar
et al. [116], Fink *et al.* [117], Peri and Hensley [118],
Kol'tsov *et al.* [119].

In attaching organometallic complexes to the silica surface,

$$n\text{Si–OH} + MR_m \rightarrow (SiO)_m\text{–}MR_{m-n} + nRH, \qquad (28)$$

when $M = Ti, Zr, Nb, Hf, V$, or Cr, the surface catalyzes olefin polymerization; when $M = Ni, Cr, Ti, Zr$, or Mo, the surface catalyzes diene polymerization; and when $M = Mo$ or W, the surface catalyzes olefin metathesis. Subsequent oxidation or reduction of the attached complexes can be used to form hydrogenation, oxidation, or supported metal catalysts according to the following scheme. When $M = Cr, Mo, W$, hydrogenation catalysts are obtained according to

$$(SiO)_n\text{–}MR_{m-n} + \frac{(m-n)}{2}H_2 \rightarrow (SiO)_nM + (m-n)RH. \qquad (29)$$

When $M = Mo$ or W, oxidation catalysts are obtained from

$$(SiO)_n\text{–}MR_{m-n} + O_2 \rightarrow SiO_nMO_1 + \text{oxidation products of } R, \qquad (30)$$

and when $M = Ni, Pd$ or Pt, supported metallic catalysts are obtained from

$$(SiO)_n\text{–}MR_{m-n} + \frac{m}{2}H_2 \rightarrow n\text{Si–OH} + M^0 + (m-n)RH. \qquad (31)$$

Thus, a wide range of surface chemistries are possible. These classes of reactions are not limited to catalyst formation, however; they also represent potential routes for composite (diphasic) structures, optical and electronic materials, and sensors. This is obviously a promising area for future research. Additional information concerning these types of reactions is found in refs. [113–120].

6.

CONSEQUENCES OF SURFACE MORPHOLOGY

Many of the topics discussed in the previous sections rely on reactions between mutually physisorbed or chemisorbed surface species, or reactions between a diffusing gas phase species and physisorbed or chemisorbed surface species. We expect that the kinetics of such reactions will depend on the details of the geometry of the surface. In the first case, the reactant species are adsorbed on the surface of the porous solids; thus their locations are dictated by the underlying geometry, and reaction kinetics may depend upon diffusion on the irregular surface. In the second case, one of the reactant species must diffuse within the pore network, and thus the diffusion coefficient depends on scattering from the pore walls, again introducing a geometrical contribution. Geometrical effects are in general of importance in catalytic reactions, chemisorption, and surface derivatization.

De Gennes [121] considered the problem of "anomalous" diffusion on fractal networks in an attempt to understand the conductivity threshold of a percolation cluster. In normal diffusion the mean-square displacement, δ, is related to the diffusion coefficient, D, according to:

$$\langle\delta\rangle^2 = Dt \tag{32}$$

where the time interval t is related to the number of steps taken by the diffusing species. In anomalous diffusion, time is related to distance by

$$t \propto \delta^{\psi+2}. \tag{33}$$

ψ is an exponent characterizing the anomalous part of diffusion in the following sense: $\psi = 0$ for normal random walks, but $\psi > 0$ for random walks constrained to a fractal. By analogy with the problems of random walk and self-avoiding random walk, the fractal dimension of the path of the walker is given by the exponent in the relationship between steps taken and displacement [122]. Thus the path of a species performing a random walk on the fractal surface has the factal dimension, $d_w = \psi + 2$. For a percolation cluster ($d_f = 2.5$), d_w has been estimated to have the value of 3.5 in three-dimensions [122]. The corresponding diffusion distance R in time t is given by $R \propto t^{1/(\psi+2)}$ compared to $t^{1/2}$ for normal diffusion. Thus the kinetics of diffusion-limited reactions on irregular surfaces characterized by a mass or surface fractal dimension (see, e.g., Figs. 13 and 14 in Chapter 9) are expected to be retarded compared to reactions carried out on regular surfaces.

Several groups have studied electron energy transfer (ET) between adsorbed donors (DO), e.g., rhodamine 6G, and adsorbed or diffusing acceptors (AC), e.g., malachite green in an attempt to understand how surface–molecule interactions are influenced by the underlying geometry [123–128]. The idea is to utilize the effect that environmental geometry has on the rate law of nonradiative one-step ET from excited DO molecules to AC molecules supported by or diffusing within the porous matrix. The geometry is characterized by a fractal dimension (fractal geometry) or pore size distribution, mean pore size, and tortuosity (Euclidean geometry).

Pines and coworkers [123–128] argue that when the procedure for embedding the material with DO and AC is adsorption, the ET experiment probes the structural properties of the surface of the material as seen by the adsorbent–adsorbate interactions. Based on this assumption they have analyzed the ET process for several porous and nonporous silicas using a single parameter fit (d_s) to the Klafter–Blumen [129] equation describing the survivability of DO in the presence of AC. Their results for three DO–AC pairs adsorbed on porous silica gels with pore diameters ranging from 60 to 200 Å give d_s values between 2.3 and 2.8, whereas results on fumed silica give $d_s = 2.08$. The d_s values reflect the spatial distribution of the acceptors around the donors. Pines *et al.* take this interpretation one step further and suggest that d_s also reflects the geometry of the underlying silica support as seen by molecular–material interactions. Based on this interpretation they conclude that the surfaces of the silicas are fractal in the range from the molecule size up to the critical (Förster) radius over which ET can occur. A trend of increasing d_s with increasing surface area for the porous silicas is observed that agrees qualitatively with results by Hurd and coworkers [130]. The d_s value obtained for fumed silica produced by oxidation of $SiCl_4$ above 2000°C, 2.08, is arguably different from the value, $d_s = 2$, that is indicative of a smooth (uniform) surface.

This fractal interpretation has been questioned by Drake and coworkers [131,132], who performed ET experiments on two homologous series of silica gels morphologically similar to each other on the scale of the pore size R_p. One series was composed of randomly packed monodisperse particles with partially sintered interfaces. The other series was similar but exhibited a hierarchical morphology. Both BET measurements and small-angle-scattering indicated that the gel surfaces were essentially smooth on length scales greater than ~1.0 nm. (BET area = geometric area, and Porod slope, P = −4.) ET experiments performed with adsorbed DO and AC molecules were interpreted on the basis of Euclidean geometry by the relationship between R_p and a critical radius R_{max} which defines the length scale over which ET is sensed. When $2R_p/R_{max} \gg 1$, ET probes length scales less than R_p. In this case an effective d_s of about 2 is observed indicative of a smooth surface. When $2R_p/R_{max} \ll 1$, ET probes length scales well above R_p and is sensitive to the pore network. In this case, the local surface appears space filling with an effective d_s near 3. These results show that for this series of gels, the ET experiment can be explained adequately in terms of classical pore models in which the dynamics of the donor relaxation are dominated by R_p.

A second experiment performed by Drake and coworkers [131] involved adsorbed donors and diffusing acceptors: the decay of excited triplet benzophenone by O_2. Even for Euclidean materials, pore morphology is expected to play a major role in the Knudsen regime [133] where the mean free path due to gas-phase molecule–molecule scattering is larger than the mean pore size, R_p, so that the diffusing AC molecules collide more often with the pore walls than with each other. Drake *et al.* observed that the decay rate exhibited a linear dependence on R_p over almost a decade of mean pore sizes (34–286 Å). This scaling behavior supports an interpretation based on classical Knudsen diffusion where R_p emerges as the relevant geometrical length. Thus the results of Drake and coworkers cast some doubt as to the uniqueness of the fractal interpretation. Drake *et al.* emphasize that the applicability of the fractal concept to porous surfaces must be tested in each case and confronted with experimental facts.

In summary, we should expect that reactions carried out on the surfaces of gels will often be affected by the underlying surface geometry. Whether the geometrical effects are properly interpreted on the basis of Euclidean or fractal geometry apparently depends on the gel-synthesis procedure, which is known to affect the surface irregularity. To date, effects of surface geometry for reactions of interest in gel processing (dehydroxylation, rehydration, ammonolysis, etc.) remain virtually unexplored.

7.

SUMMARY

This chapter has shown that the chemistry and structure of xerogels is dominated by their high surface areas. High surface area is problematic when preparing ultrahigh-purity glasses for fiber optics, since the surface is terminated with hydroxyl groups (4.9 OH/nm²) that must be removed prior to sintering. A typical high surface area xerogel (BET surface area = 900 m²/g) may contain over 6 wt% H_2O! Surfaces may be dehydroxylated by thermal or chemical means (or both). Thermal dehydroxylation is aided by high sintering temperatures and large pore sizes. Chemical dehydroxylation of the silica surface with halogens appears to be quantitative, but it replaces OH with halogen atoms that may cause bloating or dealkalization. Due to the high strength of the Si–F bond, fluorine treatments minimize the bloating problem. Thermally dehydrated surfaces contain large concentrations of strained rings, mainly cyclic trisiloxanes, that are significantly more reactive toward dissociative chemisorption than bulk vitreous silica.

High surface area is advantageous from the standpoint of chemical modification, since high surface area implies accessibility of reactant species to the solid phase. Reactions with ammonia to form oxynitride glasses have been discussed in some detail. Other important reactions used to chemically modify surfaces include those with metal halides, metal alkyls, metal alkoxides, or silylating reagents. Due to the irregular (fractal) geometry of xerogel surfaces, we expect that the kinetics of diffusion-limited reactions will be retarded compared to smooth (nonfractal) surfaces.

REFERENCES

1. R.K. Iler, *The Chemistry of Silica* (Wiley, New York, 1979).
2. H. Knözinger and P. Ratonasanny, *Catal. Rev.-Sci. Eng.*, **17** (1978) 31–70.
3. P.P. Mardilovich, V.M. Zelenkovskii, G.N. Lysenko, A.I. Trokhimets, and G.M. Zhidomirov, *React. Kinet. Catal. Lett.*, **36** (1988) 107–112.
4. G.N. Lysenko, P.P. Mardilovich, and A.I. Trokhimets, *Zh. Prikl. Spektrosk*, **43** (1985) 110–114.
5. Y. Oosawa and M. Graetzel, *J. Chem. Soc. Chem. Commun.* (1984) 1629–1630.
6. S.S. D'yakonov, V.I. Lygin, B.Z. Shalumov, Z.G. Khlopova, and N.A. Shimicheva, *Kolloidn, Zh.*, **47** (1985) 146–150.
7. J.C. Lavalley and M. Benaissa, *J. Chem. Soc. Commun.* (1984) 908–909.
8. F. Cavani, G. Centi, E. Foresti, F. Trifiro, and G. Busca, *J. Chem. Soc. Faraday Trans. I*, **84** (1988) 237–254.
9. W.C. Conner, E.L. Weist, and L.A. Pederson, *Stud. Surf. Sci. Catal.*, **31** (Prep. Catal. 4) (1987) 323–332.
10. P. Burtin, J.P. Brunelle, M. Pijolat, and M. Soustelle, *Appl. Catal.*, **34** (1987) 239–254.
11. J.D. Mackenzie, I.W.M. Brown, C.M. Cardile, and R.H. Meinhold, *J. Mat. Sci.*, **22** (1987) 2645–2654.
12. S. Brunauer, P.H. Emmett, and E. Teller, *J. Am. Chem. Soc.*, **60** (1938) 309.
13. D. Farin and D. Avnir, *J. Chromatog.*, **406** (1987) 317–324.
14. D.M. Smith, private communication.
15. L.D. Belyakova, O.M. Dzhigit, A.V. Kiselev, G.G. Muttk, and K.D. Scherbakova, *Russ. J. Phys. Chem. (Eng. trans.)*, **33** (1959) 551.
16. D.P. Gallegos, D.M. Smith, and C.J. Brinker., *J. of Colloid and Interface Science*, **124** (1988) 186–198.
17. A.V. Kiselev, *Kolloidn. Zh.*, **2** (1936) 17.
18. P.C. Carman, *Trans. Faraday Soc.*, **36** (1940) 964.
19. A.V. Kiselev, *Trans. Faraday Soc. Disc.*, **52** (1971) 14.
20. D. Barby, "Silicas" in *Characterization of Powder Surfaces*, eds. G.D. Parfitt and K.S.W. Sing (Academic Press, New York, 1976), chapter 8, p. 353.
21. C. Okkerse, "Porous Silica" in *Phys. and Chem. Aspects of Adsorbents and Catalysts*, ed. B.G. Linsen (Academic Press, New York, London, 1970), chapter 5, p. 214.
22. M.L. Hair, *Infrared Spectroscopy in Surface Chemistry* (Dekker, New York, 1967), p. 79.

23. H.P. Boehm, *Angew. Chem.*, **5** (1966) 533; *Adv. Catal.*, **16** (1966) 226.
24. F.A. Cotton and G. Wilkinson, *Adv. Inorg. Chem.* (Wiley Interscience, New York, 1972), p. 113.
25. C.A. Fyfe, G.C. Gobbi, and G.J. Kennedy, *J. Phys. Chem.*, **89** (1985) 277–281.
26. D.W. Sindorf and G.E. Maciel, *J. Am. Chem. Soc.*, **105** (1983) 1487–1493.
27. W.G. Klemperer, V.V. Mainz, and D.M. Millar in *Better Ceramics Through Chemistry II*, eds. C.J. Brinker, D.E. Clark, and D.R. Ulrich (Mat. Res. Soc., Pittsburgh, Pa., 1986), pp. 15–26.
28. C.J. Brinker, R.J. Kirkpatrick, D.R. Tallant, B.C. Bunker, and B. Montez, *J. Non-Cryst. Solids*, **99** (1988) 418–428.
29. T.A. Michalske and B.C. Bunker, *J. Appl. Phys.*, **56** (1984) 2686.
30. B.C. Bunker, D.M. Haaland, T.A. Michalske, and W.L. Smith, *Surf. Sci.*, **222** (1989) 95–118.
31. P. Hoffman and E. Knozinger, *Surf. Sci.*, **188** (1987) 181.
32. I. Shapiro and H.G. Weiss, *J. Phys. Chem.*, **57** (1953) 219.
33. W.K. Lowen and E.C. Broge, *J. Phys. Chem.*, **65** (1961) 16.
34. R.K. Iler, personal observation.
35. J. Erkelens and B.G. Linsen, *J. Colloid Interface Sci.*, **29** (1969) 464.
36. V. Ya. Davydov, A.V. Kiselev, and L.T. Zhuravlev, *Trans. Faraday Soc.*, **60** (1964) 2254.
37. L.T. Zhuravlev, *Langmuir*, **3** (1987) 316–318.
38. L.T. Zhuravlev, R.L. Gorelik, and L.T. Zhuravlev in *Experimental Methods in Adsorption and Molecular Chromatography*, eds. A.V. Kiselev and V.P. Dreving (MGU, Moscow, 1973), p. 250.
39. L.T. Zhuravlev, R.L. Gorelik, and L.T. Zhuravlev in *Experimental Methods in Adsorption and Molecular Chromatography*, eds. A.V. Kiselev and V.P. Dreving (MGU, Moscow, 1973), p. 266.
40. C.G. Armistead, A.J. Tyler, F.H. Hambleton, S.A. Mitchell, and J.A. Hockey, *J. Phys. Chem.*, **73** (1969) 3947.
41. J.B. Peri and A.L. Hensley, *J. Phys. Chem.*, **72** (1968) 2936.
42. D.L. Wood and E.M. Rabinovich, *J. Non-Cryst. Solids*, **82** (1986) 171–176.
43. W. Hertl and M.L. Hair, *Nature*, **223** (1969) 1151.
44. K. Klier and A.C. Zettlemoyer, *J. Colloid Interface Sci.*, **58** (1977) 216.
45. W. Stöber, G. Bauer, and K. Thomas, *Ann. Chem.*, **604** (1957) 104; *Kolloid Z.*, **149** (1956) 39.
46. K. Nassau, E.M. Rabinovich, A.E. Miller, and P.K. Gallagher, *J. Non-Cryst. Solids*, **82** (1986) 78–85.
47. J.A.G. Taylor, J.A. Hockey, and B.A. Pethica, *Proc. Brit. Ceram. Soc.*, **5** (1965) 133.
48. J. Uytterhoeven, M. Sleex *et al.*, *Bull. Soc. Chim. Fr.* (1965) 1800.
49. J.J. Fripiat, J. Uytterhoeven *et al.*, *J. Phys. Chem.*, **66** (1962) 800; *Bull. Soc. Chim. Fr.* (1965) 1800.
50. L.T. Zhuravlev and A.V. Kiselev, *Russ. J. Phys. Chem.*, **39** (1965) 236.
51. J.A.G. Taylor and J.A. Hockey, *J. Phys. Chem.*, **70** (1966) 2169.
52. I. Matsuyama, K. Susa, S. Satoh, and T. Suganuma, *Ceramic Bull.*, **63** (1984) 1408–1411.
53. V.A. Dzis'ko, A.A. Vishnevskaya, and V.S. Chesalova, *Zh. Fiz. Khim.*, **24** (1950) 1416.
54. J.D.F. Ramsay and R.G. Avery, *Bri. Ceram. Proc.*, **38** (1986) 275–283.
55. D.M. Krol, C.A.M. Mulder, and J.G. van Lierop, *J. Non-Cryst. Solids*, submitted.
56. E.M. Rabinovich, D.W. Johnson, Jr., J.B. MacChesney, and E.M. Vogel, *J. Am. Ceram. Soc.*, **66** (1983) 683.
57. T.A. Gallo, C.J. Brinker, L.C. Klein, and G.W. Scherer in *Better Ceramics Through Chemistry*, eds. C.J. Brinker, D.E. Clark, and D.R. Ulrich (Elsevier, New York, 1984), pp. 85–90.

58. C.J. Brinker and D.M. Haaland, unpublished.
59. J.P. Williams, Y.-S. Su, W.R. Strzegowski, B.L. Butler, H.L. Hoover, and V.O. Altemose, *Am. Ceram. Soc. Bull.*, **55** (1976) 524–527.
60. A.J. Downs and C.J. Adams in *Comprehensive Inorganic Chemistry*, **vol. 2**, eds. J.C. Bailar, Jr., H.J. Emeleus, R. Nyholm, and A.F. Trotman-Dickenson (Pergamon, Oxford, England, 1973), p. 1107.
61. V.A. Tertykh, V.M. Mashchenko, A.A. Chuiko, and V.V. Pavlov, *Fiz. Khim. Mekh. Liofil'nost Dispersnykh Sist.*, **4** (1973) 37.
62. T.H. Elmer, I.D. Chapman, and M.E. Nordberg, *J. Phys. Chem.*, **67** (1963) 2219.
63. G.D. Chukin, *Zh. Prikl. Spektrosh*, **21** (1974) 879 [*Chem. Abstr.*, **82**, 64779m].
64. J. Uytterhoeven and H. Naveau, *Bull. Soc. Chim. Fr.* (1962) 27.
65. M.L. Hair and W. Hertl, *J. Phys. Chem.*, **77** (1973) 2070.
66. K. Susa, I. Matsuyama, S. Satoh, and T. Suganuma, *J. Non-Cryst. Solids*, **79** (1986) 165–176.
67. E.M. Rabinovich, D.L. Wood, D.W. Johnson, Jr., D.A. Fleming, S.M. Vincent, and J.B. MacChesney, *J. Non-Cryst. Solids*, **82** (1986) 42–49.
68. C.A.M. Mulder, *J. Non-Cryst. Solids*, **95 and 96** (1987) 303–310.
69. D.M. Krol and E.M. Rabinovich, *J. Non-Cryst. Solids*, **82** (1986) 143–147.
70. B.A. Morrow and I.A. Cody, *J. Phys. Chem.*, **80** (1976) 1995.
71. B.A. Morrow and I.A. Cody, *J. Phys. Chem.*, **80** (1976) 1998.
72. B.A. Morrow, I.A. Cody, and L.S.M. Lee, *J. Phys. Chem.*, **80** (1976) 2761.
73. B.C. Bunker, D.M. Haaland, K.J. Ward, T.A. Michalske, J.S. Binkley, C.F. Melius, and C.A. Balfe, *Surface Sci.*, **210** (1989) 406–428.
74. C.J. Brinker, B.C. Bunker, D.R. Tallant, and K.J. Ward, *J. de Chimie Physique*, **83** (1986) 851.
75. J.H. DeBoer and J.M. Vleeskins, *Proc. K. Ned. Akad. Wet. Ser. B: Palaentol, Geol. Phys. Chem.*, **b61** (1958) 85–93.
76. J.A. Hockey, *Chem. Ind. (London)* (1965) 57–63.
77. R.K. Iler, *Colloid Chemistry of Silica and Silicates* (Cornell Univ. Press, Ithaca, New York, 1955).
78. R. Oestrieke, W.H. Yang, R.J. Kirkpatrick, R.L. Hervig, A. Navrotsky, and B. Montez, *Geochim. Cosmochim. Acta.*, **51** (1987) 2199.
79. C.J. Brinker, R.K. Brow, D.R. Tallant, and R.J. Kirkpatrick, *Mat. Res. Soc. Symp. Proc.*, **105** (1988) 289–294.
80. C.J. Brinker, R.K. Brown, D.R. Tallant, and R.J. Kirkpatrick, *J. Non-Cryst. Solids*, in press.
81. T. Kudo and S. Nagase, *J. Am. Chem. Soc.*, **107** (1985) 2589.
82. C.J. Brinker, D.R. Tallant, E.P. Roth, and C.S. Ashley, *J. Non-Cryst. Solids*, **82** (1986) 117.
83. C.J. Brinker, D.R. Tallant, E.P. Roth, and C.S. Ashley in *Defects in Glasses*, eds. F.L. Galeener, D.L. Griscom, and M.J. Weber (Mat. Res. Soc., Pittsburgh, Pa., 1986), pp. 387–411.
84. F.L. Galeener in *The Structure of Non-Crystalline Materials 1982*, eds. P.H. Gaskell, J.M. Parker, and E.A. Davis (Taylor & Francis, London, 1983).
85. M. O'Keeffe and G.V. Gibbs, *J. Chem. Phys.*, **81** (1984) 876.
86. G.J. Young, *J. Colloid Sci.*, **13** (1958) 67.
87. A.V. Volkov, A.V. Kiselev, and V.I. Lygin, *Zh. Fiz. Khim.*, **48** (1974) 1214 [*Chem. Abstr.*, **81** 82656e].
88. C.A. Balfe, K.J. Ward, D.R. Tallant, and S.L. Martinez in *Better Ceramics Through Chemistry II*, eds. C.J. Brinker, D.E. Clark, and D.R. Ulrich (Mat. Res. Soc., Pittsburgh, Pa., 1986), pp. 619–626.

89. G.S. Wirth and J.M. Gieskes, *J. Colloid Interface Sci.*, **68** (1979) 492.
90. A.J. Moulson and J.P. Roberts, *Trans. Farad. Soc.*, **57** (1961) 1208.
91. G.V. Gibbs, private communication.
92. R.N. Nucho and A. Madhukar, *Phys. Rev.*, **B21** (1980) 1576.
93. G. Hollinger, E. Berginat, H. Chermette, F. Himpsel, D. Lohez, M. Lannoo, and M. Bensoussan, *Phil. Mag.*, **B55** (1987) 735.
94. P. DeMayo, L.V. Natarasan, and W.R. Ware, *Am. Chem. Soc. Symp. Series*, **278** (1989) 1.
95. R.E. Loehman, *Glass IV*, eds. M. Tomozawa and R. Doremus (Academic Press, New York, 1983).
96. C.J. Brinker and D.M. Haaland, *J. Am. Ceram. Soc.*, **66** (1983) 758-765.
97. H.O. Mulfinger, *J. Am. Ceram. Soc.*, **49** [9] (1966) 462-467.
98. B.A. Morrow, I.A. Cody, and L.S.M. Lee, *J. Phys. Chem.*, **79** [22] (1975) 2405-2408.
99. V. Bazant, V. Chvalovsky, and J. Rathousky, *Organosilicon Compounds* (Academic Press, New York, 1965), pp. 85-86.
100. D.C. Bradley, R.C. Mehrotra, and D.P. Gaur, *Metal Alkoxides* (Academic Press, London, 1978), p. 231.
101. D.M. Haaland and C.J. Brinker in *Better Ceramics Through Chemistry*, eds. C.J. Brinker, D.E. Clark, and D.R. Ulrich (Elsevier, New York, 1984), pp. 267-274.
102. M.J.D. Low and N. Ramasubramanian, *J. Phys. Chem.*, **71** [9] (1967) 3077.
103. R.K. Brow and C.G. Pantano in *Better Ceramics Through Chemistry*, eds. C.J. Brinker, D.E. Clark, and D.R. Ulrich (Elsevier, New York, 1984), pp. 361-368.
104. R.K. Brow and C.G. Pantano, *J. Am. Ceram. Soc.*, **70** (1987) 9-14.
105. C.G. Pantano, R.K. Brow, and L.A. Carmen in *Sol-Gel Technology*, ed. L.C. Klein (Noyes, Park Ridge, NJ, 1988), pp. 110-138.
106. K. Kamiya, M. Ohya, and T. Yoko, *J. Non-Cryst. Solids*, **83** (1986) 208-222.
107. E.P. Plueddemann in *Silanes, Surfaces, and Interfaces*, ed. D.E. Leyden (Gordon and Breach, Amsterdam, 1986), pp. 1-24.
108. M.L. Hair in *Silanes, Surfaces, and Interfaces*, ed. D.E. Leyden (Gordon and Breach, Amsterdam, 1986), pp. 25-42.
109. M.L. Hair and W. Hertl, *J. Phys. Chem.*, **73** (1969) 2372; **75** (1971) 2181.
110. *Silanes, Surfaces, and Interfaces*, ed. D.E. Leyden (Gordon and Breach, Amsterdam, 1986).
111. *Silylated Surfaces*, eds. D.E. Leyden and W.T. Collins (Gordon and Breach, New York, 1980).
112. E.P. Plueddemann, *Silane Coupling Agents* (Pleunum, New York, 1982).
113. Yu. I. Yermakov, B.N. Kuznetsov, and V.A. Zakharov, *Catalysis by Supported Complexes* (Elsevier, Amsterdam, 1981), p. 13.
114. M. Liefländer and W. Stöber, *Z. Naturforsch.*, **15** (1960) 411.
115. H.W. Kochlschütter and U. Bögel, *Fortschrittsber. Kolloide Polym.*, **55** (1971) 29.
116. R.J. Peglar, F.H. Hambleton, and J.A. Hockey, *J. Catal.*, **20** (1971) 309.
117. P. Fink, G. Pforr, and B. Rackow, *Z. Chem.*, **10** (1970) 424; **12** (1972) 451.
118. J.B. Peri and A.L. Hensley, *J. Phys. Chem.*, **72** (1968) 2936.
119. S.I. Kol'tsov, V.B. Kopylov, A.N. Volkova, and V.B. Aleskovskii, *Izv. Vyssh. Uchebn. Zaved. Khim. Khim. Tekhnol.*, **16** (1973) 1475.
120. Yu. I. Yermakov, *Catal. Rev.-Sci. Eng.*, **13** (1976) 77-120.
121. P.G. de Gennes, *Recherche*, **7** (1976) 919.
122. S.H. Lin in *Solid State Physics*, vol. 39, eds. H.E. Ehrenreich and D. Turnbull (Academic Press, 1986), pp. 207-273.
123. D. Pines, D. Huppert, and D. Avnir, *J. Chem. Phys.*, **89** [2] (1988) 1177-1180.

124. U. Even, K. Rademan, J. Jortner, N. Manor, and R. Resifeld, *Phys. Rev. Lett.*, **52** (1984) 2164; **58** (1987) 285.

125. D. Rojanski, D. Huppert, H.D. Bale, X. Dacai, P.W. Schmidt, D. Farin, A. Seri-Levy, and D. Avnir, *Phys. Rev. Lett.*, **56** (1986) 2505.

126. A. Takami and N. Mataga, *J. Phys. Chem.*, **91** (1987) 618.

127. Y. Lin, M.C. Nelson, and D.M. Hanson, *J. Chem. Phys.*, **86** (1987) 158.

128. D. Pines-Rojanski, D. Huppert, and D. Avnir, *Chem. Phys. Lett.*, **139** (1987) 109.

129. J. Klafter and A. Blumen, *J. Chem. Phys.*, **80** (1984) 879.

130. A.J. Hurd, D.W. Schaefer, and J.E. Martin, *Phys. Rev. A.*, **35** (1987) 2361.

131. J.M. Drake, P. Levitz, S.K. Sinha, and J. Klafter, *Chem. Phys. Lett.*, in press.

132. P. Levitz, and J.M. Drake, *Phys. Rev. Lett.*, **58** (1987) 686.

133. E.L. Cussler in *Diffusion: Mass Transfer in Fluid Systems* (Cambridge University Press, 1984).

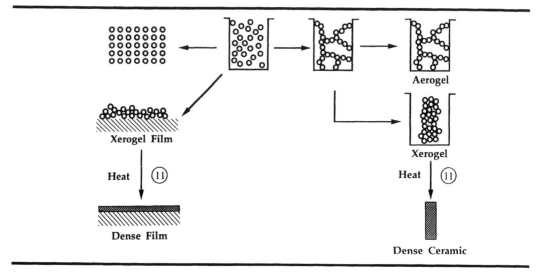

Xerogel Film

Heat | ⑪

Dense Film

Aerogel

Xerogel

Heat | ⑪

Dense Ceramic

CHAPTER 11

Sintering

Sintering is a process of densification driven by interfacial energy. Material moves by viscous flow or diffusion in such a way as to eliminate porosity and thereby reduce the solid–vapor interfacial area. In gels, that area is enormous, so the driving force is great enough to produce sintering at exceptionally low temperatures, where the transport processes are relatively slow. In this chapter, we review the theories of sintering for amorphous and crystalline materials, and we compare the predicted and experimentally observed behavior. The kinetics of densification of gels are complicated by the concurrent processes of dehydroxylation and structural relaxation. This leads to the remarkable result that faster heating permits complete densification at a lower temperature. For crystalline gels there are the further complications of grain growth and phase transformations. A general conclusion that emerges from the data is that it is advantageous to complete sintering before crystallization of the gel. This is particularly true if the gel is the matrix of a composite.

1. *Theories of Viscous Sintering* describes the concept of energy balance by which Frenkel first analyzed sintering, and it reviews the microstructural models that have been developed. Effects of a distribution of pore sizes and of gas trapped in the pores are considered.

2. *Experimental Studies of Viscous Sintering* shows that conventional materials obey the theories closely, but gels exhibit time-dependent viscosity, because of continual loss of water and structural rearrangement. This makes it difficult to extract useful information from studies of densification during constant heating.

3. *Theory of Sintering of Crystalline Materials* describes models based on transport by diffusion. The low initial density of gels and the strong tendency toward grain growth make it difficult to sinter to full density.

4. *Experimental Studies of Diffusive Sintering* presents the limited data available on the densification kinetics of crystalline gels. A given composition is shown to sinter much more slowly if it is allowed to crystallize first.

5. *Competition between Sintering and Crystallization* compares the kinetics of the processes and shows how time–temperature transformation curves can be used to establish optimal sintering schedules.

6. *Composites* briefly examines the problem of densifying a composite in which the matrix is a gel.

1.

THEORIES OF VISCOUS SINTERING

Amorphous materials sinter by viscous flow and crystalline materials sinter by diffusion, so the paths along which material moves, and the relationship between the rate of transport and the driving force, are quite different [2]. Analysis of viscous sintering is relatively simple in principle, but exact treatments are prevented by the complex geometry of the porous body. Fortunately, simple approximations, while not strictly realistic, yield satisfactory results.[†]

1.1. Physical Principles

1.1.1. ENERGY BALANCE

Viscous sintering is driven by the energy gained by reduction in surface area of the porous body. Given a microstructural model, it is possible to relate the change in surface area to the overall change in dimensions (i.e., the strain). The energy gained when this strain occurs is the product of the specific surface energy and the change in surface area. When a viscous body flows, energy is expended, and the rate of this dissipation of energy is proportional to the square of the strain rate. Frenkel [3] suggested that the rate of strain (or densification) could be found by equating the rate of change in the surface energy to the rate of energy dissipation. This insight is the basis for all analyses of viscous sintering, which differ only in the models they

---

[†]The material in Sections 1 and 2 is drawn in large part from ref. [1].

adopt to represent the geometry of the body. As an example of this type of analysis, consider a simple cylinder with radius a and length l; for the sake of simplicity, we ignore the end surfaces and assume that the body retains its cylindrical shape. The rate of dissipation of energy during flow is [4]

$$\dot{E}_f = V(\sigma_r \dot{\varepsilon}_r + \sigma_\theta \dot{\varepsilon}_\theta + \sigma_z \dot{\varepsilon}_z) \tag{1}$$

where V is the volume of the cylinder, and σ_r, σ_θ, σ_z, and $\dot{\varepsilon}_r$, $\dot{\varepsilon}_\theta$, $\dot{\varepsilon}_z$ are the stresses and strain rates in the radial, circumferential, and axial directions, respectively. (See Fig. 1.) In a solid cylinder [4], $\sigma_r = \sigma_\theta$ and $\dot{\varepsilon}_r = \dot{\varepsilon}_\theta$, so Eq. 1 becomes

$$\dot{E}_f = (2\sigma_r \dot{\varepsilon}_r + \sigma_z \dot{\varepsilon}_z)\pi a^2 l. \tag{2}$$

The rate of change of the volume of the cylinder is

$$dV/dt = V(\dot{\varepsilon}_r + \dot{\varepsilon}_\theta + \dot{\varepsilon}_z) \tag{3}$$

but the viscous cylinder is incompressible ($dV/dt = 0$), so Eq. 3 implies

$$\dot{\varepsilon}_r = \dot{\varepsilon}_\theta = -\dot{\varepsilon}_z/2. \tag{4}$$

The axial strain rate is related to the stresses by

$$3\eta\dot{\varepsilon}_z = \sigma_z - \frac{\sigma_r + \sigma_\theta}{2} = \sigma_z - \sigma_r \tag{5}$$

Fig. 1.

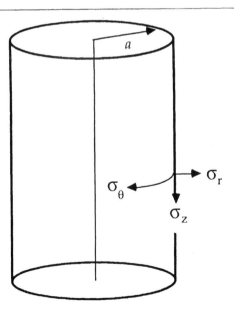

Stresses and strains in cylinder.

where η is the viscosity. Equations 4 and 5 can be used to substitute for $\dot{\varepsilon}_r$ and σ_r in Eq. 2 with the result

$$\dot{E}_f = (3\eta\dot{\varepsilon}_z^2)\pi a^2 l. \tag{6}$$

The surface area of the cylinder is $S = 2\pi a l$, so the rate of change of S is

$$\frac{dS}{dt} = 2\pi a l\left(\frac{1}{a}\frac{da}{dt} + \frac{1}{l}\frac{dl}{dt}\right) = 2\pi a l(\dot{\varepsilon}_r + \dot{\varepsilon}_z) \tag{7}$$

and the rate of energy gained by the change in surface area is

$$\dot{E}_s = \gamma_{sv}\, dS/dt = \gamma_{sv}\pi a l\dot{\varepsilon}_z \tag{8}$$

where γ_{sv} is the solid–vapor interfacial energy. According to Frenkel's postulate, these energies must balance:

$$\dot{E}_f + \dot{E}_s = 0. \tag{9}$$

Using Eqs. 6 and 8 in Eq. 9 leads to a prediction of the axial strain rate of the viscous cylinder:

$$\dot{\varepsilon}_z = -\gamma_{sv}/3\eta a. \tag{10}$$

This means that the cylinder will contract axially at a rate that is directly proportional to the interfacial energy and inversely proportional to the viscosity and the radius. The same result could have been obtained from a conventional stress analysis, using Laplace's equation for the radial stress ($\sigma_r = -\gamma_{sv}/a$) and recognizing that there is an axial membrane stress of $\sigma_z = -2\gamma_{sv}/a$. The energy method is easier for some more complicated geometries.

This general method of analysis has been applied to several geometrical models, which are discussed in the next subsection. In every case, however, the same dimensionless group appears: $\tau = \gamma_{sv}t/d\eta$, where d is a characteristic dimension of the structure. Densification is complete when $\tau \approx 1$ or $t \approx d\eta/\gamma_{sv}$. The sintering rate is therefore faster for bodies with smaller particles or pores. In gels, the pore diameters are typically on the order of 2 to 10 nm, so they densify orders of magnitude faster than bodies made using coventionally crushed powders.

1.1.2. VISCOUS FLOW

The variable over which one has the greatest control is the viscosity, because of its strong dependence on temperature. For a silicate material, η changes by an order of magnitude for every 20–40°C change in temperature, with the temperature dependence being steepest in the vicinity of the glass transition temperature, T_g. This typical behavior is shown in Fig. 2a for a soda-lime silicate glass, whose viscosity is well described by an equation of the Vogel–Fulcher form:

$$\eta = \eta_0 \exp[A/(T - T_0)] \tag{11}$$

where A, T_0, and η_0 are constants and T is temperature; η_0 has units of Pascal·seconds (Pa·s) or poises (1 Pa·s = 10 poises). Over a narrow range of temperature (~ 50–$100°C$), η is adequately represented by the Arrhenius equation,

$$\eta = \eta_0 \exp(Q/RT) \tag{12}$$

where Q is an activation energy and R is the ideal gas constant. The viscosity

Fig. 2a.

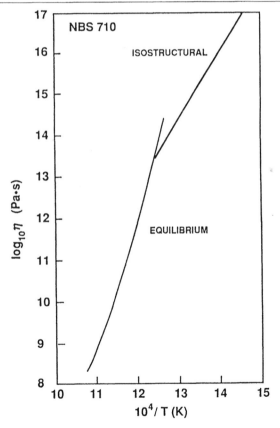

Viscosity of NBS710 soda-lime silicate glass, based on equilibrium data of Napolitano *et al.* [5] and isostructural data of Mazurin *et al.* [6]; from Scherer [7]. Along the isostructural curve the atomic configuration is fixed (see discussion on following pages).

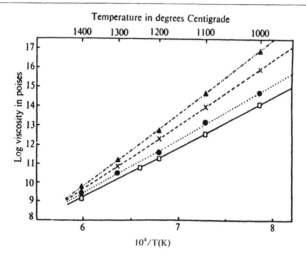

Viscosity of silica as function of temperature for glasses with hydroxyl contents (wt%) of 0.0003 (▲), 0.027 (×), 0.04 (●), and 0.12 (□); from Hetherington *et al.* [8]. Activation energy Q is proportional to slopes of lines.

of silica is anomalous in that it obeys the Arrhenius equation all the way from the liquidus to T_g. As shown in Fig. 2b, the activation energy decreases as the hydroxyl content increases; for silica, $120 \leq Q \,(\text{kcal/mol}) \leq 170$. The reduction in viscosity and Q results from the creation of nonbridging bonds as *silanol* (Si–OH) groups replace *siloxane* (Si–O–Si) bonds. This effect is seen in multicomponent glasses (e.g., [9]), as well as in silica. Additions of alkali or halogens reduce the viscosity in the same way.

The glass transition is in the temperature range where the viscosity is $\sim\!10^{12}\text{–}10^{13}$ Pa·s. As a liquid is cooled through this temperature range, the movement of the atoms becomes so slow that the structure of the liquid drifts out of equilibrium. If the cooling is stopped in the glass transition range, the properties of the liquid can be observed to change with time as the structure approaches its equilibrium configuration. This process is called *structural relaxation* [10]. Figure 2a shows the *isostructural viscosity* measured below T_g by Mazurin *et al.* [6]: the liquid is equilibrated near $10^{13.5}$ Pa·s, then quickly cooled; the viscosity is measured before the structure has time to change significantly. The viscosity is *lower* than its equilibrium value, because the structure is frozen into a configuration with high *free volume*,[†]

[†] Free volume roughly means the volume available to an atom in excess of the volume required for thermal vibrations. The mobility of an atom generally increases with its free volume. Although this parameter alone is not adequate to describe the state of the glass, it provides a conceptually appealing rationalization for the results of quenching.

characteristic of the higher temperature from which it was quenched. The *fictive temperature*, T_f, is generally understood[†] to be the temperature where the current structure would be in equilibrium. In Mazurin's experiment, $T_f \approx 522°C$, where the equilibrium viscosity is $10^{13.5}$ Pa·s. After the sample is quenched to a lower temperature, its structure relaxes toward equilibrium and its viscosity rises with time. An extreme example of structural relaxation was provided by Hara and Suetoshi [11], who quenched a hot glass into a salt bath at 320°C and produced a fictive temperature ~65°C higher than T_g. The density of the glass at room temperature was ~0.4% lower than that of a slowly cooled glass; if the quenched glass were *annealed* (equilibrated) near T_g, its viscosity would rise by a factor of ~25. Gels may be regarded as equivalent to severely quenched glasses, since they have very low skeletal densities (~10–25% less than the corresponding melted glass). In fact, Puyané *et al.* [12] found that the glass transition behavior of a gel was just like that of a quenched melt of the same composition. Therefore we may expect the viscosity of a gel to rise during an isothermal hold near T_g as excess free volume is eliminated from the oxide network. The kinetics of densification indicate that this does indeed occur.

1.1.3. SURFACE ENERGY

The specific surface energy (γ_{sv}) of glass is weakly dependent on both temperature and composition (see Fig. 3), with the lowest values being ~0.06 J/m² for P_2O_5 and ~0.08 J/m² for B_2O_3 [14], and typical values being ~0.3 J/m² (e.g., for SiO_2 [14] and soda-lime silicate glass [15]). The surface energy of silica does change significantly (~2×) with the concentration of hydroxyl groups [16], but the loss of OH occurs upon heating over a wide range of temperature. Over the same range the viscosity changes by many orders of magnitude; since the sintering rate is proportional to γ_{sv}/η, practically speaking, the temperature dependence of the sintering rate is governed entirely by $\eta(T)$.

1.2. Comparison of Models

In Frenkel's classic paper [3], he presented an analysis of the coalescence of a pair of spheres, which is representative of the sintering of a body of packed powder. As shown in Fig. 4 (from an excellent monograph on sintering by

[†]Strictly speaking, one cannot always assign a fictive temperature according to this definition, because the structure produced by an arbitrary thermal cycle may not correspond to equilibrium at *any* temperature. Therefore, the more rigorous definition is that T_f is the temperature from which one would have to quench in order to obtain the observed property [10].

Fig. 3.

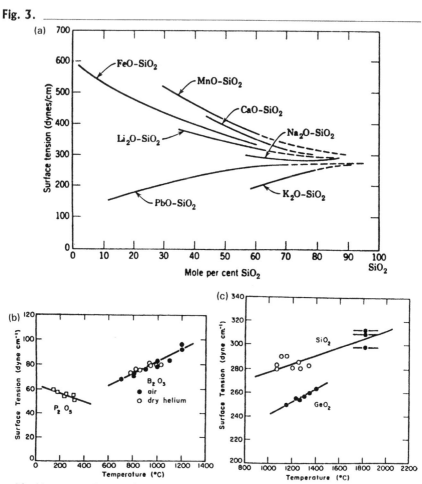

Liquid–vapor surface tension γ_{sv} versus composition in some silicate systems (a) and versus temperature for P_2O_5 and B_2O_3 (b) and SiO_2 and GeO_2 (c). Fig. 3a from Kingery *et al.* [13]; Figs. 3b and 3c from Kingery [14], reprinted by permission of the American Ceramic Society.

Exner [17]), the centers of the spheres approach one another as the neck between them widens; the change in distance between the centers of the two spheres is assumed to equal the linear contraction of a compact of such particles. This geometry is not as simple as it may seem, because the shape of the neck between the particles changes considerably by the time the centers move significantly. The mathematical description of the shape of the neck is therefore difficult, and is handled with severe simplifying assumptions. In addition, the flow pattern in the spheres is expected [18] to resemble that

Fig. 4.

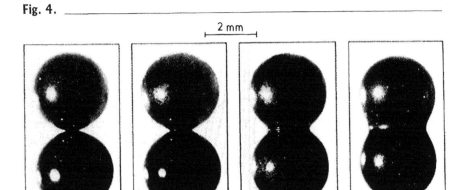

2 mm

2 min	9 min	14 min	30 min
x/a = 0.16	x/a = 0.28	x/a = 0.53	x/a = 0.75

Two-particle model made of glass spheres (3 mm diameter) sintered at 1000°C. x and a are, respectively, the radii of the neck (denoted by z in the text) and the sphere. From Exner [17].

on the left side of Fig. 5, but the analysis assumes the simpler pattern shown on the right side of that figure. With these assumptions, Frenkel[†] obtained a simple relationship for the growth in radius z, of the neck between spheres of radius a:

$$(z/a)^2 = 3\gamma_{sv}t/8\eta a. \tag{13}$$

This result has been verified experimentally in several studies [19,20]. When the neck is relatively small, there is a simple geometrical relationship between z and the distance between centers of the spheres, which is used to obtain an expression for the change in linear dimension L of the sintering body [21]:

$$\frac{L(t)}{L(0)} = 1 - \frac{3\gamma_{sv}t}{8\eta a}. \tag{14}$$

This equation is widely used, because it is simple and it works rather well. However, the assumptions employed in the derivation limit its applicability to the first few percent of linear shrinkage. Frenkel's model thus describes the initial stage of sintering, during which there is considerable neck growth but little densification. The success of Eq. 14 in describing much larger shrinkages has been attributed [21] to offsetting errors contributed by the assumed shape of the interparticle neck and neglect of particle rearrangement. Indeed, it has been shown [21,22,23] that there is significant particle

[†] Frenkel's original solution contained an erroneous factor of π.

Fig. 5.

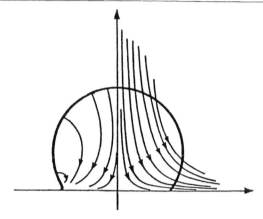

Flow fields for viscous sintering of spheres (schematic). Right-hand side, uniaxial contraction assumed in Frenkel's model. Left-hand side, form expected in real situations. From Uhlmann *et al.* [18]. *Note added in proof*: Recent simulations using finite element analysis (A. Jagota and P.R. Dawson, to be published in *J. Am. Ceram. Soc.*) show that the flow field on the right is, indeed, correct.

rearrangement during densification of two-dimensional arrays, principally because of nonuniform packing [22]. It is not clear how important that process is in three-dimensional bodies, where movement of the particles is more constrained.

Mackenzie and Shuttleworth (MS) analyzed [24] the shrinkage rate of a spherical shell according to Frenkel's method. The shell, shown in Fig. 6, can be used to represent the densification of a body containing spherical pores. The dimensions of the shell are chosen so that the central void occupies the same volume fraction as the pores in the sintering body. This is a much more elegant treatment than the analysis of the coalescence of spheres, because the shell remains spherical as it shrinks. Exact expressions can be written for the change in surface area and the energy dissipated in viscous flow as the shell contracts. The result is

$$\int_0^t \left(\frac{\gamma_{sv} n^{1/3}}{\eta}\right) dt = \frac{2}{3}\left(\frac{3}{4\pi}\right)^{1/3} \int_{\rho_0}^{\rho} \frac{d\rho}{(1-\rho)^{2/3}\rho^{1/3}} \tag{15}$$

or

$$\frac{\gamma_{sv} n^{1/3} t}{\eta} = f_{MS}(\rho) - f_{MS}(\rho_0) \tag{16}$$

where

$$f_{MS}(\rho) = \frac{2}{3}\left(\frac{3}{4\pi}\right)^{1/3}\left[\frac{1}{2}\ln\left(\frac{1+\rho^3}{(1+\rho)^3}\right) - \sqrt{3}\tan^{-1}\left(\frac{2\rho-1}{\sqrt{3}}\right)\right]. \tag{17}$$

Fig. 6. _____

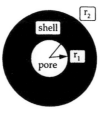

The Mackenzie–Shuttleworth model describes the shrinkage of a spherical shell with inner and outer radii of r_1 and r_2, respectively. ρ_s is the skeletal density (i.e., density of solid phase). V = volume of solid phase = $4\pi(r_2^3 - r_1^3)/3$; ρ_b = bulk density = $\rho_s V/(4\pi r_2^3/3)$; ρ = relative density = $\rho/\rho_s = 1 - (r_1/r_2)^3$; $n = 1/V$, $n^{1/3} = 0.62(1/\rho - 1)/r_1$.

Figure 7 shows the predicted change in relative density, $\rho = \rho_b/\rho_s$, with time, where ρ_b is the bulk density of the porous body and ρ_s is the density of the solid phase (i.e., the skeletal density); in Eq. 15, ρ_0 (generally called the *green density*) is the initial value of ρ. The time scale uses the dimensionless variable (*reduced time*) $\gamma_{sv} n^{1/3} t/\eta$, where n is the number of pores per unit volume of solid (see Fig. 6). For a given relative density, the larger n, the smaller the pores; therefore, n is related to the reciprocal of the pore size. This solution is exact, and applies ideally to the final stage of sintering, when the pores can be modelled as isolated spherical voids. Note that Eq. 16 applies when sintering is performed isothermally. If the temperature and, consequently, the viscosity, change with time, then the integral on the left side of Eq. 15 must be retained.

Mackenzie and Shuttleworth [24] extended their analysis to include the effects of applied pressure or gas trapped in the pores. This was done by including another term in the energy balance representing the mechanical work done by the pressure. They considered the case of an insoluble gas and of a gas that dissolves into the solid phase at such a rate that the pressure in the bubble remains constant. If the gas is insoluble, the bubble shrinks until the pressure of the gas equals $P_c = 2\gamma_{sv}/r$, where P_c is the capillary pressure driving sintering and r is the radius of the bubble. If, at the moment when the pore becomes a closed sphere, the radius of the bubble is r_0 and the pressure of the gas is p_0, then the equilibrium size of the bubble will be [24] $r_{eqm}^2 = p_0 r_0^3/2\gamma_{sv}$. This phenomenon can prevent sintered bodies from reaching full density, so it is advisable to sinter in a vacuum or an atmosphere of soluble gas; for example, for silica it is best to sinter in a helium atmosphere. In some cases, gas is evolved by the sintering body (e.g., water may be released from a gel upon heating), and this may cause the pores to expand, instead of collapse. Phallipou *et al.* [25] found this to be a

Fig. 7.

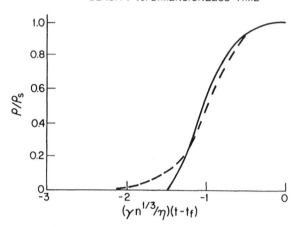

DENSITY vs. DIMENSIONLESS TIME

Relative density (ρ/ρ_s) versus reduced time for Mackenzie–Shuttleworth model (solid curve) and cylinder model (dashed curve). For $\rho/\rho_s > 0.942$, the MS model applies. The curves have been shifted to coincide at the time t_f when sintering is complete. From Scherer [28].

problem in silica aerogels: samples containing high amounts of hydroxyl (~ 3000 ppm OH) shrank first, then expanded (after the pores closed and the evolving gas could not escape), whereas samples with less hydroxyl (~ 500 ppm OH) would sinter to full density.

If pressure is applied to the outside of the sintering body, instead of inside the pore, the shrinkage rate is increased. Murray et al. [26] showed that when the applied pressure, P_A, is very large, as in hot pressing, the capillary pressure, P_c, can be ignored, and the MS solution is simplified considerably:

$$\ln[1 - \rho(t)] = \ln(1 - \rho_0) + 3P_A t/4\eta. \tag{18}$$

According to Eq. 18, the relative density approaches $\rho = 1$ asymptotically, whereas the complete MS analysis shows that densification is completed in a finite time. The reason for the discrepancy is that P_c becomes infinite when the radius of the pore approaches zero. Therefore the capillary pressure cannot be ignored and Eq. 18 does not apply at the very end of densification.

The structure shown in Fig. 8 (hereafter called the *cylinder model*) has been proposed [28] as a model for the intermediate stage of sintering, during which the pores are interconnected (open to the atmosphere). The structure is represented by a unit cell with edge length l. The relative density is

$$\rho = x^2(3\pi - 8\sqrt{2}x) \tag{19}$$

Fig. 8.

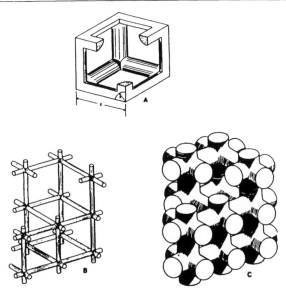

Microstructural model used to represent gels in sintering analysis. (A) Unit cell of structure with edge length l and cylinder radius a. (B) Microstructure with relative density of ~ 0.1. (C) Microstructure with relative density of ~ 0.5. From Brinker and Scherer [27].

where $x = a/l$ and a is the radius of each cylinder. The inverse of Eq. 19 is [29]

$$x = \left(\frac{\pi \sqrt{2}}{8}\right)\left[\cos\left(\theta + \frac{4\pi}{3}\right) + \frac{1}{2}\right] \tag{20}$$

where

$$\theta = \frac{1}{3}\cos^{-1}\left[1 - \left(\frac{4}{\pi}\right)^3 \rho\right]. \tag{21}$$

The volume of solid phase in the unit cell is $l^3\rho$, and each cell contains one pore, so the number of pores per unit volume of solid is

$$n = 1/l_0^3\rho_0 \tag{22}$$

where l_0 and ρ_0, the initial values of the cell size and relative density, are used, since the volume of solid does not change. The analysis of the sintering kinetics of the cylinder model is quite similar to that presented in Section 1.1. The result is [28]

$$\int_0^t \left(\frac{\gamma_{SV} n^{1/3}}{\eta}\right) dt = \int_{x_0}^x \frac{2dx}{(3\pi - 8\sqrt{2}x)^{1/3}x^{2/3}} \tag{23}$$

or

$$\frac{\gamma_{SV} n^{1/3} t}{\eta} = f_S(y) - f_S(y_0) \tag{24}$$

where

$$f_S(y) = -\left(\frac{2}{\alpha}\right)\left[\frac{1}{2}\ln\left(\frac{\alpha^2 - \alpha y + y^2}{(\alpha + y)^2}\right) + \sqrt{3}\tan^{-1}\left(\frac{2y - \alpha}{\alpha\sqrt{3}}\right)\right] \tag{25}$$

and

$$y = \left(\frac{3\pi}{x} - 8\sqrt{2}\right)^{1/3}. \tag{26}$$

In Eqs. 23 and 24, x_0 and y_0 are initial values; in Eq. 25, $\alpha \equiv (8\sqrt{2})^{1/3}$. Equation 24 is plotted as the dashed curve in Fig. 7. In view of the profound difference in the structure, the densification curve is surprisingly close to that found by Mackenzie and Shuttleworth [24]. As the cylinder model densifies, the cylinders become shorter and thicker and, when the relative density reaches $\rho \sim 0.94$, neighboring cylinders touch. At that point, the unit cell contains an isolated pore, so the remainder of the densification is described by the MS model.[†]

It is enlightening to compare the densification kinetics of the cylinder model with the predictions of Frenkel's equation, Eq. 14. Figure 9 shows unit cells with relative densities of 0.26 and 0.52 composed of spheres. If the unit cell is regarded as containing one pore, then the densification of these structures can be calculated using Frenkel's equation and the results can be plotted on the same dimensionless time scale as the cylinder model [30], as in Fig. 10. Over the range of shrinkage ($<5\%$) for which Frenkel's equation applies (and beyond!), the densification kinetics are identical to those of the cylinder model. Evidently the geometric details of the model have a minor effect on the calculated sintering rate; it is only necessary to provide an accurate measure of the mean pore size or number of pores per unit volume of solid. The practical implication of the similarity of these curves is that the cylinder model can be applied to predict densification without regard to the microstructural details of the material under study. A body composed of loosely packed spheres will densify initially according to Frenkel's equation. After the radii of the necks have grown to approach the radii of the spheres, the microstructure will resemble the cylinder model. Eventually, the pores will become isolated, and the MS model will apply. Thanks to the similarity of the theoretical curves, the entire densification (up to $\rho \approx 0.94$) can be adequately described using the cylinder model. It should be recognized,

[†]Strictly speaking, the pore formed by contact of the cylinders would not be spherical, although it would rapidly approach that shape. At any rate, the cylindrical geometry is merely a convenient approximation and does not deserve to be taken too seriously.

Fig. 9. _____

PORES/PARTICLE

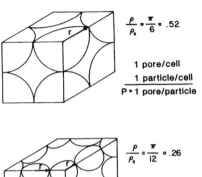

$\dfrac{\rho}{\rho_s} = \dfrac{\pi}{6} = .52$

1 pore/cell
1 particle/cell
P = 1 pore/particle

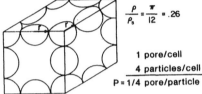

$\dfrac{\rho}{\rho_s} = \dfrac{\pi}{12} = .26$

1 pore/cell
4 particles/cell
P = 1/4 pore/particle

Unit cells of microstructures of monosized spheres with relative densities of 0.52 and 0.26. From Scherer [30], reprinted by permission of the American Ceramic Society.

Fig. 10. _____

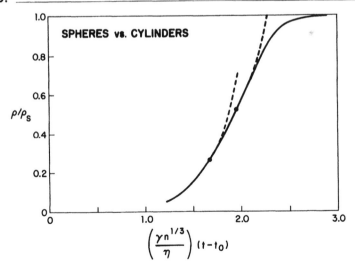

Relative density versus reduced time for Frenkel's model (dashed curve) and cylinder model (solid curve) for microstructures shown in Fig. 9. (Cylinder model used up to $\rho = 0.94$, MS model thereafter). From Scherer [30], reprinted by permission of the American Ceramic Society.

however, that this happy situation does not apply to the surface area of the
body, which evolves at different rates according to these models. Consider
the structures in Fig. 8c and Fig. 9, both of which have $\rho_0 \simeq 0.5$. Slight
coalescence of the spheres produces a large change in surface area, but a small
change in relative density; the same densification of the cylinders would
require a much smaller change in area (but less transport of matter and hence
less dissipation of energy). It is coincidental that $\rho(t)$ is so similar for these
two models. To predict the evolution of surface area, more realistic models
are required; it is necessary to take account of the distribution of pore sizes,
because the smaller pores dominate the surface area, and the larger pores
control the final shrinkage rate.

The cylinder model was first extended to describe the behavior of a body
containing two different pore sizes [30] and then to allow for an arbitrary
distribution of pore sizes [31]. Since the larger pores tend to shrink more
slowly, they are subjected to compressive stresses by the smaller pores. Conversely, the shrinkage of the smaller pores is inhibited by the larger pores,
so the former experience tensile stresses. This interaction affects the densification rate of both sets of pores, as shown in Fig. 11, where V_L is the volume
fraction of large pores at the outset; the pore diameters in this calculation
differ by a factor of 4. When $V_L = 0$, there are no large pores, and the
densification represents the free contraction rate of the small pores; the curve
for $V_L = 1$ illustrates the densification rate of the large pores alone. When
$0 < V_L < 1$, the model [30] predicts that the initial densification rate is rapid

Fig. 11. _____

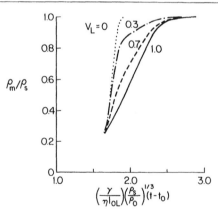

Relative density versus reduced time for bimodal pore size distribution. (ρ_m = bulk
density of matrix with both pore sizes; ρ_s = theoretical density of glass; V_L = volume
fraction of large cells; l_{0L} = cell size corresponding to large pore.) From Scherer, *Journal
of the American Ceramic Society*, **67** [11] (1984) 709–715. Reprinted by permission of
the American Ceramic Society.

until the small pores have collapsed; at that point, the large pores remain open and are separated by dense glass where the small pores had been. As the large pores shrink, they must drag along this extra glass, so their densification rate is lower than it would have been if the small pores had never been present. Consequently, the last part of the curve has a small slope, and the time required to reach full density is predicted to be almost independent of V_L.

The evolution of the pore size distribution in a sintering body is illustrated in Fig. 12. These calculations are based on the cylinder model [31], assuming that the pore volume is initially distributed among pores with diameters (d) from 0.01 to 1 (arbitrary units). In case II, equal volumes are assigned to groups of pores that are equally spaced in $\log d$, so the cumulative pore volume plot (Fig. 12a) is linear. In case I, the pore volume increases linearly from zero for the smallest pores to a maximum for the largest pores; in case III, the pore volume varies in the opposite direction, so that most of the volume is occupied by small pores. As the density increases from $\rho_0 = 0.5$, the average pore diameter ($\bar{d}$) increases as the smallest pores close first (Fig. 12c). The change is so small for case I that the measured value of $\bar{d}$ would appear to be nearly constant for $0.5 \leq \rho \leq 0.9$; this reflects a balance between the loss of the smallest pores (which are few) and the shrinkage of the larger pores. This sort of distribution could account for Iler's observation [16] that the pore size of silica gel does not change during sintering. As the proportion of small pores increases (I → III), the apparent increase in $\bar{d}$ is greater. Note that *all* of the pores shrink in this simulation, but the average size (as reflected in BET or Hg penetration data) increases because the smaller pores disappear faster. This sometimes leads investigators to the erroneous conclusion that pore growth is occurring during sintering. If growth does occur, the peak of the differential curve (Fig. 12b) will not only shift toward larger sizes, but successive curves will cross as the volume of the larger pores increases. Pore coarsening would be expected during hydrothermal treatment (Chapter 6) or low-temperature treatment when surface diffusion is predominant (see Section 3.1), but not during sintering of gels at ambient pressure in the vicinity of T_g. For example, Fig. 13 shows the evolution of the pore size distribution during sintering of a series of silica xerogels, indicating that pores of all sizes shrink.

The cylinder model has also been applied to the sintering of a film on a rigid substrate [33]. As shown in Fig. 14, the substrate retards densification of the film by preventing contraction in the plane. The film can only contract normal to the substrate. For example, the model predicts that a free body with an initial density of $\rho_0 = 0.5$ will require about one unit of reduced time to reach full density, whereas a film of the same material would require ~1.3 units to reach $\rho = 1$.

Fig. 12.

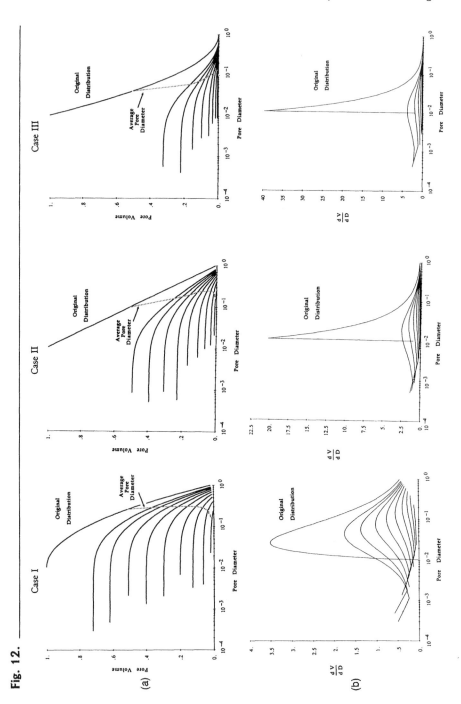

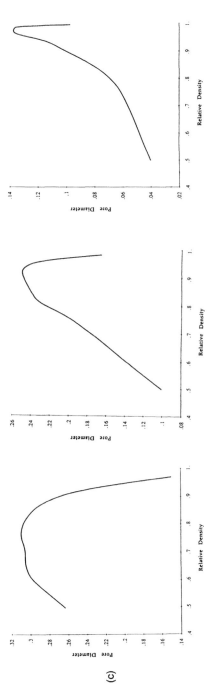

Calculated variation in pore size distribution during sintering, assuming pore diameters (*d*) uniformly spaced in log *d*, with pore volumes initially equal (II) for each pore size, or linearly increasing toward large (I) or small (III) pores. The cumulative pore volume (a) decreases continuously during densification; successive curves show the distribution after decrements of 10% (by number) in the number of open pores, and the dashed curve traces the average pore diameter. (50 vol% of the open pores lie on each side of the average.) Derivative curves (b) indicate a drift toward larger sizes as the smallest pores close first. The average diameter (c) increases initially, then drops as the largest pores finally close. Analysis based on cylinder model, according to Scherer [31].

Fig. 13.

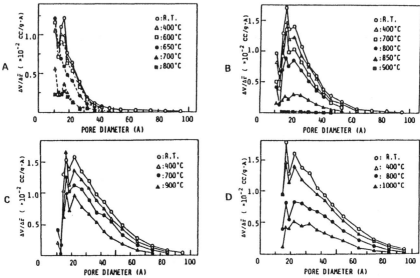

Change in pore size distribution in various silica xerogels with heat treatment: (A) gelled at 54°C; (B) 65°C; (C) 68°C; (D) 70°C. From Yamane and Okano [32].

Fig. 14.

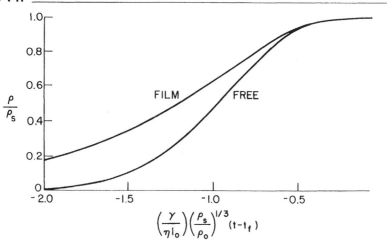

Relative density versus reduced time for film on rigid substrate and for unconstrained (free) material. Film sinters more slowly, because shrinkage cannot occur in the plane of the substrate. From Scherer and Garino [33], reprinted by permission of the American Ceramic Society.

1.3. Viscoelasticity

The analyses just discussed are appropriate for Newtonian viscous materials. However, as we shall see, gels sinter at temperatures near the glass transition region, so they are expected to exhibit viscoelastic behavior. In principle, the sintering of a viscoelastic material can be analyzed in a simple manner, by calculating the stresses and strains expected for an elastic material and then using the viscoelastic analogy [10]. However, there is generally no point in doing so. The viscoelastic response consists of three parts: an instantaneous elastic strain, a delayed elastic strain, and a viscous strain; the two elastic components are comparable in magnitude. Even in a gel, the capillary stress is less than 50 MPa and the modulus is ≥ 6 GPa [34], so the elastic strain is less than 1%. Since gels shrink by 20% (linearly) or more during sintering, the observed shrinkage is almost entirely caused by viscous flow, and the delayed elastic strain can be legitimately neglected.

2.

EXPERIMENTAL STUDIES OF VISCOUS SINTERING

This section describes studies of densification kinetics that illustrate the principles discussed in Section 1. A broad discussion of viscous sintering in the preparation of glass bodies is provided in a review by Rabinovich [35].

2.1. Particulate Gels

The cylinder model [28] was originally developed to describe the densification kinetics of optical waveguide preforms made by vapor deposition, and has been successfully applied to those materials [36–38]. Vapor phase processes are used in optical fiber fabrication because the resulting particles are pure and are small enough (<100 nm) to allow sintering at convenient temperatures. Instead of depositing the particles directly, it is possible to collect the particles without agglomeration and to disperse them in a liquid. The dispersion can be molded, gelled, dried, and sintered to produce glass of high optical quality. Scherer and Luong, [39] used Aerosol OX-50®, a commercial silica powder made by flame oxidation of $SiCl_4$ vapor, to produce silica glass of quality sufficient to make optical telecommunications fiber, as discussed in Chapter 4. The sintering kinetics of the dried gel are compared with the cylinder model as follows. Samples are treated isothermally for various periods of time, and their densities are measured. For each sample, the theoretical curve from Fig. 7 is used to find the reduced time, $\gamma_{sv} n^{1/3} t/\eta$, corresponding to its density. A plot is constructed of the reduced time versus the duration of heat treatment, as illustrated in Fig. 15. If the model applies, we should find straight lines with slopes related to the viscosity at the temperature of heat treatment, since the slope of the line is $\gamma_{sv} n^{1/3}/\eta$; the lines should have a common point of intersection corresponding to the initial density of the samples. For these gels, the model evidently works quite well. Using the known value for γ_{sv} and the measured value of the geometric parameter, the viscosity was calculated from the fit to the sintering data, and was found to have reasonable values and an appropriate activation energy (141 kcal/mol). These gels also contained a small quantity of large pores, so the sintering kinetics needed to be analyzed using the model for a bimodal pore size distribution [30], and this was done (unpublished work). The amount of large pores was small enough that the calculated curves for bimodal and monodisperse pores were barely distinguishable, and the resulting viscosities were negligibly different.

Similar gels were prepared by Rabinovich *et al.* [40,41] using Cab-O-Sil®, another flame-generated silica soot, in aqueous suspension. The small

Fig. 15. _____

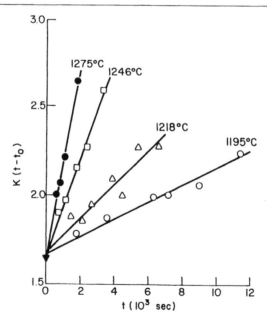

Reduced time versus elapsed time for sintering of particulate gel at indicated temperatures; $K = \gamma_{sv} n^{1/3}/\eta$. Linearity indicates agreement with cylinder model. From Scherer and Luong [39].

particle size, together with the high surface tension of water, caused these gels to crack upon drying. The pieces were pulverized in a blender, and the resulting agglomerates were redispersed and dried. No cracking occurred on the second drying, because of the relatively large pores between agglomerates. The sintering kinetics were studied by Johnson *et al.* [42], and found to deviate from the cylinder model as shown in Fig. 16. This most likely reflects the bimodal pore size distribution that results from the method of preparation: small pores exist between the primary particles and large pores between the aggregates. In fact the trend of the data in Fig. 16 is similar to the shape of the curve for $V_L = 0.3$ in Fig. 11. Orgaz and Corral [43] examined silica gels made in the same way and found very good agreement with the cylinder model for relative densities between 0.3 and 0.9. It may be that one pore size dominated the sintering behavior in their gels in that density range. They found some deviation from the model at short times (in a direction indicating a lower viscosity) which was attributed to the effect of OH.

Fig. 16.

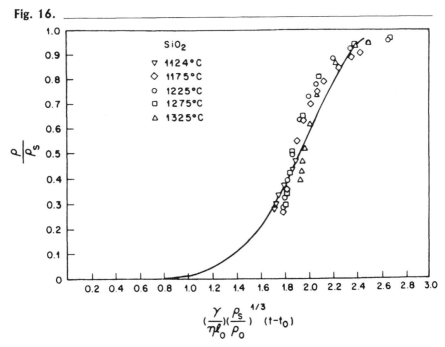

Relative density versus reduced time for particulate gels. Data (symbols) disagree with cylinder model (solid curve), possibly because gel contains bimodal pore size distribution and data are compared to model for unimodal pores. From Johnson *et al.* [42], reprinted by permission of the American Ceramic Society.

Shoup [44] prepared silica gels from solutions of potassium silicate that were seeded with Ludox®, a colloidal silica consisting of ~5 nm particles. The particles grow as silica from the solution deposits on them, until the particles impinge and the solution gels. The resulting dried gels have very narrow pore size distributions, and the size can be varied over a range from ~300 nm down to ~60 nm. Gels with smaller pores crack during drying, because of the large capillary pressure. The solid phase of these gels consists of dense silica, as revealed by the small weight loss on heating. The microstructure consists of an open network of particles with a relative density of $\rho \approx 0.25$, and, as shown in Fig. 16 of Chapter 4, looks rather like the cylinder model. The sintering kinetics of Shoup's gels are well described by that model [36].

Stöber et al. [45] showed that monodisperse particles of silica could be formed in solution by hydrolysis of tetraethyl orthosilicate (TEOS) under basic conditions, as discussed in Chapter 4. Shimohira et al. [46] found that such particles could be arranged in close-packed arrays by settling from solution (Fig. 17). Several studies [46–48] have shown that such arrays sinter much more rapidly than random packings of the same particles. The randomly packed particles may undergo rotation and rearrangement that can cause some pores to grow initially, instead of shrink. For the close-packed particles, rearrangement is unlikely; moreover, the large number of nearest neighbors (high coordination number) facilitates densification, since each particle grows ~12 necks. Sacks and Tseng [48] analyzed the sintering kinetics of such arrays using the MS and cylinder models, and found good agreement with both, as shown in Fig. 18. The adjustable parameter used to fit the data to the models was the viscosity. They found that the viscosity obtained from the fit was smaller than expected, but had a reasonable activation energy. The quality of the fit to the models shows that the viscosity did not change with time during sintering, as it might if the hydroxyl content of the particles changed significantly with time. Uniform packing of the particles was shown to have a profound effect on the rate of densification: ordered arrays reached full density in 24 h at 1000°C, while the relative density of random arrays increased from 0.45 to 0.58. Tseng and Yu [49] extended this work by examining the effect of atmosphere on the sintering rate of the ordered arrays of silica spheres. They found the rate at 1000°C to decrease in the order $H_2O + N_2 > air > N_2 > HCl + N_2$, with about one order of magnitude difference at each step. This remarkably large effect is attributed to the influence of water on the viscosity of silica. Tseng and Yu argue that HCl lowers the surface tension of the silica, but the effect seems too large to be explained in that way. More probably, the HCl removes hydroxyl from the silica; the chlorine must then escape (i.e., Cl cannot substitute for OH in the silica), since Cl and OH have a similar effect on viscosity.

Fig. 17.

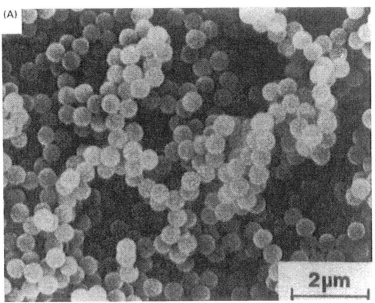

(A)

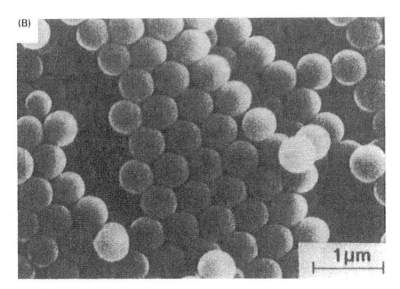

(B)

Scanning electron micrographs of compacts made from (A) flocculated suspension of silica spheres and (B) compact containing close-packed particles (sedimented from stable dispersion). From Sacks and Tseng [48].

Fig. 18.

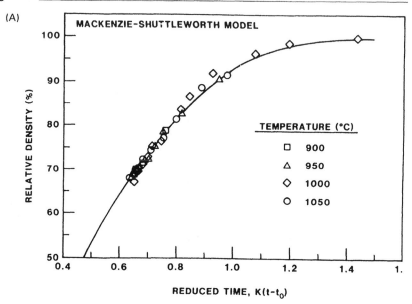

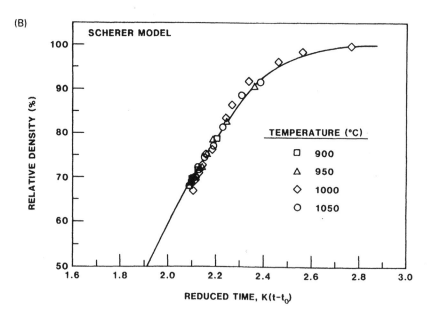

Relative density versus reduced time for sintering of close-packed spheres shown in Fig. 17B, using MS model (A) and cylinder model (B). From Sacks and Tseng [48], reprinted by permission of the American Ceramic Society.

2.2. Aerogels

As explained in Chapter 8, aerogels are dried under supercritical conditions where the capillary forces are absent, so very little shrinkage occurs as the solvent is removed. These gels therefore have very low green density ($0.03 \leq \rho_0 \leq 0.2$). Relatively few studies of the sintering kinetics of such materials have been performed.

Phalippou *et al.* [25] studied the densification behavior of silica aerogels and found that densification occurred rapidly at $\sim 1100°C$. The gels contained ~ 3000 ppm of hydroxyl after supercritical drying, which caused them to foam when heated to 1200°C. A chlorine treatment (discussed in Chapter 10) was used to reduce the OH content and thereby prevent foaming. The viscosity of silica increases as the OH content decreases (see Fig. 2b), so the temperature of rapid sintering was raised by the chlorine treatment. However, all of the gels densified at a viscosity of $\sim 10^{10}$ Pa·s. Woignier *et al.* [50] prepared aerogels in the $SiO_2-B_2O_3$, $SiO_2-P_2O_5$, and $SiO_2-B_2O_3-P_2O_5$ systems. The gels sintered in the same range of viscosity as silica aerogel; the phosphorus-containing systems tended to crystallize.

Onorato [51] studied the sintering kinetics of silica aerogels made from TMOS [52], and found good agreement with the cylinder model, as shown in Fig. 19. Acid-catalyzed gels sintered at lower temperatures, but bloated immediately after densifying; base-catalyzed gels could be sintered without bloating, because the large pore diameters (~ 30 nm) facilitated escape of OH. The densification rate was found to be atmosphere-dependent, because of the effect of atmospheric water on the viscosity of silica. However, the viscosity showed no time dependence, evidently because there is not much retained organic material or water in the solid phase of these gels.[†] Samples were also sintered during heating at a constant rate, and the data were fit to the model by substituting Eq. 12 into the left side of Eq. 23. The results were found to be in good agreement with the model, and Fig. 20 shows that the viscosity found from the isothermal experiments agreed well with that found using constant heating. Woignier *et al.* [53] have also found good agreement between the cylinder model and the isothermal densification kinetics of silica aerogels. In spite of the low initial densities of aerogels (≥ 0.03 g/cm^3) [54], they can be sintered to glass of high enough quality to make optical fibers with losses as low as 2.5 dB/km [55].

[†] This is not surprising, since the gels were subjected to a high temperature and pressure in the autoclave. The situation might be different for gels prepared by the low-temperature CO_2 process (see Chapter 8). Unfortunately, sintering kinetics have not been reported for such gels.

Fig. 19.

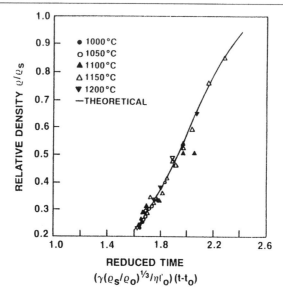

Relative density versus reduced time for silica aerogel analyzed using cylinder model. From Onorato [51].

2.3. Xerogels

The structure of xerogels is strongly dependent on the conditions prevailing during hydrolysis. Base-catalyzed silicates produce gels that are granular in texture and retain less organic material. Acid-catalysis leads to gels with a finer, denser structure that is not particulate, but consists of comparatively linear, lightly cross linked polymeric clusters. Even though acid-catalyzed gels are completely hydrolyzed in solution, the dried gels may contain a large number of chemically bound alkoxy groups, because of reesterification during the drying process. When the gels are heated, the alkoxy and hydroxyl groups are removed by condensation reactions that cause the large weight loss illustrated in Fig. 27b of Chapter 9. These reactions produce new cross-links and stiffen the structure. The instability of the structure is illustrated in Fig. 21, which shows that the alkoxide-derived silica gels exhibit a large thermal expansion coefficient ($\sim$1–4 $\times$ 10^{-5}°C^{-1}) below $\sim$300°C, although the expansion coefficient of vitreous silica is very small (5 $\times$ 10^{-7}°C^{-1}). As discussed in Chapter 9, the behavior of the gel results from the large number of nonbridging oxygens. If the gel is heated to 500°C, to allow the condensation reactions to occur, and then cooled, the thermal expansion coefficient of the gel will be the same as that of silica glass when reheated [57]. The

Fig. 20.

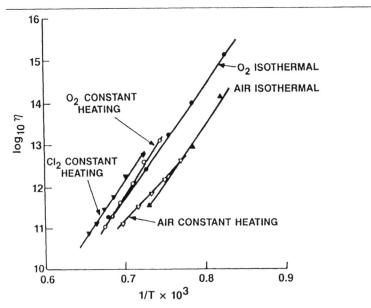

Temperature-dependence of viscosity (η) found by fitting sintering data for silica aerogel to cylinder model. Data obtained under isothermal conditions and at constant heating rate are in good agreement. Lower viscosity in air attributed to atmospheric moisture; chlorine reduces OH content and raises η. From Onorato [51].

particulate gels have only a small concentration of hydroxyl groups on the surfaces of the flame-generated particles, so the weight loss is small [39] and does not significantly affect the structure of the glass.

The condensation and pyrolysis reactions that occur during heating of a xerogel liberate a large volume of gas that can generate a high pressure, because of the low permeability of the small pores in the network. Consequently, xerogels may crack when heated between room temperature and ~400°C; typically damage is avoided with slow heating rates or by heating in a series of steps. Much less trouble is encountered in heating aerogels and particulate gels made from fumed silica, because the smaller volume of gas produced and the relative ease of flow through the larger pores.

As explained in Chapter 9, the density of the solid phase in an alkoxide-derived gel is lower than that of a melted glass of the same composition, even after the gel is heated to 500°C to promote the condensation reactions. The solid phase densifies completely only when heated near to the glass transition temperature, T_g. The densification near T_g is associated with a very small weight loss, so it is evidently caused by breaking and reforming bonds in the silicate network. This process is believed to be an extreme example of the

Fig. 21.

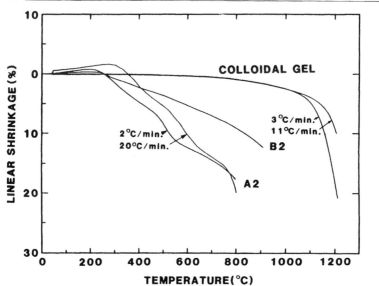

Linear thermal shrinkage at constant heating rates for particulate gel made using flame-generated particles (Colloidal Gel), base-catalyzed gel from TEOS (B2), and acid-catalyzed gel from TEOS (A2). (Reprinted by permission of publisher from Brinker *et al.* [56], copyright 1984, Elsevier Science Publ. Co., Inc.)

sort of structural relaxation that is observed in melted glass during annealing near T_g [10]. The density of the glass increases as the structure approaches that of the equilibrium liquid at the annealing temperature. The structural rearrangements (relaxation) that allow the solid phase to densify are expected to increase the viscosity, since much smaller density changes in melted glasses are associated with large increases in viscosity. For gels, viscous sintering begins in a temperature range in which structural relaxation, as well as some condensation reactions, may occur. This can cause the viscosity to vary as a function of time, even during isothermal sintering, as we shall see.

2.3.1. SILICA GELS

Nogami and Moriya [58] showed that the chemical and microstructural differences between acid- and base-catalyzed gels were reflected in their densification behavior. The initial shrinkage of the base-catalyzed gel could be described using Frenkel's equation, Eq. 14. The acid-catalyzed gels sintered at lower temperatures, and the kinetics of densification did not follow the same model. A more detailed comparison was performed by Brinker *et al.* [56] using the three gels described in Fig. 21. The isothermal sintering kinetics of the gels were analyzed using the cylinder model, but plots of reduced time versus elapsed time (as in Fig. 15) were not linear. The slope of such a plot is $K = \gamma_{sv} n^{1/3}/\eta$, so a change in η with time will cause a continuous change in slope. To investigate the change in viscosity, the slope was determined as a function of time. Figure 22 shows that $1/K$, a quantity proportional to viscosity, increases by as much as three orders of magnitude during an isothermal hold. This increase can only be attributed to a change in viscosity, because the surface energy is not subject to large variations, and there was neither crystallization nor growth in pore size. The increase in η is attributed to two factors: loss of hydroxyls and structural relaxation. Condensation reactions increase the cross link density, and structural relaxation increases the skeletal density toward the value characteristic of vitreous silica, creating a stiffer structure. These changes in viscosity were

Fig. 22.

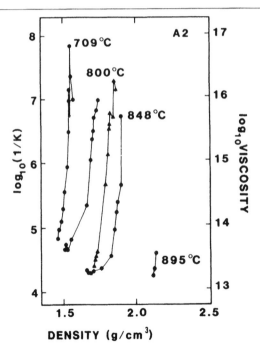

Isothermal plots of $1/K$ and viscosity for acid-catalyzed gel (A2) after heating at $2°C/min$ (●) or $20°C/min$ (▲) to indicated temperature; $K = \gamma_{sv} n^{1/3}/\eta$. (Reprinted by permission of publisher from Brinker *et al.* [56], copyright 1984, Elsevier Science Publ. Co., Inc.)

directly verified by Gallo and Klein [59], who compared the viscosity inferred
from sintering experiments with viscosities obtained by beam-bending
measurements on porous gels under identical conditions. The viscosity values
were found to have the same time-dependence and to agree in magnitude
within a factor of 2 to 4. This is reasonable agreement, considering the
difficulty of accurately measuring the bending of a rod that is also shrinking,
and especially the uncertainty in estimating the load-bearing fraction of the
cross-sectional area of the porous gel rod.

Note that the viscosity at the start of the isothermal hold is weakly
dependent on temperature (Fig. 22). The reason is that the hydroxyl content
decreases during heating to the temperature of the hold, so there is less
OH in the samples heated to higher temperatures, and this difference in
composition offsets the difference in temperature. The viscosity at the start
of sintering is $\sim 10^{12}$ Pa·s, which is characteristic of the glass transition
temperature. The small pores provide such a large driving force for densification that appreciable shrinkage is observable even at such enormous

Fig. 23. _____

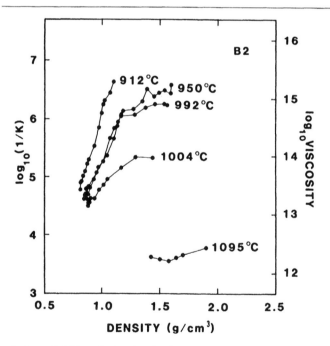

Isothermal plots of $1/K$ and viscosity for base-catalyzed gel (B2) after heating at
2°C/min (●) to indicated temperature; $K = \gamma_{sv} n^{1/3}/\eta$. (Reprinted by permission of
publisher from Brinker *et al.* [56], copyright 1984, Elsevier Science Publ. Co., Inc.)

viscosities. The increase in η arrests the sintering eventually, so that full density is not reached at the lower hold temperatures. Figure 23 shows that the base-catalyzed gels also exhibit an increase in viscosity during an isothermal hold, but η eventually reaches a plateau. The activation energy calculated from the plateau values is 155 kcal/mol, which is a typical value for vitreous silica. Evidently, the viscosity becomes constant because the skeleton of the gel has achieved the (metastable) equilibrium structure of silica. In the particulate gel, since the particles are dense silica, no structural changes occur during sintering, so, as shown in Fig. 24, the viscosity is constant within experimental error.

Gallo and Klein [60] showed that the effects of structural relaxation and OH content could be separated. Samples of gel were heated using two different schedules: direct heating to the hold temperature, or heating in several steps with 4 h holds at the intermediate temperatures. Following these treatments, gels were sintered at 750, 800, and 850°C, and the kinetics were analyzed using the cylinder model. Since the kinetics of dehydroxylation and structural relaxation are not identical, these treatments can produce gels with similar OH contents, but different degrees of relaxation. The opportunity for relaxation is considerable because the skeletal density of the gel is only 1.99 g/cm³ after heating to 550°C, and does not approach the density of vitreous silica (2.2 g/cm³) until heating above 800°C. The viscosity values obtained from the data are shown in Fig. 25. Higher viscosities are found for the samples subjected to stepwise heating. Since the OH contents of the samples overlap for each temperature, the differences must result from structural relaxation.

Fig. 24.

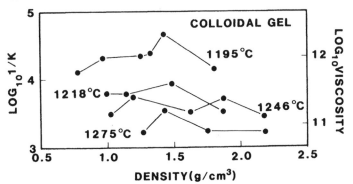

Isothermal plots of $1/K$ and viscosity for particulate gel made using flame-generated particles (same data as in Fig. 15); $K = \gamma_{sv} n^{1/3}/\eta$. (Reprinted by permission of publisher from Brinker *et al.* [56], copyright 1984, Elsevier Science Publ. Co., Inc.)

Fig. 25.

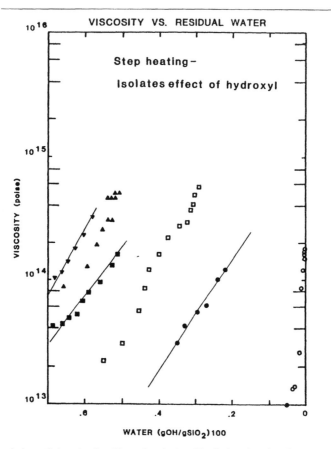

Isothermal plots of viscosity for silica gels calculated by fitting sintering data to cylinder model. Data joined by lines were heated in steps to (750 (▼), 800 (■), or 850°C (●), and unconnected data were heated continuously to 750 (▲), 800 (□), or 850°C (○). At a given OH content, different viscosities are obtained, because of different degrees of structural relaxation produced by the heat treatments. From Gallo and Klein [60].

2.3.2. MULTICOMPONENT GELS

Brinker and Scherer [61] studied the sintering kinetics of gels with the nominal composition (wt%): $71SiO_2$, $18B_2O_3$, $7Al_2O_3$, $4BaO$. The gels were made from TEOS, aluminum *sec*-butoxide, trimethyl borate, and barium acetate; batches were hydrolyzed under various conditions of pH and water concentration in order to produce a range of gel structures. (See Table 1.)

Table 1.

Matrix of Gelation Conditions and Measured Bulk
Densities (g/cm^3) of Desiccated Gels.

H$_2$O	pH = 2.5	pH = 6.8	pH = 9.2
5	LpH5	MpH5	—
	1.27	1.17	
8	LpH8	MpH8	HpH8
	1.28	1.28	0.75
12	LpH12	MpH12	HpH12
		1.15	0.79

Source: Brinker and Scherer [61].
Note: LpH, MpH, and HpH refer to low, medium, and
high pH, respectively. The numbers 5, 8, and 12 refer to
the water addition in moles of water/mole alkoxides.

The densification behavior of two of these gels is compared in Fig. 54 of Chapter 9, where the cross-hatched area indicates the change attributable to the increase in skeletal density. The skeleton changes more in gels prepared at low pH or with a small amount of water, because such gels have a lower cross link density. This structural difference is reflected in the sintering kinetics, as shown in Fig. 26. The sample prepared using lower pH and less water exhibits a much larger increase in η during densification, and reaches a plateau only at high temperatures or very long times. The apparent rise in η at the end of densification is attributed to the presence of air trapped in isolated pores. Satoh *et al.* [62,63] showed that such bubbles formed in their gels when the relative density exceeded ~0.6. The narrower the initial pore size distribution, the higher the density at which closed pores should form. The cylinder model [28] predicts that monodisperse pores become isolated when $\rho > 0.94$.

A more detailed study of a similar gel was reported by Scherer *et al.* [64]. Typical data, shown in Fig. 27, indicate that the sintering kinetics follow the cylinder model after a long period during which the viscosity increases. Data for several temperatures are summarized in Fig. 28. As for silica, the viscosity is similar (~10^{12} Pa·s) at the start of each hold, and rises eventually to a plateau. Figure 29 shows that the initial viscosities suggest a ridiculously low activation energy for viscous flow, whereas the plateau values give a reasonable value. As explained previously, the initial values apply to gels with different OH contents and different degrees of structural relaxation, so the activation energy is meaningless. Measurements of the weight loss and residual OH content of the gels indicated that the sintered gels all had

Fig. 26.

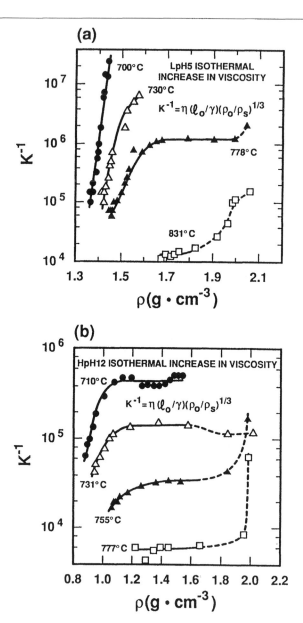

Isothermal plots of $1/K$ for gel LpH5 (a) and HpH12 (b) (see Table 1) for gelation conditions); $K = \gamma_{sv} n^{1/3}/\eta$. From Brinker and Scherer [61], pp. 43–59 in *Ultrastructure Processing of Ceramics, Glasses, and Composites*, eds. L.L. Hench and D.R. Ulrich (Wiley, New York, 1984).

Fig. 27.

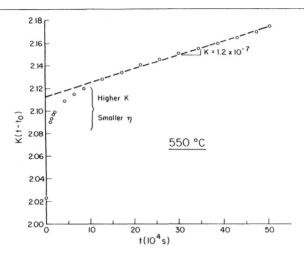

Plot of reduced time (from cylinder model) versus elapsed time at 550°C for multi-component gel; $K = \gamma_{sv} n^{1/3}/\eta$. Viscosity increases continuously during initial 10^5 s. From Scherer *et al.* [64].

similar OH contents, so the plateau values apply to essentially identical glasses, and the activation energy is valid. As an indirect test of the viscosity values found from the sintering experiments, the sintered glasses were examined in a differential scanning calorimeter (DSC). The location of the glass transition and the kinetics of structural relaxation are dependent on the viscosity of the glass [10]; the viscosity values inferred from the DSC data were found to be consistent with those found from sintering.

The DSC results [64] also permitted an estimate of the dependence of the viscosity on the OH content of the gel. It was found that structural relaxation must be responsible for a significant fraction of the increase in viscosity at low temperatures. Gallo *et al.* [65] used thermogravimetric analysis and infrared spectroscopy to establish the relative importance of structural relaxation and OH content on the viscosity changes during sintering of a similar gel. Their results are summarized in Fig. 30. Note that the sample sintered at 595°C shows an increase in η of an order of magnitude with no change in hydroxyl content; all of this loss is attributed to structural relaxation. At lower temperatures, both factors contribute to the rise in η. As in Fig. 25, the lines seem to converge at low residual OH content, which would mean that η is a weak function of temperature when the OH content is low. In fact, the opposite is true. (See Fig. 2b.) This anomaly reflects the influence of structural relaxation, which increases the slopes of the curves, and is more influential at higher temperatures (lower hydroxyl contents).

Fig. 28.

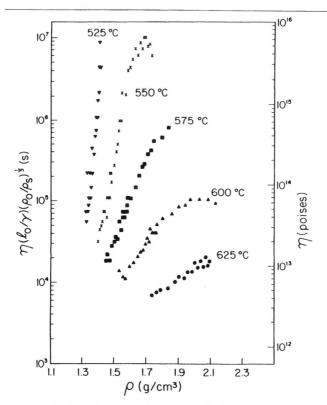

Isothermal plots of $1/K$ and viscosity (η) versus bulk density, obtained by analyzing sintering kinetics of multicomponent gel in terms of cylinder model; $K = \gamma_{sv} n^{1/3}/\eta$. From Scherer *et al.* [64].

2.3.3. SINTERING AT A CONSTANT HEATING RATE

A variety of models [66–69] have been proposed for analysis of densification data obtained while heating a sample at a constant rate, $q = dT/dt$. The starting point is the general equation for isothermal initial-stage sintering [69]:

$$\frac{d}{dt}\left(\frac{\Delta L}{L_0}\right) = \frac{K_0 \exp(-Q/RT)}{(\Delta L/L_0)^n} \tag{27}$$

where $\Delta L = L_0 - L(t)$, $L_0 = L(0)$, and $L(t)$ is the linear dimension of the sample at time t. The constants K_0 and n (not to be confused with the geometric parameter, n, introduced in Eq. 15) depend on the mechanism of

Fig. 29.

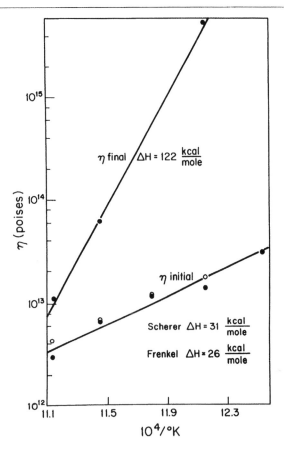

Viscosity versus reciprocal temperature from data in Fig. 28, using η at the start of isothermal hold (η_{initial}) and from plateaus at long times (η_{final}); η_{initial} found using both cylinder model (●) and Frenkel's model (○). From Scherer *et al.* [64].

densification. Equation 27 is identical to Eq. 14 if the viscosity is given by Eq. 12, $K_0 = 3\gamma_{\text{sv}}/8\eta_0 a$, and n = 0; that is, this model reduces to Frenkel's equation when n = 0. To predict the densification rate during heating, it is necessary to assume that the sintering rate depends only on the current value of the temperature and strain ($\Delta L/L_0$); then if K_0, n, Q, and q are constant, it is easy to integrate Eq. 27. Recognizing that $Q \gg RT$, the integral can be approximated by

$$\left(\frac{\Delta L}{L_0}\right)^{n+1} = \left[\frac{K_0 RT^2(n+1)}{qQ}\right]\exp(-Q/RT). \tag{28}$$

Fig. 30.

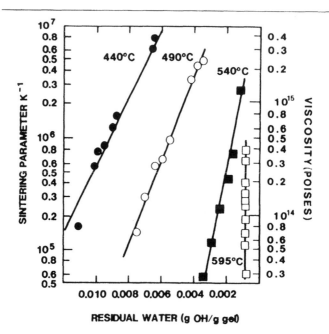

Isothermal plots of $1/K$ and viscosity versus residual OH content, found by analyzing sintering kinetics of multicomponent gel using cylinder model. (Reprinted by permission of the publisher from Gallo *et al.* [65], copyright 1984, Elsevier Science Publ. Co., Inc.)

If this equation applies, then a plot of $\ln(\Delta L/L_0)$ versus $1/T$ should yield a straight line with a slope of $-Q/(n + 1)R$. The parameter n is found by performing experiments at several heating rates and plotting $\ln(\Delta L/L_0)$ versus $\ln(q)$ at constant temperature; the result should be a line with a slope of $-1/(n + 1)$.

For viscous sintering it is clear that this theory is limited to the early stages (first ~5%) of shrinkage, since it is based on Frenkel's equation. To cover a broader range of density it is necessary to integrate Eq. 15 or 23. As shown in Fig. 31, if the sintering kinetics actually obeyed the cylinder model, then plotting the data according to Eq. 28 would indicate than n varies from -0.5 to $+0.7$, instead to being zero [70]. Unfortunately, regardless of the model one chooses, this approach will fail for gels, because the viscosity is *not a single-valued function of temperature.* The viscosity changes with time at constant temperature, and evolves constantly during heating, so the densification kinetics depend on thermal history, not just the current temperature. (This point is directly illustrated by the work of Gallo and Klein presented in Fig. 25.) The calculated curves in Fig. 31 assume that the

Fig. 31.

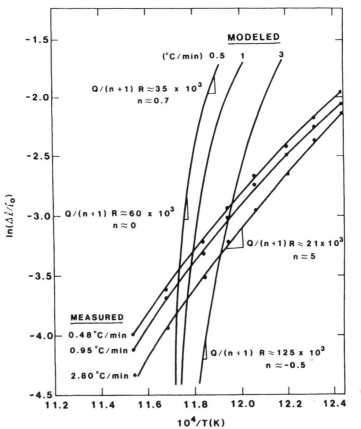

ln(shrinkage) versus $1/T(K)$ during heating at constant rate: curves calculated from cylinder model (MODELED) and measured data (●) for multicomponent gel (MEASURED). From Brinker *et al.* [70].

activation energy for viscous flow equals ~120 kcal/mol throughout densification. However, as shown in Fig. 29, the apparent activation energy depends on hydroxyl content, which varies with temperature and time. This is reflected in the data in Fig. 31, which imply a high value of n; depending on the temperature at which the slope is measured, n ranges from ~8–13 [70].

Although the constant heating rate models do not provide useful information for xerogels, the effect of heating rate on densification is of considerable interest and importance. The particulate gels in Fig. 21 exhibit "normal" behavior: when the heating rate increases, the shrinkage occurs at higher temperatures. This is to be expected, since the gel has less time at a given temperature when the heating rate is greater. However, the A2 gel shows the opposite behavior: the shrinkage curve for the sample heated at 20°C/min *crosses* that for 2°C/min at ~750°C! This is possible because the more rapidly heated sample retains more hydroxyls, which depress its viscosity, and this offsets the shorter time available for sintering. During an isothermal hold, as we have seen, the viscosity of a xerogel can rise by orders of magnitude, causing the sintering to grind to a halt. Therefore, it is advantageous to continue heating during sintering, so that the rising temperature compensates for the structural relaxation and loss of OH that raise the viscosity. The faster the heating, the lower the temperature at which densification will be completed. *However*, it is essential to recognize that excessively rapid heating can cause trapping of evolved gases that can crack or bloat the gel. The maximum safe heating rate for a particular gel must be established empirically.

2.4. Hot Pressing

Hot pressing is sintering under a high pressure. Typically, the sample is placed in a cylinder and the load is applied uniaxially by a piston. When the applied pressure, P_A, is large compared to the capillary pressure, $P_c = 2\gamma_{sv}/r$, then the densification kinetics are expected to obey Eq. 18. As noted earlier, this equation does not predict complete densification in a finite time, because of neglect of P_c, which becomes infinite as $r \to 0$. Vasilos [71] sintered crushed vitreous silica with $P_A = 6.9$ and 17.2 MPa and found that his data were accurately described by Eq. 18 for $0.55 < \rho < 0.99$. Similar agreement was reported by Decottignies *et al.* [72] for crushed vitreous silica, but gel powders showed some deviation from the model. A linear relation between $\ln(1 - \rho)$ and t was found only for $\rho > 0.9$. The discrepancy was attributed to the open porosity of the gel, as opposed to the closed pores assumed in the model. That explanation is dubious, given the insensitivity of the models to geometric details and, more importantly, the good results for equally porous crushed glass. The observed deviations most probably reflect the changing viscosity, as in the gels discussed previously. This interpretation is supported by the low activation energy (111 kcal/mol) for flow that was found from the hot-pressing experiments. As explained in connection with Fig. 29, since more OH is retained at lower temperatures, the measured viscosity and activation energy can be unrealistically low.

Decottignies *et al* [72] performed hot-pressing experiments with a constant heating rate, $q = dT/dt$. Subject to the assumptions discussed in Section 2.3, they showed that Eq. 18 could be written as

$$\frac{d \ln(1 - \rho)}{dT} = \frac{3P_A}{4q\eta}.$$ (29)

They found $\eta(T)$ from the slope of a plot of $\ln(1 - \rho)$ versus T. At higher temperatures, the results of these experiments agreed with the viscosity found from isothermal tests, but lower viscosities were found at lower temperatures. This is to be expected, since rapid heating would trap a higher OH content in the gel, thereby reducing its viscosity. Similar studies have been reported by Jabra *et al.* [73] and Pancrazi *et al.* [74].

2.5. Films

Very little information is available on the kinetics of sintering of gel films. The data of Brinker and Mukherjee [75], shown in Fig. 32, indicate that a film densifies faster than a bulk gel made from the same preparation, even though the film has lower surface area. This is contrary to the theory illustrated in Fig. 14, which predicts that a film will sinter more slowly than an unconstrained gel, since a film can shrink only in the direction perpendicular to the substrate. The experimental results suggest that the films

Fig. 32. _____

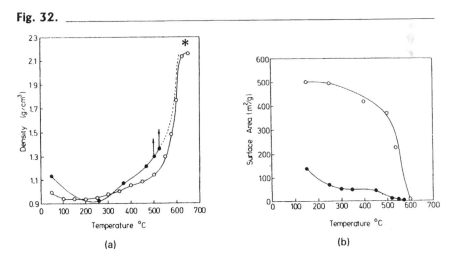

(a)   (b)

(a) Densification of monoliths (O) and films (●) heated at 1°C/min to indicated temperatures and quenched; the symbol * indicates density of conventionally melted glass of this composition, and ↑ indicates rapid densification with time. (b) BET surface area for monoliths (O) and films (●). From Brinker and Mukherjee [75].

differ from bulk gels, either chemically or microstructurally. As explained in Chapter 13, films are in fact denser than bulk gels, because rapid evaporation causes films to remain compliant, so that the pores are collapsed by the capillary pressure during drying.

3.

THEORY OF SINTERING OF CRYSTALLINE MATERIALS

In addition to the specific references cited in the following section, there are many excellent reviews on this subject, including the text by Kingery *et al.* [13]. Coble and Burke [76] describe the physical principles and the early theoretical and experimental work. Theoretical and practical aspects of microstructural control are discussed by Yan [77,78]. In the models discussed next, rough approximations to the geometry are used, so that simple expressions can be found for the mass transport kinetics. To follow the evolution of the geometry of the sintering body, more detailed models are required. A variety of approaches, including topological descriptions (which provide no predictions of kinetics), and computer models that describe both geometry and kinetics, are briefly reviewed by Johnson [79].

3.1.   Physical Principles

As for viscous sintering, the driving force for densification of crystalline materials is surface energy. The surface energies are often higher than for glasses, typical values being ~1.0–1.2 J/m^2 [13], but are equally weakly dependent on temperature and composition. Material tends to move from regions of positive (convex) curvature to regions of negative (concave) curvature, as shown in Fig. 1 of Chapter 4, and this leads to the filling of necks between particles. One important respect in which crystalline and amorphous materials differ is that the plane of contact between crystals, called the *grain boundary*, has a specific interfacial energy, γ_{gb}. This energy reflects the fact that the crystal planes in the respective particles do not match perfectly at the boundary. The existence of the grain boundary energy means that the energy gained by eliminating porosity is partially offset by the energy invested in creating necks between the grains. It also means that grains have a tendency to grow to reduce their energy

Fig. 33.

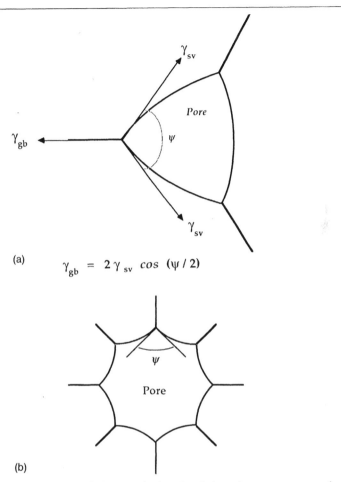

(a)

$$\gamma_{gb} = 2\gamma_{sv} \cos(\psi/2)$$

(b)

Dihedral angle determined by relative magnitudes of grain-boundary energy, γ_{gb}, and solid–vapor surface energy, γ_{sv}. Pore formed at junction of three grains (a) will shrink, but pore surrounded by many grains (b) will grow.

by decreasing their surface-to-volume ratio. The grain-boundary energy dictates the shapes of the pores, as indicated in Fig. 33a: balance of forces requires that the surfaces of the grains meet at the grain boundary at the *dihedral angle*, ψ, where

$$\psi = 2\cos^{-1}(\gamma_{gb}/2\gamma_{sv}). \tag{30}$$

The pore in Fig. 33a will shrink, because the surface of the grain is concave toward the pore. The tension created by the curvature produces a higher concentration of vacancies, so atoms tend to diffuse from the grain boundary to the surface of the pore. When many grains surround a pore (in which case the pore is said to have a high *coordination number*), the requirement of preserving ψ at the grain boundary forces the grains to have positive curvature, as in Fig. 33b. This reduces the vacancy concentration at the surface of the pore, inducing diffusion away from the pore, causing it to grow. Whether a pore will shrink or grow thus depends on the value of ψ and the number of grains surrounding the pore [80,81]. (The fewer neighbors, the more likely it is to shrink.) Neither stable nor growing pores are thermodynamically favorable in amorphous systems, and this accounts in part for the relative ease of sintering glasses.

Several mechanisms and paths by which material can be transported are illustrated in Fig. 34. The easiest paths are (1) along the surface of the particle and (3) through the vapor phase. Both *surface diffusion* and *evaporation–condensation* carry material to the neck and thereby reduce the energy of the system, but neither mechanism produces densification. The same is true of path (2), which brings material from the convex surface to the neck by diffusion through the crystal lattice. Such processes *coarsen* the body by making the necks, and possibly the pores, larger. (See Fig 35.) Shrinkage can only occur if material is removed from the point of contact (i.e., from the grain boundary), so that the centers of the particles can move toward one another. Paths (4) and (5) carry material from the grain boundary to the perimeter of the neck, so these mechanisms produce shrinkage. The mobility of the atoms is high within the grain boundary, but the flux of atoms is small because the grain boundary is very thin (~ 1 nm). Diffusion through the lattice is much slower, but the flux can pass through a much larger area, so the net rate of transport by lattice diffusion may be greater. The size

Fig. 34. _____

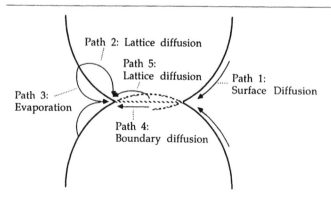

Paths for material transport in sintering of crystalline materials.

Fig. 35.

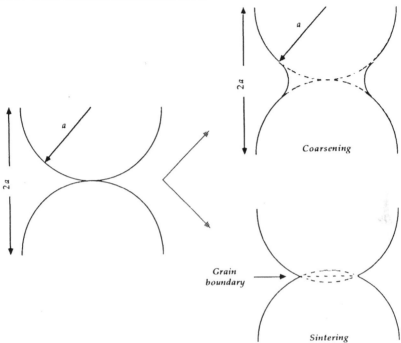

Coarsening is a growth in the neck between particles without any decrease in the distance between their centers. Densification requires removal of material from the grain boundary between the particles.

of the particles influences whether *grain boundary diffusion* or *lattice diffusion* (also called *volume diffusion*) is more important: the smaller the grain, the larger the volume fraction occupied by the grain boundary. The activation energy for the lattice diffusion coefficient (D_L) is higher than for the grain boundary diffusion coefficient (D_{gb}), so D_L increases faster with T and tends to be more important at higher temperatures. The surface-diffusion coefficient (D_s) also has a relatively low activation energy, so coarsening often occurs in preference to densification at low temperatures, because of the relative importance of D_s. The kinetics of sintering depend on which of these paths and transport mechanisms is dominant. The importance of grain boundaries in sintering of crystalline materials can hardly be overestimated. A pore that is not intersected by a grain boundary is, for all practical purposes, trapped. That is, a pore *inside* a grain cannot shrink without diffusion of atoms through the grain from some distant boundary.

3.2. Models

The sintering of crystalline materials is conceptually divided into stages similar to those described in Section 1.2. In the *initial stage*, the pores are open to the exterior; the necks grow, producing a few percent of shrinkage, but there is no growth of the particles. In the *intermediate stage*, the pores remain interconnected and all pores are intersected by grain boundaries; grain growth may occur. The *final stage* begins when the pores become isolated; they may become detached from grain boundaries, in which case further shrinkage is very slow. The final stage typically occurs when the relative density is ≥ 0.95, so most of the shrinkage occurs during the intermediate stage.

Models for the initial stage are based on the two-sphere geometry analyzed by Frenkel. The surface energy produces an excess of vacancies in the crystal lattice in regions of negative curvature, and this results in a flux of vacancies away from the perimeter of the neck (or, equivalently, a flux of atoms toward the perimeter). It is difficult to calculate accurately the concentration gradient or to predict the path along which transport will occur, and this has given rise to a multitude of models based on different (though generally similar) assumptions. For example, Kingery and Berg [20] calculated the rate of neck growth between spheres of radius a, assuming that transport followed path (5) in Fig. 34, and found

$$(z/a)^2 = (80t/\tau_L)^{2/5} \tag{31}$$

where τ_L is the characteristic time for sintering by lattice diffusion,

$$\tau_L \equiv \frac{kTa^3}{D_L \gamma_{sv} \delta^3} \tag{32}$$

where k is Boltzmann's constant, T is temperature, D_L is the lattice diffusion coefficient, and δ is the radius of an atom. In the initial stage the shrinkage, $\Delta L/L$, is proportional to $(z/a)^2$, so $\Delta L/L \propto t^{2/5}$; this is to be compared to the linear time-dependence predicted by Eq. 14 for viscous sintering.

In the intermediate stage, the structure is assumed to consist of polyhedral grains with cylindrical pores along the edges. The density is predicted to increase linearly with time when mass transport occurs by lattice diffusion with the grain boundary as a source of atoms [82]:

$$(d\rho/dt)_L = 336/\tau_L. \tag{33}$$

It is clear from Fig. 7 that the rate of viscous sintering is also approximately constant for $0.3 < \rho < 0.9$; for the cylinder model, $\rho \approx K(t - t_0) - 1.43$, so

$$(d\rho/dt)_{VF} \approx \gamma_{sv} n^{1/3}/\eta. \tag{34}$$

To compare the rates of densification by viscous flow and lattice diffusion, we estimate that there is ~1 pore/particle, so $n \approx 1/(4\pi a^3/3)$, and we use the Stokes–Einstein equation to relate the viscosity and lattice diffusion coefficient,

$$D_L = kT/3\pi \, \delta\eta. \tag{35}$$

With these substitutions, we find from Eqs. 33 and 34 that $(d\rho/dt)_{VF} >$ $(d\rho/dt)_L$ when $a > 6\delta$. Thus, an amorphous material will sinter by viscous flow, rather than lattice diffusion whenever the particle is larger than several angstroms. Seigle [83] reached the same conclusion by comparison of Frenkel's equation, Eq. 13, with an expression similar to Eq. 31. Obviously these are order-of-magnitude estimates, because neither the geometric assumptions nor the thermodynamic relationships underlying these equations are valid for such small particles.

From Eqs. 32 and 33 we find that the densification rate varies as $1/a^3$ when the mass transport path is by lattice diffusion from the grain boundary; the same result is found for the final stage. When diffusion occurs along the grain boundary, rather than through the lattice, the densification rate varies as $1/a^4$, so this path is favored for very small particles. The rate of coarsening by surface diffusion also varies as $1/a^4$, so the latter two processes are expected to be in competition in gels, because of their small particle size. Which process will dominate depends on the relative magnitudes of the grain boundary and surface diffusion coefficients. Since those properties have different activation energies, either one may dominate depending on the temperature of sintering.

Evidently there are several mechanisms that could be operating concurrently during the sintering of a crystalline material. Models have been proposed to allow for multiple mechanisms [84,85] and Ashby [86] has developed "sintering maps" that indicate the dominant mechanisms as functions of temperature and particle size. Unfortunately, one usually needs more data than are available to calculate maps or to predict the principal mechanisms for a given set of conditions. The situation is further complicated by coarsening processes, including grain growth and pore growth, which compete with densification. The grains tend to grow to reduce the total area of the grain boundaries. The rate of increase of the grain diameter, G, is found to obey [82]

$$G^3(t) - G^3(0) \propto t \tag{36}$$

so

$$dG/dt \propto 1/G^2, \tag{37}$$

which means that grain growth is relatively rapid for small particles. Thus gels are subject to rapid grain growth and rapid surface diffusion, both of which tend to inhibit densification.

One common method for facilitating sintering of crystalline materials is to introduce a liquid phase. The liquid tends to invade the grain boundary between particles, where it helps to dissolve the solid phase, then provides an easy path for transport of atoms away from the point of contact [87,88]. The liquid may also help to "lubricate" the particles, allowing them to slip into a denser packing. In the initial stage, this process is predicted to produce shrinkage proportional to $t^{1/3}$, with a rate proportional to $a^{4/3}$. Another method for accelerating densification is application of pressure, P_A. In the initial stage [89], the pressure increases the driving force from $\gamma_{sv}\kappa$ to $\gamma_{sv}\kappa + P_A$, where κ is the curvature of the neck. If the pressure is large enough, the surface energy is negligible and the microstructure has relatively little effect on the rate of densification.

4.

EXPERIMENTAL STUDIES OF DIFFUSIVE SINTERING

The complex behavior described in the preceding section discourages direct tests of sintering models for crystalline materials. Nevertheless, the theory clearly indicates the parameters that must be controlled to get rapid densification. Small particles provide short diffusion paths from the pore to the point of contact between particles, so the sintering rate increases as a^{-3} or a^{-4}. Most importantly, the particles must be densely and uniformly packed to minimize the pore volume while maximizing the number of particle contacts. This creates a situation in which there are many grain boundaries feeding atoms into a relatively small pore space along short diffusion paths. The most dramatic illustration of these principles was provided by Rhodes [90] in a classic study of the sintering of yttria-stabilized zirconia. He prepared a suspension of alkoxide-derived powder (8–12 nm crystallite size) and allowed the agglomerates to settle, then used the fine particles remaining in the supernatant to prepare dense compacts by gravitational settling. The dried compact had a relative density of 0.74, equal to that of close-packed uniform spheres. As shown in Fig. 36, that material sintered to theoretical density in 1 h at 1100°C, whereas cold-pressed pellets of the as-received powder only reached $\rho \approx 0.95$ at 1500°C. The densification kinetics of the well-packed powder suggested that grain boundary diffusion was the controlling mechanism, as expected for such fine powders.

Another clear illustration of the importance of particle packing is provided by studies of sintering of monospheres of titania. The spheres are made by hydrolysis of titanium tetraethoxide and, as discussed in Chapter 4, they grow by aggregation of much smaller particles. Edelson and Glaeser [91,92]

Fig. 36.

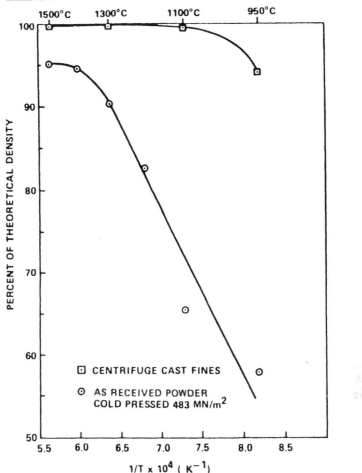

Effect of temperature (1 h cycle) on sintered density for sample (cold-pressed) with 2.9 μm agglomerates compared with agglomerate-free powder (centrifuge cast). From Rhodes [90], reprinted by permission of the American Ceramic Society.

showed that amorphous titania spheres with diameters of 0.35 μm consist of ~10 nm particles, giving the spheres a surface area of ~310 m^2/g. The spheres contain ~35 vol% porosity that is eliminated by heating to 800°C, during which time the spheres sinter and convert to rutile. Each sphere consists of 1 to 3 crystallites by the time that the spheres start to sinter together. Barringer and Bowen [93] examined the densification of compacts made by settling such spheres. Rapid settling produced bodies with green

Fig. 37.

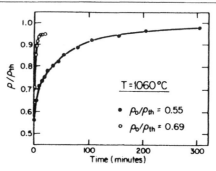

Relative density as a function of time at 1060°C for compacts with $\rho_0 = 0.55$ and 0.69. Improved packing has significantly enhanced densification. From Barringer and Bowen [93].

density of $\rho_0 = 0.55$, while slow settling gave $\rho_0 = 0.69$. As shown in Fig. 37, the more densely packed body sinters much more rapidly. Note that the densification stops at $\rho \approx 0.94$, because there are ~6 vol% large voids created by packing faults in the green body; i.e., there were vacancies (missing spheres) in the original compact that could close. Such pores correspond to the stable or growing pore depicted in Fig. 33b.

These studies show that alkoxide-derived powders permit sintering at exceptionally low temperatures, *if* the powders are processed in such a way as to produce dense and uniform packing. On the other hand, monolithic gels have relatively low green densities, so crystalline gels do not sinter well in spite of their very small pore size. Moreover, the gels are usually amorphous after drying and may develop very large crystallites when heated. The most thoroughly studied example is alumina. Yoldas showed that clear monolithic alumina gels with 2.5 nm pores and 7 nm crystallites [94] could be made by hydrolysis of alkoxides [95–96]. However, the relative density of the dried gel is 0.37, and when it is sintered at 1200°C it shrinks only to $\rho = 0.62$; the sintering is accompanied by a growth in pore size to ~10 nm, and transformation of the amorphous gel to α-alumina. The gel turns white as it sinters, because of the large grain size of the crystalline phase. Kumagai and Messing [97,98] showed that the sintering kinetics of alumina gels are dominated by the microstructure produced during crystallization. Boehmite (AlOOH) goes through a complicated series of phase transformations on heating, converting to γ-, δ-, θ-, and finally α-alumina (see discussion in section 2.6 of Chapter 9); Yoldas' sols convert to boehmite when aged at 80°C [95]. To control this phase evolution, Kumagai and Messing prepared samples of colloidal boehmite (AlOOH) mixed with varying amounts of "seeds" of α-alumina particles (0.06–0.2 μm diameter) to nucleate the

transformation. The relative density of the gels was ~0.45 times that of
α-alumina, and the surface area was ~268 m²/g. Gels containing 1.5 wt%
seeds reached $\rho > 0.98$ at 1200°C, while unseeded samples only reached
$\rho \approx 0.95$ at 1600°C. The reason for this profound difference is revealed by
the microstructures in Fig. 38: the unseeded sample develops large *vermicular*
(i.e., wormlike) grains with internal porosity, while the seeded samples have
small uniform grains with pores only at grain boundaries. The vermicular
grains densify slowly, because they are single crystals, so the pore volume is
not intersected by many grain boundaries. The seeded samples contain
approximately one crystal of α-alumina per nucleating particle [99], so
effective seeding requires a high number density of seeds ($\sim 5 \times 10^{13}/cm^3$).
That is, each nucleus generates a crystal that sweeps through the matrix until
it encounters another α-alumina crystal, so the grain size after the transformation is directly dependent on the number of nuclei. The transformation
can also be facilitated by seeding with a γ-alumina powder that transforms
to α before the gel [100], but this is less effective than using α-alumina
directly. McArdle and Messing [101] showed that α-Fe$_2$O$_3$ could be used as
a nucleating agent, because of its small lattice mismatch (<5%) with
α-alumina. As indicated in Table 2, the transformation temperature is
reduced from 1215 to 1085°C by high concentrations of seeds (15–90 nm) of
α-Fe$_2$O$_3$, and to 968°C by addition of iron nitrate to the sol. The salt
apparently produces a large number of small, uniformly distributed seeds,
making it more effective (on a weight basis) than direct addition of α-Fe$_2$O$_3$
particles. The importance of seed concentration is further illustrated by the
failure of Suwa *et al.* [102] to reduce the transformation of alumina with
seeds of α-Fe$_2$O$_3$, apparently because the particles they added were too few
and too large.

The elegant work of Messing *et al.* shows that dense ceramics can be made
from monolithic gels at modest temperatures, if the phase transformations
are carefully controlled. However, even under these ideal conditions,
monolithic gels offer little advantage over conventionally processed
powders, as illustrated by Fig. 39. Yeh and Sacks [103] have shown that use
of a suitable distribution of particle sizes, elimination of aggregates, and
careful dispersion allow preparation of powder compacts with very high
green density ($\rho_0 \approx 0.7$) that sinter *better* than the seeded gels. They used
commercial alkoxide-derived powder with an average particle size of
~0.06 μm. This is very fine compared to conventional ground powders, but
still has a surface area ~1/10 of that of the boehmite used by Messing *et al.*
The coarser powders can be suspended in higher concentrations and are less
subject to cracking during drying. A critical advantage of this material is that
it is already the equilibrium phase, α-alumina, rather than the amorphous
or nonequilibrium phases generally produced by alkoxide gels. Further

Fig. 38.

(a)

(b)

Structural evolution in boehmite made by hydrolysis of aluminum *sec*-butoxide.
(a) Large vermicular grains of α-alumina in unseeded boehmite sintered 2 h at 1200°C.
(b) Uniform fine grains produced in gel with 10^{14} seeds/cm^3 of α-Fe$_2$O$_3$, sintered 2 h at
1100°C. From J.L. McArdle and G.L. Messing, private communication.

Table 2.

Effect of Fe_2O_3 Additions on the θ- to α-Al_2O_3 Transformation.

Sample Designation	Form of Fe_2O_3 Addition	Equivalent wt% Fe_2O_3	DTA Transformation Temperature (°C)[a]
Unseeded	—	0.0	1215
$10^{12}/cm^3$	particle	0.1	1168
$10^{13}/cm^3$	particle	0.3	1144
$10^{14}/cm^3$	particle	4.2	1085
0.1 M	salt	1.6	1005
0.5 M	salt	8.1	968

Source: McArdle and Messing [101].

[a] Corresponds to peak of differential thermal analysis (DTA) data.

reductions in sintering temperature could be obtained by using finer powders, so alkoxide-derived *powders* offer greater advantages for the preparation of crystalline ceramics than *monolithic* gels. Clasen has shown [104] that powders of α-alumina (average diameter ~50 nm) can be prepared by flame oxidation of $AlCl_3$. The relative merits (properties and cost) of solution- and vapor-derived powders remain to be established.

Fig. 39.

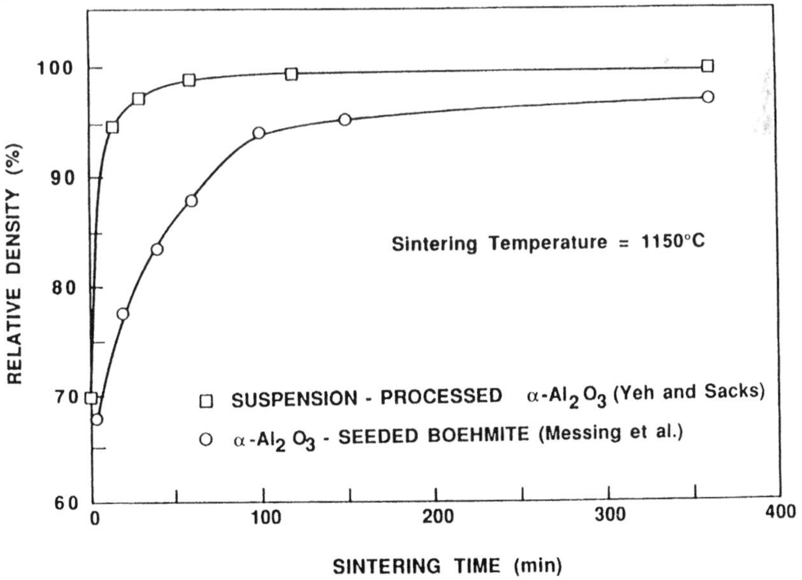

Comparison of densification kinetics of conventionally processed alumina powder and seeded alumina gels. From Yeh and Sacks [103].

The *worst* prospects for densification are offered by crystalline aerogels, because of their extremely low relative density. For example, Rahaman *et al.* [105] made gels with the composition of mullite ($3Al_2O_3 \cdot 2SiO_2$) from TEOS and Al *sec*-butoxide. The gels were hypercritically dried using the low temperature CO_2 process [108], resulting in amorphous bodies with $\rho_0 \approx 0.04$. When heated at a constant rate, the gels began to crystallize at ~1000°C, and the density stopped increasing after reaching only $\rho \approx 0.51$; the gels were heated to 1350°C, but densification stopped at ~1250°C. The same material, when crushed and pressed into a pellet, reached $\rho \approx 0.97$ at <1200°C; this is better than the sintering behavior of xerogels of the same composition [107]. Similarly, alkoxide-derived alumina powders sinter well if seeded with α-alumina powder, but sintering of aerogels made from the same sol leads to cracking and little densification [108].

In summary, the most rapid densification is obtained for amorphous materials, even aerogels. The next best case is well-packed fine powders, but the phase transformation behavior must be controlled. Crystalline monolithic gels will sinter relatively slowly, even if phase transformations are controlled, because of their low green density and crystalline aerogels will not reach full density under any circumstances.

5.

COMPETITION BETWEEN SINTERING AND CRYSTALLIZATION

It was shown in Section 3.2 that the densification rate is orders of magnitude faster when transport occurs by viscous flow, rather than diffusion. This point is illustrated in Fig. 40 for the case of cordierite: amorphous cordierite (even in compacts made from crushed glass powder [110,111]) sinters readily in the range of 800 to 860°C, but much higher temperatures are required for the crystalline form. Even an aerogel with $\rho_0 \approx 0.13$ sinters to a fully dense clear glass at 900°C [112]. This suggests the most favorable course for densification of monolithic gels: sinter to full density *before* crystallization. In fact, there is a patent by Ian Thomas [113] that teaches that dense multicomponent glass ceramics can be made from aerogels only if the amount of crystallization during sintering is no more than 5–10 vol%. For some compositions, such as GeO_2 [114] and alkali silicates [115], crystallization of the gels is so fast that it is easier to prepare dense amorphous materials by melting. The problem is illustrated in Fig. 41 for the case of soda silicate gels: at low soda concentrations the gels crystallize at low temperatures and

Fig. 40.

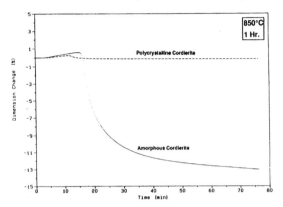

Comparison of densification kinetics of polycrystalline and amorphous cordierite. The crystalline material is stoichiometric and has mean particle size of 8.6 μm; the amorphous material is melted and crushed glass with some B_2O_3 and P_2O_5 and slight excess of SiO_2 and MgO relative to stoichiometric cordierite and has a mean size of 10 μm. From Ewsuk *et al* [109].

sintering does not go to completion.[†] In this section, we examine the relationship between the kinetics of viscous sintering and crystallization, and see how thermal cycles can be optimized to obtain dense glasses.

Time-temperature transformation (TTT) diagrams have long been used by metallurgists to show the influence of thermal history on phase transformations. Uhlmann [117] introduced this approach as a means of establishing the critical cooling rate needed to produce a glass by quenching a melt. Using standard theories (or measured values) for the rates of crystal nucleation (I_v) and growth (u), the volume fraction V of crystals produced by a given thermal history can be calculated using Avrami's equation [118]:

$$V = 1 - \exp(-\pi I_v u^3 t^4/3) \approx \pi I_v u^3 t^4/3. \qquad (38)$$

In this equation the growth and nucleation rates are assumed to be constant; the approximation applies when the volume crystallized is small. According to the classical theory, the nucleation rate (nuclei per unit volume per second) is given by [119]

$$I_v \propto \exp[-(W^* + \Delta G_n)/kT] \qquad (39)$$

[†] A different sort of problem occurs at high soda contents, where viscous sintering occurs at such low temperatures that organics may be trapped. This must be addressed by minimizing the amount of volatiles (e.g., by using excess water for hydrolysis and/or prolonged treatments at low temperatures); however, the problem cannot always be entirely avoided.

Fig. 41.

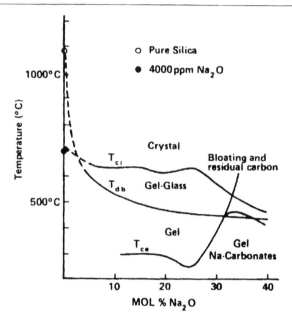

Gel-processing diagram for Na_2O–SiO_2 system: T_{ce} = carbonates eliminated, T_{db} = densification begins, T_{cl} = crystallization begins. At low Na_2O concentrations, crystallization precedes densification. At high concentrations, sintering begins before the carbonates are eliminated, causing bloating and trapping of carbon. From Prassas and Hench [116], pp. 100-125 in *Ultrastructure Processing of Ceramics, Glasses, and Composites*, eds. L.L. Hench and D.R. Ulrich (Wiley, New York, 1984).

where W^* is the thermodynamic barrier to formation of a nucleus and ΔG_n is the kinetic barrier to diffusion across the liquid–nucleus interface. The crystal growth rate increases with undercooling (i.e., as T decreases below the liquidus temperature), because the thermodynamic driving force increases; then the rate decreases because of the falling atomic mobility at low temperatures, so u goes through a maximum at temperature T_{max}. For gels that crystallize rapidly during heating, the transformation will generally be complete before that maximum is reached. Below T_{max} the crystal growth rate is approximated by

$$u \propto T \exp(-\Delta G_c/kT) \qquad (40)$$

where ΔG_c is the kinetic barrier to transport at the liquid–crystal interface. It is generally assumed that $\Delta G_c \approx \Delta G_n$, and that both are equal to Q, the activation for viscous flow (see Eq. 12). In that case, Eq. 38 becomes

$$V \propto (t/\eta)^4 T^3 \exp(-W^*/kT). \qquad (41)$$

For a spherical nucleus, the thermodynamic barrier is given by

$$W^* = (16\pi/3)\gamma_{SL}{}^3/(\Delta G_v)^2 \tag{42}$$

where γ_{SL} is the crystal–liquid interfacial energy and ΔG_v is the free energy change per unit volume crystallized. Thus the temperature-dependence of V is determined by Q and ΔG_v. The analysis can be extended to nonisothermal conditions [120,121] by assuming that the rates of nucleation and crystallization are single-valued functions of temperature (i.e., they are dependent on the current temperature, but not the thermal history). This may not be valid for gels, as we saw in Section 2.3.3, because the viscosity is sensitive to the changing hydroxyl content during heating.

A typical example of a TTT curve is presented in Fig. 42, where I_v and u have been calculated using the viscosity values obtained from sintering studies by Sacks and Tseng [48], as well as viscosity values for conventional dry silica (i.e., having a low OH content). The curves represent a volume fraction of crystals of $V = 10^{-6}$: for times to the left of the curve (e.g., <10 min at 1575 K for the "wet" material), $V < 10^{-6}$. The viscosity is so much lower in the glass with the high hydroxyl content that the time to reach that degree of crystallization is reduced by ~ 3.5 orders of magnitude. Note that the curves are in reasonable agreement with the crystallization behavior observed by Sacks and Tseng.

Fig. 42.

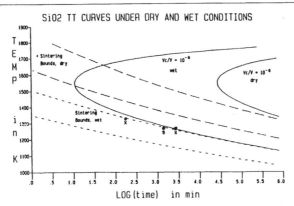

SiO2 TT CURVES UNDER DRY AND WET CONDITIONS

Silica sintering (dashed curves) and crystallization (solid curves) under dry and "wet" conditions; upper and lower sintering bounds correspond to assumed pore diameters of 500 and 5 nm, respectively. The lower viscosity observed by Sacks and Tseng [48] shifts all curves to lower times at any temperature. Their data for sol-gel derived glasses are also included: X marks treatments that resulted in crystallization with little sintering and S marks treatments that produced sintered glasses free of crystallinity. From Uhlmann et al, [122], pp. 173–183 in *Science of Ceramic Chemical Processing*, eds. L.L. Hench and D.R. Ulrich (Wiley, New York, 1986).

This type of analysis was further extended by Uhlmann *et al.* [123] by combining the transformation curves with densification curves calculated from the theory of viscous sintering. The dashed curves in Fig. 42 represent the time to reach full density at each temperature according to the cylinder model discussed in Section 1.2; pairs of curves correspond to assumed initial pore diameters of 5 and 500 nm. If the sintering curve does not cross the crystallization curve, then densification can be completed before detectable crystallization occurs. Gels with smaller pores (lower dashed curves) sinter faster, so they are less likely to crystallize before sintering. This reflects an important concept: since the goal is to minimize the amount of crystallization, we need to identify those parameters that affect *to different extents* the rates of sintering and crystallization. The crystallization rate does not depend on the microstructure of the gel, so manipulation of the pore size (e.g., by control of pH during hydrolysis and condensation) permits decoupling of the rates of transformation and densification. Zarzycki [124] has examined this question in detail. He points out that the transport rates of both processes are determined by the viscosity, so the extent of transformation, V, is a function of t/η (as indicated in Eq. 41). Anything that changes the viscosity, such as the hydroxyl content, influences both processes in such a way that the volume fraction of crystals is exactly the same by the time a given density is reached. During heating, the amount of sintering depends only on $\int dt/\eta$, but V is complex if homogeneous nucleation occurs [119,121]. With a fixed number of nuclei (e.g., heterogeneities or crystals formed during aging), V also depends only on $\int dt/\eta$, and the thermal history does not matter. This reasoning is based on the assumption that the thermodynamic barrier is not affected by changes in hydroxyl content or thermal history. There is some indication that W^* may be influenced by hydroxyl content [9,124], but the effect seems to be relatively small.

The most important factors that can be used to control the degree of crystallization during sintering are microstructure, applied pressure, impurities, and heterogeneous nuclei. Reducing the pore size speeds sintering without significantly affecting crystal nucleation or growth [124,125]. For example, acid-catalyzed silicate gels have finer pores and higher green densities, so they are more likely to sinter without crystallizing than are base-catalyzed gels. The sintering rate increases in proportion to the applied pressure, as shown by Eq. 18, so hot-pressing is one solution to the problem. Unfortunately, this process makes it difficult to sinter gels with complex shapes and it raises the likelihood of contamination. Some gels crystallize so fast that even hot-pressing cannot prevent devitrification during sintering. For example, Zarzycki [124] examined silica gels made from Ludox® and found that dense amorphous pieces could not be produced by hot-pressing. Applied pressure generally does more to accelerate sintering than crystallization, but in some systems the effect of pressure on the phase-transformation kinetics cannot be ignored. Since crystallization usually results in a decrease in specific volume, applied pressure can accelerate the transformation. Different phases may be obtained by hot-pressing than by free sintering, or densification may be completed in a lower temperature range where the desired crystal phase is unfavored (e.g., [126]).

The crystallization rate of the Ludox® gels [124] was too high to be explained by the OH content. Alkali impurities were present, and these can reduce the viscosity and influence the crystal phases that appear [127]. However, when compared to purer gels made from TEOS, the difference in viscosity produced by OH and alkali was not great enough to account for the faster crystallization. It was concluded that the gels made from Ludox® contained heterogeneous nucleation sites, so that reduction of the thermodynamic barrier, rather than acceleration of the transport kinetics, controlled the result. In alkoxide-derived gels doped with small (<0.2 mol%) amounts of alkali oxides, the increase in V was also greater than could be explained on the basis of viscosity alone. Zhu *et al.* noted that the alkali raised the pH of the solution and thereby coarsened the gel, and they suggested that this allowed heterogeneous nuclei to enter the pores [127].

In summary, the volume fraction of crystals that develops during the time required to sinter a gel depends on several factors. The thermal cycle is not important if growth occurs from pre-existing nuclei. However, the maximum in I_v is generally near T_g, in the same temperature range where gels are typically sintered; therefore the choice of sintering temperature or heating rate can have a strong effect on V if homogeneous nucleation is dominant. Impurities, such as OH, that affect only the viscosity do not change the extent of crystallization during sintering. Microstructural changes should only affect the sintering behavior, and the degree of control over pore size and green density in gels is so great that this parameter should be the experimentalists' most powerful tool. Of course, care must be taken to avoid accidental introduction of heterogeneous nucleation sites. Contamination (e.g., by dust in glassware used to prepare sols) must be avoided, because the processing temperatures of gels are so low that particles will not dissolve, as they may in conventional ceramics.

6.

COMPOSITES

Spheroidal particles and whiskers are known to impart strength and toughness to ceramic matrices, but the presence of these inclusions interferes with sintering. It has been shown that as little as 3 vol% of SiC particles in ZnO [128], or of Al_2O_3 particles in TiO_2 [129], can significantly retard densification of the matrix. The same effect is seen in whisker composites, and the retardation increases with the aspect ratio of the whiskers [130]. The matrix is inhibited when it tries to contract around such inclusions, and it has been argued that this leads to large stresses that prevent sintering. However, it has been shown that theories predicting stresses large enough to account for the observed sintering rates are erroneous [131]. The behavior of crystalline matrices may indicate that coarsening mechanisms (e.g., surface diffusion) are favored when inclusions inhibit densification. In addition, as illustrated in Fig. 43, the layer of particles surrounding an inclusion cannot shrink in the circumferential direction, so material must diffuse ''around the corner'' from A to grow the neck at B. Such problems do not exist in glass matrices, which are relatively insensitive to the presence of inclusions [132]. The small retardation seen in glass matrices is in quantitative agreement with the prediction of a realistic model for the sintering stresses [133]. However, that model breaks down when the solids loading exceeds the percolation threshold (~ 16 vol% for monospheres), so that the inclusions form a contiguous network. Such high loadings cause a much greater reduction in shrinkage rate, but the effect is still minor compared to that seen in crystalline systems.

Continuous fibers and woven mats provide the greatest improvements in the mechanical properties of ceramic matrices, but they inhibit shrinkage to such an extent that hot-pressing is required even for glass matrices [134]. The same is true when the matrix is produced from an alkoxide-derived sol, but the pressing temperature may be lower [135]. The disadvantage of gels as

Fig. 43.

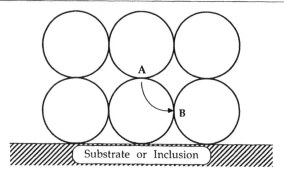

Change in diffusion path for sintering when particles sit on a substrate or surround an inclusion. To fill in the neck at B, material must diffuse from the neck at A. From Bordia and Scherer [131].

matrix materials is that they shrink so much on drying that cracking is likely and voids between the fibers are inevitable. It is necessary to dip the composites repeatedly to achieve an acceptable volume fraction of matrix [135,136]. Clark and colleagues [136–138] have studied the preparation of composites containing whiskers or fibers in a matrix of alumina made by the Yoldas [94–96] process. They have shown that cracking during drying can be prevented by adding a "plasticizer" (viz., glycerol) to the sol, and controlling the relative humidity [139]. However, all of the efforts of Clark *et al.* to densify these composites without pressure have produced porous bodies with modest strength. The same result is obtained with a gel-derived silica matrix containing 60 vol% SiC whiskers and fibers [140]. In contrast, strong and tough composites have been made by Fitzer and Gadow [135] by hot-pressing gel-derived matrices of alumina, silica, and various silicates with inclusions of C, SiC, and alumina; the pressing condition for alumina, for example, was ~1700 K and ~20 MPa.

The advantage of glass matrices, especially those derived from gels, is that less damage is done to the fibers during pressing, because viscous flow allows the matrix to infiltrate the fibers at relatively low temperatures. Glasses also offer a broad range of compositions, and therefore of thermal expansion coefficient. This is important because stresses from thermal expansion mismatch between the matrix and the inclusion can weaken the composite. The greatest advantage comes from glass-ceramic compositions that can be processed in the amorphous state and then crystallized to obtain refractory matrices that can be used at temperatures comparable to the pressing temperature [134]. The high-temperature performance of such composites is actually limited by degradation of the fibers, not the matrices.

7.

SUMMARY

Viscous sintering is a relatively simple process, and the observed densification kinetics are in good agreement with theory for a variety of materials, from melted and crushed glass to gels made from pyrogenic particles. However, in alkoxide-derived xerogels the sintering process is concurrent with changes in structure and hydroxyl content, so the kinetics are more complex. The good performance of the sintering theories for conventional glasses gives us confidence to turn the problem around: we assume that the theory is correct, then fit the kinetics to the theory to determine the time-dependence of the viscosity of the gel. For alkoxide-derived silicate xerogels the viscosity may increase by three orders of magnitude during an isothermal hold in the vicinity of the glass transition temperature, with greater increases observed for acid-catalyzed than for base-catalyzed gels.

The kinetics of densification of crystalline materials are slower and much more dependent on microstructure. The difficulty of reaching theoretical density is exacerbated in gels by the fact that the dried gel is rarely the equilibrium crystal phase. The low green density of the gel, together with the extensive grain growth produced during phase changes, can lead to arrested sintering. It is advantageous to sinter the gel while it is amorphous and then to crystallize after complete densification. This is easy for some compositions and impossible for others. Some insight into the optimal firing history can be obtained through the construction of TTT diagrams, but experience indicates that the most important factors to control are the microstructure (viz., the finer the better) and the concentration of heterogeneous nuclei (viz., the fewer the better).

Sintering of composites is most difficult, even for glass matrices, because the inclusions prevent shrinkage. This encourages cracking during drying and prevents contraction during sintering. Full density can be achieved in glass matrices containing low (<16 vol%) inclusions, but higher loadings or crystalline matrices demand hot-pressing.

REFERENCES

1. G.W. Scherer in *Surface and Colloid Science*, **vol. 14**, ed. E. Matijević (Plenum, New York, 1987), pp. 265–300.
2. R.W. Hopper and D.R. Uhlmann, *Mater. Sci. and Eng.*, **15** (1974) 137–144.
3. J. Frenkel, *J. Phys.* (Moscow), **9** [5] (1945) 385–391.
4. A.E.H. Love, *Mathematical Theory of Elasticity*, 4th ed. (Dover, New York, 1944), p. 145.
5. A. Napolitano, J.H. Simmons, D.H. Blackburn, and R.E. Chidester, *J. Res. Natl. Bur. Stand.*, **78A** [3] (1974) 323–329.
6. O.V. Mazurin, Yu. K. Startsev, and L.N. Potselueva, *Sov. J. Phys. Chem. Glass*, **5** [1] (1979) 68–79 (Eng. trans.).
7. G.W. Scherer, *J. Am. Ceram. Soc.*, **67** [7] (1984) 504–511.
8. G. Hetherington, K.H. Jack, and J.C. Kennedy, *Phys. Chem. Glasses*, **5** [5] (1964) 130–136.
9. C.J.R. Gonzalez-Oliver, P.S. Johnson, and P.F. James, *J. Mater. Sci.*, **14** (1979) 1159–1169.
10. G.W. Scherer, *Relaxation in Glass and Composites* (Wiley, New York, 1986).
11. M. Hara and S. Suetoshi, *Rep. Res. Lab. Asahi Glass Co.*, **5** (1955) 126–135.
12. R. Puyané, P.F. James, and H. Rawson, *J. Non-Cryst. Solids*, **41** (1980) 105–115.
13. W.D. Kingery, H.K. Bowen, and D.R. Uhlmann, *Introduction to Ceramics*, 2d ed. (Wiley, New York, 1976), chapter 5.
14. W.D. Kingery, *J. Am. Ceram. Soc.*, **42** [1] (1959) 6–10.
15. N.M. Parikh, *J. Am. Ceram. Soc.*, **41** [1] (1958) 18–22.

16. R.K. Iler, *The Chemistry of Silica* (Wiley, New York, 1979), p. 645.
17. H.E. Exner, *Reviews on Powder Metallurgy and Physical Ceramics*, **1** [1-4] (1979) 1-251.
18. D.R. Uhlmann, L. Klein, and R.W. Hopper in *The Moon* (D. Reidel, Dordrecht, Holland, 1975), pp. 277-284.
19. G.C. Kuczynski, *J. Appl. Phys.*, **20** [10] (1949) 1160-1163.
20. W.D. Kingery and M. Berg, *J. Appl. Phys.*, **26** [10] (1955) 1205-1212.
21. H.E. Exner and G. Petzow in *Sintering and Catalysis, Materials Science Research*, vol. **10**, ed. G.C. Kuczynski (Plenum, New York, 1975), pp. 279-293.
22. M.W. Weiser and L.C. de Jonghe, *J. Am. Ceram. Soc.*, **69** [11] (1986) 822-826.
23. E. Liniger and R. Raj, *J. Am. Ceram. Soc.*, **70** [11] (1987) 843-849.
24. J.K. Mackenzie and R. Shuttleworth, *Proc. Phys. Soc.*, **62** [12-B] (1949) 838-852.
25. J. Phalippou, T. Woignier, and J. Zarzycki in *Ultrastructure Processing of Ceramics, Glasses, and Composites*, eds. L.L. Hench and D.R. Ulrich (Wiley, New York, 1984), pp. 70-87.
26. P. Murray, E.P. Rodgers, and A.E. Williams, *Trans. Brit. Cer. Soc.*, **53** (1954) 474-510.
27. C.J. Brinker and G.W. Scherer, *J. Non-Cryst. Solids*, **70** (1985) 301-322.
28. G.W. Scherer, *J. Am. Ceram. Soc.*, **60** [5-6] (1977) 236-239.
29. G.W. Scherer, *J. Non-Cryst. Solids*, **34** (1979) 239-256.
30. G.W. Scherer, *J. Am. Ceram. Soc.*, **67** [11] (1984) 709-715.
31. G.W. Scherer, *J. Am. Ceram. Soc.*, **71** (1988) C447-C448.
32. M. Yamane and S. Okano, *Yogyo-Kyokai Shi*, **87** (1979) 434-438.
33. G.W. Scherer and T.A. Garino, *J. Am. Ceram. Soc.*, **68** [4] (1985) 216-220.
34. M.J. Murtagh, E.K. Graham, and C.G. Pantano, *J. Am. Ceram. Soc.*, **69** [11] (1986) 775-779.
35. E.M. Rabinovich, *J. Mater. Sci.*, **20** (1985) 4259-4297.
36. G.W. Scherer and D.L. Bachman, *J. Am. Ceram. Soc.*, **60** [5-6] (1977) 239-243.
37. M.F. Yan, J.B. MacChesney, S.R. Nagel, and W.W. Rhodes, *J. Mater. Sci.*, **15** (1980) 1371-1378.
38. K.L. Walker, J.W. Harvey, F.T. Geyling, and S.R. Nagel, *J. Am. Ceram. Soc.*, **63** (1980) 96-102.
39. G.W. Scherer and J.C. Luong, *J. Non-Cryst. Solids*, **63** (1984) 163-172.
40. E.M. Rabinovich, D.W. Johnson, Jr., J.B. MacChesney, and E.M. Vogel, *J. Am. Ceram. Soc.*, **66** [10] (1983) 683-688.
41. E.M. Rabinovich in *Sol-Gel Technology for Thin Films, Fibers, Preforms, Electronics, and Speciality Shapes*, ed. L.C. Klein (Noyes, Park Ridge, N.J., 1988), pp. 260-294.
42. D.W. Johnson, Jr., E.M. Rabinovich, J.B. MacChesney, and E.M. Vogel, *J. Am. Ceram. Soc.*, **66** [10] (1983) 688-693.
43. F. Orgaz and M.P. Corral, *Bol. Soc. Esp. Ceram. Vidr.*, **26** [5] (1987) 291-297.
44. R.D. Shoup in *Colloid and Interface Science*, vol. **III** (Academic Press, New York, 1976), pp. 63-69.
45. W. Stöber, A. Fink, and E. Bohn, *J. Colloid Interface Sci.*, **26** (1968) 62-69.
46. T. Shimohira, A. Makishima, K. Kotani, and M. Wakakuwa in *Proc. Int. Symp. Factors in Densification and Sintering of Oxide and Non-Oxide Ceramics*, Hakone, Japan, eds. S. Somiya and S. Saito (Gakiyutsu Bunken Fokyu-Kai, Assoc. for Sci. Doc. Info., Tokyo Inst. of Tech., Tokyo, 1978), pp. 119-127.
47. E.A. Barringer and H.K. Bowen, *J. Am. Ceram. Soc.*, **65** [12] (1982) C199-C201.
48. M.D. Sacks and T.-Y. Tseng, *J. Am. Ceram. Soc.*, **67** [8] (1984) 526-532, 532-537.
49. T.Y. Tseng and J.J. Yu, *J. Mater. Sci.*, **21** (1986) 3615-3624.
50. T. Woignier, J. Phalippou, and J. Zarzycki, *J. Non-Cryst. Solids*, **63** (1984) 117-130.
51. P.I.K. Onorato, unpublished.

52. S.R. Su and P.I.K. Onorato in *Better Ceramics Through Chemistry II*, eds. C.J. Brinker, D.E. Clark, and D.R. Ulrich (North-Holland, New York, 1986), pp. 237–244.

53. T. Woignier, J. Phalippou, and M. Prassas, "Glasses from aerogels: II. The aerogel–glass transformation," *J. Mater. Sci.*, to be published.

54. C.A.M. Mulder, J.G. van Lierop, and G. Frens, *J. Non-Cryst. Solids*, **82** (1986) 92–96.

55. E. Papanikolau, W.C.P.M. Meerman, R. Aerts, T.L. van Rooy, J.G. van Lierop, and T.P.M. Meeuwsen, *J. Non-Cryst. Solids*, **100** (1988) 247–249.

56. C.J. Brinker, W.D. Drotning, and G.W. Scherer in *Better Ceramics Through Chemistry*, eds. C.J. Brinker, D.E. Clark, D.R. Ulrich (North-Holland, New York, 1984), pp. 25–32.

57. T. Kawaguchi, J. Iura, N. Taneda, H. Hishikura, and Y. Kokubu, *J. Non-Cryst. Solids*, **82** (1986) 50–56.

58. M. Nogami and Y. Moriya, *J. Non-Cryst. Solids*, **37** (1980) 191–201.

59. T.A. Gallo and L.C. Klein in *Better Ceramics Through Chemistry II*, eds. C.J. Brinker, D.E. Clark, and D.R. Ulrich (North-Holland, New York, 1986), pp. 245–250.

60. T.A. Gallo and L.C. Klein, *J. Non-Cryst Solids*, **82** (1986) 198–204.

61. C.J. Brinker and G.W. Scherer in *Ultrastructure Processing of Ceramics, Glasses, and Composites*, eds. L.L. Hench and D.R. Ulrich (Wiley, New York, 1984), pp. 43–59.

62. S. Satoh, K. Susa, I. Matsuyama, and T. Suganuma, *J. Non-Cryst. Solids*, **55** (1983) 455–457.

63. S. Satoh, K. Susa, I. Matsuyama, and T. Suganuma, *J. Am. Ceram. Soc.*, **68** [7] (1985) 399–402.

64. G.W. Scherer, C.J. Brinker, and E.P. Roth, *J. Non-Cryst. Solids*, **72** (1985) 369–389.

65. T.A. Gallo, C.J. Brinker, L.C. Klein, and G.W. Scherer in *Better Ceramics Through Chemistry*, eds. C.J. Brinker, D.E. Clark, and D.R. Ulrich (North-Holland, New York, 1984), pp. 85–90.

66. W.S. Young, S.T. Rasmussen, and I.B. Cutler in *Ultra-Fine Grain Ceramics*, eds. J.J. Burke, N.L. Reed, and V. Weiss (Syracuse Univ. Press, Syracuse, New York, 1970), pp. 185–202.

67. W.S. Young and I.B. Cutler, *J. Am. Ceram. Soc.*, **53** [12] (1970) 659–663.

68. J.L. Woolfrey and M.J. Bannister, *J. Am. Ceram. Soc.*, **55** [8] (1972) 390–394.

69. M.J. Bannister, *J. Am. Ceram. Soc.*, **51** (1968) 548–553.

70. C.J. Brinker, G.W. Scherer, and E.P. Roth, *J. Non-Cryst. Solids*, **72** (1985) 345–368.

71. T. Vasilos, *J. Am. Ceram. Soc.*, **43** [10] (1960) 517–519.

72. M. Decottignies, J. Phalippou, F. Sorrentino, and J. Zarzycki, *J. Non-Cryst. Solids*, **63** (1984) 81–93.

73. R. Jabra, J. Phalippou, and J. Zarzycki, *J. Non-Cryst. Solids*, **42** (1980) 489–498.

74. F. Pancrazi, J. Phalippou, F. Sorrentino, and J. Zarzycki, *J. Non-Cryst. Solids*, **63** (1984) 81–93.

75. C.J. Brinker and S.P. Mukherjee, *Thin Solid Films*, **77** (1981) 141–148.

76. R.L. Coble and J.E. Burke in *Progress in Ceramic Science*, **vol. 3**, ed. J.E. Burke (Pergamon, New York, 1963), pp. 197–258.

77. M.F. Yan, *Mater. Sci. Eng.*, **48** (1981) 53–72.

78. M.F. Yan in *Advances in Powder Technology*, ed. G.Y. Chin (Am. Soc. Metals, Metals Park, Ohio, 1982), pp. 99–133.

79. D.L. Johnson in *Sintering–Theory and Practice*, eds. K. Kolar, S. Pejovnik, and M.M. Ristic (Elsevier, Amsterdam, 1982), pp. 17–26.

80. W.D. Kingery and B. Francois in *Sintering and Related Phenomena*, eds. G.C. Kuczynski, N.A. Hooton, and G.F. Gibbon (Gordon Breach, New York, 1967), pp. 471–498.

81. F. Lange, *J. Am. Ceram. Soc.*, **67** [2] (1984) 83–89.
82. R.L. Coble, *J. Appl. Phys.*, **32** [5] (1961) 787–792; 793–799.
83. L.L. Seigle, *Prog. Powder Metall.*, **20** (1964) 221–238.
84. D.L. Johnson, *J. Am. Ceram. Soc.*, **53** [10] (1970) 574–577.
85. D.L. Johnson and T.M. Clarke, *Acta Metall.*, **12** (1964) 1173–1179.
86. M.F. Ashby, *Acta Metall.*, **22** (1974) 275–289.
87. W.D. Kingery, *J. Appl. Phys.*, **30** [3] (1959) 301–306.
88. R.M. German, *Liquid Phase Sintering* (Plenum, New York, 1985).
89. R.L. Coble, *J. Appl. Phys.*, **41** [12] (1970) 4798–4807.
90. W.H. Rhodes, *J. Am. Ceram. Soc.*, **64** [1] (1981) 19–22.
91. L.H. Edelson and A.M. Glaeser, *J. Am. Ceram. Soc.*, **71** [4] (1988) 225–235.
92. L.H. Edelson and A.M. Glaeser, *J. Am. Ceram. Soc.*, **71** [4] (1988) C198–C201.
93. E.A. Barringer and H.K. Bowen, *Appl. Phys. A*, **45** (1988) 271–275.
94. B.E. Yoldas, *Ceram. Bull.*, **54** [3] (1975) 286–290.
95. B.E. Yoldas, *J. Appl. Chem. Biotechnol.*, **23** (1973) 803–809.
96. B.E. Yoldas, *J. Mater. Sci.*, **10** (1975) 1856–1860.
97. M. Kumagai and G.L. Messing, *J. Am. Ceram. Soc.*, **67** [11] (1984) C230–C231.
98. M. Kumagai and G.L. Messing, *J. Am. Ceram. Soc.*, **68** [9] (1985) 500–505.
99. R.A. Shelleman, G.L. Messing, and M. Kumagai, *J. Non-Cryst. Solids*, **82** (1986) 277–285.
100. J.L. McArdle and G.L. Messing, *J. Am. Ceram. Soc.*, **69** [5] (1986) C98–C101.
101. J.L. McArdle and G.L. Messing, *Adv. Ceram. Mater.*, **3** [4] (1988) 387–392.
102. Y. Suwa, R. Roy, and S. Komarneni, *J. Am. Ceram. Soc.*, **68** [9] (1985) C238–C240.
103. T.S. Yeh and M. Sacks, *J. Am. Ceram. Soc.*, **71** [10] (1988) 841–844.
104. R. Clasen, *Glastech. Ber.*, **61** [5] (1988) 119–126.
105. M.N. Rahaman, L.C. de Jonghe, S.L. Shinde, and P.H. Tewari, *J. Am. Ceram. Soc.*, **71** [7] (1988) C338–C341.
106. P.H. Tewari, A.J. Hunt, and K.D. Lofftus, *Mater. Lett.*, **3** [9–10] (1985) 363–367.
107. S. Komarneni, Y. Suwa, and R. Roy, *J. Am. Ceram. Soc.*, **69** [7] (1986) C155–C156.
108. S.M. Wolfrum, *J. Mater. Sci. Lett.*, **6** (1987) 706–708.
109. K.G. Ewsuk, G.W. Scherer, L.W. Harrison, W.G. Peet, S.L. Bors, and U. Chowdhry, presented at Annual Meeting, Am. Ceram. Soc., Cincinnati, Ohio, 1988.
110. E.A. Giess, J.P. Fletcher, and L.W. Herron, *J. Am. Ceram. Soc.*, **67** [8] (1984) 349–352.
111. E.A. Giess, C.F. Guerci, G.F. Walker, and S.H. Wen, *J. Am. Ceram. Soc.*, **68** [12] (1985) C326–C329.
112. H. Vesteghem, A.R. di Giampaolo, and A. Dauger, *J. Mater. Sci. Lett.*, **6** (1987) 1187–1189.
113. I.M. Thomas, U.S. Patent 3,791,808 (February 12, 1974).
114. S.P. Mukherjee, *J. Non-Cryst. Solids*, **82** (1986) 293–300.
115. I. Schwartz, P. Anderson, H. de Lambilly, and L.C. Klein, *J. Non-Cryst. Solids*, **83** (1986) 391–399.
116. M. Prassas and L.L. Hench in *Ultrastructure Processing of Ceramic, Glasses, and Composites*, eds. L.L. Hench and D.R. Ulrich (Wiley, New York, 1984), pp. 100–125.
117. D.R. Uhlmann, *J. Non-Cryst. Solids*, **7** (1972) 337–348.
118. M. Avrami, *J. Chem. Phys.*, **7** (1939) 1103–1112; **8** (1940) 212–224; **9** (1941) 177–184.
119. J.W. Christian *The Theory of Transformations in Metals and Alloys*, 2d ed. (Pergamon, New York, 1975).
120. P.I.K. Onorato and D.R. Uhlmann, *J. Non-Cryst. Solids*, **22** (1976) 367–378.
121. R.W. Hopper, G. Scherer, and D.R. Uhlmann, *J. Non-Cryst. Solids*, **15** (1974) 45–62.

122. D.R. Uhlmann, B.J. Zelinski, L. Silverman, S.B. Warner, B.D. Fabes, and W.F. Doyle in *Science of Ceramic Chemical Processing*, eds. L.L. Hench and D.R. Ulrich (Wiley, New York, 1986), pp. 173–183.

123. D.R. Uhlmann, L. Klein, P.I.K. Onorato, and R.W. Hopper, *Proc. Lunar Sci. Conf. 6th* (1975) 693–705.

124. J. Zarzycki in *Advances in Ceramics*, **vol. 4** (Am. Ceram. Soc., Columbus, Ohio, 1982), pp. 204–216.

125. J. Zarzycki, *J. Non-Cryst. Solids*, **48** (1982) 105–116.

126. H. Perthius and Ph. Colomban, *Ceram. Int.*, **12** (1986) 39–52.

127. C. Zhu, J. Phalippou, and J. Zarzycki, *J. Non-Cryst. Solids*, **82** (1986) 321–328.

128. L.C. de Jonghe, M.N. Rahaman, and C.H. Hsueh, *Acta Metall.*, **34** [7] (1986) 1467–1471.

129. R.K. Bordia and R. Raj, *J. Am. Ceram. Soc.*, **71** [4] (1988) 302–310.

130. M.D. Sacks, H.-W. Lee, and O.E. Rojas, *J. Am. Ceram. Soc.*, **71** [5] (1988) 370–379.

131. R.K. Bordia and G.W. Scherer, *Acta Metall.*, **36** [9] (1988) 2411–2416.

132. M.N. Rahaman and L.C. de Jonghe, *J. Am. Ceram. Soc.*, **70** [12] (1987) C348–C351.

133. G.W. Scherer, *J. Am. Ceram. Soc.*, **70** [10] (1987) 719–725.

134. K.M. Prewo in *Tailoring Multiphase and Composite Ceramics*, Mater. Sci. Res., **vol. 20**, eds. R.E. Tressler, G.L. Messing, C.G. Pantano, and R.E. Newnham (Plenum, New York, 1986), pp. 529–547.

135. E. Fitzer and R. Gadow in *Tailoring Multiphase and Composite Ceramics*, Mater. Sci. Res., **vol. 20**, eds. R.E. Tressler, G.L. Messing, C.G. Pantano, and R.E. Newnham (Plenum, New York, 1986), pp. 571–607.

136. D.E. Clark in *Science of Ceramic Chemical Processing*, eds. L.L. Hench and D.R. Ulrich (Wiley, New York, 1986), pp. 237–246.

137. J.J. Lannuti and D.E. Clark in *Better Ceramics through Chemistry*, eds. C.J. Brinker, D.E. Clark, and D.R. Ulrich (North-Holland, New York, 1984), pp. 369–374.

138. J.J. Lannuti and D.E. Clark in *Better Ceramics through Chemistry*, eds. C.J. Brinker, D.E. Clark, and D.R. Ulrich (North-Holland, New York, 1984), pp. 375–381.

139. R.H. Krabill and D.E. Clark in *Better Ceramics through Chemistry II*, eds. C.J. Brinker, D.E. Clark, and D.R. Ulrich (North-Holland, New York, 1986), pp. 641–646.

140. B.I. Lee and L.L. Hench in *Science of Ceramic Chemical Processing*, eds. L.L. Hench and D.R. Ulrich (Wiley, New York, 1986), pp. 231–236.

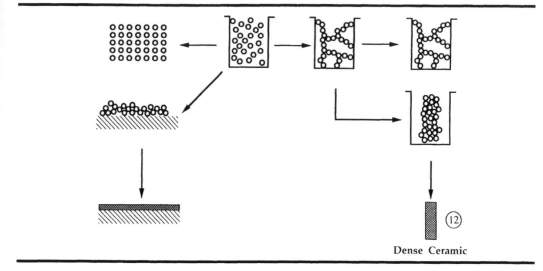

Dense Ceramic

CHAPTER 12

Comparison of Gel-derived and Conventional Ceramics

In this chapter we explore the unique features of gel-derived ceramic materials and contrast their properties with those of conventional ceramics. The latter category includes melted glasses and crystalline ceramics made by high-temperature processing of crushed powders. The most extraordinary properties are possessed by unfired gels, which retain the structure created in solution. The higher the temperature to which the gel is exposed, the more nearly it approaches the behavior of an ordinary ceramic. It has occasionally been claimed that the properties of gels reflect the conditions of polymerization even after the gel has been melted, but in almost every case it can be shown that such reports are erroneous. On the other hand, unique materials of many kinds can be made from gels by exploiting low temperature processing. For example, homogeneous glasses have been prepared in the $SrO–SiO_2$ system, which exhibits stable liquid–liquid immiscibility, and nonequilibrium crystal phases that cannot be obtained from glasses can be made from gels. Novel materials and processes made possible by sol-gel processing are discussed in Chapter 14. The outline of this chapter is as follows.

1. *Thermodynamics and Kinetics* presents background information for subsequent discussion.

    1.1. *Sols, Gels, and Glasses* provides definitions of these states of matter and discusses their relative free energies.

    1.2. *Phase Transformations* discusses the thermodynamics and kinetics of crystallization and phase separation.

2. *Unique Properties of Sols* describes what may be the most important features of the sol-gel process, namely the ability to make coatings, fibers, and powders.

3. *Unique Properties of Gels* reviews briefly the physical properties of unfired gels. Their high porosity, permeability, reactivity, and conductivity offer many potential advantages for applications.

4. *Glasses* presents a detailed comparison of the properties and phase-transformation behavior of gel-derived and conventional glasses. A distinction is drawn between glasses that have not been heated above the glass transition temperature, which may not be in thermodynamic equilibrium, and those that have been melted.

5. *Crystalline Systems* examines the new phases and superior microstructures that can be obtained through use of sol-gel processing.

1.

THERMODYNAMICS AND KINETICS

Many of the studies that draw comparisons between gel-derived and conventional ceramics examine phase changes, including crystallization and liquid–liquid phase separation. This section provides a brief review of the thermodynamics and kinetics of those processes.

1.1. Sols, Gels, and Glasses

A *sol* is a dispersion of solid particles or polymers in a liquid. It is possible to precipitate particles that are amorphous or crystalline, or to make amorphous particles that become crystalline through dissolution and reprecipitation. The latter process can produce particles that differ little from ordinary ceramic oxides, except that the sol particles are small (submicron). If the solubility of the solid phase in the liquid is limited, monomers may attach irreversibly to a growing cluster, so that rearrangement into the equilibrium structure is impossible. In that case, polymeric clusters appear with fractal structures that are quite different from ceramics; they typically have much lower connectivity (i.e., fewer bridging oxygen bonds) and consequently contain many hydroxyl and organic ligands.

A *gel* consists of a continuous solid network surrounding and supporting a continuous liquid phase. The network can result from destabilization of a particulate sol by reduction of a repulsive double layer or removal of a steric barrier, or it can be built by crosslinking of polymeric clusters. In either case, the free energy of the gel is similar to that of the sol, because relatively few bonds have to be formed to produce the spanning cluster. Therefore, there is no thermal event when gelation occurs. The *free energy, g,*

$$g = h - Ts \tag{1}$$

is shown schematically in Fig. 1; h is the *enthalpy,* T the temperature, and s the *entropy.* Physical changes occur in such a way as to reduce the free energy of the system. For example, the entropy of the sol is decreased when particles stick together, and this tends to increase g; yet the particles aggregate spontaneously when the repulsive barrier in the sol is reduced, because the enthalpy gained on contact overcomes the loss of entropy, resulting in a net reduction in g. It is important to recognize that a system may rest in a state of *metastable equilibrium* with a free energy higher than that of the equilibrium state, if a barrier (such as the electrostatic double layer) to rearrangement exists. The gel has a higher free energy than a melted glass of the same nominal composition, because the gel has a higher interfacial area and lower crosslink density (with terminal OH and OR groups). The gel is *unstable,* not metastable, because there is no energy barrier to prevent the gel from relaxing toward the glassy state; the structure of the gel is preserved only because the mobility of the atoms is limited at low temperatures. Aging, which changes the structure in the direction of a denser amorphous or crystalline state, occurs whenever mobility is provided by elevated temperatures and/or the presence of a solvent. In Chapter 9 it was

Fig. 1.

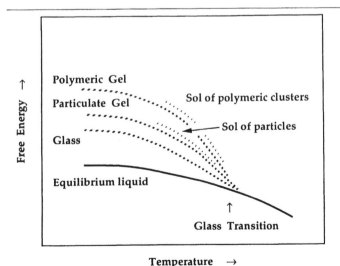

Schematic illustration of free energy of sol, gel, and glass phases as functions of temperature. Polymeric gels and sols are less densely crosslinked, so they are more energetic than the particulate variety.

shown that the skeletal density of a gel increases during heating as soon as enough thermal energy becomes available to allow formation of new cross-links; at higher temperatures the gel sinters, eliminating the solid–vapor interface. All of these processes reduce the free energy of the gel, and all occur spontaneously without the need to surmount an energy barrier.

A *glass* is a noncrystalline solid. Historically, a glass has been defined as a supercooled liquid, because that used to be the only known way to make one. Now amorphous solids are routinely made by vapor deposition, by applying shock waves or irradiation to crystals, and by a variety of other tricks including sol-gel processing [1]. There are occasional arguments as to whether all of these products should be called glasses, or whether materials not quenched from liquids should be distinguished as noncrystalline solids or amorphous solids. If the properties of the material were dependent on the method of fabrication, such a distinction might make sense; however, when there are no significant differences, distinguishing between silica glass and noncrystalline silica is as impractical as calling Albuquerque by different names, according to which road you took to get there. Glass is an unstable material, so there is no "equilibrium glass" to use as a standard for comparison. The properties of a glass depend on thermal history (e.g., cooling rate, annealing time) and can vary over a wide range, so the properties obtained by vapor deposition (for example) can generally be obtained from a melted glass following a suitable heat treatment. Consequently we shall use the terms *glassy, amorphous*, and *noncrystalline* interchangeably. Further, we regard the solid phase of a non-crystalline gel as a glass, usually with a composition high in OH and OR.

To examine the thermodynamics of the glassy state, it is convenient to follow the changes that occur during cooling of a liquid. As shown in Fig. 2, the thermal expansion coefficient decreases from that of the liquid (α_L) to that of the glass (α_g) in the vicinity of the glass transition temperature, T_g. Below T_g, the atoms are fixed in position and the thermal expansion results from the increase in the amplitude of their vibrations around those positions. Above T_g the atoms become increasingly free to break bonds and move about, so the volume can increase more quickly as the temperature rises (making $\alpha_L > \alpha_g$). If the liquid were cooled infinitely slowly, the atoms would be able to rearrange into the equilibrium configuration and the volume would follow the equilibrium curve. However, the rate of rearrangement decreases as the viscosity (η) increases, and η rises rapidly as the temperature drops. At any realistic cooling rate ($q = dT/dt$) the liquid will fall out of equilibrium and become a glass, frozen into a structure with a volume greater than that of the equilibrium liquid. The liquid is in metastable equilibrium below the liquidus temperature: it has greater free energy than the equilibrium crystal phase, but must overcome a nucleation barrier (discussed in

Fig. 2.

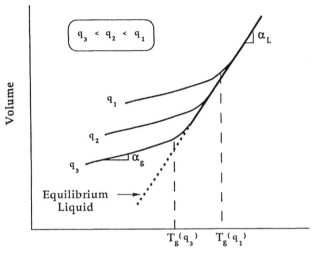

a)

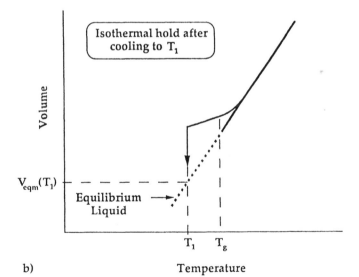

b)

(a) Change in the volume of a glass-forming liquid during cooling. The higher the cooling rate, $q = dT/dt$, the higher the temperature at which the volume departs from that of the equilibrium liquid. The thermal expansion coefficients of the glass and liquid are α_g and α_L, respectively. (b) If a liquid is cooled to temperature, T_1, below the glass transition temperature T_g, and then held isothermally, it relaxes toward the volume of the equilibrium liquid.

Section 1.2) to crystallize. Note that the glass is *unstable* with respect to the equilibrium liquid, so it relaxes spontaneously toward that structure at a rate limited only by the viscosity.

If the liquid is held isothermally in the vicinity of T_g (say, at T_1 in Fig. 2), the volume gradually relaxes toward its equilibrium value, V_{eqm}. The rate of *structural relaxation* can be crudely represented by[†] [3]

$$\frac{dV}{dt} = \frac{V_{eqm} - V(t)}{\tau_R} \tag{2}$$

where τ_R is the *structural relaxation time*, which is proportional to η:

$$\tau_R = \eta/K_R \tag{3}$$

where K_R is a constant with a value of $\sim 2.5 \times 10^9$ Pa for oxide liquids [4]. Cooling can be regarded as a sequence of temperature steps of size ΔT followed by relaxation (according to Eq. 2) for time $\Delta t = \Delta T/q$, which means that less time is available in each temperature interval when the cooling rate is greater. Relaxation is arrested when T drops to the point that τ_R increases to Δt, so the liquid falls out of equilibrium at a higher temperature when q is larger. For a typical oxide glass, an increase in q of one order of magnitude raises T_g by $\sim 25°C$.

When a glass is reheated, its structure responds sluggishly to the increasing temperature [2]. The atoms remain in a densely packed configuration so (as indicated in Fig. 3) the volume of the glass crosses the equilibrium liquid curve and falls *below* that of the liquid on reheating; then when the temperature becomes high enough to provide sufficient mobility, the structure rapidly approaches that of the equilibrium liquid. The slope of $V(t)$ during the sudden approach to equilibrium is higher than α_L, as shown in Fig. 3b; this *overshoot* in the slope is most pronounced when the heating rate is much higher than the preceding cooling rate. If the cooling rate is much faster than the heating rate, then the volume of the glass is relatively high and the atoms are so mobile that the structure decreases toward equilibrium at low temperatures; instead of an overshoot, there is an *undershoot* of the thermal expansion coefficient before T_g. The enthalpy of the glass exhibits the same behavior, including the overshoot and undershoot in the slope of the curve; that slope is the *isobaric heat capacity* ($c_p = dh/dT$). This property is readily measured using a differential scanning calorimeter (DSC), so it is widely used

---

[†]Structural relaxation is not a simple exponential process, as seems to be implied by Eq. 2, in part because η and τ_R also change with the structure [2]. When the free volume of the glass is large, the mobility of the atoms is greater and η is lower than at equilibrium; during relaxation η increases and V decreases. In addition, actual relaxation data reveal a distribution of relaxation times; τ_R can be regarded as an average of that distribution.

Fig. 3.

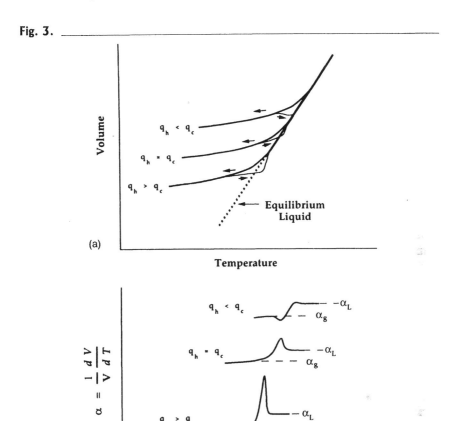

(a)

(b)

Change in volume during cooling and reheating of glass. (a) On reheating, the volume of the glass crosses the equilibrium liquid curve (falling below that of the liquid), then rapidly rises to equilibrium. (b) Slope $\alpha = d \ln V/dT$ rises above α_L (overshoots) at T_g before it approaches equilibrium. When the cooling rate (q_c) is greater than the heating rate (q_h), there is an exotherm (undershoot) before the overshoot; when $q_h > q_c$ there is no undershoot, and the overshoot is higher.

to examine the kinetics of relaxation. (See Chapter 12 of ref. [2].) A study of structural relaxation in gels is discussed in Section 4.2.

If a glass is allowed to relax isothermally, it is difficult to be sure that its final volume represents structural equilibrium, rather than being an unstable state in which it is trapped by its high viscosity. Since glasses can be prepared with volumes above or below the equilibrium value, it is possible to establish that they approach the same state from both directions, as illustrated in

Fig. 4.

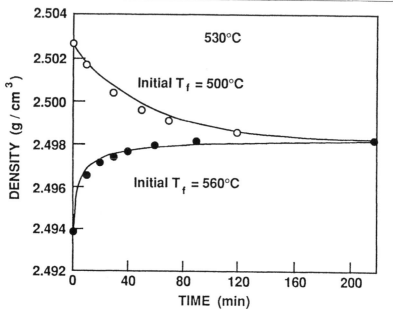

Change in density with time at 530°C for soda-lime-silicate glass previously equilibrated at 560°C (●) and 500°C (○); the same equilibrium state is approached from above and below. Data from Hara and Suetoshi [5].

Fig. 4. If there were an equilibrium state with lower energy, the denser glass would be expected to approach that state rather than rising to meet the sample relaxing from high volume. This is a strong indication that the metastable equilibrium liquid represents the configuration with the lowest free energy; that is, there is no indication of the existence of more than one metastable liquid structure. If there were, then the glass would sometimes fall into one of those states rather than relaxing continuously toward the equilibrium liquid, and that has never been observed.

1.2.   Phase Transformations

The liquid is in a state of *metastable equilibrium* with respect to the crystal when the temperature is below the liquidus, T_L. There is an energy barrier to the *nucleation* of crystals, since that requires the creation of a liquid–solid interface with a specific energy of γ_{SL}. A crystal will not form unless the energy invested in forming the interface is balanced by the free energy gained by forming a unit volume of the new phase, Δg_c. Consider a spherical crystal (called a *nucleus*) of radius r. The free energy change when it forms is [6]

$$\Delta g = 4\pi r^2 \gamma_{SL} + \tfrac{4}{3}\pi r^3 \Delta g_c. \tag{4}$$

The first term on the right is positive, as energy is expended to form the surface of the crystal, and the second term is negative. An increase in the size of the crystal will decrease the free energy of the system only if $d\,\Delta g/dr < 0$, so the *critical radius of the nucleus* (i.e., the size beyond which it will grow spontaneously) is found by setting $d\,\Delta g/dr = 0$ in Eq. 4. The result is

$$r_c = -2\gamma_{SL}/\Delta g_c \tag{5}$$

(which is positive, since $\Delta g_c < 0$). Crystals smaller than r_c tend to melt (or dissolve), so crystallization will proceed only after a statistical "accident" produces a nucleus larger than the critical size. Once this barrier is surmounted, the system can move directly toward equilibrium by growth of the crystal. The barrier diminishes as the temperature drops, because

$$\Delta g_c \approx \Delta h_c (T_L - T)/T_L \tag{6}$$

where Δh_c is the *heat of fusion*. At the melting point $(T = T_L)$ the free energies of the liquid and crystal are equal, so no energy is gained by forming a crystal, and the nucleation barrier is infinite. At low temperatures the energy gain (Δg_c) is large, and the critical nucleus becomes small enough to be formed with ease; however the *rate* of formation of nuclei, I_v, depends on the viscosity [7]

$$I_v = \frac{k_1}{\eta} \exp\left(-\frac{k_2 \gamma_{SL}^{\;3}}{T(\Delta g_c)^2}\right) \tag{7}$$

where k_1 and k_2 are constants. Thus, nucleation becomes energetically favorable at lower temperatures, but may be inhibited kinetically by the high viscosity.

The process of formation of a critical nucleus is called *homogeneous nucleation*, when only the liquid and corresponding crystal are involved. If another substance facilitates nucleation, the process is called *heterogeneous nucleation*. For example, suppose that a piece of crystalline "dirt" is embedded in the glass, as in Fig. 5. If the energy of the solid–solid interface between the inclusion and a growing crystal is small, then the energy balance in Eq. 4 tips in favor of crystal growth: the volume of crystal can form with less energy consumed by the solid–liquid interface. This allows crystals to form even at temperatures very near T_L. It is often observed that crystals appear first on the exterior surface of a body, which leads to the erroneous impression that surfaces are heterogeneous nucleation sites. However, free surfaces actually impede nucleation, because a crystal–vapor interface has a

Fig. 5.

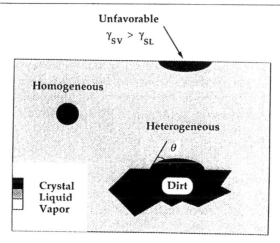

Schematic illustration of heterogeneous nucleation. If the contact angle, θ, between the crystal, liquid, and inclusion is small, less energy is expended in creating the surface of the nucleus. Nucleation at the exterior surface is unfavorable, because it requires formation of a solid–vapor interface with a high specific energy (γ_{LV}).

higher energy than a crystal–liquid interface; therefore, it is easier for a crystal to form inside the liquid than at the liquid–vapor interface. The reason for the prevalance of surface nucleation is that surfaces are generally contaminated, and the dirt provides heterogeneous nucleation sites.

Once a nucleus has formed, the crystal grows at the rate u, which can be written as [7]

$$u = \frac{f}{\eta}\left[1 - \exp\left(-\frac{\Delta g_c}{R_g T}\right)\right] \tag{8}$$

where R_g is the ideal gas constant and f is the *interface site factor*, which represents the fraction of sites on the interface that are available for attachment of atoms. For example, growth might occur by joining of atoms to a screw dislocation, in which case surface sites not adjacent to the ledge of the dislocation would be inactive. The quantity in brackets represents the probability that an atom that has jumped from the liquid to the surface of the growing crystal will remain there, rather than jumping back into the liquid. As the temperature decreases, that probability approaches unity and so does f, so the growth rate depends only on the viscosity, $u \propto 1/\eta$. At any given temperature, Eq. 8 indicates that the growth rate is constant in time.

Equation 8 applies when the composition of the liquid is the same as that of the crystal, which is often not the case; for example, cristobalite grows in alkali silicate glasses containing $\leq 10\,\text{mol}\%$ alkali. As indicated in Fig. 6,

Fig. 6. _____

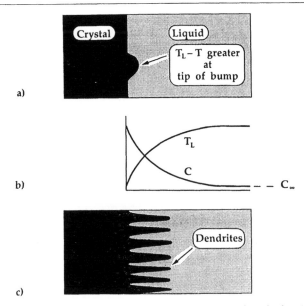

a)

b)

c)

When the crystal differs in composition from the liquid (a), the rejected solute (e.g., alkali oxide excluded from a cristobalite crystal) increases in concentration (c) at the interface where it reduces T_L (b). This can result in instability of the planar interface, leading to the appearance of dendrites (c).

the rejected alkali builds up at the crystal–liquid interface and the rate of growth becomes controlled by the rate of diffusion of solute (alkali, in this example) away from the crystal. In that case, u decreases as $t^{-1/2}$, because the layer of solute becomes thicker as the crystal grows, so transport of silica to the interface becomes increasingly difficult. The elevated concentration of alkali not only affects transport, it causes a reduction in T_L, as shown in Fig. 6b. This situation, where T_L increases with distance from the interface, is called *constitutional supercooling* [8] and it is important because it can destabilize the planar interface: if a bump forms on the crystal, it projects into a region where it has a greater driving force for growth ($T_L - T$), so u increases and the bump races away from the planar interface. This is often the origin of the formation of *dendritic* (treelike) crystals or *spherulites* (spherical clusters of dendrites growing from a common point). An interesting feature of dendritic growth is that it proceeds at a constant rate, rather than decreasing as $t^{-1/2}$, because the tip of the dendrite is constantly advancing into "fresh" liquid with a low solute content. Therefore, the rate of crystal growth is usually constant, even when the compositions of the crystal and liquid are different.

The amount of liquid that is transformed into crystal in a given time depends on the rate of growth of existing crystals and the rate of nucleation of new ones. If a spherical crystal is nucleated at time t' and grows at a constant rate u, its volume increases as $(4\pi/3)[u(t - t')]^3$. It was shown by Avrami (see discussion in Chapter 12 of ref. [6]) that the volume fraction of such crystals at time t is

$$x = 1 - \exp\left(-\frac{4\pi}{3}u^3 \int_0^t I_v(t - t')^3 \, dt'\right) \tag{9a}$$

$$= 1 - \exp\left(-\frac{\pi}{3}u^3 I_v t^4\right) \tag{9b}$$

where Eq. 9b applies when the nucleation rate is constant. In general, allowing for nonspherical crystals and time-dependent nucleation rates, the *Avrami equation* can be written as

$$x = 1 - \exp\left[-\left(\int_0^t k \, dt'\right)^n\right] \tag{10a}$$

$$= 1 - \exp[-(kt)^n] \tag{10b}$$

where k is a temperature-dependent function related to u and I_v, and n is a constant. It is important to remember that Eqs. 9 and 10b apply only for *isothermal* phase transformations. If the experiment is done at a constant rate of heating or cooling, some relatively simple approximations to Eq. 10a can be obtained [9,10]; nevertheless one often sees Eq. 10b erroneously applied to such data, in spite of frequent criticism (e.g., [11–13]) of the practice.

A common method of analysis of nonisothermal crystallization data is to assume that k in Eq. 10 can be represented by the Arrhenius equation,

$$k = k_0 \exp(-Q/R_g T) \tag{11}$$

where k_0 is a constant and Q is an "activation energy for crystallization." This is a severe approximation, as is revealed by a glance at Eqs. 7 and 8, since k represents the temperature-dependences of both nucleation and growth. In fact, the growth rate is zero at T_L, rises to a maximum with undercooling as the driving force increases, and then decreases with further cooling as the rising viscosity inhibits growth. The nucleation rate also passes through a maximum (typically at a much lower temperature than the maximum in the growth rate). For such a function the "activation energy" is negative at high temperatures (since the rate decreases with increasing temperature), passes through zero, and becomes positive at low temperatures. Therefore, the use of Eq. 11 is appropriate only when crystallization occurs with a fixed number of growth sites (zero nucleation rate) and when growth is completed before T rises to the temperature of the maximum growth rate. In that case, the activation energy is related to that of the viscosity, η; that is, the crystallization kinetics are controlled by the atomic mobility.

Another type of transformation is *phase separation*, where one liquid separates into two immiscible liquid phases [14–16]. The two liquids may be stable phases (occurring above the liquidus), as in the alkaline earth silicate systems shown in Fig. 7a, or metastable as in the Na_2O–SiO_2 system shown in Fig. 7b. The dashed curve in Fig. 8 is called the *spinodal*. Liquids whose compositions fall between the miscibility boundary and the spinodal must overcome an energy barrier to separate into two phases, so phase separation proceeds by a process of nucleation and growth. Compositions within the spinodal are unstable, so they separate freely by a process called *spinodal decomposition*. It is sometimes stated that the mechanism of separation can be determined from the morphology of the phases, because nucleation and growth produces droplets, while spinodal decomposition leads to two continuous interconnected phases (reminiscent of the structure of a gel). However, the morphology is not a reliable indicator, because the continuous phases can break up and the droplets can form necks, so either process can produce either structure.

2.

UNIQUE PROPERTIES OF SOLS

Here we briefly note some of the features of inorganic sols that are of technological and scientific importance, and that are not obtainable with conventional ceramics:

(1) *Film deposition* is probably the most important new capability offered by sol-gel processing, as discussed at length in Chapters 13 and 14.

(2) *Fiber drawing* is a related capability that is also of commercial importance [18–20]. (See discussion in Chapters 3 and 14.)

(3) *Growth of particles* from sols allows the preparation of fine powders to be used in low-temperature sintering (as discussed in Chapter 11). Sols are also used in photochemistry, with the particles acting as floating substrates [21]. A particularly intriguing opportunity is the preparation of monodisperse spheres (discussed in Chapters 3 and 4). They can be used not only for low-temperature sintering, but for fundamental studies in colloid chemistry, including measurement of interparticle forces, studies of the kinetics of aggregation and the geometry of aggregates, and tests of theories of scattering of radiation.

Fig. 7.

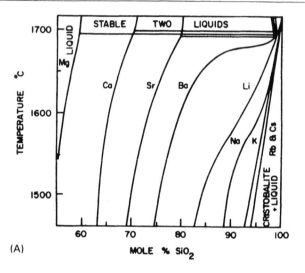

(A)

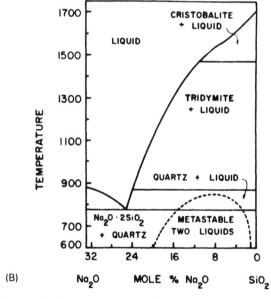

(B)

Liquid–liquid immiscibility in silicates. (A) Stable immiscibility (above the liquidus) in the alkaline earth silicate systems. (B) Metastable immiscibility in the Na_2O–SiO_2 system. From Seward [16].

Fig. 8.

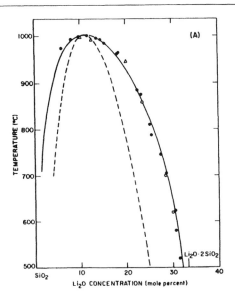

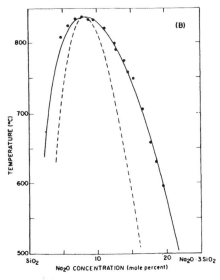

Metastable miscibility gaps in (A) Li_2O-SiO_2 and (B) Na_2O-SiO_2 systems. Phase separation occurs below the solid curves; the dashed curves are the spinodal boundaries. Between the dashed and solid curves, phase separation occurs by nucleation and growth of liquid droplets. Within the dashed curve, spinodal decomposition occurs. From Haller *et al.* [17].

3.

UNIQUE PROPERTIES OF GELS

The most obvious characteristic of a dried gel is its *porosity*, and this feature provides gels with properties that cannot be obtained with conventional ceramics. The low refractive index of xerogels is of interest in optics [22], and the spectacularly low index of aerogels (~ 1.007–1.24) makes them useful for Cherenkov radiation detectors [23]. Their high surface area makes gels attractive as catalytic substrates, especially in the case of aerogels, which have unusually high chemical reactivity [24]. The reactive surface also permits chemical modifications, such as nitridation, as discussed in Chapter 10. The negative aspect of this property is that gels may be attacked by the atmosphere; for example, sodium silicate compositions react with ambient CO_2 to form carbonates [25,26] that interfere with sintering. The high porosity of aerogels provides both transparency and low thermal conductivity that makes them valuable for thermally insulating windows [27,28]. The ability to prepare bodies with macropores and/or micropores makes it possible to use gels as hosts to contain optically active molecules [29] or as filters for separation of species [30]. The pores can also be filled with polymers to produce strong light-weight composites [31].

The high proportion of *nonbridging bonds* in gels produces a very high coefficient of thermal expansion [32,33] as well as providing reaction sites for chemical modification. The open structure of gels also provides exceptional ionic conductivity [34] as well as electronic conductivity and electrochromic performance much superior to crystalline ceramics [35].

A particularly remarkable feature of gels is their capacity to incorporate both *organic and inorganic components*, as demonstrated by the excellent work of Schmidt and colleagues [36–39]. These materials can be formulated to have high or low permeability to gases, toughness, and scratch-resistance, and they can be applied as coatings on plastics, because of their low curing temperatures.

4.

GLASSES

There has been some controversy over the possible differences between glasses made by melting of conventional raw materials (such as sand, minerals, and salts) and those derived by sol-gel processing. The most lucid discussions of the problem are by Cooper [40] and Uhlmann *et al.* [7], who distinguish between glasses melted at high temperature from conventional batch materials (MB) or from gels (MG), and glasses obtained by sintering gels at temperatures in the vicinity of the glass transition temperature (SG). In this section we follow their line of argument.

It is clear that MB are true equilibrium liquids at high temperatures, because (as explained in Section 1.1) they relax to the same state from above or below (i.e., after being displaced from equilibrium by heating or cooling). If MG liquids are different from MB liquids, then the MG liquids must have *higher* free energy; otherwise we would have to conclude that liquids (which always relax to the state of MB) relax toward a state of higher free energy (in violation of the concept of equilibrium), or that an activation barrier exists between these liquid states. It is hard to conceive of such a barrier, but Cooper [40] notes that its existence could be tested by "seeding" (nucleating) an MB melt with MG liquid; if the MG liquid were lower in free energy, there would be a phase change in that direction. Spectroscopic studies (e.g., [32,41]) show that the local bonding in a gel approaches that of a melted glass at temperatures below T_g; since the structure changes in that direction, it is safe to assume that MB represents the equilibrium liquid state. This means that g must be higher for the MG liquid, if it is not identical to MB; in the latter case, it can be shown [40] that it must have a higher liquidus temperature, T_L, so any crystal that formed at $T_L(MG)$ would be above $T_L(MB)$. Such a crystal would be unstable with respect to the MB liquid and (since there is no barrier to melting) would immediately melt; thus we should never expect to see crystals in an MG melt at temperatures above $T_L(MB)$.

One might argue that the MG melt is indeed higher in free energy than MB, and the relaxation of the structure toward equilibrium is so slow that differences in the properties of the MG liquid persist. According to Eq. 3, the characteristic time for structural relaxation is <1 s when $\eta < 2.5 \times 10^9$ Pa·s, which is at least four orders of magnitude greater than the viscosity of a silicate liquid at its liquidus temperature. Even at T_g the relaxation time is $\tau_R \sim 1000$ s, so a brief heat treatment would permit a homogeneous liquid to reach equilibrium, However, this refers only to relaxation of the local bonding arrangement. Differences in structure produced by inhomogeneity of the batch or growth of second phases are governed by a different characteristic time, τ_D:

$$\tau_D \approx L^2/D_c \tag{12}$$

where D_c is the chemical diffusion coefficient and L is the size of the dissolving heterogeneity or growing phase. The diffusion coefficient can be related to the viscosity by the Stokes–Einstein equation [42],

$$D_c = \frac{k_B T}{6\pi a_0 \eta} \tag{13}$$

where k_B is Boltzmann's constant and a_0 is the radius of the diffusing species. With $a_0 = 0.1$ nm, Eq. 12 becomes

$$\tau_D(s) \approx 150L^2\eta/T \qquad (14)$$

where L is in microns, η in Pascal·seconds, and T in degrees Kelvin. Batch is typically melted at a viscosity of $\sim 10^3$ Pa·s, so a 1 μm heterogeneity would dissolve in 100 s (assuming $T = 1500$ K); under the same conditions, the structural relaxation time is on the order of *micro*seconds. This means that relaxation is immeasurably fast in a melt, and homogeneity is achieved in minutes to hours, depending on the scale of the heterogeneity. Note, however, that near T_g (say, $\eta \approx 10^{13}$ Pa·s, $T = 1000$ K), the same heterogeneity would survive for $\sim 10^5$ years! This has obvious implications for the properties of sintered gels (discussed in Section 4.2). Clearly we can expect equilibrium values for properties that depend only on local bonding (e.g., density and modulus); however, in sintered gels that have never been heated above T_g (of the corresponding melted glass) differences are possible in transport-controlled properties, such as homogeneity or degree of crystallization.

Although phase changes can be inhibited by low temperature processing of gels, the rate of transformation at a given temperature is likely to be faster for MG or SG than for MB [7,40,43]. The higher free energy of the gel-derived material increases the driving force for the transformation, and the (typically) higher hydroxyl content reduces η and thereby accelerates both the nucleation and growth rates. Equation 7 indicates that the nucleation rate may be more affected, because of its stronger dependence on g and especially its exponential dependence on γ_{SL}^3. Unfortunately, it is difficult to predict the effect of OH content or subtle structural differences on the interfacial energy. Apart from differences in the material, it is important to consider the effect of differences in thermal history between MB and SG, as emphasized by Uhlmann *et al.* [7]. When the melt is cooled it passes through the range of maximum crystal growth rate before reaching T_g, so nuclei (particularly heterogeneous nuclei) can grow extensively. On the other hand, SG are sintered in the vicinity of T_g, where u is negligible and I_v is small. On heating to a somewhat higher temperature, however, SG passes through the range of the maximum homogeneous nucleation rate; if brought to a temperature where the growth rate is appreciable, SG may transform faster than MB, because of the larger number of nuclei and lower viscosity. It is often suggested that SG contain fewer heterogeneous nuclei, because they are not contaminated by contact with a crucible at high temperatures. However, it should be remembered that lab dirt is largely composed of aluminosilicates, so gels may be full of potent nuclei if they are not prepared from carefully cleaned glassware and filtered reagents.

In the following subsections we examine the experimental evidence for differences in properties, starting with comparisons between MB and MG. As expected, the differences are generally attributable to variations in hydroxyl content. More interesting results are found in comparisons of MB and SG, where the enormous difference in thermal history permits the preparation of truly novel materials.

4.1. Melted Gels

One of the earliest applications of sol-gel technology was to facilitate the preparation of homogeneous multicomponent oxides for studies of phase relations (see discussion in Edgar [44]), and this capability has been exploited for reasons as diverse as the preparation of optical glasses with exceptionally low scattering [45] and solder fluxes allowing improved arc stability [46]. It should be recognized, however, that some systems develop chemical and structural inhomogeneity as a result of crystallization during gelation and aging (e.g., [47,48]); even in borosilicate gels, which do sinter to homogeneous glasses, the Si–O–B bonds form during sintering, not during gelation [49]. Nevertheless, it has been amply demonstrated that gels do provide exceptionally homogeneous melts. The gels' advantage comes from the fact that the scale of the heterogeneity, L^2 in Eq. 14, is smaller in the gel than in conventional batch, so that dissolution is faster at a given temperature (or can be achieved at a lower temperature in a given time).

Relatively few comparisons have been made of the crystallization behavior of MB and MG (comparisons with SG are discussed in Section 4.2). Mukherjee *et al.* [50] prepared refractory glasses in the lanthanum silicate family, including some containing alumina or zirconia. Both MB and MG glasses were made by melting at ~2000°C, during which some preferential evaporation of silica occurred. Differences were seen in the kinetics of crystallization at 1150°C and in the phases that appeared, but no chemical analyses were performed after melting, so it is quite possible that the MB and MG were not identical in oxide content. Neilson and Weinberg [51] attempted to address this uncertainty by comparing the crystallization behavior of MB and MG glasses of 19 wt% Na_2O-81 wt% SiO_2, where the composition was analyzed by atomic absorption after melting. The glasses were melted at 1565°C and crystallized during an isothermal hold at 720°C; the phases and their relative amounts are given in Table 1. The same phases appeared in both glasses, but the volume fraction crystallized from the MG was slightly greater, a result attributable to the higher OH content of the gel-derived glass. In fact, the hydroxyl

Table 1.

Crystalline Species Formed by Heating Sodium Silicate Gel and Glasses.

		Heating Time (h)								
		5		7		9		16		66	
Phase	Form[a]	SS	GP	SS	GP	SS	GP	SS	GP	SS	GP
α-Cristobalite	MB	2	3	3	—	3	—	3	4	2	4
	MG	2	3	—	—	3	—	4	5	3	5
	XG	—	5	—	—	—	—	—	5	—	5
α-Na$_2$O·2SiO$_2$	MB	0	0	0	—	0	—	0	0	0	0
	MG	0	0	—	—	—	—	0	0	0	0
	XG	—	4	—	—	—	—	—	4	—	4
3Na$_2$O·8SiO$_2$	MB	0	0	2	—	1	—	2	3	2	3
	MG	2	1	1	—	1	—	3	3	2	3
	XG	—	0	—	—	—	—	—	1	—	2
Na$_2$O·3SiO$_2$	MB	—	4	3	—	3	—	3	5	4	4
	MG	1	4	—	—	3	—	4	0	4	0
	XG	—	0	—	—	—	—	—	0	—	0
Na$_2$O·4SiO$_2$	MB	0	0	3	—	3	—	5	0	2	0
	MG	1	0	—	—	4	—	5	0	4	0
	XG	—	0	—	—	—	—	—	0	—	0

Source: Data from Weinberg and Nielson [53].

[a] SS = surface slice from bulk sample. GP = ground powder. MB = melted batch. MG = melted gel. XG = xerogel powder (no prior heat-treatment). Numbers 0–5 indicate relative amount of crystal phase (0 = none, 5 = large amount).

content can sometimes affect the phases obtained, as well as the kinetics: Low and McPherson [52] found that different phases appeared in the same gel when heated in various atmospheres, including steam, air, and argon. The higher OH content obtained by firing in Ar was the most important factor; molecular water from steam had little effect on the crystallization kinetics.

The kinetics of phase separation in MB and MG glasses have been compared in several studies that are reviewed by Weinberg and Neilson [53]. Faber and Rindone [54] prepared a soda-lime-silicate glass by three methods: MB from sand and carbonates, MG(I) using Ludox® colloidal silica plus nitrates of Na and Ca, and MG(II) using TEOS plus nitrates. Chemical analysis showed that MB and MG(I) were very close in composition, but MG(II) was low in silica because of vaporization of TEOS. The glasses were melted at 1320°C for 48 h under flowing atmospheres of dry oxygen, water-saturated oxygen, or steam. Phase separation was induced by heating at 600°C for 1 h, then at 650°C for 4 h, and the resulting microstructure was examined by small angle X-ray scattering (SAXS) and TEM. The OH content of MB was strongly dependent on the melting atmosphere (increasing from <10 ppm in dry O_2 to 150 ppm in steam), and the size of the droplets of second phase decreased as OH increased. For glasses with comparable concentrations of OH, there was little difference in behavior between MG and MB. The smaller droplets in the "wet" MB and the MG indicate that OH speeds nucleation. A startlingly different result was obtained by Weinberg and Neilson [55] in a study of a soda–silica glass reportedly containing 18.56 mol% Na_2O: not only were the kinetics of separation different for MB and MG, but the temperature of the miscibility boundary was shifted by more than 100°C! Similar results were obtained in an expanded study in the same system [54]. However, it was subsequently demonstrated [53] that the chemical analyses of those glasses were incorrect. For glasses of the same composition, negligible differences in behavior were found.

Two studies report differences in MB and MG at elevated temperatures. Weinberg and Neilson [57] found that the liquidus temperature was higher for MG, which is impossible according to Cooper's argument [40]: a crystal forming at T_L(MG) would immediately melt, because it would be above T_L(MB). Again, this surprising result was a consequence of faulty chemical analysis. Another remarkable report is less easy to explain: Yoldas [58] found that the high-temperature viscosity of a soda-lime-silicate glass was different for MB and MG, even though the oxide and hydroxyl contents were the same. His data, shown in Fig. 9, indicate that the liquids were different in viscosity even after 150 h at 1375°C! Chemical analysis at the end of that treatment showed the glasses to have the same composition to within 0.1 wt% of the oxides; MB had 388 ppm OH and MG had 436 ppm. It is curious that the difference was greater at higher temperatures, whereas changes in alkali content usually have a greater effect on the low-temperature viscosity. It seems inescapable that there was some error in the chemical analysis of the glasses, but the reported difference in viscosity would require a change in oxide content of ~1 wt% Na_2O or ~3 wt% CaO, which should be easily detected [59]. Yoldas [60] notes that considerable oxygen substoichiometry can exist in a gel, and that could have an important effect on properties. However, in the viscosity study the oxygen deficiency was measured to decrease from 8 to 2.3 ppm during the 150 h of melting; the melted oxide had a deficiency <1 ppm. Since the original glass has ~1 nonbridging oxygen per Si (because of the Na and Ca content), it is hard to imagine how that degree of oxygen deficiency could affect the viscosity.

Fig. 9.

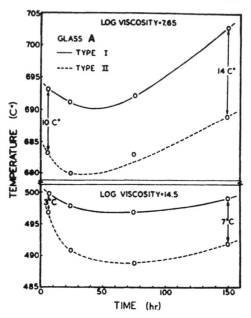

Glass with composition (wt%) 8.25 CaO-17.75 Na$_2$O-74.00 SiO$_2$ prepared from conventional batch (type I) or from alkoxides (type II). Differences in viscosity remain even after melting at high temperature (1375°C) for 150 h. From Yoldas [58].

4.2. Sintered Gels

Mackenzie [61] surveyed the literature in 1982 and concluded that the properties of gels became indistinguishable from those of ordinary melted glasses after the gels had been heated to T_g. For silica gel made from TEOS, Bertoluzza *et al.* [41] found that the infrared and Raman spectra of the gel evolved continuously toward that of silica, and the transformation was complete at ~800°C; the gel differed from fused silica only in that it had a higher OH content. This similarity in structure accounts for the observation by Yamane *et al.* [62] that silica gels heated only to 900°C were identical to silica in density, refractive index, hardness, thermal expansion, and modulus. Of course, in comparing properties it is necessary to take account of the effects of impurities. For example, Hench *et al.* [22,63] found that chlorine-dried silica gels after densification had lower thermal expansion and higher refractive index than fused silica (see Table 2), possibly because of retained Cl. Chandrashekhar and Shafer [64] found that silica gels made

Table 2.

Comparison of Properties of Silica Gel and Fused Silica.

Type of Silica	Tradenames	Total Cation Impurity (ppm)	OH Impurity (ppm)	Cl Impurity (ppm)	UV Trans. (50% Trans.) (nm)	Thermal Exp. Coefficient (cm/cm×10^7)	Bubbles and Inclusions (max #/cubic in.)	Strain (nm/cm)	Refractive[h] Index n(d)	Dispersion[i] v(d)	Index Homogeneity[i] (×10^7)	Density (g/cc)
(I) Electromelted quartz	Vitreosil-IR[a] Intrasil[b] Puropsil A[f]	30–200	<5	0	212–223	0.54	0–8	5–10	1.458	67.8±0.5	3–100	2.21
(II) Flame-fused quartz	Homosil[b] NSG-OX[e] Vitreosil-O55[a] Optosil[b]	10–30	150–1500	0	210–220	0.55	0–5	5–10	1.458	67.8±0.5	3–4	2.21
(III) Hydrolyzed SiCl₄	7940[c] Dynasil 1000[d] Spectrosil-A&B[g] Suprasil-1[b] NSG-ES[e] Tetrasil[f]	1–2	600–1000	100	165–188	0.55–0.57	0–3	5–10	1.458	67.8±0.5	1–300	2.20
(IV) Oxidized SiCl₄	Spectrosil WF[a] 7943[c] Suprasil-W[b]	1–2	0.4–5	up to 200	165–180	0.55	0–2	10–40	1.458	67.8±0.5	10–40	2.20
(V) Dense gel-silica	GELSIL[g]	1–2	<1	0–>1000	165–168	0.2	0	5	1.458–1.463	67.8–66.4±0.5	15	2.20
(VI) Porous optical gel-silica	Poro-GELSIL[g]	1–2	>2000	0	250–300	—	0	—	1.28–1.45	—	—	1.30–2.10

Source: Hench et al. [22].

[a] Thermal American Fused Quartz; Montville, N.J.   [b] Heraus Amersil; Sayreville, N.J.   [c] Corning Glass Works; Corning, N.Y.   [d] Dynasil; Berlin, N.J.
[e] NSG Quartz; Japan.   [f] Electro Quartz; Saint-Pierre-Lès-Nemours, France.   [g] GELTECH, Inc.; Alachua, Fla.   [h] At sodium d-line.
[i] Abbé number at sodium d-line

from TMOS or fumed silica had higher dielectric constants and dielectric loss than fused silica. The properties did not correlate with alkali or OH impurities, but were found to result from residual carbon; the dielectric constant was linearly proportional to the elemental C content of the gel. Weimer et al. [65] later showed that gel-derived films could be prepared with electrical properties comparable to thermally oxidized silicon. Of course, it is also essential to account for the effect of porosity in any comparison. Partially sintered gels are lower in refractive index [22], chemical durability [66,67], strength [68], and modulus [69]. In the following, we use SG to refer to fully densified gels.

The relaxation behavior of SG was found by Puyané et al. [70] to resemble that of a rapidly quenched glass. (See Fig. 10.) On heating the gel in the differential thermal analyzer (DTA), they found an exotherm preceding the endothermic glass transition; when the SG was cooled and reheated the exotherm did not reappear. An MB sample with the same composition that was rapidly quenched showed remarkably similar behavior (Fig. 10b). Similar behavior was seen by Brinker et al. [71] in an acid-catalyzed silica gel, which showed an exotherm before the glass transition when the gel was first heated, but when the SG was cooled and reheated only an endotherm appeared. (See Fig. 34b in Chapter 9.) This pattern of relaxation is shown schematically in Fig. 3b: the rapidly quenched glass is frozen into an open structure (a state of high free energy), which relaxes exothermically at a relatively low temperature on reheating. The gel is expected to have excess free energy as well, because its bonds are formed in solution and the connectivity of the network (as revealed by NMR) is less than in a melted glass. The structure relaxes abruptly as soon as the temperature is raised toward T_g.

Scherer et al. [72] made a detailed comparison of the relaxation kinetics of an SG and MB glass with the composition given in Table 3. The gel was prepared from TEOS, aluminum sec-butoxide, boron trimethoxide, and sodium acetate using a procedure of sequential addition in that order [73]. The gels were sintered by heating to 650°C at 1°C/min in air. To reduce the OH content of the glass, other samples of gel (called "desiccated") were heated at 1°C/min to 400°C in air and held for 10 h, then heated to 500°C and held under a vacuum of 10^{-7} torr for 24 h. They were then heated at 1°C/min in helium to 650°C, by which time they were completely sintered. One sample of the gel was melted at 1325°C for 30 min (MG). The MB was made from sand, B_2O_3, Al_2O_3, Na_2CO_3, and NaCl, which was ball-milled and then melted at 1600°C for 6 h. The OH contents were found from the IR absorption at 2.7 μm, assuming an extinction coefficient of 56 1/mol·cm [74]. Heat capacity measurements were made by heating the samples to 980 K in the DSC, then cooling at rates of 0.62, 2.5, or 10°C/min, and reheating at 20°C/min. The dependence of the shape of the reheating curve on the

Fig. 10.

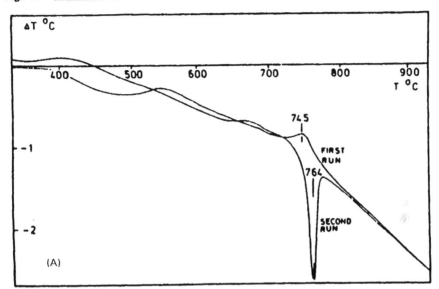

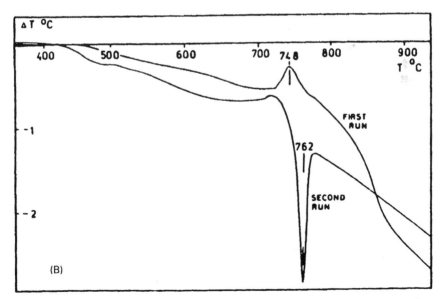

DTA traces for $13Na_2O \cdot 87SiO_2$ (wt%) gel (A) and conventionally melted and quenched glass (B). The exotherm in the first run indicates rapid relaxation of a structure with high free energy. (Compare with Fig. 3.) From Puyané *et al.* [70].

Table 3.

Analyzed Composition (wt%) of Samples Used in Relaxation Study.

Sample	Na$_2$O	B$_2$O$_3$	Al$_2$O$_3$	SiO$_2$	[OH] (ppm)
Sintered gel (SG)	7.71, 6.5	10.0, 10.7	1.0, 0.92	Balance	2200
Melted gel (MG)	7.01	8.9	1.08	83	620
Desiccated gel (SG)	—	—	—	—	440
Melted oxide A (MB)	7.09	10.0	1.01	81.9	770
Melted oxide B (MB)	12	15	1	82	490

Source: Scherer et al. [72].

Note: Analyses were duplicated at Corning Glass Works and Sandia National Laboratories (using inductively coupled plasma atomic emission spectroscopy).

prior cooling rate is a reflection of the spectrum of relaxation times controlling structural relaxation, and can be analyzed using a technique developed by Moynihan and his students [75], based on a theory by Narayanaswamy [76]. The details of the theory are discussed in ref. [2]. The result is that a single set of four parameters gives a very good fit to the data for the desiccated gel (SG), as shown in Fig. 11; more importantly, the *same* parameters provide an equally good fit for the MB. The curves for the SG that was not desiccated had the same shape, but were shifted to lower temperatures because of the lower viscosity produced by their higher OH content. Note that the gels were never heated more than ~60°C above T_g, yet their local structure (as revealed by the relaxation behavior) is indistinguishable from that of oxides melted at 1600°C. When the gel was melted at 1325°C, its subsequent relaxation behavior was unchanged (i.e., MG = SG). When this work was originally presented [77], it was reported that there was a considerable difference in behavior between the SG and MB (glass B in Table 3). However, it was later discovered that there was a difference in the oxide contents of those glasses (because of loss of Na and B during preparation of the gel), so another oxide was melted with a closer match to the gel (glass A). The difference in relaxation behavior of those melted oxides is shown in Fig. 12. This illustrates the importance of insuring that comparisons are made on identical compositions.

There are few studies of the kinetics of phase separation in sintered gels. Villegas et al. [78] compared SG and MB with the nominal composition 5.8Na$_2$O·15.6B$_2$O$_3$·78.6SiO$_2$ (mol%), but it is not clear whether the compositions were analyzed. The glasses were treated at 600°C for 12 h or 700°C for 4 h to induce phase separation, and it was found that the microstructure developed faster in the MB. This is surprising in light of the high hydroxyl content of the SG; in fact, the authors found that the SG crystallized upon heating to ~830°C, and attributed that fact to the mineralizing effect of

Fig. 11.

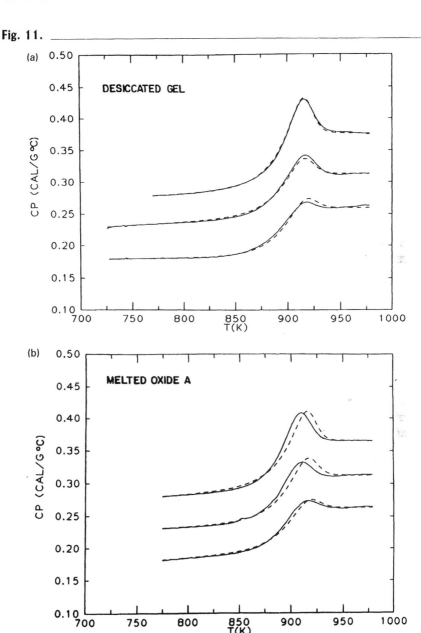

Reheating curves (20°C/min) following cooling at 0.62 (top curve), 2.5, and 10 (bottom curve) °C/min for: (a) desiccated gel and (b) melted oxide A (see Table 3). Curves shifted downward by 0.05 cal/g °C for clarity. Dashed curves in both figures calculated using best-fit parameters for gel. From Scherer *et al.* [72].

Fig. 12.

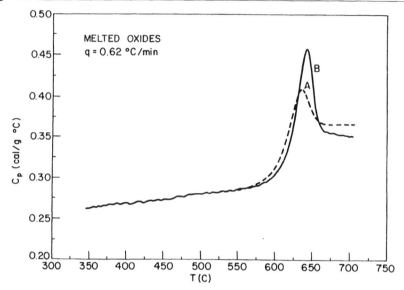

Reheating curves (20°C/min) for melted oxides A and B (see Table 3) following cooling at 0.62°C/min. From Scherer *et al.* [72].

OH. The slow development of phase separation may reflect inhomogeneity in the sintered gel, as Si–O–B bonds were found to appear only during sintering at 550 to 600°C; some heterogeneity was evidently present on a larger scale, since isolated regions of the gel contained crystals.

Yasumori *et al.* [79] found the opposite result in a glass with the nominal composition $10Na_2O \cdot 90SiO_2$ (mol%, analysis not reported): the SG exhibited much faster phase separation. The gel was prepared from sodium acetate and TMOS under acidic conditions, then heated slowly to 460°C, and held for 24 h, at which point it was clear but porous. When heated to 515°C, the gel became translucent as its surface area decreased from 160 to 50 m²/g; the IR spectrum after 48 h at 515°C was the same as for the MB, except for a higher OH content and a trace of retained carboxyl groups. The SAXS intensity (indicating phase separation) rose much faster than for the corresponding MB. This could be attributed to the higher OH content, but Yasumori *et al.* argue that it also reflects the higher mobility of sodium ions, which fail to bind to the network in the gel.

Chen and James [80] examined a glass containing $10Li_2O \cdot 90SiO_2$ (mol%, analyzed) and also found that the SG exhibited much faster phase separation than the corresponding MB. It is remarkable that the SG contained an interconnected microstructure on a scale of ~50 nm, yet it contained pores

~4 nm in diameter (revealed by TEM as well as mercury porosimetry). This is unexpected, because the characteristic time for sintering is approximately (see Eq. 13 of Chapter 11)

$$\tau_s \sim \frac{8\eta a}{3\gamma_{LV}} \tag{15}$$

where a is the radius of the pore or particle[†] and γ_{LV} is the liquid–vapor interfacial energy. Comparing the time to produce a structure with $L \approx 50$ nm at ~1000 K with the time to collapse a pore with $a \approx 4$ nm, assuming $\gamma_{LV} \approx 0.3$ J/m^2, Eqs. 12 and 15 indicate that $\tau_s/\tau_D \sim 10^{-4}$. Therefore, densification should be done long before phase separation, *unless* the pores are exclusively in the silica-rich phase; in that case, the viscosity controlling sintering would be much higher than that controlling phase separation. There is clearly a need for more research to establish the impact of hydroxyl content and network structure on the kinetics of phase separation in sintered gels. It is clear, however, as shown shortly, that homogeneous SG can be prepared in ranges of composition where MB are always phase separated.

The kinetics of crystallization have been much more thoroughly examined. The work has been motivated by the fact that some compositions are quite easily made from the melt, but are difficult to prepare from gels without devitrification. For example, Schwartz *et al.* [81] prepared a gel containing $15Li_2O \cdot 85SiO_2$ that turned opaque (phase separated) then crystallized on heating from 600 to 800°C, and was still not fully dense at 900°C. Branda *et al.* [82] made a more detailed study of $Li_2O \cdot 2SiO_2$ glasses made from alkoxides or melted from oxides (chemical analyses not reported). The gels were dried and heated for 2 h at 420 to 450°C, at which point they were white and opaque, but amorphous. The differential thermal analysis traces showed that the SG and MB had the same T_g (~480°C) and T_L (~1046°C), and both crystallized to lithium disilicate, but the SG crystallized at a lower temperature. The kinetics of crystallization during heating at a constant rate were erroneously analyzed using Eq. 10b. This means that the parameters have doubtful significance, but since the analysis was performed consistently, the comparison between SG and MB is valid. The result was that the same activation energy was found for both materials. Thus the only difference was in the rate of crystallization, which was presumably accelerated by the higher OH content of the SG. It is particularly interesting that the gel yielded lithium metasilicate if it was not preheated at 420–450°C; that treatment may have allowed the lithium to diffuse into the skeleton, making the gel structure more like that of the melted glass.

[†] Strictly, it is the particle size, since this relation comes from Frenkel's equation; but it doesn't really matter for an order-of-magnitude estimate.

As noted earlier, the soda silicates react with the atmosphere to form carbonates [25], and the consequences are indicated in Fig. 41 of Chapter 11. At high Na_2O contents, densification begins before the carbonates decompose, causing bloating; at low alkali contents, crystallization precedes densification. The $Na_2O \cdot 2SiO_2$ composition was examined by Grassi et al. [83], who prepared gels from alkoxides and a melted glass from oxides (chemical analyses not reported). Powders of gel were preheated to 425°C for 24 h, then heated at a constant rate in the DTA. The SG and MB were found to have the same T_g (~ 475°C) and to crystallize to the same phases. Again, the nonisothermal kinetics were analyzed inappropriately by application of Eq. 10b; nevertheless, both materials exhibited the same activation energy for crystallization (~ 60 kcal/mol), which is about equal to the activation energy for viscous flow in this glass. Neilson and Weinberg [51] compared SG, MG, and MB with the composition $19Na_2O \cdot 81SiO_2$ (mol%, analyzed); whereas the MB and MG yielded the same phases (as discussed in Section 4.1), the SG produced different phases and crystallized much faster. In this study, the gel was not pretreated, but was heated directly to 720°C for crystallization, so the differences in crystallization behavior may reflect inhomogeneity in the sodium distribution in the gel. Perhaps a pretreatment of the sort used by Branda et al. [82] would produce a different result.

Gels with very low soda contents were examined by Phalippou et al. [84], who compared silica gels made from TMOS with others made from Ludox® colloidal silica, which contains ~ 4000 ppm Na_2O. The colloidal (particulate) gels crystallized to cristobalite between 900 and 1250°C, and therefore could not be sintered to full density. Above 800°C the infrared spectra of both types of gel were the same, and they had similar OH contents, so the difference in behavior is attributed to the alkali content. The effect of traces of alkali was directly demonstrated by Zhu et al. [85], who added small amounts to otherwise pure silica gels made from TMOS or TEOS. The temperature at which crystallization began was reduced by 250°C by additions of 200 ppm Na_2O or K_2O, and by 400°C by that amount of Li_2O. In contrast, the temperature of crystallization was reduced only ~ 40°C by changes in gel structure and OH content resulting from changing the water : alkoxide ratio from 4 to 16. Zarzycki [43] concludes that the difference in crystallization rate between gel-derived and melted silica is so large, even allowing for differences in alkali and OH impurities, that the gels must contain heterogeneous nuclei.

Zelinski et al. [86] prepared anorthite ($CaO \cdot Al_2O_3 \cdot 2SiO_2$) from TEOS and the double alkoxide $Ca(Al(OC_2H_5)_4)_2$, and cordierite ($2MgO \cdot 2Al_2O_3 \cdot 5SiO_2$) from TEOS and $Mg(Al(OC_2H_5)_4)_2$. Dried gel powders were calcined at 700°C, at which point they remained amorphous to X-rays; sintering becomes appreciable in these gels only above 800°C [87]. The anorthite powders were then heated at rates (q_H) of 5 to 100°C/min in the DTA, with the results shown in Fig. 13: the peak of the exotherm (T_{cr}) shifts to higher temperatures as q_H increases. The data were fit to a sophisticated computer model [88] that takes proper account of the temperature dependences of the nucleation and growth rates. The dependence of T_{cr} on q_H for the SG was similar to that for MB, but the gel crystallized at a lower temperature. The difference between SG and MB could be explained by a lower barrier to nucleation in the gel ($\sim 61kT_L$ versus $\sim 64kT_L$ for the MB). An alternative

Fig. 13. _____

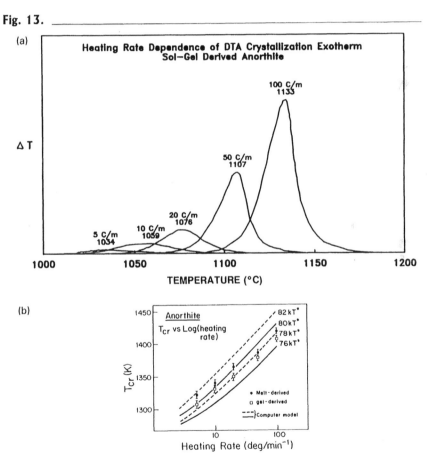

(a) Heating rate dependence of crystallization in gel-derived anorthite powder. Peak of exotherm, T_{cr}, shifts to higher temperatures as the heating rate (°C/min) increases. (b) T_{cr} versus log(heating rate) for MB and SG anorthite glass powders; $T^* = 0.8T_L$, so $80kT^* = 64kT_L$. From Zelinski *et al.* [86].

explanation, favored by the authors, is that the viscosity of SG is lower by a factor of ~1.5. The crystallization kinetics of the cordierite glass were not examined in the same detail, because of the complicated series of phase changes occurring in that material. (For a detailed discussion, see Gensse and Chowdhry [89].) It was found, however, that the sequence of crystallization (beta quartz converting to cordierite) in SG was the same as for MB.

It is often claimed that sol-gel processing produces materials with exceptional homogeneity, and we have seen that this is true for *melted* gels. However, in the gel state the situation is different. The Si–O–B linkage is easily hydrolyzed (see discussion in Chapter 3), so it is generally not present in the gel until it is sintered [49,90]. Sen and Thiagarajan [91] found no evidence for Si–O–Al bonds in a gel with the composition of mullite ($3Al_2O_3 \cdot 2SiO_2$), presumably because of the difference in reactivity of the alkoxides. (Pouxviel *et al.* [92] avoided this problem by using a metalorganic precursor already containing Si–O–Al bonds. Irwin *et al.* [93] found some Si–O–Al in gels made from separate alkoxides, and some Al not bonded to Si even when a double alkoxide was used. Clearly the outcome depends on the details of the synthesis.) Dauger *et al.* [47] found that heterogeneities, such as zirconia crystals, appeared in their gels. (Fortunately, they dissolved during sintering so it was possible to make homogeneous superionic conducting glasses from gels outside the usual range of glass formation.) Alkali silicate gels are notoriously difficult to prepare, because the alkali are not incorporated into the network, whereas the glasses are easily melted by conventional methods; rather than homogenizing, the gels tend to crystallize when heated. However, the sol-gel method has been used successfully to prepare homogeneous glasses from network-forming oxides that are difficult or impossible to melt. For example, TiO_2–SiO_2 glasses, which are important for their exceptionally low thermal expansion coefficients [94], are commercially prepared by a vapor deposition process operating at ~1800°C. Kamiya and Sakka [95] used alkoxides to make titania-silica glasses with the same low expansion without exceeding a temperature of 900°C. Kamiya *et al.* [96] used alkoxides to prepare ZrO_2–SiO_2 glasses, which are very difficult to melt, and showed them to have outstanding chemical durability. The densities of the glasses, densified at only 700°C, were in agreement with values interpolated from related melted glasses.

Perhaps the most impressive achievement is the preparation of homogeneous glasses in the alkaline-earth silicate systems. As shown in Fig. 7a, these liquids exhibit stable immiscibility extending to extremely high temperatures (>2000°C), so they are virtually impossible to melt; even if homogeneous melts were prepared, they would phase separate during cooling to the glassy state. However, several of them have been made as homogeneous glasses from gels, by taking advantage of low-temperature processing. Although alkaline earths are not network-forming elements, the divalent ions evidently bind more tenaciously to the network than the alkalis, so that segregation is less of a problem, and crystallization does not precede densification.

Hayashi and Saito [97] prepared gels containing 10, 20, and 40 wt% CaO from calcium ethoxide and TEOS; when hydrolyzed slowly by exposure to atmospheric moisture, homogeneous gels were obtained that sintered to clear glass at 800°C. Heating to higher temperature caused crystallization to wollastonite. Yamane and Kojima [98] used strontium nitrate and TMOS to make gels containing 1, 5, 10, and 20 mol% SrO; nitrate crystallized from the 20% composition during drying, but the others remained homogeneous. The gels fractured, but became transparent when fully dense (at 930, 845, and 800°C, respectively, for 1, 5, and 10% SrO), and then crystallized at higher temperatures. (See Fig. 14.) Quartz crystals appeared in the SG containing 1–10% SrO; the 20% composition did not become clear, but crystallized to $SrO \cdot 2SiO_2$. Melts were prepared from strontium carbonate and sand containing 31–45 mol% SrO (outside of the miscibility gap shown in Fig. 7A), and the refractive indices of the glasses were measured. The indices of the SG fell on the line extrapolated from the MB to pure silica, as shown in Fig. 15. Bansal [99] prepared gels containing 10–20 mol% MgO from magnesium nitrate and TEOS; gels made with magnesium acetate were opaque powders, so that precursor was abandoned. Nitrate crystals formed on the gels with 15 and 20% MgO if dried at room temperature, but not if drying was performed at 76°C. Monolithic gels were obtained if glycerol was included, but the densification behavior of such gels was not reported; retention of carbonates might be expected [100]. At 900°C the SG with 10% MgO was "partly clear" (evidently the gel was not macroscopically homogeneous), while the other compositions were white and opaque; all were amorphous to X-rays. Quartz crystals appeared in the SG with 10% MgO when it was heated to 1000°C. Thus, all of these alkaline earth silicate systems yield clear SG glasses that cannot be made by melting.

5.

CRYSTALLINE SYSTEMS

For crystalline systems, the novel features of gel-derived materials include novel forms (e.g., nanometer-sized particles, film, and fibers), low-temperature fabrication, and new phases. As noted in Section 2, the capacity to produce films and fibers from oxides, including crystalline systems, is one of the most important features of sol-gel processing. Oxide fibers drawn from solution are used for reinforcement in matrices including metals and

Fig. 14.

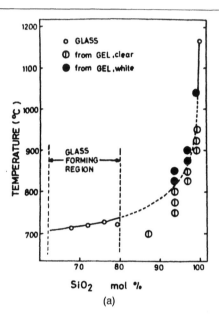

(a)

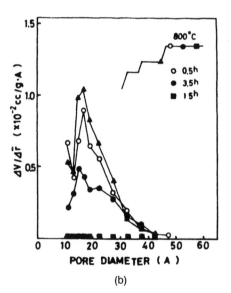

(b)

(a) The dashed line separates SG that are clear and dense from those that are partially crystallized. (b) Pore size distributions for the gel containing 10 mol% SrO demonstrate that the clear gel is fully dense at 800°C. Samples heated to 750°C for 8 h (▲), then to 800°C for 0.5 (O), 3.5 (●), or 15 h (■). From Yamane and Kojima [98].

Fig. 15.

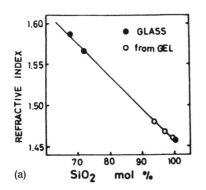

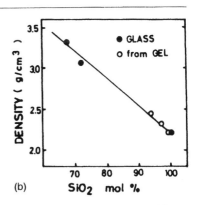

(a)   (b)

Comparison of the refractive index (a) and density (b) of MB and SG in the $SrO-SiO_2$ system; MB compositions outside miscibility gap shown in Fig. 7A. From Yamane and Kojima [98].

polymers, while films are used for a wide range of purposes, from protection against corrosion or abrasion to control of reflectivity. Sol-gel films can be used for coating fibers and whiskers to provide control over matrix-inclusion interactions. As described in Chapter 4, the nuclear-fuel industry uses sol-gel processing to make small crystalline spheres of refractory oxides, without the risk of generating the hazardous dust that would result from conventional processing. Many such applications, taking advantage of the ability to form gels directly into useful shapes, are discussed in Chapters 13 and 14.

Mineralogists were among the first users of solution methods for preparation of oxides [44,101,102], because of the relative ease of producing equilibrium crystal phases from such homogeneous powders. The effect of homogeneity is illustrated by the intensity of the exotherm upon crystallization of mullite [103], shown in Fig. 16. When the alkoxide solution was hydrolyzed with excess water (A), the oxide was inhomogeneous because of the great difference in the rates of hydrolysis of TEOS and aluminum sec-butoxide; more homogeneous products were made by gentler methods of hydrolysis. When the powders were heated, the most homogeneous powder (C) transformed completely at 980°C, whereas A remained amorphous to much higher temperatures. Similarly, Keller et al. [104] used nitrates to produce a powder with the composition $ZnAl_2O_4$ that was fully crystallized to that spinel phase by 600°C; mixed oxides did not convert completely to spinel until 1200°C.

In the preceding examples, equilibrium is rapidly achieved because there is no other crystal phase with the same composition. In general, however, glasses crystallize to the phase with the lowest density, which presumably has

Fig. 16.

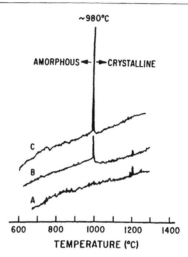

Intensity of crystallization peak in mullite gel is related to the conditions of hydrolysis, which control the homogeneity of the powder. (A) Solution hydrolyzed with excess water. (B) Thin layer of solution hydrolyzed by wet air. (C) Solution stirred and slowly hydrolyzed over 90 days. From Yoldas [103].

a structure most resembling that of the glass. For example, cristobalite (density 2.32 g/cm^3) grows in alkali silicate glasses at temperatures where quartz (density 2.635 g/cm^3) is the stable phase. In vapor-phase oxidation, oxide particles are formed by rapid quenching of liquid droplets, but the high-temperature phases do not appear: alumina crystallizes in the gamma or theta phase (3.5–3.9 g/cm^3), rather than alpha (3.97 g/cm^3) [105]; titania forms anatase (3.90 g/cm^3), rather than rutile (4.27 g/cm^3) [106]. This tendency can be exploited to make disordered phases from gels; for example, Pentinghaus [102] found that reedmergnerite ($NaBSi_3O_8$) could only be formed from gel, not from melted glass. Heistand et al. [107] used alkoxides to make amorphous titania that crystallized into anatase, then took advantage of the fine particle size to sinter it at such a low temperature that it did not convert into rutile. Livage [21] points out that sol-gel processing permits the formation of dense (i.e., nonporous) anatase films, which are expected to be superior to rutile as photoanodes.

The phase evolution can be profoundly affected by the conditions of preparation of the gel. For example, aluminum alkoxides yield bayerite ($Al(OH)_3$) if hydrolyzed in cold water, and boehmite ($AlOOH$) in hot water [108]. The conversion to the alpha alumina phase occurs at lower temperatures when a higher ratio of water to alkoxide is used in the hydrolysis; the same is true for the transition to the monoclinic phase in zirconia gels [109].

It should be noted that the occurrence of phase transformation *during* sintering can be detrimental, because of the tendency toward excessive grain growth of the new phase. This problem is discussed in Chapters 9 and 11. The best method for control of phase transformations is seeding of the gel with the equilibrium phase, as demonstrated by Messing and students [110,111].

6.

SUMMARY

Although homogeneity is frequently cited as a feature of sol-gel processing, it is often not achieved until the gel is sintered or even melted. Generally the scale of heterogeneity is small enough so that homogeneous ceramics are readily prepared at relatively low temperatures and short firing times. In some cases, such as alkali silicates, the mobility of the ions is so great that large scale segregation can occur, preventing the body from sintering into a dense single phase. On the other hand, where less mobile ions are present (as in the alkaline earth silicates), low processing temperature may permit fabrication of homogeneous glasses that cannot be made by melting.

After melting, the properties of a homogeneous glass must be independent of prior thermal history, because the relaxation time for the structure is so fast in the liquid state. Differences that appear between gel-derived and oxide glasses after melting must result from differences in composition (including OH, Cl, and other impurities). Even after sintering in the vicinity of the glass transition temperature, differences in properties are not observed. However, although the local bonding structure relaxes quickly near T_g, the development of microstructure by phase separation or crystallization may be arrested. The difference is in the scale of the process: phase separation requires diffusion over distances of several nanometers (at least) and is likely to take longer than viscous sintering of nanometer-sized pores. Consequently, dense glasses with metastable structures can be prepared. The same is true of crystalline systems, which can sometimes be sintered without transforming into the equilibrium phase.

REFERENCES

1. G.W. Scherer and P.C. Schultz in *Glass: Science and Technology*, **vol. 1**, eds. D.R. Uhlmann and N.J. Kreidl (Academic Press, New York, 1983), pp. 49–103.
2. G.W. Scherer, *Relaxation in Glass and Composites* (Wiley, New York, 1986).
3. A.Q. Tool, *J. Res.*, **34** (1945) 199–211.

4. S.M. Rekhson, A.V. Bulaeva, and O.V. Mazurin, *Sov. J. Inorg. Mater.*, **7** [4] (1971) 622–623 (Eng. trans.).
5. M. Hara and S. Suetoshi, *Rep. Res. Lab. Asahi Glass Co.*, **5** (1955) 126–135.
6. J.W. Christian, *The Theory of Transformations in Metals and Alloys, Part I* (Pergamon, New York, 1975), pp. 420–422.
7. D.R. Uhlmann, M.C. Weinberg, and G. Teowee, *J. Non-Cryst. Solids*, **100** (1988) 154–161.
8. B. Chalmers, *Principles of Solidification* (R.E. Krieger Publ., Huntington, N.Y., 1977).
9. T.J.W. de Bruijn, W.A. de Jong, and P.J. van den Berg, *Thermochimica Acta*, **45** (1981) 315–325.
10. D.W. Henderson, *J. Non-Cryst. Solids*, **30** (1979) 301–315.
11. T.B. Tang, *Thermochimica Acta*, **58** (1982) 373–377.
12. T.B. Tang and M.M. Chaudhri, *J. Am. Ceram. Soc.*, **66** [11] (1983) C218.
13. T. Kemény and L. Gránásy, *J. Non-Cryst. Solids*, **68** (1984) 193–202.
14. *Phase Separation in Glass*, eds. O.V. Mazurin and E.A. Porai-Koshits (North-Holland, New York, 1984).
15. D.R. Uhlmann and A.G. Kolbeck, *Phys. Chem. Glasses*, **17** [5] (1976) 146–158.
16. T.P. Seward in *Phase Diagrams, Materials Science and Technology*, ed. A.M. Alper (Academic Press, New York, 1970), pp. 295–338.
17. W. Haller, D.H. Blackburn, and J.H. Simmons, *J. Am. Ceram. Soc.*, **57** [3] (1974) 120–126.
18. S. Sakka in *Sol-Gel Technology for Thin Films, Fibers, Preforms, Electronics, and Specialty Shapes*, ed. L.C. Klein (Noyes, Park Ridge, N.J., 1988), pp. 140–161.
19. H.G. Sowman in *Sol-Gel Technology for Thin Films, Fibers, Preforms, Electronics, and Specialty Shapes*, ed. L.C. Klein (Noyes, Park Ridge, N.J., 1988), pp. 162–183.
20. W.C. LaCourse in *Sol-Gel Technology for Thin Films, Fibers, Preforms, Electronics, and Specialty Shapes*, ed. L.C. Klein (Noyes, Park Ridge, N.J., 1988), pp. 184–198.
21. J. Livage in *Better Ceramics Through Chemistry II*, eds. C.J. Brinker, D.E. Clark, and D.R. Ulrich (Mater. Res. Soc., Pittsburgh, Pa., 1986), pp. 717–724.
22. L.L. Hench, S.H. Wang, and J.L. Nogues, *SPIE Multifunctional Materials*, **878** (1988) 76–85.
23. G. Poelz in *Aerogels*, ed. J. Fricke (Springer-Verlag, New York, 1986), pp. 176–187.
24. S.J. Teichner in *Aerogels*, ed. J. Fricke (Springer-Verlag, New York, 1986), pp. 22–30.
25. M. Prassas and L.L. Hench in *Ultrastructure Processing of Ceramics, Glasses, and Composites*, eds. L.L. Hench and D.R. Ulrich (Wiley, New York, 1984), pp. 100–125.
26. L.L. Hench, S. Wallace, S. Wang, and M. Prassas, *Ceram. Eng. Sci. Proc.*, **4** [9–10] (1983) 732–739.
27. J. Fricke in *Aerogels*, ed. J. Fricke (Springer-Verlag, New York, 1986), pp. 94–103.
28. P.H. Tewari, A.J. Hunt, and K.D. Lofftus in *Aerogels*, ed. J. Fricke (Springer-Verlag, New York, 1986), pp. 31–37.
29. B. Dunn, E. Knobbe, J.M. McKiernan, J.C. Pouxviel, and J.I. Zink in *Better Ceramics Through Chemistry III*, eds. C.J. Brinker, D.E. Clark, and D.R. Ulrich (Mat. Res. Soc., Pittsburgh, Pa., 1988), pp. 331–342.
30. L.C. Klein in *Sol-Gel Technology for Thin Films, Fibers, Preforms, Electronics, and Specialty Shapes*, ed. L.C. Klein (Noyes, Park Ridge, N.J., 1988), pp. 382–399.
31. E.J.A. Pope and J.D. Mackenzie in *Better Ceramics Through Chemistry II*, eds. C.J. Brinker, D.E. Clark, and D.R. Ulrich (Mat. Res. Soc., Pittsburgh, Pa., 1986), pp. 809–814.
32. N. Tohge, G.S. Moore, and J.D. Mackenzie, *J. Non-Cryst. Solids*, **63** (1984) 95–103.
33. T. Kawaguchi, J. Iura, N. Taneda, H. Hishikura, and Y. Kokubu, *J. Non-Cryst. Solids*, **82** (1986) 50–56.

34. J.P. Boilot and P. Colomban in *Sol-Gel Technology for Thin Films, Fibers, Preforms, Electronics, and Speciality Shapes*, ed. L.C. Klein (Noyes, Park Ridge, N.J., 1988), pp. 303–329.
35. J. Livage, *Solid State Chemistry 1982*, eds. R. Metselaar, H.J.M. Heifligers, and J. Schoonman (Elsevier, Amsterdam, 1983), pp. 17–32.
36. H. Schmidt in *Better Ceramics Through Chemistry*, eds. C.J. Brinker, D.E. Clark, and D.R. Ulrich (North-Holland, New York, 1984), pp. 327–335.
37. H. Schmidt and B. Seiferling in *Better Ceramics Through Chemistry II*, eds. C.J. Brinker, D.E. Clark, and D.R. Ulrich (Mat. Res. Soc., Pittsburgh, Pa., 1986), pp. 739–750.
38. H. Schmidt, G. Rinn, R. Nass, and D. Sporn in *Better Ceramics Through Chemistry III*, eds. C.J. Brinker, D.E. Clark, D.R. Ulrich (Mat. Res. Soc., Pittsburgh, Pa., 1988), pp. 743–754.
39. H. Schmidt, B. Seiferling, G. Philipp, and K. Deichmann in *Ultrastructure Processing of Advanced Ceramics*, ed. J.D. Mackenzie and D.R. Ulrich (Wiley, New York, 1988), pp. 651–660.
40. A.R. Cooper in *Better Ceramics Through Chemistry II*, eds. C.J. Brinker, D.E. Clark, and D.R. Ulrich (Mater. Res. Soc., Pittsburgh, Pa., 1986), pp. 421–430.
41. A. Bertoluzza, C. Fagnano, M.A. Morelli, V. Gottardi, and M. Guglielmi, *J. Non-Cryst. Solids*, **48** (1982) 117–128.
42. R.B. Bird, W.E. Stewart, and E.N. Lightfoot, *Transport Phenomena* (Wiley, New York, 1960).
43. J. Zarzycki in *Advances in Ceramics*, **vol. 4** (Am. Ceram. Soc., Columbus, Ohio, 1982), pp. 204–216.
44. A.D. Edgar, *Experimental Petrology* (Clarendon, Oxford, England, 1973), chapter 3.
45. S.P. Mukherjee and R.K. Mohr, *J. Non-Cryst. Solids*, **66** (1984) 523–527.
46. P.S. Dunn, C.A. Natalie, and D.L. Olson, *J. Mater. Energy Systems*, **8** [2] (1986) 176–184.
47. A. Dauger, F. Chaput, J.C. Pouxviel, and J.P. Boilot, *J. de Phys.*, **46** [12] (1985) C8-455–C8-459.
48. S.P. Mukherjee and S.K. Sharma, *J. Non-Cryst. Solids*, **71** (1985) 317–325.
49. M.A. Villegas and J.M. Fernandez Navarro, *J. Mater. Sci.*, **23** (1988) 2464–2478.
50. S.P. Mukherjee, J. Zarzycki, and J.P. Traverse, *J. Mater. Sci.*, **11** (1976) 341–355.
51. G.F. Neilson and M.C. Weinberg, *J. Non-Cryst. Solids*, **63** (1984) 365–374.
52. I.M. Low and R. McPherson, *J. Mater. Sci.*, **23** (1988) 3544–3549.
53. M.C. Weinberg and G.F. Neilson in *Sol-Gel Technology for Thin Films, Fibers, Preforms, Electronics, and Speciality Shapes*, ed. L.C. Klein (Noyes, Park Ridge, N.J., 1988), pp. 28–48.
54. K.T. Faber and G.E. Rindone, *Phys. Chem. Glasses*, **21** [5] (1980) 171–177.
55. M.C. Weinberg and G.F. Neilson, *J. Mater. Sci.*, **13** (1978) 1206–1216.
56. G.F. Neilson and M.C. Weinberg in *Materials Processing in the Reduced Gravity Environment of Space*, ed. G.E. Rindone (Elsevier, New York, 1982), pp. 333–342.
57. M.C. Weinberg and G.F. Neilson, *J. Am. Ceram. Soc.*, **66** [2] (1983) 132–134.
58. B.E. Yoldas, *J. Non-Cryst. Solids*, **51** (1982) 105–121.
59. P. Danielson, Corning Inc., Corning, NY, private communication. Estimate based on experience with similar compositions.
60. B.E. Yoldas, *Diffusion and Defect Data*, **53–54** (1987) 351–362.
61. J.D. Mackenzie, *J. Non-Cryst. Solids*, **48** (1982) 1–10.
62. M. Yamane, S. Aso, S. Okano, and T. Sakaino, *J. Mater. Sci.*, **14** (1979) 607–611.
63. S.H. Wang, C. Campbell, and L.L. Hench in *Ultrastructure Processing of Advanced Ceramics*, eds. J.D. Mackenzie and D.R. Ulrich (Wiley, New York, 1988), pp. 145–157.

64. G.V. Chandrashekhar and M.W. Shafer in *Better Ceramics Through Chemistry II*, eds. C.J. Brinker, D.E. Clark, and D.R. Ulrich (Mater. Res. Soc., Pittsburgh, Pa., 1986), pp. 705–710.

65. R.A. Weimer, P.M. Lenahan, T.A. Marchione, and C.J. Brinker, *Appl. Phys. Lett.*, **51** (1987) 1179–1181.

66. L.L. Hench, M. Prassas, and J. Phalippou, *Ceram. Eng. Sci. Proc.*, **3** [9–10] (1982) 477–483.

67. L.L. Hench, S. Wallace, S. Wang, and M.Prassas, *Ceram. Eng. Sci. Proc.*, **4** [9–10] (1983) 732–739.

68. S.C. Park and L.L. Hench in *Science of Ceramic Chemical Processing*, eds. L.L. Hench and D.R. Ulrich (Wiley, New York, 1986), pp. 168–172.

69. T. Woignier, J. Phalippou, and R. Vacher in *Better Ceramics Through Chemistry III*, eds. C.J. Brinker, D.E. Clark, and D.R. Ulrich (Mat. Res. Soc., Pittsburgh, Pa., 1988), pp. 697–702.

70. R. Puyané, P.F. James, and H. Rawson, *J. Non-Cryst. Solids*, **41** (1980) 105–115.

71. C.J. Brinker, E.P. Roth, D.R. Tallant, and G.W. Scherer in *Science of Ceramic Chemical Processing*, eds. L.L. Hench and D.R. Ulrich (Wiley, New York, 1986), pp. 37–51.

72. G.W. Scherer, C.J. Brinker, and E.P. Roth, *J. Non-Cryst. Solids*, **82** (1986) 191–197.

73. C.J. Brinker and G.W. Scherer, *J. Non-Cryst. Solids*, **70** (1985) 301–322.

74. J.P. Williams, Y.S. Su, W.R. Strzegowski, B.L. Butler, H.L. Hoover, and V.O. Altemose, *Am. Ceram. Soc. Bull.*, **55** (1976) 524–527.

75. C.T. Moynihan, A.J. Easteal, M.A. DeBolt, and J. Tucker, *J. Am. Ceram. Soc.*, **59** (1976) 12–16, 16–21.

76. O.S. Narayanaswamy, *J. Am. Ceram. Soc.*, **54** (1971) 491–498.

77. C.J. Brinker, E.P. Roth, and G.W. Scherer, *Am. Ceram. Soc. Bull.*, **64** (1985) 476.

78. M.A. Villegas and J.M. Fernandez Navarro, *J. Mater. Sci.*, **23** (1988) 2142–2152.

79. A. Yasumori, S. Inoue, and M. Yamane, *J. Non-Cryst. Solids*, **82** (1986) 177–182.

80. A. Chen and P.F. James, *J. Non-Cryst. Solids*, **100** (1988) 353–358.

81. I. Schwartz, P. Anderson, H. de Lambilly, and L.C. Klein, *J. Non-Cryst. Solids*, **83** (1986) 391–399.

82. F. Branda, A. Aronne, A. Marotta, and A. Buri, *J. Mater. Sci. Lett.*, **6** (1987) 203–206.

83. A. Grassi, F. Branda, S. Saiello, A. Buri, and A. Marotta, *Thermochimica Acta*, **76** (1984) 133–138.

84. J. Phalippou, J. Zarzycki, and J.F. Lalanne, *Ann. Chim. Fr.*, **3** (1978) 99–105.

85. C. Zhu, J. Phalippou, and J. Zarzycki, *J. Non-Cryst. Solids*, **82** (1986) 321–328.

86. B.J.J. Zelinski, B.D. Fabes, and D.R. Uhlmann, *J. Non-Cryst. Solids*, **82** (1986) 307–313.

87. B.J.J. Zelinski, M.L. Galiano, and D.R. Uhlmann in *Ultrastructure Processing of Advanced Ceramics*, eds. J.D. Mackenzie and D.R. Ulrich (Wiley, New York, 1988), pp. 855–864.

88. D. Crammer, R. Salomaa, H. Yinnon, and D.R. Uhlmann, *J. Non-Cryst. Solids*, **45** (1981) 127–136.

89. C. Gensse and U. Chowdhry in *Better Ceramics Through Chemistry II*, eds. C.J. Brinker, D.E. Clark, and D.R. Ulrich (Mat. Res. Soc., Pittsburgh, Pa., 1986), pp. 693–703.

90. D.M. Haaland and C.J. Brinker in *Better Ceramics Through Chemistry*, eds. C.J. Brinker, D.E. Clark, and D.R. Ulrich (North-Holland, New York, 1984), pp. 267–273.

91. S. Sen and S. Thiagarajan, *Ceram. Int.*, **14** (1988) 77–86.

92. J.C. Pouxviel, J.P. Boilot, A. Dauger, and L. Huber in *Better Ceramics Through Chemistry II*, eds. C.J. Brinker, D.E. Clark, and D.R. Ulrich (Mat. Res. Soc., Pittsburgh, Pa., 1986), pp. 269–274.

93. A.D. Irwin, J.S. Holmgren, and J. Jonas, *J. Mater. Sci.*, **23** (1988) 2908-2912.
94. P.C. Schultz and H.T. Smyth in *Amorphous Materials*, eds. R.W. Douglas and B. Ellis (Wiley, New York, 1972), pp. 453-461.
95. K. Kamiya and S. Sakka, *J. Mater. Sci.*, **15** (1980) 2937-2939.
96. K. Kamiya, S. Sakka, and Y. Tatemichi, *J. Mater. Sci.*, **15** (1980) 1765-1771.
97. T. Hayashi and H. Saito, *J. Mater. Sci.*, **15** (1980) 1971-1977.
98. M. Yamane and T. Kojima, *J. Non-Cryst. Solids*, **44** (1981) 181-190.
99. N.P. Bansal, *J. Am. Ceram. Soc.*, **71** [8] (1988) 666-672.
100. L.L. Hench in *Better Ceramics Through Chemistry*, eds. C.J. Brinker, D.E. Clark, and D.R. Ulrich (North-Holland, New York, 1984), pp. 101-110.
101. R. Roy, *J. Am. Ceram. Soc.*, **39** [4] (1956) 145-146.
102. H. Pentinghaus, *J. Non-Cryst. Solids*, **63** (1984) 193-199.
103. B.E. Yoldas in *Ultrastructure Processing of Advanced Ceramics*, eds. J.D. Mackenzie and D.R. Ulrich (Wiley, New York, 1988), pp. 333-345.
104. J.T. Keller, D.K. Agrawal, and H.A. McKinstry, *Adv. Ceram. Mater.*, **3** [4] (1988) 420-422.
105. R. Clasen, *Glastech. Ber.*, **61** [5] (1988) 119-126.
106. G.W. Scherer in *Better Ceramics Through Chemistry*, eds. C.J. Brinker, D.E. Clark, and D.R. Ulrich (North-Holland, New York, 1984), pp. 205-211.
107. R.H. Heistand II, Y. Oguri, H. Okamura, W.C. Moffatt, B. Novich, E.A. Barringer, and H.K. Bowen in *Science of Ceramic Chemical Processing*, eds. L.L. Hench and D.R. Ulrich (Wiley, New York, 1986), pp. 482-496.
108. B. Yoldas, *J. Appl. Chem. Biotechnol.*, **23** (1973) 803-809.
109. B.E. Yoldas, *J. Am. Ceram. Soc.*, **65** [8] (1982) 387-393.
110. M. Kumagai and G.L. Messing, *J. Am. Ceram. Soc.*, **68** [9] (1985) 500-505.
111. J.L. McArdle and G.L. Messing, *Adv. Ceram. Mater.*, **3** [4] (1988) 387-392.

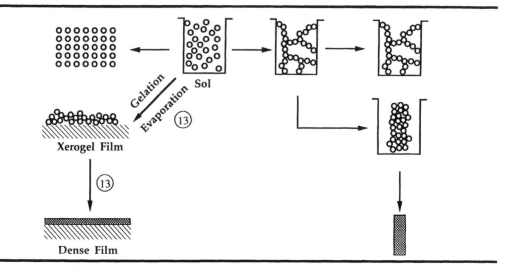

CHAPTER 13

Film Formation

Certainly one of the most technologically important aspects of sol-gel processing is that, prior to gelation, the fluid sol or solution is ideal for preparing thin films by such common processes as dipping, spinning, or spraying. Electrophoretic and thermophoretic film deposition have also been investigated in several cases. Compared to conventional thin film forming processes such as CVD, evaporation, or sputtering, sol-gel film formation requires considerably less equipment and is potentially less expensive; however the most important advantage of sol-gel processing over conventional coating methods is the ability to control precisely the microstructure of the deposited film, i.e., the pore volume, pore size, and surface area.

This chapter examines the fundamental physics and chemistry underlying thin film formation from solution-derived precursors.[†,††] Applications of sol-gel–derived thin films are described in Chapter 14, and constrained one-dimensional drying and sintering of thin films are discussed in Chapters 8 and 11, respectively. In this chapter we show that in dipping or spinning, the film microstructure depends on the size and extent of branching (or aggregation) of the solution species prior to film deposition and the relative rates of condensation and evaporation during film deposition. Control of these factors enables the technologist to tailor the film porosity. For example, the pore volume may be varied from 0 to 65%; the pore size from <0.4 nm to >5.0 nm; and the surface area from <1 to >250 m^2/g. The outline of

[†] In this chapter *precursor* refers to the inorganic species present in the dilute coating sol prior to the deposition process.

[††] We do not discuss adhesion, since it is generally not a problem in thin (50–200 nm) sol-gel-derived films.

this chapter is as follows:

1.   *Physics of Film Formation* examines the dipping and spinning processes with respect to such parameters as withdrawal rate, spin speed, viscosity, surface tension, and evaporation rate. Specific attention is given to the consequences of the overlap of the deposition and drying stages.
2.   *Precursor Structure* discusses the effects of precursor size, extent of branching or aggregation, and fractal dimension on the microstructure of the deposited film.
3.   *Deposition Conditions* describes how the relative rates of condensation and evaporation during deposition affect film microstructure and presents evidence for shear-induced restructuring, alignment, or ordering.
4.   *Other Coating Methods* investigates some less common film-forming processes including electrophoresis, thermophoresis, and settling, as well as hybrid methods combining several processes.

1.

PHYSICS OF FILM FORMATION

1.1.   Dip Coating

In an excellent review of dip and spin coating, Scriven [1] divided the batch *dip coating* process into five stages: immersion, start-up, deposition, drainage, and evaporation. (See Fig. 1.) With volatile solvents, such as alcohol, evaporation normally accompanies the start-up, deposition, and drainage steps. The continuous dip coating process (shown in Fig. 1f) is simpler because it separates immersion from the other stages, essentially eliminates start-up, and "hides" drainage in the deposited film.

The moving substrate entrains liquid in a fluid mechanical boundary layer carrying some of the liquid toward the deposition region, 3, where the boundary layer splits in two (Fig. 1b). The inner layer moves upward with the substrate, while the outer layer is returned to the bath. The thickness of the deposited film is related to the position of the streamline dividing the upward- and downward-moving layers. A competition between as many as six forces in the film deposition region governs the film thickness and position of the streamline: (1) viscous drag upward on the liquid by the moving substrate, (2) force of gravity, (3) resultant force of surface tension in the concavely curved meniscus, (4) inertial force of the boundary layer liquid arriving at the deposition region, (5) surface tension gradient, and (6) the disjoining or conjoining pressure (important for films less than 1 μm thick) [1].

Fig. 1.

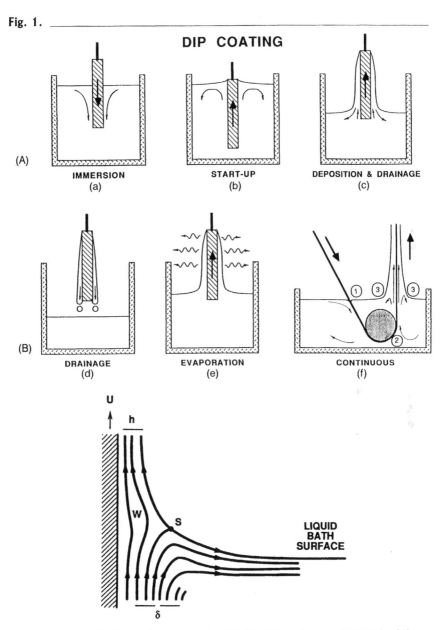

(A) Stages of the dip coating process: (a–e) batch; (f) continuous. (B) Detail of the liquid flow patterns in area 3 of the continuous process. U is the withdrawal speed, S is the stagnation point, δ is the boundary layer, and h is the thickness of the fluid film. From Scriven [1].

When the liquid viscosity (η) and substrate speed (U) are high enough to lower the curvature of the meniscus, then the deposited film thickness (h) is the thickness that balances the viscous drag ($\propto \eta U/h$) and gravity force ($\rho g h$) [1,2]:

$$h = c_1(\eta U/\rho g)^{1/2} \tag{1}$$

where the proportionality constant, c_1, is about 0.8 for Newtonian liquids [2]. When the substrate speed and liquid viscosity are not high enough, as is often the case in sol-gel processing, this balance is modulated by the ratio of viscous drag to liquid–vapor surface tension (γ_{LV}) according to the following relationship derived by Landau and Levich [3]:

$$h = 0.94(\eta U/\gamma_{LV})^{1/6}(\eta U/\rho g)^{1/2} \tag{2}$$

or rearranging terms:

$$h = 0.94(\eta U)^{2/3}/\gamma_{LV}^{1/6}(\rho g)^{1/2}. \tag{3}$$

The applicability of Eqs. 1–3 to sol-gel film formation has been examined in a limited number of cases [4–6]. Strawbridge and James [4] determined the relationship between coating thickness and viscosity for an acid-catalyzed silicate solution ($r = 1.74$) deposited on glass substrates at speeds, U, ranging from 1 to 15 cm/min. Their results are presented in Fig. 2 where they are compared to predicted values based on a modification of Eq. 1 that takes into account film shrinkage due to evaporation and partial sintering. Rather significant discrepancies with the model are observed as the substrate speed is increased. Brinker and Ashley [5] investigated the relationship between film thickness and withdrawal speed for a variety of silicate sols in which the precursor structures ranged from rather weakly branched "polymers" to highly condensed particles. Their results plotted in Fig. 3a show that, for polymeric systems, h varies approximately as $U^{2/3}$ in accordance with Eqs. 2 and 3 and with previous results of Dislich and Hussmann [6]. Plotting the logarithm of the product of h and refractive index ($n \propto \rho$) versus ln U (Fig. 3b), provides a better fit to Eqs. 2 and 3, indicating that the (mass of film)/(unit area) varies as $U^{2/3}$ for most of the investigated systems. This appears not to be the case for sols composed of mutually repulsive, monosized particles. As discussed in Section 3, repulsive particles tend to order as U increases, causing deviations from the predicted behavior.

Several other factors may be responsible for deviations from the predicted behavior in various pH, viscosity regimes. Equations 1–3 assume constant Newtonian viscosity and ignore the effects of evaporation. Since extended polymers or aggregates exhibit a strong concentration dependence of the viscosity and Guinier radius R (see Section 2.6.6 in Chapter 3), the concentrating effect of evaporation is expected to progressively increase the

Fig. 2.

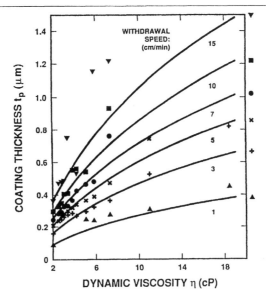

Coating thickness versus viscosity at different withdrawal speeds. The curves are predictions according to Eq. 1. From Strawbridge and James [4].

viscosity resulting finally in network formation and non-Newtonian behavior. In addition, differences between theory and experiment may be caused by nonideal free surface behavior [7] where, for example, the surface is nonextensible so that the shear strain, rather than the shear stress is zero. In both cases the expected result is thicker films than predicted by theory.

In sol-gel film deposition, evaporation is generally relied upon to solidify the coating. The most significant factor in the rate of evaporation is the rate of diffusion of the vapor away from the film surface [1]. This in turn depends on the movement of the gas within a very thin layer ($l \sim 1$ mm), because even a tiny bit of convection can greatly enhance diffusion. The rate of evaporation, m, is generally expressed in terms of an empirical mass transfer coefficient, k, according to [8]:

$$m = k(p_e - p_i) \tag{4}$$

where p_e is the partial pressure of the volatile species in local equilibrium with the surface and p_i is its partial pressure a distance, l, away. To a first approximation the evaporation rate is independent of the liquid depth. During dip coating, the movement of the substrate can strongly influence the evaporation rate, but in practice k and $p_e - p_i$ are probably dominated by large-scale, uncontrolled currents above the bath [1].

Fig. 3.

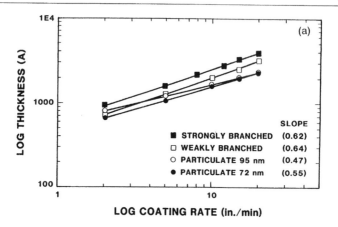

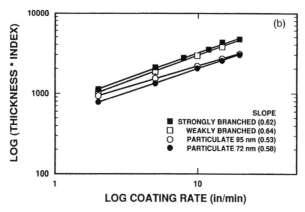

(a) Coating thickness versus substrate speed for a variety of "polymeric" and particulate silicate sols plotted according to Eq. 3. From Brinker and Ashley, ref. [5]. (b) Product of thickness and refractive index (proportional to mass) versus substrate speed. From Brinker and Ashley, ref. [5].

Although the composition of the liquid bath may be relatively unaffected by evaporation, the much thinner film experiences a substantial increase in concentration. The slower the substrate speed, the thinner the film and the greater the overlap of the deposition and drying stages. Since condensation continues to occur during sol-gel film formation, the relative condensation and evaporation rates will dictate the extent of further cross-linking that accompanies the deposition and drainage stages. The evaporation rate can, of course, be controlled by the deposition ambient. We shall see that the condensation rate may be controlled by varying the pH of the coating bath.

Fig. 4.

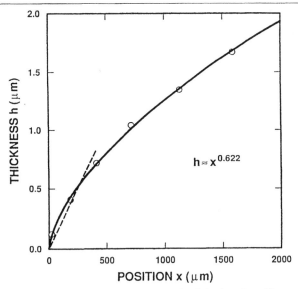

Thickness profile of drying titanate film determined by imaging ellipsometry. The displacement along the film was measured from a fiducial point near the drying front. The dashed line represents a constant evaporation profile with no surface tension. The dotted line is the profile for gravitational draining with a non-constant evaporating rate. Thickness varies as $h \sim x^{0.62}$. From Hurd (unpublished).

The overlap of the deposition and drainage stages has been investigated by Hurd and Brinker [9,10] using an *ellipsometric imaging* method that allows determination of the steady-state thickness and refractive index profile. Figure 4 shows the thickness profile of a film deposited from a titania sol at 10 cm/min [9]. The solid line represents the wedge profile expected in general for constant evaporation assuming that the rate of mass loss due to evaporation is proportional to the surface area of the film. Figure 5 shows the thickness profile expected for steady-state *gravitational draining* with constant evaporation and no curvature effects according to ref. [10]:

$$Sx = h(1 - h^2/3\lambda^2) \tag{5}$$

where the dynamic contact angle is $S = m/\rho U_0 \approx 10^{-3}$, m is the evaporation rate which was measured to be about 10^{-5} g/sec/cm^2, and $\lambda = (U_0 \eta / \rho g)^{1/2}$ is a length of order 10 μm. (See Eq. 1.) The thickness profile in Fig. 5 predicts the expected wedge profile seen in Fig. 4, a dynamic contact angle of about 1 milliradian, and a height of the drying front $X_0 = 2\lambda/3S \approx 1$ cm where h diverges to meet the reservoir—all consistent with experiment. However, the curvature is unexpected near the drying front where it has a parabolic profile

Fig. 5.

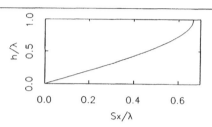

Thickness profile for steady-state gravitational draining with constant evaporation and no curvature effects according to Eq. 5. From Hurd and Brinker [10].

(as expected for gravitational draining with no evaporation). Since the shapes of thickness profiles for pure solvents (ethanol, propanol, etc.) are qualitatively similar to Fig. 4,[†] the difference between theory and experiment is probably not due to effects of the gel network. The most likely explanation is that the evaporation rate is not constant along the film due to different rates of diffusion of the solvent vapor away from the film's surface owing to the shape of the liquid profile [10,11]. Using a solution of the steady-state diffusion equation analogous to Laplace's equation for electrostatics [13], Hurd [11] obtains a parabolic thickness profile in reasonable agreement with experiment (see Fig. 4).

According to Scriven [1], overlap of the deposition and drying stages can cause flow driven by a gradient in surface tension (Marangoni effect [14]) that significantly affects the drying profile. This phenomenon results from differential evaporation of the more volatile components causing a gradient in composition and therefore a potential gradient in surface tension. Hurd [11] has studied this phenomenon in water–alcohol systems using imaging ellipsometry. As the film is formed from an initially homogeneous solution (28 vol% H_2O and 72 vol% n-propanol), n-propanol evaporates preferentially leaving a water-rich component that is concentrated at the leading edge of the drying film.[††] (See Fig. 6.) Alternatively, in systems that form lower–boiling-point azeotropes, the water–alcohol azeotrope may be removed first, concentrating an alcohol-rich component at the leading edge. It is presently unclear what effect these phenomena have on the structure of the depositing films, but the mixed solvent thickness profiles are wrong to explain the profile seen in Fig. 4. Other factors that may influence the curvature of the drying profile are Van der Waals forces [15] and capillary pressure [10].

[†] Although MeOH appeared to be an exception [12], its negative curvature and zero contact angle are attributable to traces of H_2O [11].

[††] A similar thickness profile might be generated if one of the liquid components has a much stronger affinity for the substrate surface [11].

Fig. 6.

Ellipsometric image of a 28 vol% H_2O and 72 vol% n-propanol solution during dip coating at 10 cm/min. A water-rich region grows near the drying front due to the preferential evaporation of propanol. From Hurd [11].

1.2. Spin Coating

Bornside *et al.* [8] divide *spin coating* into four stages: *deposition, spin-up, spin-off,* and *evaporation* (see Fig. 7), although, as discussed for dip coating, evaporation may accompany the other stages. An excess of liquid is dispensed on the surface during the deposition stage. In the spin-up, stage, the liquid flows radially outward, driven by centrifugal force. In the spin-off stage, excess liquid flows to the perimeter and leaves as droplets. As the film thins, the rate of removal of excess liquid by spin-off slows down, because the thinner the film, the greater the resistance to flow, and because the concentration of the nonvolatile components increases raising the viscosity. In the fourth stage, evaporation takes over as the primary mechanism of thinning.

An advantage of spin coating is that a film of liquid tends to become uniform in thickness during spin-off and, once uniform, tends to remain so provided that the viscosity is not shear dependent and does not vary over the substrate [1]. This tendency arises due to the balance between the two main forces: centrifugal force, which drives flow radially outward, and viscous force (friction), which acts radially inward [16]. During spin-up, centrifugal force overwhelms the force of gravity, and the rapid thinning quickly squelches all inertial forces other than centrifugal force.

Fig. 7.

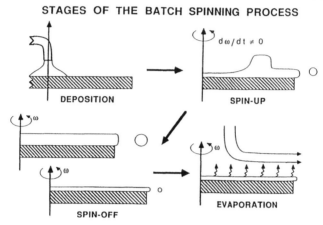

Stages of the spin-coating process. From Bornside *et al.* [8].

The thickness of an initially uniform film during spin-off is described by [1,17]:

$$h(t) = h_0/(1 + 4\rho\omega^2 h_0^2 t/3\eta)^{1/2} \tag{6}$$

where h_0 is the initial thickness, t is time, and ω is the angular velocity: ρ and ω assumed constant. Even films that are not initially uniform tend monotonically toward uniformity, sooner or later following Eq. 6 [1].

The spinning procedure creates a steady forced convection in the vapor above the substrate that causes the mass transfer coefficient, k, to be quite uniform. Thus the evaporation rate in spin coating tends to be quite uniform also. A spun film arrives at its final thickness by evaporation after the film becomes so thin and viscous that its flow stops. According to a model of spin coating by Meyerhofer [18] that separates the spin-off and evaporation stages, the final thickness and total elapsed time to achieve this thickness are

$$h_{\text{final}} = (1 - \rho_{\text{A}}/\rho_{\text{A}})\left(\frac{3\eta m}{2\rho_{\text{A}}\,\omega^2}\right)^{1/3} \tag{7}$$

and

$$t_{\text{final}} = t_{\text{spin-off}} + h_{\text{spin-off}}\rho_{\text{A}}/m\rho_{\text{A}} \tag{8}$$

where ρ_{A} is the mass of volatile solvent per unit volume, ρ_{A} is its initial value, and e is the evaporation rate that depends on the mass transfer coefficient. Bornside and coworkers [8,19] are developing models pertinent to sol-gel film formation that account for overlap of spin-off and evaporation, progressively increasing viscosity, etc.

Equations 6–8 pertain to Newtonian liquids that do not exhibit a shear rate dependence of the viscosity during the spin-off stage. If the liquid is shear thinning, the lower shear rate experienced near the center of the substrate causes the viscosity to be higher there and the film to be thicker. This is often the case for particulate sols and gel-forming systems. In order to overcome this situation, commercial spin-coating apparatus employs sophisticated dispensing systems that meter out liquid from a radially moving arm.

2.

PRECURSOR STRUCTURE

The previous section ignores for the most part how the structure of the inorganic precursor influences the structure of the deposited film. In this section we shall show that the size and extent of branching of the solution precursors prior to film deposition and the relative rates of evaporation and condensation during film deposition control the pore volume, pore size, and surface area of the final film.

Figure 8 schematically represents the steady-state deposition stage of the dip coating process. Dilute, noninteracting, polymeric species that make up the coating bath are concentrated on the substrate surface by gravitational draining accompanied by vigorous evaporation and further condensation reactions. The solution concentration increases by a factor of 20 or 30 forcing the initially dilute precursors into close proximity. Correspondingly, the viscosity progressively increases due both to the increasing concentration and further condensation reactions promoted by the increasing concentration.

Polymer growth during the deposition stage probably occurs by a process similar to *cluster–cluster aggregation* [21] (see Chapters 3 and 4) with trajectories ranging from Brownian (dilute conditions) to ballistic in the latter part of this stage where strong convective motions exist due to evaporation. We refer to gelation as the moment when the condensing network is sufficiently stiff to withstand flow due to gravity yet is still filled with solvent. From this point, further evaporation may collapse the film or generate porosity within the film.

Although the underlying physics and chemistry that govern polymer growth and gelation are essentially the same for films as bulk gels, several factors distinguish structural evolution in films [20]. (1) The overlap of the deposition and evaporation stages establishes a competition between evaporation (which compacts the structure) and continuing condensation reactions (which stiffen the structure, thereby increasing the resistance to compaction). In bulk systems, the gelation and drying stages are normally

Fig. 8.

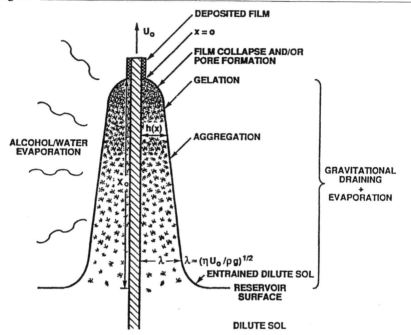

Schematic of the steady state sol gel dip coating process, showing the sequential stages of structural development that result from draining accompanied by solvent evaporation and continued condensation reactions. From Brinker *et al.* [20].

separated. (2) Compared to bulk systems, aggregation, gelation, and drying occur in seconds to minutes during dipping or spinning rather than days or weeks.[†] (3) The short duration of the deposition and drying stages causes films to experience considerably less *aging* (crosslinking) than bulk gels. This generally results in more compact dried structures. (4) Fluid flow, due to draining, evaporation, or spin-off, combined with attachment of the precursor species to the substrate, impose a shear stress within the film during deposition. After gelation, continued shrinkage due to drying and further condensation reactions creates a tensile stress within the film. Bulk gels are not constrained in any dimension.

[†] Nevertheless, precursor transport by Brownian diffusion, as opposed to evaporation-driven convection, remains dominant in the film up to about the last 10% of the drying time (assuming gelation to occur by random bond percolation at a critical volume fraction of about 16% [22]). During that time the precursors have ample time to test new configurations unless precluded by irreversible condensation. In the last 10% of the drying time, the concentration and viscosity rise dramatically, and the system has less chance to equilibrate.

2.1. Effects of Branching, Size, and Condensation Rate

The possible structures of the solution precursors range from weakly branched polymeric species characterized by a *mass fractal dimension*, d_f, to uniform (nonfractal) particles that may or may not be aggregated. (See Chapters 2 to 4.) During the deposition and drying stages, these various species are rapidly concentrated on the substrate surface. How efficiently they pack (i.e., the volume fraction solids) depends on the extent of branching or aggregation and on the condensation rate. For branched clusters or aggregates, the precursor or aggregate size is also important.

The extent of branching dictates steric constraints. Mandelbrot [23] has shown that if two structures of radius R are placed independently in the same region of space, the number of intersections, $M_{1,2}$, is expressed as

$$M_{1,2} \propto R^{d_{f,1}+d_{f,2}-d} \qquad (9)$$

where $d_{f,1}$ and $d_{f,2}$ are the respective fractal (or Euclidean) dimensions and d is the dimension of space ($d = 3$). Thus if each structure has a fractal dimension less than 1.5, the probability of intersection decreases indefinitely as R increases. The structures are *mutually transparent*: they freely interpenetrate one another as they are forced into close proximity by the increasing concentration [24]. De Gennes visualized such networks as "entangled worms" (Fig. 9) [25]. Alternatively, if the fractal dimensions of both objects are greater than 1.5, the probability of intersection increases algebraically with R: the structures, even though porous, are *mutually opaque*, much like an assemblage of tumbleweeds (Fig. 10).

The concepts of transparency and opacity based on Eq. 9 pertain to perfectly rigid structures that *stick* irreversibly at each point of intersection. In reality, the precursor structures are more or less compliant; for example, the elastic moduli of mass fractal objects decrease with R [24]. In addition, the *sticking probability* is normally $\ll 1$ and depends on the condensation rate. In silicate systems, the sticking probability is highest at intermediate pH, where the condensation rate is greatest [26]. (See Fig. 11.) The sticking probability is low near pH 2 and (for high–molecular-weight species) above about pH 10 where silanols tend to be deprotonated causing mutual repulsion of silicate particles.

The preceding factors mitigate the criterion for mutual transparency: under conditions where the condensation rate is minimized, few intersections result in sticking, and even highly branched structures interpenetrate as we shall show in a following example. Still, for any particular condensation rate, the observed trends are consistent with Eq. 9.

Fig. 9. _____

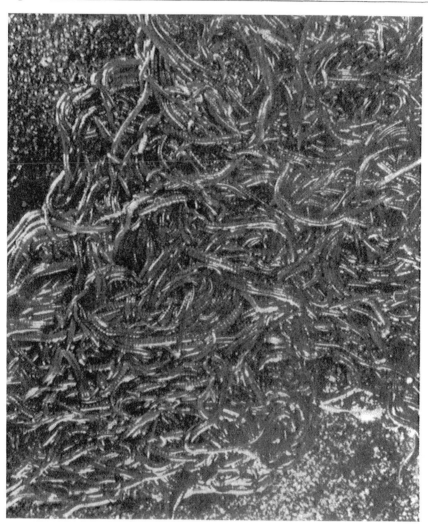

A highly entangled and densely packed network composed of "Red Wigglers"
(*Lumbricus rubellus*). From de Gennes [25].

The following subsections examine the interplay between physical and
chemical phenomena in rapidly concentrating systems during film formation. We present examples of how the structure of the precursors and the
rates of condensation, evaporation, and shear during deposition affect the
structure of the depositing film.

Fig. 10. _____

A porous layer formed by "ballistic" deposition of highly branched, fractal "Tumbleweeds" on a fence.

Fig. 11. _____

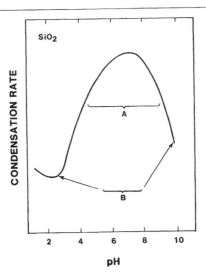

Condensation rate versus pH for aqueous silicates. A and B correspond to regions of higher and lower sticking probability, respectively. From Brinker *et al.* [20].

2.1.1. WEAKLY BRANCHED SYSTEMS

As discussed in Chapter 3, acid-catalyzed conditions combined with low H_2O : Si ratios, r, result in rather weakly branched, extended polymers; the average local environment of silicon is described by a Q^n distribution dominated by incompletely condensed Q^2 and Q^3 species [27,28]. These sols exhibit a concentration dependence of the Guinier radius [29] and the reduced viscosity, η_{sp}/C [30], a molecular weight dependence of the intrinsic viscosity, $[\eta]$ [30], and a mass fractal dimension, $d_f \leq 2$ [29]. These factors are consistent with the criteria for mutual transparency as discussed in conjunction with Eq. 9: weakly branched systems prepared under conditions that inhibit condensation are predicted to interpenetrate reversibly, leading to the observed molecular weight and concentration dependencies listed previously. Interpenetration should allow the precursors to become densely packed during film formation as their concentration is increased by evaporation. The combined effects of draining and evaporation during film deposition allow only a brief time period for interpenetration to occur; however, there is an equally brief time for condensation reactions to occur that inhibit interpenetration.

Brinker and coworkers [20] investigated film formation in weakly branched systems prepared by two-step acid-catalyzed hydrolysis of TEOS with $r = 5$ or ~1.5 (W1 and W2, respectively), deposited by dip coating near pH 2. System W2 was a *spinnable* sol (see Chapter 3) that was diluted with solvent to achieve a viscosity suitable for dip coating. The ^{29}Si NMR spectra [31] and Porod plots [29] shown in Figs. 12a and 12b are consistent with rather weakly branched structures described by $d_f < 2$. Spinnability is achieved in general under conditions in which the condensation rate and extent of branching are low enough that the polymers can be highly concentrated (causing the viscosity to increase) without premature gelation [32]. Thus the criteria for spinnability are also those expected to promote dense packing during film deposition.

System W1 exhibits a concentration-dependence of the Guinier radius [29];[†] i.e., dilution with solvent during the growth stage causes an increase in the measured Guinier radius. (See Fig. 13.) This behavior suggests that dilution causes disentanglement or swelling of the polymers; conversely, concentration during deposition is expected to cause interpenetration or collapse of the polymers.

Figure 14 shows the refractive index of films deposited from the W1 system after different stages of growth under concentrated conditions

---

[†] It is likely that the spinnable system W2 also shows a concentration dependence of the Guinier radius, but this has not been verified experimentally.

(defined in Fig. 13) followed by a fivefold dilution [20]. Also shown is the refractive index of a film deposited from the spinnable precursor W2 after dilution. Pore volumes of the films were evaluated from the refractive index according to the Lorentz–Lorenz relationship [33]:

$$(n_f^2 - 1)/(n_f^2 + 2) = V_s(n_s^2 - 1)/(n_s^2 + 2) \tag{10}$$

where n_f is the film refractive index, V_s is the volume fraction solids, and n_s is the refractive index of the solid skeleton. The porosities are less than 5%, regardless of the size of the silicate precursors, which varied in the W1 system by over an order of magnitude [29]. The corresponding nitrogen adsorption-desorption isotherm (Fig. 15) [34] is of Type II [35] indicative of adsorption on a nonporous surface. Thus any porosity that exists must have a throat diameter less than 0.4 nm, the kinetic diameter of the nitrogen molecule. The BET surface area (~ 1 cm^2/cm^2) equals the geometric area of the exterior film surface, as expected for a nonporous film.

The absence of measurable porosity or surface area is explained by the following hypotheses [20]: (a) weak branching combined with limited condensation during film formation allows the precursors to interpenetrate

Fig. 12a. ⎯⎯⎯⎯⎯⎯⎯⎯⎯⎯⎯⎯⎯⎯⎯⎯⎯⎯⎯⎯⎯⎯⎯⎯

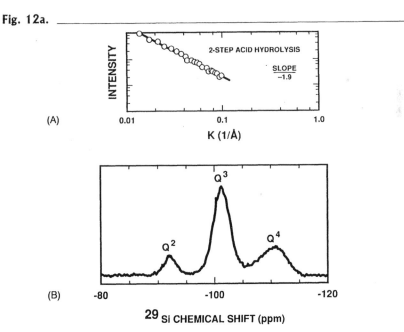

(A) Porod plot for the two-step acid-catalyzed silicate precursor (W1). K is the scattering vector $4\pi/\lambda(\sin\theta/2)$. From Schaefer and Keefer [29]. (B) Corresponding ^{29}Si NMR spectrum. From Assink [31].

Fig. 12b. _____

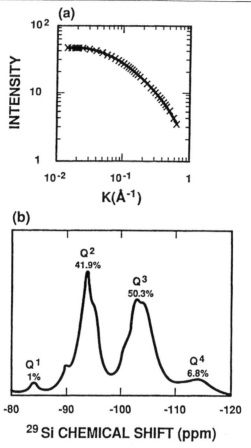

(A) Slit-smeared Porod plot for spinnable silicate precursor (W2) measured by SAXS. $K = 4\pi / \lambda(\sin\theta/2)$. (B) Corresponding ^{29}Si NMR spectrum. From Brinker and Assink [32].

in response to the decreasing solvent concentration, promoting dense packing and low pore volume; (b) the weakly branched, compliant precursors are collapsed by the decreasing solvent quality (good to θ conditions as the polymer concentration increases [36]) and by the high *capillary pressures* (up to 200 MPa) attained at the final stages of drying (see Chapters 7 and 8) when the liquid-vapor menisci recede into the film interior.[†] These mechanisms

_____

[†] By analogy to titanate films deposited from tiny precursors under conditions where the condensation rate was low [9,10], collapse of the film at the final stage of drying due to the capillary pressure is expected to play an important role (see discussion in following subsection), but both interpenetration and collapse probably contribute to the low porosities.

Fig. 13. _____

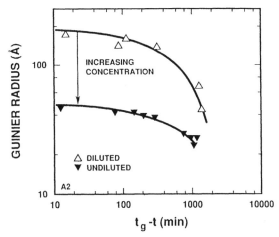

Concentration dependence of the Guinier radius for the two-step acid-catalyzed precursor (W1) as a function of the time from gelation, $t_g - t$. From Schaefer and Keefer [29]. Arrow suggests the compaction in the structure expected from solvent evaporation during dipcoating.

Fig. 14. _____

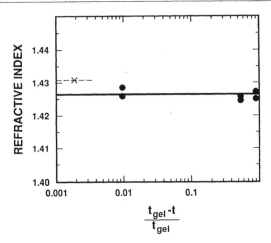

Refractive index of films prepared from the two-step acid-catalyzed precursor (W1) as a function of normalized times from gelation (defined in Fig. 13). The silicate species were grown under more concentrated conditions (undiluted) and then diluted with solvent immediately before dip coating. X denotes the refractive index of a film made from the spinnable precursor (W2). From Brinker *et al.* [20].

Fig. 15.

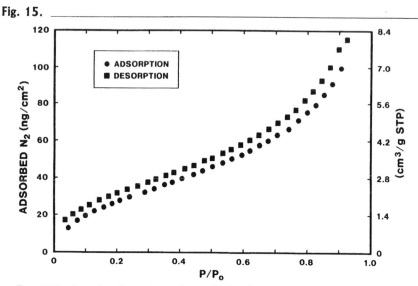

Type II N_2 adsorption-desorption isotherm obtained by a surface acoustic wave method for a non-porous film deposited from the W1 precursor. From Frye *et al.* [34].

are in fact synergistic, since interpenetration reduces the radii of the emptying liquid-filled channels, increasing the magnitude of the capillary pressure and, consequently, increasing the extent of film collapse. Only the strong concentration dependence of the viscosity inhibits interpenetration, but since little further condensation occurs during the deposition step, the precursor stiffness remains low.

2.1.2. HIGHLY BRANCHED SYSTEMS

A second example demonstrating the effect of precursor branching on the structure of films utilizes more highly branched, rigid borosilicate species synthesized [37] and deposited near pH 3. Figure 16 shows the ^{29}Si NMR spectrum [31] and SAXS profile of the borosilicate precursor solution after 16 days of aging. Compared to the silicate systems described previously, the borosilicate polymers are more highly branched, as illustrated by the greater proportion of Q^4 Si species in the NMR spectrum, and the power law scattering behavior, though limited, indicative of branched, compact fractal clusters with $d_f \approx 2.4$. The reduction in slope at low K indicates the beginning of the Guinier regime where $KR \sim 1$ (or $R \sim 5.0\,nm$) [20].

In contrast to the weakly branched precursors previously described, the more highly branched borosilicate precursors exhibit an inverse dependence

Fig. 16.

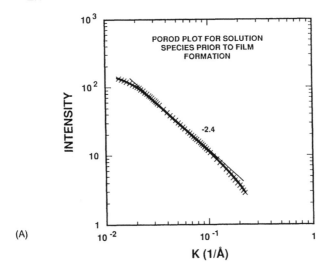

(A)

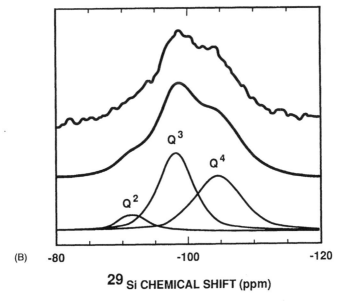

(B)

(A) Porod plot for multicomponent borosilicate precursors after aging for two weeks at 50°C and pH 3. $K = 4\pi/\lambda(\sin\theta/2)$. Limited power law region corresponds to $d_f \approx 2.4$. From Brinker *et al.* [20]. (B) Corresponding ^{29}Si NMR spectrum. From Assink [31].

Fig. 17.

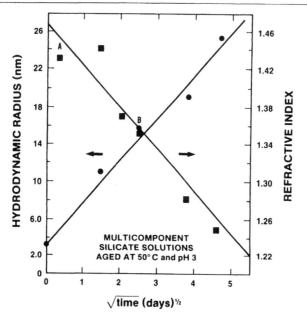

Reciprocal relationship between the hydrodynamic radius of multicomponent precursors as a function of root aging time at 50°C and pH 3 and corresponding refractive indices of films prepared from the multicomponent borosilicate precursors after different aging times. The strong mutual screening of branched polymeric precursors leads to a reciprocal relationship between polymer size and pore volume (inversely related to refractive index). From Brinker *et al.* [20].

of the pore volume on the precursor size as shown in Fig. 17. The density, ρ_f, of a mass fractal object decreases with distance, r, from the center of mass according to

$$\rho_f \propto 1/r^{(3-d_f)} \tag{11}$$

so the porosity increases as $r^{(3-d_f)}$. Thus the reciprocal relationship (Fig. 17) is consistent with a system of noninterpenetrating (mutually opaque) or weakly interpenetrating fractal clusters.

Figure 18 shows the SAXS profile of a film deposited from a precursor solution aged at 50°C for 14 days prior to dipping [20]. The Porod plot is composed of two limited power law regions corresponding to $d_f = 1.7$ and $d_f = 2.8$ separated by a "break" at $K^{-1} \sim 1.4$ nm. The corresponding pore size distribution obtained from a nitrogen desorption isotherm [38] (Fig. 18) indicates that the break corresponds closely to the pore radius, ~ 1.9 nm. The SAXS and pore size data suggest that during the deposition and drying stages, the individual clusters ($d_f = 2.4$) are compacted due to the capillary

Fig. 18.

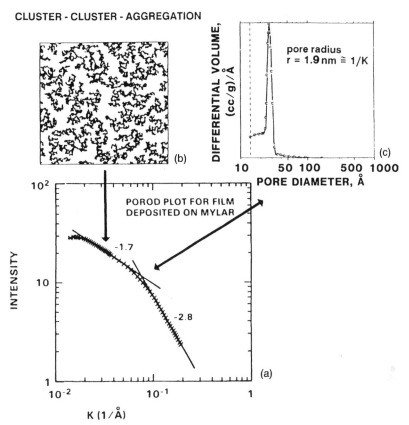

(a) SAXS Porod plot for film deposited on MYLAR® from the multicomponent borosilicate precursor after about two weeks of aging at 50°C and pH 3. $K=4\pi/\lambda(\sin\theta/2)$. From Brinker *et al.* [20]. (b) Simulation of diffusion-limited cluster–cluster aggregation from Meakin [21]. (c) Corresponding pore size distribution determined from the desorption branch of the N_2 isotherm from Brinker *et al.* [38]. Pore radius $r = 1.4$ nm $\approx 1/K$.

pressure, causing d_f to increase to 2.8 on length scales <1.5 nm.[†] On length scales >1.5 nm $d_f = 1.7$ is consistent with simulations of diffusion limited cluster–cluster aggregation [21]. ($d_f = 1.75$; see Fig. 18b.) This suggests that on length scales greater than 1.5 nm, the film structure is composed of aggregated, compacted clusters that create interconnected pores with throat diameters ~3.0 nm. Thus the porosity is attributed both to mutually opaque fractal clusters on short-length scales and interstitial

[†] The compacted clusters should have radii $R_2 = R_1{}^{d_{f_1}/d_{f_2}} = 5.0^{2.4/2.8} \sim 2.9$ nm ± 0.07 nm).

porosity created by cluster aggregation on long length scales. It should be noted that it is necessary to invoke fractal geometry to explain the reciprocal relationship between precursor size and porosity, since porosity created by conventional (Euclidean) particle packings (e.g., random close packing) is scale-invariant, i.e., the volume fraction solids of random close packing of uniform monosized particles is independent of particle size.

The refractive index of the film prepared from small, unaged precursors (point A in Fig. 17, hydrodynamic radius ~3.0 nm) indicates that the porosity is less than 5% according to Eq. 10. The corresponding nitrogen adsorption–desorption isotherm (Fig. 19) [38] is of Type II and the BET surface area is about $1 \text{ cm}^2/\text{cm}^2$ indicative of adsorption on a nonporous surface.

Based on ellipsometric imaging of titania films prepared from small (~0.5 nm diameter) precursors [9,10], it is likely that the elimination of porosity in the deposited film (A in Fig. 17) occurs primarily at the final stage of drying. Figures 20a and 20b show the profiles of refractive index and volume fraction solids, ϕ, as a function of distance from an arbitrary

Fig. 19.

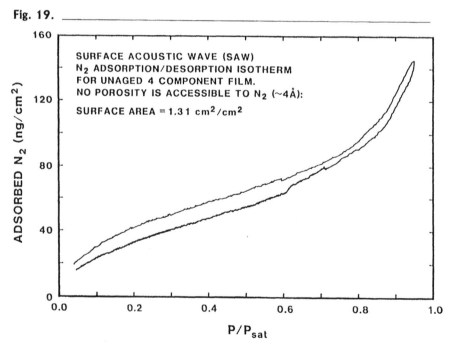

Type II N_2 adsorption–desorption isotherm of dip-coated film prepared from unaged multicomponent precursors. From Brinker *et al.* [34].

Fig. 20. _____

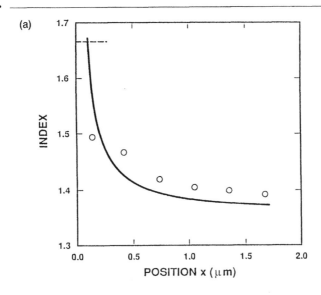

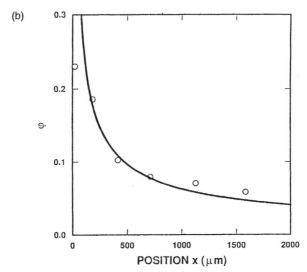

(a) Steady-state index-thickness curve of a drying titanate film prepared by dip coating. The solid line represents the index expected for steady-state conditions where hϕ is a constant. The dashed line represents the refractive index of the dried film, ~1.67. From Hurd and Brinker [9]. (b) Profile of steady-state volume fraction of solids, ϕ, of drying titanate film prepared by dip coating. The solid line represents $\phi \approx x^{-0.62}$ according to the non-constant evaporation model (see Fig. 4). From C.J. Brinker. A.J. Hurd, G.C. Frye, K.J. Ward, and C.S. Ashley, *J. Non-Cryst. Solids*, in press.

position ($0\,\mu$m) close to the drying front. (See Fig. 21 for details.) The refractive index—and ϕ—increase gradually as the drying front is approached, lagging the expectation for the non-constant evaporation model, and then increase dramatically within $100\,\mu$m of the drying front. The initial slowing down of the change in index and ϕ near the drying front has been observed in many films [9,10]. It is attributed either to partial drying, which reduces the index and invalidates Eq. 10, or to network formation that exerts counterpressure against the solvent–air interface. The subsequent dramatic compaction in the vicinity of the drying front (thickness or position $\rightarrow 0$) is attributed to the response of the network to the capillary pressure created as solvent–air menisci recede into the film interior. The curvature of the menisci, κ, and therefore the pressure, P, increase as the radius, r, of the emptying liquid channel decreases:

$$\kappa = 2\cos(\theta/r) \tag{12}$$

$$P = 2\gamma_{LV}\cos(\theta/r) \tag{13}$$

where θ is the contact angle. Assuming $\gamma_{LV} = 0.3$ J/m, $\theta = 0°$, and $r = 1.0$ nm, the maximum capillary pressure according to Eq. 13 is estimated to exceed 60 MPa (see Chapter 7) due to the tiny pores, which explains the remarkable compaction of films prepared from small precursors. Smaller pores (see Table 1) cause even greater pressures.

The film pore volume increases with the size of the precursor species prior to deposition. A typical adsorption–desorption isotherm for a porous film

Fig. 21a. _____

TiO$_2$

$\varDelta = 180°$ $\psi = 0°$ $\varDelta = 180°$ $\psi = 39°$ $\varDelta = 180°$ $\psi = 57°$ $\varDelta = 180°$ $\psi = 90°$

Ellipsometric images of drying titanate film during dip coating for four values of the ellipsometric angle, ψ, corresponding to points A–D in Fig. 21b. Orientation is such that coating bath is on left and dried film on right. From Hurd and Brinker [20].

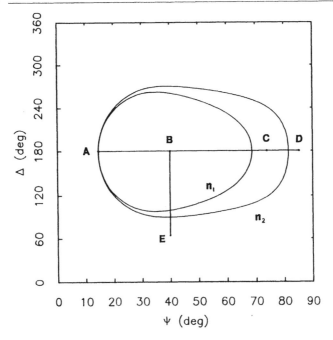

Locus of ellipsometric extinctions for null detection for two values of the refractive index, n, where $n_1 < n_2$. During ellipsometric imaging, motorized polarizers allow searching along paths ABCD or BE that guarantee crossing an extinction condition between the bounding indices, n_1 and n_2. From Hurd and Brinker [10].

(point B in Fig. 17) is shown in Fig. 22 [38]. It is of type IV [35] corresponding to a *mesoporous* solid (2–50 nm pore diameters). The pore size distribution calculated from the desorption isotherm [38] (see inset, Fig. 22) is quite narrow. Both the average pore sizes and surface areas of the films increase with the size of the precursor species prior to deposition. Table 1 [38] lists pore sizes and surface areas for films prepared from precursors aged from 0 to 25 days, corresponding to hydrodynamic radii ranging from about 2 to 26 nm (Fig. 17). This table emphasizes that the growth process prior to film deposition allows the film microstructure to be carefully tailored.

For this more highly branched system, the progressive increase in porosity with precursor size is attributed to the structure produced by transport-limited aggregation of mutually opaque, fractal clusters. Presumably this fragile structure is at least partially preserved during the drying process because the average radius of the liquid-filled channels between clusters increases with cluster size, causing the capillary pressure (and hence extent

Fig. 22. _____

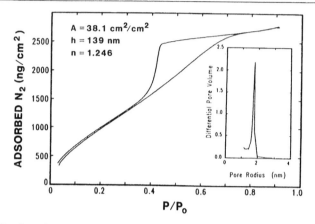

Type IV N_2 adsorption–desorption isotherm for a film prepared from multicomponent precursors after aging for two weeks at 50°C and pH 3. Inset Corresponding pore size distribution determined from the desorption branch. From Brinker *et al.* [38].

of film collapse) to decrease. In Section 3 we point out that the extent of collapse also depends on the relative rates of condensation and evaporation during film deposition.

2.1.3.   PARTICULATE SYSTEMS

Figure 23 shows a representative scattering curve for a particulate film prepared from a sol composed of unaggregated ~10 nm diameter silica spheres at pH 11.5 [20]. The important features of the scattering curve are:

Table 1.

Porosity versus Aging Conditions.

Sample Aging Times	Refractive Index	Porosity[a] (%)	Median Pore[a] Radius (nm)	Surface Area[a] (m²/g)
Unaged	1.45	0	<0.2	1.2–1.9
3 Day	1.31	16	1.5	146
1 Week	1.25	25	1.6	220
2 Week	1.21	33	1.9	263
3 Week[b]	1.18	52	3.0	245

[a] Determined by N_2 adsorption–desorption using a surface acoustic wave technique [34].
[b] The 3-week sample gelled. It was re-liquified at high shear rates and diluted with ethanol prior to film deposition.

Fig. 23.

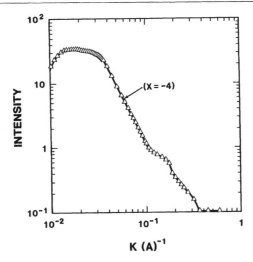

Porod plot of film prepared from monosized particulate silicate precursors (~10-nm diameters). From Brinker *et al.* [20].

the Porod slope of -4; the local maximum at $K = 1.5\,\mathrm{nm}^{-1}$ $(0.15\,\text{Å}^{-1})$; the absence of a second power law regime for dimensions $>3.5\,\mathrm{nm}$; and an apparent liquid-like peak at $K \approx 0.02$. The first two features are consistent with scattering from smooth, monodisperse 10 nm diameter particles. The Porod slope of -4 (for $K > 0.2\,\mathrm{nm}^{-1}$) is characteristic of scattering from smooth surfaces [38] for dimensions greater than about 35 nm. The local maximum represents the first single sphere scattering peak for particles about 8.4 nm in diameter $(R_g = 3.3\,\mathrm{nm})$ [39]. The absence of a second power law regime indicates that the film structure is quite uniform for length scales greater than the particle size. This shows that the particles remain highly dispersed as the concentration is increased during deposition: aggregation would cause a crossover to a Porod slope ≥ -2 [40]. Instead, the appearance of a liquid-like peak suggests that the repulsive double layer surrounding the particles (Chapter 4) lets them continue to rearrange in response to the increasing concentration. Finally, a liquid-like structure [41] is frozen in by the increasing viscosity.[†]

The final film is uniformly porous with a refractive index, $n = 1.24$, corresponding to about 45% porosity. This high porosity indicates that, although the system remains unaggregated, the capillary pressure that results from emptying the relatively large channels created between the particles

---

[†] We shall show in the following section that the particle packing can vary from liquid-like to ordered depending on the withdrawal rate. (See also discussion on pages 254–256.)

Fig. 24.

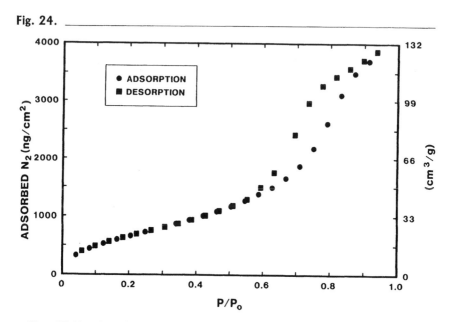

Type IV N$_2$ adsorption-desorption isotherm for a film prepared from monosized particulate silicate precursors (~20-nm diameters). From Frye *et al.* [34].

is insufficient to achieve close packing during the final stages of drying. The adsorption–desorption isotherm shown in Fig. 24 [31] is of Type IV, corresponding to a mesoporous film [35]. The BET surface area is 21.5 cm^2/cm$^2_{film}$, and the average pore radius determined from the desorption isotherm is about 3 nm [34]. This physical description inferred from the SAXS, ellipsometry, and adsorption results appears consistent with regions of random dense packing observed in the TEM micrograph in Figure 25 [20].

Aggregation of colloidal silica particles at intermediate pH (4–11) occurs by reaction-limited cluster aggregation [42], producing mass fractal aggregates described by $d_f = 2.1$. Based on the concept of mutual opacity (Eq. 9), film deposition from these aggregated precursors should yield porous films that exhibit a reciprocal relationship between refractive index and aggregate size similar to the strongly branched systems described in Subsection 2.1.2. Brinker *et al.* [20] deposited films from silica sols after one-hour aggregation steps at pH 2 to 11. Figure 26 shows the relationship between refractive index and the pH employed for aggregation and deposition. pH 7 corresponds to the highest aggregation rate (see inset, Fig. 26) and therefore the largest aggregates (~100 nm radii as measured by quasi-elastic light scattering (QELS)), resulting in the greatest film porosity (~65% according to Eq. 10).

Fig. 25.

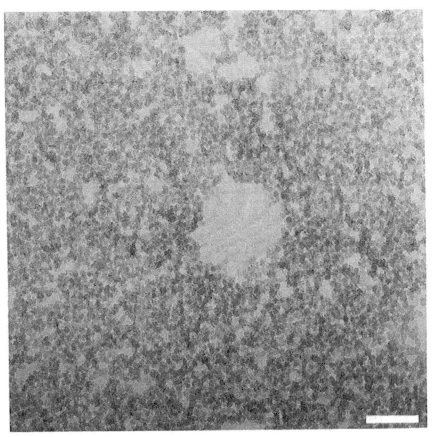

TEM micrograph of silicate film prepared on a holey carbon grid from spherical, particulate precursors at pH 11.5. Bar = 100 nm. From Brinker *et al.* [20]. Large voids result from sample preparation.

The high porosities are a consequence of both the fractal properties of the aggregate structure and the low capillary pressures that exist at the final stage of drying due to the aggregated particles. Figure 27 shows an ellipsometric image of a drying film prepared from silica aggregates [9]. Compared to images of films prepared from unaggregated precursors (see, e.g., Fig. 21), there is evidence of an extensive, partially wetted state indicated by a bright "plume." The plume represents a slowly changing plateau where the film is gradually thinning down rather than abruptly collapsing. Due to the low capillary pressure, little compaction occurs, so a rather thick, porous film is produced.

Fig. 26. _____

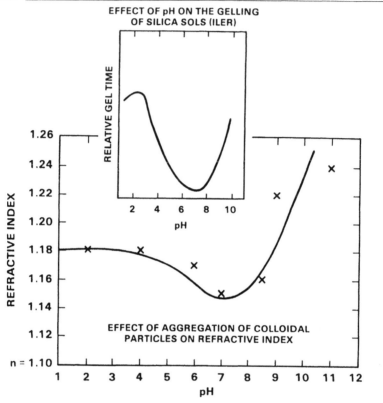

**REFRACTIVE INDEX OF DEPOSITED FILMS AFTER
DIFFERENT EXTENTS OF AGGREGATION.**

Refractive index of particulate silica films as a function of the pH of the aggregation step prior to deposition (~2 hour aggregation times). From Brinker *et al*. [20]. Inset: Gel times for aqueous colloidal silica as a function of pH. From Iler [26].

3. _____

DEPOSITION CONDITIONS

3.1.   Relative Rates of Condensation and Evaporation

The overlap of the deposition and drying stages in dipping and spinning establishes a competition between evaporation (which compacts the film) and continued condensation reactions (which stiffen the film, increasing its resistance to compaction). Thus the porosity of the film depends on the

Fig. 27. _____

Aggregated SiO₂

Δ = 180°   ψ = 75°

Ellipsometric image of a drying silicate film prepared from pre-aggregated, particulate precursors by dipping. Orientation is such that the coating bath is on the left and the dry film on the extreme right. Bright "plume" near the drying front on right is a long-lived partially wet area where the film is slowly thinning down by evaporation and collapse due to capillary pressure. Ellipsometric angles, $\Delta = 180°$; $\psi = 75°$. From Hurd and Brinker [9].

relative rates of condensation and evaporation. This is illustrated by comparing the porosities of films and bulk gels. Figure 28 shows nitrogen adsorption–desorption isotherms for a film and bulk gel prepared from identical acid-catalyzed silicate precursors [34]. The isotherm for the bulk gel is of Type I indicative of a _microporous_ solid (pore diameters <2.0 nm), whereas the film isotherm is of Type II corresponding to adsorption on a nonporous surface. This difference in porosities is attributed to the slow drying of the bulk gel (~1 month) compared to the rapid drying of the film (a few seconds). The low evaporation rate of the bulk gel allows it to become much stiffer at an earlier stage of drying (through continued condensation reactions) consequently reducing its shrinkage during the later stages of drying, and increasing its porosity.

During film formation the condensation rate can be controlled by varying the pH of the coating bath (see Fig. 11), while the evaporation rate can be controlled by varying the partial pressure of solvent in the coating ambient. Figure 29 illustrates the effect of reducing the condensation rate with respect to the evaporation rate during dip coating [38]. Films prepared from strongly branched, fractal ($d_f = 2.4$) precursors at pH 3.2 are quite porous ($n \approx 1.27$, corresponding to about 30% porosity according to Eq. 10). Reduction of the pH of the coating bath immediately prior to deposition reduces the condensation rate during deposition and drying (see Fig. 18b in Chapter 3 and Fig. 11), causing the refractive index of the deposited films to increase and the porosity to decrease ($V_p \sim 3\%$ at pH 1.2). The reduction in the condensation rate mitigates the criterion for mutual

Fig. 28.

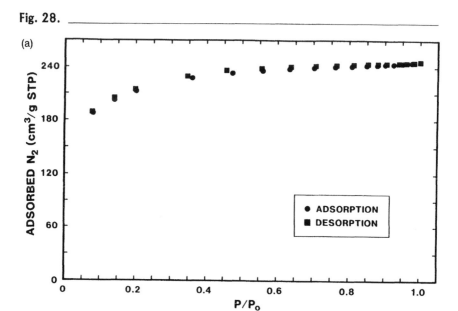

(a)

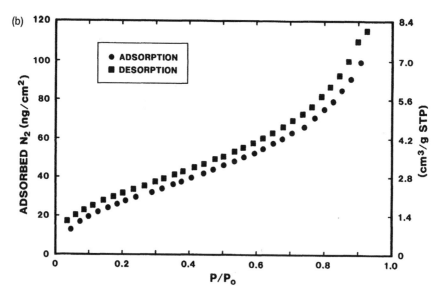

(b)

(a) Type I isotherm of a microporous bulk xerogel prepared from two-step acid-catalyzed precursors by gelation and slow drying. From Frye *et al.* [34].

(b) Type II isotherm of a non-porous film prepared from same precursors as in Fig. 28a. From Frye *et al.* [34].

Fig. 29.

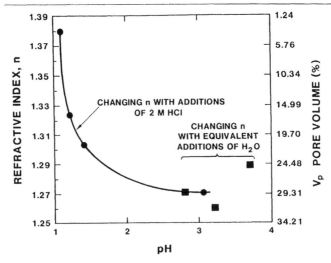

Refractive index and pore volume of films prepared from multicomponent silicate precursors (aged at 50°C and pH 3 for two weeks prior to deposition) versus the deposition pH achieved by additions of 2 M HCl. Refractive index and pore volume are very sensitive to pH. This reflects the reduction in the condensation rate with decreasing pH. Changes in index and pore volume resulting from additions of the same volume of H_2O are shown as boxes. From Brinker *et al.* [38].

transparency (Eq. 9) and retards the stiffening of the film during deposition and drying: the precursors initially interpenetrate and then are collapsed by the high capillary pressure.

The response of particulate systems to changes in the relative rates of condensation and evaporation depends on whether the particles are attractive or repulsive. If the particles are attractive, a reduction in the evaporation rate is analogous to an increase in the condensation rate: It provides greater time for aggregation to occur, increasing the film porosity. Conversely, an increase in the evaporation rate is analogous to a reduction in the relative condensation (or aggregation) rate, resulting in denser films as described previously in conjunction with Fig. 26. If the particles are repulsive, however, a reduction in the evaporation rate provides additional time for the particles to order. Figure 30 shows a densely packed, partially ordered film deposited from monosized silicate particles at pH 11.5 by very gradual draining of a reservoir. The film porosity is obviously much less than the 40% (estimated from the refractive index; obtained for similar silicate sols by dip coating at 10 cm/min, see Fig. 26).

To date the criteria for ordering (e.g., the maximum drainage or withdrawal rates, range of repulsive forces, and degree of monodispersity) are

Fig. 30.

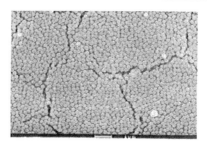

bar = 1 micron

Thick partially ordered silicate film prepared from ~200 nm spherical particles by slow draining. Bar = 1 μm. From Brinker and Ashley [43].

not well established. Figure 31 shows the effect of withdrawal speed, U, on the refractive index of films prepared from monosized particles at high pH and compares the behavior of these mutually repulsive particulate sols to the strongly branched system discussed in Section 2.1.2 (labeled "polymer" in Fig. 31). All of the particulate systems show an increase in refractive index (decrease in porosity) with U while the "polymer" sample shows the opposite trend. For the 55 nm sol, increasing U from 2 to 30 in/min (5 to 76 cm/min) causes the refractive index to increase from 1.24 to 1.35, corresponding to a reduction in porosity from 50 to about 25%.[†] This latter value is the porosity of monosized particles arranged in face centered cubic or hexagonal close-packing, indicating that the particles tend to order when deposited at high withdrawal rates. Both increased shear rates (see Section 3.2) and slower drying may contribute to the apparent ordering at high U. Molecular dynamics calculations indicate that repulsive particles tend to align in planes in a shear field [44]. The time required for alignment and registration of the sheets of repulsive particles is provided by the slower drying of thick films compared to thin ones.[††] By comparison, slower drying of the strongly branched precursors deposited at pH 3.5, where the condensation rate is high, results in an increase in porosity, since slower drying provides a longer time for condensation reactions to occur. In addition to shear rate, it is expected that the simultaneous application of an electric field [45] or, for magnetic particles, a magnetic field [46] should enhance the ordering kinetics.

[†] Corresponding N_2 adsorption-desorption, studies showed that the porosity was reduced from ~36% (random close packing) to ~25% (face centered cubic) as U was varied from 12.7 to 45.7 cm/min. G.C. Frye and C.J. Brinker, unpublished.

[††] Since drying is limited by evaporation from the liquid–vapor interface (the area of which is independent of thickness), thicker films deposited at higher U necessarily take longer to dry than thinner films.

Fig. 31.

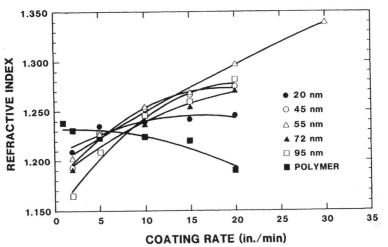

Refractive index versus withdrawal speed for films deposited from repulsive monosized particulate sols at pH ≥ 11 compared to a "polymer" sol prepared from attractive, strongly branched precursors (see Section 2.1.2) deposited at pH ~ 3. From Brinker *et al.* [38].

3.2. Effects of Shear

Both dipping and spinning establish shear fields that could influence the structure of the depositing film. In dipping, a gentle velocity gradient is established normal to the plane of the substrate. In spinning, a stronger gradient normal to the substrate is established in which the shear stress increases with both radius and angular velocity (spin frequency). Comparison of films prepared by dipping with films prepared by spinning at different frequencies provides a means of evaluating the effects of shear.

Figure 32 shows the refractive index and pore volume of films prepared from strongly branched precursors (described in Section 2.1.2) by either dipping or spinning. For all spin frequencies (1000–6000 RPM), the spun films exhibit lower pore volumes than the dipped film. The density of the spin-coated films increases with spin frequency and with proximity to the edge. Since the porosity of the dipped film was attributed to the fragile structure resulting from transport-limited aggregation of fractal clusters, the continuous decrease in pore volume with shear rate suggests breakdown or restructuring of the aggregates, shear-induced interpenetration, or alignment [20]. Alternatively, these results may reflect a change in the relative rates

Fig. 32.

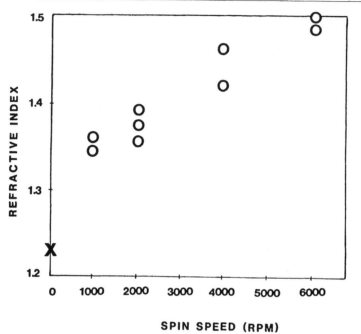

Refractive index versus spinning frequency for spun films prepared from multi-component silicate precursors aged at 50°C and pH 3 for two weeks prior to deposition. X denotes refractive index of corresponding dipped films. From Brinker *et al.* [20].

of evaporation and condensation: increased shear rates cause an increase in the evaporation rate (analogous to a decrease in the condensation rate), resulting in denser films as illustrated in Fig. 29.[†]

A second shear-induced effect is alignment of precursor species parallel to the direction of the applied shear. (See also discussion in Section 3.1.) Shear-induced alignment is expected to be maximized in systems composed of linear or rigid-rod polymers and minimized in systems composed of unaggregated equiaxed particles. Brinker *et al.* [20] tested this hypothesis by comparing the structures of spin-coated films prepared from monosized, spherical, silicate particles with weakly branched, spinnable, silicate polymers [47]. Figure 33 shows polarized IR spectra of the centers and edges of films prepared by spinning at 8000 RPM. If the precursors were aligned parallel to the direction of applied shear, effects due to polarized vibrational modes would cancel

---

[†] In the case of Fig. 32, the results are very sensitive to the deposition conditions, which indicates that the relative rates of evaporation and condensation rather than shear-induced restructuring are responsible for the behavior observed in Fig. 32.

Fig. 33.

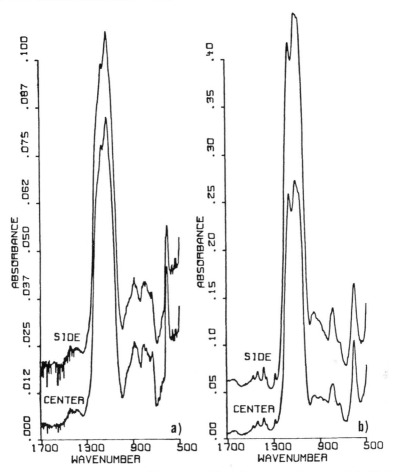

Polarized FTIR spectra of spun films prepared from (a) unaggregated, spherical, silicate particles and (b) spinnable polysilicate precursors at 8000 rpm. Spectra collected near edge (side) are vertically offset for clarity. From Brinker *et al.* [20].

at the center of the film due to radial symmetry, whereas near the edges alignment effects would be maximized. Figure 33 shows that the relative intensities of the silica vibrational modes are the same for the center and edge of the film prepared from spherical silicate particles. However, for films prepared from spinnable precursors, the relative intensities of the 1118 and 1172 cm^{-1} Si–O stretching modes are quite different at the center compared to the edge, indicating some shear-induced alignment of the siloxane backbone. Correspondingly, optical analysis [48] showed these films to be birefringent. Thus even though the spinnable precursors are not composed of rigid rod or linear polymers (see discussion in Section 2.6.6 of Chapter 3), some alignment or restructuring occurs at high shear rates (perhaps due to the asymmetry of fractal aggregates [21]).

4.

OTHER COATING METHODS

Although dipping and spinning are the most extensively investigated sol-gel coating methods, they suffer from the general requirements of axially or radially symmetric substrates and the difficulty in achieving thick layers with a single coating. For these reasons, other coating methods such as electrophoresis, thermophoresis, sedimentation (also called settling), and spraying have been explored. This section briefly reviews the first three of these methods as well as hybrid methods.

4.1. Electrophoresis

The phenomenon of *electrophoresis* (EP) is the movement of charged particles through a liquid under the influence of an external electric field applied across the suspension [49]. The particles or polymers move in a direction opposite or parallel to the external current, depending on their charge, and deposit on either the cathode or anode. (See Fig. 34.) EP is therefore limited to conductive substrates that can serve as an electrode, but complicated shapes are easily accommodated.

Unlike dip coating, in EP the particles move in linear trajectories and impact the stationary substrate with a maximum velocity, v, that depends on the applied electric field, E, and the particle charge, q. By analogy to settling (but replacing gravity with the electric field), Foss derived the following expression for v [50]:

$$v = \left(\frac{qE}{6\pi\eta r} \right) \times 10^9 \qquad (14)$$

where q is in Coulombs, E in volts/cm, the viscosity (η) in centipoise (m Pa·s), the hydrodynamic radius (r) in cm, and v in cm/sec. However, unlike settling, where the largest particles settle first, causing a deposit whose particle size decreases from bottom to top, in EP the compensating effects of the particle charge and hydrodynamic radius cause all the particles to have approximately the same velocity [51]. Therefore, the particle size distribution is the same at all levels of the deposit.

Fig. 34.

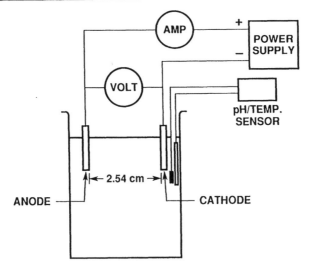

Schematic of electrophoretic coating cell. From Clark *et al.* [49].

Normally electrophoresis is performed in either a constant voltage or constant current mode. Under constant voltage conditions, the current decreases with time because the deposited film continually increases the cell resistance. Consequently, the deposition rate n (particles/unit area) decreases with time as the potential loss across the film, V_f, increases [49]:

$$dn/dt = NZV/D = NZ(V_{appl} - V_f)/D \qquad (15)$$

where N is the particle density of the sol, Z is the electrophoretic mobility (which depends on the dielectric constant and viscosity of the solvent, a geometric shape factor, and the zeta potential), V is the applied voltage, and D is the separation between the electrodes.

The constant current mode supplies a constant voltage drop across the suspension, so all particles arrive with the same velocity over the deposition time and the deposition rate is uniform. The number of deposited particles/area varies linearly with time [52]:

$$n(t) = NZI_0 R_s t/D \qquad (16)$$

where I_0 is the current and R_s the resistance of the sol. The voltage required to maintain this constant current also increases linearly with time [52]:

$$V_t = V_0(1 + \nu NZI_0 t/\sigma AD) \qquad (17)$$

where ν is the particle volume, σ the film conductance, and A the film area.

The exact mechanism of deposition is not well understood [52]. The charged species may be neutralized at the liquid–solid interface via oxidation (anode) or reduction (cathode) reactions, reducing the repulsive barrier created by the electrical double layer. Or the high concentration in the vicinity of the electrode could sufficiently reduce the particle separation so that aggregation occurs by London–van der Waals attractions. Both Hamaker [53] and Krishna *et al.* [54] conclude that EP coatings are most adherent when the sol is unstable in the vicinity of the electrode, resulting in an aggregated coating, not simply a compact viscous layer.

Dalzell [55] compared boehmite (AlO(OH)) films deposited by EP under constant voltage conditions to boehmite films deposited by dipping. Boehmite sols were prepared by a modification of the *Yoldas process* [56,57] in which the peptization step was designed to minimize the gelation volume.

Below about pH 9 boehmite acquires a positive charge by protonation [55]:

$$AlO(OH) + H^+ \rightarrow AlO(OH_2)^+. \tag{18}$$

During EP, the pertinent cathode and anode reactions are [55,58]

$$2AlO(OH_2)^+ + 2e^- \rightarrow 2AlO(OH) + H_2 \tag{19}$$
$$\text{colloid surface} \qquad\quad \text{cathode} \qquad \text{gas}$$

$$2H_2O \rightarrow H_2 + 2OH^- \tag{20}$$
$$\text{solution} \qquad \text{cathode}$$

$$NO_3^- + 2OH^- \rightarrow NO_2^- + O_2 + H_2O + 2e^- \tag{21}$$
$$\text{solution} \qquad\qquad \text{anode} \qquad\quad \text{solution}$$

$$2H_2O \rightarrow O_2 + 4H^- + 4e^-. \tag{22}$$
$$\text{solution} \qquad \text{anode}$$

Equation 19 is the reduction reaction that leads to the deposition of boehmite on the cathode. The coating yield (weight per unit area) as a function of time is shown in Fig. 35. [49]. The initial low rate of yield (region A) is due to the formation of dipoles and inertial effects accompanying their orientation in the electric field. At long times (region C) the rate approaches zero due to the increase in voltage drop across the deposited film (Eq. 15).

Compared to films prepared by dipping, the most significant differences observed for EP Films are the following: (1) whereas the yield of dipped films did not depend on the immersion time prior to the withdrawal step, the EP film yield increased with immersion time as illustrated in Fig. 35; (2) the maximum bulk density of the EP films was 2.85 g/cm^3 [55,49], corresponding to a relative density, $\rho_{film}/\rho_{AlOOH} = 0.95$, while the bulk density of the dipped films was 1.13 g/cm^3 [55,49], corresponding to a relative density of only 0.38; (3) for films of equivalent thickness, the EP films were harder than the dipped films.

Fig. 35. _____

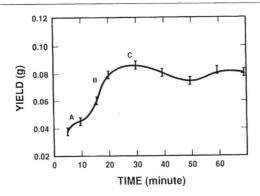

TIME (minute)

Electrophoretic coating yield per unit area versus deposition time at 5 volts for an aqueous, boehmite sol. From Clark *et al.* [49].

Since the boehmite particles are fibrillar or plate-like [59], the remarkable difference in bulk densities suggests that EP causes smectic or nematic alignment of the platelets (Fig. 62, Ch. 9), leading to efficient packing. During dipping the platelets tend to align face to edge due to their charges [59] creating open networks with low bulk densities. The EP process also tends to be self-healing, because the field is greatest at a hole or pore, causing particles to deposit there preferentially. It is not clear, however, just how dense the film becomes during EP, since the EP film is withdrawn from the coating bath and allowed to dry and is thus subjected to capillary pressure. For the aligned films the spacing between platelets could be quite small, resulting in large capillary pressures that further compact the film during drying.

The nitrogen adsorption-desorption isotherm of an EP boehmite film deposited at 5 volts is shown in Fig. 36. It is a type IV isotherm indicative of mesoporosity, but the enhanced adsorption at low pressure, $P/P_0 < 0.1$, indicates considerable microporosity [35]. This isotherm is consistent with a structure comprising regions of densely packed platelets (that create microporosity) separated by larger channels responsible for the mesoporosity. The larger mesoporous channels might be created by H_2 evolution (Eqs. 19, 20) which is observed during deposition above about 3 volts [55]. At higher voltages (>15 V), H_2 evolution creates large macroporous channels that dominate the film structure and preclude the use of high-voltage EP films for most applications.

Apart from increasing the number and size of channels produced by H_2 evolution, the increasing voltage above about 3 volts produces a very gradual compaction of the deposited film, manifested as a reduction in surface

Fig. 36.

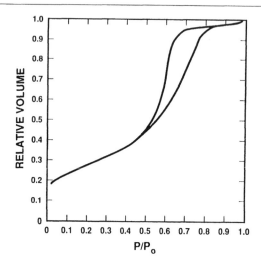

Type IV N_2 adsorption–desorption isotherm for an electrophoretic coating deposited at
5 volts from an aqueous boehmite sol. From Dalzell [55].

area, pore volume, and pore size (excluding macroporosity which is not
accounted for by nitrogen adsorption). This suggests that increasing the
particle velocity causes a denser deposit to form. This same reasoning leads
to the hypothesis that constant-voltage conditions should cause a gradual
reduction in film density with time as the voltage drop across the film
increases and the depositing particle velocity decreases (Eq. 15). From this
standpoint, constant-current conditions are expected to produce more
uniform films.

4.2.   Thermophoresis

Thermophoresis is the movement of suspended particles through a fluid
under the influence of an applied thermal gradient [60]. The thermal gradient
causes the particles to experience a net force in the direction of decreasing
temperature (positive thermophoresis), because molecules impacting the
particle on opposite sides through thermal motion have different average
velocities due to their differences in temperature [61–63]. The thermophoretic velocity acquired by a particle suspended in a liquid in the direction
of decreasing temperature gradient, ∇T, has the general form [58]:

$$v_t = -K_t(\eta/T)\,\nabla T \qquad (23)$$

where K_t is a constant that depends on the molecular flow regime of the sol.

Particles suspended in a temperature gradient should move in more or less ballistic trajectories and impact the coating surface with a velocity normal to the surface proportional to the temperature gradient. It is expected therefore that, like EP coatings, thermophoretic coatings may be denser than dip coatings. Unlike EP coatings, thermophoretic coatings have the advantage that electrically conductive substrates are not a requirement. A thorough comparison between the structure and properties of thermophoretic and dip coatings has not been performed.

4.3. Settling

A final coating method that is amenable to particulate systems is a *settling* technique [64] adapted by Garino [65–67] for deposition of particulate sols. The sol is spread onto a horizontal substrate with a moving rod much like doctor-blading. The particles deposit on the substrate surface under the influence of gravity accompanied by the convective motion resulting from solvent evaporation. For large particles that settle within the time constraints established by evaporation, this method is similar to EP since the particles impact the surface with a finite velocity component normal to the surface. According to Stoke's law [68] the steady-state velocity of a spherical particle settling in a viscous medium is:

$$v = g(\rho_p - \rho_0)D_s^2/18\eta_0 \qquad (24)$$

where ρ_p is the particle density, ρ_0 is the density of the fluid, D_s is the diameter, and η_0 is the viscosity. Unlike EP the largest particles arrive first, potentially causing a gradation in particle size, and the deposition and evaporation stages may overlap, depending on the relative rates of settling and evaporation. For smaller particles that do not settle quickly, this method differs from dip or spin coating in that the depositing film is not thinned by draining and no shear field is established. Unlike any of the previously discussed methods, all the particles in the sol are deposited on the substrate surface.

Figures 37a and 37b show two perspectives of a film deposited by settling an aqueous sol composed of electrostatically stabilized (repulsive) 0.5-μm silica spheres formed by the *Stöber process* [69]. Numerous ordered (face-centered cubic or hexagonal close-packed) colloidal crystalline regions are evident, separated by random close-packed regions serving as domain boundaries. According to Eq. 24, the settling velocity, $v \approx 10\,\mu$m/min. For coatings that dry over a period of 5 to 30 min (depending on temperature), the settling distance varies from 0.50 to 0.3 mm, which represents a

Fig. 37.

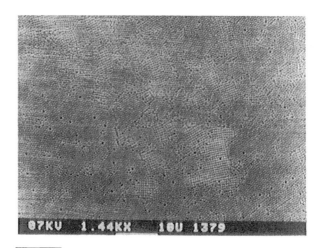

Edge and top view of an ordered silicate film prepared from 0.5-μm particles by settling. Edge view: bar = 5 μm. Top-view: bar = 10 μm. From Garino and Bowen [66,67].

significant fraction of the original sol thickness, ~0.5 mm. Thus, for this film, both settling and convection due to drying contribute significantly to the deposition process. It is presently unclear whether settling is necessary for ordering or whether monodispersity and long-range forces are sufficient. During sedimentation without evaporation, repulsive, monosized particles are known to order [70,71]: the gradually increasing concentration allows particles sufficient time to crystallize in response to uniform, long-range, repulsive forces. (See refs. [60,61] of Chapter 4.) By analogy to rapidly quenched melts, rapid concentration via evaporation without settling may "freeze-in" an amorphous liquid-like structure. Presumably an investigation of the dependence of ordering on the relative rates of settling and evaporation would clarify this issue.

4.4. Hybrid Methods

It is important to note that all of the coating methods described thus far are suitable for multiple deposition steps employing the original solution

Fig. 38.

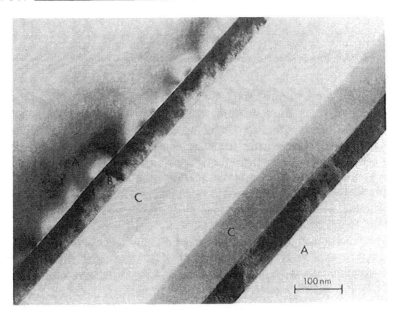

Triple-layer $TiO_2/SiO_2/SiO_2$ film deposited on Si by spinning. SiO_2/SiO_2 interface is unresolvable. Bar = 100 nm. From Pettit and Brinker [73].

precursors to build up thicker coatings or different precursors (structurally or compositionally) to obtain layered or graded coatings [72]. When each layer of a multilayer coating is quite dense as deposited, a very sharp interface is established as illustrated in Fig. 38 for a triple layer, $TiO_2/SiO_2/SiO_2$ film deposited on Si [73]. When the underlying layer is porous, intermixing of the two layers may occur. Based on the concept of mutual transparency or opacity defined by Eq. 9, the extent of intermixing may depend on the characteristic sizes and fractal dimensions associated with the depositing precursor and the underlying porous substrate. This interpenetration process may be aided by the capillary pressure created by solvent flow into the underlying porous media. To the best of our knowledge, this topic has not been addressed in conjunction with film formation.

It is also possible to deposit sequential layers by different coating methods resulting in hybrid films with unique properties unattainable by a single method. For example, dip coating a conductive substrate to obtain a porous layer of composition A followed by electrophoretic deposition of composition B may be a viable route to a dense, molecular composite (AB) film. The possibilities are practically unlimited and virtually unexplored.

5.

SUMMARY

The structures of films prepared from polymeric or particulate precursors by liquid-based coating methods depend on geometric factors such as precursor size and extent of branching or aggregation prior to deposition and concurrent phenomena (such as continued condensation reactions, evaporation, settling, electrophoresis, and shear) during deposition. Unlike bulk gels, for most film formation methods the aggregation, gelation, and drying stages significantly overlap, establishing a time scale for aggregation or ordering, gelation, and aging that depends on the evaporation rate of the solvent, typically alcohol or water. Generally, this has the effect of producing denser structures than the corresponding bulk (monolithic) xerogels, because little aging can occur in the brief time span of film formation. Control of both the precursor structure and the deposition conditions results in precise tailoring of the film microstructure and properties: porosity, pore size, surface area, and refractive index. In the following chapter we discuss how these tailored properties have been exploited in numerous applications of sol-gel–derived thin films.

REFERENCES

1. L.E. Scriven in *Better Ceramics Through Chemistry III* eds. C.J. Brinker, D.E. Clark, and D.R. Ulrich (Mat. Res. Soc., Pittsburgh, Pa., 1988), pp. 717–729.
2. R.P. Spiers, C.V. Subaraman, and W.L. Wilkinson, *Chem. Eng. Sci.*, **29** (1974) 389–396.
3. L.D. Landau and B.G. Levich, *Acta Physiochim*, U.R.S.S., **17** (1942) 42–54.
4. I. Strawbridge and P.F. James, *J. Non-Cryst. Solids*, **82** (1986) 366–372.
5. C.J. Brinker and C.S. Ashley, unpublished results.
6. H. Dislich and E. Hussmann, *Thin Solid Films*, **98** (1981) 129.
7. N.E. Bixler, private communication.
8. D.E. Bornside, C.W. Macosko, and L.E. Scriven, *J. Imaging Tech.*, **13** (1987) 122–129.
9. A.J. Hurd and C.J. Brinker, *J. de Phys.*, **49** (1988) 1017–1025.
10. A.J. Hurd and C.J. Brinker in *Better Ceramics Through Chemistry III* eds. C.J. Brinker, D.E. Clark, and D.R. Ulrich (Mat. Res. Soc., Pittsburgh, Pa., 1988), pp. 731–742.
11. A.J. Hurd, to be published, Proc. of AIChE Symp. on Emerging Technologies, San Francisco, 1989.
12. A. Servida and L.E. Scriven, private communication.
13. J.D. Jackson, *Classical Electrodynamics*, 2d ed. (Wiley, New York, 1979), chapter 2.
14. A.W. Adamson, *Physical Chemistry of Surfaces*, 4th ed. (Wiley, New York, 1982).
15. P.G. de Gennes, *Rev. Mod. Phys.*, **57** (1985) 827.
16. B. Higgins, *Phys. Fluids*, **29** (1986) 3522–3529.
17. A.G. Emslie, F.T. Bonner, and L.G. Peck, *J. Appl. Phys.*, **29** (1958) 858–862.
18. D. Meyerhofer, *J. Appl. Phys.*, **49** (1978) 3993–3997.
19. D.E. Bornside, C.W. Macosko, and L.E. Scriven, *J. Imaging Tech.* **13** (1987), 122–130.
20. C.J. Brinker, A.J. Hurd, and K.J. Ward in *Ultrastructure Processing of Advanced Ceramics*, eds. J.D. Mackenzie and D.R. Ulrich (Wiley, New York, 1988), p. 223.
21. P. Meakin, *Phys. Rev. Lett.*, **51** (1983) 1119.
22. R. Zallen in *The Physics of Amorphous Solids* (Wiley-Interscience, New York, 1983), chapter 4.
23. B.B. Mandelbrot, *The Fractal Geometry of Nature* (Freeman, San Francisco, 1982).
24. T.A. Witten and M.E. Gates, *Science*, **232** (1986) 1607.
25. P.G. de Gennes, *Physics Today*, **36** (1983) 33–39.
26. R.K. Iler in *The Chemistry of Silica* (Wiley, New York, 1979).
27. W.D. Klemperer and S.D. Ramamurthi in *Better Ceramics Through Chemistry III*, eds. C.J. Brinker, D.E. Clark, and D.R. Ulrich (Mat. Res. Soc., Pittsburgh, Pa., 1988), pp. 1–13.
28. L.W. Kelts and N.J. Armstrong in *Better Ceramics Through Chemistry III*, eds. C.J. Brinker, D.E. Clark, and D.R. Ulrich (Mat. Res. Soc., Pittsburgh, Pa., 1988), pp. 519–522.
29. D.W. Schaefer and K.D. Keefer in *Better Ceramics Through Chemistry*, eds. C.J. Brinker, D.E. Clark, and D.R. Ulrich (Elsevier, New York, 1984), pp. 1–14.
30. S. Sakka and H. Kozuka, *J. Non-Cryst. Solids*, **100** (1988) 142–153.
31. R.A. Assink, unpublished data.
32. C.J. Brinker and R.A. Assink, *J. Non-Cryst. Solids*, **111** (1989) 48–54.
33. M. Born and E. Wolf in *Principles of Optics* (Pergamon, New York, 1975), p. 87.
34. G.C. Frye, A.J. Ricco, S.J. Martin, and C.J. Brinker in *Better Ceramics Through Chemistry III*, eds. C.J. Brinker, D.E. Clark, and D.R. Ulrich (Mat. Res. Soc., Pittsburgh, Pa., 1988), pp. 349–354.
35. The types of isotherms and their significance are discussed in S.J. Gregg and K.S.W. Sing, *Adsorption, Surface Area and Porosity* (Academic Press, New York, 1982).

36. P.J. Flory, *Principles of Polymer Chemistry* (Cornell University Press, Ithaca, New York, 1978).
37. C.S. Ashley and S.T. Reed, Sandia National Laboratories Report SAND 84-0662 (1984) (available from NTIS).
38. C.J. Brinker, G.C. Frye, A.J. Hurd, K.J. Ward, and T. Bein, to be published, Proc. of IVth Int'l. Conference on Ultrastructure Processing of Glasses, Ceramics, and Composites, eds. D.R. Uhlmann and D.R. Ulrich (Wiley, New York, 1990).
39. G. Porod, *Kolloid Z.*, **124** (1951) 83.
40. D.W. Schaefer, J.E. Martin, P. Wiltzius, and D.S. Cannell in *Kinetics of Aggregation and Gelation*, ed. F. Family and D.P. Laudau (Elsevier, New York, 1984), p. 71.
41. D.W. Schaefer and A.J. Hurd in *Chemistry and Physics of Composite Media*, Electrochemical Society Proceedings, **vol. 85-8**, eds. M. Tomkiewics and P.N. Sen (Electrochemical Society, N.J., 1986).
42. C. Aubert and D.S. Cannell, *Phys. Rev. Lett.*, **56** (1986) 738.
43. C.J. Brinker and C.S. Ashley, unpublished.
44. J.H. Simmons, R.K. Mohr, and C.J. Montrose, *J. Appl. Phys.*, **53** (1982) 4075.
45. A.J. Hurd in *Better Ceramics Through Chemistry II*, eds. C.J. Brinker, D.E. Clark, and D.R. Ulrich (Mat. Res. Soc., Pittsburgh, Pa., 1986), pp. 345–350.
46. A.J. Hurd and B.C. Bunker, unpublished.
47. S. Sakka and K. Kamiya, *J. Non-Cryst. Solids*, **48** (1982) 31.
48. A.J. Hurd and C.J. Brinker, unpublished.
49. D.E. Clark, W.J. Dalzell, and D.C. Folz, *Ceram. Eng. Sci. Proc.*, **9** (1988) 1111–1118.
50. L. Foss, M.S. thesis, Massachusetts Institute of Technology, Cambridge, Mass. (1982).
51. S. Storz, *J. Colloid and Int. Sci.*, **65** (1978) 118.
52. A. Sussman and T.J. Ward, *RCA Review*, **42** (1981) 178–197.
53. H.C. Hamaker, *Trans. Faraday Soc.*, **36** (1940) 279.
54. D.U. Krishna Rao and E.C. Subbarao, *Ceramic Bulletin*, **58** (1979) 467–469.
55. W.J. Dalzell, M.S. thesis, University of Florida, Gainesville (1988).
56. B.E. Yoldas, *J. Applied Chem. Biotechnology*, **23** (1973) 803–809.
57. B.E. Yoldas, *J. Mater. Sci.*, **10** (1975) 1856–1860.
58. W.J. Dalzell and D.E. Clark, *Cer. Eng. and Sci. Proc.*, **7** (1986) 1014–1026.
59. A.C. Pierre and D.R. Uhlmann in *Better Ceramics Through Chemistry II*, eds. C.J. Brinker, D.E. Clark, and D.R. Ulrich (Mat. Res. Soc., Pittsburgh, Pa., 1986), pp. 481–488.
60. B. Derjaguin, Ya. I. Rabinovich, A.I. Storozhilova, and G.I. Shcherbing, *J. Colloid Interface Sci.*, **57** (1976) 451–461.
61. K.L. Walker, F.T. Geyling, and S.R. Nagel, *J. Am. Ceram. Soc.*, **63** (1980) 552–558.
62. E.A. Mason and S. Chapman, *J. Chem. Phys.*, **36** (1962) 627.
63. J.R. Sellers, M. Tribus, and J.S. Klein, *Trans. ASME*, **78** (1956) 441.
64. W.A. Pilskin and E.E. Conrad, *Electro Chem. Tech.*, **2** (1964) 196–199.
65. T.J. Garino, Ph.D. thesis, Massachusetts Institute of Technology, Cambridge, Mass. (1986).
66. T.J. Garino and H.K. Bowen, *J. Am. Ceram. Soc.*, **70** (1987) 315–317.
67. T.J. Garino and H.K. Bowen, *J. Am. Ceram. Soc.*, **70** (1987) 311–314.
68. L.G. Bunville in *Modern Methods of Particle Size Analysis*, ed. H.G. Barth (Wiley, New York, 1984), p. 6.
69. W. Stöber, A. Fink, and E. Bohn, *J. Colloid Interface Sci.*, **26** (1968) 62.
70. E.A. Barringer, N. Jubb, B. Fegley, R.L. Pober, and H.K. Bowen in *Ultrastructure Processing of Ceramics, Glasses, and Composites*, eds. L.L. Hench and D.R. Ulrich (Wiley, New York, 1984), pp. 315–333.

71. E.A. Barringer and H.K. Bowen, *J. Am. Ceram. Soc.*, **65** (1982) 199–201.
72. H. Schröder in *Physics of Thin Films: Advances in Research and Development*, **vol. 5**, eds. G. Hass and R.E. Thun (Academic Press, NY, 1969), p.87.
73. R.B. Pettit, C.J. Brinker, and C.S. Ashley, *Solar Cells*, **15** (1985) 267–278.

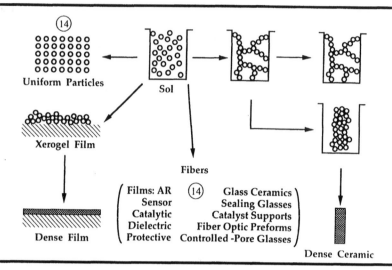

Uniform Particles

Sol

Xerogel Film

Dense Film

Fibers

Films: AR
Sensor
Catalytic
Dielectric
Protective

Glass Ceramics
Sealing Glasses
Catalyst Supports
Fiber Optic Preforms
Controlled -Pore Glasses

Dense Ceramic

CHAPTER 14

Applications

Applications for sol-gel processing derive from the various special shapes obtained directly from the gel state (e.g., monoliths, films, fibers, and monosized powders) combined with compositional and microstructural control and low processing temperatures. Compared to conventional sources of ceramic raw materials, often minerals dug from the earth, synthetic chemical precursors are a uniform and reproducible source of raw materials that can be made extremely pure through various synthetic means. Low processing temperatures, which result from microstructural control (e.g., high surface areas and small pore sizes), expand glass-forming regions by avoiding crystallization or phase separation, making new materials available to the technologist. In the opposite sense, microstructural control results from low processing temperatures, since metastable, porous structures created in solution are preserved, leading to applications in filtration, insulation, separations, sensors, and antireflective surfaces. The advantages of the sol-gel process (for preparing glass) are summarized in Table 1 [1].

The disadvantages of sol-gel processing include the cost of the raw materials, shrinkage that accompanies drying and sintering, and processing times. (See Table 2 [1].) Thin films benefit from most of the advantages of sol-gel processing just cited while avoiding these disadvantages; as of 1989, thin films are one of the few successful commercial applications. However even films suffer from cracking problems associated with attempts to prepare thick films ($>1 \mu$m).

Successful applications should be distinguished from applications in general. According to Livage [2] for an application to be commercially successful, it must significantly improve upon a current product or process

Table 1.

Some Advantages of the Sol-Gel Method over Conventional Melting for Glass.

1. Better homogeneity from raw materials.
2. Better purity from raw materials.
3. Lower temperature of preparation:
    a. Save energy;
    b. Minimize evaporation losses;
    c. Minimize air pollution;
    d. No reactions with containers, thus purity;
    e. Bypass phase separation;
    f. Bypass crystallization.
4. New noncrystalline solids outside the range of normal glass formation.
5. New crystalline phases from new noncrystalline solids.
6. Better glass products from special properties of gel.
7. Special products such as films.

Source: Mackenzie [1].

or create an entirely new commodity. It is unlikely, for example, that sol-gel processing will be successful as an alternative to conventional glass melting unless new glasses with unique properties are the result, or sol-gel processing solves problems brought about by the conventional high temperature approach.

This chapter presents a brief overview of current and potential applications of sol-gel processing. Because two books published in 1988 [3,4] and a number of review papers of the 1980s [1,2,5–12] discuss this topic in detail, the primary purpose of this chapter is to provide a source of references to current technology and to discuss critical issues associated with various classes of applications that must be addressed in order to advance sol-gel technology. An outline of this chapter is as follows:

1. *Thin Films and Coatings* discusses applications for optical, electronic, protective, and porous thin films or coatings.

Table 2.

Some Disadvantages of the Sol-Gel Method.

1. High cost of raw materials.
2. Large shrinkage during processing.
3. Residual fine pores.
4. Residual hydroxyl.
5. Residual carbon.
6. Health hazards of organic solutions.
7. Long processing times.

Source: Mackenzie [1].

2. *Monoliths* reviews applications for cast bulk shapes dried without cracking in such areas as optical components, transparent superinsulation, and ultralow-expansion glasses.

3. *Powders, Grains, and Spheres* addresses uses for powders as ceramic precursors or abrasive grains and applications of dense or hollow ceramic or glass spheres.

4. *Fibers* drawn directly from viscous sols are used primarily for reinforcement or fabrication of refractory textiles.

5. *Composites* discusses uses for gels as matrices for fiber-, whisker-, or particle-reinforced composites and as hosts for organic, ceramic, or metallic phases.

6. *Porous Gels and Membranes* describes applications that result from the ability to tailor the porosity of thin free-standing membranes, as well as bulk xerogels or aerogels.

1.

THIN FILMS AND COATINGS

Films and coatings represent the earliest commercial application of sol-gel technology [13]. Thin films (normally $<1 \mu$m in thickness) formed by dipping or spinning use little in the way of raw materials and may be processed quickly without cracking, overcoming most of the disadvantages of sol-gel processing. In addition, large substrates may be accommodated and it is possible to uniformly coat both sides of planar and axially symmetric substrates such as pipes, tubes, rods, and fibers not easily handled by more conventional coating processes. The early applications for sol-gel films were in optical coatings as reviewed by Schroeder [14] in 1969. Since then, many new uses for sol-gel films have appeared in electronic, protective, membrane, and sensor applications. Current (late 1980s) and potential applications for sol-gel thin films and coatings were reviewed in references [15–21]. This section briefly summarizes these reports with respect to the most promising applications and the most severe problems. Table 3 provides a representative set of references to the various applications of sol-gel thin films and coatings.

1.1. Optical Coatings

Optical coatings alter the reflectance, transmission, or absorption of the substrate. IROX® (TiO_2/Pd) coated architectural glass (Fig. 1) is the best example of the current state of the art [22]: TiO_2 controls the reflectivity

Table 3.

Applications of Films and Coatings.

Topic	References
General review	13–21
Optical coatings	
Colored	22–24
Antireflective coatings	25–37
Optoelectronic	38–40
Optical memory	41–42
Electronic coatings	
Photo anodes	43–47
High-temperature superconductors	48–51
Conductive	52–57
Ferroelectric–Electro-optic	58–67
Protective coatings	
Corrosion-resistant	68–74
Mechanical	
Planarization	75–76
Scratch- and wear-resistant	77–81
Strengthening	82–83
Adhesion promoting	84–89
Passivation (electronic)	90–97
Porous coatings	98–100
Miscellaneous coatings	101–105

and the Pd content provides the desired absorption. In this manner buildings appear outwardly uniformly reflective, while light transmission is controlled in accordance with sun exposure to minimize summer cooling costs. (See Figs. 1b and 2.) Schott Glaswerke produces millions of square meters of optical oxide coatings on glass each year [21]. Most of these coatings are single or multilayer interference films produced in the SiO_2–TiO_2 binary system where the refractive index for films consolidated at 500°C decreases continuously with SiO_2 content from about 2.2 to 1.4. (See Fig. 3a.) Controlled introduction of porosity can further reduce the refractive index of SiO_2 films to below 1.2 [17] (Fig. 3b), either uniformly or in a graded manner. The absorption of thin films has also been modified by incorporation of transition metals to produce a variety of colored coatings on various substrates [23,24].

In addition to reflective or colored coatings, oxide coatings on glass and silicon substrates (single layers, multilayers, and porous layers) have been used extensively as antireflective (AR) surfaces in solar-related applications to improve device efficiency and as laser-damage–resistant AR coatings for laser optics, especially in inertial confinement fusion applications [25–37].

Fig. 1.

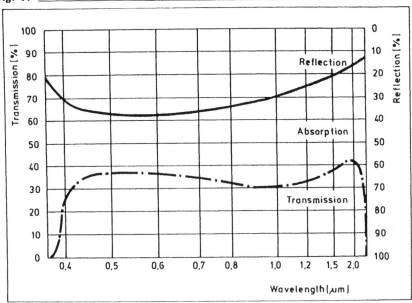

(a)

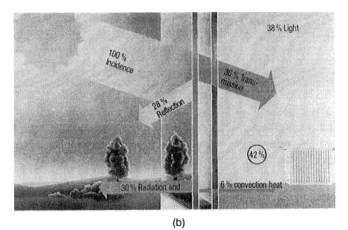

(b)

(a) Percentages of transmission, reflection, and absorption for IROX® A1 architectural glass coating (percent transmission + percent reflection + percent absorption = 100). From Dislich [21].

(b) Schematic illustration of optical properties of IROX® A1 coatings particularly well suited for all-glass facades. From [21], Dislich, in *Sol-Gel Technology for Thin Films, Preforms, Electronics, and Speciality Shapes*, ed. L.C. Klein (Noyes Publications, Park Ridge, New Jersey, 1988).

Fig. 2. _____

Photo of Euro-House in Frankfurt, West Germany illustrating the uniformity in reflection of IROX®-coated glass panels, each of which exceeds $6\,m^2$ in area. From Dislich [15].

Using a single coating of the correct thickness and refractive index, the reflectance at a single wavelength can be reduced to a value near zero. In air at normal incidence, this condition is met when the film has a real part of the refractive index, n_f, given by

$$n_f = (n_s)^{1/2} \tag{1}$$

where n_s is the real part of the refractive index of the substrate ($n_s \approx 1.5$ for glass and 4.0 for silicon). The film thickness, d (generally, 50–150 nm), is adjusted to give minimum reflectance at a wavelength, λ_0, defined through the equation

$$\lambda_0 = 4n_f d \tag{2}$$

where for solar energy applications, λ_0 is normally centered in the solar spectrum. From Eq. 1, porous layers ($n \approx 1.2$) are required to antireflect glass, while silicon substrates used in solar cells require a thin film with $n_f \approx 2$ achieved, for example, with a binary $80TiO_2 \cdot 20SiO_2$ composition (Fig. 3a).

Since for a single, uniform layer the reflectance is minimized at only one wavelength defined by Eq. 2, multilayer films, comprising alternating layers of high and low index (refer to Fig. 38 in Chapter 13), and graded refractive index films have been developed. The reflectance of a solar cell coated with

Fig. 3.

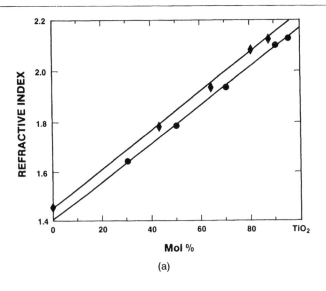

(a)

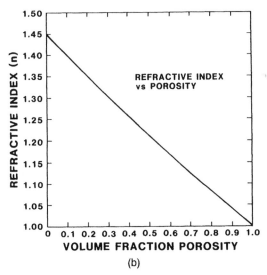

(b)

(a) Refractive index as a function of x (mol%) for gel-derived films in the binary system $x\text{TiO}_2 \cdot (100 - x)\,\text{SiO}_2$. Diamonds are data of Schroeder [14] for films densified at 500°C. Filled circles are data of Brinker and Harrington [32] for films densified at 450°C. From Pettit et al. [25], in Sol-Gel Technology for Thin Films, Preforms, Electronics, and Specialty Shapes, ed. L.C. Klein (Noyes Publications, Park Ridge, New Jersey, 1988). (b) Refractive index of silicate films as a function of volume fraction porosity according to the Lorentz–Lorenz relationship. (Equation 10 in Chapter 13.)

Fig. 4a.

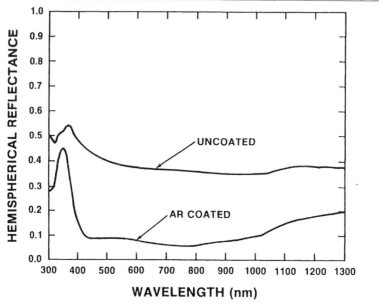

Measured spectral hemispherical reflectance of uncoated (as received) and TiO_2/SiO_2 double layer coated silicon solar cells. The approximate 7% reflectance at 700 nm results from silver metallization ($\sim$7% surface coverage). From Pettit *et al.* [33].

a double layer (TiO_2/SiO_2) AR film is compared with that of an uncoated solar cell in Fig. 4a. The double layer AR coating caused the measured cell efficiency to increase by 44% [33]. The reflectance of a surface can almost completely be eliminated when the refractive index of the surface varies smoothly from the value of air (1.0) to the value of the substrate. Yoldas and Partlow achieved an excellent graded index surface on glass by controlled etching of a porous SiO_2 film (see Fig. 4b) [31].

Critical issues associated with reflective or antireflective coatings are the precise control of thickness and refractive index and, for multilayer films, development of a discrete, step-change in refractive index between layers. As reflected by the impressive production by Schott Glaswerke, the technology is sufficiently mature that these issues no longer constitute a barrier to the successful application of sol-gel technology.

In addition to reflective, colored, and AR coatings, more recent optical applications include contrast enhancement filters [15,21], cold mirrors [15-21], patternable thick films for diffraction gratings [41] and optical memory disks [41,42], and ferroelectric films [38-40,58-67] (PLZT, $KNbO_3$, or $LiNbO_3$) for optoelectronic and integrated optics applications [38-40].

Fig. 4b.

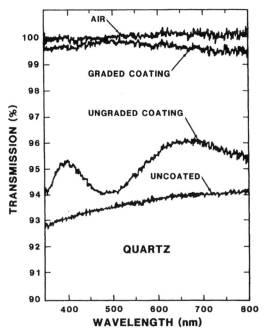

Spectral transmission curves for a porous SiO_2-coated fused silica substrate before and after grading of the refractive index by acid-leaching. Reference curves for silica and air are also included. From Yoldas and Partlow [31].

PLZT and potassium or lithium niobate exhibit a change in refractive index at a particular applied voltage defined by the electro-optic coefficient. For use in integrated optics as modulators or switches, it is necessary to control stoichiometry precisely, to develop a fine-grained, well-crystallized microstructure that is optically transparent, and to deposit relatively thick (>0.5-μm), crack-free films with dielectric strengths exceeding $300\,kV/cm$ [106]. These criteria constitute a much more challenging set of demands than mere thickness and refractive index control. Compared to conventional thin film deposition processes such as sputtering, sol-gel processing has the advantages that (1) stoichiometry is better controlled and (2) the development of oriented films to overcome clamping (inability to reorient domain boundaries, that reduce the effective electro-optic coefficient) may be possible. However the deposition of thick films is often a tedius, time-consuming process requiring repeated dipping and firing operations, which, in addition to increasing cost, increase the likelihood of contamination. To date there are few commercial sources of sol-gel optoelectronic films.

1.2. Electronic Films

Active electronic thin films include high temperature superconductors [48–51], conductive indium tin oxide (ITO) and vanadium pentoxide [52–57], ferroelectric barium and lead titanates and PLZT [58–67], electrochromic tungsten (VI) oxide [65], and titania films used as photoanodes [43–47]. (See Table 3.) Livage and coworkers [2,47,52,53] have reviewed the electronic properties of transition metal oxide gels, including titania, vanadia, and tungsten oxide. An interesting commercial application is a conductive vanadia coating deposited on photographic film to reduce static electricity. The conductive properties of transition metal oxide gels are a consequence of electronic properties arising from hopping of unpaired electrons between metal ions in different valence states in the solid phase and ionic properties arising from proton diffusion through the liquid phase when in water [2]. For V_2O_5 films, a rather high conductivity of about 10^{-2} ohm^{-1} cm^{-1} is observed under ambient conditions. Livage also postulates development of "all-gel devices" such as an electrochromic display comprising an indium tin oxide (ITO) or doped SnO_2 transparent conductive layer, an electrochromic WO_3 layer, and an electrolytic layer made, e.g., from a hydrous oxide [2]. Such devices have been made in the laboratory. They open up a wide range of new possibilities in microionics.

Ferroelectric $BaTiO_3$ films prepared by sol-gel methods were first reported by Fukushima in 1975 [58], followed by lead titanate [60], lead zirconate titanate (PZT) [62,63] and lead, lanthanum, zirconate, titanate (PLZT) [64]. Superconducting $YBa_2Cu_3O_{7-x}$ and thallium-based films have been prepared using sol-gel methods by several groups [48–51]. Both ferroelectric and superconducting films share similar processing criteria with optoelectronic films described in the previous section, viz. precisely controlled stoichiometry combined with well-crystallized crack-free microstructures and absence of impurities. The possibility of orientation through physico-

Fig. 5. _____

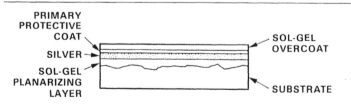

Schematic illustration of flexible metal mirror design. Sol-gel coatings serve several protective roles: the planarization layer prevents galvanic corrosion of the Fe–Ag couple in humid environments; the protective film prevents abrasion or corrosion of the underlying silver reflective surface. From Ashley and Reed [71,72].

chemical procedures (see Chapter 13), epitaxy, or seeding is especially attractive for increasing critical currents in superconducting thin films. Sol-gel thin film superconductors have not exhibited properties comparable to the best sputtered films. Current processing problems include interactions with substrate materials during heat treatments employed for crystallization and removal of impurities, small grain size [51], and the insolubility of copper alkoxides of lower alcohols. The technical challenge is to develop a truly low temperature, chemical route to well-crystallized superconducting phases.

1.3. Protective Films

Protective films [68–97] impart corrosion or abrasion resistance, promote adhesion, increase strength, or provide passivation or planarization. (See Table 3.) Figure 5 shows the design of a flexible metal-mirror developed for terrestrial and low earth orbit solar applications. Sol-gel films perform several protective functions: the planarization layer prevents corrosion of the Fe–Ag galvanic couple while planarizing the stainless steel substrate, significantly improving the specularity of mirror; the overlayer prevents corrosion and abrasion of the underlying silver reflective surface.

In addition to optical applications, surface planarization is essential to multilayer microelectronics processing. Spin-on glasses (SOG) show promise as interlayer dielectrics capable of planarizing complex surface topologies [76]. (See Fig. 6.) Since sol-gel films appear to planarize features up to several micrometers in size (Fig. 6), it is not surprising that sol-gel films applied to abraded substrates increase their tensile strength presumably by reducing the sizes of the largest surface flaws [81–83]. For both strengthening and planarization applications, sol-gel systems that exhibit a strong concentration dependence of the viscosity are desirable, because only a small change in concentration is necessary to "freeze-in" the pristine, flat surface created by surface tension during deposition.

Electronic passivation [90,97] of a surface requires a dense, high-purity, pin-hole free film exhibiting high dielectric strength, low conductivity, and (for microelectronics applications) low surface state densities. In silicon-based microelectronics applications, sol-gel films are a low-temperature alternative to thermal SiO_2 and compete with CVD silica. Although superior to CVD films, thus far sol-gel SiO_2 films processed at or below 1000°C have electronic properties inferior to device-quality thermal SiO_2 (dielectric strengths less than 8–10 MV/cm and surface-state densities generally greater than $10^{10}/cm^2$ eV) [94,95] and cannot be considered for gate oxides in metal-oxide-Si (MOS) architectures. Interlayer dielectrics have less severe

Fig. 6.

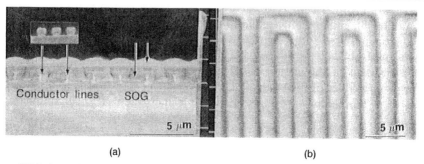

(a)   (b)

SEM photomicrographs of the edge (a) and top surface (b) of a metallized Si substrate coated with a spin on glass. From Davison *et al.* [51].

electronic requirements, and the planarization properties of dipped films and SOG (see Fig. 6) make sol-gel films quite attractive. If adequate electronic properties can be achieved at temperatures compatible with metallizations (~450°C for Al), many applications are envisioned.

Oxidation of III–V compound semiconductors such as InP and GaAs creates mixed oxides with poor electronic properties. SOG and dipped sol-gel silicate films have been evaluated in metal-insulator-semiconductor (MIS) devices by several groups [90-92,95]. Based on capacitance versus voltage measurements (Fig. 7), Warren *et al.* [96] concluded that silicate films processed at 300°C exhibit low surface-state densities ($<10^{11}/cm^2$ eV) in the upper half of the band gap and very few slow interface states. MIS structures subjected to rapid thermal anneals at 400 to 550°C have high breakdown strengths (up to 6 MV/cm); surface states increased presumably due to P depletion. Using sol-gel processing, it is expected that P-doped silicate films can be deposited that preserve the InP interface stoichiometry, leading to high quality MIS devices useful in high speed logic applications.

In general, sol-gel films appear well suited to many protective/passivation applications, since relatively dense pin-hole–free layers can be prepared at low temperatures compatible with the substrate, and the film chemistry can be precisely tailored. However, three major drawbacks of sol-gel films, from the standpoint of abrasion- or corrosion-resistant layers are (1) thick coatings ($>1\,\mu$m) are difficult to achieve without cracking; (2) sol-gel films are in general quite brittle; and (3) relatively high temperatures are required to achieve good properties. In this regard organically modified systems are of interest. Schmidt and coworkers [20,79,80] can deposit thick dense layers at low temperatures (~120°C) without cracking that can potentially combine the hardness of ceramics with the toughness of organic polymers. Organically modified silicate (ORMOSIL) films are currently being used as antiscratch

Fig. 7.

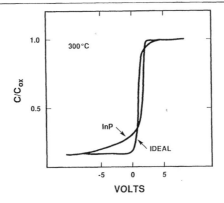

Capacitance versus voltage (CV) curve for an InP/sol-gel SiO$_2$ capacitor structure compared to an "ideal" CV curve. The gel film was consolidated at 300°C. From Warren *et al.* [96].

coatings on acrylic and polycarbonate windows and lenses and as protective coatings to protect medieval glass from corrosion. Properties of ORMOSIL coated and uncoated eyeglass lens materials are summarized in Tables 4 and 5. In addition, preliminary studies suggest that thick organically modified coatings can be heated to produce thick, crack-free oxide layers [102]. If this is generally true, many protective applications could be realized.

Table 4.

Scratch and Abrasion Tests of Different Coatings
(Composition (mol%): Metal Oxide: 20; Epoxysilane: 50; SiO$_2$: 30).

	Coatings and Polymers				
Test	CR 39 (Uncoated)	PMMA (Uncoated)	Coating 1 M = Zr	Coating 2 M = Ti	Coating 3 M = Al
1 (load in g)	1–2	<1	10	20–30	50
2 (haze in %) (200 rev.)	12–13	>20	—	1.5	—
3 (haze in %)	4[a]	—	—	1.2	—
	15[b]	—	—	6	—

Test 1: Minimum Vickers diamond indent load in grams required to detect a scratch optically by microscopy.
Test 2: Percentage of scattered light after Taber abrasion compared to a nonabraded surface.
Test 3: Special abrasive test simulating cleaning under dust-contaminated conditions: [a] diamond powder, [b] boron carbide powder.

Source: Schmidt *et al.* [80].

Table 5.

Summary of Important Tests of ORMOSIL Coated Eyeglass Lenses.

Test	Result
3-min ultrasonic treatment (10 wt% tartratic acid in water)	Coating unaffected
Antireflective coating	Applied without problems
Coloring by dye diffusion	Applied without problems
3-min ultrasonic treatment in 0.1N NaOH	Minor cracks on lab products around coating defects
Temperature change in water baths (+90 to +15°C), 5 cycles	Coating unaffected
Temperature change in lab air (+80 to −20°C)	Coating unaffected
16-hr physiological NaCl solution (room temperature)	Coating unaffected
Xenotest (ultraviolet lamp 180 klux)	>80 hr, unaffected (with appropriate ultraviolet absorber)
Adhesion	Cross-cut test value similar before and after tape test

Source: Schmidt *et al.* [80].

1.4. Porous Films

Pore volume, pore size, surface area, or surface reactivity of porous films can be tailored to achieve specific goals. Using methods developed by Brinker and coworkers [17], it is possible to vary these properties by aging the precursor sols and/or by controlling the relative evaporation and condensation rates. (For details refer to Chapter 13.) Pore volume control permits the refractive index to be optimized for AR applications as described in Section 1.1. (See Fig. 3.) Control of pore size and surface area lead to applications as sensor or catalytic surfaces. Frye and co-workers [99] deposited controlled pore silicate films on the active surface of surface acoustic wave (SAW) devices in order to distinguish between various environmental molecules on the basis of size. Bein and coworkers [100] embedded small zeolite molecular sieves (pore diameters ≈ 0.5–1.0 nm) in silicate sol-gel matrices with pore sizes less than 0.4 nm forcing adsorption of all molecules larger than nitrogen to occur exclusively within the zeolite channels. A ZSM-5-zeolite/silicate composite film (ZSM-5 pore size ≈ 0.6 nm) deposited on a SAW device easily distinguished propanol (kinetic diameter = 0.47 nm) from iso-octane (kinetic diameter = 0.62 nm). (See Table 6.) FTIR studies with Lewis base probe

Table 6.

Gas Adsorption by a ZSM-5/Silicate Film on a
Surface Acoustic Wave (SAW) Device.

	SAW Response to Vapor		
Species	Kinetic Diameter ($\AA$)	Frequency Shift (Hz)	Mass Change (ng/cm^2)
Methanol	~3.8	−6530	540
Propanol	~4.7	−10200	840
Iso-octane	~6.2	74	−6.1

Conditions
  Vapor: 0.1% saturated.
  Temperature: 23°C.
  Pore size: $5.5 \times 6.0\,A^2$.

Source: T. Bein, G.C. Frye, and C.J. Brinker, *J. Am. Chem. Soc.*
111 (1989) 7640–7641.

molecules verified that all of the porosity within the embedded zeolite crystals
was accessible to gaseous probes smaller than the zeolite channel size [100].

Surface reactivity (e.g., acid–base properties, ion-exchange behavior, and
ability to selectively complex metal cations) may be modified by appropriate
choice of precursor metals, by derivitization with organic ligands, or by
reactions of the surface with metal halides, metal alkyls, etc. (See Chapter
10.) Combining microstructure control (demonstrated for silicates) with
control of surface reactivity should lead to many applications for porous
films in the areas of sensors and catalysis.

2.

MONOLITHS

Monoliths are defined as bulk gels (smallest dimension ≥ 1 mm) cast to
shape and processed without cracking. Monolithic gels are potentially of
interest because complex shapes may be formed at room temperature and
consolidated at rather low temperatures without melting. The principle
applications for monolithic gels are optical ones: fiber optic preforms,
lenses and other near-net-shape optical components, graded refractive
index (GRIN) glasses, and transparent foams (aerogels) used as Cherenkov
detectors and as superinsulation. (See Table 7.)

Table 7.

Applications for Monolithic Gels.

Topic	References
Review	[107–108]
Optical glasses and fiber-optic preforms	[109–131]
Near net shape optical components	[132–135]
Aerogel transparent insulation	[136–140]
Substrates	[141–148]
Graded refractive index (GRIN) glass	[149–156]
Ultralow expansion glass	[157–158]
Miscellaneous	[159–162]

    Optical fiber fabrication from monolithic preforms involves casting a cyclindrical shape followed by drying, sintering under conditions that reduce the OH concentration to ppb levels (for details, refer to Chapter 10), and conventional fiber drawing above the glass softening temperature. Compared to fibers drawn from the melt, monolithic gel processing has the advantage that melting is not required, avoiding the introduction of impurities from the crucible. However, the gel approach appears to have no apparent technological or commercial advantage over current CVD processes. In fact, cracking problems and slow processing times associated with shrinkage during drying are a detriment. Clever schemes such as double dispersion processing [108] or the use of fumed silica or colloidal silica fillers [110] apparently permit more rapid processing without cracking. Molecular stuffing the interior of gel cylinders with GeO_2 provides a method of fabricating step-index preforms [110]. Unfortunately, fiber quality is inferior to that prepared from CVD preforms [108]: the best gel-derived fibers have optical losses greater than 3 dB/km [128,129] (see, e.g., Fig. 8), whereas state-of-the-art CVD-derived fibers exhibit losses of <1 dB/km. It is not well understood whether the greater optical losses in gel-derived fibers result from scattering or absorption due to OH or other impurities. Even if optical losses are reduced, CVD methods appear better suited for accurate doping to achieve tailored refractive index profiles.

    A related broad area of applications for monoliths is the formation of high-purity optical components without melting or polishing [132–135]. Shaping is accomplished at room temperature by casting in a precision mold. Drying and sintering (near T_g) preserve the shape and surface finish even though the volume may change considerably. A wide variety of shapes have been made [133,135] including lenses and circular, concentric, hollow cylinders formed by casting in a rotating mold [126]. The technical challenge is to avoid cracking and to reproducibly account for volume changes accompanying drying and sintering. Incorporation of particulate

Fig. 8.

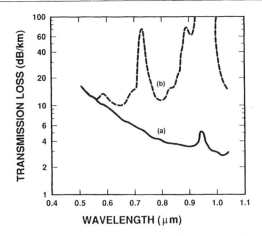

Typical transmission loss spectra of SiO_2-core optical fibers prepared from sintered SiO_2 gel preforms. (a) Chlorine-treated low-OH sintered glass preform. (b) Untreated high-OH glass preform. From Matsuyama *et al.* [129].

fillers in alkoxide-based solutions (e.g., [110]) appears to be a promising new method that allows large optical shapes to be prepared quickly. (See Fig. 9.) The fillers increase the permeability and reduce the capillary pressure during drying, lowering the tendency for cracking and increasing the drying rates. Fillers also reduce the drying and sintering shrinkage, so volume changes are more easily accounted for.

GRIN glasses made by gel methods [149–156] take advantage of the high diffusivity of ions in the liquid phase at the gel point. A multicomponent monolithic gel is prepared containing one or more index-modifying ions and soaked in a solution containing different modifier ions. Interdiffusion and ion exchange create a GRIN gel. Drying and sintering produce a GRIN glass. (See Fig. 10.) The advantages of the gel approach over conventional ion-exchange processes are that large quantities of index-modifying ions such as lead can deeply interdiffuse quite quickly (hours compared to months for conventional ion-exchange methods), and almost any two ions can be exchanged regardless of valence [149]. Compared with CVD methods used to manufacture optical fibers, this process is envisioned to be less expensive and better suited for optical glasses that do not require ultralow optical losses.

Applications for GRIN glasses are principally as lenses for compact photocopiers, endoscopes, and compact disk players [149]. Problems that need to be addressed are ion redistribution and cracking during drying and uniform consolidation throughout a steep compositional gradient.

Fig. 9. _____

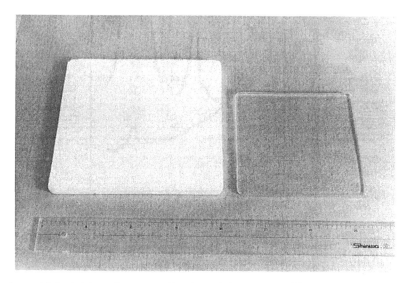

Large dried gel plates and corresponding sintered glasses prepared from particulate silica-filled gels. From Toki *et al.* [110,126].

One of the first applications of monolithic gels was as low refractive index elements in Cherenkov detectors used in high energy physics [131]. Instead of using pressurized gases, aerogels provided a convenient means of achieving refractive indices below 1.1 under ambient conditions. Following this rather esoteric application, the low thermal conductivities of aerogels have led to more mundane uses as transparent or translucent superinsulations to improve energy utilization [136–140]. Figure 11a depicts a superinsulated window system in which an evacuated SiO_2 aerogel layer of 15-20 mm provides a thermal loss coefficient of about 0.5 W/(m²K), corresponding to an R-value of 11 [140]! The obvious practical disadvantage of this concept is that autoclaves would need to be large enough to accommodate the window. For this reason translucent, granular aerogel insulations have been proposed for use in trombe walls where optical clarity is not an issue [140] (Fig. 11b).

In general, good applications for monolithic gels are those that take advantage of purity, process simplification (e.g., elimination of polishing), or the inherent porosity of gels (aerogels and GRIN glasses). Monolithic processing also reduces processing temperatures and expands glass-forming regions leading to possible applications in synthesis of very refractory compositions such as SiO_2–TiO_2 ultralow-expansion (ULE®) glasses [147,148]

Fig. 10. _____

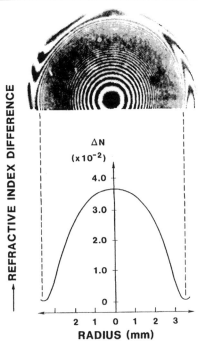

REFRACTIVE INDEX DIFFERENCE

ΔN
$(\times 10^{-2})$

4.0

3.0

2.0

1.0

0

2   1   0   1   2   3
RADIUS (mm)

Interferogram and refractive index profile of a GRIN glass made by ion exchange of a porous gel followed by sintering. From Yamane *et al.* [149].

and metastable oxynitride compositions [159]. The principal critical issues remain rapid processing without cracking and achieving reproducible shrinkage from the cast shape to the final consolidated gel.

3.

POWDERS, GRAINS, AND SPHERES

Powders are the starting point for most polycrystalline ceramic processing schemes. Ceramic powders and grains are also used as catalysts, pigments, abrasives, and fillers, and they are employed in electro-optical and magnetic devices. Uranyl spheres are used as nuclear fuels, porous beads are used in chromatography, and hollow spheres are used as targets in inertial confinement fusion. (See Table 8.) As such the potential applications for sol-gel–derived powders in high-tech, high value added products are extensive.

Fig. 11.

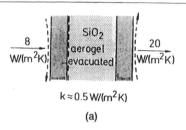

$$k \approx 0.5 \, W/(m^2 K)$$

(a)

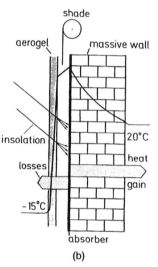

(b)

(a) Schematic of superinsulated window system consisting of an evacuated aerogel layer 15–20 mm thick between two panes of glass. This design provides a thermal loss coefficient of about 0.5 W/m²K corresponding to an *R* value of 11. From Fricke [137], in *Sol-Gel Technology for Thin Films, Preforms, Electronics, and Speciality Shapes.* ed. L.C. Klein (Noyes Publications, Park Ridge, New Jersey, 1988). (b) Trombe wall design with translucent aerogel insulation. The absorbed solar radiation is partly available for room heating. From Fricke [137].

Potential advantages of sol-gel powders over conventional powders (often physical mixtures of minerals and chemicals) are controlled size and shape, molecular scale homogeneity, and enhanced reactivity (lower processing temperatures). For successful commercial applications these advantages must outweigh inherent disadvantages such as cost, lengthy processing times, and low yields.

The diversity of sizes and shapes of ceramic powders prepared by homogeneous precipitation, phase transformation, or aerosol techniques are reviewed by Matijevic and Gherardi [163]. (See, e.g., Fig. 12.) The

Table 8.

Applications for Gel-Derived Powders, Grains, and Spheres.

Topic	References
Review	[163–167]
High-temperature superconductors	[168–172]
Electronic	[173–177]
Waste immobilization	[178–181]
Refractory compositions	[182–187]
Abrasive grains	[188–193]
Nuclear fuel	[194–196]
Beads and microspheres	[197–201]
Miscellaneous	[202–204]

reported advantages of monosized spherical particles as precursors for polycrystalline ceramics is that they are easier to process into uniform green microstructures, which results in easier control of the microstructure during densification [166] (see Fig. 13), although higher green density is achieved with a particle size distribution. For other applications that utilize optical, magnetic, electronic, or mechanical properties of dispersed particles, desirable shapes include rods (whiskers), ellipsoids, platelets, and oblate or prolate disks. There does not appear to be a general synthetic

Fig. 12a,b. _____

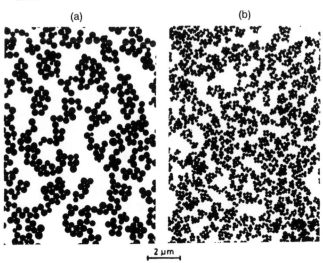

(a)   (b)

2 μm

TEM of barium titanate precursor obtained by aging a solution containing Ti(IV) iso-propoxide at (a) 60°C for 2 hours or (b) 40°C for 2 hours. From Matijevic and Gherardi [163]. See also discussions in Chapters 2 and 4.

Fig. 12c,d. _____

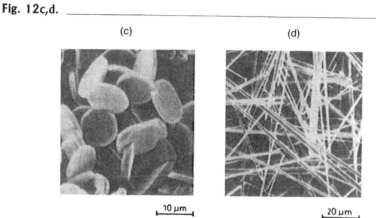

(c)   (d)

(c) SEM of hematite particles prepared from $Fe(NO_3)_3$. From Matijević and Gherardi [163]. (d) SEM of zinc oxide particles prepared from $Zn(NO_3)_2$. From Matijević and Gherardi [163].

approach to obtain unusually shaped particles in a variety of different compositional systems, so each new composition must be investigated separately [163]. Thus widespread applications for shaped particles await continued basic research.

High temperature superconductors [168–172] represent an application where particles with high-purity, molecular-scale mixing and controlled shape are desirable for low temperature processing of dense, oriented microstructures to achieve low critical temperatures and high critical currents. The difficulty in achieving molecular-scale mixing in multicomponent systems is matching the rates of hydrolysis, condensation, or precipitation of the various precursor species [164]. From this standpoint, two-phase aerosol [163] or emulsion techniques [172,187] appear promising: compositional inhomogeneities are limited to the aerosol or emulsion droplet sizes, which may be accurately controlled by nebulizers or appropriate use of surfactants. A potential disadvantage of this approach with respect to superconductors is that it is limited to spherical shapes, but these techniques are eminently well suited to meet most electronic materials processing requirements where purity, homogeneity, and reproducibility are key concerns.

Sol-gel abrasive processing [188–193] takes advantage of the often detrimental tendency for gels to crack into pieces during drying. Crushing and sizing the friable gel prior to sintering is certainly less energy-intensive than conventional comminution of calcined or fused material. The lower processing temperatures of gels compared to conventional polycrystalline

Fig. 13.

Fig. 13.

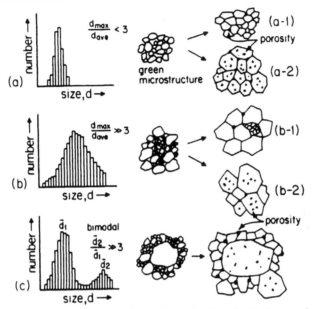

Schematic representation of the effects of powder size distribution on the sintered microstructure. For example, in (a-1), a narrow size distribution and control of the grain-boundary pore interaction result in a dense sintered ceramic with a grain size slightly larger than the original particle size. From Barringer *et al.* [166]. A *distribution* of particle sizes may result in better sintering, however. See Fig. 39 in Chapter 11.

ceramics also represents a cost savings. Alumina abrasive grains have been produced commercially by 3M and Norton since about 1982. Bauer's patents [188–189] represent the first commercial use of seeding to control the phase evolution of alumina gels. (See discussion in Chapter 11.) Abrasive grains combining oxides and nitrides should also be possible by sol-gel methods. Such composite systems may represent the future trend in sol-gel abrasives.

Uranyl spheres made by either the internal or external gelation of emulsions [194–196] were one of the first applications of sol-gel processing. Free-flowing uranyl beads up to 1 mm in diameter are easy to process by remote methods and, due to liquid processing, eliminate environmental hazards associated with radioactive dusts. Porous beads and hollow microspheres [197–201] take advantage of inherent gel porosity and the potential for forming shaped particles, either in the gel state (beads) or by use of surface hydroxyls as blowing agents (hollow spheres). According to Livage's criteria, these applications make sense since they represent new products difficult to make by conventional ceramic processing.

4.

FIBERS

In addition to conventional high temperature fiber formation from sol-gel–derived preforms (discussed in Section 2), two other sol-gel methods of fiber formation are drawing fibers directly from viscous sols at room temperature and unidirectional freezing of gels. (For details, refer to the reviews of sol-gel fiber technology [205–207].) The conditions for preparing viscous sols from alkoxides are generally acid-catalyzed hydrolyses employing low $H_2O:M$ ratios, r. (See Chapter 3 for details.) Consequently, as-drawn fibers tend to be microporous and to contain relatively high quantities of residual organics—unsuitable microstructures for achieving ultrahigh-purity, sintered glass fibers. (See Chapter 10.) In addition, many of the synthesis conditions produce fibers with noncircular cross sections [205]. For these reasons, fibers drawn from viscous sols are not appropriate for optical fiber applications. Potential applications include reinforcement, refractory textiles, and high-temperature superconductors. (See Table 9.)

Three critical concerns for reinforcement applications are strength, temperature stability, and reactivity with the matrix phase. Tensile strengths of amorphous sol-gel fibers consolidated below T_g are considerably less than those of conventional glass fibers drawn from the melt. (See Fig. 14.) Most likely residual dirt, porosity, entrapped water, or organics constitute surface defects larger in size than defects present on a pristine melt-formed surface. The advantage of the sol-gel process is that very refractory and chemically durable fibers can be formed at room temperature that would be extremely difficult to prepare by the conventional high temperature approach. A good example is alkali-resistant zirconia-containing fibers for reinforcement of concrete [205]: glasses containing more than 20% ZrO_2 are difficult or impossible to prepare conventionally, because of the high melting temperatures and the tendency to crystallize on cooling.

Sol-gel methods may also be used to prepare continuous, refractory, polycrystalline fibers that exhibit high strength and stiffness in addition

Table 9.

Applications for Gel-Derived Fibers.

Topic	References
Review	[202–208]
Reinforcement	[209–214]
Superconducting	[215]
Electrolysis	[216]
Optical	[217]

Fig. 14.

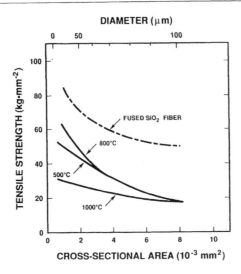

Tensile strength versus cross-sectional area for SiO_2 fibers prepared from viscous sols and heated at 500, 800, and 1000°C compared to conventional fused silica fibers. From Sakka [205], in *Sol-Gel Technology for Thin Films, Preforms, Electronics, and Speciality Shapes*, ed. L.C. Klein (Noyes Publications, Park Ridge, New Jersey, 1988).

to chemical durability. Properties of commercial aluminate, aluminosilicate, and aluminoborosilicate polycrystalline fibers are listed in Table 10. Nextel® aluminoborosilicate fibers are drawn from viscous sols prepared by concentrating an aqueous dispersion of a basic aluminum acetate, $(Al(OH_2(OOCCH_3)\cdot\frac{1}{3}H_3BO_3)$, and a particulate silica sol to a syrupy consistency suitable for spinning [210]. Subsequent heat treatments above 900°C produce aluminum borosilicate and mullite crystalline phases [210]. At 1100°C Nextel® fibers are unaffected by metals such as aluminum, boron, chromium, iron, nickel, and silicon leading to applications in ceramic-reinforced metal-matrix composites. Other applications of ceramic fibers for reinforcement are fiber-reinforced polymer and ceramic matrix composites. Ceramic fibers having intermediate values of modulus of elasticity (138–152 GPa), achieved by different compositions and heat-treatment schedules, are interesting as hybrids with organic Kevlar® fibers [210]. Because the stiffness of the two fiber systems can be matched, hybrid fiber-reinforced polymer matrix composites can be produced that are lighter due to the Kevlar fibers and exhibit greater compressive strengths due to the ceramic fibers.

Table 10.

Properties of Commercial Gel-Derived Ceramic Fibers.

Producer	Name	Composition	Tensile Strength (MPa)	Tensile Modulus (GPa)	Density (g/cm^3)
3M	Nextel 312	Al_2O_3, SiO_2, B_2O_3	1750	154	2.70
3M	Nextel 440	Al_2O_3, SiO_2, B_2O_3	2100	189	3.05
3M	Nextel 480	Al_2O_3, SiO_2, B_2O_3	2275	224	3.05
du Pont	PRD-166	Al_2O_3, ZrO_2	2100	385	4.20
du Pont	FP	alpha Al_2O_3	1400	3853	3.90
Sumitomo		Al_2O_3, SiO_2	ave. 2200	ave. 230	3.20

Ceramic fibers can be coated with ceramic slurries and hot-pressed to make dense fiber reinforced ceramic- or glass-matrix composites. Alternatively, continuous ceramic fibers can be combined with other fibers, whiskers, or powders and formed into porous shapes used for insulation (e.g., space-shuttle tiles) or subsequently infiltrated by CVD techniques to form fiber reinforced ceramic matrix composites without hot-pressing. For example, Nextel-SiC composites made by infiltration are being considered for applications such as heat exchangers and radiant gas burner tubes [210].

The suppleness and strength of glass and ceramic fibers formed by sol-gel methods allow refractory fabrics to be woven using conventional textile processes. Applications for refractory textiles include high temperature particle filtration (where resistance to corrosive gases such as sulfuric acid is important), furnace belts, flame curtains, gaskets, and cables.

Unidirectional freezing of gels [202,208] produces discontinuous fibers with polygonal cross sections (see Fig. 15) and high surface areas (240–900 m^2/g). Potential applications of these fibers are as supports for catalysts and enzymes.

5.
COMPOSITES

Composites combine different types of materials to obtain synergistic properties unattainable by one material alone. Sol-gel processing can be used to form the matrix phase, the reinforcement phase or phases (fibers, particles, etc.), or both in ceramic–ceramic composites. Because of inherently low processing temperatures, mixed organic–inorganic and ceramic–metal composites are also possible. Rice reviews the benefits of ceramic composites

Fig. 15. _____

(a)

(b)

(c)

Fibers prepared by unidirectional freezing of a TiO$_2$ gel. From Sakka [205], in _Sol-Gel Technology for Thin Films, Preforms, Electronics, and Speciality Shapes_, ed. L.C. Klein (Noyes Publications, Park Ridge, New Jersey, 1988).

and the associated processing problems [218] Applications of sol-gel composites are listed in Table 11.

High-performance ceramic–ceramic composites such as SiC-reinforced alumina are candidate materials for turbine blades as well as highly efficient ceramic diesel engines. To date these materials have been prepared principally by slurry-coating ceramic fibers or textiles followed by composite assembly and hot-pressing. Not only is this process cost-intensive, but fiber strengths are degraded due to abrasion. Ceramic matrices prepared by sol-gel

Table 11.

Applications for Sol-Gel Composites.

Topic	References
Review	[218]
Fiber-reinforced sol-gel matrix	[219–225]
Ceramic–ceramic or ceramic–metal	[226–230]
Glass or ceramic–organic	[231–242]

methods are potentially attractive alternatives, because low viscosity sols can be cast to shape around fibers contained in a precision mold, simplifying assemblage and reducing abrasion. In addition, the high surface area and small pore size of the gel matrix may establish a sufficiently high sintering force to obviate hot-pressing. To date, the large shrinkage of the sol-gel matrix phase has hindered the realization of fully dense ceramic–ceramic composites without hot-pressing, at least for three-dimensional fiber-reinforced composites. The fibers reinforce the matrix against shrinkage during drying and sintering, resulting most often in cracking or fiber-reinforced porous ceramics. These problems are somewhat mitigated in unidirectional fiber-reinforced composites, although the ceramic matrix must "slide" along the fiber interface without cracking to accommodate the drying and sintering strains. To overcome these processing problems, it will be necessary either to greatly reduce gel matrix shrinkage during drying and sintering (perhaps by using particle-filled systems or by multiple infiltration steps) or to coat the fibers with a rubbery, lubricating layer that could accommodate drying strain by deformation and then burn off at elevated temperatures leaving gaps to accommodate the sintering strain. Short-fiber (whisker) [219] or particle-reinforced systems also mitigate these processing problems but at the expense of composite performance.

Because sol-gel–derived materials may be cast to shape at room temperature, they are excellent low-temperature hosts for organic molecules, polymers, and fibers [231–242]. ORMOSILS [20,241] and CERAMERS [242] are names coined to describe mixed organic–inorganic composites prepared by sol-gel methods. In a review of this topic, Schmidt and Seiferling [20] describe how the organic components, either network formers or network modifiers, are tailored for specific applications: epoxysilanes for scratch-resistant materials; thermoplastic and photocurable groups based on diphenylsilanes and photocurable ligands such as methacryl, vinyl, or allyl groups, combined with a variety of polymerizable monomers, for coatings, adhesive films, and bulk materials, respectively. Pope and Mackenzie [231,232] and Lee and Hench [233–235] have made novel organic–inorganic composites by impregnating porous xerogels with organic monomers that could be cured in situ. Silica-polymethylmethacrylate composites were transparent and exhibited good abrasion resistance [231]. Organosilane impregnated silica gels were pyrolyzed to form SiC-reinforced silica two to three times harder than the silica matrix alone [233].

Incorporation of organic molecules or network-modifying ligands leads to applications in optics [237,240], catalysis [80], and sensors [80]. Organic dye molecules (e.g., rhodamine 6G) dispersed in transparent silicate gels result in lasing and optical gain [237]. Compared to organic polymer matrices, greater laser pumping can be sustained without photobleaching the dye [237].

Monofunctional organosilanes terminating the surfaces of porous xerogels or coatings can be used as catalyst supports, membranes, or sensors [241]. For example, Fig. 16a shows that incorporation of aminosilanes in a porous silicate coating results in significant CO_2 adsorption, despite rather low surface areas. In order to reduce the influence of water vapor, hydrophobic components can be incorporated. Figure 16b shows the effect of a propylsilane addition on the water uptake of an aminosilane–silicate film.

Fig. 16.

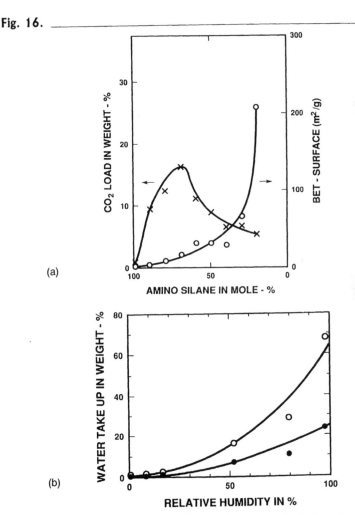

(a)

(b)

(a) CO_2 adsorption at 40°C and 10^3 kPa CO_2 pressure for SiO_2–aminosilane films.
(b) Water uptake from the atmosphere for a pure aminosilane condensate (O) and a propylsilane–aminosilane (1:1) condensate (●). From Schmidt and Seiferling [20].

Because a wide range of organic functionalities is possible, potential applications of organic–inorganic coatings for sensors appear limitless.

Ceramic–metal composites containing a dispersed metallic phase have applications in optics, catalysis, and electronics. IROX® TiO_2/Pd coatings discussed in Section 1 are perhaps the first commercial use of such cermet composite materials in an optics application: the Pd additions control the absorptivity of the coating. Roy and coworkers [230] describe the incorporation of metal halides in gels (so-called di-phasic gels) to produce photochromic glasses and the incorporation of a variety of metals, including Ni, Cu, Sn, and Pt, introduced as salts, that produce highly dispersed metallic phases when reduced between 200 and 700°C. Since at these temperatures the gel matrix phase is generally still porous with a surface area exceeding $100 \, m^2/g$, these cermet composites appear promising for applications in catalysis.

Nanao and Eguchi [228] describe a general scheme whereby metal alkoxides and metal chelates are combined to form a gel. Thermal decomposition of the chelates produces composite multilayer films with applications as sensors, conductive or piezoelectric layers, or opto-electronic materials. Adair *et al.* [226] discuss a novel process based on Liesegang periodic precipitation to produce a layered SiO_2/Cu composite potentially useful for fabrication of insulator–conductor multilayers for electronic packaging.

6.

POROUS GELS AND MEMBRANES

High surface areas and small pore sizes characteristic of inorganic gels are properties unattainable by conventional ceramic processing methods. These unique properties may be exploited in applications such as filtration, separations, catalysis, and chromatography. (See Table 12.) The use of porous gels in separations and filtrations was reviewed by Klein [243].

Table 12.

Applications of Porous Gels and Membranes.

Topic	References
Review	[243]
Membranes	[244–252]
Porous glass substrates	[253–255]
Catalyst supports	[256–269]

The preparation of controlled pore films supported on nonporous substrates for AR and sensor surfaces was discussed in Section 1. When supported on macroporous substrates, controlled pore films may be used as microfilters (50 nm to 1 μm pore sizes), ultrafilters, or membranes (<50 nm pore sizes). Compared to conventional organic polymer membranes, sol-gel membranes offer several advantages [243]: (1) they can be operated or sterilized at high temperatures; (2) they do not swell or shrink in contact with liquids; and (3) they are much more abrasion-resistant. Figure 17 shows the cross-section of an ultrafilter membrane prepared by Uhlhorn [249]. It comprises three successive layers exhibiting progressively smaller pore diameters. The final layer has a pore diameter of about 2 nm and is prepared by partially filling the pores in a titania layer with alumina, vanadia, or silver. The small pore size permits applications in ultrafiltrations and reverse osmosis. Similar schemes have been developed by Cot *et al.* [246–248], Leenaars *et al.* [244,245], and Gillot *et al.* [250]. Applications include microfiltration of water, wine, and beverages, and ultrafiltration of milk. Critical processing parameters are reproducibly attaining a desired pore size and, most importantly, avoidance of pin holes and cracks. Even smaller

Fig. 17. _____

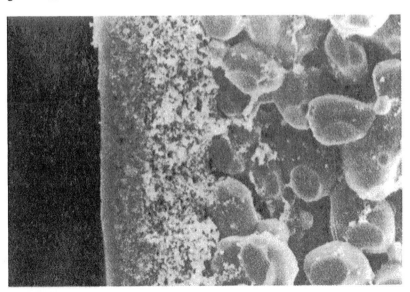

SEM photomicrograph of a four-layer TiO₂ membrane. Thin layer on left is a ceramic "skin" with ultrafiltration and reverse osmosis properties. The other three layers form a microfiltration system. From Haggin [249] reporting on membranes prepared by Uhlhorn.

pore sizes useful for gas separations may be obtained by deposition of branched, fractal, silicate species after various stages of growth using procedures discussed in Chapter 13 [17].

We have explained that porous gels, either xerogels or aerogels, in the form of beads, granules, or plates have applications in chromatography and thermal or acoustical insulation (see Sections 2 and 4). Porous gel granules, powders, and films are also used as desiccants and as catalyst supports. The use of high surface area aerogels for catalyst supports was described in several publications by Teichner [257–263]. Stephens *et al.* [264] and more recently Bunker *et al.* [265] describe novel catalysts prepared from porous sodium titanate gels by ion exchange. These catalysts show improved catalytic activity for processes such as coal liquifaction compared to catalysts prepared on traditional suppports such as alumina. One reason for the improved catalytic activity may be that metals can be atomically dispersed on the sodium titanate via ion exchange. This latter application takes advantage of both controlled porosity and controlled chemistry achievable via sol-gel techniques. As described in the previous section, interesting catalyst materials may also be prepared by reduction or decomposition of porous gels containing metal salts or chelates.

7.

SUMMARY

There are many potential applications of sol-gel–derived materials in the form of films, fibers, monoliths, powders, composites, and porous media. The most successful applications are those that utilize the potential advantages of sol-gel processing such as purity, homogeneity, and controlled porosity combined with the ability to form shaped objects at low temperatures, while avoiding inherent disadvantages such as cost of raw materials, slow processing times, and high shrinkage. For these reasons, dense or porous thin films or membranes, fine abrasive grains, refractory, corrosion-resistant fibers and near-net-shape optics emerge as the best applications.

REFERENCES

1. J.D. Mackenzie in *Ultrastructure Processing of Glasses, Ceramics, and Composites*, eds. L.L. Hench and D.R. Ulrich (Wiley, New York, 1984), p. 15.
2. J. Livage in *Transformation of Organometallics into Common and Exotic Materials: Design and Activation*, NATO ASI series E, no. **141**, ed. R.M. Laine (Martinus Nijhoff, Dordrecht, 1988), pp. 250–255.

3. *Sol-Gel Technology for Thin Films, Fibers, Preforms, Electronics, and Specialty Shapes*, ed. L.C. Klein (Noyes, Park Ridge, N.J., 1988).

4. *Transformation of Organometallics into Common and Exotic Materials: Design and Activation*, NATO ASI series E, no. **141**, ed. R.M. Laine (Martinus Nijhoff, Dordrecht, 1988).

5. L.C. Klein, *The Glass Industry* (January, 1981), pp. 14–17.

6. L.C. Klein and G.J. Garvey in *Soluble Silicates*, ACS Symposium Series, no. **194**, ed. J.S. Falcone, Jr. (Am. Chem. Soc., Washington DC, 1982), pp. 293–304.

7. D.R. Ulrich in *Transformation of Organometallics into Common and Exotic Materials: Design and Activation*, NATO ASI series E, no. **141**, ed. R.M. Laine (Martinus Nijhoff, Dordrecht, 1988), p. 207.

8. D.R. Uhlmann, B.J.J. Zelinski, and G.E. Wnek in *Better Ceramics Through Chemistry*, eds. C.J. Brinker, D.E. Clark, and D.R. Ulrich (Elsevier, New York, 1984), p. 59.

9. H. Dislich, *J. Non-Cryst. Solids*, **80** (1986) 115–121.

10. D.R. Ulrich, *J. Non-Cryst. Solids*, **100** (1988) 174.

11. J.D. Mackenzie, *J. Non-Cryst. Solids*, **100** (1988) 162.

12. D.W. Johnson, Jr., *Amer. Ceram. Soc. Bull.*, **64** (1985) 1597–1602.

13. W. Geffcken and E. Berger, Deutsches Reichspatent 736411 (May 6, 1939), assigned to Jenaer Glaswerke Schott & Gen., Jena.

14. H. Schroeder in *Physics of Thin Films*, ed. G. Hass, **5** (Academic Press, New York, 1969), pp. 87–141.

15. H. Dislich in *Transformation of Organometallics into Common and Exotic Materials: Design and Activation*, NATO ASI series E, no. **141**, ed. R.M. Laine (Martinus Nijhoff, Dordrecht, 1988), p. 236.

16. D.R. Uhlmann and G.P. Rajendran in *Ultrastructure Processing of Advanced Ceramics*, eds. J.D. Mackenzie and D.R. Ulrich (Wiley, New York, 1988), p. 241.

17a. C.J. Brinker in *Transformation of Organometallics into Common and Exotic Materials: Design and Activation*, NATO ASI series E, no. **141**, ed. R.M. Laine (Martinus Nijhoff, Dordrecht, 1988), p. 261.

17b. C.J. Brinker, A.J. Hurd, and K.J. Ward in *Ultrastructure Processing of Advanced Ceramics*, eds. J.D. Mackenzie and D.R. Ulrich (Wiley, New York, 1988), p. 223.

18. H. Dislich, *J. Non-Cryst. Solids*, **73** (1985) 599–612.

19. D.L. Segal and J.L. Woodhead in "Novel Ceramic Fabrication Processes and Applications," ed. R.W. Davidge, *Br. Ceram. Proc.*, **38** (1986) 245.

20. H. Schmidt and B. Seiferling in *Better Ceramics Through Chemistry II*, eds. C.J. Brinker, D.E. Clark, and D.R. Ulrich (Mat. Res. Soc., Pittsburgh, Pa., 1986), p. 739.

21. H. Dislich in *Sol-Gel Technology for Thin Films, Fibers, Preforms, Electronics, and Specialty Shapes*, ed. L.C. Klein (Noyes, Park Ridge, N.J., 1988), p. 50.

22. H. Dislich and E. Hussmann, *Thin Solid Films*, **77** (1981) 129.

23. A. Duran, J.M. Fernandez Navarro, P. Mazon, and A. Joglar, *J. Non-Cryst. Solids*, **82** (1986) 391.

24. F. Orgaz and H. Rawson, *J. Non-Cryst. Solids*, **82** (1986) 378.

25. R.B. Pettit, C.S. Ashley, and C.J. Brinker in *Sol-Gel Technology for Thin Films, Fibers, Preforms, Electronics, and Specialty Shapes*, ed. L.C. Klein (Noyes, Park Ridge, N.J., 1988), pp. 80–109.

26. R.B. Pettit and C.J. Brinker, *Solar Energy Materials*, **14** (1986) 269–287.

27. B.E. Yoldas and T.W. O'Keefe, *Applied Optics*, **18** (1979) 3133.

28. B.E. Yoldas, *Applied Optics*, **19** (1980) 1425–1429.

29. B.E. Yoldas, *Applied Optics*, **23** (1984) 1418–1424.

30. S.P. Mukherjee and W.H. Lowdermilk, *Applied Optics*, **21** (1982) 293–296.

31. B.E. Yoldas and D.P. Partlow, *Applied Optics*, **23** (1984) 1418–1424.
32. C.J. Brinker and M.S. Harrington, *Solar Energy Materials*, **5** (1981) 159–172.
33. R.B. Pettit, C.J. Brinker, and C.S. Ashley, *Solar Cells*, **15** (1985) 267–278.
34. C.S. Ashley and S.T. Reed in *Better Ceramics Through Chemistry II*, eds. C.J. Brinker, D.E. Clark, and D.R. Ulrich (Mat. Res. Soc., Pittsburgh, Pa., 1986), p. 671.
35. H.M. McCollister and N.L. Boling, U.S. Patent 4273826, June 1981.
36. P. Hinz and H. Dislich, *J. Non-Cryst. Solids*, **82** (1986) 411.
37. K. Hara, T. Inazumi, and T. Izumatani, *J. Non-Cryst. Solids*, **100** (1988) 490–493.
38. S. Hirano and K. Kato, *J. Non-Cryst. Solids*, **100** (1988) 538.
39. D.J. Eichorst and D.A. Payne in *Better Ceramics Through Chemistry III*, eds. C.J. Brinker, D.E. Clark, and D.R. Ulrich (Mat. Res. Soc., Pittsburgh, Pa., 1988), p. 773.
40. R.V. Ramaswamy, T. Chia, R. Srivastava, A. Miliou, and J. West, *SPIE*, **878** "Multifunctional Materials" (1988) 86–93.
41. N. Tohge, A. Matsuda, T. Minami, Y. Matsuno, S. Katayama, and Y. Ikeda, *J. Non-Cryst. Solids*, **100** (1988) 501–505.
42. S. Masuda, A. Inubushi, M. Okubo, A. Matsumoto, H. Sadamura, and K. Suzuki, *Mater. Sci. Monogr.*, **38C** (1987) 2093–2104.
43. T. Yokoo, K. Kamiya, and S.Sakka, *Denki Kagaku Oyobi Kogyo Butsuri Kagaku*, **54** (1986) 284–285.
44. T. Yoko, K. Kamiya, A. Yuasa, K. Tanaka, and S. Sakka, *J. Non-Cryst. Solids*, **100** (1988) 483–489.
45. T. Yoko, K. Kamiya, and S. Sakka, *Yogyo-Kyokai-Shi*, **95** (1987) 150–155.
46. S. Doeff, M. Henry, and C. Sanchez in *Better Ceramics Through Chemistry II*, eds. C.J. Brinker, D.E. Clark, and D.R. Ulrich (Mat. Res. Soc., Pittsburgh, Pa., 1986), p. 653.
47. J. Livage in *Better Ceramics Through Chemistry II*, eds. C.J. Brinker, D.E. Clark, and D.R. Ulrich (Mat. Res. Soc., Pittsburgh, Pa., 1986), p. 717.
48. P. Strehlow, H. Schmidt, and M. Birkhahn in *Better Ceramics Through Chemistry III*, eds. C.J. Brinker, D.E. Clark, and D.R. Ulrich (Mat. Res. Soc., Pittsburgh, Pa., 1988), p. 791.
49. M. Tatsumisago, H. Sago, and T. Minami, *Chem. Express*, **3** (1988) 311–314.
50. S.A. Kramer, G. Kordas, J. McMillan, G.C. Hilton, and D.J. Van Harligen, *Appl. Phys. Lett.*, **53** (1988) 156–158.
51. W.W. Davison, S.G. Shyu, R.D. Roseman, and R.C. Buchanan in *Better Ceramics Through Chemistry III*, eds. C.J. Brinker, D.E. Clark, and D.R. Ulrich (Mat. Res. Soc., Pittsburgh, Pa., 1988), p. 797.
52. J. Livage and J. Lemerle, *Ann. Rev. Mater. Sci.*, **12** (1982) 103–122.
53. J. Livage in *Better Ceramics Through Chemistry*, eds. C.J. Brinker, D.E. Clark, and D.R. Ulrich (Elsevier, New York, 1984), p. 125.
54. C.J.R. Gonzalez-Oliver and I. Kato, *J. Non-Cryst. Solids*, **82** (1986) 400.
55. N.J. Arfsten, *J. Non-Cryst. Solids*, **63** (1984) 243.
56. C. Sanchez, F. Babonneau, R. Morineau, J. Livage, and J. Bullot, *Phil. Mag. B.*, **47** (1983) 279.
57. H. Dislich and P. Hinz, *J. Non-Cryst. Solids*, **48** (1982) 11.
58. J. Fukushima, *Yogyo Kyoaishi*, **83** (1975) 204.
59. R.G. Dosch in *Better Ceramics Through Chemistry*, eds. C.J. Brinker, D.E. Clark, and D.R. Ulrich (Elsevier, New York, 1984), p. 157.
60. K.D. Budd, S.K. Dey, and D.A. Payne in *Better Ceramics Through Chemistry II*, eds. C.J. Brinker, D.E. Clark, and D.R. Ulrich (Mat. Res. Soc., Pittsburgh, Pa., 1986), p. 711.
61. J. Fukushima, K. Kodaira, and T. Matsushita, *J. Mat. Sci.*, **19** (1984) 595–598.

62. R.A. Lipeles, D.J. Coleman, and M.S. Leung in *Better Ceramics Through Chemistry II*, eds. C.J. Brinker, D.E. Clark, and D.R. Ulrich (Mat. Res. Soc., Pittsburgh, Pa., 1986), p. 665.

63. R.A. Lipeles and D.J. Coleman in *Ultrastructure Processing of Advanced Ceramics*, eds. J.D. Mackenzie and D.R. Ulrich (Wiley, New York, 1988), p. 919.

64. K.D. Budd, S.K. Dey, and D.A. Payne, *Brit. Ceram. Proc.*, **36** (1985) 107–121.

65. A. Chemseddine, R. Morineau, and J. Livage. *Solid State Ionics*, **9–10** (1983) 357.

66. J. Livage, P. Barboux, J.C. Badot, and N. Baffier in *Better Ceramics Through Chemistry III*, eds. C.J. Brinker, D.E. Clark, and D.R. Ulrich (Mat. Res. Soc., Pittsburgh, Pa., 1988), p. 167.

67. S. Hirano and K. Kato, *Adv. Ceram. Mater.*, **3** (1988) 503–506.

68. C.G. Pantano, R.K. Brow, and L.A. Carman in *Sol-Gel Technology for Thin Films, Fibers, Preforms, Electronics, and Specialty Shapes*, ed. L.C. Klein (Noyes, Park Ridge, N.J., 1988), p. 110.

69. M. Masashi, B. Yamagishi, M. Noshiro, Y. Jitsurgiri, and K. Ohnishi, European Patent Appl. 85107552.3, June 19, 1985.

70. R.B. Pettit and C.J. Brinker, "*SPIE* Optical Coatings for Energy Efficiency & Solar Applications", **324** (1982) 176.

71. S.T. Reed and C.S. Ashley in *Better Ceramics Through Chemistry III*, eds. C.J. Brinker, D.E. Clark, and D.R. Ulrich (Mat. Res. Soc., Pittsburgh, Pa., 1988), p. 631.

72. C.S. Ashley, S.T. Reed, and A.R. Mahoney in *Better Ceramics Through Chemistry III*, eds. C.J. Brinker, D.E. Clark, and D.R. Ulrich (Mat. Res. Soc., Pittsburgh, Pa., 1988), p. 635.

73. M. Guglielmi and A. Maddalena, *J. Mat. Sci. Lett.*, **4** (1985) 123.

74. A. Maddalena, M. Guglielmi, V. Gottardi, and A. Raccanelli., *J. Non-Cryst. Solids*, **82** (1986) 356.

75. S.M. Sim, P.-Y. Chu, R.H. Krabill, and D.E. Clark in *Ultrastructure Processing of Advanced Ceramics*, eds. J.D. Mackenzie and D.R. Ulrich (Wiley, New York, 1988), p. 1011.

76. S.G. Shyu, T.J. Smith, S. Baskaran, and R.C. Buchanan in *Better Ceramics Through Chemistry III*, eds. C.J. Brinker, D.E. Clark, and D.R. Ulrich (Mat. Res. Soc., Pittsburgh, Pa., 1988), p. 767.

77. G.R. Dunton, III, and L.Q. Green, U.S. Patent 3387994, June 11, 1968.

78. J. Martinsen, R.A. Figat, and M.W. Shafer in *Better Ceramics Through Chemistry*, eds. C.J. Brinker, D.E. Clark, and D.R. Ulrich (Elsevier, New York, 1984), p. 145.

79. H. Schmidt in *Better Ceramics Through Chemistry*, eds. C.J. Brinker, D.E. Clark, and D.R. Ulrich (Elsevier, New York, 1984), p. 327.

80. H. Schmidt, B. Seiferling, G. Phillipp, and K. Deichmann in *Ultrastructure Processing of Advanced Ceramics*, eds. J.D. Mackenzie and D.R. Ulrich (Wiley, New York, 1988), p. 651.

81. M.F. Gruninger, J. B. Wachtman, Jr., and R.A. Haber, *Mat. Res. Soc. Symp. Proc.*, **54** (1986) 823–828.

82. B.D. Fabes, W.F. Doyle, B.J.J. Zelinski, L.A. Silverman, and D.R. Uhlmann, *J. Non-Cryst. Solids*, **82** (1986) 349.

83. F. Orgaz and F. Capel, *Riv. Stn. Sper. Vetro*, **16** (1986) 147–152.

84. E.P. Plueddemmann, *Silane Coupling Agents* (Plenum, New York, 1982).

85. T.J. Davies, H.G. Emblem, K. Jones, and P. Parkes, U.K. Patent App; GB 2143809 A1, February 20, 1985.

86. I.R. McKeer, K. Jones, H.G. Emblem, J.M. McCullough, and R.D. Shaw, U.K. Patent App. GB 2063848 A, November 11, 1980.

87. R. Feagin, European Patent App. 0202674 A2, June 4, 1986.
88. W.C. Stevens, U.S. Patent 4738896 A, April 19, 1988.
89. T.A. Michalske and K.D. Keefer in *Better Ceramics Through Chemistry III*, eds. C.J. Brinker, D.E. Clark, and D.R. Ulrich (Mat. Res. Soc., Pittsburgh, Pa., 1988), p. 187.
90. Y.W. Lam and H.C. Lam, *J. Phys. D. Appl. Phys.*, **9** (1976) 1677.
91. T.A. Ma and K. Miyanchi, *Appl. Phys. Lett.*, **34** (1979) 88.
92. S.K. Gupta and C.G. Audain, *SPIE*, **469** (1984) 179.
93. W.A. Yarbrough, T.R. Gururaja, and L.E. Cross, *Am. Ceram. Soc. Bull.*, **66** (1987) 692–698.
94. W.L. Warren, P.M. Lenahan, and C.J. Brinker in *Better Ceramics Through Chemistry III*, eds. C.J. Brinker, D.E. Clark, and D.R. Ulrich (Mat. Res. Soc., Pittsburgh, Pa., 1988), p. 803.
95. R.A. Weimer, P.M. Lenahan, T.A. Marchione, and C.J. Brinker, *Appl. Phys. Lett.*, **51** (1987) 1179.
96. W.L. Warren, P.M. Lenahan, and C.J. Brinker, *Thin Solid Films*, submitted.
97. A. Nazeri-Eshghi, J.D. Mackenzie, and J-M. Yang in *Better Ceramics Through Chemistry III*, eds. C.J. Brinker, D.E. Clark, and D.R. Ulrich (Mat. Res. Soc., Pittsburgh, Pa., 1988), p. 561.
98. I.M. Thomas, UCRL-PREPRINT, Lawrence Livermore National Laboratories, Livermore CA, December 2, 1985.
99. G.C. Frye, A.J. Ricco, S.J. Martin, and C.J. Brinker in *Better Ceramics Through Chemistry III*, eds. C.J. Brinker, D.E. Clark, and D.R. Ulrich (Mat. Res. Soc., Pittsburgh, Pa., 1988), p. 349.
100. T. Bein, K. Brown, P. Enzel, and C.J. Brinker in *Better Ceramics Through Chemistry III*, eds. C.J. Brinker, D.E. Clark, and D.R. Ulrich (Mat. Res. Soc., Pittsburgh, Pa., 1988), p. 761.
101. G. Kordas in *Better Ceramics Through Chemistry II*, eds. C.J. Brinker, D.E. Clark, and D.R. Ulrich (Mat. Res. Soc., Pittsburgh, Pa., 1986), p. 685.
102. H. Schmidt, G. Rinn, R. Naß, and D. Sporn in *Better Ceramics Through Chemistry III*, eds. C.J. Brinker, D.E. Clark, and D.R. Ulrich (Mat. Res. Soc., Pittsburgh, Pa., 1988), p. 743.
103. R.R. Chianelli and M.B. Dines, *Inorg. Chem.*, **17** (1978) 2758.
104. B.B. Nayak, N.H. Acharya, T.K. Chandhuri, and G.B. Mitra, *Thin Solid Films*, **9** (1982) 309–314.
105. H. Schmidt, O. von Stetten, G. Kellermann, H. Patzelt, and W. Naegele, IAEA-SM-259/67, IAEA, Vienna, Austria, 1982, p. 111–120.
106. B.A. Tuttle, *MRS Bulletin* (October–November 1987) 40–45.
107. M. Yamane in *Sol-Gel Technology for Thin Films, Fibers, Preforms, Electronics, and Specialty Shapes*, ed. L.C. Klein (Noyes, Park Ridge, N.J., 1988), p. 200.
108. E.M. Rabinovich in *Sol-Gel Technology for Thin Films, Fibers, Preforms, Electronics, and Specialty Shapes*, ed. L.C. Klein (Noyes, Park Ridge, N.J., 1988), p. 260.
109. F. Hayashi and A. Iwai, Japanese Patent 88182222 A2: JP 63182222, July 27, 1988.
110. M. Toki, S. Miyashita, T. Takeuchi, S. Kanbe, and A. Kochi, *J. Non-Cryst. Solids*, **100** (1988) 479.
111. M. Matsuo, Japanese Patent 88100032 A2: JP 63100032, May 2, 1988.
112. A. Yajima, O. Horibata, and S. Sakai, Japanese Patent 88134525 A2: JP 63134525, June 7, 1988.
113. T. Takeuchi, Japanese Patent 88117919 A2: JP 63117919, May 21, 1988.
114. T. Takeuchi, Japanese Patent 88107821 A2: JP 63107821, May 12, 1988.

115. S. Sakai, O. Horibata, and A. Yajima, Japanese Patent 88147836 A2: JP 63147836, June 20, 1988.

116. A. Yajima, O. Horibata, and S. Sakai, Japanese Patent 88147832 A2: JP 63147832, June 20, 1988.

117. S. Sakai, O. Horibata, and A. Yajima, Japanese Patent 88144127 A2: JP 63144127, June 16, 1988.

118. A. Yajima,O. Horibata, and S. Sakai, Japanese Patent 88129028 A2: JP 63129028, June 1, 1988.

119. E.M. Rabinovich, European Patent App. EP 281282 A1, August 7, 1988.

120. E.M. Rabinovich, J.B. Machesney, D.W. Johnson Jr., J.R. Simpson, B.W. Meagher, F.V. Dimarcello, D.L. Wood, and E.A. Sigety, *J. Non-Cryst. Solids*, **63** (1984) 155-161.

121. I.M. Thomas, U.S. Patent 4028085, June 7, 1977.

122. S. Shibata and N. Nakahara, IOOC-ECOC (1985).

123. S. Shibata, T. Kitagawa, F. Hanawa, and M. Horiguchi, *J. Non-Cryst. Solids*, **88** (1986) 345-365.

124. R. Puyané, C. Gonzalez-Oliver, and A.L. Harmer, European Patent App. 0099440 A1, July 26, 1982.

125. B. Lintner, N. Arfsten, H. Dislich, H. Schmidt, G. Philipp, and B. Seiferling, *J. Non-Cryst. Solids*, **100** (1988) 378.

126. T. Mori, M. Toki, M. Ikejiri, M. Takei, M. Aoki, S. Uchiyama, and S. Kanbe, *J. Non-Cryst. Solids*, **100** (1988) 523-525.

127. D.W. Johnson, J.B. MacChesney, and E.M. Rabinovich, U.S. Patent 4605428, August 12, 1986.

128. G.W. Scherer, U.S. Patent 4574063, March 4, 1986.

129. I. Matsuyama, K. Susa, S. Satoh, and T. Suganuma, *Ceramic Bull.*, **63** (1984) 1408-1411.

130. K. Nassau, E.M. Rabinovich, and D.L. Wood, U.S. Patent App. 774666, September 11, 1985.

131. M. Cantin, M. Casse, L. Koch, R. Jouan, P. Mestran, D. Roussel, F. Bonnin, J. Moutel, and S.J. Teichner, *Nucl. Instrum. Methods*, **118** (1974) 177.

132. H. Schmidt, *J. Non-Cryst. Solids*, **73** (1985) 681-691.

133. S.-H. Wang, C. Campbell, and L.L. Hench in *Ultrastructure Processing of Advanced Ceramics*, eds. J.D. Mackenzie and D.R. Ulrich (Wiley, New York, 1988), p. 145.

134. L.L. Hench and J.-L. Nogues in *Proceedings of IVth Int'l. Conf. on Ultrastructure Processing of Ceramics, Glasses, and Composites*, eds. D.R. Uhlmann and D.R. Ulrich (Wiley, New York, to be published 1990).

135. R.D. Shoup in *Ultrastructure Processing of Advanced Ceramics* eds. J.D. Mackenzie and D.R. Ulrich (Wiley, New York, 1988), p. 347.

136. *Aerogels*, ed. J. Fricke (Springer-Verlag, Berlin, 1987).

137. J. Fricke in *Sol-Gel Technology for Thin Films, Fibers, Preforms, Electronics, and Specialty Shapes*, ed L.C. Klein (Noyes, Park Ridge, N.J., 1988), p. 226.

138. J. Fricke, *J. Non-Cryst. Solids*, **100** (1988) 169.

139. J. Fricke and R. Caps in *Ultrastructure Processing of Advanced Ceramics*, eds. J.D. Mackenzie and D.R. Ulrich (Wiley, New York, 1988), p. 613.

140. J. Fricke and G. Reichenauer in *Better Ceramics Through Chemistry II*, eds. C.J. Brinker, D.E. Clark, and D.R. Ulrich (Mat. Res. Soc., Pittsburgh, Pa., 1986), p. 775.

141. B.E. Yoldas, U.S. Patent 4,286,024, August 25, 1981.

142. J.J. Lannutti and D.E. Clark, *Ceramics International*, **11** (1985) 91-96.

143. M. Nogami and M. Tomozawa, *J. Am. Ceram. Soc.*, **69** (1986) 99.

144. M. Nogami, K. Nagasaka, K. Kadono, and T. Kishimoto, *J. Non-Cryst. Solids*, **100** (1988) 198.

145. G.V. Chandrashekar and M.W. Shafer in *Better Ceramics Through Chemistry II*, eds. C.J. Brinker, D.E. Clark, and D.R. Ulrich (Mat. Res. Soc., Pittsburgh, Pa., 1986), p. 711.

146. J.E. Mark and C.-C. Sun, *Polym. Bull.*, **18** (1987) 259–264.

147. R.T. Paine, J.F. Janik, and C. Narula in *Better Ceramics Through Chemistry III*, eds. C.J. Brinker, D.E. Clark, and D.R. Ulrich (Mat. Res. Soc., Pittsburgh, Pa., 1988), p. 461.

148. L.V. Interrante, C.L. Czekaj, J.L.J. Hackney, G.A. Sigel, P.J. Schields, and G.A. Slack in *Better Ceramics Through Chemistry III*, eds. C.J. Brinker, D.E. Clark, and D.R. Ulrich (Mat. Res. Soc., Pittsburgh, Pa., 1988), p. 465.

149. M. Yamane, J.B. Caldwell, and D.T. Moore in *Better Ceramics Through Chemistry II*, eds. C.J. Brinker, D.E. Clark, and D.R. Ulrich (Mat. Res. Soc., Pittsburgh, Pa., 1986), p. 765.

150. M. Yamane, J.B. Caldwell, and D.T. Moore, *J. Non-Cryst. Solids*, **85** (1986) 244–246.

151. S. Kurosaki and M. Watanbe, U.K. Patent App. GB 2086877 A, September 10, 1981.

152. M. Yamane, H. Kawazoe, A. Yasumori, and T. Takahashi, *J. Non-Cryst. Solids*, **100** (1988) 506–510.

153. L.J. Rysdale, *Electron Lett.*, **22** (1986) 99–102.

154. K. Shingyochi, S. Konishi, and K. Susa, Japanese Patent JP 87119120 A2: JP 62119120, May 30, 1987.

155. S. Konishi, K. Shingyouchi, and A. Makishima, *J. Non-Cryst. Solids*, **100** (1988) 511–513.

156. S. Shibata, T. Kitagawa, and M. Horiguchi, *J. Non-Cryst. Solids*, **100** (1988) 269.

157. C.P. Scherer and C.G. Pantano, *J. Non-Cryst. Solids*, **82** (1986) 246.

158. Z. Deng, E. Breval, and C.G. Pantano, *J. Non-Cryst. Solids*, **100** (1988) 364.

159. C.J. Brinker and D.M. Haaland, *J. Am. Ceram. Soc.*, **66** (1983) 754.

160. B.E. Yoldas and I.K. Lloyd, *Ceram. Bull.*, **18** (1983) 1171–1177.

161. E.A. Hayri and M. Greenblatt, *Diff. and Defect Data*, **53–54** (1987) 433–438.

162. A. Salomoni, E.H. Toscano, A. Caniero, A. Montenero, and G. Ondracel, *Mat. Chem. and Phys.*, **17** (1987) 475–484.

163. E. Matijević and P. Gherardi, *Transformation of Organometallics into Common and Exotic Materials: Design and Activation*, NATO ASI series E, no. **141**, ed. R.M. Laine (Martinus Nijhoff, Dordrecht, 1988), p. 279.

164. I.M. Thomas in *Sol-Gel Technology for Thin Films, Fibers, Preforms, Electronics, and Speciality Shapes*, ed. L.C. Klein (Noyes, Park Ridge, N.J., 1988), p. 2.

165. J.B. Blum in *Sol-Gel Technology for Thin Films, Fibers, Preforms, Electronics, and Specialty Shapes*, ed. L.C. Klein (Noyes, Park Ridge, N.J., 1988), p. 296.

166. E. Barringer, N. Jubb, B. Fegley, R.L. Pober, and H.K. Bowen in *Ultrastructure Processing of Glasses, Ceramics, and Composites*, eds. L.L. Hench and D.R. Ulrich (Wiley, New York, 1984), p. 315.

167. R.L. Downs, M.A. Ebner, and W.J. Miller in *Sol-Gel Technology for Thin Films, Fibers, Preforms, Electronics, and Speciality Shapes*, ed. L.C. Klein (Noyes, Park Ridge, N.J., 1988), p. 330.

168. B.C. Bunker, J.A. Voigt, D.L. Lamppa, D.H. Doughty, E.L. Venturini, J.F. Kwak, D.S. Ginley, T.J. Headley, M.S. Harrington, M.O. Eatough, R.G. Tissot, Jr., and W.F. Hammetter in *Better Ceramics Through Chemistry III*, eds. C.J. Brinker, D.E. Clark, and D.R. Ulrich (Mat. Res. Soc., Pittsburgh, Pa., 1988), p. 373.

169. F. Mahloojchi, F.R. Sale, N.J. Shah, and J.W. Ross, *Br. Ceram. Proc.*, **40** (1988) 1–14.

170. M.F. Yan, H.C. Ling, H.M. O'Bryan, P.K. Gallagher, and W.W. Rhodes in *Better Ceramics Through Chemistry III*, eds. C.J. Brinker, D.E. Clark, and D.R. Ulrich (Mat. Res. Soc., Pittsburgh, Pa., 1988), p. 385.

171. J.C. Bernier, S. Vilminot, S. El Hadigui, C. His, J. Guille, T. Dupin, R. Barral, and G. Bouzat in *Better Ceramics Through Chemistry III*, eds. C.J. Brinker, D.E. Clark, and D.R. Ulrich (Mat. Res. Soc., Pittsburgh, Pa., 1988), p. 831.

172. M.J. Cima, R. Chiu, and W.E. Rhine, *Mater. Res. Soc. Symp. Proc.*, **9** (1988) 241–244.

173. K. Suzuki, M. Naito, and S. Shirasaki, Japanese Patent 88151675 A2: JP 63151675, June 24, 1988.

174. J.P. Boilot, A. Gay, P. Colomban, and M. Lejeune, French Patent FT 2545003 A1, November 2, 1984.

175. C. Gensse and U. Chowdry in *Better Ceramics Through Chemistry II*, eds. C.J. Brinker, D.E. Clark, and D.R. Ulrich (Mat. Res. Soc., Pittsburgh, Pa., 1986), p. 693.

176. J. Ravez, N. Puyoo-Castaings, and F. Duboudin, *Lab. Chim. Solide*, **81** (1988) 313–316.

177. R.W. Schwartz, D.A. Payne, P.M. Eccles, and D.J. Eichorst in *Ultrastructure Processing of Advanced Ceramics*, eds. J.D. Mackenzie and D.R. Ulrich (Wiley, New York, 1988), p. 487.

178. E. Crispino, P. Gerontopoulos, G. Arcangeli, S. Cao, M. Forno, and W. Muller, *Radiochimica Acta*, **36** (1984) 69–74.

179. J.M. Pope and D.E. Harrison, *Waste Management* (February 23–26, 1981) 237–248.

180. E.R. Vance, *J. Mater. Sci.*, **21** (1986) 1413–1416.

181. L.J. Yang, S. Komarneni, and R. Roy, *Mater. Res. Soc. Proc.*, **26** (1984) 567–574.

182. V. Gottardi, M. Guglielmi, and A. Tiziani, *J. Non-Cryst. Solids*, **43** (1981) 105–114.

183. V. Gottardi, M. Guglielmi, and R. Sechi, *Rivista della Staz. Sper. Vetro*, **12** (1982) 132–139.

184. J.S. Sparks and D.S. Tucker, *Adv. Ceram. Mater.*, **3** (1988) 509–510.

185. S. Chakrabarti and A. Paul, *Trans. Indian Cer. Soc.*, **45** (1986) 7–13.

186. A.I. Kingon, A. Van Zyl, and P.M. Smit, U.K. Patent Appl. GB 2168334 A1. June 18, 1986.

187. H. Saize, P. Odier, and B. Cales, *J. Non-Cryst. Solids* **82** (1986) 314–320.

188. R. Bauer, European Patent App. 85106752.0, 0168606 A2, May 31, 1985.

189. R. Bauer, European Patent App. 85100506.6, January 18, 1985.

190. A.P. Gerk and M.G. Schwabel, European Patent App. 86303121.7, 0200487 A2, April 24, 1986.

191. A.P. Gerk, U.S. Patent 4574003, March 4, 1986.

192. E. Breval, G.C. Dodds, and N.H. Macmillan, *Mat. Res. Bull.*, **20** (1985) 413–429.

193. T.E. Cottringer, R.H. Van de Merwe, and R. Bauer, European Patent App. EP 152768 A2, August 28, 1985.

194. V. Baran, K. Stamberg, M. Tympl, and J. Kinzelova, *J. Nucl. Matl.*, **58** (1975) 59–66.

195. *Proceedings of the Symposium on Sol-Gel Processes and Reactor Fuel Cycles*, Gatlinburg, TN, May 1970, Oak Ridge National Laboratory Report CONF-700502 (1970).

196. *Proceedings of the Conference on Sol-Gel Processes for Ceramic Nuclear Fuels*, May 1968 (Int'l. Atomic Energy Authority, Vienna, Austria, 1968).

197. I.M. Thomas, U.S. Patent 3709833, January 9, 1973.

198. G. Gonzalez-Oliver, European Patent App. 86810344.1, August 4, 1986.

199. C.J.R. Gonzalez-Oliver, M. Schneider, K. Nawata, and H. Kusano, *J. Non-Cryst. Solids*, **100** (1988) 274.

200. R.L. Downs, M.A. Ebner, B.D. Homyk, and R.L. Nolen, *J. Vac. Sci. Technol.*, **18** (1981) 1272–1275.

201. R.L. Downs and W.J. Miller, U.S. Patent 4336338, June 22, 1982.
202. N. Yamamoto, T. Goto, and Y. Horiguchi, European Patent App. EP 261593 A1, March 30, 1988.
203. W.A. Zdaniewski, P.M. Shah, and H.P. Kirchner, *Adv. Ceram. Mat.*, **2** [3A] (1987) 204-208.
204. P.S. Dunn, C.A. Natalie, and D.L. Olson, *J. Mat. for Energy Syst.*, **8** (1986) 176-184.
205. S. Sakka in *Sol-Gel Technology for Thin Films, Fibers, Preforms, Electronics, and Speciality Shapes*, ed. L.C. Klein (Noyes, Park Ridge, N.J., 1988), p. 140.
206. H.G. Sowman in *Sol-Gel Technology for Thin Films, Fibers, Preforms, Electronics, and Specialty Shapes*, ed. L.C. Klein (Noyes, Park Ridge, N.J., 1988), p. 162.
207. W.C. LaCourse in *Sol-Gel Technology for Thin Films, Fibers, Preforms, Electronics, and Specialty Shapes*, ed. L.C. Klein (Noyes, Park Ridge, N.J., 1988), p. 184.
208. W. Mahler and M.F. Bechtold, *Nature*, **285** (1980) 27.
209. S. Sakka, *Hyomen*, **19** (1981) 430-437.
210. H.G. Sowman, European Patent App. 87301350.2, February 17, 1987.
211. W.C. LaCourse in *Better Ceramics Through Chemistry*, eds. C.J. Brinker, D.E. Clark, and D.R. Ulrich (Elsevier, New York, 1984), p. 53.
212. G.F. Everitt in *Ultrastructure Processing of Advanced Ceramics*, eds. J.D. Mackenzie and D.R. Ulrich (Wiley, New York, 1988), p. 463.
213. H. Mizuguchi, M. Ota, K. Katsuhiki, and J. Kobayashi, Japanese Patent 88112437 A2: JP 63112437, May 17, 1988.
214. F.I. Hurwitz, L.H. Hyatt, J.P. Gorecki, and L.A. D'Armore in *Ultrastructure Processing of Advanced Ceramics*, eds. J.D. Mackenzie and D.R. Ulrich (Wiley, New York, 1988), p. 973.
215. H. Kozuka, T. Umeda, J.S. Jin, and S. Sakka in *Better Ceramics Through Chemistry III*, eds. C.J. Brinker, D.E. Clark, and D.R. Ulrich (Mat. Res. Soc., Pittsburgh, Pa., 1988), p. 639.
216. E. Leroy, C. Robin-Brosse, and J.P. Torre, Comm. Eur. Communities, *Ceram. Adv. Energy Technol.*, Report No. 9210 (1984) 501-517.
217. E. Papanikolau, W.C.P.M. Meermann, R. Aerts, T.L. van Rooij, J.G. van Lierop, and T.P.M. Meeuwsen, *J. Non-Cryst. Solids*, **100** (1988) 247.
218. R. Rice in *Better Ceramics Through Chemistry*, eds. C.J. Brinker, D.E. Clark, and D.R. Ulrich (Elsevier, New York, 1984), p. 337.
219. J.J. Lannutti and D.E. Clark in *Better Ceramics Through Chemistry*, eds. C.J. Brinker, D.E. Clark, and D.R. Ulrich (Elsevier, New York, 1984), p. 369.
220. J.J. Lannutti and D.E. Clark in *Better Ceramics Through Chemistry*, eds. C.J. Brinker, D.E. Clark, and D.R. Ulrich (Elsevier, New York, 1984), p. 375.
221. F.K. Chi and G.L. Stark, European Patent App. EP 125005 A1, November 14, 1984.
222. C.C. Payne, U.S. Patent 4430369, February 7, 1984.
223. D. Qi and C.G. Pantano in *Ultrastructure Processing of Advanced Ceramics*, eds. J.D. Mackenzie and D.R. Ulrich (Wiley, New York, 1988), p. 635.
224. C.G. Pantano, G.L. Messing, D. Qi, and W. Minehan, Final Report, AFWL-TN-86-59, November, 1987.
225. E. Fitzer and R. Gadow in *Tailoring Multiphase and Composite Ceramics*, eds. R.E. Tressler, G.L. Messing, C.G. Pantano, and R.E. Newnham (Plenum, New York, 1986), p. 571.
226. J.H. Adair, S.A. Touse, and P.J. Melling, *Am. Ceram. Soc. Bull.*, **66** [10] (1978) 1490-1494.
227. J.C. Withers, R.O. Loufty, and K.L. Stuffle, European Patent App. EP 379673 A2, August 24, 1988.

228. T. Nanao and T. Eguchi, European Patent App. 84104194.0, April 13, 1984.
229. G. Larnac and J. Phalippou, Technical Report, DRET-86/1369 Dir. Rech., Etud. Tech., Paris, Fr. (1987).
230. R. Roy, S. Komarneni, and W. Yarbrough in *Ultrastructure Processing of Advanced Ceramics*, eds. J.D. Mackenzie and D.R. Ulrich (Wiley, New York, 1988), p. 571.
231. E.J.A. Pope and J.D. Mackenzie in *Better Ceramics Through Chemistry II*, eds. C.J. Brinker, D.E. Clark, and D.R. Ulrich (Mat. Res. Soc., Pittsburgh, Pa., 1986), p. 809.
232. E.J.A. Pope and J.D. Mackenzie, *Mat. Sci. Research*, **20** (1986) 187–194.
233. B.I. Lee and L.L. Hench in *Better Ceramics Through Chemistry II*, eds. C.J. Brinker, D.E. Clark, and D.R. Ulrich (Mat. Res. Soc., Pittsburgh, Pa., 1986), p. 815.
234. B.I. Lee and L.L. Hench, *Ceram. Eng. Sci. Proc.*, **8** (1987) 685–692.
235. B.I. Lee and L.L. Hench, *Cer. Eng. and Sci. Proc.*, **7** (1986) 994–1000.
236. R.R. Haghighat, R.F. Kovar, and R.W. Lusignea in *Better Ceramics Through Chemistry III*, eds. C.J. Brinker, D.E. Clark, and D.R. Ulrich (Mat. Res. Soc., Pittsburgh, Pa., 1988), p. 755.
237. B.Dunn, E. Knobbe, J.M. McKiernan, J.C. Pouxviel, and J.I. Zink in *Better Ceramics Through Chemistry III*, eds. C.J. Brinker, D.E. Clark, and D.R. Ulrich (Mat. Res. Soc., Pittsburgh, Pa., 1988), p. 331.
238. R.F. Kovar and R.W. Lusignea in *Ultrastructure Processing of Advanced Ceramics*, eds. J.D. Mackenzie and D.R. Ulrich (Wiley, New York, 1988), p. 715.
239. J.E. Mark in *Ultrastructure Processing of Advanced Ceramics*, ed. J.D. Mackenzie and D.R. Ulrich (Wiley, New York, 1988), p. 623.
240. A. Makishima and T. Tani, *J. of Am. Ceram. Soc.*, **69** (1986) C-72–74.
241. W.F. Doyle, B.D. Fabes, J.C. Root, K.D. Simmons, Y.M. Chiang, and D.R. Uhlmann in *Ultrastructure Processing of Advanced Ceramics*, eds. J.D. Mackenzie and D.R. Ulrich (Wiley, New York, 1988), p. 953.
242. G.L. Wilkes, B. Orler, and H. Huang, *Polymer Prepr.*, **26** (1985) 300–302.
243. L.C. Klein in *Sol-Gel Technology for Thin Films, Fibers, Preforms, Electronics, and Specialty Shapes*, ed. L.C. Klein (Noyes, Park Ridge, N.J., 1988), p. 382.
244. A.F.M. Leenaars, K. Keizer, and A.J. Burgraaf in *Studies in Inorganic Chemistry*, **vol. 3**, eds. R. Metselaar, H.J.M. Heijligers, and J. Schoonman (Elsevier, Amsterdam, 1983), pp. 401–404.
245. A.F.M. Leenaars, K. Keizer, and A.J. Burgraaf, *J. Mat. Sci.*, **10** (1984) 1077–1088.
246. A. Larbot, J.A. Alary, J.P. Fabre, C. Guizard, and L. Cot in *Better Ceramics Through Chemistry II*, eds. C.J. Brinker, D.E. Clark, and D.R. Ulrich (Mat. Res. Soc., Pittsburgh, Pa., 1986), p. 659.
247. L. Cot, A. Larbot, and C. Guizard in *Ultrastructure Processing of Advanced Ceramics*, eds. J.D. Mackenzie and D.R. Ulrich (Wiley, New York, 1988), p. 211.
248. A. Larbot, J.A. Alary, C. Guizard, J.P. Fabre, N. Idrissi, and L. Cot in *High Tech Ceramics*, **38c**, ed. P. Vincenzini (Elsevier, Amsterdam, 1987), pp. 2259–2263.
249. J. Haggin, *Chem. Eng. News*, **66** (1988) 25.
250. J. Gillot, G. Brinkman, and D. Garcera, "New Ceramic Filter Media for Cross-Flow Microfiltration and Ultrafiltration," S.C.T. Ceramic Membranes, Tarbes, France.
251. M.A. Anderson, M.J. Geiselmann, and Q. Xu, *J. Membrane Sci.*, **39** (1988) 243–258.
252. M.J. Gieselmann, M.A. Anderson, M.D. Mogsemiller, and G.G. Hill, Jr., *Sep. Sci. Tech.*, **23** (1988) 1695–1714.
253. S.A. Bonis, *Electronic Packaging and Production*, **14** (1974). 46–56.
254. M. Nogami, *Yogyo Kyokaishi*, **93** (1985) 195–200.
255. A. Suprynowicz, B. Buszewski, R. Lodkowski, and A.L. Dawidowicz, *J. Chromatogr.*, **446** (1988) 347–357.

256. T. Hanada, H. Fukaya, and O. Sugawa, *Fire Sci. and Tech.*, **3** (1983) 1–12.
257. S.J. Teichner in *Aerogels*, ed. J. Fricke (Springer-Verlag, Berlin, Heidelberg, 1986), p. 22.
258. M. Astier, A. Bertrand, D. Bianchi, A Chenard, G. Pajonk, M.B. Taghavi, S.J. Teichner, and B.L. Villemin in *Preparation of Catalysts*, eds. B. Delmon, P.A. Jacobs, and G. Poncelet (Elsevier, Amsterdam, 1976), p. 315.
259. M. Formenti, J. Juillet, P. Mériaudeau, and S.J. Teichner, *Bull. Soc. Chim. France* (1972) 69.
260. M.B. Taghavi, G. Pajonk, and S.J. Teichner, *Bull. Soc. Chim. France* (1978) 302.
261. M.B. Taghavi, G. Pajonk, and S.J. Teichner, *J. Coll. Interf. Sc.,* **71** (1979) 451.
262. A. Muller, F. Juillet, and S.J. Teichner, *Bull. Soc. Chim. France* (1976) 1361.
263. G. Matis, F. Juillet, and S.J. Teichner, *Bull. Soc. Chim. France* (1976) 1633, 1637.
264. H.P. Stephens, R.G. Dosch, and F.V. Stohl, *Ind. I and D. Eng. Chem. Prod. Res. and Dev.*, **24** (1985) 15.
265. B.C. Bunker, C.H.F. Peden, D.R. Tallant, S.L. Martinez, and G.L. Turner in *Better Ceramics Through Chemistry III*, eds. C.J. Brinker, D.E. Clark, and D.R. Ulrich (Mat. Res. Soc., Pittsburgh, Pa., 1988), p. 105.
266. J.A. Cairns, D.L. Segal, and J.L. Woodhead in *Better Ceramics Through Chemistry*, eds. C.J. Brinker, D.E. Clark, and D.R. Ulrich (Elsevier, New York, 1984), p. 135.
267. A.N. Speca and R.D. Laib, U.S. Patent 4536489A, August 20, 1985.
268. M. Carbini, E. Baretter, G. Navazio, and M. Guglielmi, *Riv. Stn. Spe. Vetro*, **14** (1984) 161–165.
269. G. Caturan, G. Facchin, V. Gottardi, M. Guglielmi, and G. Navazio, *J. Non-Cryst. Solids*, **48** (1982) 219.

INDEX

D

D1 defect. *See* Cyclic silicates
D2 defect. *See* Cyclic silicates
Darcy averaging, 421, 428, 477
Darcy's law, 420–425, 427, 423–424, 438
  drying and, 470, 478
  versus Fick's law, 426–428
DC arc plasma, 287–288
DCCA. *See* Drying control chemical additive
Dealkalization, by chlorine, 641–643
Debye forces, 239
Debye–Hückel screening length, 242
Decomposition of solution, evaporative, 284
Defect
  paramagnetic, 588–590
  silicate, 575
  See also Cyclic silicates
Deformation, theory, 437–448
Degree of reaction, 305–306, 316–317,
    343–346. *See also* Order of reaction
Dehydroxylation, 582–588, 592, 628–645
  See also Chlorine drying
  aerogel, 633
  borosilicate, 634
  by chlorine, 637–643
  by condensation reaction, 628–637
  curvature effect, 633, 634
  effect of pore size, 632, 633
  by fluorine, 643–645
  strategy, 632–635
  thermal, 554, 629–637
  in vacuum, 636
Delayed elastic strain, 384, 396
Delta alumina, 600, 602–604
Dendrite, 755
Densification, skeletal, 562–567
Density
  aerogel, 538–540
  bulk, of silica xerogels, 581
  gel, 477
  gradient, 477, 492–493
  green, 685, 701, 726–727
  packing, 263, 267–269
  particle, 275
  relaxation, 681, 750–752
  of silica, 766–767
  of silicon alkoxides, 113, 114
  skeletal, *See* Skeletal density

of strontium silicate, 777, 778
of xerogel, 537
Deposition, film. *See* Film formation
Desiccant, 870
Deuterium isotope effect, 133–134
Devitrification, 773. *See also* Crystallization
Dialysis, 409, 489
Diborate, lithium, 81, 88
Dielectric constant, 242, 768
Dielectric loss, 768
Dielectric strength, 849, 850
Diesel engine, 865
Differential scanning calorimetry (DSC)
  of gel-to-glass transition, 560–561, 570–571
  of glass transition, 711, 750, 768–772
  of silicate dehydroxylation, 578
Differential shrinkage. *See* Differential strain
Differential strain, 418, 445, 476, 484, 490,
    493–494
Differential thermal analysis (DTA), 547, 552,
    562
  of glass transition, 562, 768–769
  of phase transformation, 773–775, 780
Diffusion, 426–428, 439
  drying and, 463, 489–492, 500
  characteristic time, 761–762, 773
  coefficient, 279, 313, 335–340, 426–427,
    490, 565
  control, 279, 565
  on fractal surface, 665
  grain boundary, 720–721, 723–724
  lattice, 720–723
  osmotic, 408–410, 418, 426, 463–464
  path, 720–721
  sintering and, *See* Diffusive sintering
  surface, 691, 721–722, 723
  vapor, 455–456, 475, 477
Diffusion-limited aggregation, 36, 195, 196,
    202, 252–254, 259, 290, 335–343
Diffusive sintering, 718–730
  diffusion path, 720–721
  experimental studies, 724–730
  theory, 718–724
Dihedral angle, 719–720
Dilational symmetry, 194
Dimensional, fractal. *See* Fractal dimension
Dimensionless time, 678, 685, 688–689, 696
Dimethyl formamide, 500
Dip coating process, 797, 798, 841
  See also Film formation